MW01614480

Advanced Tactical and Strategic Missile Guidance

Seventh Edition, Volume 2

Advanced Tactical and Strategic Missile Guidance

Seventh Edition, Volume 2

Paul Zarchan

Volume 259
Progress in Astronautics and Aeronautics

Timothy C. Lieuwen, Editor-in-Chief
Georgia Institute of Technology
Atlanta, Georgia

Published by
American Institute of Aeronautics and Astronautics, Inc.
12700 Sunrise Valley Drive, Suite 200, Reston, VA 20191-5807

Cover image credit: U.S. Air Force image, Minuteman III missile postboost vehicle/Graphic by Staff Sgt. Mark A. Schurman

American Institute of Aeronautics and Astronautics, Inc., Reston, VA.

ISBN 978-1-62410-538-8

To
Maxine, Adina, Ari, Ronit, Lila, Nathan and Zachary

TABLE OF CONTENTS

Preface to the Seventh Edition .... **xiii**

Acknowledgments .... **xv**

Introduction .... **xvii**

**Chapter 1 Unstable Airframe and the Three-Loop
Autopilot** **1**

Introduction 1
Review and Demonstration of Airframe Instability 1
Simulation to Demonstrate Airframe Center of Gravity Location
can Cause Instability 5
Reexamining the Previous Linearization 10
Modifying the Three-Loop Autopilot Gain Algorithm for an
Unstable Airframe 14
Open-Loop Analysis 15
Closed-Loop Analysis...................................... 17
Summary 29
References 29

Chapter 2 Flexible Body Dynamics **31**

Introduction 31
Rigid and Flexible Body Dynamics 31
Methods for Alleviating Instability Due to Rate Gyro Sensing
Body Bending 43
Summary 53
References 53

Chapter 3 Modern Control and Autopilot Design **55**

Introduction 55
Modern Control Applied to Missile Roll Autopilot................ 55
Classical Analysis of Modern Control Techniques 78
How to Use Modern Control for Practical System Design 85
Summary 88
References 88

Chapter 4   Advanced Adjoint Applications **89**

Introduction 89
Multiple Sampling Rate Adjoint 89
Adjoint of Discrete Linear Kalman Filter 101
Simulating an Impulse 126
Summary 129
References 130

**Chapter 5   Shaping Filters and Their Use in Miss
              Distance Analysis** **131**

Background and Introduction 131
Shaping Filter Representation of Uniformly Distributed
Target Maneuver 131
Summary 145

Chapter 6   Weaving Targets **147**

Introduction and Background 147
Weave Maneuver in Single Time Constant Guidance System 147
Closed-Form Solutions for Miss Distance 155
Higher-Order Guidance System Dynamics 161
Acceleration Saturation 165
Reducing the Time Constant to Improve Performance 168
Advanced Guidance Techniques to Improve Performance 171
Using the Discrete Fourier Transform to Find Miss
Caused by Weaving Target 179
Summary 182
References 183

Chapter 7   Filtering and Weaving Targets **185**

Introduction 185
Review of Original Three-State Linear Kalman Filter 185
Four-State Weave Kalman Filter 194
Miss Distance Analysis 208
Extended Kalman Filter 211
Summary 229
References 229

Chapter 8   Filter Bank Approach to Weaving Target Problem **231**

Introduction 231
Review of Five-State Extended-Kalman-Filter Performance 231
Review of Four-State Linear Weave Kalman-Filter Performance 233

Filter Bank Methodology 237
Three-Filter Bank Example 239
Summary 254
References 254

Chapter 9   Predictor–Corrector Guidance **255**

Introduction 255
Surface-to-Surface Missile Problem 255
Predictor-Corrector Guidance with Surface-to-Surface
Intercept Problem 263
Roll Angle or Rate Guidance Problem 272
Predictor–Corrector Guidance with Roll-Rate Problem 277
Summary 288
References 288

**Chapter 10   Optimal Guidance Law to Minimize
Interceptor Jerk** **289**

Introduction 289
Derivation of the Jerk Guidance Law............................ 289
Preliminary Evaluation of Jerk Guidance Law 294
Alternate Form of Jerk Guidance 305
Summary 306
References 306

**Chapter 11   Use of Poisson/Singer Target Maneuver Model
for Kalman Filter Design** **307**

Introduction 307
Shaping Networks for Poisson and Singer Target Maneuver Models 308
Modeling the Poisson Target Maneuver 312
Kalman Filter Comparison 319
Increasing the Process Noise of the Normal Kalman Filter 326
Summary 331
References 331

**Chapter 12   Alternative Approaches to Guidance Law
Development** **333**

Introduction 333
Optimal Control 333
Using Optimal Control to Derive Guidance Law for Single-Lag
Flight Control System 336

Deriving Guidance Law for Weaving Target Using Optimal Control 343
Guidance Portion Due to Maneuvering Targets 349
Alternative Numerical Approach as a Result of
Flight-control System Dynamics 351
Deriving New Guidance Law for Cubic Flight-control System 356
Alternative Approach to Cubic Flight-control-system
Guidance Law 362
Performance Comparison of Guidance Laws in Presence of Cubic
Flight-control System 366
Summary 374
References 374

Chapter 13   Trajectory Shaping Guidance **375**

Introduction 375
Problem Setup 375
Using the Schwartz Inequality for Trajectory Shaping Guidance 377
Alternate Form of Trajectory Shaping Guidance Law 382
Testing Trajectory Shaping Guidance in the Linear World 383
Closed-Form Solutions 391
Nonlinear Results 397
Summary 407
References 407

**Chapter 14   Guidance Methods Against
                Stationary Targets** **409**

Introduction 409
Biased Proportional Navigation for Controlling Impact Angle
Against Stationary Targets 409
Polynomial Guidance 421
Summary 437
References 437

Chapter 15   The Vertical-S Target Maneuver **439**

Introduction 439
Shaping Filter Approach for Vertical-S Target Maneuver 439
Kalman Filter for Vertical-S Target Maneuver 452
Summary 464
References 464

Chapter 16   Improving the Performance of a Tail-Controlled Missile at High Altitudes **465**

Introduction 465
Zero Over Pole Guidance Law 465
Integrating Optimal Control Equation to Confirm Zero Over
Pole Guidance Law 467
Performance of Guidance Law When Real Autopilot is
Zero Over Pole 475
Zero over Pole Guidance and Three-Loop Autopilot 482
Summary 490
References 490

Chapter 17   Engagement Simulations in Three Dimensions **491**

Introduction 491
Weaving Targets in Three Dimensions 491
Ballistic Target Trajectory Generator in Three Dimensions 503
Intercept Point Prediction for Ballistic Targets 512
Strategic Missile-target Engagement Simulation 517
Summary 535
References 535

Chapter 18   Smallest Possible Miss Distance for a Radar Homing Missile **537**

Introduction 537
Smallest Possible Miss Distance for a Radar Homing Missile ... 537
Summary 553
References 553

Chapter 19   Comparison of Differential Game Guidance with Optimal Guidance **555**

Introduction 555
Traditional Guidance Law Review 556
Differential Game Guidance Law 558
Target Maneuvers 559
Guidance Law Comparison 562
Making Differential Game Guidance More Practical 575
Target Dynamics 578
Summary 580
References 580

Chapter 20   Boost-Phase Filtering Options **583**

Introduction 583
ICBM Model 584
ICBM Guidance 584
Filtering Options 592
Two-State Template-Based Filter 593
Three-State Filter 594
Summary 610
References 610

Chapter 21   Kinematics of Intercepting a Ballistic Target **611**

Operational Area 623
Launch Area Denied 629
Defended Area 633
Summary 637

**Chapter 22   Kill Vehicle Guidance and Control Sizing
                      for Boost-Phase Intercept** **639**

Introduction 639
Background 639
Air-Launched Interceptor Approach 640
Guidance and Control Issues 641
One-Dimensional Model for Understanding Guidance 641
Developing Formulas for Divert Due to Boosting Target and PIP Errors . . 647
Interceptor–IRBM Engagements 649
Interceptor–ICBM Engagements 668
Noise and Filtering 673
Interceptor Engagements with Noise and Filtering 690
Summary 692
References 693

Index **695**

Supporting Materials **707**

PREFACE TO THE SEVENTH EDITION

PREFACE TO TACTICAL AND STRATEGIC MISSILE GUIDANCE: AN INTRODUCTION, SEVENTH EDITION AND ADVANCED TACTICAL AND STRATEGIC MISSILE GUIDANCE

Tactical and Strategic Missile Guidance, Sixth Edition contained more than 1000 pages and so it was decided to split the book into two volumes so that existing material could be enhanced and new material could be added without having to worry about page constraints. The first volume of this two volume set is entitled *Tactical and Strategic Missile Guidance: An Introduction, Seventh Edition* and contains twenty-three of the chapters from the sixth edition. Nine of the chapters of this volume have major enhancements and the chapters have now been placed in a more logical sequence. The second volume of the two-volume set entitled *Advanced Tactical and Strategic Missile Guidance* has twenty-two chapters of which nine are completely new. These nine new chapters are based on material from the AIAA continuing education short course "Fundamentals of Tactical and Strategic Missile Guidance" that I have been teaching since 1990. The new topics are treated from both an analytical and a simulation point of view so that readers with different backgrounds and learning styles can benefit from the new material. The rest of the chapters in this volume are enhanced and taken from either the sixth or fifth editions of the original *Tactical and Strategic Missile Guidance*.

Both volumes contain chapters on important topics related to improving missile guidance system performance and understanding key design concepts and tradeoffs. In addition, at the request of many readers, all of the FORTRAN source code that appeared in the text of the first five editions of *Tactical and Strategic Missile Guidance* has been converted to MATLAB. Interested readers can obtain all MATLAB source code—and the equivalent FORTRAN versions—in electronic form on the AIAA Web site as mentioned at the back of both volumes on the Supporting Materials page. Also at the request of some readers a Listing Index has been added to the end of both volumes to make it easier for interested readers to quickly find the source code without having to search both volumes.

On a personal note, it continues to be very gratifying for me to learn that many people working with or needing to learn about missile guidance have found *Tactical and Strategic Missile Guidance* useful. Over the years, many readers have contacted me and have asked questions when the book's material was not clear to them. Material in the existing chapters has been clarified so that all will benefit from my interaction with the readers. It is still my hope that this two-volume set will be of value not only to new readers, but will also be worthwhile to those who have read previous editions.

Paul Zarchan
February 2019

ACKNOWLEDGMENTS

Although many of the devices discussed in *Tactical and Strategic Missile Guidance* work in a vacuum, I do not. In this section I would like to take the opportunity to acknowledge those individuals who contributed in some way to the content and publication of this text.

First, I would like to thank Dr. Donald C. Fraser, former Executive Vice President of The Charles Stark Draper Laboratory, Inc. (CSDL) for recommending me as an author for AIAA's Progress in Astronautics and Aeronautics book series. He knew of my desire to write a book on missile guidance and through his efforts the dream is now a reality. During my 16 years at CSDL, Mr. John M. Elwell, former Guidance, Navigation & Control Director, provided me with numerous technical opportunities that expanded my missile guidance background. The technical thread running throughout the text is based on those technical opportunities. The highly competent staff and superb technical environment of CSDL provided the atmosphere that not only made this project possible but also fun. In particular, I would like to thank CSDL technical staff members Dr. Owen Deutsch, Mr. Steven L. Nelson, and Dr. Richard E. Phillips for our many valuable technical discussions. Much of the material in the strategic interceptor chapters was a consequence of our valuable technical interchanges. Some of the material in several of the chapters pertaining to tactical missile guidance was based on knowledge gained during my employment at Raytheon. Here, I not only learned tactical interceptor guidance and control fundamentals from the people who first made radar homing missiles a reality, but also learned of and developed an appreciation for the hard work involved in converting theory to something that works.

Next, I would like to acknowledge those associated with AIAA responsible for the inception and completion of this project. I would like to thank Dr. Allen E. Fuhs, former AIAA Progress in Astronautics and Aeronautics Editorial Board Member, for inviting me to participate in this series and then helping me formulate an outline from which a manuscript could be written. His kind and constructive review of the initial draft led to numerous changes, all of which improved the overall quality of the text. Mr. John Newbauer, former AIAA Administrator of Scientific Publications, provided me with excellent examples of the standards expected of a Progress series volume and, in addition, helped outline the preface. Mrs. Jeanne Godette, former AIAA Director of Book Publications, and her very competent staff not only taught me a great deal about grammar but, in addition, moved this project forward very rapidly. Throughout later years Rodger Williams and David Arthur of AIAA helped with the many editions of the text that followed. In addition, I would like to thank Pat DuMoulin of AIAA for making valuable suggestions and converting my manuscripts for past editions of both *Tactical and Strategic Missile Guidance* and *Fundamentals of Kalman Filtering: A Practical Approach* to actual books.

Finally, I wish to express my love and gratitude to my wife Maxine for having patience and understanding, while our home fell into a state of disrepair, for not complaining while I put in the many hours required in writing this text.

INTRODUCTION

The requirements for tactical missile guidance systems were born at the end of World War II as a result of the highly effective kamikaze attacks on U.S. vessels. After the war it was clear that naval guns using unguided shells were not adequate for shooting down hostile aircraft making suicidal attacks against U.S. ships. To counter kamikaze-like threats, the U.S. Navy initiated the development of the Lark guided missile in 1944. Approximately six years of hard work was required before the first successful intercept, made by the Lark against an unmanned aircraft, occurred on Dec. 2, 1950. A very readable history of this significant event, told by one of the engineering participants, can be found in a work by M. W. Fossier ("The Development of Radar Homing Missiles," *Journal of Guidance, Control, and Dynamics*, Vol. 7, No. 6, 1984, pp. 641–651).

Operational endoatmospheric tactical missile systems, using the same guidance principles as those used by the Lark, have now been in existence for approximately seven decades. These missiles use guidance concepts that work well not only against stationary or predictable targets but also are effective against responsive threats (such as aircraft executing evasive maneuvers) whose future position is highly uncertain. In the tactical arena, current guidance law technology appears to be adequate if the flight time is long compared to the effective time constant of the guidance system and if the missile enjoys a considerable acceleration advantage over the target. In fact, it is not uncommon for a tactical missile to have an acceleration advantage of more than five against an aircraft target. This advantage is more than adequate for a successful intercept with current guidance law technology.

Strategic ballistic missiles are different from tactical guided missiles because they travel much longer distances and are designed to intercept stationary targets whose location is known precisely. A tactical homing missile acquires the target with a seeker and then guides all the way to intercept, whereas a strategic interceptor does not require a seeker because all of its guidance is in the boost phase. Because the boost phase represents only a small fraction of the total flight of a strategic ballistic missile, the interceptor glides most of the way toward the stationary target. In this type of strategic application, precise instrumentation is necessary so that the interceptor can steer to the correct position and velocity states at the end of the boost phase. With the correct states the missile will be able to glide ballistically, without further corrective maneuvers, toward the target.

The concept of long-range strategic ballistic missiles was also born during World War II with the highly effective V-2 rocket developed by German scientists at Peenemunde. The first combat round, launched on September 6, 1944, represented the culmination of nearly 14 years of work (G. P. Kennedy, Rockets, Missiles, and Spacecraft of the National Air and Space Museum, Smithsonian Institution Press, Washington, DC, 1983). The range of the V-2 was only a few hundred miles, and a 2000-lb warhead compensated somewhat for its lack of accuracy. After World War II some of the German scientists from Peenemunde were brought to the United States, along with 80 missiles, in order to advance the V-2 technology. Guidance accuracy was improved by more than an order of magnitude by incorporating precise inertial sensors into the guidance system design. The short-range Redstone missile became the first U.S. ballistic missile with a highly accurate inertial guidance system. More precise and smaller inertial sensor hardware and improved guidance software were achieved with the U.S. medium-range Jupiter missile. An eyewitness account of this advance in strategic missile technology can be found in a work by W. Haeussermann ("Developments in the Field of Automatic Guidance and Control of Rockets," *Journal of Guidance, Control, and Dynamics*, Vol. 4, No. 3, 1981, pp. 225–239). Today's highly accurate strategic ballistic missiles travel at intercontinental ranges at near- orbital speeds.

In newer systems interceptors will have to make use of and improve on yesterday's tactical and strategic missile technology. Tomorrow's interceptors will have to fly strategic distances against moving and possibly accelerating targets whose future position is highly uncertain. In these applications it is not sufficient to apply only ballistic missile technology. Some type of homing guidance system is required after the interceptor boost phase to take out the inevitable errors. However, it is not clear whether tactical guidance law technology is appropriate for the homing phase of these strategic interceptors. Unlike tactical endoatmospheric interceptors, which maneuver by generating lift, the newer exoatmospheric interceptors require fuel for the missile divert engines in order to maneuver or divert in response to guidance commands. If all of the divert fuel is consumed, the interceptor cannot maneuver. In addition, fuel is at a premium, because, according to the rocket equation, interceptor weight grows exponentially with fuel weight. Therefore, unlike tactical endoatmospheric missiles, the newer exoatmospheric interceptors require guidance laws that minimize fuel consumption. In addition, because of practical limits on achievable divert engine thrust-to-weight ratios; the newer interceptors may no longer enjoy the tactical missile's acceleration advantage over the target. In fact, strategic exoatmospheric interceptors may be working at an *acceleration disadvantage* against a boosting target!

With this perspective, these two volumes attempt to lay the foundation for meeting today's new challenges. The principles of both tactical and strategic missile guidance are presented in a common language, notation, and perspective. The mathematics, arguments, and examples presented in the text are intended to be nonintimidating so that the designers working in the tactical world will be able to understand and appreciate the difficulty of the strategic problem and vice versa. Numerous examples are presented to illustrate all of the concepts presented in these two volumes. In addition, examples are presented showing elementary ways in which tactical and strategic guidance principles can be combined. In this way the potential for a cross-fertilization of ideas, which is necessary for today's challenges in guidance technology, is increased.

The book is written with both the expert and novice in mind. Proven methods of guidance in both the tactical and strategic world are presented from several points of view. The guidance laws are usually first derived mathematically, then explained from a heuristic point of view, and finally a numerical example, along with a MATLAB listing, is included to prove that the guidance law performs as expected. The numerical example is usually chosen to highlight the strengths and weaknesses of the guidance approach. The expert or more interested reader using the MATLAB listing may not only get a deeper understanding of the principles involved, but may also explore issues beyond the scope of the text.

It might appear from glancing at the many block diagrams scattered throughout the text that the subject matter requires a control system background. *This is not true!* The block diagrams are offered as a pictorial description of differential equations. The reader can always ignore the block diagrams and just read the differential equations. The only assumption made in the text is that the reader has either an engineering, mathematics or physics background. All of the numerical and mathematical techniques used in the text are explained in earlier chapters of the first volume. A synopsis of the chapters of each of the two volumes *Tactical and Strategic Missile Guidance: An Introduction, Seventh Edition* and *Advanced Tactical and Strategic Missile Guidance* follows.

TACTICAL AND STRATEGIC MISSILE GUIDANCE: AN INTRODUCTION, SEVENTH EDITION

Chapter 1 presents the basis of most of the numerical techniques used throughout the text. The utility of Laplace transforms in solving and representing differential equations is explained. It is then shown how differential equations can be solved using the second-order numerical Runge–Kutta integration technique. MATLAB code for a representative example is presented so that the reader can better understand how numerical integration is actually applied. The code is written so that it is easy to apply to problems involving other differential equations without changing more than a few lines of code. In fact, most of this sample code is repeated throughout the text when the differential equations representing the system under consideration change. Introductory material on state space notation and the importance of the fundamental matrix are also included in this chapter. This material will help the reader understand Kalman filtering and advanced guidance techniques which will appear later in both volumes.

Chapter 2 introduces the reader to the most important and widely used tactical missile guidance law: proportional navigation. A simplified engagement simulation is developed in order to show how proportional navigation operates and why it is an effective guidance law. Next, linearization techniques are used to get a deeper understanding of how proportional navigation works and to form an analytical foundation from which guidance theory can be advanced. Closed-form solutions for the required missile acceleration due to heading error and target maneuver disturbances are derived so that the reader can see how these important errors influence total system performance. It is then demonstrated that the derived closed-form solutions are not only analytically convenient but are in fact accurate indicators of the expected system performance. The concept of zero effort miss is introduced as a way of further understanding both proportional navigation and more advanced guidance concepts that will appear later in both volumes.

Chapter 3 introduces one of the most important methods used in analyzing tactical missile guidance systems: the method of adjoints. The rules for constructing an adjoint are presented, and the necessary mathematics required to understand why adjoints are useful are included in this chapter. A numerical example involving a missile guidance system is used both as a practical application of the theory and to ensure that the reader fully understands how to apply the adjoint technique. The numerical results of the adjoint example are compared to traditional simulation results so that the reader can appreciate both the power and utility of this very elegant technique. New closed-form solutions are derived using adjoints so that the reader can also begin to appreciate the relationship between miss distance, the guidance system time constant, and the error source.

Chapter 4 reviews all of the necessary theory so that the reader can understand how measurement noise and random phenomenon influence the performance of a missile guidance system. Basic definitions are reviewed, and numerical examples are presented showing the reader how theory is used in practice. The method of adjoints, which was introduced in Chapter 3 for deterministic systems, is extended so that a missile guidance system can be analyzed if the error sources are random. A numerical example is presented showing how the method of adjoints can be used to obtain statistical performance projections in only *one computer run*. These results are compared to multiple-run Monte Carlo projections (that is, repeated simulation trials with ensemble averaging of the resultant output data). Chapter 4 also includes examples of the miss due to various noise error sources for a single time constant guidance system and, in addition, shows how the adjoint techniques can be used to make an error budget.

Another popular computerized analytical technique, known as covariance analysis, is introduced in Chapter 5. It is shown that covariance analysis can also be used to yield exact statistical performance projections of a missile guidance system in the presence of random error sources in one computer run. The numerical requirements for the successful implementation of covariance analysis are discussed and compared to the adjoint method. Chapter 5 also shows how the adjoint technique can be extended to yield acceleration as well as miss distance information.

Chapter 6 presents most of the important properties of a proportional navigation guidance system. It is shown that an accurate dynamic model of the guidance system is crucial in obtaining accurate miss distance performance projections. Normalized design curves are developed for a high-order canonical guidance system so that an engineer can quickly estimate system performance given a minimum amount of information. The concept of an optimal target evasive maneuver is developed. It is shown how one can compute the maximum miss distance that a maneuvering target can induce. This type of information is important in determining the vulnerability of a guidance system. Finally, the influence of saturation and parasitic effects on system performance are considered. Design curves are presented showing how to size the missile to target acceleration advantage so that the interceptor will be effective. It is also shown in this chapter that parasitic effects such as radome place fundamental limits on the attainable speed and gain of the guidance system.

The missile seeker provides a noisy measurement of the line-of-sight angle. Chapter 7 demonstrates how simple digital fading memory filters can be used as part of a missile guidance system to provide an estimate of the line-of-sight angle and rate from the noisy measurement. Adjoint theory is again extended to handle systems with continuous and digital parts so that a mixed continuous-discrete missile guidance system can be analyzed efficiently. Some of the more important properties of fading memory filters are illustrated via numerical examples. This chapter also shows that, if range measurements are available, it is possible to estimate target acceleration with a fading memory filter.

Chapter 8 reviews proportional navigation concepts and sets up the mathematical foundation so that more advanced tactical guidance laws can be derived. Augmented proportional navigation is derived from some properties of the Schwartz inequality. It is then shown, via a numerical example, how guidance system time constants can degrade miss distance performance of both proportional and augmented proportional navigation. Recognizing the cause of the miss distance degradation, a new guidance law is derived. It is demonstrated that, as long as the missile guidance system dynamics is known, the new guidance law can effectively eliminate miss distance degradation due to a guidance system time constant.

Chapter 9 introduces a class of optimal digital noise filters known as Kalman filters. Using the theoretical Kalman filtering equations, the chapter develops, in detail, a digital Kalman filter that is very useful for missile guidance system applications. It is shown, in detail, how a Kalman filter can be used in conjunction with an optimal guidance law to improve system performance and to relax missile acceleration requirements. Various experiments are conducted in order to illustrate important filtering and guidance concepts. If the measurement noise and sampling time can be decreased, the estimation robustness of the Kalman filter to different types of target maneuvers is presented. In addition, Chapter 9 shows the interaction of optimal guidance with radome effects.

Chapter 10 wraps up the discussion of tactical interceptors with consideration of the missile's operational zone. Reach considerations, based on the rocket equation and drag effects, are presented. Gravitational effects, which were previously neglected, are offered as another phenomenon that further limits the zone. Numerical examples are used to illustrate important effects considered in this chapter.

In Chapter 11, the focus of the text switches to endoatmospheric ballistic targets. Closed-form solutions are derived and validated, based on the properties of ballistic targets, showing how the magnitude of endoatmospheric ballistic target deceleration varies with speed, altitude, and re-entry angle. It is demonstrated that unless advanced guidance techniques are used, the high deceleration levels of a ballistic target make it difficult to hit under many engagement conditions.

Advanced guidance techniques and fire control logic for endoatmospheric intercepts require knowledge of the target's ballistic coefficient. Using a simplified extended Kalman filter as an example, the challenges of estimating a target's ballistic coefficient are demonstrated in Chapter 12. Common filter design pitfalls and their engineering fixes are illustrated in easy-to-understand examples. Guidelines for making an extended Kalman filter robust to large initialization errors are presented. In addition, an example showing how the performance of an extended Kalman filter can be improved when Kalman filter states are changed is presented.

Chapter 13 integrates many of the text's concepts to further explain why endoatmospheric ballistic targets are challenging. Formulas are derived showing the geometry dependence of noise-induced miss distance. The dependence of the minimum achievable guidance system time constant on radome slope, geometry, and missile aerodynamic properties is shown. Numerical examples are presented in order to highlight missile performance differences for both aircraft and ballistic target threats.

Chapter 14 introduces other forms of tactical guidance. A command guidance implementation of proportional navigation is compared to a homing guidance implementation in terms of system noise propagation. Beam riding and command to line-of-sight methods of guidance are introduced and compared to proportional navigation. It is shown that, although the performance of these new methods of guidance are geometry-dependent, they can be made to work rather effectively.

Chapter 15 introduces strategic interceptor concepts from a tactical point of view. A gravitational model, based on Newton's law of universal gravitation, is developed for strategic flight. Comparisons between a flat-Earth gravitational model and a strategic gravitational model are made. Although strategic engagement simulation models are presented in the text in a Cartesian Earth-centered coordinate system, a polar coordinate system is also introduced so that important closed-form solutions can be derived. Key formulas for velocity and flight time for an impulsive ballistic missile to travel a fixed distance, given an initial flight-path angle, are developed using the polar coordinate system. A Cartesian Earth- centered simulation is used to confirm the analytical results. Finally, Chapter 15 shows that a strategic interceptor has two ways of reaching the target: the short way and the long way. Tradeoffs between the velocity required and time of travel for each possible method of travel is presented.

Chapter 16 shows how preliminary strategic booster sizing can be done with the rocket equation. Simplified booster sizing examples are presented in order to clarify the concepts. The rocket equation is extended so that the virtues of staging can be illustrated via a numerical example. Finally, the gravity turn maneuver is introduced as the simplest possible steering method a booster can employ in traveling from its launch point to a desired destination.

Starting from the closed-form solutions derived in Chapter 15, the concepts of Lambert steering, which is fundamental to booster and spacecraft steering, are developed in Chapter 17. A simple to understand but numerically inefficient way of solving Lambert's problem is derived. A numerical example is presented showing how the numerical solution to Lambert's problem can be implemented. A novel use of the secant method is demonstrated to speed up the solution to Lambert's problem by more than two orders of magnitude! It is then shown how the implemented solution can be modified with a simple feedback scheme to steer an interceptor, during its boost phase, to its intended target. Another subset of Lambert steering, known as general energy management (GEM) steering, is also derived and demonstrated. A numerical example highlighting the similarities and differences between Lambert and GEM steering is presented. Chapter 17 also shows how the solution to Lambert's problem can be used to solve Kepler's problem.

Chapter 18 shows elementary but fundamental methods of combining the tactical missile guidance concepts of Chapters 2–9 with the strategic notions of Chapters 15–17. Unifying numerical examples are used to illustrate the strengths and weaknesses of the combined approach. Previously derived closed-form solutions for the required missile acceleration to hit a target by a tactical interceptor are converted to strategic lateral divert formulas. Nonlinear strategic engagement simulation results are used to show that the divert formulas for prediction error, apparent target maneuver, and guidance law are not only useful because of their simplicity but are in fact accurate indicators of strategic interceptor requirements.

Chapter 19 presents some additional concepts that are very important to the strategic world. It is shown how compensating for known gravity effects in the guidance law can considerably reduce interceptor lateral divert requirements. Next, predictive guidance is introduced as the ultimate guidance law. It is shown that, if accurate a *priori* information exists concerning the target, then predictive guidance can be used to substantially alleviate interceptor lateral divert requirements. All of the homing guidance concepts introduced have assumed that strategic interceptor divert engines were effectively throttleable. A pulsed guidance law is developed assuming that guidance commands can only be issued with a few discrete burns. The performance of the pulsed guidance law is compared to that of proportional navigation.

When two aircraft are flying in close formation, they both appear within the pursuing interceptor's seeker beam. When the missile is close enough to the aircraft, one of the targets will drop out of the seeker beam, and it will appear to the missile as if the target has moved instantaneously. Large miss distances may result because of insufficient remaining homing time. Chapter 20 develops normalized design curves to both illustrate and quantify the multiple target problem for cases in which the seeker can and cannot move instantaneously. Rules of thumb are developed relating the necessary ratio of the time left for homing after resolution has taken place to the guidance system time constant.

To design a flight-control system the interceptor airframe equations must first be linearized. Chapter 21 shows how the nonlinear missile force and moment equations are related to the geometry of the missile airframe. A simple method for linearizing the force and moment equations is introduced so that transfer functions can be derived for the missile airframe. It is shown that the transfer function approximation to the airframe is an excellent match to reality for small angles of attack.

Chapter 22 illustrates how the flight-control system interacts with the guidance system. Both the open-loop and rate gyro flight-control systems are discussed in detail. It is shown that the open-loop flight-control system has the dynamics of the bare airframe and is therefore not usually acceptable in radar homing applications because of its low damping. It is demonstrated that the rate gyro flight-control system improves the system damping by using a sensor and the principles of feedback. Chapter 22 also introduces a brute force method that can be used as a check on the analytically derived open-loop frequency response of the rate gyro flight control system.

Chapter 23 demonstrates that by using an accelerometer and rate gyro with the flight-control system, the system damping, time constant, and open-loop crossover frequency can be controlled independently. Controlling the system damping ensures that the guidance system is not overly sensitive to radome slope effects at the high altitudes. Selecting the system time constant means that there will be adequate performance against maneuvering targets. Controlling the open-loop crossover frequency ensures that the design will be robust and not overly sensitive to unmodeled high frequency dynamics. A unifying example is presented to demonstrate how typical guidance system tradeoffs are conducted. The method of adjoints is used to show how miss distance error budgets can be generated, actuator requirements set, and missile acceleration requirements derived.

ADVANCED TACTICAL AND STRATEGIC MISSILE GUIDANCE

In the derivation of the three-loop autopilot gain algorithm of Chapter 23 of *Tactical and Strategic Missile Guidance: An Introduction, Seventh* Edition it was assumed that the airframe was stable. The missile airframe can actually be unstable during the missile boost phase since the center of gravity can be behind the center of pressure as fuel is being consumed. Chapter 1 of *Advanced Tactical and Strategic Missile Guidance* shows how the three-loop autopilot gain algorithm can be re-derived when the airframe is unstable. An example is presented showing that the flight control step responses are virtually identical to the case when the airframe is stable.

Chapter 2 demonstrates the importance of considering body bending effects for autopilot design in order to avoid potential stability problems. The use of notch filters is shown to be one way of stabilizing the flight control system in the presence of body bending. Reducing the autopilot gain and actuator bandwidth are shown to be additional ways of stabilizing the flight control system. The effectiveness of all the techniques discussed are demonstrated by many numerical examples.

In Chapter 3 a variety of numerical examples are presented showing that if a control system is designed using modern control techniques, a classical frequency response analysis must be performed in order to determine the crossover frequency and stability margins of the open-loop system. Numerous examples are presented in this chapter showing that if the crossover frequency is too high the system will go unstable – even if the phase and gain margins are excellent. Chapter 3 also demonstrates that the crossover frequency can be adjusted by adjusting the weighting terms in the performance index. An example is presented showing how a low order controller can stabilize a high order plant by adjusting the weighting terms in the performance index in order to lower the open-loop crossover frequency.

Chapter 4 presents three new applications of the method of adjoints for a mixed continuous-discrete system. The first application involves multiple samplers used in a system in which measurement data is first preprocessed at one sampling rate and then used for input to a fading memory filter operating at another sampling rate. The second application involves taking the adjoint of a three-state discrete Kalman filter in the homing loop. Finally, the third example shows shows how the blind range phenomenon can be analyzed using the method of adjoints.

Chapter 5 emphasizes that the shaping filter approach for representing random target maneuvers can be very useful when missile saturation does not occur. However, in evaluating system performance when missile saturation occurs, the use of shaping filters to represent the actual target maneuver can lead to erroneously pessimistic predictions of the rms miss distance. For miss distance studies involving missile acceleration saturation it is demonstrated that one should model the actual target maneuver so that realistic performance projections can be obtained.

It is well-known that the barrel roll or weave maneuver can cause large miss distances. Because tactical ballistic missiles can spiral or weave into resonance as they re-enter the atmosphere due to either mass or configurational asymmetries, the weave maneuver is of particular interest to the guidance system designer. Chapter 6 studies the influence of the target weave maneuver on a proportional navigation guidance system. This chapter demonstrates how the target weave frequency and amplitude, the missile guidance system time constant, effective navigation ratio, and acceleration capability all play an important role in determining system performance. Chapter 6 demonstrates that, in general, speeding up a missile guidance system and increasing the missile-to-target acceleration advantage will help reduce the miss distance due to a weaving target. It is also shown how special guidance laws that require more information than proportional navigation can be used to improve system performance.

Chapter 6 has shown that advanced guidance laws can be used to improve system performance against spiraling targets. Chapter 7 explores the various filtering options that can be used to estimate the weaving target states required for advanced guidance techniques. After a brief review of existing filtering techniques for the weaving target problem, it is assumed that the target weave frequency is known and an optimal linear four-state weave Kalman filter, which estimates both target acceleration and jerk, is both derived and evaluated. It is then assumed that the target weave frequency is not known but must be estimated along with target acceleration and jerk using an extended five-state Kalman filter. The chapter compares the various candidate Kalman filter options and appropriate guidance laws in terms of both performance and robustness.

Chapter 7 has shown that the states required for advanced guidance laws against weaving targets can be obtained from a four-state linear Kalman filter if the weave frequency of the target is known. It was shown that the combination of such a Kalman filter and advanced guidance law could improve the rms miss distance. However, Chapter 7 also showed that if the target weave frequency is unknown it can be estimated by using a five-state extended Kalman filter. However this approach only works if the extended Kalman filter is initialized close to the actual target weave frequency. Chapter 8 presents a linear Kalman-filter bank approach, originally introduced in the 1960s, for accurately estimating the target weave frequency. As with other material in the text, this chapter uses a step-by-step approach in order to explain how the filter bank approach works, and when used in conjunction with advanced guidance laws, can significantly improve performnce when the weave frequency of the target is unknown.

For completeness, an alternative guidance approach used by the space community is developed and compared to the more traditional guidance approaches in Chapter 9. It is first shown how the predictor-corrector method can be applied to a problem in which other more conventional guidance laws are also applicable. The various guidance approaches are compared in terms of performance and robustness. A problem is also selected in which the direct application of a proportional-navigation-type guidance law is not possible. The strengths and weaknesses of predictor-corrector guidance approach are demonstrated for an unusual example.

Chapter 10 introduces a new guidance law known as jerk guidance. It is shown that for the constant target maneuver and heading error cases the new guidance law minimizes the integral of the missile jerk squared. It is also shown that for a very stressing Vertical-S target maneuver the jerk guidance law could yield smaller miss distances than proportional navigation implemented as a missile jerk command when the missile jerk was limited.

Chapter 11 shows that Kalman filters based on either the Poisson or Singer representation of random target maneuvers are identical. When a normal three-state Kalman filter, based on a random step target maneuver, is compared to a Kalman filter based on based on the Poisson target maneuver model—the Poisson Kalman filter yields much smaller rms miss distances. However, it is also demonstrated that if the process noise of the normal Kalman filter is increased, it can yield comparable rms miss distances to the Poisson Kalman filter—except in the case when the actual target maneuver is a Poisson target maneuver.

So far, all of the guidance laws presented in the text were derived using an analytical technique based on the Schwartz inequality. Chapter 12 presents two alternative ways of developing guidance laws numerically. These techniques can either be used to check existing guidance laws or to derive more advanced guidance laws when the Schwartz inequality technique becomes either too cumbersome or impossible to use. An example is presented showing the advantages of a new advanced guidance law, based on a more realistic model of the flight control system, can improve performance at very high altitude.

In ChapterIIn Chapter 13 the trajectory shaping guidance law is derived using the Schwartz inequality. It is demonstrated that with this new guidance law we could not only hit the target but can also control the final line-of-sight angle. The price paid for trajectory shaping is that more acceleration is required to hit the target. Formulas are also derived in this chapter that can be used to predict the missile acceleration requirements for the trajectory shaping guidance law under a variety of circumstances. It is demonstrated that these formulas are also an accurate indicator of performance in the nonlinear world.

Chapter 14 deals with guidance laws **that do not require time-to-go information** for hitting stationary targets. The first guidance law is biased proportional navigation which both hits the target and controls the final impact angle. Biased proportional navigation compares favorably to the optimal trajectory shaping guidance law derived using the Schwartz inequality (in Chapter 13) in terms of the acceleration requirements. A polynomial guidance law, develped for constant missile velocity, is also presented in Chapter 14. The polynomial guidance law enables a missile to hit the target at a desired time and, in addition, also drives the seeker look angle to zero. This new guidance law leads to some unusual trajectories and does not require excessive missile acceleration.

Chapter 15 derives a two-term Fourier series approximation to represent the Vertical-S target maneuver so that a six-state Kalman filter can be developed. It is shown that the six-state Kalman filter out performes either a three-state Kalman filter (derived assuming a random constant target maneuver) or a four-state weave Kalman filter against the Vertical-S target maneuver.

Chapter 16 introduces a new closed-form guidance law that also has the potential of improving the performance of a tail-controlled missile operating at very high altitude. It is shown that the performance of the new closed-form guidance law is better than optimal guidance for the example chosen. However, the computerized guidance law of Chapter 12, which was specifically derived for the complex model of the flight control system transfer function, still offers the most benefits for a tail-controlled missile operating at very high altitudes.

Chapter 17 shows how three-dimensional simulations can be used to convey important information. This chapter provides new insights against spiraling targets which can only be gained in three dimensions. For example, the miss caused by a weaving target does not oscillate in three dimensions as it does in one or two dimensions but in fact approaches steady state as the flight time increases. Code has been provided in Chapter 17 to extend the Lambert routine to three dimensions in our strategic engagement simulations. New code for a Kepler subroutine has also been provided so that we can do intercept point prediction for three-dimensional ballistic targets without having to resort to numerical integration.

Formulas are developed in Chapter 18 for the minimum possible rms miss distance for the case in where there is white glint noise, a uniformly distributed target maneuver, and infinite missile acceleration capability. As these error sources are up to the target rather than the missile designer, they place a lower limit on the best achievable performance of a radar homing missile that has infinite acceleration capability. Finite missile acceleration capability plus other sources of error will only serve to increase the miss distance.

Chapter 19 introduces the differential game guidance law with bounded controls and demonstrates its performance improvements over conventional guidance under extreme circumstances in which the missile-to-target acceleration advantage is very low. The chapter also demonstrates how the chattering caused by the bang-bang nature of differential game guidance with bounded controls can be dramatically reduced without significant loss of performance.

Chapter 20 examines two filtering options for a boost phase intercept problem. The two filters compared are a template-based Kalman filter that has perfect a *priori* information and a linear three-state polynomial Kalman filter that does not require such information. The filters are compared under ideal conditions and then are compared in a more realistic environment.

Chapter 21 introduces techniques for graphically presenting strategic information on successful intercepts of an impulsive ballistic target being pursued by an impulsive interceptor. The three techniques discussed are the operational area method, the launch area denied method, and the defended area method. Examples are presented illustrating the three methods of presentation. Cases are examined showing how the size of the successful engagement area is influenced by both the interceptor velocity and the target trajectory type.

Finally, Chapter 22 addresses some of the guidance and control issues involved in enabling an air-launched interceptor carrying a highly maneuverable kinetic kill vehicle to perform an exoatmospheric intercept of a boosting threat missile capable of traveling many thousands of kilometers. The chapter takes the reader through part of the first iteration of the multi-iteration design process in order to show how much divert and acceleration are required by the kinetic kill vehicle to hit the target. Simplified examples are presented to indicate how conventional guidance and filtering techniques can be used as a starting point in the iterative design process for this important problem in missile defense.

These two volumes attempt to present many of the important guidance principles involved in enabling an interceptor to hit its intended target. The utilities of these principles are explained and demonstrated with pictures, equations, and computer code. However, missing from these volumes is the intensity, challenge of the unknown, and plain hard work that it takes in going from a paper design to something that not only flies but also meets the system objectives. In light of the new challenges facing today's guidance engineer, it is very appropriate to quote from the work by M. W. Fossier:

> On joining Raytheon as a young engineer, I found an intensity of spirit that I had never experienced before. I felt myself carried along in what seemed almost a crusade, sharing a burning commitment to succeed against an immense challenge.
>
> As a result of this shared feeling, each triumph led to a broad-based feeling of great elation. I still vividly recall the first flight test by Lark against a low-flying drone aircraft in 1951 over the ocean at Point Mugu. The intercept was a relatively short range and was the first to be in full view of the handful of observers permitted on the beach. When the missile homed unerringly to a spectacular direct hit on the drone, the human explosion matched the one in the air. The formal celebration ended in the wee hours of the morning, but the emotional wave lasted for months.
>
> On the other hand, I can still feel the despair that resulted when a technician inadvertently connected $B+$ (250V) to the filament string, blowing out every tube in an early Sparrow being prepared for flight test. At that time, the best flight test engineers we had took about a month of 12-hour days to check out a missile. The resulting delay was felt at a personal level in every corner of the organization.
>
> For years I attributed this intensity to the relative youth of most of the participants. However, I found that age has served mainly to mute their exuberance, but not their intensity, and I was forced to seek another source. My current view is that the intensity was (and is) a result of the intellectual challenge of the unknown. It is the characteristic of the engineering profession that there are always new problems waiting in the wings to replace the old ones as they are solved. The constant element is the challenge itself, which demands that we do our utmost and rewards us only when we do.

CHAPTER 1

Unstable Airframe and the Three-Loop Autopilot

INTRODUCTION

In Chapter 23 of the previous volume, the three-loop autopilot gain algorithm was derived and tested in both the time and frequency domains. The basic assumption in the derivation of the three-loop autopilot gain algorithm was that the missile airframe was stable; however, in reality, for part of a missile's flight the airframe of a tail-controlled missile can be unstable. For example, during the boost phase of flight, the center of gravity (CG) can be to the left (towards tail) of the center of pressure (CP), which makes the airframe unstable. As the fuel is consumed, the center of gravity moves to the right of the center of pressure (towards the nose), which stabilizes the airframe as shown in Fig. 1.1. We may want to start guidance during the boost phase of flight, so it is critical that the flight control system (autopilot plus airframe) be stable. In this chapter we shall rederive the three-loop autopilot gain algorithm, assuming that the missile airframe might be unstable.

REVIEW AND DEMONSTRATION OF AIRFRAME INSTABILITY

A typical tail-controlled, aerodynamic, winged missile, as shown in Fig. 1.1, generates lift by moving control surfaces. In this endoatmospheric missile, the movable control surface or tail can be deflected about the hinge line through a fin angle δ in order to help the missile develop an angle of attack α.

The fixed surface or wing plus the missile body help the missile develop additional acceleration. The normal force acts through the center of pressure (CP). We can express the normal force equation as

$$F_N = QS_{ref}C_N$$

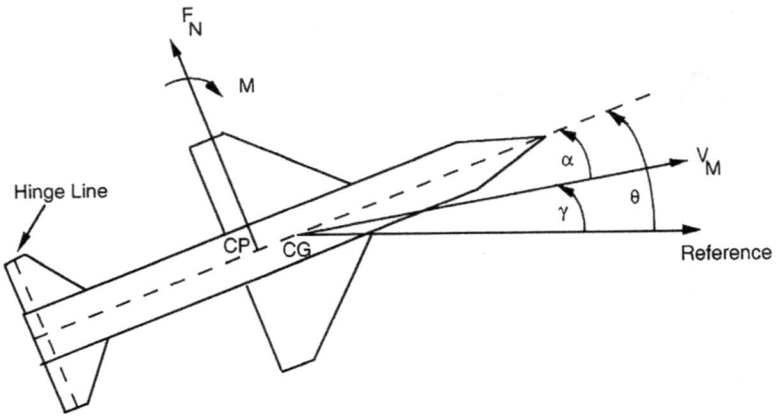

Fig. 1.1   Tail-controlled missile with stable airframe.

where C_N is the normal force coefficient, Q is the dynamic pressure, and S_{ref} is the reference area. The dynamic pressure and reference area are given by

$$Q = 0.5\rho V_M^2$$

$$S_{ref} = \frac{\pi d^2}{4}$$

where ρ is the air density in units of slug/ft^3, V_M is the missile velocity in units of ft/s, and d is the missile diameter in units of ft.

The total force acting on the missile body consists of component forces on the body, wing tail, and nose, as shown in Fig. 1.2. In this simplified diagram, all interference effects are neglected, and the total force is simply the sum of the individual forces [1]. Each of the component forces acts through its own center of pressure. The centers of pressure for the body, wing, and nose are denoted X_{CPB}, X_{CPW}, and X_{CPN}, respectively. The force acting on the tail acts through the hinge line X_{HL}.

It was shown in Chapter 21 of the previous volume that, neglecting interference effects, the normal force coefficient can be approximated as

$$C_N = \underbrace{2\alpha}_{\text{Nose}} + \underbrace{\frac{1.5 S_{PLAN}\alpha^2}{S_{ref}}}_{\text{Body}} + \underbrace{\frac{8 S_W \alpha}{\beta S_{ref}}}_{\text{Wing}} + \underbrace{\frac{8 S_T(\alpha + \delta)}{\beta S_{ref}}}_{\text{Tail}}$$

where α is the angle of attack, δ is the control surface deflection, and S_W, S_T, and S_{PLAN} are per panel wing, tail, and planform areas, respectively. Because the wing and tail are approximated by trapezoids in Fig. 1.2, their panel areas are given by

$$S_W = 0.5h_W(C_{TW} + C_{RW})$$
$$S_T = 0.5h_T(C_{TT} + C_{RT})$$

where the subscript T denotes a tip chord and the subscript R denotes a root chord. For a cylindrical missile body with a parabolic nose (radome), the planform area can be approximated as

$$S_{\text{PLAN}} = (L - L')d + 0.67L'd \approx Ld$$

where L is the missile length and L' is the radome length. The parameter β in the normal force coefficient equation is a normalized speed; for supersonic travel it is

$$\beta = \sqrt{\text{Mach}^2 - 1}$$

The missile Mach number is simply the missile speed divided by the speed of sound. Although the speed of sound is altitude dependent, we shall assume for simplicity that the speed of sound is always 1000 ft/s. Multiplying the force by

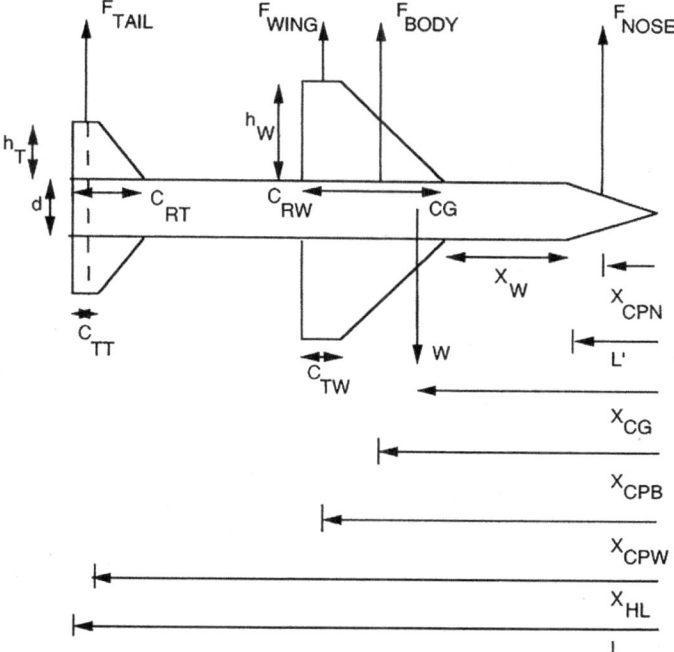

Fig. 1.2   Forces on a tail-controlled missile.

its moment arm yields the developed moment, or

$$M = F_N * \text{Moment Arm}$$

Therefore, the moment coefficient can be approximated as

$$C_M = 2\alpha \frac{(X_{CG} - X_{CPN})}{d} + \frac{1.5 S_{PLAN} \alpha^2}{S_{ref}} \frac{(X_{CG} - X_{CPB})}{d}$$

$$+ \frac{8 S_W \alpha}{\beta S_{ref}} \frac{(X_{CG} - X_{CPW})}{d} + \frac{8 S_T (\alpha + \delta)}{\beta S_{ref}} \frac{(X_{CG} - X_{HL})}{d}$$

where X_{CG} is the distance from the nose to the missile center of gravity and X_{CPN}, X_{CPB}, and X_{CPW} are the distances from the nose to the centers of pressure for the nose, body, and wing, respectively. The preceding expression assumes a tail-controlled missile, and X_{HL} is the distance from the nose to the missile hinge line. The nose, body, and wing centers of pressure (referenced with respect to the nose) can be analytically approximated because of their geometrical shape and can be shown to be

$$X_{CPN} = 0.67 L'$$

$$X_{CPW} = L' + X_W + 0.7 C_{RW} - 0.2 C_{TW}$$

$$X_{CPB} = \frac{0.67 A_N L' + A_B [L' + 0.5 (L - L')]}{A_N + A_B}$$

where X_W is the distance from the wing to the radome tangency point, as shown in Fig. 1.2. The nose and body areas are given by

$$A_N = 0.67 L' d$$

$$A_B = (L - L') d$$

Again for simplicity, at the end of the boost phase of flight, we will assume that the missile center of gravity is approximately in the center of the missile, or

$$X_{CG} = 0.5 L$$

The total moment M on the missile can be expressed in terms of the moment coefficient according to

$$M = Q S_{ref} d C_M$$

We now have enough information to express the normal and angular accelerations acting on the missile in terms of the geometry of the missile configuration. The acceleration normal to the missile body can be expressed in terms of the normal force according to

$$n_B = \frac{F_N g}{W} = \frac{g Q S_{ref} C_M}{W}$$

where W is the missile weight. The angular acceleration acting on the missile can be expressed in terms of the moment according to

$$\ddot{\theta} = \frac{M}{I_{yy}} = \frac{QS_{ref}dC_M}{I_{yy}}$$

where I_{yy} is the missile moment of inertia. If the missile body is approximated as a cylinder, the formula for the moment of inertia is given by [2]

$$I_{yy} = \frac{W[3(0.5d)^2 + L^2]}{12g} \approx \frac{WL^2}{12g}$$

Finally, from Fig. 1.1 we can see that the angle of attack can be expressed in terms of the missile body and flight-path angles according to

$$\alpha = \theta - \gamma$$

Taking derivatives of both sides of the equation and recognizing that the flight-path rate can also be expressed in terms of the missile acceleration yields

$$\dot{\alpha} = \dot{\theta} - \dot{\gamma} = \dot{\theta} - (n_L/V_M)$$

If we assume that the angle of attack is small, the missile acceleration perpendicular to the velocity n_L is approximately the same as the missile acceleration perpendicular to the body n_B. Therefore, the derivative of the angle of attack can be expressed as

$$\dot{\alpha} = \dot{\theta} - \dot{\gamma} = \dot{\theta} - (n_B/V_M)$$

In this chapter, as was done in the previous volume, we shall assume that the acceleration perpendicular to the body is approximately the same as the missile acceleration perpendicular to the velocity vector.

SIMULATION TO DEMONSTRATE AIRFRAME CENTER OF GRAVITY LOCATION CAN CAUSE INSTABILITY

We now have enough information to simulate the force and moment equations and thus find out how the missile airframe responds when the tail is deflected. Consider the hypothetical 1000-lb tail-controlled missile, shown in Fig. 1.3, similar to the one first considered by Jerger [3] and the one that was used in the previous volume. In this example both the wing and tail are triangular in shape. The locations of the wing, hinge line, and center of gravity along with all the other airframe dimensions are indicated in Fig. 1.3.

A simulation was written utilizing the preceding nonlinear force and moment equations for the hypothetical missile of Fig. 1.3. The angle of attack rate and missile angular acceleration differential equations, which involve the force and

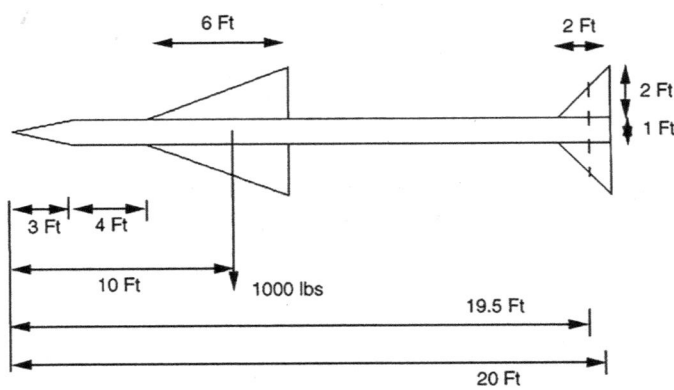

Fig. 1.3   Hypothetical missile.

moment equations, are integrated and the simulation appears in Listing 1.1. We can see from the inputs at the beginning of the simulation that the airframe inputs and data of Fig. 1.3 are consistent. The simulation assumes that the speed of sound is always equal to 1000 ft/s. As was mentioned in the previous volume, this is an approximation because the speed of sound is altitude dependent and can be as much as 10% different than the number used in the simulation. We can also see that we are using the exponential approximation to the atmosphere that was first introduced in Chapter 10 of the previous volume. The airframe differential equations for angle of attack rate and angular body acceleration appear before the FLAG = 1 statement.

LISTING 1.1   AIRFRAME SIMULATION

```
clear
n=0;
VM=3000.;
DEL=5./57.3;
ALT=0.;
A=1000.;
DIAM=1.;
FR=3.;
XL=20.;
CTW=0.;
CRW=6.;
HW=2.;
CTT=0.;
CRT=2.;
HT=2.;
```

```
XN=4.;
XCG=10.;
XHL=19.5;
WGT=1000.;
if ALT<=30000.
    RHO=.002378*exp(-ALT/30000.);
else
    RHO=.0034*exp(-ALT/22000.);
end
SWING=.5*HW*(CTW+CRW);
STAIL=.5*HT*(CTT+CRT);
SREF=3.1416*DIAM*DIAM/4.;
XLP=FR*DIAM;
SPLAN=(XL-XLP)*DIAM+1.33*XLP*DIAM/2.;
XCPN=2*XLP/3;
AN=.67*XLP*DIAM;
AB=(XL-XLP)*DIAM;
XCPB=(.67*AN*XLP+AB*(XLP+.5*(XL-XLP)))/(AN+AB);
XCPW=XLP+XN+.7*CRW-.2*CTW;
XMACH=VM/A;
XIYY=WGT*(3*((DIAM/2)^2)+XL*XL)/(12*32.2);
TMP1=(XCG-XCPW)/DIAM;
TMP2=(XCG-XHL)/DIAM;
TMP3=(XCG-XCPB)/DIAM;
TMP4=(XCG-XCPN)/DIAM;
B=sqrt(XMACH^2-1);
Q=.5*RHO*VM*VM;
THD=0;
ALF=0;
T=0;
H=.0025;
S=0.;
while T<1.99999
    THDOLD=THD;
    ALFOLD=ALF;
    STEP=1;
    FLAG=0;
    while STEP<=1
      if FLAG==1
          STEP=2;
          THD=THD+H*THDD;
          ALF=ALF+H*ALFD;
          T=T+H;
      end
      CN=2*ALF+1.5*SPLAN*ALF*ALF/SREF+8*SWING*ALF/(B*SREF)+...
                 8*STAIL*(ALF+DEL)/(B*SREF);
```

```
        CM=2*ALF*TMP4+1.5*SPLAN*ALF*ALF*TMP3/SREF+...
            8*SWING*ALF*TMP1/(B*SREF)...
                +8*STAIL*(ALF+DEL)*TMP2/(B*SREF);
        THDD=Q*SREF*DIAM*CM/XIYY;
        XNL=32.2*Q*SREF*CN/WGT;
        ALFD=THD−XNL/VM;
                FLAG=1;
    end
    FLAG=0;
    THD=.5*(THDOLD+THD+H*THDD);
    ALF=.5*(ALFOLD+ALF+H*ALFD);
    S=S+H;
    if S>=.0099999
        S=0.;
        n=n+1;
        ArrayT(n)=T;
        ArrayXNLG(n)=XNL/32.2;
        ArrayALFDEG(n)=ALF*57.3;
    end
end
figure
plot(ArrayT,ArrayXNLG),grid
xlabel('Time (Sec)')
ylabel('Missile Acceleration (G)')
figure
plot(ArrayT,ArrayALFDEG),grid
xlabel('Time (Sec)')
ylabel('Angle of Attack (Deg)')
clc
output=[ArrayT',ArrayXNLG',ArrayALFDEG'];
save datfil.txt output − ascii
disp 'simulation finished'
```

In this simulation the center of gravity is nominally to the right of the center of pressure (i.e., right is towards the nose in this example), so the airframe is stable. A case was run for the hypothetical missile of Fig. 1.3 in which the missile fin was deflected 5 deg when the missile was at sea level traveling at 3000 ft/s. We can see from Fig. 1.4 that a 5-deg fin deflection in the positive direction causes the missile to accelerate. Figure 1.4 shows that the steady-state acceleration due to a 5-deg fin deflection is approximately 13 g at this flight condition. We can see that the achieved acceleration due to a fixed fin deflection is oscillatory but stable when the center of gravity is 10 ft from the nose. However, if we move the center of gravity towards the tail (i.e., XCG gets larger), we can see from Fig. 1.5 that the missile airframe eventually goes unstable.

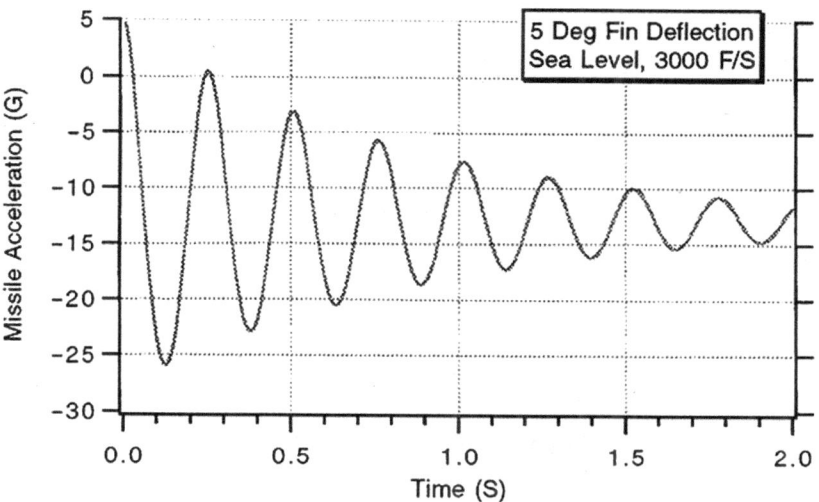

Fig. 1.4   A 5-deg fin deflection results in an oscillatory but stable response when the center of gravity is 10 ft from the nose.

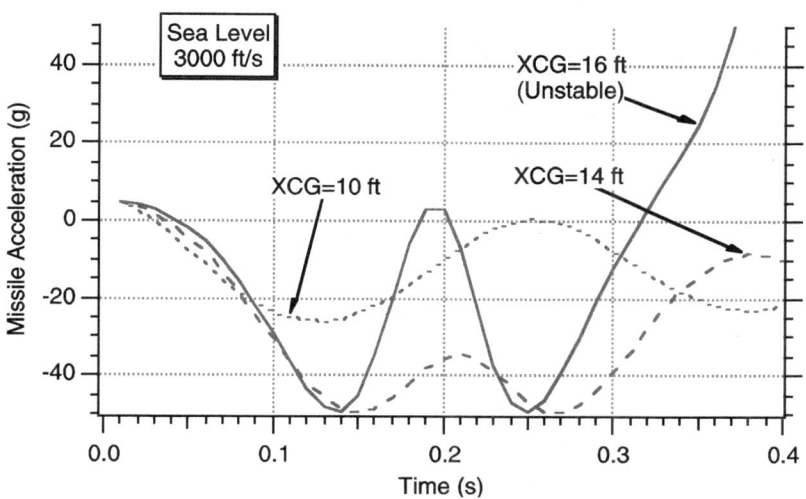

Fig. 1.5   Moving the center of gravity towards the tail (increasing XCG) eventually results in an unstable airframe.

REEXAMINING THE PREVIOUS LINEARIZATION

In Chapter 21 of the previous volume we were able to linearize the nonlinear force and moment equations. In this section we will repeat the linearization up to a point because the previous linearization did not take into account that the airframe might be unstable. An examination of the force and moment coefficients reveals that if we assume that the missile speed and altitude are constant, the equations are mostly linear except for the angle of attack squared term in each of the equations. For the constant speed, constant altitude flight condition we can linearize by assuming that each equation is linear in angle of attack and fin deflection. This means that the normal force coefficient is approximated as

$$C_N = f(\alpha,\delta) \approx C_{N\alpha}\alpha + C_{N\delta}\delta$$

One method of finding $C_{N\alpha}$ and $C_{N\delta}$ is to simply divide the angle of attack terms in C_N by α to get $C_{N\alpha}$ and then divide the fin deflection term by δ to get $C_{N\delta}$, yielding

$$C_{N\alpha} = 2 + \frac{1.5S_{\text{PLAN}}\alpha}{S_{\text{ref}}} + \frac{8S_W}{\beta S_{\text{ref}}} + \frac{8S_T}{\beta S_{\text{ref}}}$$

$$C_{N\delta} = \frac{8S_T}{\beta S_{\text{ref}}}$$

Note that $C_{N\alpha}$ depends on the angle of attack.

Because we are assuming that the acceleration normal to the body is nearly equal to the missile acceleration perpendicular to the velocity vector (the angle of attack is small), we can express the missile turning rate in terms of $C_{N\alpha}$ and $C_{N\delta}$, or

$$\dot{\gamma} \approx \frac{n_L}{V_M} = \frac{gQS_{\text{ref}}}{WV_M}[C_{N\alpha}\alpha + C_{N\delta}\delta] = -Z_\alpha\alpha - Z_\delta\delta$$

where Z_α and Z_δ are defined as

$$Z_\alpha = \frac{-gQS_{\text{ref}}C_{N\alpha}}{WV_M}$$

$$Z_\delta = \frac{-gQS_{\text{ref}}C_{N\delta}}{WV_M}$$

Therefore, we have expressed the missile turning rate or acceleration as a linear function of angle of attack or fin deflection. In a similar way the moment coefficient can be linearized as

$$C_M = f(\alpha,\delta) \approx C_{M\alpha}\alpha + C_{M\delta}\delta$$

As before, we can find $C_{M\alpha}$ and $C_{M\delta}$ by simply first dividing the angle of attack terms of C_M by α and then dividing the fin deflection term by δ, yielding

$$C_{M\alpha} = \frac{2(X_{CG} - X_{CPN})}{d} + \frac{1.5S_{PLAN}\alpha}{S_{ref}}\frac{(X_{CG} - X_{CPB})}{d}$$

$$+ \frac{8S_W}{\beta S_{ref}}\frac{(X_{CG} - X_{CPW})}{d} + \frac{8S_T}{\beta S_{ref}}\frac{(X_{CG} - X_{HL})}{d}$$

$$C_{M\delta} = \frac{8S_T}{\beta S_{ref}}\frac{(X_{CG} - X_{HL})}{d}$$

Note that $C_{M\alpha}$ is not a constant for a given speed and altitude but depends on the angle of attack.

We can now express the linearized missile angular acceleration as

$$\ddot{\theta} = \frac{M}{I_{yy}} = \frac{QS_{ref}d}{I_{yy}}[C_{M\alpha}\alpha + C_{M\delta}\delta] = M_\alpha\alpha + M_\alpha\delta$$

where M_α and M_δ are defined as

$$M_\alpha = \frac{QS_{ref}dC_{M\alpha}}{I_{yy}}$$

$$M_\delta = \frac{QS_{ref}dC_{M\delta}}{I_{yy}}$$

At this point it is important to note that M_α is negative for a stable airframe. Because the derivative of the angle of attack is given by

$$\dot{\alpha} = \dot{\theta} - \dot{y}$$

we can say that

$$\dot{\alpha} = \dot{\theta} + Z_\alpha\alpha + Z_\delta\delta$$

As mentioned previously, M_α and Z_α are not constants in our linearized model but vary with angle of attack. These aerodynamic parameters are usually evaluated at a trim angle of attack. The vehicle is considered to be at trim when the moment is zero ($C_M = 0$). At the trim condition one solves for the angle of attack and uses that value to evaluate M_α and Z_α.

The linearized airframe equations can also be represented in block diagram form, as shown in Fig. 1.6 The two integrators shown in the block diagram indicate that the airframe can be considered to be a second-order system. It is important to note that this diagram assumes that the input fin deflection δ is in units of degrees and that the output acceleration n_L is in units of g. All internal angles and rates are in units of either degrees or degrees per second.

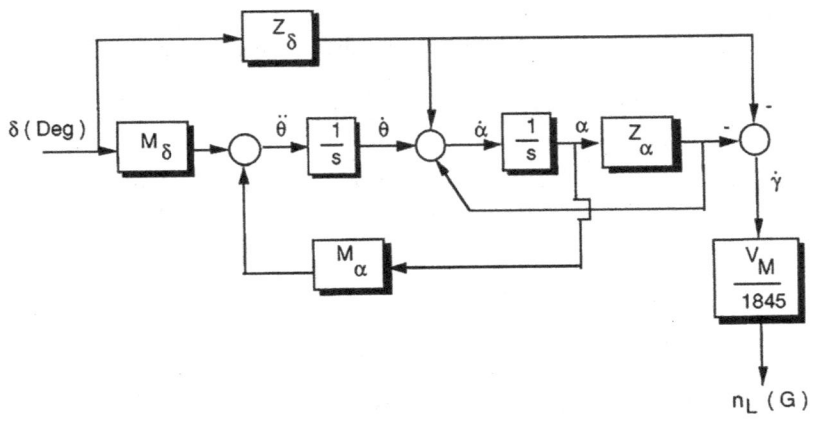

Fig. 1.6   Linearized airframe.

Often it is convenient to have a transfer function representation of the airframe. Strictly speaking, the transfer function is only valid when the missile is at a fixed speed, altitude, and trim angle of attack. After some algebra we can find the transfer function relating the achieved missile acceleration to the fin deflection from Fig. 1.6 as

$$\frac{n_L}{\delta} = \frac{-V_M[M_\alpha Z_\delta - Z_\alpha M_\delta]}{1845 M_\alpha} \left[1 - \frac{Z_\delta s^2}{M_\alpha Z_\delta - Z_\alpha M_\delta}\right] \bigg/ \left(1 + \frac{Z_\alpha}{M_\alpha}s - \frac{s^2}{M_\alpha}\right)$$

From the previous linearization of Chapter 21 of the previous volume, the preceding transfer function can be simplified to

$$\frac{n_L}{\delta} = K_1\left(1 - \frac{s^2}{\omega_Z^2}\right) \bigg/ \left(1 + \frac{2\zeta_{AF}}{\omega_{AF}}s + \frac{s^2}{\omega_{AF}^2}\right)$$

where

$$K_1 = \frac{-V_M[M_\alpha Z_\delta - Z_\alpha M_\delta]}{1845 M_\alpha}$$

and

$$\omega_Z = \frac{M_\alpha Z_\delta - Z_\alpha M_\delta}{Z_\delta}$$

$$\omega_{AF} = \sqrt{-M_\alpha}$$

$$\zeta_{AF} = \frac{Z_\alpha \omega_{AF}}{2 M_\alpha}$$

The problem with the preceding linearization is the definition of the airframe frequency ω_{AF}, *which assumes* M_α *is negative (stable airframe) and is not defined for a positive* M_α. The linearization can easily be corrected by using the more general airframe transfer function based on the definitions of K_1 and ω_z

$$\frac{n_L}{\delta} = \frac{K_1\left(1 - \dfrac{s^2}{\omega_z^2}\right)}{1 + \dfrac{Z_\alpha}{M_\alpha}s + \dfrac{s^2}{M_\alpha}}$$

The preceding transfer function is valid for both negative and positive M_α.

Similarly, the transfer function from missile pitch rate to fin deflection can also be written from Fig. 1.6 as

$$\frac{\dot{\theta}}{\delta} = \frac{-[M_\alpha Z_\delta - Z_\alpha M_\delta]}{M_\alpha}\left[1 + \frac{M_\delta s}{M_\alpha Z_\delta - Z_\alpha M_\delta}\right] \bigg/ \left(1 + \frac{Z_\alpha}{M_\alpha}s - \frac{s^2}{M_\alpha}\right)$$

which in Chapter 21 of the previous volume simplifies to

$$\frac{\dot{\theta}}{\delta} = K_3(1 + T_\alpha s) \bigg/ \left(1 + \frac{2\zeta_{AF}}{\omega_{AF}}s + \frac{s^2}{\omega_{AF}^2}\right)$$

where

$$K_3 = \frac{-[M_\alpha Z_\delta - Z_\alpha M_\delta]}{M_\alpha} = \frac{1845K_1}{V_M}$$

$$T_\alpha = \frac{M_\delta}{M_\alpha Z_\delta - Z_\alpha M_\delta}$$

Again, the problem with this linearization is the definition of ω_{AF}, which assumes M_α is negative (stable airframe). The linearization can easily be corrected by using the more general airframe transfer function

$$\frac{\dot{\theta}}{\delta} = \frac{K_3(1 + T_\alpha s)}{1 + \dfrac{Z_\alpha}{M_\alpha}s + \dfrac{s^2}{M_\alpha}}$$

If an accelerometer is used in a flight-control system to measure the achieved missile acceleration, it will probably not be located at the center of gravity. Its measurement of the actual acceleration will be corrupted by the body angular acceleration according to

$$n_A = n_L + \frac{(X_{CG} - X_{ACC})\ddot{\theta}}{1845}$$

where X_{ACC} is the accelerometer location with respect to the nose, X_{CG} is the center of gravity of the missile, and n_A is acceleration measured by the accelerometer. However, we shall neglect this effect and assume that the measured and achieved accelerations are identical in order to simplify the ensuing analysis.

MODIFYING THE THREE-LOOP AUTOPILOT GAIN ALGORITHM FOR AN UNSTABLE AIRFRAME

We shall now express the more general airframe transfer functions as

$$\frac{n_L}{\delta} = \frac{K_1\left(1 - \dfrac{s^2}{\omega_z^2}\right)}{1 + \dfrac{Z_\alpha}{M_\alpha}s + \dfrac{s^2}{M_\alpha}} = \frac{K_1\left(1 - \dfrac{s^2}{\omega_z^2}\right)}{1 + b_{11}s + b_{12}s^2}$$

$$\frac{\dot{\theta}}{\delta} = \frac{K_3(1 + T_\alpha s)}{1 + \dfrac{Z_\alpha}{M_\alpha}s + \dfrac{s^2}{M_\alpha}} = \frac{K_3(1 + T_\alpha s)}{1 + b_{11}s + b_{12}s^2}$$

Therefore, we can see that

$$b_{11} = \frac{Z_\alpha}{M_\alpha}$$

$$b_{12} = \frac{1}{M_\alpha}$$

Now we have enough information to rederive the three-loop autopilot gain algorithm of Chapter 23 of Volume 1.

The flight-control system with the three-loop autopilot appears in Fig. 1.7. In this system the rate gyro feeds body rate information into the autopilot while the accelerometer feeds back achieved acceleration information. For simplicity, it has been assumed that the accelerometer location is at the missile center of gravity so that the acceleration sensed is the true acceleration. The three autopilot gains K_A, ω_I, and K_R must be chosen to satisfy some designer-chosen criteria, and the gain K_{DC} is computed from the other gains so that the achieved acceleration will match the commanded acceleration.

As was discussed in Chapter 23 of the previous volume, a particularly useful methodology in gain selection is to choose the open-loop crossover frequency so that many stability problems can be avoided. In addition, the dominant flight-control system time constant can be selected so that rapid speed of response can be achieved in order to hit maneuvering targets. Finally, adequate damping can also be chosen by the designer to alleviate potential radome coupling problems [4].

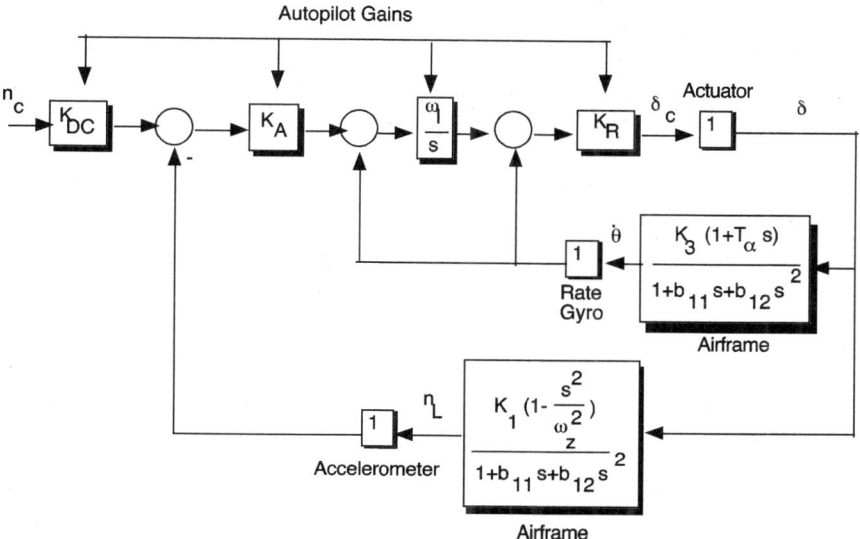

Fig. 1.7   Flight-control system with three-loop autopilot.

OPEN-LOOP ANALYSIS

Because the open-loop crossover frequency has no meaning in the time domain, we must first shift to the frequency domain to see how the autopilot gains influence the crossover frequency. Figure 1.8 shows the three-loop autopilot with the loop broken right before the actuator, as was done in Chapter 23 of the previous volume.

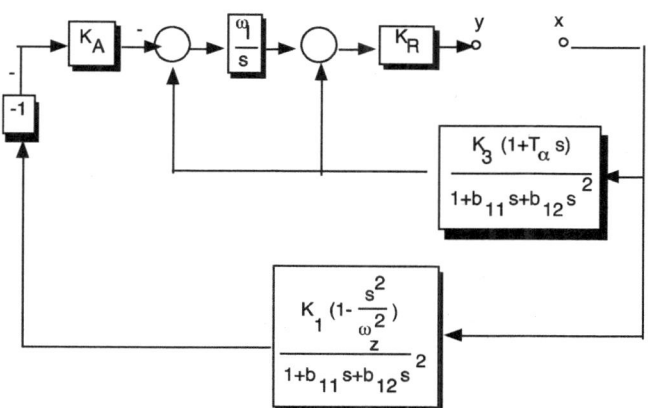

Fig. 1.8   Open-loop representation of three-loop autopilot.

By inspection of Fig. 1.8, we can write an expression for the open-loop transfer function $HG(s)$ as

$$HG(s) = -\frac{y}{x} = -K_R \left[G_3 + \frac{G_3 \omega_I}{s} + \frac{G_1 K_A \omega_I}{s} \right]$$

where G_1 and G_3 are shorthand notation for the airframe transfer functions and are given by

$$G_1 = \frac{n_L}{\delta} = \frac{K_1 \left(1 - \frac{s^2}{\omega_z^2} \right)}{1 + b_{11}s + b_{12}s^2}$$

$$G_2 = \frac{\dot{\theta}}{\delta} = \frac{K_3(1 + T_\alpha s)}{1 + b_{11}s + b_{12}s^2}$$

After much algebra, the open-loop transfer function can be rewritten as

$$HG = -K_R \omega_I K_A \left(\frac{K_3}{K_A} + K_1 \right) \bigg/ \left[s(1 + b_{11}s + b_{12}s^2) \right]$$

$$\times \left[1 + \frac{s(K_3 + \omega_I K_3 T_\alpha)}{\omega_I K_3 + K_A \omega_I K_1} + \frac{s^2 [K_3 T_\alpha - (K_A \omega_I K_1 / \omega_z^2)]}{\omega_I K_3 + K_A \omega_I K_1} \right]$$

which simplifies to

$$HG = -K_0 \left(1 + \frac{2\zeta_0}{\omega_0}s + \frac{s^2}{\omega_0^2} \right) \bigg/ \left[s\left(1 + b_{11}s + b_{12}s^2 \right) \right]$$

where the gain K_0 is given by

$$K_0 = K_R \omega_I K_A \left(\frac{K_3}{K_A} + K_1 \right)$$

and the numerator coefficients can be expressed in terms of the autopilot gains and aerodynamic parameters according to

$$\frac{2\zeta_0}{\omega_0} = \frac{(K_3 + \omega_I K_3 T_\alpha)}{\omega_I K_3 + K_A \omega_I K_1}$$

$$\frac{1}{\omega_0^2} = \frac{[K_3 T_\alpha - (K_A \omega_I K_1 / \omega_z^2)]}{\omega_I K_3 + K_A \omega_I K_1}$$

If we define

$$K_c = \frac{K_3}{K_A K_1}$$

then

$$\frac{2\zeta_0}{\omega_0} = \frac{(K_c/\omega_I) + K_c T_\alpha}{1 + K_c}$$

$$\frac{1}{\omega_0^2} = \frac{-(1/\omega_z^2) + (K_c T_\alpha/\omega_I)}{1 + K_c}$$

In addition, if we define the intermediate gain K by

$$K = K_R \omega_I K_A$$

we can say that

$$K_0 = K_R \omega_I K_A [(K_3/K_A) + K_1] = K(K_1 K_c + K_1) = KK_1(1 + K_c)$$

If we assume that the crossover frequency is beyond the airframe dynamics, we can set the magnitude of the open-loop crossover frequency to unity, as we did in Chapter 23 of the previous volume, in order to obtain

$$1 \approx \frac{-(K_0 \omega_{CR}^2/\omega_0^2)}{(b_{12}\omega_{CR}\omega_{CR}^2)} = \frac{-K_0}{b_{12}\omega_{CR}\omega_0^2}$$

Solving for the open-loop crossover frequency yields

$$\omega_{CR} = \frac{-K_0}{b_{12}\omega_0^2}$$

From the preceding equation we can see that the open-loop crossover frequency is a function of both the aerodynamics and the autopilot gains.

CLOSED-LOOP ANALYSIS

Now we can go back to the time domain to complete the autopilot design. By inspection of Fig. 1.7, we can write an expression for the relationship between the output control surface deflection and the input acceleration command as

$$\frac{\delta}{n'_c} = \frac{-K_A K_R \omega_I/s}{1 - K_R G_3 - (K_R \omega_I G_3/s) - (K_R \omega_I K_A G_1/s)}$$

Rewriting the preceding expression in terms of the open-loop transfer function yields

$$\frac{\delta}{n'_c} = \frac{-K_A K_R \omega_I/s}{1 + HG}$$

The relationship between the flight-control system output acceleration and the input command can then be obtained from the chain rule as

$$\frac{n_L}{n_c'} = \frac{\delta}{n_c'} * \frac{n_L}{\delta} = \left[\frac{-K_A K_R \omega_I / s}{1 + HG} \right]$$

$$* \left[K_1 \left(1 - \frac{s^2}{\omega_z^2} \right) \middle/ (1 + b_{11}s + b_{12}s^2) \right]$$

After much algebra one can show that the flight-control system transfer function becomes

$$\frac{n_L}{n_c'} = \frac{K_A K_R \omega_I K_1 \left(1 - \dfrac{s^2}{\omega_z^2} \right) / K_0}{1 + s \left[\dfrac{2\zeta_0}{\omega_0} - \dfrac{1}{K_0} \right] + s^2 \left[\dfrac{1}{\omega_0^2} - \dfrac{b_{11}}{K_0} \right] - \dfrac{b_{12}s^3}{K_0}}$$

We are not interested in controlling the airframe zeros (that is, the numerator in the flight-control system transfer function) but would like to have the preceding closed-loop transfer function to have the form

$$\frac{n_L}{n_c'} = \frac{K_A K_R \omega_I K_1 \left(1 - \dfrac{s^2}{\omega_z^2} \right) / K_0}{(1 + \tau s) \left[1 + \dfrac{2\zeta s}{\omega} + \dfrac{s^2}{\omega^2} \right]}$$

where we have made the denominator of the flight-control system transfer function a real pole times a quadratic. The two preceding flight-control system transfer functions are equivalent if the denominators are the same, or

$$\frac{2\zeta}{\omega} + \tau = \frac{2\zeta_0}{\omega_0} - \frac{1}{K_0}$$

$$\frac{1}{\omega^2} + \frac{2\zeta\tau}{\omega} = \frac{1}{\omega_0^2} - \frac{b_{11}}{K_0}$$

$$\frac{\tau}{\omega^2} = -\frac{b_{12}}{K_0}$$

In addition, we have already derived an expression for the open-loop crossover frequency to be

$$\omega_{CR} = \frac{-K_0}{b_{12}\omega_0^2}$$

The preceding four equations have four unknowns. At a given flight condition we know the aerodynamics or b_{11} and b_{12}. If we specify the desired time constant

τ, damping ζ, and open-loop crossover frequency ω_{CR} of the flight-control system, the remaining four unknowns are ω, ζ_0, ω_0, and K_0. Because we have four equations, there is sufficient information to solve for these four unknowns. *However, because the preceding four equations are not linear, some work is involved in determining the autopilot gains.*

If we solve the fourth equation for K_0 and substitute it into the second equation

$$\frac{1}{\omega^2} + \frac{2\zeta\tau}{\omega} = \frac{1}{\omega_0^2} - \frac{b_{11}}{K_0}$$

we get

$$\frac{1}{\omega^2} + \frac{2\zeta\tau}{\omega} = \frac{1}{\omega_0^2} + \frac{b_{11}}{\omega_{CR} b_{12} \omega_0^2} = \frac{1}{\omega_0^2}\left(1 + \frac{b_{11}}{\omega_{CR} b_{12}}\right)$$

We can also substitute K_0 into

$$\frac{\tau}{\omega^2} = -\frac{b_{12}}{K_0}$$

yielding

$$\frac{\tau}{\omega^2} = -\frac{b_{12}}{K_0} = \frac{-b_{12}}{-\omega_{CR} b_{12} \omega_0^2} = \frac{1}{\omega_{CR} \omega_0^2}$$

Substituting the preceding equation into

$$\frac{1}{\omega^2} + \frac{2\zeta\tau}{\omega} = \frac{1}{\omega_0^2} + \frac{b_{11}}{\omega_{CR} b_{12} \omega_0^2} = \frac{1}{\omega_0^2}\left(1 + \frac{b_{11}}{\omega_{CR} b_{12}}\right)$$

yields

$$\frac{1}{\omega^2} + \frac{2\zeta\tau}{\omega} = \frac{\omega_{CR}\tau}{\omega^2}\left(1 + \frac{b_{11}}{\omega_{CR} b_{12}}\right)$$

Note that all of the terms in the preceding equation are known except for ω. We can solve the preceding equation for ω yielding

$$\omega = \left[\tau\omega_{CR}\left(1 + \frac{b_{11}}{\omega_{CR} b_{12}}\right) - 1\right]/(2\zeta\tau)$$

Because

$$\frac{\tau}{\omega^2} = \frac{1}{\omega_{CR} \omega_0^2}$$

we can solve for ω_0 and get

$$\omega_0 = \frac{\omega}{\sqrt{\tau \omega_{CR}}}$$

Because we have already solved for ω, ω_0, and K_0 we can invert

$$\frac{2\zeta}{\omega} + \tau = \frac{2\zeta_0}{\omega_0} - \frac{1}{K_0}$$

to get

$$\zeta_0 = 0.5\omega_0 \left(\frac{2\zeta}{\omega} + \tau - \frac{1}{\omega_{CR} b_{12} \omega_0^2} \right)$$

Recall that from the open-loop definitions of the previous section,

$$\frac{2\zeta_0}{\omega_0} = \frac{\dfrac{K_c}{\omega_I} + K_c T_\alpha}{1 + K_c} = a_1$$

$$\frac{1}{\omega_0^2} = \frac{\dfrac{-1}{\omega_z^2} + \dfrac{K_c T_\alpha}{\omega_I}}{1 + K_c} = a_2$$

We would like to solve the preceding two equations for K_c. Therefore,

$$a_1(1 + K_c) = \frac{K_c}{\omega_I} + K_c T_\alpha$$

$$a_2(1 + K_c) = \frac{-1}{\omega_z^2} + \frac{K_c T_\alpha}{\omega_I}$$

Multiplying the first of the preceding two equations by T_α on both sides and subtracting it from the preceding equation yields

$$a_2(1 + K_c) - a_1 T_\alpha(1 + K_c) = \frac{-1}{\omega_z^2} - T_\alpha^2 K_c$$

Therefore

$$K_c(a_2 - a_1 T_\alpha + T_\alpha^2) = \frac{-1}{\omega_z^2} - a_2 + a_1 T_\alpha$$

Solving for K_c yields

$$K_c = \frac{\dfrac{-1}{\omega_z^2} - a_2 + a_1 T_\alpha}{a_2 - a_1 T_\alpha + T_\alpha^2}$$

Substituting the expressions for a_1 and a_2 into the preceding equation yields

$$K_c = \frac{\dfrac{-\omega_0^2}{\omega_z^2} - 1 + 2\zeta_0\omega_0 T_\alpha}{1 - 2\zeta_0\omega_0 T_\alpha + \omega_0^2 T_\alpha^2}$$

We already know from our open-loop definitions that

$$K_c = \frac{K_3}{K_A K_1}$$

Solving the preceding equation for the autopilot gain K_A yields

$$K_A = \frac{K_3}{K_c K_1}$$

We already know from the definition of a_2 that

$$\frac{1}{\omega_0^2} = \frac{\dfrac{-1}{\omega_z^2} + \dfrac{K_c T_\alpha}{\omega_I}}{1 + K_c} = a_2$$

Inverting the preceding equation and solving for the autopilot gain ω_I yields

$$\omega_I = \frac{T_\alpha K_c \omega_0^2}{1 + K_c + \dfrac{\omega_0^2}{\omega_z^2}}$$

Recall from

$$\frac{\tau}{\omega^2} = -\frac{b_{12}}{K_0}$$

we can solve for K_0, yielding

$$K_0 = \frac{-b_{12}\omega^2}{\tau}$$

Recall from the open-loop definitions that

$$K_0 = K K_1 (1 + K_c)$$

Solving the preceding equation for K yields

$$K = \frac{K_0}{K_1(1 + K_c)}$$

From the open-loop definitions, recall that

$$K = K_R \omega_I K_A$$

Solving the preceding equation for the autopilot gain K_R yields

$$K_R = \frac{K}{K_A \omega_I}$$

To get unity gain in the closed-loop transfer function,

$$\frac{n_L}{n'_c} = \frac{K_A K_R \omega_I K_1 \left(1 - \frac{s^2}{\omega_z^2}\right) \Big/ K_0}{(1 + \tau s)\left[1 + \frac{2\zeta s}{\omega} + \frac{s^2}{\omega^2}\right]}$$

we can see that the steady-state gain K_{DC} must be equal to unity, or

$$\frac{K_{DC} K_A K_R \omega_I K_1}{K_0} = 1$$

Solving for the autopilot gain K_{DC} yields

$$K_{DC} = \frac{K_0}{K_A K_R \omega_I K_1} = \frac{K_0}{K K_1} = \frac{K K_1 (1 + K_c)}{K K_1} = 1 + K_c$$

Using the open loop definition for K_C we can simplify the expression for the steady-state autopilot gain to be

$$K_{DC} = 1 + \frac{1}{K_A V_M}$$

We now have enough information to simulate the three-loop flight-control system with the autopilot gain algorithm we just derived. Listing 1.2 presents a time domain step response simulation of the three-loop autopilot in the presence of the nonlinear airframe dynamics. We can see from the listing that the required aerodynamic parameters are derived from the geometry of the airframe. For a given flight condition, these airframe parameters are used to both describe the airframe and determine the autopilot gains using the gain algorithm we just derived. Although the original gain algorithm derivation neglected the dynamics of the actuator, these dynamics are included in the step response simulation to test the robustness of the autopilot gains. The actuator is modeled as a second-order transfer function, or

$$\frac{\delta}{\delta_c} = 1 \Bigg/ \left(1 + \frac{2\zeta_{ACT}}{\omega_{ACT}} s + \frac{s^2}{\omega_{ACT}^2}\right)$$

with a natural frequency ω_{ACT} of 150 rad/s and damping ζ_{ACT} of 0.7.

LISTING 1.2   THREE-LOOP AUTOPILOT STEP RESPONSE IN PRESENCE OF NONLINEAR AIRFRAME

```
clear
count=0;
XNC=322.;
XNCG=XNC/32.2;
ALT=0.;
TS=.01;
H=.001;
VM=3000.;
WCR=50.;
ZETA=.7;
TAU=.3;
WACT=150.;
ZACT=.7;
SLOPE=1.5;
DIAM=1.;
FR=3.;
XL=20.;
CTW=0.;
CRW=6.;
HW=2.;
CTT=0.;
CRT=2.;
HT=2.;
XN=4.;
XCG=10.;
XHL=19.5;
A=1000.;
if ALT<=30000.
    RHO=.002378*exp(−ALT/30000.);
else
    RHO=.0034*exp(−ALT/22000.);
end
WGT=1000.;
XACC=XCG;
SWING=.5*HW*(CTW+CRW);
STAIL=.5*HT*(CTT+CRT);
SREF=3.1416*DIAM*DIAM/4.;
XLP=FR*DIAM;
SPLAN=(XL−XLP)*DIAM+1.33*XLP*DIAM/2.;
XCPN=2*XLP/3;
ADEL=1.33*XLP*DIAM/2;
ASQ=(XL−XLP)*DIAM;
XCPB=(2*ADEL*XLP/3+ASQ*(XLP+.5*(XL−XLP)))/(ADEL+ASQ);
```

```
XCPW=XLP+XN+.7*CRW−.2*CTW;
XMACH=VM/A;
XIYY=WGT*(3*((DIAM/2)^2)+XL*XL)/(12*32.2);
TMP1=(XCG−XCPW)/DIAM;
TMP2=(XCG−XHL)/DIAM;
TMP3=(XCG−XCPB)/DIAM;
TMP4=(XCG−XCPN)/DIAM;
B=sqrt(XMACH^2−1);
Q=.5*RHO*VM*VM;
T=0.;
S=0.;
THD=0;
ALF=0;
XX=0.;
DELNL=0.;
DELNLD=0.;
P1=WGT*XNCG/(Q*SREF);
Y1=2+8*SWING/(B*SREF)+8*STAIL/(B*SREF);
Y2=1.5*SPLAN/SREF;
Y3=8*STAIL/(B*SREF);
Y4=2*TMP4+8*SWING*TMP1/(B*SREF)+8*STAIL*TMP2/(B*SREF);
Y5=1.5*SPLAN*TMP3/SREF;
Y6=8*STAIL*TMP2/(B*SREF);
P2=Y2−Y3*Y5/Y6;
P3=Y1−Y3*Y4/Y6;
A0=−P1*Y6;
ALFTR=(−P3+sqrt(P3*P3+4*P2*P1))/(2.*P2);
DELTR=−Y4*ALFTR/Y6−Y5*ALFTR*ALFTR/Y6;
ALFRQD=ALFTR;
CNA=2+SLOPE*SPLAN*ALFTR/SREF+8*SWING/(B*SREF)+8*STAIL/(B*SREF);
CND=8*STAIL/(B*SREF);
CMAP=2*TMP4+SLOPE*SPLAN*ALFTR*TMP3/SREF+8*SWING*TMP1/(B*SREF);
CMA=CMAP+8*STAIL*TMP2/(B*SREF);
CMD=8*STAIL*TMP2/(B*SREF);
XMA=Q*SREF*DIAM*CMA/XIYY;
XMD=Q*SREF*DIAM*CMD/XIYY;
ZA=−32.2*Q*SREF*CNA/(WGT*VM);
ZD=−32.2*Q*SREF*CND/(WGT*VM);
WZ=sqrt((XMA*ZD−ZA*XMD)/ZD);
B11=ZA/XMA;
B12=−1/XMA;
XK1=−VM*(XMA*ZD−XMD*ZA)/(1845*XMA);
XK2=XK1;
TA=XMD/(XMA*ZD−XMD*ZA);
XK3=1845*XK1/VM;
W=(TAU*WCR*(1+B11/(WCR*B12))−1)/(2*ZETA*TAU);
```

```
W0=W/sqrt(TAU*WCR);
Z0=.5*W0*(2*ZETA/W+TAU−1/(W0*W0*WCR*B12));
XKC=(−W0^2/WZ^2−1.+2.*Z0*W0*TA)/(1.−2.*Z0*W0*TA+W0*W0*TA*TA);
XKA=XK3/(XK1*XKC);
XK0=−B12*W*W/TAU;
XK=XK0/(XK1*(1+XKC));
WI=XKC*TA*W0*W0/(1.+XKC+W0^2/WZ^2);
XKR=XK/(XKA*WI);
XKDC=1.+1845./(XKA*VM);
while T<=.999999
    S=S+H;
    THDOLD=THD;
    ALFOLD=ALF;
    XXOLD=XX;
    DELNLOLD=DELNL;
    DELNLDOLD=DELNLD;
    STEP=1;
    FLAG=0;
    while STEP<=1
        if FLAG==1
            STEP=2;
            THD=THD+H*THDD;
            ALF=ALF+H*ALFD;
            XX=XX+H*XXD;
            DELNL=DELNL+H*DELNLD;
            DELNLD=DELNLD+H*DELNLDD;
            T=T+H;
        end
        CN=2*ALF+1.5*SPLAN*ALF*ALF/SREF+8*SWING*ALF/(B*SREF)...
        +8*STAIL*(ALF+DELNL/57.3)/(B*SREF);
        CM=2*ALF*TMP4+1.5*SPLAN*ALF*ALF*TMP3/SREF...
            +8*SWING*ALF*TMP1/(B*SREF)+8*STAIL*(ALF+DELNL/57.3)...
            *TMP2/(B*SREF);
        THDD=Q*SREF*DIAM*CM/XIYY;
        THDDEG=THD*57.3;
        XNL=32.2*Q*SREF*CN/WGT;
        XNLG=XNL/32.2;
        ALFD=THD−XNL/VM;
        XNANL=XNLG;
        XXD=WI*(THDDEG+XKA*(XNANL−XNCG*XKDC));
        DELCNL=XKR*(XX+THDDEG);
        DELNLDD=WACT*WACT*(DELCNL−DELNL−2*ZACT*DELNLD/WACT);
        FLAG=1;
    end
    FLAG=0;
    THD=.5*(THDOLD+THD+H*THDD);
```

```
    ALF=.5*(ALFOLD+ALF+H*ALFD);
    XX=.5*(XXOLD+XX+H*XXD);
    DELNL=.5*(DELNLOLD+DELNL+H*DELNLD);
    DELNLD=.5*(DELNLDOLD+DELNLD+H*DELNLDD);
    if S>=(TS−.00001)
       S=0.;
       XNCG=XNC/32.2;
       XNLG=XNL/32.2;
       count=count+1;
       ArrayT(count)=T;
       ArrayXNCG(count)=XNCG;
       ArrayXNLG(count)=XNLG;
       ArrayDELNL(count)=DELNL;
       ArrayALF(count)=ALF*57.3;
    end
end
output=[ArrayT',ArrayXNCG',ArrayXNLG',ArrayDELNL',ArrayALF'];
save datfil.txt output −ascii
disp 'simulation finished'
clc
figure
plot(ArrayT,ArrayXNCG,ArrayT,ArrayXNLG),grid
xlabel('Time (s)')
ylabel('Acceleration (g)')
axis([0 1−1 11])
figure
plot(ArrayT,ArrayDELNL),grid
xlabel('Time (s) ')
ylabel('Fin Deflection (deg')
figure
plot(ArrayT,ArrayALF),grid
xlabel('Time (s)')
ylabel('Angle of Attack (deg')
```

Using the same airframe characteristics of Listing 23.1 of the previous volume for the flight condition in which the missile was at sea level and traveling at 3000 ft/s, a 10-g command was issued to the autopilot. We can see from Listing 23.2 that the nominal design goals given the autopilot gain algorithm were to achieve a time constant of 0.3 s, an open-loop crossover frequency of 50 rad/s, and a damping of 0.7. The three autopilot gains for these requirements at this flight condition turn out to be $K_A = 1.15$ deg/g-s, $\omega_I = 12.9$ rad/s, and $K_R = 0.0928$ s. In this example the missile center of gravity is 10 ft (XCG = 10), which yields a stable airframe response. We can see from Fig. 1.9 that the overall time constant of the flight-control system is slightly in excess of 0.3 s. This is not in disagreement with theory, because the overall time constant of the third-order flight-control

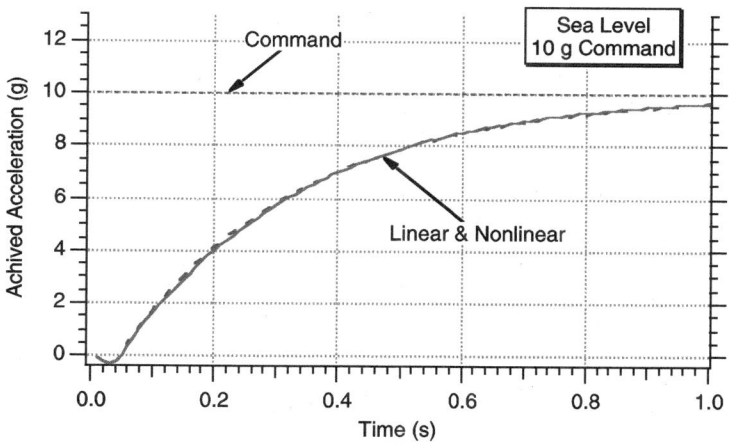

Fig. 1.9   Three-loop autopilot yields stable response at sea level when airframe is stable.

system can be approximated as

$$\tau_{TOT} = \tau + \frac{2\zeta}{\omega}$$

Because the damping is 0.7 and the natural frequency is approximately 35 rad/s, the overall or total time constant is 0.34 s, which is consistent with Fig. 1.9. In other words, the response reaches 63% of the steady-state value in 0.34 s. The altitude of the response was increased to 50,000 ft, and we can see from Fig. 1.10

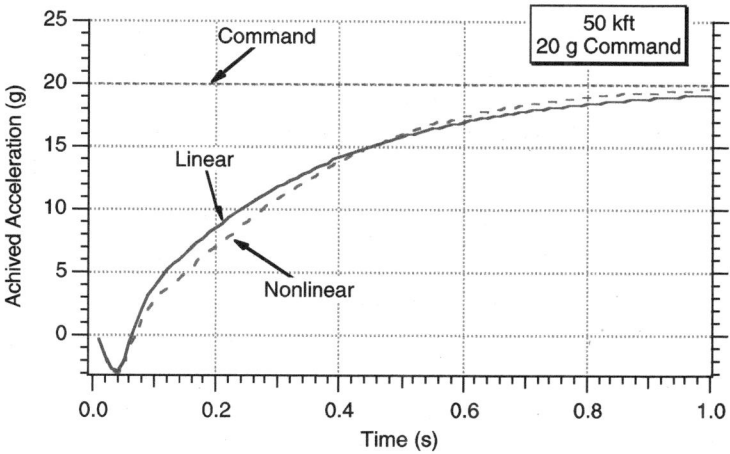

Fig. 1.10   Three-loop autopilot yields stable response at 50,000 ft altitude when airframe is stable.

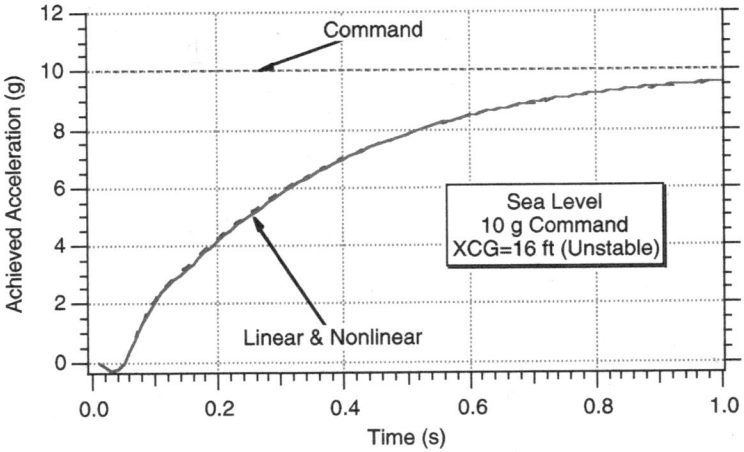

Fig. 1.11   Three-loop autopilot yields stable response at sea level when airframe is unstable.

that the speed of response of the flight-control system is still maintained with the new autopilot gain algorithm.

Next the center of gravity was moved towards the tail (XCG = 16). We know that in this case, as was shown at the beginning of this chapter, the airframe is unstable. We can see from Figs. 1.11 and 1.12 that the flight control system

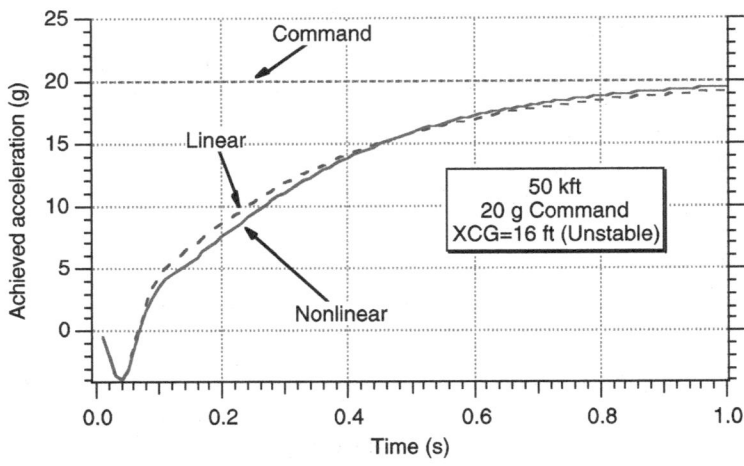

Fig. 1.12   Three-loop autopilot yields stable response at 50,000 ft altitude when airframe is unstable.

response at sea level and 50,000 ft altitude is virtually identical to the response when the airframe was stable.

Thus we can see that a slight modification of the original three-loop gain algorithm leads to a flight-control system design that meets the system objectives even when the airframe is unstable.

SUMMARY

In this chapter we have seen how the three-loop autopilot gain algorithm can be rederived when the airframe is unstable. An example was presented showing that the flight-control step responses were virtually identical to the case when the airframe was stable.

REFERENCES

[1] Giragosian, P., "Aerodynamic Considerations in the Design of a Vertically Launched Advanced Interdiction Missile," *Proceedings of AIAA Atmospheric Flight Mechanics Conference*, AIAA, San Diego, CA, 9–11 Aug. 1982.

[2] Selby, S. M., *Standard Mathematical Tables*, 20th ed., Chemical Rubber Co., Cleveland, OH, 1972.

[3] Jerger, J. J., *System Preliminary Design*, Van Nostrand, Princeton, NJ, 1960.

[4] Nesline, F. W., and Nabbefeld, N. C., "Design of Digital Autopilots for Homing Missiles," *Proceedings of AGARD Flight Mechanics Panel Symposium*, London, May 1979.

CHAPTER 2

Flexible Body Dynamics

INTRODUCTION

We saw in Chapter 21 of the previous volume that the bare airframe has a low damped oscillatory response. It becomes apparent that as we strive to make the time constant of the flight-control system small, the bandwidth of the flight-control system increases. As the bandwidth increases, a parasitic path from the control surface through the missile structure to the rate gyro output of the autopilot becomes very important. The rate gyro senses the missile body structural vibrations and feeds them back to the actuator, thus introducing the possibility of destructive instability. This chapter deals with techniques for alleviating the possibility of the body bending instability. The rate gyro flight-control system topology of Chapter 22 of the previous volume will be used to illustrate the importance of flexible body dynamics.

RIGID AND FLEXIBLE BODY DYNAMICS

When the missile is in flight, aerodynamic forces are created that excite the natural vibrational modes of the missile. Although there are many modes, an example of the first structural mode shape [1] is shown in Fig. 2.1.

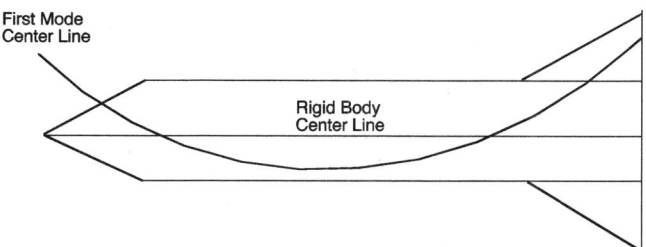

Fig. 2.1 An example of the first vibrational mode shape.

The general transfer function from the control surface angle to the body rate for the first vibrational or bending mode can be shown to be [1]

$$\frac{\dot{\theta}_{FB}}{\delta} = \frac{K_{FB}\left(1 + \dfrac{s^2}{\omega_{ZFB}^2}\right)s}{\left(1 + \dfrac{2\zeta_1}{\omega_1}s + \dfrac{s^2}{\omega_1^2}\right)}$$

where K_{FB} is the gain for the structural path of the first mode, ω_{ZFB} is the quadratic zero of the first mode, ζ_1 is the damping of the first mode, and ω_1 is the vibrational frequency of the first mode. The parameters involved in the derivation of the preceding transfer function are functions of the location of the rate gyro, the control surface hinge line, and the location of the fin center of gravity. Recall from Chapter 21 of the previous volume and Chapter 1 of this volume that the rigid body rate dynamics of the airframe is given by the transfer function

$$\frac{\dot{\theta}_{RB}}{\delta} = \frac{K_3(1 + T_\alpha s)}{1 + \dfrac{Z_\alpha}{M_\alpha}s - \dfrac{s^2}{M_\alpha}} = \frac{K_3(1 + T_\alpha s)}{1 + b_{11}s + b_{12}s^2}$$

The preceding equations can be put in block diagram form to show how the rigid and flexible body dynamics combine, as shown in Fig. 2.2.

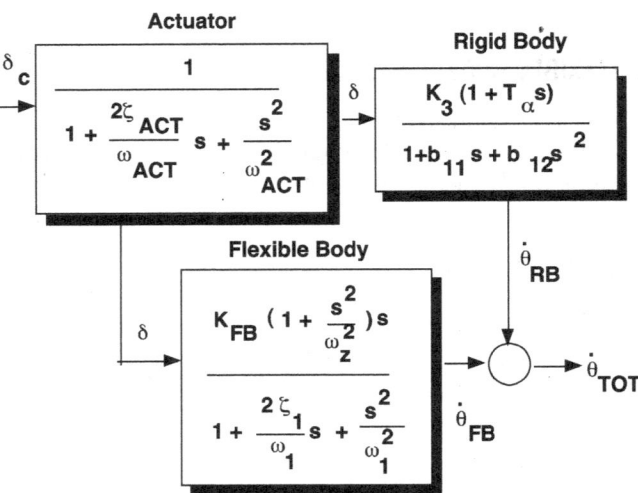

Fig. 2.2   How rigid and flexible body rate dynamics combine.

**TABLE 2.1   RIGID AND FLEXIBLE BODY RATE
AIRFRAME PARAMETERS**

Symbol	Value
b_{11}	0.00461
b_{12}	0.00136
K_3	-0.362
$T\alpha$	0.831 s
K_{FB}	1.34×10^{-3}
ω_{ZFB}	395 rad/s
ζ_1	0.015
ω_1	259 rad/s

In order to demonstrate the importance of the flexible body dynamics, let's consider a numerical example in Table 2.1, taken from [1], with values for all the parameters.

From Table 2.1 and the preceding equation for the rigid body dynamics, we can express the rigid body transfer function as

$$\frac{\dot{\theta}_{RB}}{\delta} = \frac{K_3(1 + T_\alpha s)}{1 + b_{11}s + b_{12}s^2} = \frac{-0.362(1 + 0.831s)}{1 + \dfrac{2 * 0.0625}{27.1}s + \dfrac{s^2}{27.1^2}}$$

In order to simulate the effects of rigid and flexible body dynamics in the time domain, it is first necessary to convert the preceding equation in the Laplace transform domain to the time domain. Starting with the preceding equation for the rigid body dynamics, we can see from the chain rule in calculus that

$$\frac{\dot{\theta}_{RB}}{\delta} = \frac{e_1}{\delta} * \frac{\dot{\theta}_{RB}}{e_1}$$

Defining

$$\frac{e_1}{\delta} = \frac{1}{1 + b_{11}s + b_{12}s^2}$$

$$\frac{\dot{\theta}_{RB}}{e_1} = K_3(1 + T_\alpha s)$$

we can cross multiply in the preceding two equations and convert to the time domain to obtain

$$e_1 + b_{11}\dot{e}_1 + b_{12}\ddot{e}_1 = \delta$$

$$\dot{\theta}_{RB} = K_3(e_1 + T_\alpha \dot{e}_1)$$

Therefore, the differential equations to be integrated for a simulation are

$$\ddot{e}_1 = \frac{1}{b_{12}}(\delta - e_1 - b_{11}\dot{e}_1)$$

$$\dot{\theta}_{RB} = K_3(e_1 + T_\alpha \dot{e}_1)$$

Similarly, the flexible body transfer function, using the values of Table 2.1, can be evaluated as

$$\frac{\dot{\theta}_{FB}}{\delta} = \frac{K_{FB}\left(1 + \frac{s^2}{\omega_{ZFB}^2}\right)s}{\left(1 + \frac{2\zeta_1}{\omega_1}s + \frac{s^2}{\omega_1^2}\right)} = \frac{1.34 * 10^{-3}\left(1 + \frac{s^2}{395^2}\right)s}{\left(1 + \frac{2 * 0.015}{259}s + \frac{s^2}{259^2}\right)}$$

Again, in order to simulate the effects of the flexible body dynamics in the time domain it is necessary to convert the preceding equation in the Laplace transform domain to the time domain. Starting with the preceding equation for the flexible body dynamics, we can see from the chain rule in calculus that

$$\frac{\dot{\theta}_{FB}}{\delta} = \frac{e}{\delta} * \frac{\dot{\theta}_{FB}}{e}$$

Defining

$$\frac{e}{\delta} = \frac{1}{\left(1 + \frac{2\zeta_1}{\omega_1}s + \frac{s^2}{\omega_1^2}\right)}$$

$$\frac{\dot{\theta}_{FB}}{e} = K_{FB}\left(1 + \frac{s^2}{\omega_{ZFB}^2}\right)s$$

we can cross multiply in the preceding two equations and convert from the Laplace transform to the time domain to obtain

$$e + \frac{2\zeta_1}{\omega_1}\dot{e} + \frac{s^2}{\omega_1^2}\ddot{e} = \delta$$

$$\dot{\theta}_{FB} = K_{FB}\left(\dot{e} + \frac{\ddot{e}}{\omega_{ZFB}^2}\right)$$

Therefore, one of the differential equations for simulation of the flexible body dynamics is given by

$$\ddot{e} = \omega_1^2\left(\delta - e - \frac{2\zeta_1}{\omega_1}\dot{e}\right)$$

To find the third derivative of e, which is required for the flexible body rate equation, we take the derivative of both sides of the preceding equation, yielding

$$\dddot{e} = \omega_1^2 \left(\dot{\delta} - \dot{e} - \frac{2\zeta_1}{\omega_1} \ddot{e} \right)$$

A simulation of the rigid and flexible body rate equations due to a 1-deg step command in fin deflection (DELC=1) can be found in Listing 2.1.

LISTING 2.1   RIGID AND FLEXIBLE BODY RESPONSE TO A FIN DEFLECTION COMMAND

```
clear all
close all
count=0;
TS=.001
XK3=-.362;
TA=.831;
B11=.00461;
B12=.00136;
XKFB=.00134;
WZFB=395.;
Z1=.015;
W1=259.;
TF=1.;
DELC=1.;
WACT=100.;
ZACT=.7;
T=0.;
S=0.;
H=.00001;
E=0.;
ED=0.;
E1=0.;
E1D=0.;
E1DD=0.;
DEL=0.;
DELD=0.;
while T<=TF
    EOLD=E;
    EDOLD=ED;
    E1OLD=E1;
    E1DOLD=E1D;
    E1DDOLD=E1DD;
```

```
DELOLD=DEL;
DELDOLD=DELD;
DELDD=WACT*WACT*(DELC-DEL-2.*ZACT*DELD/WACT);
EDD=(DEL-E-B11*ED)/B12;
E1DDD=W1*W1*(DELD-E1D-2.*Z1*E1DD/W1);
THDRB=XK3*(E+TA*ED);
THDFB=XKFB*(E1D+E1DDD/WZFB^2);
THDTOT=THDRB+THDFB;
E=E+H*ED;
ED=ED+H*EDD;
E1=E1+H*E1D;
E1D=E1D+H*E1DD;
E1DD=E1DD+H*E1DDD;
DEL=DEL+H*DELD;
DELD=DELD+H*DELDD;
T=T+H;
DELDD=WACT*WACT*(DELC-DEL-2.*ZACT*DELD/WACT);
EDD=(DEL-E-B11*ED)/B12;
E1DDD=W1*W1*(DELD-E1D-2.*Z1*E1DD/W1);
THDRB=XK3*(E+TA*ED);
THDFB=XKFB*(E1D+E1DDD/WZFB^2);
THDTOT=THDRB+THDFB;
E=.5*(EOLD+E+H*ED);
ED=.5*(EDOLD+ED+H*EDD);
E1=.5*(E1OLD+E1+H*E1D);
E1D=.5*(E1DOLD+E1D+H*E1DD);
E1DD=.5*(E1DDOLD+E1DD+H*E1DDD);
DEL=.5*(DELOLD+DEL+H*DELD);
DELD=.5*(DELDOLD+DELD+H*DELDD);
S=S+H;
if S>=(TS-.00001)
   S=0.;
   count=count+1;
   ArrayT(count)=T;
   ArrayTHDRB(count)=THDRB;
   ArrayTHDFB(count)=THDFB;
   ArrayTHDTOT(count)=THDTOT;
  end
end
figure
plot(ArrayT,ArrayTHDRB),grid
xlabel('Time (Sec)')
ylabel('Rigid Body Rate (deg/s)')
figure
plot(ArrayT,ArrayTHDFB),grid
xlabel('Time (Sec)')
```

```
ylabel('Flexible Body Rate (deg/s)')
figure
plot(ArrayT,ArrayTHDTOT),grid
xlabel('Time (Sec)')
ylabel('Total Body Rate (deg/s)')
clc
output=[ArrayT',ArrayTHDRB',ArrayTHDFB',ArrayTHDTOT'];
save datfil.txt output -ascii
disp 'simulation finished'
```

Listing 2.1 was run for the nominal case, and the rigid body rate, flexible body rate, and total body rate responses appear in Figs. 2.3 to 2.5, respectively. We can see from Fig. 2.3 that the rigid body rate oscillates at the natural frequency and damping of the bare airframe (27.1 rad/s frequency and 0.0625 damping). Figure 2.4 shows that the flexible body rate response oscillates at 259 rad/s with a damping of 0.015. Comparing the amplitudes of the responses of Figs. 2.4 and 2.3, we can see that the rigid body rate amplitude is two orders of magnitude larger than the flexible body rate amplitude. Figure 2.5 shows that from a time domain point of view, the total body rate response (i.e., sum of rigid and flexible body rate responses) is not distinguishable from the rigid body rate response. One might think that because of the small amplitude of the flexible body rate response its effects on flight-control system design are not important. *We shall soon see that considering flexible body dynamics is critical in preventing instabilities within the flight-control system.*

Next the flexible and rigid body rate dynamics were incorporated into the rate gyro flight-control system of Chapter 22 of the previous volume in Fig. 2.6. Recall from Chapter 22 that the gain K_R controlled both the closed-loop

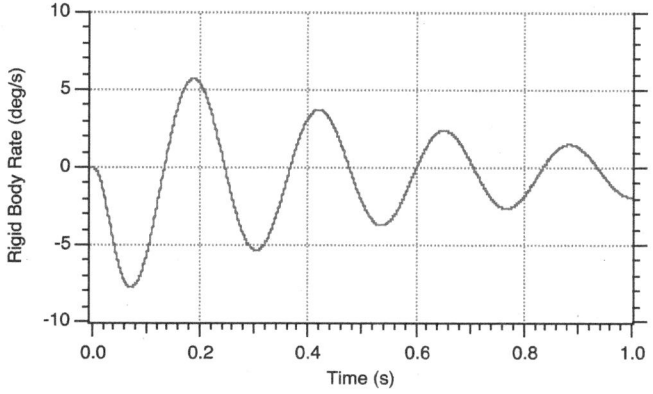

Fig. 2.3   Rigid body response has low damping and a natural frequency of 27.1 rad/s.

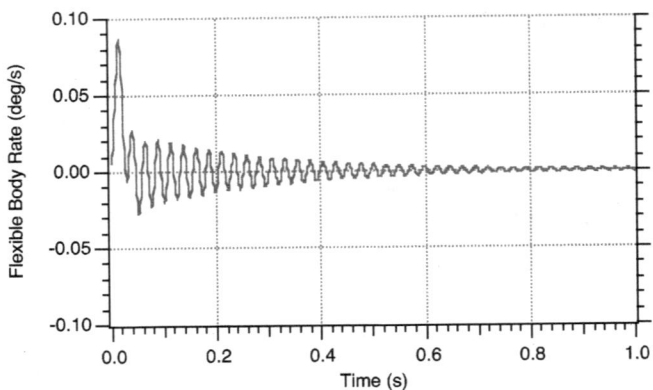

Fig. 2.4   Flexible body response is higher frequency but an order of magnitude smaller than rigid body response.

damping and open-loop crossover frequency of the rate gyro flight-control system. We saw in Chapter 22 of the previous volume that increasing the flight-control system gain K_R increases both the damping and the open-loop crossover frequency. Listing 2.2 programs the differential equations describing both the rigid and flexible body dynamics of the rate gyro flight-control system. Note that actuator dynamics (ω_{ACT} and ζ_{ACT}) are included in the block diagram of the rate gyro flight-control system. From Fig. 2.6 we can see that the actuator bandwidth ω_{ACT} can be made a parameter for study. We can also see from Listing 2.2 that the simulation can be run with rigid body dynamics only when FB=0 and with both rigid and flexible body dynamics when FB=1.

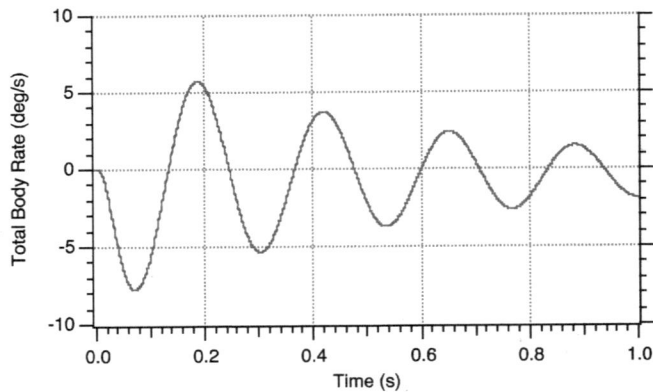

Fig. 2.5   Total body rate response appears to be identical to rigid body response.

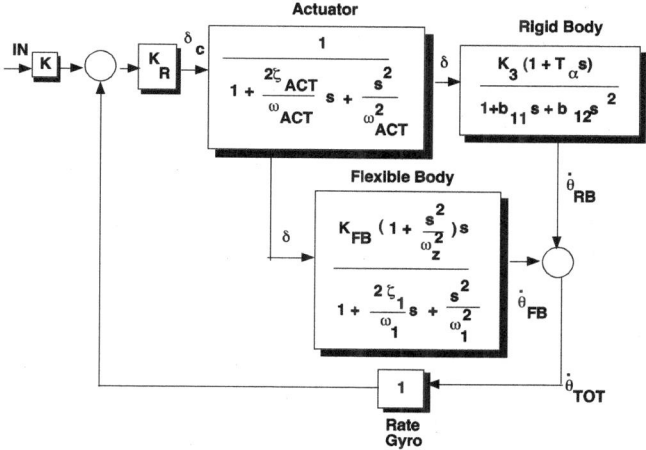

Fig. 2.6   Rate gyro autopilot with flexible body mode.

LISTING 2.2   RATE GYRO FLIGHT CONTROL SYSTEM WITH BOTH RIGID AND FLEXIBLE BODY DYNAMICS

```
clear all
close all
count=0;
TS=.001;
XK3=-.362;
TA=.831;
B11=.00461;
B12=.00136;
XKFB=.00134;
WZFB=395.;
Z1=.015;
W1=259.;
WT=W1;
ZT=Z1;
WB=W1;
ZB=.5;
TF=1.;
XIN=1.;
WACT=100.;
WACT=400.;
ZACT=.7;
XKR=.1;
```

```
XKR=.2;
XKR=.3;
XK=(1.-XKR*XK3)/(XKR*XK3)
FB=0.;
% FB=1;
T=0.;
S=0.;
H=.00001;
E=0.;
ED=0.;
E1=0.;
E1D=0.;
E1DD=0.;
DEL=0.;
DELD=0.;
DELC=0;
while T<=TF
    EOLD=E;
    EDOLD=ED;
    E1OLD=E1;
    E1DOLD=E1D;
    E1DDOLD=E1DD;
    DELOLD=DEL;
    DELDOLD=DELD;
    DELDD=WACT*WACT*(DELC-DEL-2.*ZACT*DELD/WACT);
    EDD=(DEL-E-B11*ED)/B12;
    E1DDD=W1*W1*(DELD-E1D-2.*Z1*E1DD/W1);
    THDRB=XK3*(E+TA*ED);
    THDFB=XKFB*(E1D+E1DDD/WZFB^2);
    THDTOT=THDRB+FB*THDFB;
    DELC=XKR*(XK*XIN+THDTOT);
    E=E+H*ED;
    ED=ED+H*EDD;
    E1=E1+H*E1D;
    E1D=E1D+H*E1DD;
    E1DD=E1DD+H*E1DDD;
    DEL=DEL+H*DELD;
    DELD=DELD+H*DELDD;
    T=T+H;
    DELDD=WACT*WACT*(DELC-DEL-2.*ZACT*DELD/WACT);
    EDD=(DEL-E-B11*ED)/B12;
    E1DDD=W1*W1*(DELD-E1D-2.*Z1*E1DD/W1);
    THDRB=XK3*(E+TA*ED);
    THDFB=XKFB*(E1D+E1DDD/WZFB^2);
    THDTOT=THDRB+FB*THDFB;
    DELC=XKR*(XK*XIN+THDTOT);
```

```
E=.5*(EOLD+E+H*ED);
ED=.5*(EDOLD+ED+H*EDD);
E1=.5*(E1OLD+E1+H*E1D);
E1D=.5*(E1DOLD+E1D+H*E1DD);
E1DD=.5*(E1DDOLD+E1DD+H*E1DDD);
DEL=.5*(DELOLD+DEL+H*DELD);
DELD=.5*(DELDOLD+DELD+H*DELDD);
S=S+H;
if S>=(TS-.00001)
    S=0.;
    count=count+1;
    ArrayT(count)=T;
    ArrayTHDTOT(count)=THDTOT;
end
end
figure
plot(ArrayT,ArrayTHDTOT),grid
xlabel('Time (Sec)')
ylabel('Total Body Rate (deg/s)')
clc
output=[ArrayT',ArrayTHDTOT'];
save datfil.txt output -ascii
disp 'simulation finished'
```

In order to understand the simulation results that follow, we must first remind ourselves how the open-loop crossover frequency ω_{CR} is related to the rigid body airframe parameters. Recall from Chapter 22 of the previous volume that the expression for the open-loop crossover frequency was given by

$$\omega_{CR} = \frac{-K_R K_3 T_\alpha}{b_{12}}$$

Using the parameters of Listing 2.2 when the flight-control system gain K_R is 0.3, the crossover frequency becomes

$$\omega_{CR} = \frac{-K_R K_3 T_\alpha}{b_{12}} = \frac{-0.3 * (-0.362) * 0.831}{0.00136} = 66 \, \text{rad/s}$$

Because the open-loop crossover frequency is more than one-third the actuator bandwidth, we expect a stability problem. The nominal case (no flexible body dynamics and actuator bandwidth at 100 rad/s) was run, and we can see from Fig. 2.7 that increasing the autopilot gain K_R to 0.3 indeed destabilizes the system just due to the rigid body dynamics (response starts to oscillate). The rigid body response can be stabilized when the autopilot gain is 0.3 ($K_R = 0.3$) by increasing the bandwidth of the actuator to 400 rad/s, as can be seen from

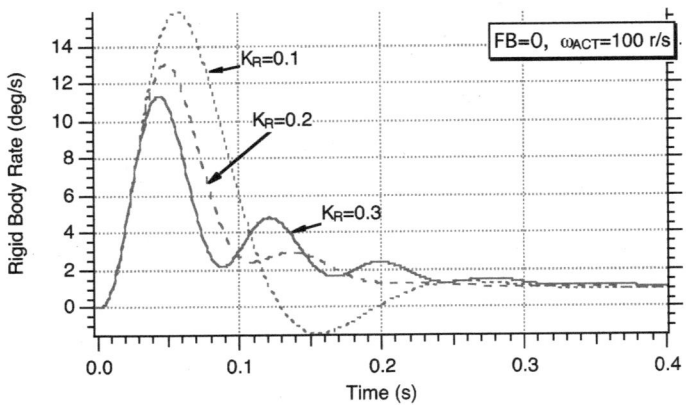

Fig. 2.7   Increasing autopilot gain K_R without flexible body dynamics decreases stability of rate gyro flight-control system.

Fig. 2.8. Now the crossover frequency is well below one-third of the actuator bandwidth (i.e., $66 < 400/3$). However, Fig. 2.9 shows that when the flexible body dynamics are included (FB=1) the system goes unstable—even for a very high actuator bandwidth of 400 rad/s. Therefore, we can see that consideration of flexible body dynamics in flight control system design is very important. The next section will consider methods for alleviating the destabilizing effects of flexible body dynamics.

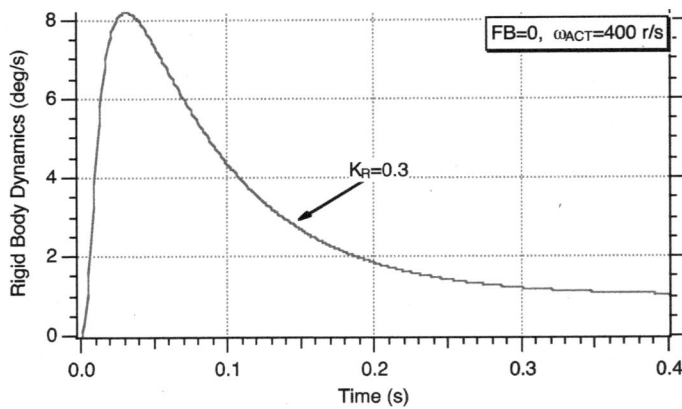

Fig. 2.8   Increasing bandwidth of actuator stabilizes rigid body response for high autopilot gain ($K_R = 0.3$).

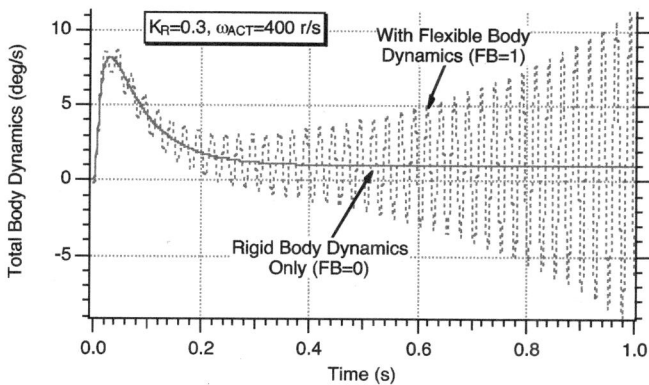

Fig. 2.9   Flexible body dynamics can cause rate gyro autopilot to go unstable.

METHODS FOR ALLEVIATING INSTABILITY DUE TO RATE GYRO SENSING BODY BENDING

One method for alleviating body bending effects is to reduce the autopilot gain K_R. We can see from Fig. 2.10 that reducing K_R from 0.3, where the response is unstable (see Fig. 2.9), to 0.1 stabilizes the response. However, reducing the autopilot gain K_R is not always desirable because it may also decrease the crossover frequency so that the resultant autopilot response may not be acceptable. Figure 2.11 shows that another method for stabilizing the response is to keep K_R at 0.3 but reduce the actuator bandwidth from 400 rad/s to 200 rad/s. This method may not always be acceptable because of possible interference with another body bending mode.

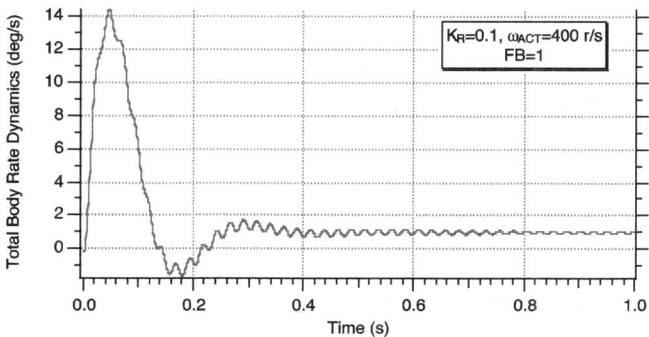

Fig. 2.10   Reducing autopilot gain has a stabilizing effect when flexible body dynamics are considered.

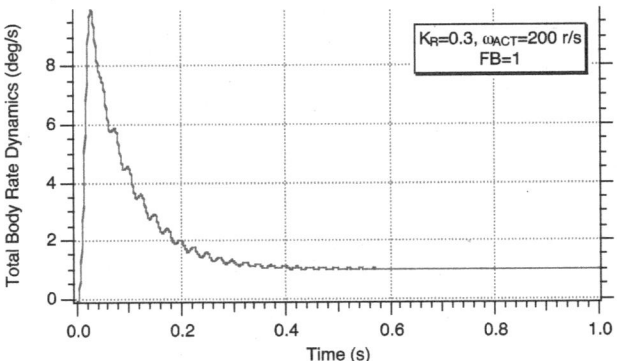

Fig. 2.11   Reducing actuator bandwidth has a stabilizing effect when flexible body dynamics are considered.

Another approach for stabilizing the response due to flexible body dynamics while keeping the gain high and actuator bandwidth high is the use of a notch filter, as shown in Fig. 2.12. As can be seen from Fig. 2.12, the notch filter is

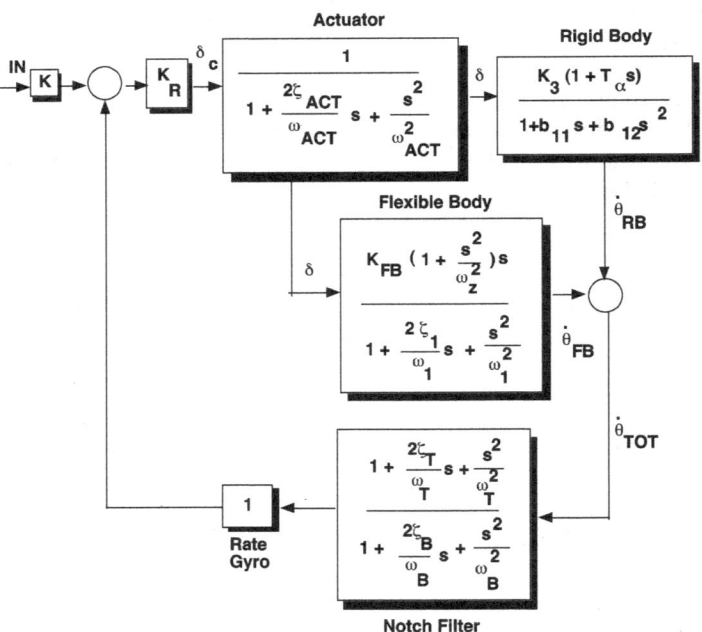

Fig. 2.12   Using a notch filter to cancel out first bending mode.

placed before the input to the rate gyro. The transfer function of the notch filter is given by

$$\frac{\dot{\theta}}{\dot{\theta}_{\text{TOT}}} = \frac{1 + \dfrac{2\zeta_T}{\omega_T}s + \dfrac{s^2}{\omega_T^2}}{1 + \dfrac{2\zeta_B}{\omega_B}s + \dfrac{s^2}{\omega_B^2}}$$

where

$$\omega_T \approx \omega_1$$
$$\zeta_T \approx \zeta_1$$
$$\omega_B \approx \omega_1$$
$$\zeta_B \approx 0.5$$

We can see from the preceding equations that the notch filter is trying to increase the low damping of the bending mode from ζ_1 to 0.5 in this example by using a cancellation technique, assuming that the frequency and damping of the bending mode is known in advance from testing [2]. A multiplication of the notch filter transfer function with the transfer function of the bending mode, by making the preceding substitutions, yields

$$\frac{1 + \dfrac{2\zeta_T}{\omega_T}s + \dfrac{s^2}{\omega_T^2}}{\left(1 + \dfrac{2\zeta_B}{\omega_B}s + \dfrac{s^2}{\omega_B^2}\right)} * \frac{K_{FB}\left(1 + \dfrac{s^2}{\omega_{ZFB}^2}\right)s}{\left(1 + \dfrac{2\zeta_1}{\omega_1}s + \dfrac{s^2}{\omega_1^2}\right)}$$

$$\approx \frac{1 + \dfrac{2\zeta_1}{\omega_1}s + \dfrac{s^2}{\omega_1^2}}{\left(1 + \dfrac{2*.5}{\omega_1}s + \dfrac{s^2}{\omega_1^2}\right)} * \frac{K_{FB}\left(1 + \dfrac{s^2}{\omega_{ZFB}^2}\right)s}{\left(1 + \dfrac{2\zeta_1}{\omega_1}s + \dfrac{s^2}{\omega_1^2}\right)} = \frac{K_{FB}\left(1 + \dfrac{s^2}{\omega_{ZFB}^2}\right)s}{\left(1 + \dfrac{2*.5}{\omega_1}s + \dfrac{s^2}{\omega_1^2}\right)}$$

In other words, using the notch filter effectively increases the damping of the bending mode from 0.015 to 0.5.

Similarly, if two bending modes are considered, then a notch filter is required for each bending mode, as shown in Fig. 2.13.

To illustrate the effectiveness of the notch filtering technique, a simulation was set up where there was the possibility of two bending modes, one bending mode, or no bending nodes through the parameters FB1 and FB2. In addition there are options for using zero, one, or two notch filters through the parameter NOTCH. Let's first concentrate on the first bending mode (FB1=1, FB2=0). From Listing

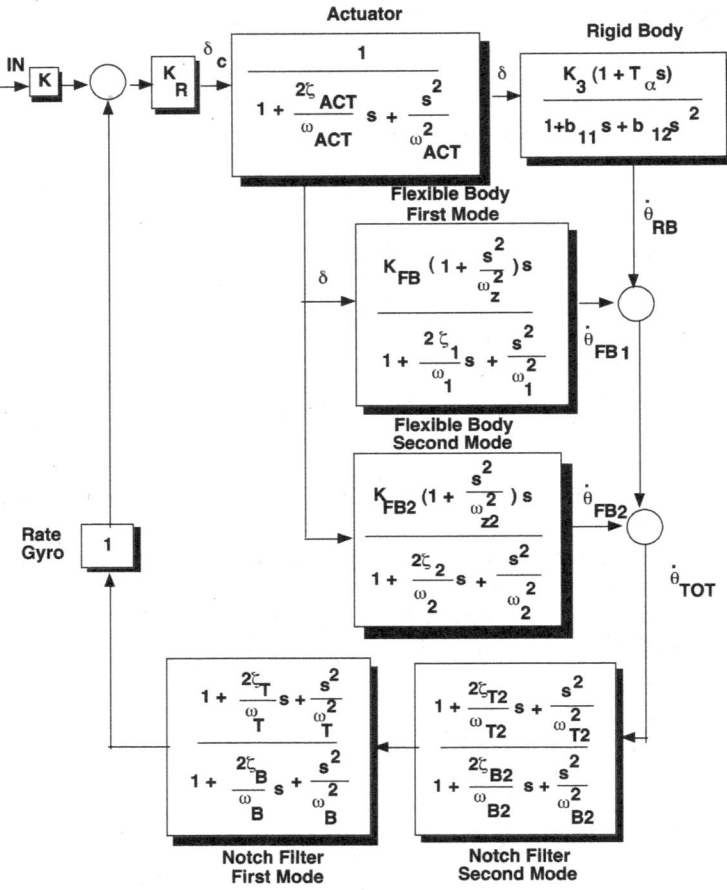

Fig. 2.13   Two flexible body modes and two notch filters.

2.3 we can see that if there is a notch filter for the first mode, then NOTCH=1. If there are notch filters for the two bending modes, then NOTCH=2. Finally, if NOTCH=3 then there are no notch filters. We have already shown the transfer functions for the first mode and its notch filter.

LISTING 2.3   NOTCH FILTER SIMULATION

```
clear all
close all
count=0;
TS=.001;
```

```
XK3=-.362;
TA=.831;
B11=.00461;
B12=.00136;
XKFB=.00134;
WZFB=395.;
Z1=.015;
W1=259.;
XKFB2=.000664;
WZFB2=255.;
Z2=.022;
W2=649.;
WT=W1;
ZT=Z1;
WT2=W2;
ZT2=Z2;
WB=W1;
ZB=.5;
WB2=W2;
ZB2=.5;
TF=1.;
XIN=1.;
WACT=100.;
WACT=400.
ZACT=.7;
XKR=.1
XKR=.3
% Notch=1 (first mode filter), Notch=2 (both filters), Notch=3 (no filters)
NOTCH=3;
NOTCH=1;
NOTCH=2;
XK=(1.-XKR*XK3)/(XKR*XK3)
% FB1=1 (First mode) FB2=1 (Second mode)
FB1=1.;
FB2=1.;
T=0.;
S=0.;
H=.00001;
E=0.;
ED=0.;
E1=0.;
E1D=0.;
E1DD=0.;
DEL=0.;
DELD=0.;
DELC=0;
```

```
E2=0.;
E2D=0.;
E3=0.;
E3D=0.;
E3DD=0.;
E4=0.;
E4D=0.;
THDFIL1=0;
THDFIL=0;
while T<=TF
    EOLD=E;
    EDOLD=ED;
    E1OLD=E1;
    E1DOLD=E1D;
    E1DDOLD=E1DD;
    DELOLD=DEL;
    DELDOLD=DELD;
    E2OLD=E2;
    E2DOLD=E2D;
    E3OLD=E3;
    E3DOLD=E3D;
    E3DDOLD=E3DD;
    E4OLD=E4;
    E4DOLD=E4D;

    DELDD=WACT*WACT*(DELC-DEL-2.*ZACT*DELD/WACT);
    EDD=(DEL-E-B11*ED)/B12;
    E1DDD=W1*W1*(DELD-E1D-2.*Z1*E1DD/W1);
    E3DDD=W2*W2*(DELD-E3D-2.*Z2*E3DD/W2);
    THDRB=XK3*(E+TA*ED);
    THDFB=XKFB*(E1D+E1DDD/WZFB2^2);
    THDFB2=XKFB2*(E3D+E3DDD/WZFB2^2);
    THDTOT=THDRB+FB1*THDFB+FB2*THDFB2;
    if NOTCH==1
        DELC=XKR*(XK*XIN+THDFIL1);
    elseif NOTCH==2
        DELC=XKR*(XK*XIN+THDFIL);
    else
        DELC=XKR*(XK*XIN+THDTOT);
    end
    E2DD=WB*WB*(THDTOT-E2-2*ZB*E2D/WB);
    THDFIL1=E2+2*ZT*E2D/WT+E2DD/WT^2;
    E4DD=WB2*WB2*(THDFIL1-E4-2*ZB2*E4D/WB2);
    THDFIL=E4+2*ZT2*E4D/WT2+E4DD/WT2^2;

    E=E+H*ED;
```

```
ED=ED+H*EDD;
E1=E1+H*E1D;
E1D=E1D+H*E1DD;
E1DD=E1DD+H*E1DDD;
DEL=DEL+H*DELD;
DELD=DELD+H*DELDD;
E2=E2+H*E2D;
E2D=E2D+H*E2DD;
E3=E3+H*E3D;
E3D=E3D+H*E3DD;
E3DD=E3DD+H*E3DDD;
E4=E4+H*E4D;
E4D=E4D+H*E4DD;
T=T+H;

DELDD=WACT*WACT*(DELC-DEL-2.*ZACT*DELD/WACT);
EDD=(DEL-E-B11*ED)/B12;
E1DDD=W1*W1*(DELD-E1D-2.*Z1*E1DD/W1);
E3DDD=W2*W2*(DELD-E3D-2.*Z2*E3DD/W2);
THDRB=XK3*(E+TA*ED);
THDFB=XKFB*(E1D+E1DDD/WZFB^2);
THDFB2=XKFB2*(E3D+E3DDD/WZFB2^2);
THDTOT=THDRB+FB1*THDFB+FB2*THDFB2;
if  NOTCH==1
    DELC=XKR*(XK*XIN+THDFIL1);
elseif NOTCH==2
    DELC=XKR*(XK*XIN+THDFIL);
else
    DELC=XKR*(XK*XIN+THDTOT);
end
E2DD=WB*WB*(THDTOT-E2-2*ZB*E2D/WB);
THDFIL1=E2+2*ZT*E2D/WT+E2DD/WT^2;
E4DD=WB2*WB2*(THDFIL1-E4-2*ZB2*E4D/WB2);
THDFIL=E4+2*ZT2*E4D/WT2+E4DD/WT2^2;

E=.5*(EOLD+E+H*ED);
ED=.5*(EDOLD+ED+H*EDD);
E1=.5*(E1OLD+E1+H*E1D);
E1D=.5*(E1DOLD+E1D+H*E1DD);
E1DD=.5*(E1DDOLD+E1DD+H*E1DDD);
DEL=.5*(DELOLD+DEL+H*DELD);
DELD=.5*(DELDOLD+DELD+H*DELDD);
E2=.5*(E2OLD+E2+H*E2D);
E2D=.5*(E2DOLD+E2D+H*E2DD);
E3=.5*(E3OLD+E3+H*E3D);
E3D=.5*(E3DOLD+E3D+H*E3DD);
```

```
E3DD=.5*(E3DDOLD+E3DD+H*E3DDD);
E4=.5*(E4OLD+E4+H*E4D);
E4D=.5*(E4DOLD+E4D+H*E4DD);

S=S+H;
if S>=(TS-.00001)
    S=0.;
    count=count+1;
    ArrayT(count)=T;
    ArrayTHDTOT(count)=THDTOT;
  end
end
figure
plot(ArrayT,ArrayTHDTOT),grid
xlabel('Time (Sec)')
ylabel('Total Body Rate (deg/s)')
clc
output=[ArrayT',ArrayTHDTOT'];
save datfil.txt output -ascii
disp 'simulation finished'
```

The simulation of Listing 2.3 was run, and we can see from Fig. 2.14 that the first mode notch filter stabilizes the response of the rate gyro flight-control system to the first bending mode (FB1=1, FB2=0, NOTCH=1).

If, on the other hand, there were two bending modes (FB1=1, FB2=1) and we only had a notch filter for the first mode (NOTCH=1), we can see from Fig. 2.15 that the rate gyro flight-control system becomes unstable.

If, on the other hand, there were two bending modes (FB1=1, FB2=1) and we had a notch filter for the two modes (NOTCH=2), we can see from Fig. 2.16 that the rate gyro flight-control system is stabilized.

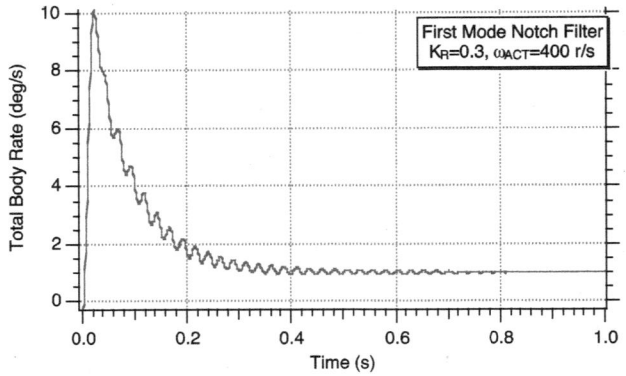

Fig. 2.14   Using a notch filter is another method for stabilizing total body rate response.

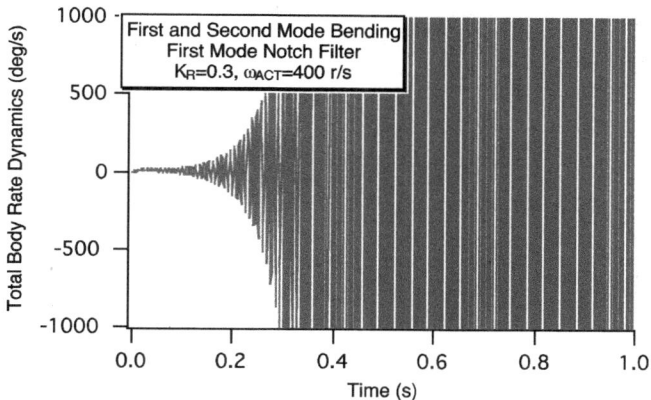

Fig. 2.15   First mode notch filter not adequate for handling second bending mode.

Let's see if there is another way of stabilizing the rate gyro flight-control system in the presence of two bending modes without the use of a notch filter for each body bending mode. Figure 2.17 considers just reducing the flight-control system gain K_R from 0.3 to 0.1 while keeping the actuator bandwidth at 400 rad/s. We can see from Fig. 2.17 that reducing the flight-control system gain alone is not sufficient for eliminating the destabilizing effects of the first two body bending modes.

Again let's see if there is another way of stabilizing the rate gyro flight-control system without the use of a notch filter for each body bending mode. Figure 2.18

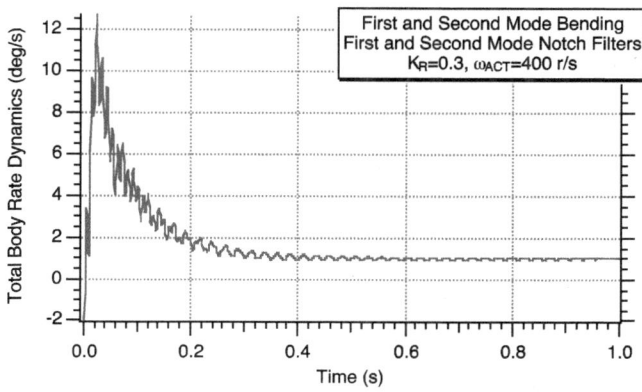

Fig. 2.16   First and second mode notch filters are required for stabilizing first and second bending modes.

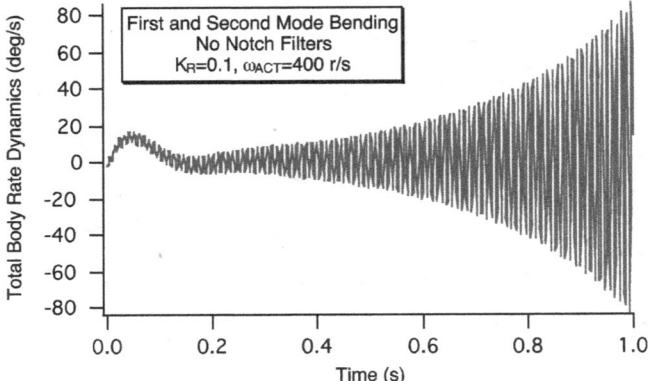

Fig. 2.17   Reducing flight-control system gain alone in presence of first and second bending modes is not sufficient for stabilizing the flight-control system.

considers reducing the actuator bandwidth from 400 rad/s to 200 rad/s while keeping the flight-control system gain at 0.3. We can see from Fig. 2.18 that reducing the actuator bandwidth alone is also not sufficient for eliminating the destabilizing effects of the first two body bending modes.

Finally let's see if there is another way of stabilizing the rate gyro flight-control system without the use of a notch filter for each body bending mode. Figure 2.19 considers reducing both the flight-control system gain K_R from 0.3 to 0.1 and the actuator bandwidth ω_{ACT} from 400 rad/s to 200 rad/s. We can see from Fig. 2.19 that reducing both the flight-control system gain and actuator bandwidth is sufficient for eliminating the destabilizing effects of the first two body bending modes.

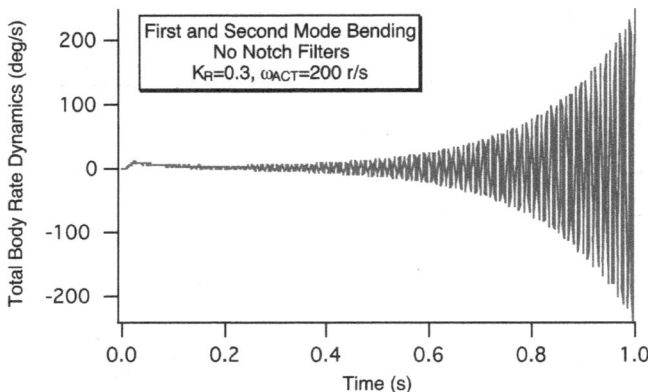

Fig. 2.18   Reducing actuator bandwidth alone in presence of first and second bending modes is not sufficient for stabilizing the flight-control system

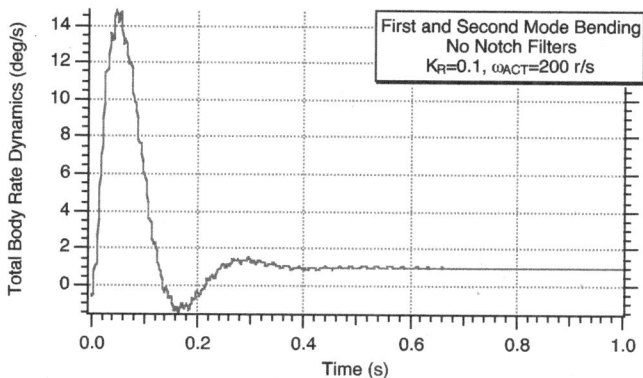

Fig. 2.19   Reducing gain and actuator bandwidth in presence of first and second bending modes stabilizes the flight control system.

SUMMARY

In this chapter we have shown the importance of considering body bending effects in order to avoid potential stability problems. The use of notch filters was shown to be one way of stabilizing the flight-control system in the presence of body bending. Reducing the autopilot gain and actuator bandwidth were shown to be additional ways of stabilizing the flight-control system. The effectiveness of all the techniques discussed was demonstrated by many numerical examples.

REFERENCES

[1]   Nesline, F. W., and Nesline, L. M., "Phase and Gain Stabilization of Structural Feedback Oscillations in Homing Missile Autopilots," American Control Conference, Boston, MA, 1985, pp. 323–329.

[2]   Wells, B. H., "Tactical Missile Structural Testing and Model Verification for Autopilot Design," AIAA Guidance, Navigation and Control Conference, AIAA Paper 91-2616, 1991.

CHAPTER 3

Modern Control and Autopilot Design

INTRODUCTION

Modern control system design techniques are powerful tools for control system engineers. In Chapter 12 of this volume we shall show how these techniques can also be used to design missile guidance laws. However, when it comes to autopilot design, where high frequency effects are extremely important, the direct application of modern control techniques can lead to a flight-control system that is unduly complex and extremely sensitive to high frequency effects that are not modeled. The purpose of this chapter is to illustrate, via numerical examples taken from Ref. [1], how things can go terribly wrong if modern control is directly applied to autopilot design without investigating the resultant design in the frequency domain. Examples of increasing complexity are presented, showing how modern roll autopilot designs with apparently adequate stability characteristics (i.e., good phase and gain margins) can easily go unstable. As was the case in Chapters 22–23 of the previous volume and Chapter 1 of this volume, the importance of the open-loop crossover frequency is again demonstrated, and its relationship to robustness is highlighted though examples.

MODERN CONTROL APPLIED TO MISSILE ROLL AUTOPILOT

Modern control theory can be applied if our model of the real world can be expressed in state space form as

$$\dot{x} = Fx + Gu$$

and the control u is linearly related to the states x according to

$$u = -Cx$$

In the preceding equation, C is a set of control gains. With modern control techniques we are trying to minimize a performance index J that can be expressed in matrix form and is given by

$$J = x^T(t_F)S_F x(t_F) + \int_0^{t_F} (x^T A x + u^T B u)\,dt$$

where $x(t_F)$ is the state vector at the final time.

Minimizing the performance index involves solving the nonlinear matrix Riccati differential equation

$$\dot{S} = -SF - F^T S + SGB^{-1}G^T S - A$$

with boundary value

$$S(t_F) = S_F$$

The control gains C are related to the solution of the preceding differential equation for S and can be obtained from

$$C = B^{-1}G^T S$$

There can be significant numerical difficulty in solving the matrix Riccati differential equation numerically because it is a boundary-value problem. In fact, in this text and the previous volume we have not yet solved a boundary-value problem numerically. However, this difficulty can be easily avoided by simply changing variables in the Riccati equation in order to convert the nonlinear matrix differential equation with a boundary value to an alternate matrix differential equation with an initial condition. Recall that from the overdot notation we know that

$$\dot{S} = \frac{dS}{dt}$$

and if we define

$$\tau = t_F - t$$

we can say that

$$\frac{dS}{dt} = -\frac{dS}{d\tau}$$

Therefore, by changing variables the nonlinear matrix Riccati differential equation becomes

$$\dot{S} = SF + F^T S - SGB^{-1}G^T S + A$$

with initial value

$$S(0) = S_F$$

In autopilot design we are *not* trying to control the final state, and so

$$S_F = 0$$

Because we have changed variables, S is now a function of τ or time to go rather than t or time. The control gains for the optimal control law are still obtained from

$$C = B^{-1}G^T S$$

but again the control gains C are now functions of time to go rather than a function of time. A very small integration step size is often required for the successful integration of the matrix Riccati differential equation because of its numerical fragility. In autopilot design we will integrate the matrix Riccati differential equation until a steady-state condition is reached (i.e., S approaches a matrix of constants, and therefore the control gains C will approach a set of constants).

The transfer function that forms the basis of the roll autopilot design is the following rigid body equation:

$$\frac{\dot{\phi}}{\delta_c} = \frac{K_\delta}{\omega_{RR}} \frac{1}{1 + s/\omega_{RR}}$$

where $\dot{\phi}$ is the roll rate, δ_c is the fin deflection command, K_δ is the fin effectiveness, and ω_{RR} is the roll rate bandwidth. Typically the roll rate response, determined by the preceding equation, is too slow and must be made faster by the autopilot. The maximum roll angle and rate must be kept within specified limits. The preceding transfer function appears in block diagram form in Fig. 3.1 with an extra state added so that we can control the roll angle ϕ.

Cross multiplying terms of the transfer function for the second-order plant model, as was done in Chapter 1 of the previous volume, and converting from the Laplace transform domain to the time domain yields the second-order differential equation

$$\ddot{\phi} = -\omega_{RR}\dot{\phi} + K_\delta\delta_c$$

The preceding differential equation can be put into state space form as

$$\begin{bmatrix} \dot{\phi} \\ \ddot{\phi} \end{bmatrix} = \begin{bmatrix} 0 & 1 \\ 0 & -\omega_{RR} \end{bmatrix} \begin{bmatrix} \phi \\ \dot{\phi} \end{bmatrix} + \begin{bmatrix} 0 \\ K_\delta \end{bmatrix} \delta_c$$

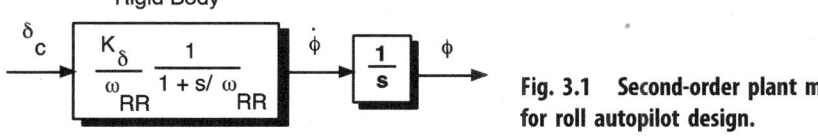

Fig. 3.1   Second-order plant model for roll autopilot design.

The performance index that we would like to minimize for this example is

$$J = \int_0^\infty \left[\left(\frac{\delta_{cMX}}{\phi_{MX}} \right)^2 \phi^2 + \left(\frac{\delta_{cMX}}{\dot{\phi}_{MX}} \right)^2 \dot{\phi}^2 + \delta_c^2 \right]$$

where ϕ_{MX} is the maximum desired value for the roll angle ϕ, $\dot{\phi}_{MX}$ is the maximum desired value for the roll rate $\dot{\phi}$, and δ_{cMX} is the maximum desired value for the fin deflection command δ_c. Therefore, the states of the system and the A matrix are given by

$$x = \begin{bmatrix} \phi \\ \dot{\phi} \end{bmatrix}$$

$$A = \begin{bmatrix} \left(\dfrac{\delta_{cMX}}{\phi_{MX}} \right)^2 & 0 \\ 0 & \left(\dfrac{\delta_{cMX}}{\dot{\phi}_{MX}} \right)^2 \end{bmatrix}$$

and the matrices required by the Riccati equation are

$$F = \begin{bmatrix} 0 & 1 \\ 0 & -\omega_{RR} \end{bmatrix}$$

$$G = \begin{bmatrix} 0 \\ K_\delta \end{bmatrix}$$

$$S(0) = \begin{bmatrix} 0 & 0 \\ 0 & 0 \end{bmatrix}$$

$$B = 1$$

Because the control is linearly related to the states, the feedback control law is of the form

$$\delta_c = -C_1 \phi - C_2 \dot{\phi}$$

Listing 3.1 numerically integrates the matrix differential Riccati equation to solve for the control gains C_1 and C_2. Notice that a small integration step size is used (H=.0002). We can see from Listing 3.1 that it is assumed that the roll rate bandwidth ω_{RR} is 2 rad/s and the fin effectiveness K_δ is 9000 $1/s^2$.

LISTING 3.1   SOLVING FOR CONTROL GAINS IN TWO-STATE SYSTEM

```
clear
count=0;
ORDER=2;
KDEL=9000.;
```

```
WRR=2.;
DELCMX=30.;
PHIMX=10.;
PHIDMX=300.;
G=zeros([ORDER,1]);
A=zeros([ORDER,ORDER]);
F=zeros([ORDER,ORDER]);
S=zeros([ORDER,ORDER]);
G(2,1)=KDEL;
F(1,2)=1;
F(2,2)=-WRR;
A(1,1)=(DELCMX/PHIMX)^2;
A(2,2)=(DELCMX/PHIDMX)^2;
T=0;
H=.0002;
S1=0;
TF=5.;
while ~(T >= (TF-.0001))
    S1=S1+H;
    SOLD=S;
    STEP=1;
    FLAG=0;
    while STEP<=1
        if FLAG==1
            STEP=2;
            HSD=H*SD;
            S=S+HSD;
            T=T+H;
        end
        SD=S*F+F'*S-S*G*G'*S+A;
        C=G'*S;
        FLAG=1;
    end
    FLAG=0;
    H2=.5*H;
    HSDP=H2*SD;
    SS=SOLD+S;
    SSP=.5*SS;
    S=SSP+HSDP;
    if S1>=.0499999
        S1=0;
        C1=C(1,1);
        C2=C(1,2);
        count=count+1;
        ArrayT(count)=T;
        ArrayC1(count)=C1;
        ArrayC2(count)=C2;
```

```
        end
end
output=[ArrayT',ArrayC1',ArrayC2'];
save datfil.txt output -ascii
disp 'simulation finished'
clc
figure
plot(ArrayT,ArrayC1),grid
xlabel('Time (s) ')
ylabel('C1')
figure
plot(ArrayT,ArrayC2),grid
xlabel('Time (s) ')
ylabel('C2')
C1
C2
```

Listing 3.1 was run, and the two resultant control gains in the steady state turned out to be

$$C_1 = 3$$
$$C_2 = 0.103$$

Next a simulation was written to perform a transient response of the roll autopilot using the two preceding control gains. The simulation appears in Listing 3.2. The simulation has the possibility of running when the plant model matches the two-state controller (i.e., no actuator dynamics, QACT=0) or when actuator dynamics are included in the plant model (i.e., QACT=1). It is important to emphasize that the actuator dynamics were not considered in the derivation of the two control gains. When actuator dynamics are considered, the actuator model is given by the transfer function

$$\frac{\delta}{\delta_c} = \frac{1}{1 + \dfrac{2\zeta_{ACT}}{\omega_{ACT}}s + \dfrac{s^2}{\omega_{ACT}^2}}$$

where the damping of the actuator is 0.65 (ZACT=.65) and the bandwidth of the actuator is 100 rad/s (WACT=100).

LISTING 3.2   CHECKING TWO-STATE AND FOUR-STATE PLANT WITH TWO-STATE CONTROL

```
count=0;
QACT=1;
QACT=0;
WACT=100.;
ZACT=.65;
XKD=9000.;
```

```
WRR=2.;
C1=3.;
C2=.103;
T=0;
H=.0002;
S=0.;
PHI=10.;
PHID=0.;
DEL=0.;
DELD=0.
while ~(T>.5)
    S=S+H;
    PHIOLD=PHI;
    PHIDOLD=PHID;
    DELOLD=DEL;
    DELDOLD=DELD;
    STEP=1;
    FLAG=0;
    while STEP <=1
      if FLAG==1
          STEP=2;
          PHI=PHI+H*PHID;
          PHID=PHID+H*PHIDD;
          DEL=DEL+H*DELD;
          DELD=DELD+H*DELDD;
          T=T+H;
      end
      DELC=-C1*PHI-C2*PHID;
      DELDD=WACT*WACT*(DELC-DEL-2.*ZACT*DELD/WACT);
          if QACT==1
              E=DEL;
          else
              E=DELC
          end
      PHIDD=XKD*E-WRR*PHID;
      FLAG=1;
    end
    FLAG=0;
    PHI=.5*(PHIOLD+PHI+H*PHID);
    PHID=.5*(PHIDOLD+PHID+H*PHIDD);
    DEL=.5*(DELOLD+DEL+H*DELD);
    DELD=.5*(DELDOLD+DELD+H*DELDD);
    if S>=.0-9999
      S=0.;
      count=count+1;
      ArrayT(count)=T;
      ArrayPHI(count)=PHI;
```

```
        ArrayPHID(count)=PHID;
        ArrayDEL(count)=DEL;
     end
end
figure
plot(ArrayT,ArrayPHI),grid
xlabel('Time (s) ')
ylabel('PHI (deg)')
clc
figure
plot(ArrayT,ArrayPHID),grid
xlabel('Time (s) ')
ylabel('PHID (deg/s)')
figure
plot(ArrayT,ArrayDEL),grid
xlabel('Time (s) ')
ylabel('DEL (deg)')
clc
clc
output=[ArrayT',ArrayPHI',ArrayPHID',ArrayDEL'];
save datfil.txt output -ascii
disp '*** Simulation Complete'
```

Listing 3.2 was first run without the actuator dynamics (QACT=0), and the well-behaved roll angle response is shown in Fig. 3.2. The response is stable, and all indications are that the use of optimal control in this application was a

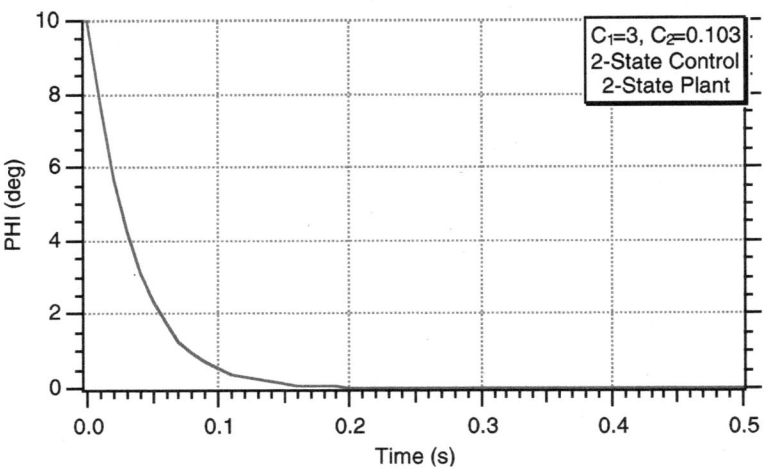

Fig. 3.2   Well-behaved response when control law is matched to the real world.

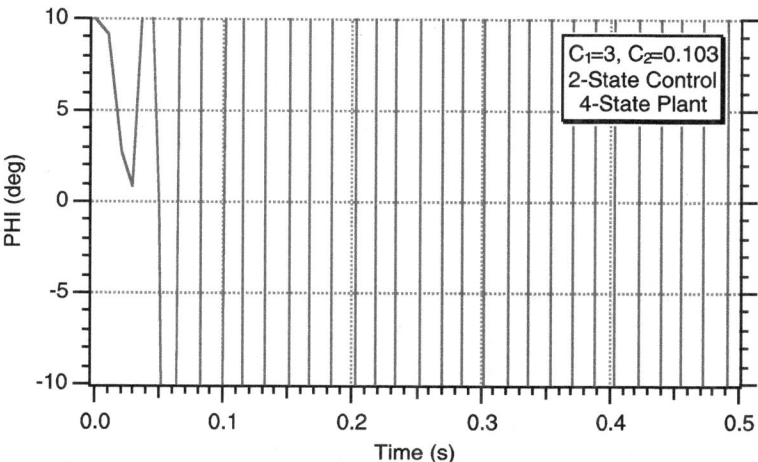

Fig. 3.3   Unstable response when control law is not matched to the real world.

good choice. However, if we include the dynamics of the actuator, Fig. 3.3 indicates an unstable response. The indications are that in this case, because the control law is not matched to the real world, there is a problem with the direct application of modern control to autopilot design.

In order to address the stability problem we just observed, the higher order plant model that includes actuator dynamics, shown in Fig. 3.4, will be used in the derivation of the control gains. It is recognized in advance that this fourth-order model will require four control gains rather than the two control gains of our two-state model.

The two transfer functions for the fourth-order plant model, shown in Fig. 3.4, can be rewritten as two second-order differential equations by cross multiplying terms in the transfer function and converting to the time domain from the Laplace transform domain (see Chapter 1 of the previous volume). The

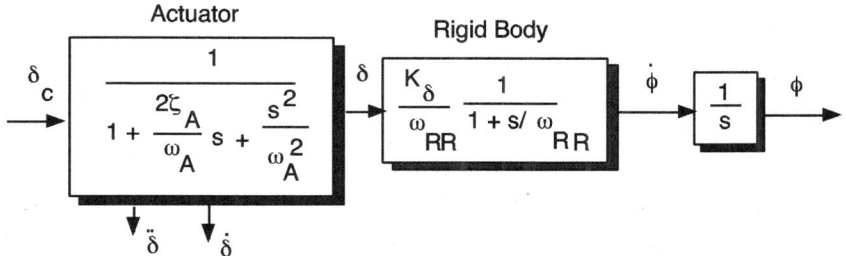

Fig. 3.4   Fourth-order plant model for roll autopilot design.

differential equation describing the rigid body dynamics is

$$\ddot{\phi} = -\omega_{RR}\dot{\phi} + K_{\delta}\delta$$

while the differential equation describing the actuator dynamics can be written as

$$\ddot{\delta} = \omega_A^2\left(\delta_c - \delta - \frac{2\zeta_A}{\omega_A}\dot{\delta}\right)$$

The preceding two differential equations can be put into state space form as

$$\begin{bmatrix} \dot{\phi} \\ \ddot{\phi} \\ \dot{\delta} \\ \ddot{\delta} \end{bmatrix} = \begin{bmatrix} 0 & 1 & 0 & 0 \\ 0 & -\omega_{RR} & K_{\delta} & 0 \\ 0 & 0 & 0 & 1 \\ 0 & 0 & -\omega_A^2 & -2\zeta_A\omega_A \end{bmatrix} \begin{bmatrix} \phi \\ \dot{\phi} \\ \delta \\ \dot{\delta} \end{bmatrix} + \begin{bmatrix} 0 \\ 0 \\ 0 \\ \omega_A^2 \end{bmatrix} \delta_c$$

Again, the performance index that we would like to minimize for this example is

$$J = \int_0^\infty \left[\left(\frac{\delta_{cMX}}{\phi_{MX}}\right)^2 \phi^2 + \left(\frac{\delta_{cMX}}{\dot{\phi}_{MX}}\right)^2 \dot{\phi}^2 + \delta_c^2 \right]$$

where ϕ_{MX} is the maximum desired value for the roll angle ϕ, $\dot{\phi}_{MX}$ is the maximum desired value for the roll rate $\dot{\phi}$, and δ_{cMX} is the maximum desired value for the fin deflection command δ_c. Therefore, the states of the system are

$$x = \begin{bmatrix} \phi \\ \dot{\phi} \\ \delta \\ \dot{\delta} \end{bmatrix}$$

and the matrices required by the Riccati equation are

$$F = \begin{bmatrix} 0 & 1 & 0 & 0 \\ 0 & -\omega_{RR} & K_{\delta} & 0 \\ 0 & 0 & 0 & 1 \\ 0 & 0 & -\omega_A^2 & -2\zeta_A\omega_A \end{bmatrix}$$

$$G = \begin{bmatrix} 0 \\ 0 \\ 0 \\ \omega_A^2 \end{bmatrix}$$

$$A = \begin{bmatrix} \left(\dfrac{\delta_{cMX}}{\phi_{MX}}\right)^2 & 0 & 0 & 0 \\ 0 & \left(\dfrac{\delta_{cMX}}{\dot{\phi}_{MX}}\right)^2 & 0 & 0 \\ 0 & 0 & 0 & 0 \\ 0 & 0 & 0 & 0 \end{bmatrix}$$

$$S = \begin{bmatrix} 0 & 0 & 0 & 0 \\ 0 & 0 & 0 & 0 \\ 0 & 0 & 0 & 0 \\ 0 & 0 & 0 & 0 \end{bmatrix}$$

$$B = 1$$

The feedback control law for the fourth-order system is of the form

$$\delta_c = -C_1\phi - C_2\dot{\phi} - C_3\delta - C_4\dot{\delta}$$

Listing 3.3 numerically integrates the fourth-order matrix differential Riccati equation to solve for the control gains using the preceding matrices as inputs to the program. Again, it is assumed that the roll rate bandwidth ω_{RR} is 2 rad/s, the fin effectiveness K_δ is 9000 $1/s^2$, the actuator damping is 0.65, and the actuator frequency is 100 rad/s.

LISTING 3.3   SOLVING FOR CONTROL GAINS IN FOUR-STATE SYSTEM

```
clear
count=0;
ORDER=4;
WACT=100.;
ZACT=.65;
KDEL=9000.;;
WR=2.;
DELCMX=30.;
PHIMX=10.;
PHIDMX=300.;
G=zeros([ORDER,1]);
A=zeros([ORDER,ORDER]);
F=zeros([ORDER,ORDER]);
S=zeros([ORDER,ORDER]);
G(4,1)=WACT*WACT;
F(1,2)=1;
F(2,2)=-WR;
F(2,3)=KDEL;
F(3,4)=1.;
```

```
F(4,3)=-WACT*WACT;
F(4,4)=-2.*ZACT*WACT;
A(1,1)=(DELCMX/PHIMX)^2;
A(2,2)=(DELCMX/PHIDMX)^2;
T=0;
H=.0002;
S1=0;
TF=5.;
while ~(T >= (TF-.0001))
     S1=S1+H;
     SOLD=S;
     STEP=1;
     FLAG=0;
     while STEP<=1
         if FLAG==1
             STEP=2;
             HSD=H*SD;
             S=S+HSD;
             T=T+H;
         end
         SD=S*F+F'*S-S*G*G'*S+A;
         C=GT*S;
         FLAG=1;
     end
     FLAG=0;
     H2=.5*H;
     HSDP=H2*SD;
     SS=SOLD+S;
     SSP=.5*SS;
     S=SSP+HSDP;
     if S1>=.0499999
             S1=0;
             C1=C(1,1);
             C2=C(1,2);
             C3=C(1,3);
             C4=C(1,4);
             count=count+1;
             ArrayT(count)=T;
             ArrayC1(count)=C1;
             ArrayC2(count)=C2;
             ArrayC3(count)=C3;
             ArrayC4(count)=C4;
     end
end
output=[ArrayT',ArrayC1',ArrayC2',ArrayC3',ArrayC4'];
save datfil.txt output -ascii
```

```
disp 'simulation finished'
clc
figure
plot(ArrayT,ArrayC1),grid
xlabel('Time (s) ')
ylabel('C1')
figure
plot(ArrayT,ArrayC2),grid
xlabel('Time (s) ')
ylabel('C2')
figure
plot(ArrayT,ArrayC3),grid
xlabel('Time (s) ')
ylabel('C3')
figure
plot(ArrayT,ArrayC4),grid
xlabel('Time (s) ')
ylabel('C4')
C1
C2
C3
C4
```

Listing 3.3 was run, and the four control gains in the steady state were found to be

$$C_1 = 3$$
$$C_2 = 0.127$$
$$C_3 = 8.81$$
$$C_4 = 0.0309$$

Next a simulation was written so the transient response of the roll autopilot could be conducted. The simulation appears in Listing 3.4. The simulation has the possibility of running when the plant model matches the four-state controller (QGYR=0) or when rate gyro dynamics are included in the plant model (QGYR=1). In this case, the rate gyro model is given by the transfer function

$$\frac{\dot{\phi}_M}{\dot{\phi}} = \frac{1}{1 + \dfrac{2\zeta_G}{\omega_G}s + \dfrac{s^2}{\omega_G^2}}$$

where the damping of the rate gyro is 0.5 (ZG=.5) and the bandwidth of the rate gyro is 200 rad/s (WG=200). The differential equation for the rate gyro can be found by cross multiplying terms in the preceding transfer function and converting to the time domain from the Laplace transform domain. The resultant differential equation for the rate gyro becomes

$$\ddot{\phi}_M = \omega_G^2\left(\dot{\phi} - \frac{2\zeta_G}{\omega_G}\ddot{\phi}_M - \dot{\phi}_M\right)$$

LISTING 3.4   CHECKING FOUR-STATE AND SIX-STATE PLANT WITH FOUR-STATE CONTROL

```
clear
count=0;
QGYR=0;
QGYR=1;
WG=200.;
ZG=.5;
WACT=100.;
ZACT=.65;
XKD=9000.;
WR=2.;
C1=3.;
C2=.127;
C3=8.81;
C4=.0309;
T=0;
H=.0002;
S=0.;
PHI=10.;
PHID=0.;
DEL=0.;
DELD=0.;
PHIM=10.;
PHIMD=0.;
PHIMDD=0.;
while ~(T>.5)
    S=S+H;
    PHIOLD=PHI;
    PHIDOLD=PHID;
    DELOLD=DEL;
    DELDOLD=DELD;
    PHIMOLD=PHIM;
    PHIMDOLD=PHIMD;
    PHIMDDOLD=PHIMDD;
    STEP=1;
    FLAG=0;
    while STEP <=1
      if FLAG==1
          STEP=2;
          PHI=PHI+H*PHID;
          PHID=PHID+H*PHIDD;
```

```
              DEL=DEL+H*DELD;
              DELD=DELD+H*DELDD;
              PHIM=PHIM+H*PHIMD;
              PHIMD=PHIMD+H*PHIMDD;
              PHIMDD=PHIMDD+H*PHIMDDD;
              T=T+H;
         end
           if QGYR==1
              DELC=-C1*PHIM-C2*PHIMD-C3*DEL-C4*DELD;
              DELDD=WACT*WACT*(DELC-DEL-2.*ZACT*DELD/WACT);
              PHIDD=XKD*DEL-WR*PHID;
              PHIMDDD=WG*WG*(PHID-PHIMD-2.*ZG*PHIMDD/WG);
           else
              DELC=-C1*PHI-C2*PHID-C3*DEL-C4*DELD;
              DELDD=WACT*WACT*(DELC-DEL-2.*ZACT*DELD/WACT);
              PHIDD=XKD*DEL-WR*PHID;
              PHIMDDD=0;
           end
       FLAG=1;
     end
     FLAG=0;
     PHI=.5*(PHIOLD+PHI+H*PHID);
     PHID=.5*(PHIDOLD+PHID+H*PHIDD);
     DEL=.5*(DELOLD+DEL+H*DELD);
     DELD=.5*(DELDOLD+DELD+H*DELDD);
     PHIM=.5*(PHIMOLD+PHIM+H*PHIMD);
     PHIMD=.5*(PHIMDOLD+PHIMD+H*PHIMDD);
     PHIMDD=.5*(PHIMDDOLD+PHIMDD+H*PHIMDDD);
     if S>=.0-9999
       S=0.;
       count=count+1;
       ArrayT(count)=T;
       ArrayPHI(count)=PHI;
       ArrayPHID(count)=PHID;
       ArrayDEL(count)=DEL;
     end
end
figure
plot(ArrayT,ArrayPHI),grid
xlabel('Time (s) ')
ylabel('PHI (deg)')
clc
figure
plot(ArrayT,ArrayPHID),grid
xlabel('Time (s) ')
ylabel('PHID (deg/s)')
figure
```

```
plot(ArrayT,ArrayDEL),grid
xlabel('Time (s) ')
ylabel('DEL (deg)')
clc
clc
output=[ArrayT',ArrayPHI',ArrayPHID',ArrayDEL'];
save datfil.txt output -ascii
disp '*** Simulation Complete'
```

Listing 3.4 was first run without the rate gyro dynamics (QGYR=0), and the resultant, well-behaved roll angle response is shown in Fig. 3.5. The response is stable, and all indications are that the use of optimal control in this application was a good choice. However, if we include the dynamics of the rate gyro (QGYR=1), Fig. 3.6 indicates an unstable response. The indications are that in this case, because the control law is not matched to the real world, again there is certainly a problem.

In order to address the stability problem we just observed, a higher order model that includes rate gyro dynamics, shown in Fig. 3.7, will be used. It is recognized in advance that this sixth-order model will require six control gains rather than the four control gains of our four-state model.

Each of the three transfer functions for the sixth-order plant model, shown in Fig. 3.7, can be rewritten as three second-order differential equations. The differential equation describing the rigid body dynamics is given by

$$\ddot{\phi} = -\omega_{RR}\dot{\phi} + K_\delta\delta$$

and the differential equation describing the actuator dynamics can be written as

$$\ddot{\delta} = \omega_A^2\left(\delta_c - \delta - \frac{2\zeta_A}{\omega_A}\dot{\delta}\right)$$

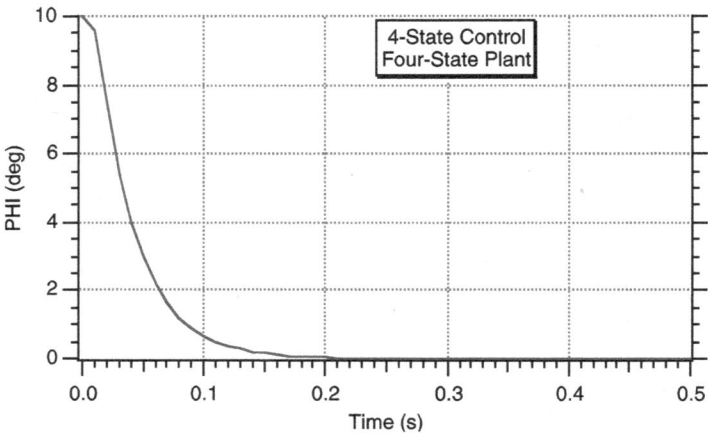

Fig. 3.5   Well-behaved response when control law is matched to the real world.

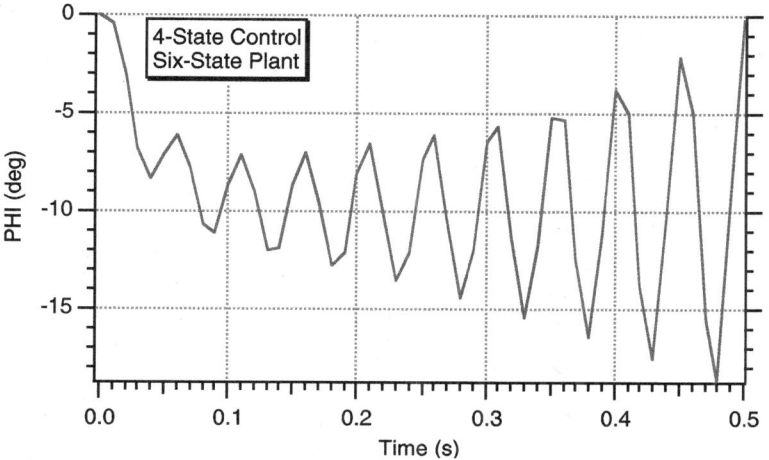

Fig. 3.6   Unstable response when control law is not matched to the real world.

The differential equation describing the rate gyro dynamics was shown to be

$$\ddot{\phi}_M = \omega_G^2\left(\dot{\phi} - \dot{\phi}_M - \frac{2\zeta_G}{\omega_G}\ddot{\phi}_M\right)$$

The preceding three second-order differential equations can be put into state space form as

$$
\begin{bmatrix}
\dot{\phi}_M \\
\ddot{\phi} \\
\dot{\delta} \\
\ddot{\delta} \\
\dot{\phi}_M \\
\ddot{\phi}_M
\end{bmatrix}
=
\begin{bmatrix}
0 & 0 & 0 & 0 & 1 & 0 \\
0 & -\omega_{RR} & K_\delta & 0 & 0 & 0 \\
0 & 0 & 0 & 1 & 0 & 0 \\
0 & 0 & -\omega_A^2 & -2\zeta_A\omega_A & 0 & 0 \\
0 & 0 & 0 & 0 & 0 & 1 \\
0 & \omega_G^2 & 0 & 0 & -\omega_G^2 & -2\zeta_G\omega_G
\end{bmatrix}
\begin{bmatrix}
\phi_M \\
\dot{\phi} \\
\delta \\
\dot{\delta} \\
\phi_M \\
\dot{\phi}_M
\end{bmatrix}
+
\begin{bmatrix}
0 \\
0 \\
0 \\
\omega_A^2 \\
0 \\
0
\end{bmatrix}
\delta_c
$$

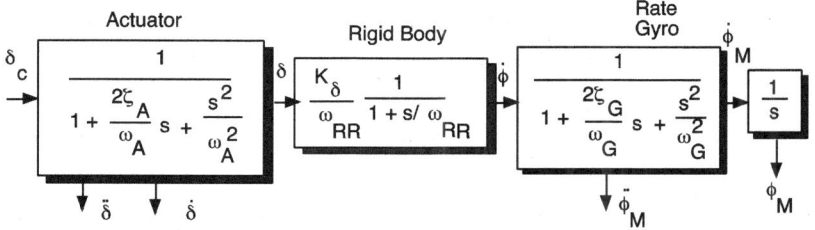

Fig. 3.7   Sixth-order plant model for roll autopilot design.

Again, the performance index that we would like to minimize for this example is still

$$J = \int_0^\infty \left[\left(\frac{\delta_{cMX}}{\phi_{MX}} \right)^2 \phi^2 + \left(\frac{\delta_{cMX}}{\dot\phi_{MX}} \right)^2 \dot\phi^2 + \delta_c^2 \right]$$

where ϕ_{MX} is the maximum desired value for the roll angle ϕ, $\dot\phi_{MX}$ is the maximum desired value for the roll rate $\dot\phi$, and δ_{cMX} is the maximum desired value for the fin deflection command δ_c. The states of the system are

$$x = \begin{bmatrix} \phi_M \\ \dot\phi \\ \delta \\ \dot\delta \\ \dot\phi_M \\ \ddot\phi_M \end{bmatrix}$$

and the matrices required by the matrix Riccati differential equation are

$$F = \begin{bmatrix} 0 & 0 & 0 & 0 & 1 & 0 \\ 0 & -\omega_{RR} & K_\delta & 0 & 0 & 0 \\ 0 & 0 & 0 & 1 & 0 & 0 \\ 0 & 0 & -\omega_A^2 & -2\zeta_A\omega_A & 0 & 0 \\ 0 & 0 & 0 & 0 & 0 & 1 \\ 0 & \omega_G^2 & 0 & 0 & -\omega_G^2 & -2\zeta_G\omega_G \end{bmatrix}$$

$$G = \begin{bmatrix} 0 \\ 0 \\ 0 \\ \omega_A^2 \\ 0 \\ 0 \end{bmatrix}$$

$$A = \begin{bmatrix} \left(\frac{\delta_{cMX}}{\phi_{MX}} \right)^2 & 0 & 0 & 0 & 0 & 0 \\ 0 & \left(\frac{\delta_{cMX}}{\dot\phi_{MX}} \right)^2 & 0 & 0 & 0 & 0 \\ 0 & 0 & 0 & 0 & 0 & 0 \\ 0 & 0 & 0 & 0 & 0 & 0 \\ 0 & 0 & 0 & 0 & 0 & 0 \\ 0 & 0 & 0 & 0 & 0 & 0 \end{bmatrix}$$

$$S = \begin{bmatrix} 0 & 0 & 0 & 0 & 0 & 0 \\ 0 & 0 & 0 & 0 & 0 & 0 \\ 0 & 0 & 0 & 0 & 0 & 0 \\ 0 & 0 & 0 & 0 & 0 & 0 \\ 0 & 0 & 0 & 0 & 0 & 0 \\ 0 & 0 & 0 & 0 & 0 & 0 \end{bmatrix}$$

$$B = 1$$

The feedback control law for the sixth-order system is of the form

$$\delta_c = -C_1\phi - C_2\dot{\phi} - C_3\delta - C_4\dot{\delta} - C_5\dot{\phi}_M - C_6\ddot{\phi}_M$$

Listing 3.5 numerically integrates the matrix differential Riccati equation to solve for the control gains using the preceding matrices as inputs to the program. Again, it is assumed that the roll rate bandwidth ω_{RR} is 2 rad/s, the fin effectiveness $K_{\delta\delta}$ is 9000 1/s^2, the actuator damping is 0.65, the actuator frequency is 100 rad/s, the rate gyro damping is 0.5, and the rate gyro frequency is 200 rad/s.

LISTING 3.5   SOLVING FOR THE CONTROL GAINS OF SIX-STATE PLANT

```
clear
count=0;
ORDER=6;
WG=200.;
ZG=.5;
WACT=100.;
ZACT=.65;
KDEL=9000.;;
WR=2.;
DELCMX=30.;
PHIMX=10.;
PHIDMX=300.;
G=zeros([ORDER,1]);
A=zeros([ORDER,ORDER]);
F=zeros([ORDER,ORDER]);
S=zeros([ORDER,ORDER]);
G(4,1)=WACT*WACT;
F(1,5)=1;
F(2,2)=-WR;
F(2,3)=KDEL;
F(3,4)=1.;
F(4,3)=-WACT*WACT;
F(4,4)=-2.*ZACT*WACT;
F(5,6)=1.;
F(6,2)=WG*WG;
```

```
F(6,5)=-WG*WG;
F(6,6)=-2.*ZG*WG;
A(1,1)=(DELCMX/PHIMX)^2;
A(2,2)=(DELCMX/PHIDMX)^2;
T=0;
H=.0002;
S1=0;
TF=5.;
while ~(T >= (TF-.0001))
      S1=S1+H;
      SOLD=S;
      STEP=1;
      FLAG=0;
      while STEP<=1
          if FLAG==1
                  STEP=2;
                  HSD=H*SD;
                  S=S+HSD;
                  T=T+H;
          end
          SD=S*F+F'*S-S*G*G'*S+A;
          C=G'*S;
          FLAG=1;
      end
      FLAG=0;
      H2=.5*H;
      HSDP=H2*SD;
      SS=SOLD+S;
      SSP=.5*SS;
      S=SSP+HSDP;
      if S1>=.0499999
              S1=0;
              C1=C(1,1);
              C2=C(1,2);
              C3=C(1,3);
              C4=C(1,4);
              C5=C(1,5);
              C6=C(1,6);
              count=count+1;
              ArrayT(count)=T;
              ArrayC1(count)=C1;
              ArrayC2(count)=C2;
              ArrayC3(count)=C3;
              ArrayC4(count)=C4;
              ArrayC5(count)=C5;
              ArrayC6(count)=C6;
      end
  end
```

```
end
output=[ArrayT',ArrayC1',ArrayC2',ArrayC3',ArrayC4',ArrayC5'...ArrayC6'];
save datfil.txt output -ascii
disp 'simulation finished'
clc
figure
plot(ArrayT,ArrayC1),grid
xlabel('Time (s) ')
ylabel('C1')
figure
plot(ArrayT,ArrayC2),grid
xlabel('Time (s) ')
ylabel('C2')
figure
plot(ArrayT,ArrayC3),grid
xlabel('Time (s) ')
ylabel('C3')
figure
plot(ArrayT,ArrayC4),grid
xlabel('Time (s) ')
ylabel('C4')
figure
plot(ArrayT,ArrayC5),grid
xlabel('Time (s) ')
ylabel('C5')
figure
plot(ArrayT,ArrayC6),grid
xlabel('Time (s) ')
ylabel('C6')
C1
C2
C3
C4
C5
C6
```

Listing 3.5 was run, and the six control gains in the steady state turn out to be

$$C_1 = 3$$
$$C_2 = 0.127$$
$$C_3 = 8.84$$
$$C_4 = 0.031$$
$$C_5 = 0.0152$$
$$C_6 = 0.0000768$$

Next a simulation was written that modified Listing 3.4 to include the rate gyro dynamics and the new control law that was aware of the six-state plant. Listing 3.6 has the transient response of the roll autopilot with the new control law.

LISTING 3.6   TRANSIENT RESPONSE OF SIX-STATE PLANT WITH SIX-STATE CONTROL LAW

```
clear
count=0;
WG=200.;
ZG=.5;
WACT=100.;
ZACT=.65;
XKD=9000.;
WR=2.;
C1=3.;
C2=.127;
C3=8.84;
C4=.031;
C5=.0152;
C6=.0000768;
T=0;
H=.0002;
S=0.;
PHI=10.;
PHID=0.;
DEL=0.;
DELD=0.;
PHIM=10.;
PHIMD=0.;
PHIMDD=0.;
while ~(T>.5)
      S=S+H;
      PHIOLD=PHI;
      PHIDOLD=PHID;
      DELOLD=DEL;
      DELDOLD=DELD;
      PHIMOLD=PHIM;
      PHIMDOLD=PHIMD;
      PHIMDDOLD=PHIMDD;
      STEP=1;
      FLAG=0;
      while STEP <=1
        if FLAG==1
            STEP=2;
            PHI=PHI+H*PHID;
            PHID=PHID+H*PHIDD;
            DEL=DEL+H*DELD;
```

```
                    DELD=DELD+H*DELDD;
                    PHIM=PHIM+H*PHIMD;
                    PHIMD=PHIMD+H*PHIMDD;
                    PHIMDD=PHIMDD+H*PHIMDDD;
                    T=T+H;
                end
                DELC=-C1*PHI-C2*PHID-C3*DEL-C4*DELD-C5*PHIMD-C6*PHIMDD;
                DELDD=WACT*WACT*(DELC-DEL-2.*ZACT*DELD/WACT);
                E=DEL;
                PHIDD=XKD*E-WR*PHID;
                PHIMDDD=WG*WG*(PHID-PHIMD-2.*ZG*PHIMDD/WG);
                FLAG=1;
            end
            FLAG=0;
            PHI=.5*(PHIOLD+PHI+H*PHID);
            PHID=.5*(PHIDOLD+PHID+H*PHIDD);
            DEL=.5*(DELOLD+DEL+H*DELD);
            DELD=.5*(DELDOLD+DELD+H*DELDD);
            PHIM=.5*(PHIMOLD+PHIM+H*PHIMD);
            PHIMD=.5*(PHIMDOLD+PHIMD+H*PHIMDD);
            PHIMDD=.5*(PHIMDDOLD+PHIMDD+H*PHIMDDD);
            if S>=.0-9999
                S=0.;
                count=count+1;
                ArrayT(count)=T;
                ArrayPHI(count)=PHI;
                ArrayPHID(count)=PHID;
                ArrayDEL(count)=DEL;
            end
end
figure
plot(ArrayT,ArrayPHI),grid
xlabel('Time (s) ')
ylabel('PHI (deg)')
clc
figure
plot(ArrayT,ArrayPHID),grid
xlabel('Time (s) ')
ylabel('PHID (deg/s)')
figure
plot(ArrayT,ArrayDEL),grid
xlabel('Time (s) ')
ylabel('DEL (deg)')
clc
clc
output=[ArrayT',ArrayPHI',ArrayPHID',ArrayDEL'];
save datfil.txt output -ascii
disp '*** Simulation Complete'
```

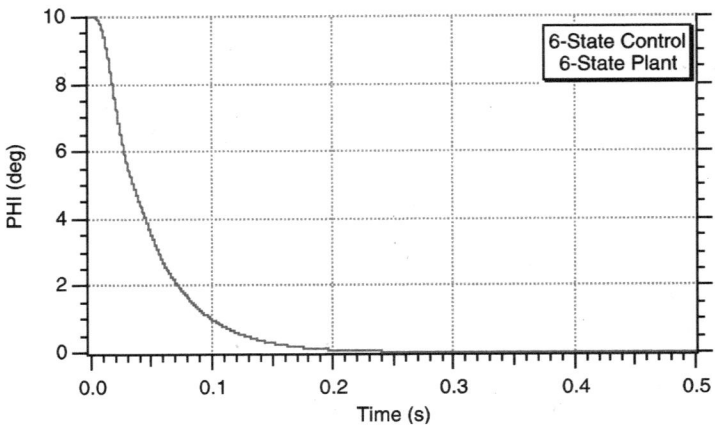

Fig. 3.8 Well-behaved response when six-state control law is matched to six-state plant.

The nominal case of Listing 3.6 was run, and the well-behaved roll angle response is shown in Fig. 3.8. The response is stable, and all indications are that the use of optimal control in this application was a good choice. However, based on past experience in this section, we know that if we include other higher order effects that are not modeled in the derivation of the control law, the system will again destabilize.

CLASSICAL ANALYSIS OF MODERN CONTROL TECHNIQUES

The series of three examples in the previous section shows that the order of the design model must match the order of the plant just to yield a stable response. In this section we will use classical control techniques to try and explain what we have observed in the previous section. We will develop in detail the open-loop transfer function for the simplest of the three examples and then state the open-loop transfer functions for the other two systems to see if this will help explain the sensitivity of the modern control techniques to unmodeled dynamics. Figure 3.9 presents the two-state plant with the two-state controller as an aid in trying to develop the open-loop transfer function of this system.

To find the open-loop response of Fig. 3.9 we can use the techniques from Chapter 22 of the previous volume. We break the loop at the input to the rigid body dynamics, zero the input, and label the output y of Fig. 3.9. The open-loop transfer function of the two-state system can then be written by inspection as

$$HG(s)_{2-\text{state}} = -\frac{y}{x} = \frac{K_\delta(C_1 + C_2 s)}{\omega_{RR}s\left(1 + \dfrac{s}{\omega_{RR}}\right)} = 13{,}500\frac{\left(1 + \dfrac{s}{29.1}\right)}{s\left(1 + \dfrac{s}{2}\right)}$$

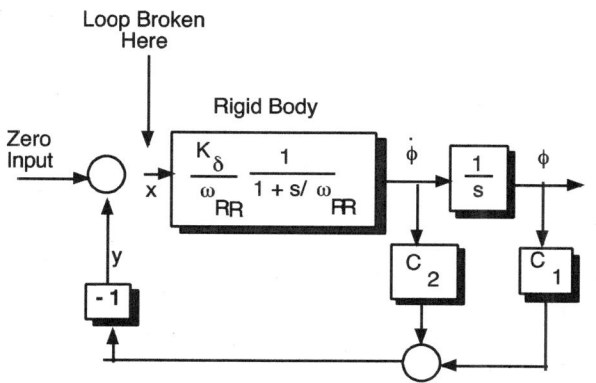

Fig. 3.9   Open-loop diagram of two-state plant and two-state control law.

The magnitude and phase of the open-loop transfer function can now be computed. As was done in Chapter 22 of the previous volume, we first let $s = j\omega$

$$HG(j\omega)_{2-\text{state}} = 13{,}500 \frac{1 + j\omega/29.1}{j\omega(1 + j\omega/2)}$$

and then we can see that the magnitude of the open-loop transfer function can be computed as

$$|HG| = 13{,}500 \frac{\sqrt{1 + (\omega/29.1)^2}}{\omega\sqrt{1 + (\omega/2)^2}} = \frac{13{,}500}{\omega} \sqrt{\frac{1 + (\omega/29.1)^2}{1 + (\omega/2)^2}}$$

The open-loop magnitude can be converted to a gain measured in dB by using the formula

$$\text{Gain} = 20 \log_{10} |HG(j\omega)|$$

and the phase of the open-loop transfer function, expressed in units of degrees, is simply

$$\text{Phase}_{HG2-\text{state}} = 57.3 \tan^{-1}(\omega/29.1) - 90 - 57.3 \tan^{-1}(\omega/2)$$

Listing 3.7 computes the gain in dB and phase in deg using the preceding equations for the open-loop transfer function in order to gain information on the phase and gain margins and the crossover frequency. As was mentioned in Chapter 22 of the previous volume, this type of information display (i.e., gain and phased vs frequency) is known as a Bode plot.

LISTING 3.7   COMPUTING THE MAGNITUDE AND PHASE OF THE OPEN-LOOP TRANSFER FUNCTION OF TWO-STATE PLANT WITH TWO-STATE CONTROL

```
clear
count=0;
for I=2:160
    W=10^(.025*I-1);
    TOP=13500*sqrt(1+(W/29.1)^2);
```

```
BOT=W*sqrt(1+(W/2.)^2);
XMAG=TOP/BOT;
GAIN=20*log10(XMAG);
PHASE=57.3*atan2(W,29.1)-90.-57.3*atan2(W,2.);
count=count+1;
ArrayW(count)=W;
ArrayGAIN(count)=GAIN;
ArrayPHASE(count)=PHASE;
end
figure
semilogx(ArrayW,ArrayGAIN),grid
xlabel('Frequency (Rad/Sec)')
ylabel('Gain (Db)')
axis([.1 1000 -60 40])
figure
semilogx(ArrayW,ArrayPHASE),grid
xlabel('Frequency (Rad/Sec)')
ylabel('Phase (Deg)')
axis([.1 1000 -400 100])
clc
output=[ArrayW',ArrayGAIN',ArrayPHASE'];
save datfil.txt output -ascii
disp 'simulation finished'
```

The nominal case of Listing 3.7 was run, and the resultant Bode plot of the open-loop transfer function appears in in Fig. 3.10. We can see from Fig. 3.10 that the stability margins are excellent (i.e., infinite gain margin and 90-deg

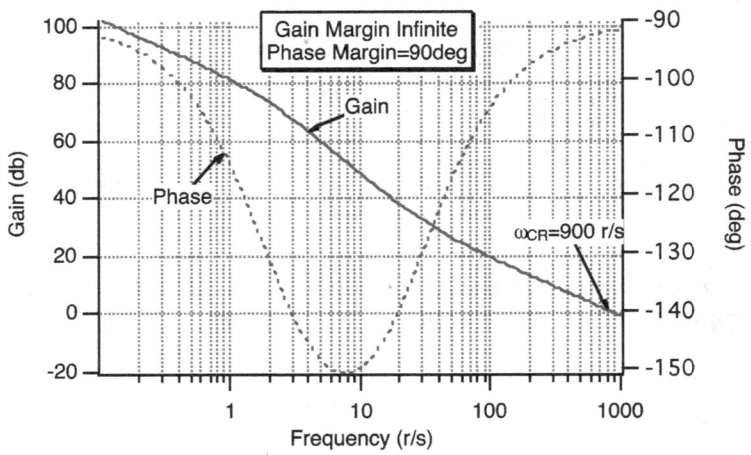

Fig. 3.10 Crossover frequency is too high for two-state plant and two-state controller.

phase margin) *but the crossover frequency is 900 rad/s*, which is extremely high. Recall our work with the three-loop autopilot in Chapter 23 of the previous volume, where we showed that the crossover frequency should be no more than one-third of the actuator bandwidth to avoid stability issues. If we allowed the crossover frequency to be 900 rad/s, it would mean that the actuator bandwidth would have to be 2700 rad/s to satisfy our previous experience. In our time domain example, when actuator dynamics were included, the system went unstable because in that example the actuator bandwidth was only 100 rad/s. In other words, the excellent phase and gain margins associated with the two-state plant and controller were only partial indicators of good stability characteristics. *The phase and gain margins only have practical meaning if we have realistic cross-over frequencies!*

In order to demonstrate that the crossover frequency is too high for an actuator with a 100-rad/s bandwidth, an experiment was conducted with Listing 3.2 to see what would happen if we could increase the actuator bandwidth. If we are looking at higher bandwidth actuators, the integration step size must be reduced in Listing 3.2 by an order of magnitude (i.e., H=.00002 rather than H=.0002). Figure 3.11 shows that if we increase the actuator bandwidth from 100 rad/s to 750 rad/s, the system is still unstable. However, Fig. 3.11 also shows that if we increase the actuator bandwidth further to 2700 rad/s (i.e., three times the crossover frequency), the system is stabilized. The preceding practical experiment illustrates the critical importance of the open-loop transfer function crossover frequency in the design process.

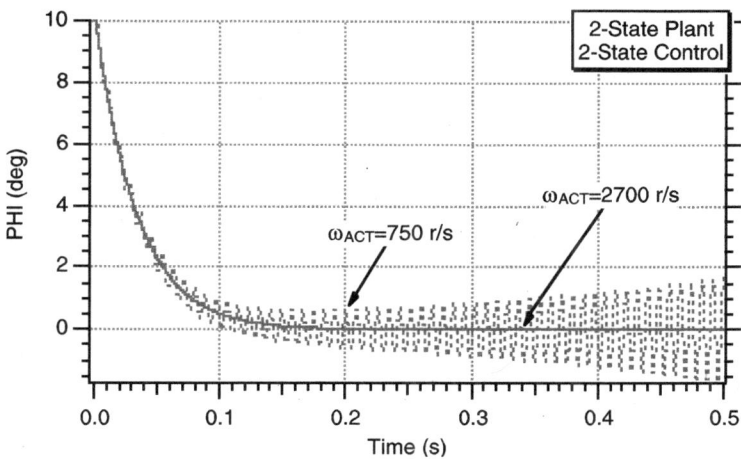

Fig. 3.11   Very high actuator bandwidths are required if crossover frequency is neglected in the autopilot design process.

Now that we have demonstrated the importance of the open-loop crossover frequency, let us now investigate the crossover frequencies of both the four-state and six-state plants with four- and six-state controllers, respectively. A great deal of extra work allows us to write the open-loop transfer functions of the four-state and six-state systems as

$$HG(s)_{4\text{-state}} = 13,500 \frac{(1 + s/29.1)\left(1 + \dfrac{2 * 0.74}{173}s + \dfrac{s^2}{173^2}\right)}{s(1 + s/2)\left(1 + \dfrac{2 * 0.65}{100}s + \dfrac{s^2}{100^2}\right)}$$

$$HG(s)_{6\text{-state}} = 13,500 \frac{(1 + s/29)\left(1 + \dfrac{2 * 0.83}{160}s + \dfrac{s^2}{160^2}\right)\left(1 + \dfrac{2 * 0.45}{216}s + \dfrac{s^2}{216^2}\right)}{s(1 + s/2)\left(1 + \dfrac{2 * 0.65}{100}s + \dfrac{s^2}{100^2}\right)\left(1 + \dfrac{2 * 0.5}{200}s + \dfrac{s^2}{200^2}\right)}$$

It is apparent from the preceding two transfer functions that modern control (i.e., full state feedback) influences only the zeroes of the open-loop transfer functions, whereas the plant determines the poles. Therefore, classical modification of the open-loop transfer function by the inclusion of compensation networks is not possible with the modern approach. In addition, full state feedback ensures that the degree of the numerator is always one less than the degree of the denominator. Therefore, there will not be a great deal of attenuation in the open-loop transfer function at the higher frequencies. We can also see that the open-loop gain constant (i.e., 13,500) is invariant with system order. This means that only the performance index weightings influence the open loop gains. The computation of the gain of the open-loop transfer function for the four-state system appears in Listing 3.8.

LISTING 3.8   COMPUTING GAIN OF FOUR-STATE OPEN-LOOP TRANSFER FUNCTION

```
clear
count=0;
for I=2:160
    W=10^(.025*I-1);
    TOP1=13500*sqrt(1+(W/29.1)^2);
    TOP2=sqrt((1-W*W/(173.*173.))^2+(2.*.74*W/173)^2);
    BOT1=W*sqrt(1+(W/2.)^2);
    BOT2=sqrt((1-W*W/(100.*100.))^2+(2.*.65*W/100)^2);
    XMAG=TOP1*TOP2/(BOT1*BOT2);
    GAIN=20*log10(XMAG);
    count=count+1;
    ArrayW(count)=W;
    ArrayGAIN(count)=GAIN;
end
```

```
figure
semilogx(ArrayW,ArrayGAIN),grid
xlabel('Frequency (Rad/Sec)')
ylabel('Gain (Db)')
clc
output=[ArrayW',ArrayGAIN'];
save datfil.txt output -ascii
```

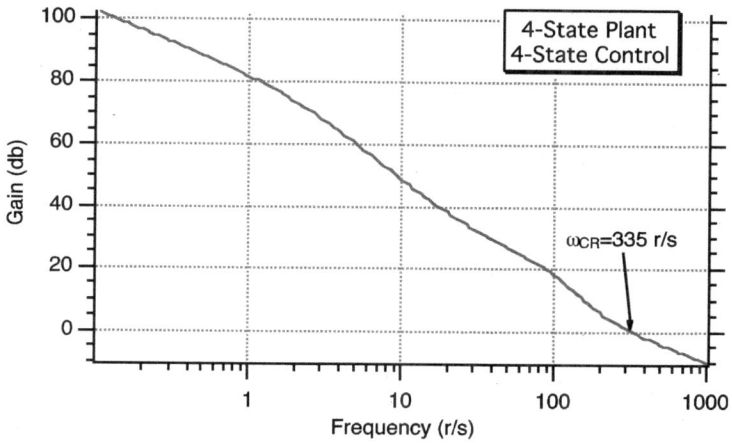

Fig. 3.12   Crossovef frequency of four-state plant with four-state controller is high.

The nominal case of Listing 3.8 was run, and the gain as a function of frequency appears in Fig. 3.12. We can see that when the gain is 0 dB (i.e., XMAG=1), *the crossover frequency is approximately 335 rad/s*, which is still high compared to the bandwidths of both the actuator and rate gyro.

The computation of the gain of the open-loop transfer function for the six-state system is performed in Listing 3.9.

LISTING 3.9   COMPUTING GAIN OF SIX-STATE OPEN-LOOP TRANSFER FUNCTION

```
clear
count=0;
for I=2:160
    W=10^(.025*I-1);
    TOP1=13500*sqrt(1+(W/29.)^2);
    TOP2=sqrt((1-W*W/(160.*160.))^2+(2.*.83*W/160)^2);
    TOP3=sqrt((1-W*W/(216.*216.))^2+(2.*.45*W/216)^2);
    BOT1=W*sqrt(1+(W/2.)^2);
    BOT2=sqrt((1-W*W/(100.*100.))^2+(2.*.65*W/100)^2);
    BOT3=sqrt((1-W*W/(200.*200.))^2+(2.*.5*W/200)^2);
```

```
XMAG=TOP1*TOP2*TOP3/(BOT1*BOT2*BOT3);
GAIN=20*log10(XMAG);
count=count+1;
ArrayW(count)=W;
ArrayGAIN(count)=GAIN;
end
figure
semilogx(ArrayW,ArrayGAIN),grid
xlabel('Frequency (Rad/Sec)')
ylabel('Gain (Db)')
axis([.1 1000 -60 40])
figure
semilogx(ArrayW,ArrayPHASE),grid
xlabel('Frequency (Rad/Sec)')
ylabel('Phase (Deg)')
axis([.1 1000 -400 100])
clc
output=[ArrayW',ArrayGAIN',ArrayPHASE'];
save datfil.txt output -ascii
disp 'simulation finished'
```

The nominal case of Listing 3.9 was run, and the gain as a function of frequency appears in Fig. 3.13. We can see that when the gain is 0 dB (i.e., XMAG=1), *the crossover frequency is approximately 334 rad/s*, which is still high compared to the bandwidths of both the actuator and rate gyro.

In the examples presented so far, we have shown that the crossover frequency is much too high. We must find a way to reduce the crossover frequency if we are

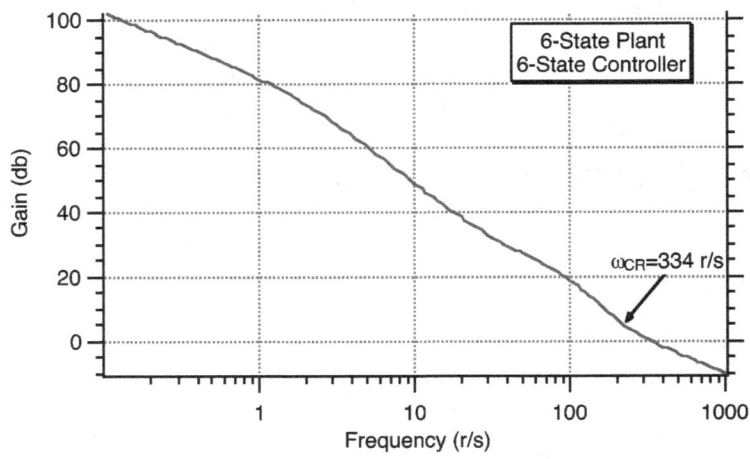

Fig. 3.13   Crossover frequency of six-state plant with six-state controller is high.

ever to get an acceptable autopilot design using the tools that modern control theory provides.

HOW TO USE MODERN CONTROL FOR PRACTICAL SYSTEM DESIGN

Most of the problems in the previous sections were caused by high crossover frequency. The open-loop crossover frequency can be controlled by adjusting parameters within the performance index. For this example, numerical experiments have shown that variations in δ_{cMX} have the most profound influence on the crossover frequency. Table 3.1 shows how the control gains vary with δ_{cMX} for the two-state controller and two-state plant. We can see from Table 3.1 that as the parameter δ_{cMX} is decreased, the two control gains also decrease.

Recall that for the two-state plant and controller, the open-loop transfer function was

$$HG(s)_{2-\text{state}} = \frac{K_\delta(C_1 + C_2 s)}{\omega_{RR}s\left(1 + \dfrac{s}{\omega_{RR}}\right)}$$

Letting $s = j\omega$ yields

$$HG(j\omega)_{2-\text{state}} = \frac{K_\delta(C_1 + C_2 j\omega)}{\omega_{RR}j\omega\left(1 + \dfrac{j\omega}{\omega_{RR}}\right)} = \frac{9000(C_1 + C_2 j\omega)}{2j\omega\left(1 + \dfrac{j\omega}{2}\right)}$$

Therefore, the magnitude of the open-loop transfer function is

$$|HG(j\omega)|_{2-\text{state}} = \frac{9000\sqrt{C_1^2 + C_2^2\omega^2}}{2\omega\sqrt{1 + \dfrac{\omega^2}{4}}}$$

TABLE 3.1 VARIATION OF TWO-STATE CONTROLLER GAINS WITH PERFORMANCE INDEX PARAMETER δ_{cMX}

δ_{cMX}	C_1	C_2
30 deg	3	0.103
20 deg	2	0.0697
10 deg	1	0.0363
1 deg	0.1	0.0056
0.5 deg	0.05	0.0035
0.1 deg	0.01	0.0013

and the gain is simply

$$\text{Gain} = 20 \log_{10} |HG(j\omega)|_{2-\text{state}}$$

Listing 3.7, which computed the magnitude of the open-loop transfer function for the two-state plant with the two-state controller, was slightly modified so that the controller gains would appear explicitly. The preceding magnitude of the open-loop transfer function formula was programmed into a slight modification of Listing 3.7, and appears in Listing 3.10. With this program we can compute the open-loop crossover frequency for different sets of the two control gains.

LISTING 3.10   COMPUTING MAGNITUDE OF OPEN-LOOP TRANSFER FUNCTION OF TWO-STATE PLANT AND CONTROLLER FOR DIFFERENT SETS OF CONTROL GAINS

```
clear
count=0;
for l=2:160
    W=10^(.025*l-1);
    C1=1.;
    C2=.0363;
    TOP=9000*sqrt(C1^2+(C2*W)^2);
    BOT=2.*W*sqrt(1+(W/2.)^2);
    XMAG=TOP/BOT;
    GAIN=20*log10(XMAG);
    PHASE=57.3*atan2(W,29.1)-90.-57.3*atan2(W,2.);
    count=count+1;
    ArrayW(count)=W;
    ArrayGAIN(count)=GAIN;
    ArrayPHASE(count)=PHASE;
end
figure
semilogx(ArrayW,ArrayGAIN),grid
xlabel('Frequency (Rad/Sec)')
ylabel('Gain (Db)')
figure
semilogx(ArrayW,ArrayPHASE),grid
xlabel('Frequency (Rad/Sec)')
ylabel('Phase (Deg)')
clc
output=[ArrayW',ArrayGAIN',ArrayPHASE'];
save datfil.txt output -ascii
disp 'simulation finished'
```

Listing 3.10 was run for different sets of control gains taken from Table 3.1, and the results appear in Fig. 3.14. We can see that as the performance index parameter δ_{cMX} is decreased, the crossover frequency is decreased until reasonable

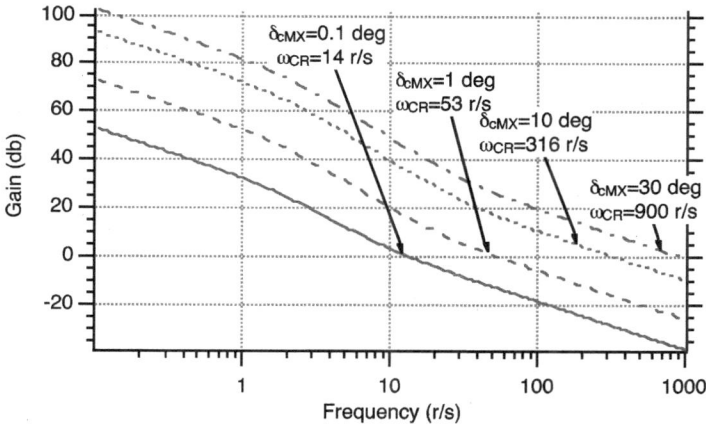

Fig. 3.14   Performance index parameter δ_{cMX} has a strong influence on open-loop crossover frequency.

values of crossover frequency are reached. Values of $\delta_{cMX} = 1$ deg and 0.1 deg yield crossover frequencies of 53 rad/s and 14 rad/s, respectively. In order to see how robust the two-state system performed with reduced crossover frequencies, a case was chosen where we evaluated the two-state controller with the six-state plant. Listing 3.6 was modified by setting $C3 = C4 = C5 = C6 = 0$ and using control gains $C1$ and $C2$ that corresponded to lower crossover frequencies. Figure 3.15 confirms that now the lower crossover frequency two-state controller can yield a stable response for the six-state plant.

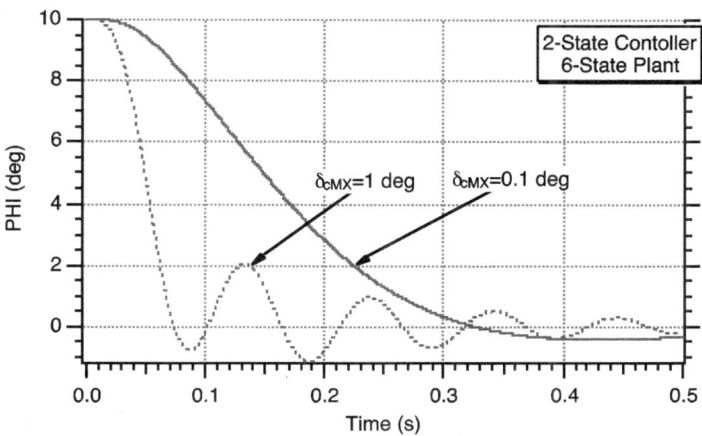

Fig. 3.15   A two-state controller can now yield a stable response for a six-state plant.

SUMMARY

In this chapter we have demonstrated by a variety of numerical examples that if a control system is designed using modern control techniques, a classical frequency response analysis must be performed in order to determine the crossover frequency and stability margins of the open-loop system. As was demonstrated in numerous examples in this chapter, if the crossover frequency is too high, the system will become unstable—even if the phase and gain margins are excellent. It was also demonstrated that the crossover frequency can be adjusted by adjusting the weighting terms in the performance index. An example was presented showing how a low-order controller could stabilize a high-order plant by lowering the crossover frequency.

REFERENCE

[1] Nesline, F. W., and Zarchan, P., "Why Modern Controllers Can Go Unstable in Practice," *Journal of Guidance, Control and Dynamics*, Vol. 7, No. 7, July-Aug. 1984, pp. 495–500.

CHAPTER 4

Advanced Adjoint Applications

INTRODUCTION

Throughout the previous volume we discussed adjoints and their application to analyzing missile guidance systems. We found that adjoints can be extremely valuable from both practical and computational points of view if the homing loop is linear (that is, without missile saturation). Single-run adjoint RMS miss distance results were shown to be equivalent to thousands of Monte Carlo runs generated from a forward engagement simulation. The value of adjoints was demonstrated not only for continuous systems that could be described by a set of linear differential equations, but also for mixed continuous discrete systems that could be described by a set of linear differential and difference equations (see chapter 7 of first volume).

In this chapter we shall demonstrate the utility of adjoints in three additional examples for the missile homing loop. In all of these examples we shall concentrate on systems that have a digital noise filter so that mixed continuous-discrete adjoints must be used. In the first example we shall consider how to implement an adjoint if there are two samplers in the missile guidance system, each operating at different sampling rates. In this example, a two-state, constant-gain digital fading memory filter is used as the noise filter. In the second example we shall consider a mixed continuous guidance system with a single sampler—but this time the digital noise filter is a three-state Kalman filter. The gains of the Kalman filter and associated guidance laws are both flight time dependent and time varying, so a new interpretation of adjoint results is required. Finally, the third example considers what to do if the seeker can no longer see the target near the end of the flight. This phenomenon is known as the *seeker blind range*.

MULTIPLE SAMPLING RATE ADJOINT

In some applications, information is available more often than can be handled by the digital noise filter because of flight computer throughput considerations. In these applications, a decision has to be made on what to do with the extra measurements that cannot be utilized by the digital noise filter. One technique for handling the extra data is to preprocess the extra data [1] as shown in Fig. 4.1. As was mentioned in [1], this technique for preprocessing the data is not necessarily the best, but is chosen in this example because it is simple and easy to understand. In this example, the normal sampling rate for the two-state digital fading memory noise filter is 10 Hz ($T_s = 0.1$ s), but data are available at 50 Hz ($T_s = 0.02$ s). In other words, we have five times as much data as can be used by the digital noise filter. In Fig. 4.1 the extra measurements are averaged by adding up five measurements and then dividing by five. The next time a measurement is available, the oldest measurement in the summation is dropped out while the new measurement is added so that the most recent five measurements are always being averaged. Note that the homing loop in Fig. 4.1 is similar to the mixed continuous discrete homing loop displayed in Fig. 7.1 of the previous volume, except for the preprocessing of the measurement data and for the single time constant representation of the flight-control system. The only source of error in the guidance system of Fig. 4.1 is a constant target maneuver n_T.

Figure 4.2 presents a more detailed block diagram of the homing loop with the two-state fading memory digital noise filter and two samplers—each operating at a different sampling rate. The input to the fading memory filter in the block diagram pz is the measured line-of-sight angle after it has been preprocessed, and the output of the fading memory filter is estimated line-of-sight angle $\hat{\lambda}$ and estimated line-of-sight rate $\dot{\lambda}$ The proportional navigation guidance law is implemented using the estimated line-of-sight rate according to

$$n_c = N' V_c \dot{\hat{\lambda}}$$

where N' is the effective navigation ratio and V_c is the closing velocity. In addition to the fading memory filter, the only other lag that exists within the guidance system in this example is the single time constant representation of the

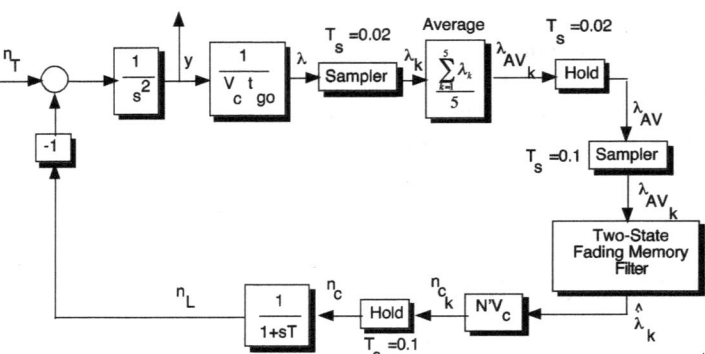

Fig. 4.1   General form of homing loop with two sampling rates and a fading memory filter.

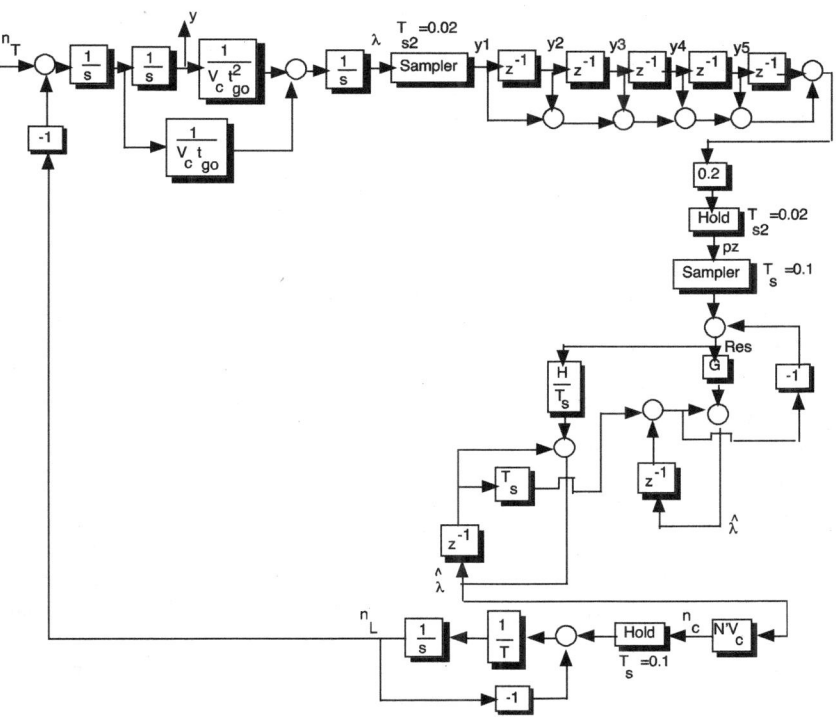

Fig. 4.2   Forward model of homing loop with two samplers operating at different rates.

flight-control system with time constant T. The two gains of the fading memory filter G and H determine the bandwidth of the filter and guidance system. As was shown in Chapter 7 of the previous volume, the two filter gains are computed from

$$G = 1 - \beta^2$$

$$H = (1 - \beta)^2$$

where β is a constant between 0 and 1 chosen by the designer. Higher values of β will result in a lower filter bandwidth or a more sluggish guidance system. Also note that in Fig. 4.2 the line-of-sight angle has been computed in a laborious way. This is done to have an extra integrator in the system so that when we take the adjoint of the sampler (which will result in a differentiator), a simple cancellation of an integrator and differentiator can take place. The adjoints of sampler and hold blocks can be found in Chapter 7 of the previous volume. To remind readers that the line-of-sight angle can be obtained in the laborious way indicated in Fig. 4.2, we start from the small angle definition of the line-of-sight angle, which is given by

$$\lambda = \frac{y}{R_{TM}}$$

Differentiation of the preceding expression using the quotient rule from calculus yields

$$\dot{\lambda} = \frac{R_{TM}\dot{y} - y\dot{R}_{TM}}{R_{TM}^2}$$

Because the range from missile to target is given by

$$R_{TM} = V_c t_{go} = V_c(t_F - t)$$

the line-of-sight rate can be expressed as

$$\dot{\lambda} = \frac{V_c t_{go}\dot{y} + yV_c}{V_c^2 t_{go}^2} = \frac{y}{V_c t_{go}^2} + \frac{\dot{y}}{V_c t_{go}}$$

Integration of the preceding equation yields the line-of-sight angle as shown in Fig. 4.2. The line-of-sight angle is sampled every 0.02 s (50-Hz rate), passed through a preprocessing section, and then passed through a zero-order hold. The output of the zero-order hold is then sampled at 0.1 s (10-Hz rate) and passed through the fading memory filter, whose gains are calculated using $\beta = 0.8$. The term z^{-1} in Fig. 4.2 represents a pure delay. Again, it is important to emphasize that the forward model of the homing loop is presented in this way to facilitate the generation of an adjoint block diagram.

The system parameters for the guidance system of Fig. 4.2 appear in Table 4.1. The forward model of Fig. 4.2 is programmed and appears in Listing 4.1. However, in the forward model simulation the line-of-sight angle does not have to be

TABLE 4.1   SYSTEM INPUTS FOR MULTIPLE SAMPLING RATE SYSTEM

Symbol	Definition	Value
V_c	Closing velocity	4000 ft/s
T	Flight-control system time constant	0.2 s
n_T	Target maneuver amplitude	96.6 ft/s^2 (3 g)
β	Fading memory filter tuning parameter	0.8
N'	Effective navigation ratio	3
T_s	Filter sampling time	0.1 s
T_{s2}	System sampling time	0.02 s

computed in the laborious way of Fig. 4.2, but instead can be computed more easily from the definition of the line-of-sight angle as

$$\lambda = \frac{y}{R_{TM}}$$

It is important to note that Listing 4.1 is nearly identical to Listing 7.2 of the previous volume except for the two samplers, the preprocessing section, and the single time constant representation of the flight-control system. In Listing 7.2 there was only one sampling time set at 0.1 s and there were two sources of error: random target maneuver and measurement noise. In Listing 4.1 we only consider a deterministic step target maneuver that always starts at the beginning of the flight and whose amplitude is always 3 g (96.6 ft/s^2). Because there is no noise disturbance in Listing 4.1, it does not have to be run in the Monte Carlo mode as was done for Listing 7.2. However, 100 runs are made in the forward model of Listing 4.1—each with different flight times. The flight times vary between 0.1 s and 10 s in steps of 0.1 s. The miss distances for each of the flight times are written to a file. It is important to note that in this forward model simulation, the high data rate sampler is encountered first. Thus the difference equations from the high data rate sampler are programmed first.

LISTING 4.1   SIMULATION OF FORWARD MODEL OF HOMING LOOP WITH TWO SAMPLERS OPERATING AT DIFFERENT RATES

```
clear
count=0;
VC=4000.;
TAP=.2;
XNT=96.6;
BETA=.8;
XNP=3.;
TS=.1;
TS2=.02;
for TF=.1:.1:10.0,
        Y=0.;
        YD=0.;
        T=0.;
        H=.001;
        S=0.;
        S2=0.;
        GFILTER=1.-BETA^2;
        HFILTER=(1.-BETA)^2;
        XLAMHOLD=0.;
        XLAMDHOLD=0.;
        Y1OLD=0.;
```

```
Y2OLD=0.;
Y3OLD=0.;
Y4OLD=0.;
Y5OLD=0.;
XNC=0.;
XNL=0.;
Y1NEW=0.;
PZ=0.;
XLAM=0.;
while T <= (TF - 1e-5)
  YOLD=Y;
  YDOLD=YD;
  XNLOLD=XNL;
  STEP=1;
  FLAG=0;
  while STEP <=1
          if FLAG==1
                  Y=Y+H*YD;
                  YD=YD+H*YDD;
                  XNL=XNL+H*XNLD;
                  T=T+H;
                  STEP=2;
          end;
          TGO=TF-T+.00001;
          RTM=VC*TGO;
          XLAM=Y/(VC*TGO);
          XLAMD=(RTM*YD+Y*VC)/(RTM^2);
          XNLD=(XNC-XNL)/TAP;
          YDD=XNT-XNL;
          FLAG=1;
  end;
  FLAG=0;
  Y=.5*(YOLD+Y+H*YD);
  YD=.5*(YDOLD+YD+H*YDD);
  XNL=.5*(XNLOLD+XNL+H*XNLD);
  S=S+H;
  S2=S2+H;
  if S2 >=(TS2 - 1e-5)
          S2=0.;
          Y1NEW=XLAM;
          Y2NEW=Y1OLD;
          Y3NEW=Y2OLD;
          Y4NEW=Y3OLD;
          Y5NEW=Y4OLD;
```

```
                    PZ=.2*(Y5OLD+Y5NEW+Y4NEW+Y3NEW+Y2NEW+XLAM);
                    Y1OLD=Y1NEW;
                    Y2OLD=Y2NEW;
                    Y3OLD=Y3NEW;
                    Y4OLD=Y4NEW;
                    Y5OLD=Y5NEW;
        end;
        if S>=(TS - 1e-5)
                    S=0.;
                    RES=PZ-(XLAMHOLD+TS*XLAMDHOLD);
                    XLAMHNEW=GFILTER*RES+XLAMHOLD+TS*XLAMDHOLD;
                    XLAMDHNEW=HFILTER*RES/TS+XLAMDHOLD;
                    XNC=XNP*VC*XLAMDHNEW;
                    XLAMHOLD=XLAMHNEW;
                    XLAMDHOLD=XLAMDHNEW;
          end;
        end;
        count=count+1;
        ArrayTF(count)=TF;
        ArrayY(count)=Y;
end;
figure
plot(ArrayTF',ArrayY'),grid
title('Standard miss for various flight times')
xlabel('Flight Time (S)')
ylabel('Miss (Ft) ')
axis([00,10,-40,100])
clc
output=[ArrayTF',ArrayY'];
save datfil.txt output -ascii
disp('Simulation Complete')
```

Figure 4.3 presents the adjoint model of the homing loop with two samplers operating at different rates derived from the forward model of Fig. 4.2. Figure 4.3 is constructed following the adjoint rules discussed in Chapters 3 and 7 of the previous volume. As was previously mentioned, the adjoints of each sampler and hold are taken according to the rules of adjoints of Chapter 7. Recall that the adjoint of a sampler yields a difference equation and a differentiator. Numerical differentiation is avoided in Fig. 4.3 because there is cancellation of the differentiator with an integrator.

The adjoint of the homing loop with two samplers operating at different rates is programmed in Listing 4.2. Because the signal flow is reversed in the adjoint model, the low data rate sampler will be encountered first. This is the opposite of what happened in the forward model. Therefore, in the adjoint listing the

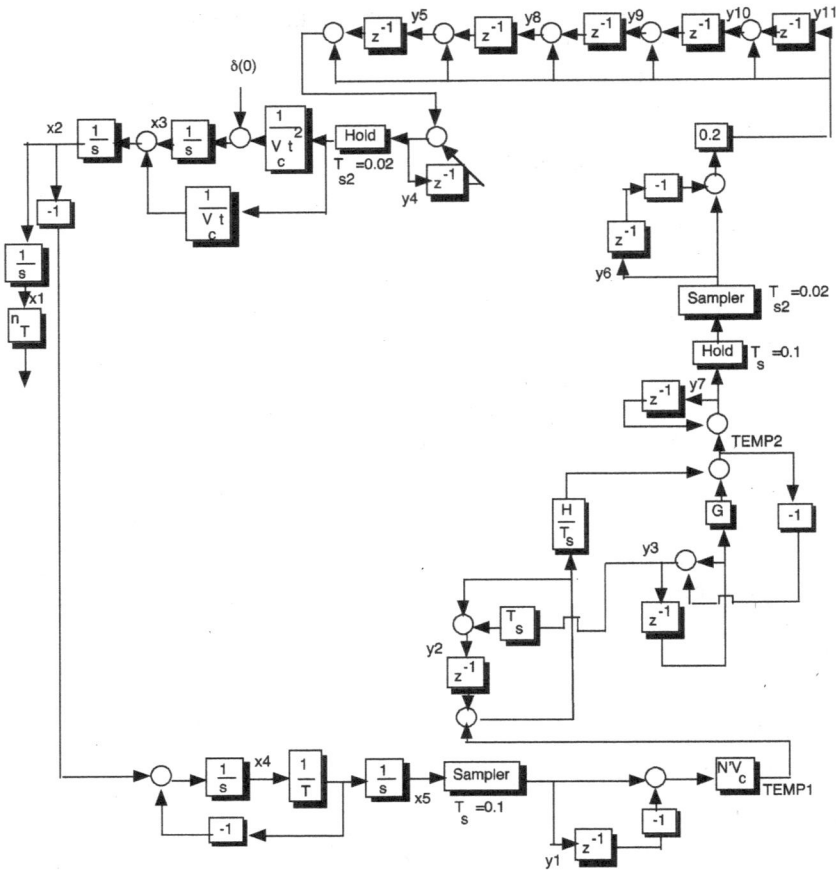

Fig. 4.3   Adjoint model of homing loop with two samplers operating at different rates.

difference equations for the low data rate sampler are programmed first. Also note that there is no for loop in the adjoint listing, so all the miss distance results are generated in one computer run.

LISTING 4.2   SIMULATION OF ADJOINT MODEL OF HOMING LOOP WITH TWO SAMPLERS OPERATING AT DIFFERENT RATES

```
clear
count=1;
XNT=96.6;
XNP=3;
TF=10.;
```

```
TS=.1;
BETA=.8;
TAP=.2;
VC=4000.;
T=0.;
S=0.;
S2=0.;
TS2=.02;
TP=T+.00001;
X1=0;
X2=0;
X3=1;
X4=0.;
X5=0.;
Y1OLD=0.;
Y2OLD=0.;
Y3OLD=0.;
Y4OLD=0.;
Y5OLD=0.;
Y6OLD=0.;
Y7OLD=0.;
Y8OLD=0.;
Y9OLD=0.;
Y10OLD=0.;
Y11OLD=0.;
Y11NEW=0.;
Y10NEW=0.;
Y9NEW=0.;
Y7NEW=0.;
Y6NEW=0.;
Y8NEW=0.;
Y4NEW=0.;
Y1NEW=0.;
Y2NEW=0.;
Y3NEW=0.;
Y5NEW=0.;
H=.001;
GFILTER=1.-BETA^2;
HFILTER=(1.-BETA)^2;
XMNT=0.;
while TP<=(TF-1e-5)
        S=S+H;
        S2=S2+H;
        X1OLD=X1;
        X2OLD=X2;
        X3OLD=X3;
```

```
X4OLD=X4;
X5OLD=X5;
STEP=1;
FLAG=0;
while STEP<=1
  if FLAG==1
  STEP=2;
        X1=X1+H*X1D;
        X2=X2+H*X2D;
        X3=X3+H*X3D;
        X4=X4+H*X4D;
        X5=X5+H*X5D;
        TP=TP+H;
  end
  TGO=TP;
  X1D=X2;
  X2D=X3+Y4NEW/(VC*TGO);
  X3D=(Y4NEW)/(VC*TGO*TGO);
  X4D=-X2-X4/TAP;
  X5D=X4/TAP;
  FLAG=1;
end
FLAG=0;
X1=(X1OLD+X1)/2+.5*H*X1D;
X2=(X2OLD+X2)/2+.5*H*X2D;
X3=(X3OLD+X3)/2+.5*H*X3D;
X4=(X4OLD+X4)/2+.5*H*X4D;
X5=(X5OLD+X5)/2+.5*H*X5D;
if S>=(TS-.0001)
  S=0.;
  Y1NEW=X5;
  TEMP1=(Y1NEW-Y1OLD)*XNP*VC;
  TEMP2=HFILTER*(Y2OLD+TEMP1)/TS+GFILTER*Y3OLD;
  Y2NEW=TEMP1+Y2OLD+TS*(Y3OLD-TEMP2);
  Y3NEW=Y3OLD-TEMP2;
  Y7NEW=TEMP2+Y7OLD;
  Y1OLD=Y1NEW;
  Y2OLD=Y2NEW;
  Y3OLD=Y3NEW;
  Y7OLD=Y7NEW;
end
if S2>=(TS2-.0001)
  S2=0.;
  Y6NEW=Y7NEW;
```

```
                Y11NEW=.2*(Y6NEW-Y6OLD);
                Y10NEW=Y11OLD+Y11NEW;
                Y9NEW=Y10OLD+Y11NEW;
                Y8NEW=Y9OLD+Y11NEW;
                Y5NEW=Y8OLD+Y11NEW;
                Y4NEW=Y4OLD+Y5OLD+Y11NEW;
                Y4OLD=Y4NEW;
                Y6OLD=Y6NEW;
                Y5OLD=Y5NEW;
                Y8OLD=Y8NEW;
                Y9OLD=Y9NEW;
                Y10OLD=Y10NEW;
                Y11OLD=Y11NEW;
                XMNT=XNT*X1;
                count=count+1;
                ArrayTP(count)=TP;
                ArrayXMNT(count)=XMNT;
            end
    end
end
figure
plot(ArrayTP,ArrayXMNT),grid
xlabel('Flight Time (Sec)')
ylabel('Target Maneuver Miss (Ft)')
clc
output=[ArrayTP',ArrayXMNT'];
save datfil.txt output -ascii
disp 'simulation finished'
```

The nominal cases of Listing 4.1 (equivalent to 100 runs) and of Listing 4.2 (1 run) were run, and the results for the miss due to a constant 3-g maneuver for different flight times are displayed in Fig. 4.4. We can see that there is exact agreement between the adjoint and forward simulations, indicating that we have correctly taken the adjoint of the homing loop with multiple samplers. We can see from Fig. 4.4 that the miss distances can be quite large due to the sluggishness of the fading memory filter ($\beta = 0.8$). Figure 4.5 shows that when the bandwidth of the fading memory filter is increased ($\beta = 0.3$), the guidance system performance improves and there is still excellent agreement between the adjoint and forward models, again indicating that the adjoint has been taken correctly.

In the forward model, the 3-g target maneuver always started at the beginning of flight and the flight time was varied from 0.1 s to 10 s in steps of 0.1 s. If we modify Listing 4.1 so that the time of flight is always 10 s but the time to go at which the 3-g maneuver occurs is varied from 0.1 s to 10 s in steps of 0.1 s, we can see from Fig. 4.6 that we get exactly the same results as those of Fig. 4.4. Thus we can see that adjoint time can be interpreted as either the time of flight

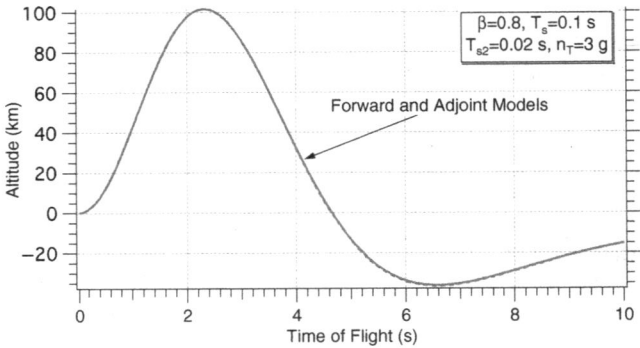

Fig. 4.4   Forward and adjoint models agree for homing loop with two samplers operating at different sampling rates.

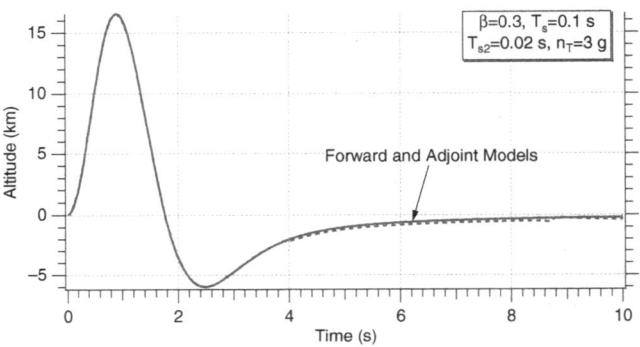

Fig. 4.5   Increasing sampling rate of first sampler causes miss distance to decrease.

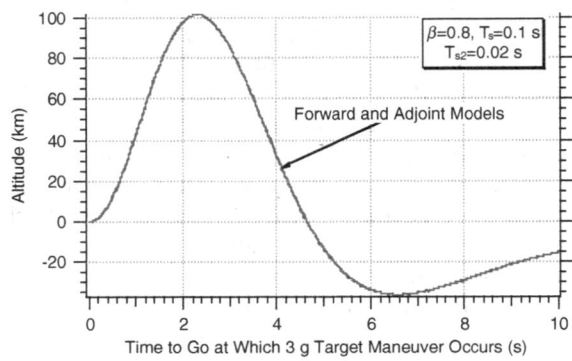

Fig. 4.6   Agreement between adjoint and forward models when target maneuver starting time is varied.

for a maneuver occurring at the beginning of flight or the time to go at which the maneuver occurs for a fixed time of flight.

ADJOINT OF DISCRETE LINEAR KALMAN FILTER

In the previous volume, whenever adjoints were taken, the noise filter was either a low pass filter, which could be represented by a differential equation, or a simple fading memory filter that could be represented by a set of difference equations. For more advanced guidance systems, a Kalman filter is normally required as the noise filter. Although the Kalman filter has the same structure as the fading memory filter, its gains are not constant and must be computed from the Riccati equations.

More advanced guidance laws can be used if a three-state Kalman filter is part of the guidance system. As the states of the three-state Kalman filter generally used in this and the previous volume are relative position, relative velocity, and target acceleration, possible guidance laws must be expressed in a slightly different way. As was shown in Chapters 8 and 9 of the previous volume, the various possible guidance laws can be expressed in terms of the Kalman filter state estimates and control gains according to

$$n_c = C_1 \hat{y}_k + C_2 \hat{\dot{y}}_k + C_3 \hat{n}_{T_k} + C_4 n_{L_k}$$

where $\hat{y}_k$ is the relative position estimate, $\hat{\dot{y}}_k$ is the relative velocity estimate, $\hat{n}_{T_k}$ is the target acceleration estimate, and n_{L_k} is the missile achieved acceleration. The gains C_1, C_2, C_3, and C_4 are known as control gains. For proportional navigation, the control gains are

$$C_1 = \frac{N'}{t_{go}^2}$$

$$C_2 = \frac{N'}{t_{go}}$$

$$C_3 = 0$$

$$C_4 = 0$$

For augmented proportional navigation, the control gains become

$$C_1 = \frac{N'}{t_{go}^2}$$

$$C_2 = \frac{N'}{t_{go}}$$

$$C_3 = 0.5N'$$

$$C_4 = 0$$

With the two preceding guidance laws, N' was a constant. If we use optimal guidance, the control gains become

$$C_1 = \frac{N'}{t_{go}^2}$$

$$C_2 = \frac{N'}{t_{go}}$$

$$C_3 = 0.5N'$$

$$C_4 = \frac{-N'T^2(e^{-x} + x - 1)}{t_{go}^2}$$

where T is the time constant of the flight-control system and N' is no longer a constant but is given by

$$N' = \frac{6x^2(e^{-x} - 1 + x)}{2x^3 + 3 + 6x - 6x^2 - 12xe^{-x} - 3e^{-2x}}$$

and x is defined as

$$x = \frac{t_{go}}{T}$$

Figure 4.7 presents a block diagram of the homing loop with a three-state linear Kalman filter as the digital noise filter. In this diagram the line-of-sight angle measurement λ_k is multiplied by range to convert the actual measurement to a pseudo measurement of relative position y_k^*. The Kalman filter estimates relative position $\hat{y}_k$, relative velocity $\hat{\dot{y}}_k$, and target acceleration $\hat{n}_{T_k}$. In this homing loop there is also a single-lag representation of the flight-control system with time constant T. Both the Kalman gains K_1, K_2, and K_3 and the control gains C_1, C_2, C_3, and C_4 are functions of time. As was previously mentioned, the control gains determine which guidance law is used in conjunction with the Kalman filter.

The forward model of the Kalman filter in the homing loop of Listing 4.3, like Listing 4.1, has only one source of error—constant target maneuver. However, here the flight time is fixed at 10 s and the time to go at which the 3-g target maneuver begins is made a parameter in the for loop. One hundred runs are made in the forward model with the maneuver starting time varying between 0.1 s and 10 s in steps of 0.1 s. This means that the time to go at which the maneuver occurs is varying between 9.9 and 0 s in steps of 0.1 s. The miss distances for each of the times to go at which the maneuver starts for the 10-s flight are written to a file. Note that in this forward model the sampling time is 0.1 s. For each run we solve for the Kalman gains via the Riccati equations, although this is not necessary because the flight time is fixed and the gains will be the same from run to run.

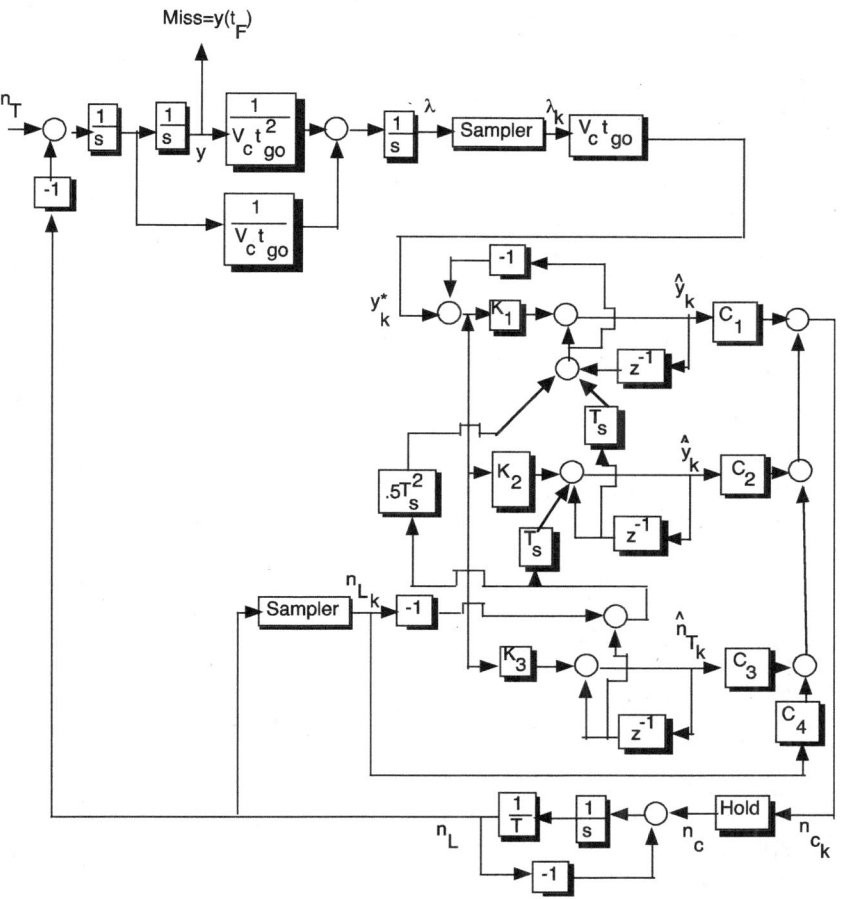

Fig. 4.7   Forward model of homing loop with discrete three-state Kalman filter.

LISTING 4.3   SIMULATION OF FORWARD MODEL OF HOMING LOOP WITH THREE-STATE KALMAN FILTER AND VARIOUS GUIDANCE LAW OPTIONS

```
clear
count=0;
VM=3000.;
VC=4000.;
XNT=96.6;
YIC=0.;
HEDEGF=20.;
XNP=3.;
SIGNOISE=.001;
```

```
TS=.1;
TAU=.5;
APN=0;
TF=10.;
TS2=TS*TS;
TS3=TS2*TS;
TS4=TS3*TS;
TS5=TS4*TS;
PHIN=XNT*XNT/TF;
for TSTART=.1:.1:10.0,
     Y=YIC;
     YD=0.;
     T=0.;
     H=.01;
     S=0.;
     RTM=VC*TF;
     SIGPOS=RTM*SIGNOISE;
     SIGN2=SIGPOS^2;
     P11=SIGN2;
     P12=0.;
     P13=0.;
     P22=(VM*HEDEGF/57.3)^2;
     P23=0.;
     P33=XNT*XNT;
     YH=0.;
     YDH=0.;
     XNTH=0.;
     XNC=0.;
     XNL=0.;
     while T <= (TF - 1e-5)
       YOLD=Y;
       YDOLD=YD;
       XNLOLD=XNL;
       STEP=1;
       FLAG=0;
       while STEP <=1
              if FLAG==1
                     Y=Y+H*YD;
                     YD=YD+H*YDD;
                     XNL=XNL+H*XNLD;
                     T=T+H;
                     STEP=2;
              end;
              TGO=TF-T+.00001;
```

```
            RTM=VC*TGO;
            XLAM=Y/(VC*TGO);
            XLAMD=(RTM*YD+Y*VC)/(RTM^2);
            XNLD=(XNC-XNL)/TAU;
            if T>TSTART
                    YDD=XNT-XNL;
            else
                    YDD=0.;
            end;
            FLAG=1;
    end;
    FLAG=0;
    Y=.5*(YOLD+Y+H*YD);
    YD=.5*(YDOLD+YD+H*YDD);
    XNL=.5*(XNLOLD+XNL+H*XNLD);
    S=S+H;
    if S>=(TS - 1e-5)
            S=0.;
            TGO=TF-T+.000001;
            RTM=VC*TGO;
            SIGPOS=RTM*SIGNOISE;
            SIGN2=SIGPOS^2;
            M11=P11+TS*P12+.5*TS2*P13+TS*(P12+TS*P22+.5*TS2*P23);
            M11=M11+.5*TS2*(P13+TS*P23+.5*TS2*P33)+TS5*PHIN/20.;
            M12=P12+TS*P22+.5*TS2*P23+TS*(P13+TS*P23+.5*TS2*P33)+TS4*
            PHIN/8.;
            M13=P13+TS*P23+.5*TS2*P33+PHIN*TS3/6.;
            M22=P22+TS*P23+TS*(P23+TS*P33)+PHIN*TS3/3.;
            M23=P23+TS*P33+.5*TS2*PHIN;
            M33=P33+PHIN*TS;
            K1=M11/(M11+SIGN2);
            K2=M12/(M11+SIGN2);
            K3=M13/(M11+SIGN2);
            P11=(1.-K1)*M11;
            P12=(1.-K1)*M12;
            P13=(1.-K1)*M13;
            P22=-K2*M12+M22;
            P23=-K2*M13+M23;
            P33=-K3*M13+M33;
            XLAMNOISE=0.;
            YSTAR=RTM*(XLAM+XLAMNOISE);
            RES=YSTAR-YH-TS*YDH-.5*TS*TS*(XNTH-XNL);
            YH=K1*RES+YH+TS*YDH+.5*TS*TS*(XNTH-XNL);
            YDH=K2*RES+YDH+TS*(XNTH-XNL);
```

```
                XNTH=K3*RES+XNTH;
                XLAMDH=(YH+YDH*TGO)/(VC*TGO*TGO);
                if APN==0
                       C1=XNP/TGO^2;
                       C2=XNP/TGO;
                       XNC=C1*YH+C2*YDH;
                elseif APN==1
                       C1=XNP/TGO^2;
                       C2=XNP/TGO;
                       C3=.5*XNP;
                       XNC=C1*YH+C2*YDH+C3*XNTH;
                else
                       X=TGO/TAU;
                       TOP=6.*X*X*(exp(-X)-1.+X);
                       BOT1=2*X*X*X+3.+6.*X-6.*X*X;
                       BOT2=-12.*X*exp(-X)-3.*exp(-2.*X);
                       XNPP=TOP/(.0001+BOT1+BOT2);
                       XNEW=XNPP*XNL*(exp(-X)+X-1.)/(X*X);
                       C1=XNPP/TGO^2;
                       C2=XNPP/TGO;
                       C3=.5*XNPP;
                       C4=-XNPP*(exp(-X)+X-1.)/(X*X);
                       XNC=C1*YH+C2*YDH+C3*XNTH+C4*XNL;
                end;
           end;
        end;
        TGOS=TF-TSTART;
        count=count+1;
        ArrayTGOS(count)=TGOS;
        ArrayY(count)=Y;
end;
figure
plot(ArrayTGOS',ArrayY'),grid
title('Miss for various tgo maneuver starting times')
xlabel('Time to go at which maneuver occurs (S)')
ylabel('Miss (Ft) ')
clc
output=[ArrayTGOS',ArrayY'];
save datfil.txt output -ascii
disp('Simulation Complete')
```

As was demonstrated in the previous section when adjoint results were generated for the fading memory filter, adjoint time can be interpreted as either missile flight time or time to go at which the target maneuver started. If we model a Kalman filter using the adjoint technique, we must reverse all time-varying gains (meaning Kalman gains and control gains) according to the rules of adjoints. Because the Kalman gains for a 10-s flight will be different than for a 5-s flight, we must fix the flight time when modeling the adjoint of a Kalman filter. Therefore, in this case adjoint time can be interpreted only as the time to go at which the disturbance occurs. Following the rules of adjoints shown in Chapter 7 of the previous volume for a mixed continuous discrete system, we obtain Fig. 4.8.

The adjoint of the homing loop with the three-state Kalman filter and various guidance law options is programmed in Listing 4.4. Note that there is no "for loop" in the adjoint listing, so all the miss distance results will be generated in one computer run. However, as was mentioned previously, adjoint time must be interpreted as time to go at which the constant maneuver starts for a 10-s

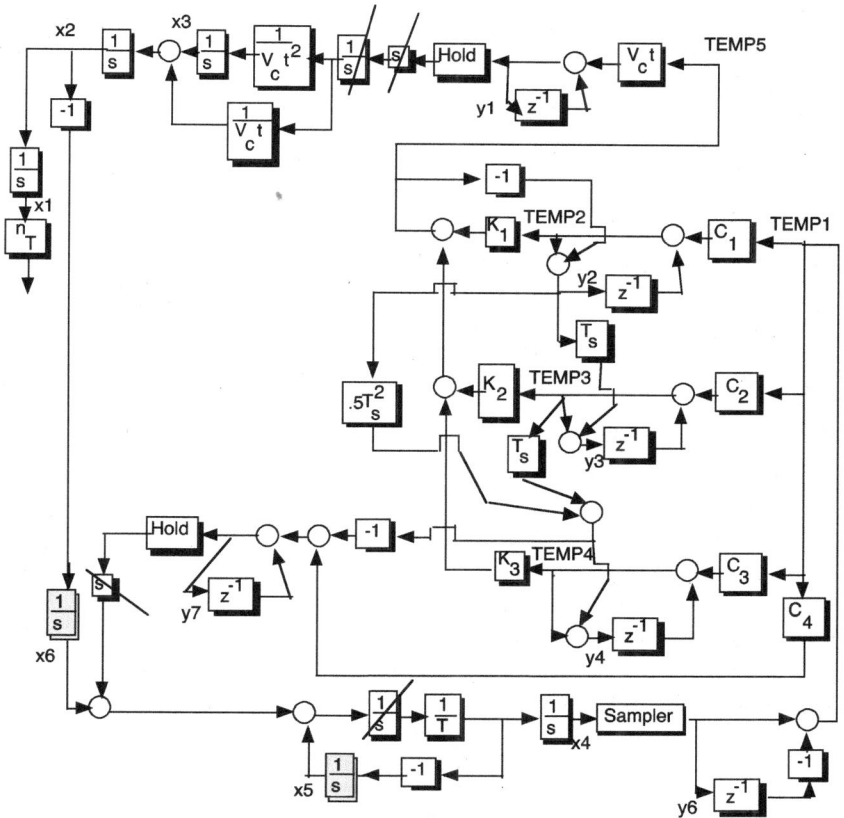

Fig. 4.8   Adjoint model of homing loop with discrete three-state Kalman filter.

flight. Also note that that the Kalman gains for a 10-s flight are first computed and stored as an array. After the Riccati equations are finished, the gains are reversed in time and also stored as an array. This array is used to provide the Kalman gains for each adjoint time.

LISTING 4.4   ADJOINT OF HOMING LOOP WITH THREE-STATE KALMAN FILTER AND VARIOUS GUIDANCE LAW OPTIONS

```
clear
count=0;
XNT=96.6;
XNP=3.;
TF=10;
TS=.1;
APN=0;
VM=3000.;
VC=4000.;
HEDEGF=20.;
SIGNOISE=.001;
TS2=TS*TS;
TS3=TS2*TS;
TS4=TS3*TS;
TS5=TS4*TS;
PHIN=XNT*XNT/TF;
RTM=VC*TF;
SIGPOS=RTM*SIGNOISE;
SIGN2=SIGPOS^2;
P11=SIGN2;
P12=0.;
P13=0.;
P22=(VM*HEDEGF/57.3)^2;
P23=0.;
P33=XNT*XNT;
C=0;
for T=TS:TS:TF

        TGO=TF-T+.000001;
        RTM=VC*TGO;
        SIGPOS=RTM*SIGNOISE;
        SIGN2=SIGPOS^2;
        M11=P11+TS*P12+.5*TS2*P13+TS*(P12+TS*P22+.5*TS2*P23);
        M11=M11+.5*TS2*(P13+TS*P23+.5*TS2*P33)+TS5*PHIN/20.;
        M12=P12+TS*P22+.5*TS2*P23+TS*(P13+TS*P23+.5*TS2*P33)+TS4*PHIN/8.;
        M13=P13+TS*P23+.5*TS2*P33+PHIN*TS3/6.;
        M22=P22+TS*P23+TS*(P23+TS*P33)+PHIN*TS3/3.;
```

```
        M23=P23+TS*P33+.5*TS2*PHIN;
        M33=P33+PHIN*TS;
        K1=M11/(M11+SIGN2);
        K2=M12/(M11+SIGN2);
        K3=M13/(M11+SIGN2);
        P11=(1.-K1)*M11;
        P12=(1.-K1)*M12;
        P13=(1.-K1)*M13;
        P22=-K2*M12+M22;
        P23=-K2*M13+M23;
        P33=-K3*M13+M33;
        C=C+1;
        U(C)=K1;
        V(C)=K2;
        W(C)=K3;
end;
% Modification
% ---------------------
ICOUNT=round((TF/TS));
E(ICOUNT)=0;
F(ICOUNT)=0;
G(ICOUNT)=0;
ICOUNT=ICOUNT-1;
% ---------------------
for I=1:1:ICOUNT
        REV=ICOUNT-I+1;
        E(REV)=U(I);
        F(REV)=V(I);
        G(REV)=W(I);
end;
TAP=.5;
VC=4000.;
T=0.;
S=0.;
TP=T+.00001;
X1=0;
X2=0;
X3=1;
X4=0.;
X5=0.;
X6=0.;
Y1OLD=0.;
Y2OLD=0.;
Y3OLD=0.;
Y4OLD=0.;
```

```
Y5OLD=0.;
Y6OLD=0.;
Y7OLD=0.;
Y7NEW=0.;
Y6NEW=0.;
Y8NEW=0.;
Y4NEW=0.;
Y1NEW=0.;
Y2NEW=0.;
Y3NEW=0.;
Y5NEW=0.;
H=.01;
XMNT=0.;
I=1;
while TP <= (TF - 1e-5)
      X1OLD=X1;
      X2OLD=X2;
      X3OLD=X3;
      X4OLD=X4;
      X5OLD=X5;
      X6OLD=X6;
      STEP=1;
      FLAG=0;
      while STEP <=1
        if FLAG==1
              X1=X1+H*X1D;
              X2=X2+H*X2D;
              X3=X3+H*X3D;
              X4=X4+H*X4D;
              X5=X5+H*X5D;
              X6=X6+H*X6D;
              TP=TP+H;
              STEP=2;
        end;
        TGO=TP;
        X1D=X2;
        X2D=X3+Y1NEW/(VC*TGO);
        X3D=(Y1NEW)/(VC*TGO*TGO);
        X4D=(X5+Y7NEW+X6)/TAP;
        X5D=-X4D;
        X6D=-X2;
        FLAG=1;
      end
      FLAG=0;
```

```
        X1=(X1OLD+X1)/2+.5*H*X1D;
        X2=(X2OLD+X2)/2+.5*H*X2D;
        X3=(X3OLD+X3)/2+.5*H*X3D;
        X4=(X4OLD+X4)/2+.5*H*X4D;
        X5=(X5OLD+X5)/2+.5*H*X5D;
        X6=(X6OLD+X6)/2+.5*H*X6D;
   S=S+H;
      if S>(TS-.0001)
        S=0.;
        K1=E(I);
        K2=F(I);
        K3=G(I);
      [TP I E(I) TS]
        I=I+1;
        if APN==0
                C1=XNP/TP^2;
                C2=XNP/TP;
                C3=0.;
                C4=0.;
        elseif APN==1
                C1=XNP/TP^2;
                C2=XNP/TP;
                C3=.5*XNP;
                C4=0.;
        else
                X=TP/TAP;
                TOP=6.*X*X*(exp(-X)-1.+X);
                BOT1=2*X*X*X+3.+6.*X-6.*X*X;
                BOT2=-12.*X*exp(-X)-3.*exp(-2.*X);
                XNPP=TOP/(.0001+BOT1+BOT2);
                C1=XNPP/TP^2;
                C2=XNPP/TP;
                C3=.5*XNPP;
                C4=-XNPP*(exp(-X)+X-1.)/(X*X);
        end
        TEMP1=X4-Y6OLD;
        TEMP2=C1*TEMP1+Y2OLD;
        TEMP3=C2*TEMP1+Y3OLD;
        TEMP4=C3*TEMP1+Y4OLD;
        TEMP5=K1*TEMP2+K2*TEMP3+K3*TEMP4;
        Y1NEW=Y1OLD+TEMP5*VC*TP;
        Y2NEW=TEMP2-TEMP5;
        Y3NEW=TEMP3+TS*Y2NEW;
        Y4NEW=TEMP4+TS*TEMP3+.5*TS*TS*Y2NEW;
```

```
            Y5=-(TS*TEMP3+.5*TS*TS*Y2NEW);
            Y7NEW=Y7OLD+C4*TEMP1+Y5;
            Y6NEW=X4;
            XMNT=XNT*X1;
            Y1OLD=Y1NEW;
            Y2OLD=Y2NEW;
            Y3OLD=Y3NEW;
            Y4OLD=Y4NEW;
            Y6OLD=Y6NEW;
            Y7OLD=Y7NEW;
        count=count+1;
        ArrayTP(count)=TP;
        ArrayXMNT(count)=XMNT;
        ArrayK1(count)=K1;
    end
end
figure
plot(ArrayTP',ArrayXMNT'),grid
title('Miss for various tgo maneuver starting times')
xlabel('Time to go at which maneuver occurs (S)')
ylabel('Miss (Ft) ')
clc
output=[ArrayTP',ArrayXMNT',ArrayK1'];
save datfil.txt output -ascii
disp('Simulation Complete')
```

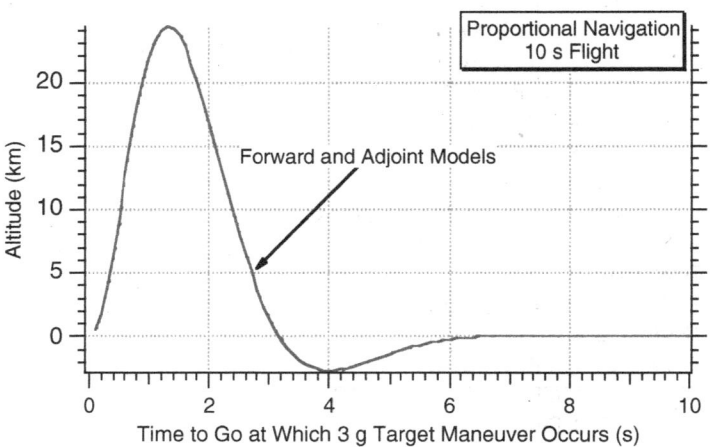

Fig. 4.9   Forward and adjoint models agree for homing loop with thee-state Kalman filter when proportional navigation is used.

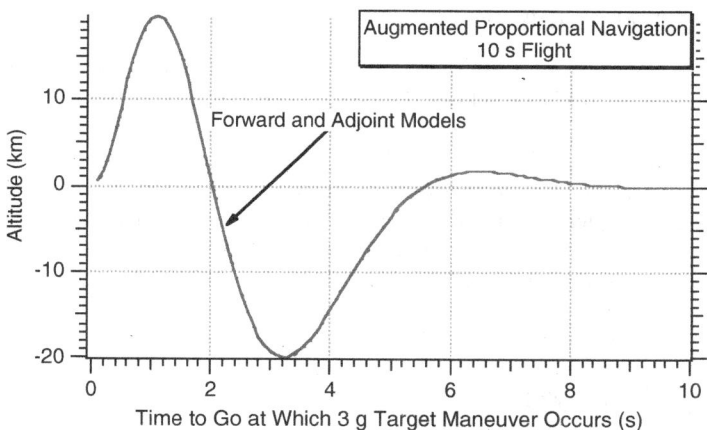

Fig. 4.10   Forward and adjoint models agree for homing loop with three-state Kalman filter when augmented proportional navigation is used.

The nominal cases of Listing 4.3 (equivalent to 100 runs) and Listing 4.4 (1 run) were run for each of the guidance law options (APN = 0, 1, and 2), and the results are displayed in Figs. 4.9 through 4.11. First we can see that there is exact agreement between the adjoint and forward simulations in all cases, indicating that we have correctly taken the adjoint of the homing loop with the discrete three-state Kalman filter. We can also see from the three cases that the miss distance results get better as the guidance law becomes more advanced.

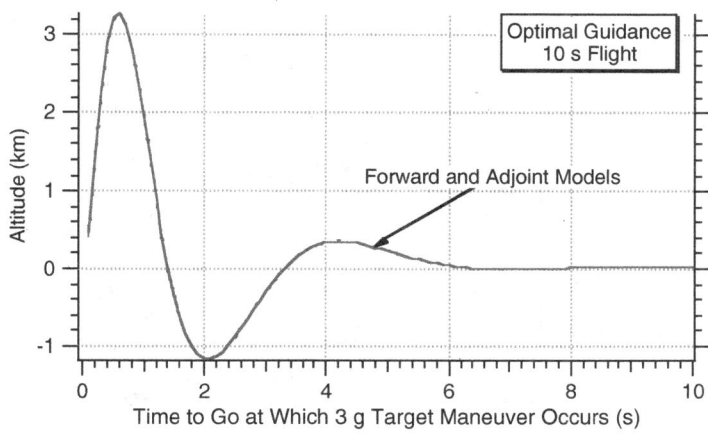

Fig. 4.11   Forward and adjoint models agree for homing loop with thee-state Kalman filter when optimal guidance is used.

ADJOINTS AND BLIND TIME

During the last portion of the missile flight, mechanical or electronic issues may cause the seeker to stop tracking the target. The range from the missile to the target at which this phenomenon occurs is known as the *blind range*, and the amount of time to go before the intercept time at which the blind range occurs is known as the *blind time*. Because no seeker information is available during the blind time, a decision must be made at what the guidance policy should be during the blind time. One potential policy might be to set the guidance command to zero during the blind time. If a Kalman filter is used on the seeker

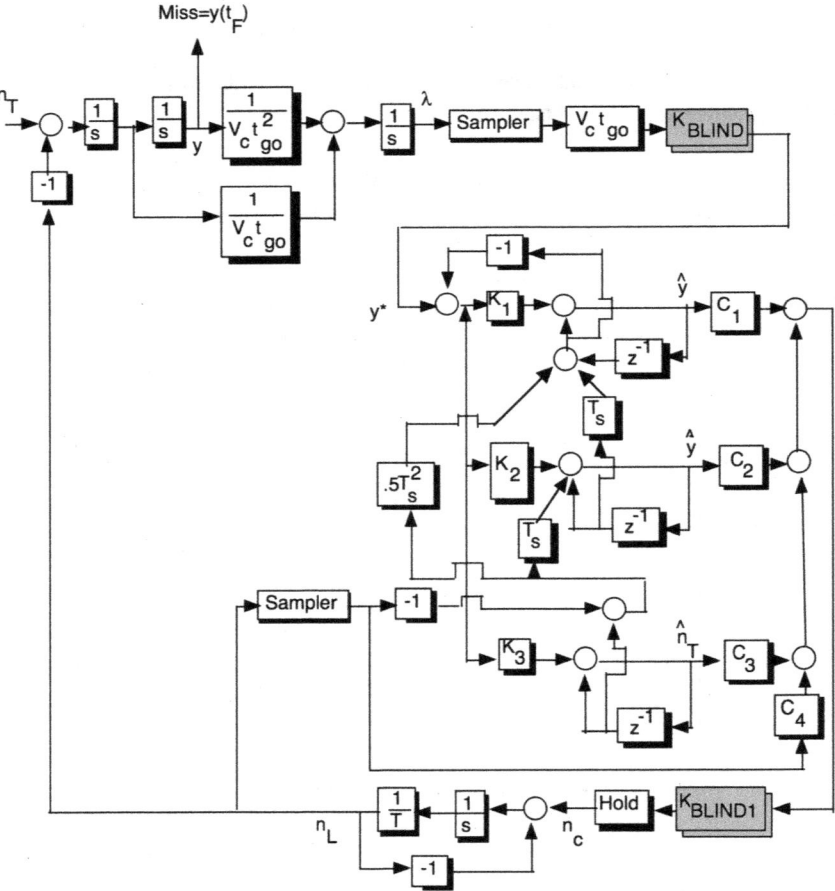

Fig. 4.12   Forward model of Kalman filter homing loop with blind time option.

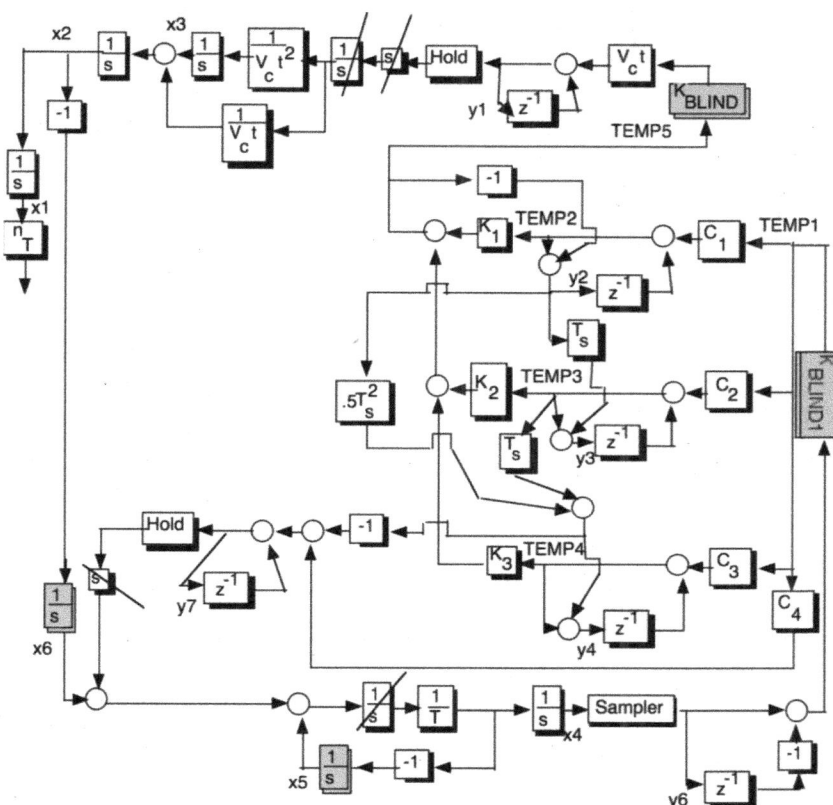

Fig. 4.13   Adjoint model of Kalman filter homing loop with blind time option.

measurements, another policy might be to let the guidance command be computed based on the Kalman filter's coasted estimates of the system states (i.e., estimates of the system states based on zero input into the Kalman filter). Figure 4.12
and Listing 4.5 show the forward model of the homing loop with the new parameter QZERO. When the time to go is less than the blind time TBLIND, the
input to the Kalman filter is essentially shut down by making SIGNOISE a large
number. This action effectively makes all the Kalman gains zero, which means
that there is no longer an input into the Kalman filter. If QZERO=1, then the guidance command after the blind time begins is zero; if QZERO=0, then the guidance
command is computed from the coasted states of the Kalman filter. Figure 4.13
and Listing 4.6 show the adjoint model of the homing loop with the new parameter QZERO.

LISTING 4.5   FORWARD MODEL OF HOMING LOOP WITH THREE-STATE KALMAN FILTER WITH TWO BLIND TIME OPTIONS

```
clear
count=0;
QZERO=0;
TBLIND=.5;
VM=3000.;
VC=4000.;
XNT=96.6;
YIC=0.;
VM=3000.;
HEDEG=20.;
XNP=3.;
SIGNOISE=.001;
TS=.1;
TAU=.5;
APN=2;
TF=10.;
TS2=TS*TS;
TS3=TS2*TS;
TS4=TS3*TS;
TS5=TS4*TS;
PHIN=XNT*XNT/TF;
for TSTART=.1:.1:10.0,
    Y=YIC;
    YD=0.;
    T=0.;
    H=.01;
    S=0.;
    RTM=VC*TF;
    SIGPOS=RTM*SIGNOISE;
    SIGN2=SIGPOS^2;
    P11=SIGN2;
    P12=0.;
    P13=0.;
    P22=(VM*HEDEG/57.3)^2;
    P23=0.;
    P33=XNT*XNT;
    YH=0.;
    YDH=0.;
    XNTH=0.;
    XNC=0.;
    XNL=0.;
    while T <= (TF - 1e-5)
```

```
YOLD=Y;
YDOLD=YD;
XNLOLD=XNL;
STEP=1;
FLAG=0;
while STEP <=1
   if FLAG==1
      Y=Y+H*YD;
      YD=YD+H*YDD;
      XNL=XNL+H*XNLD;
      T=T+H;
      STEP=2;
   end;
   TGO=TF-T+.00001;
   RTM=VC*TGO;
   XLAM=Y/(VC*TGO);
   XLAMD=(RTM*YD+Y*VC)/(RTM^2);
   XNLD=(XNC-XNL)/TAU;
   if T>TSTART
      YDD=XNT-XNL;
   else
      YDD=0.;
   end
   FLAG=1;
end
FLAG=0;
Y=.5*(YOLD+Y+H*YD);
YD=.5*(YDOLD+YD+H*YDD);
XNL=.5*(XNLOLD+XNL+H*XNLD);
S=S+H;
if S>=(TS - 1e-5)
   S=0.;
   TGO=TF-T+.000001;
   RTM=VC*TGO;
   if TGO>=TBLIND
      SIGPOS=RTM*SIGNOISE;
   else
      SIGPOS=9999999999.;
   end
   SIGN2=SIGPOS^2;
   M11=P11+TS*P12+.5*TS2*P13+TS*(P12+TS*P22+.5*TS2*P23);
   M11=M11+.5*TS2*(P13+TS*P23+.5*TS2*P33)+TS5*PHIN/20.;
   M12=P12+TS*P22+.5*TS2*P23+TS*(P13+TS*P23+.5*TS2*P33)+...
      TS4*PHIN/8.;
```

```
M13=P13+TS*P23+.5*TS2*P33+PHIN*TS3/6.;
M22=P22+TS*P23+TS*(P23+TS*P33)+PHIN*TS3/3.;
M23=P23+TS*P33+.5*TS2*PHIN;
M33=P33+PHIN*TS;
K1=M11/(M11+SIGN2);
K2=M12/(M11+SIGN2);
K3=M13/(M11+SIGN2);
P11=(1.-K1)*M11;
P12=(1.-K1)*M12;
P13=(1.-K1)*M13;
P22=-K2*M12+M22;
P23=-K2*M13+M23;
P33=-K3*M13+M33;
if TGO >=TBLIND
  XKBLIND=1.;
else
  XKBLIND=0.;
end
YSTAR=RTM*XLAM*XKBLIND;
RES=YSTAR-YH-TS*YDH-.5*TS*TS*(XNTH-XNL);
YH=K1*RES+YH+TS*YDH+.5*TS*TS*(XNTH-XNL);
YDH=K2*RES+YDH+TS*(XNTH-XNL);
XNTH=K3*RES+XNTH;
XLAMDH=(YH+YDH*TGO)/(VC*TGO*TGO);
if APN==0
  C1=XNP/TGO^2;
  C2=XNP/TGO;
  XNC=C1*YH+C2*YDH;
elseif APN==1
  XNC=XNP*VC*XLAMDH+APN*.5*XNP*XNTH;
else
  X=TGO/TAU;
  TOP=6.*X*X*(exp(-X)-1.+X);
  BOT1=2*X*X*X+3.+6.*X-6.*X*X;
  BOT2=-12.*X*exp(-X)-3.*exp(-2.*X);
  XNPP=TOP/(.0001+BOT1+BOT2);
  XNEW=XNPP*XNL*(exp(-X)+X-1.)/(X*X);
  XNC=XNPP*VC*XLAMDH+.5*XNPP*XNTH-XNEW;
end
if QZERO==1&&TGO<TBLIND
  XNC=0.;
end
    end
  end
end
```

```
  TGOS=TF-TSTART;
  count=count+1;
  ArrayTGOS(count)=TGOS;
  ArrayY(count)=Y;
end
figure
plot(ArrayTGOS',ArrayY'),grid
title('Miss for various tgo maneuver starting times')
xlabel('Time to go at which maneuver occurs (S)')
ylabel('Miss (Ft) ')
clc
output=[ArrayTGOS',ArrayY'];
save datfil.txt output -ascii
disp('Simulation Complete')
```

LISTING 4.6   FORWARD MODEL OF HOMING LOOP WITH THREE-STATE KALMAN FILTER WITH TWO BLIND TIME OPTIONS

```
clear
count=0;
XNT=96.6;
XNP=3.;
TF=10.;
TS=.1;
QZERO=0;
TBLIND=.5;
APN=2;
VM=3000.;
VC=4000.;
HEDEG=20.;
SIGNOISE=.001;
TS2=TS*TS;
TS3=TS2*TS;
TS4=TS3*TS;
TS5=TS4*TS;
PHIN=XNT*XNT/TF;
RTM=VC*TF;
SIGPOS=RTM*SIGNOISE;
SIGN2=SIGPOS^2;
P11=SIGN2;
P12=0.;
P13=0.;
P22=(VM*HEDEG/57.3)^2;
P23=0.;
P33=XNT*XNT;
```

```
C=1;
for T=TS:TS:TF
  TGO=TF-T+.000001;
  RTM=VC*TGO;
  if TGO >=TBLIND
    SIGPOS=RTM*SIGNOISE;
  else
    SIGPOS=9999999999.;
  end
  SIGN2=SIGPOS^2;
  M11=P11+TS*P12+.5*TS2*P13+TS*(P12+TS*P22+.5*TS2*P23);
  M11=M11+.5*TS2*(P13+TS*P23+.5*TS2*P33)+TS5*PHIN/20.;
  M12=P12+TS*P22+.5*TS2*P23+TS*(P13+TS*P23+.5*TS2*P33)+TS4*PHIN/8.;
  M13=P13+TS*P23+.5*TS2*P33+PHIN*TS3/6.;
  M22=P22+TS*P23+TS*(P23+TS*P33)+PHIN*TS3/3.;
  M23=P23+TS*P33+.5*TS2*PHIN;
  M33=P33+PHIN*TS;
  K1=M11/(M11+SIGN2);
  K2=M12/(M11+SIGN2);
  K3=M13/(M11+SIGN2);
  P11=(1.-K1)*M11;
  P12=(1.-K1)*M12;
  P13=(1.-K1)*M13;
  P22=-K2*M12+M22;
  P23=-K2*M13+M23;
  P33=-K3*M13+M33;
  C=C+1;
  U(C)=K1
  V(C)=K2
  W(C)=K3
  C=C+1;
  U(C)=K1;
  V(C)=K2;
  W(C)=K3;
end
% Modification
% ----------------------
ICOUNT=round((TF/TS));
E(ICOUNT)=0;
F(ICOUNT)=0;
G(ICOUNT)=0;
ICOUNT=ICOUNT-1;
% --------------------
for I=1:1:ICOUNT
REV=ICOUNT-I+1;
E(REV)=U(I);
```

```
F(REV)=V(I);
G(REV)=W(I);
end;
TAP=.5;
VC=4000.;
T=0.;
S=0.;
TP=T+.00001;
X1=0;
X2=0;
X3=1;
X4=0.;
X5=0.;
X6=0.;
Y1OLD=0.;
Y2OLD=0.;
Y3OLD=0.;
Y4OLD=0.;
Y5OLD=0.;
Y6OLD=0.;
Y7OLD=0.;
Y7NEW=0.;
Y6NEW=0.;
Y8NEW=0.;
Y4NEW=0.;
Y1NEW=0.;
Y2NEW=0.;
Y3NEW=0.;
Y5NEW=0.;
H=.01;
XMNT=0.;
I=1;
while TP <= (TF - 1e-5)
  X1OLD=X1;
  X2OLD=X2;
  X3OLD=X3;
  X4OLD=X4;
  X5OLD=X5;
  X6OLD=X6;
  STEP=1;
  FLAG=0;
  while STEP <=1
    if FLAG==1
      X1=X1+H*X1D;
      X2=X2+H*X2D;
      X3=X3+H*X3D;
```

```
        X4=X4+H*X4D;
        X5=X5+H*X5D;
        X6=X6+H*X6D;
        TP=TP+H;
        STEP=2;
      end;
      TGO=TP;
      X1D=X2;
      X2D=X3+Y1NEW/(VC*TGO);
      X3D=(Y1NEW)/(VC*TGO*TGO);
      X4D=(X5+Y7NEW+X6)/TAP;
      X5D=-X4D;
      X6D=-X2;
      FLAG=1;
end
FLAG=0;
X1=(X1OLD+X1)/2+.5*H*X1D;
X2=(X2OLD+X2)/2+.5*H*X2D;
X3=(X3OLD+X3)/2+.5*H*X3D;
X4=(X4OLD+X4)/2+.5*H*X4D;
X5=(X5OLD+X5)/2+.5*H*X5D;
X6=(X6OLD+X6)/2+.5*H*X6D;
S=S+H;
  if S > (TS-.0001)
  S=0.;
  K1=E(I);
  K2=F(I);
  K3=G(I);
  [TP I E(I) TS]
  I=I+1;
  if APN==0
    C1=XNP/TP^2;
    C2=XNP/TP;
    C3=0.;
    C4=0.;
  elseif APN==1
    C1=XNP/TP^2;
    C2=XNP/TP;
    C3=.5*XNP;
    C4=0.;
  else
    X=TP/TAP;
    TOP=6.*X*X*(exp(-X)-1.+X);
    BOT1=2*X*X*X+3.+6.*X-6.*X*X;
    BOT2=-12.*X*exp(-X)-3.*exp(-2.*X);
```

```
    XNPP=TOP/(.0001+BOT1+BOT2);
    C1=XNPP/TP^2;
    C2=XNPP/TP;
    C3=.5*XNPP;
    C4=-XNPP*(exp(-X)+X-1.)/(X*X);
  end
  TGO=TP;
    if TP < TBLIND&&QZERO==1
    XKBLIND=0.;
    else
    XKBLIND=1.;
    end
  TEMP1=XKBLIND*(X4-Y6OLD);
  TEMP2=C1*TEMP1+Y2OLD;
  TEMP3=C2*TEMP1+Y3OLD;
  TEMP4=C3*TEMP1+Y4OLD;
  TEMP5=K1*TEMP2+K2*TEMP3+K3*TEMP4;
    if TP >=TBLIND
    XKBLIND1=1.;
    else
    XKBLIND1=0.;
    end
  Y1NEW=Y1OLD+TEMP5*VC*TP*XKBLIND1;
  Y2NEW=TEMP2-TEMP5;
  Y3NEW=TEMP3+TS*Y2NEW;
  Y4NEW=TEMP4+TS*TEMP3+.5*TS*TS*Y2NEW;
  Y5=-(TS*TEMP3+.5*TS*TS*Y2NEW);
  Y7NEW=Y7OLD+C4*TEMP1+Y5;
  Y6NEW=X4;
  XMNT=XNT*X1;
  Y1OLD=Y1NEW;
  Y2OLD=Y2NEW;
  Y3OLD=Y3NEW;
  Y4OLD=Y4NEW;
  Y6OLD=Y6NEW;
  Y7OLD=Y7NEW;
  count=count+1;
  ArrayTP(count)=TP;
  ArrayXMNT(count)=XMNT;
  ArrayK1(count)=K1;
    end
end
figure
plot(ArrayTP',ArrayXMNT'),grid
```

```
title('Miss for various tgo maneuver starting times')
xlabel('Time to go at which maneuver occurs (S)')
ylabel('Miss (Ft) ')
clc
output=[ArrayTP',ArrayXMNT',ArrayK1'];
save datfil.txt output -ascii
disp('Simulation Complete')
```

Forward model and adjoint model cases were run for the case in which optimal guidance was used (APN = 2), the blind time was 0.5 s (TBLIND = .5), and a zero acceleration command was issued during the blind time (QZERO = 0). We can see from Fig. 4.14 that both the forward and adjoint models agree.

Forward model and adjoint model cases were run for the case in which optimal guidance was used (APN = 2), the blind time was 0.5 s (TBLIND = .5), and the acceleration command was coasted during the blind time (QZERO = .1). We can see from Fig. 4.15 that both the forward and adjoint models agree.

Next a simple study was done when the blind time was made a parameter for the case where optimal guidance was used and the acceleration command was coasted. We can see from Fig. 4.16 that the miss distance increases for longer blind times when the target maneuver occurs before 5 s to go before intercept.

Next we compared the guidance command policy for a 0.5-s blind time when optimal guidance was used. We can see from Fig. 4.17 that for this guidance law there is only a slight benefit in coasting the guidance command because both results are nearly identical. However, Fig. 4.18 shows that if proportional

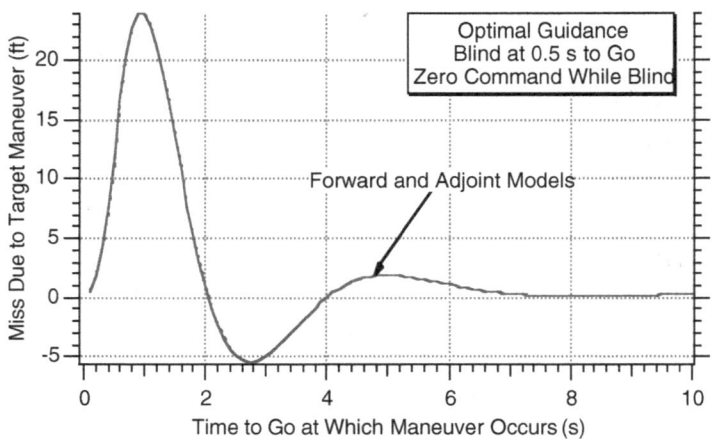

Fig. 4.14 Forward and adjoint results match for zero acceleration command during blind time.

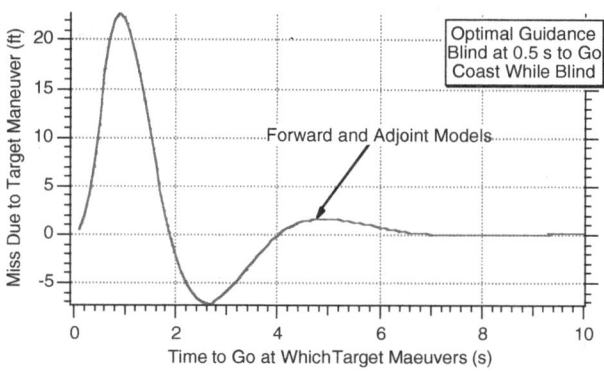

Fig. 4.15   Forward and adjoint results match for coasting command during blind time.

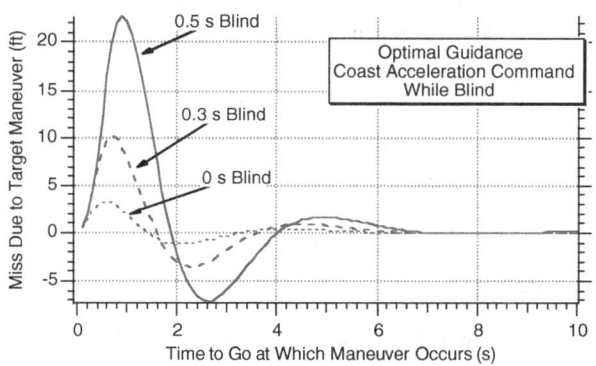

Fig. 4.16   Miss increases with increasing blind time.

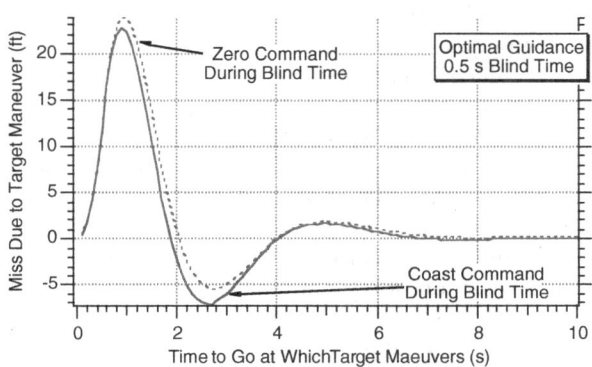

Fig. 4.17   Blind time performance is similar when optimal guidance is used.

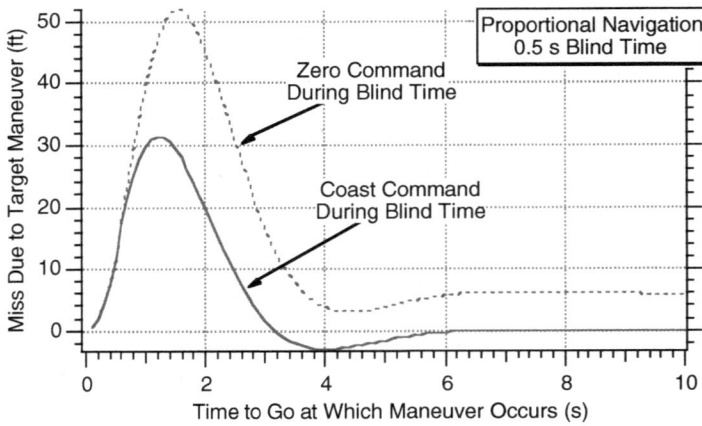

Fig. 4.18   Blind time performance is better when acceleration command is coasted during blind time when proportional navigation is used.

navigation is used instead of optimal guidance, then there is significant performance improvement in coasting the guidance command during the blind time.

SIMULATING AN IMPULSE

So far, when the adjoint simulation technique has been used, the impulse required for the implementation of the adjoint method has been simulated by finding the appropriate initial conditions on the various integrators in the system. In some applications, in which the adjoint method is being automated, it may be advantageous to actually simulate the impulse rather than trying to develop the logic in finding the appropriate initial conditions [2]. Listing 3.1 of the previous volume has been modified in Listing 4.7 to show how the impulse can be simulated. Changes to the original source code have been highlighted in bold. We can see that before the STEP=1 statement, the height of the impulse is chosen so that the simulated impulse has unit area. The width of the impulse is half an integration interval because the differential equations are called twice with the second-order Runge–Kutta integration technique. Before the STEP=1 statement, we can see that the impulse is added only once to the derivative of $x3$. When we come back to this section of code at other times, the value of the simulated impulse will be zero.

We can see from Figs. 4.19 and 4.20 that the miss due to target maneuver and heading error is identical whether we are using Listing 3.1 of the previous volume where initial conditions are used for the impulse or Listing 4.7 where the impulse

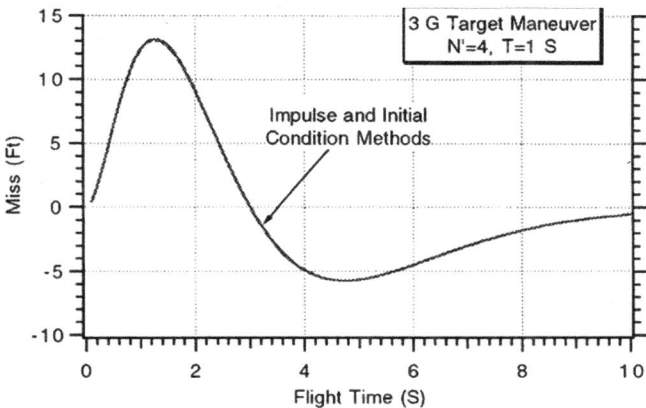

Fig. 4.19 Both impulse and initial condition methods are equivalent for target maneuver disturbance.

is actually simulated. However, it is important to note that when the impulse is simulated, the answers are more sensitive to the integration step size than when the initial condition method is used. This means that the answers in Listing 4.7 will start to diverge from the true answers sooner than the answers from Listing 3.1 if the integration interval is made larger.

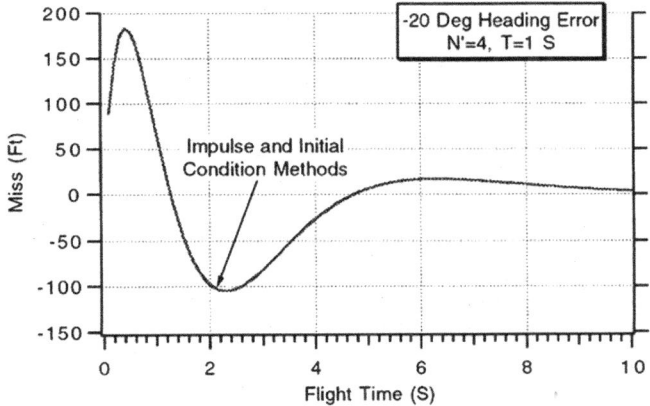

Fig. 4.20 Both impulse and initial condition methods are equivalent for heading error disturbance.

LISTING 4.7   SIMULATING AN IMPULSE RATHER THAN USING INITIAL CONDITIONS FOR USE IN THE ADJOINT METHOD (EQUIVALENT TO LISTING 3.1 IN VOLUME 1)

```
n=0.;
XNP=4;
XNT=96.6;
TAU=1;
TF=10;
VM=3000;
HEDEG=-20;
T=0.;
S=0.;
TP=T+.00001;
X1=0;
X2=0;
X3=0;
X4=0;
H=.01;
HE=HEDEG/57.3;
while TP <=(TF-1e-5)
      S=S+H;
      X1OLD=X1;
      X2OLD=X2;
      X3OLD=X3;
      X4OLD=X4;
      if TP < H/2
              IMPULSE=2./H;
      else
              IMPULSE=0;
      end
      STEP=1;
      FLAG=0;
      while STEP <=1
            if FLAG==1
                    STEP=2;
                    X1=X1+H*X1D;
                    X2=X2+H*X2D;
                    X3=X3+H*X3D;
                    X4=X4+H*X4D;
                    TP=TP+H;
            end
            X1D=X2;
            X2D=X3;
            Y1=(X4-X2)/TAU;
            TGO=TP+.00001;
```

```
        if STEP==2
                IMPULSE=0;
        end
        X3D=XNP*Y1/TGO+IMPULSE;
        X4D=-Y1;
        FLAG=1;
    end
    FLAG=0;
    X1=(X1OLD+X1)/2+.5*H*X1D;
    X2=(X2OLD+X2)/2+.5*H*X2D;
    X3=(X3OLD+X3)/2+.5*H*X3D;
    X4=(X4OLD+X4)/2+.5*H*X4D;
    if S >=.0999
            S=0.;
n=n+1;
ArrayTP(n)=TP;
            ArrayXMNT(n)=XNT*X1;
            ArrayXMHE(n)=-VM*HE*X2;
    end
end
figure
plot(ArrayTP,ArrayXMNT),grid
xlabel('Flight Time (Sec)')
ylabel('Target Maneuver Miss (Ft)')
figure
plot(ArrayTP,ArrayXMHE),grid
xlabel('Flight Time (Sec)')
ylabel('Heading Error Miss (Ft)')
clc
output=[ArrayTP',ArrayXMNT',ArrayXMHE'];
save datfil.txt output /ascii
disp 'simulation finished'
```

SUMMARY

In this chapter we have shown three new examples of the method of adjoints for mixed continuous discrete systems. The first example involved multiple samplers used in a system in which measurement data were first preprocessed at one sampling rate and then used for input to a fading memory filter operating at another sampling rate. The second example involved taking the adjoint of a three-state discrete Kalman filter in the homing loop. In both examples, experiments were conducted with a forward model to ensure that the adjoint was taken correctly. Another adjoint example introduced in this chapter included the modeling of blind time and an alternative method for simulating an impulse.

REFERENCES

[1]   Zarchan, P., *Fundamentals of Kalman Filtering: A Practical Approach*, 3rd ed, Progress in Astronautics and Aeronautics, AIAA, Reston, VA, 2009, pp. 690–701.

[2]   Bucco, D., "Adjoints Revisited: A Software Tool to Facilitate Their Application," Proceedings of 1997 AIAA Guidance and Control Conference, Washington, DC, August 1997, pp. 603–611.

CHAPTER 5

Shaping Filters and Their Use in Miss Distance Analysis

BACKGROUND AND INTRODUCTION

In the previous volume we saw how shaping filters could be used to represent random target maneuvers. It was proven analytically and demonstrated via multiple-run Monte Carlo simulations modeling that the actual random target maneuver process and single-run adjoint simulations using the shaping filter approach for the random target maneuver both yielded identical RMS miss distances. The miss distance experiments of the previous volume were conducted using linear models of the homing loop without missile acceleration saturation effects. The shaping filter representation of target maneuvers was also used in the development of both Kalman filters and guidance laws. In this chapter we shall show that in the presence of missile acceleration saturation effects, the shaping filter representation of target maneuvers is not accurate and leads to overly pessimistic RMS miss distance performance projections.

SHAPING FILTER REPRESENTATION OF UNIFORMLY DISTRIBUTED TARGET MANEUVER

In Chapter 4 of the previous volume it was demonstrated that a constant target maneuver with uniformly distributed random starting time t_{START} (i.e., t_{START} was equally likely to occur between 0 and t_F seconds) was equivalent to white noise u_s through an integrator, as shown in Fig. 5.1. The equivalence holds when the power spectral density Φ_s of the white noise is related to the target maneuver amplitude via the equation

$$\Phi_s = \frac{\beta^2}{t_F}$$

where β is the target maneuver amplitude in ft/s^2 and t_F is the flight or homing time of the engagement in units of seconds. An experiment was conducted in

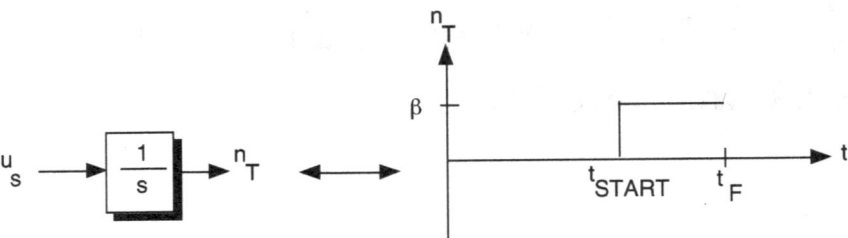

Fig. 5.1   Shaping filter approach is equivalent to constant target maneuver with random starting time that is uniformly distributed over flight time.

Chapter 4 of the previous volume showing that a multiple-run Monte Carlo simulation of the actual random maneuver yielded the same RMS miss distance as a single-run adjoint simulation using the shaping filter approach.

In order to demonstrate the power and utility of the shaping filter approach, let us consider the single-lag homing loop with time constant T depicted in Fig. 5.2. Here it is assumed that perfect continuous estimates of the line-of-sight rate $\dot{\lambda}$ are available. An acceleration command n_c is developed from the proportional navigation guidance law according to

$$n_c = N' V_c \dot{\lambda}$$

where N' is a designer-chosen gain known as the effective navigation ratio and V_c is the missile-target closing velocity. This homing loop is linear if the commanded missile acceleration is not limited (i.e., $n_{\text{LIM}}=$ infinity). Miss distance will tend to degrade as n_{LIM} decreases. The acceleration advantage that the missile has over the target is given by

$$\text{acceleration advantage} = \frac{n_{\text{LIM}}}{n_T}$$

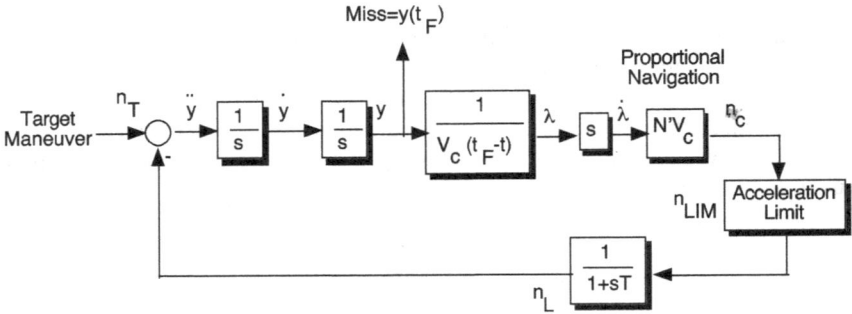

Fig. 5.2   Proportional navigation single-lag homing loop.

It was shown in Chapter 6 of the previous volume that for a proportional navigation guidance system with an effective navigation ratio of 3 against a constant target maneuver, acceleration advantages from 3 to 5 were usually required to avoid significant miss distance degradation due to a constant target maneuver in the presence of missile acceleration saturation. It is important to note that in the experiments conducted in Chapter 6 of Volume 1 where there was no acceleration limit, the actual target maneuvers (i.e., not the shaping filter equivalents) were used in the miss distance evaluation.

Listing 5.1 models the homing loop of Fig. 5.2 for a uniformly distributed constant target maneuver. From Listing 5.1 we can see that the actual random target maneuver is modeled; the statements pertaining to the target maneuver are highlighted in bold. Listing 5.1 shows that the target maneuver is equally likely to be either positive or negative (COEF equally like to be +1 or −1) and that the target maneuver starting time TSTART is taken from a uniform distribution and is a different random number between 0 and TF for each flight in the Monte Carlo set. Listing 5.1 is a Monte Carlo simulation where each run set consists of 50 different flight times TF (i.e., from 0.2 s to 10 s in steps of 0.2 s). For each flight time, 1000 runs are made to get very accurate performance projections. Therefore, for each case studied, a total of 50,000 runs are made (1000 runs times 50 flight times). From the homing loop model of Fig. 5.2 we can see that the line-of-sight angle is given by

$$\lambda = \frac{y}{V_c(t_F - t)}$$

Therefore the line-of-sight rate, required by the proportional navigation guidance law, can be calculated from the preceding equation by using the quotient rule from calculus, or

$$\dot{\lambda} = \frac{V_c(t_F - t)\dot{y} - y(-V_c)}{V_c^2(t_F - t)^2} = \frac{y + \dot{y}t_{go}}{V_c t_{go}^2}$$

Because the range from the missile to target is given by

$$R_{TM} = V_c t_{go}$$

the line-of-sight rate equation can also be rewritten as

$$\dot{\lambda} = \frac{yV_c + \dot{y}V_c t_{go}}{V_c^2 t_{go}^2} = \frac{yV_c + \dot{y}R_{TM}}{R_{TM}^2}$$

The preceding equation appears in Listing 5.1. We can also see from Listing 5.1 that the simulation is set up so that there is a 5-g target maneuver level (BETA = 161 ft/s^2), a 0.2-s guidance system time constant (TAU = 0.2), and an effective navigation ratio of 3 (XNP = 3). Initially the simulation is run in the linear mode where there is no limit on the commanded missile acceleration

(XLIM = 999999999). After every run, for each flight time, the miss distances are collected and sent to a postprocessing routine where the RMS miss distance is calculated according to

$$\text{RMS Miss} = \sqrt{\sum_{i=1}^{n} \frac{\text{Miss}_i^2}{n-1}}$$

where Miss_i represents the ith miss distance in a Monte Carlo set for a particular flight time and n is the number of runs in the Monte Carlo set (1000 in our example). Listing 5.1 prints out the RMS miss for each of the 50 flight times considered.

LISTING 5.1   MONTE CARLO SIMULATION OF SINGLE-LAG HOMING LOOP USING ACTUAL UNIFORMLY DISTRIBUTED RANDOM TARGET MANEUVER

```
clear
count=0;
VC=4000.;
BETA=161;
VM=3000.;
XNP=3.;
TAU=.2;
RUN=1000;
XLIM=999999999.;
for TF=.2:.2:10,
     Z1=0.;
     for I=1:RUN
       SUM=rand(1);
       TSTART=TF*SUM;
       PZ=rand(1);
       PZ=PZ-.5;
       if PZ > 0
              COEF=1;
       else
              COEF=-1;
       end;
       Y=0.;
       YD=0.;
       T=0.;
       H=.01;
       S=0.;
       XNC=0.;
       XNL=0.;
```

```
while T <= (TF - 1e-5)
        YOLD=Y;
        YDOLD=YD;
        XNLOLD=XNL;
        STEP=1;
        FLAG=0;
        while STEP <=1
                if FLAG==1
                        Y=Y+H*YD;
                        YD=YD+H*YDD;
                        XNL=XNL+H*XNLD;
                        T=T+H;
                        STEP=2;
                end;
                TGO=TF-T+.00001;
                RTM=VC*TGO;
                XLAM=Y/RTM;
XLAMD=(RTM*YD+Y*VC)/(RTM^2);
                XNC=XNP*VC*XLAMD;
                if XNC>XLIM
                        XNC=XLIM;
                elseif XNC <-XLIM
                        XNC=-XLIM;
                end;
                XNLD=(XNC-XNL)/TAU;
                if T<TSTART
                        XNT=0.;
                else
                        XNT=COEF*BETA;
                end
                YDD=XNT-XNL;
                FLAG=1;
        end;
        FLAG=0;
        Y=.5*(YOLD+Y+H*YD);
        YD=.5*(YDOLD+YD+H*YDD);
        XNL=.5*(XNLOLD+XNL+H*XNLD);
        S=S+H;
end
Z(I)=Y;
Z1=Z(I)+Z1;
XMEAN=Z1/I;
end
SIGMA=0.;
```

```
        Z1=0.;
        Z2=0.;
        for I=1:RUN,
          Z1=(Z(I)-XMEAN)^2+Z1;
          Z2=Z(I)^2+Z2;
          if I==1,
                    SIGMA=0;
                    RMS=0.;
          else
                    SIGMA=sqrt(Z1/(I-1));
                    RMS=sqrt(Z2/(I-1));
           end;
          end
          count=count+1;
          ArrayTF(count)=TF;
          ArrayRMS(count)=RMS;
end;
figure
plot(ArrayTF',ArrayRMS'),grid
title('RMS miss for various flight times')
xlabel('Flight Time (S)')
ylabel('RMS MISS (Ft) ')
clc
output=[ArrayTF',ArrayRMS'];
save datfil.txt output -ascii

disp('Simulation Complete')
```

Listing 5.2 also models the homing loop of Fig. 5.2; however, in Listing 5.2 the shaping filter equivalent of a uniformly distributed random starting time constant target maneuver (i.e., white noise through an integrator) is used to represent the target maneuver. The shaping filter approach requires the simulation of white noise. The approximate technique for modeling pseudo white noise was discussed in Chapter 4 of the previous volume. With this technique, zero-mean Gaussian noise with standard deviation σ is generated every integration interval h. It was demonstrated in Chapter 4 of Volume 1 that this noise appears white to the system under consideration if the integration step size h was small enough to perform accurate integration. The standard deviation of the Gaussian noise used to model white noise is given by

$$\sigma = \sqrt{\frac{\Phi}{h}}$$

where Φ is the power spectral density (in unit^2 per Hz) of the white noise we want to simulate. In Listing 5.2 the statements pertaining to the shaping filter approach to representing the uniformly distributed target maneuver are highlighted in bold. Listing 5.2 is also a Monte Carlo simulation where each run set consists of 50 different flight times TF (from 0.2 s to 10 s in steps of 0.2 s). For each flight time, 1000 runs are made to get accurate performance projections. The postprocessing routine used to calculate the RMS miss distance for each of the 50 flight times considered is identical to the one used in Listing 5.1.

LISTING 5.2   MONTE CARLO SIMULATION OF SINGLE-LAG HOMING LOOP USING SHAPING FILTER TO REPRESENT A CONSTANT TARGET MANEUVER WITH A UNIFORMLY DISTRIBUTED STARTING TIME

```
clear
count=0;
VC=4000.;
BETA=161;
VM=3000.;
XNP=3.;
TAU=.2;
RUN=1000;
XLIM=999999999.;
H=.01;
for TF=.2:.2:10,
      Z1=0.;
      for I=1:RUN
         PHI=BETA*BETA/TF;
         SIG=sqrt(PHI/H);
         Y=0.;
         YD=0.;
         T=0.;
         S=0.;
         XNC=0.;
         XNL=0.;
         XNT=0.;
         while T <= (TF - 1e-5)
                X=SIG*randn;
                YOLD=Y;
                YDOLD=YD;
                XNLOLD=XNL;
                XNTOLD=XNT;
                STEP=1;
                FLAG=0;
                while STEP <=1
                        if FLAG==1
                             Y=Y+H*YD;
```

```
                        YD=YD+H*YDD;
                        XNL=XNL+H*XNLD;
                        XNT=XNT+H*XNTD;
                        T=T+H;
                        STEP=2;
                  end;
                  TGO=TF-T+.00001;
                  RTM=VC*TGO;
                  XLAM=Y/RTM;
                  XLAMD=(RTM*YD+Y*VC)/(RTM^2);
                  XNC=XNP*VC*XLAMD;
                  if XNC>XLIM
                        XNC=XLIM;
                  elseif XNC<-XLIM
                        XNC=-XLIM;
                  end;
                  XNLD=(XNC-XNL)/TAU;
                  YDD=XNT-XNL;
                  XNTD=X;
                  FLAG=1;
            end;
            FLAG=0;
            Y=.5*(YOLD+Y+H*YD);
            YD=.5*(YDOLD+YD+H*YDD);
            XNL=.5*(XNLOLD+XNL+H*XNLD);
            XNT=.5*(XNTOLD+XNT+H*XNTD);
            S=S+H;
      end
      Z(I)=Y;
      Z1=Z(I)+Z1;
      XMEAN=Z1/I;
   end
   SIGMA=0.;
   Z1=0.;
   Z2=0.;
   for I=1:RUN,
      Z1=(Z(I)-XMEAN)^2+Z1;
      Z2=Z(I)^2+Z2;
      if I==1,
            SIGMA=0;
            RMS=0.;
      else
            SIGMA=sqrt(Z1/(I-1));
            RMS=sqrt(Z2/(I-1));
```

```
        end;
    end
        count=count+1;
        ArrayTF(count)=TF;
        ArrayRMS(count)=RMS;
end;
figure
plot(ArrayTF',ArrayRMS'),grid
title('RMS miss for various flight times')
xlabel('Flight Time (S)')
ylabel('RMS MISS (Ft) ')
clc
output=[ArrayTF',ArrayRMS'];
save datfil.txt output -ascii
disp('Simulation Complete')
```

The nominal cases of Listings 5.1 and 5.2 (i.e., no missile acceleration saturation effects) were run, and the RMS miss distances as a function of flight time are displayed in Fig. 5.3. Here we can see that that the RMS miss distances for each flight time are virtually identical for both systems. Thus Fig. 5.3 confirms that if the system under consideration is linear (i.e., infinite missile acceleration limit), then it makes no difference whether one models the actual random target maneuver or its shaping filter equivalent.

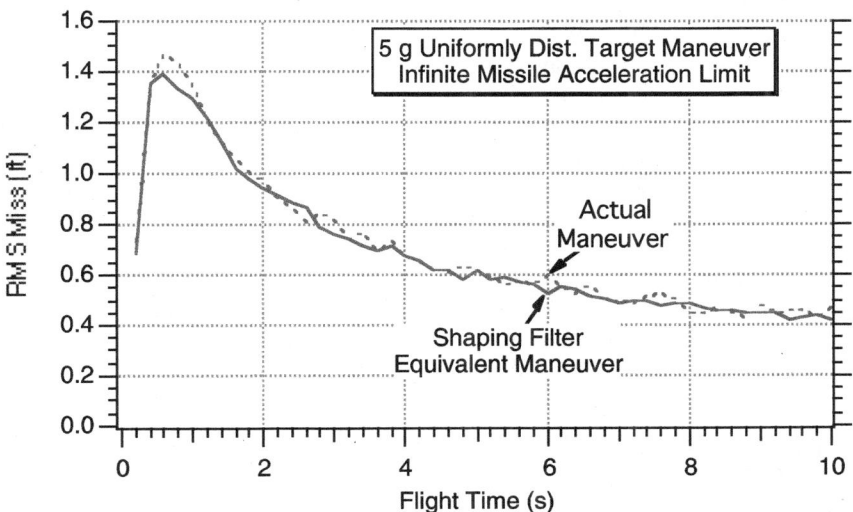

Fig. 5.3 Modeling actual target maneuver or shaping filter equivalent yield identical RMS miss distances for constant target maneuver with random starting time when missile has infinite missile acceleration capability.

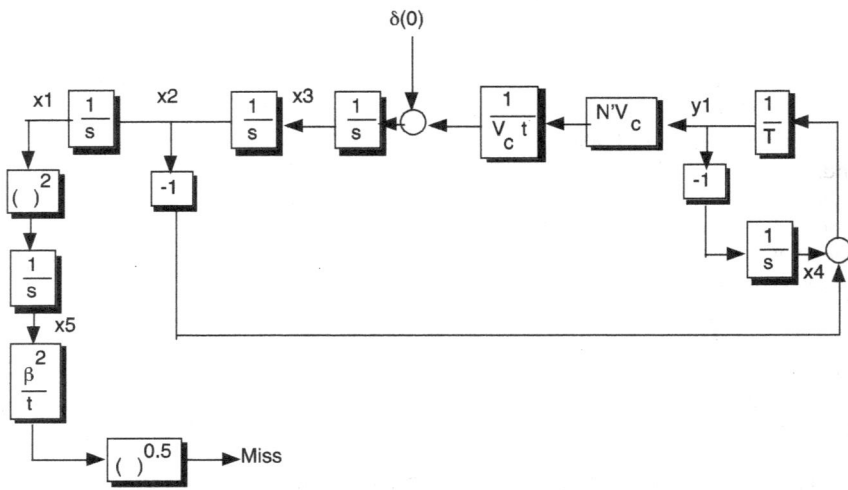

Fig. 5.4   Adjoint of single-lag homing loop with shaping filter equivalent of constant target maneuver whose starting time is uniformly distributed over the flight time.

The number of runs required to generate the results of Fig. 5.3 were identical whether the actual or shaping filter modeling techniques were utilized. One might ask why one should bother with shaping filters at all in a Monte Carlo simulation if it does not result in computational advantages over just modeling the actual target maneuver. Although the shaping filter approach does not offer computational advantages in a Monte Carlo simulation, it does offer dramatic computational advantages in an adjoint simulation when the homing loop is linear. Recall that in Fig. 5.3, *50,000 runs* were required to generate the plot. Because the single-lag homing loop under consideration is linear (i.e., no missile acceleration saturation), the adjoint of Fig. 5.2 can also be taken, as was done in Chapter 4 of the previous volume. The adjoint of the single-lag guidance system is repeated in Fig. 5.4 so that the miss due to a constant target maneuver with a starting time that is uniformly distributed over the flight time can be efficiently calculated. The adjoint simulation, based on Fig. 5.4 and using the same system parameters of Listings 5.1 and 5.2, appears in Listing 5.3. Here we can see that only *one* adjoint run is required to generate RMS miss distance results for a variety of flight times.

LISTING 5.3   ADJOINT SIMULATION OF SINGLE-LAG HOMING LOOP USING SHAPING FILTER EQUIVALENT OF CONSTANT TARGET MANEUVER WITH RANDOM STARTING TIME UNIFORMLY DISTRIBUTED OVER FLIGHT TIME

```
clear
count=0;
XNT = 161.;
```

```
XNP = 3;
TAU = .2;
TF = 10;
T=0;
S=0;
TP=T+.00001;
X1=0;
X2=0;
X3=1;
X4=0;
X5=0.;
H=.01;
while TP <= (TF - 1e-5)
      STEP=1;
      FLAG=0;
      S=S+H;
      X1OLD=X1;
      X2OLD=X2;
      X3OLD=X3;
      X4OLD=X4;
      X5OLD=X5;
      while STEP <=1
        if FLAG==1
                STEP=2;
                X1=X1+H*X1D;
                X2=X2+H*X2D;
                X3=X3+H*X3D;
                X4=X4+H*X4D;
                X5=X5+H*X5D;
                TP=TP+H;
        end;
        X1D=X2;
        X2D=X3;
        Y1=(X4-X2)/TAU;
        TGO=TP+.00001;
        X3D=XNP*Y1/TGO;
        X4D=-Y1;
        X5D=X1*X1;
        FLAG=1;
      end;
      FLAG=0;
      X1=(X1OLD+X1)/2+.5*H*X1D;
      X2=(X2OLD+X2)/2+.5*H*X2D;
      X3=(X3OLD+X3)/2+.5*H*X3D;
```

```
X4=(X4OLD+X4)/2+.5*H*X4D;
X5=(X5OLD+X5)/2+.5*H*X5D;
S=S+H;
if S>=.000999
  S=0.;
  XMUDNT=XNT*sqrt(X5/TGO);
  count=count+1;
  ArrayTP(count)=TP;
  ArrayXMUDNT(count)=XMUDNT;
end;
end
figure
plot(ArrayTP, ArrayXMUDNT),grid
title('Adjoint model using shaping filter approach')
xlabel('Flight Time (S)')
ylabel('Miss Dist Standard Deviation (Ft)')
clc
output=[ArrayTP',ArrayXMUDNT'];
save datfil.txt output -ascii
disp('Simulation Complete')
```

The nominal case of Listing 5.3 was run, and the results were compared to those generated for the actual random target maneuver of Listing 5.1. We can see from

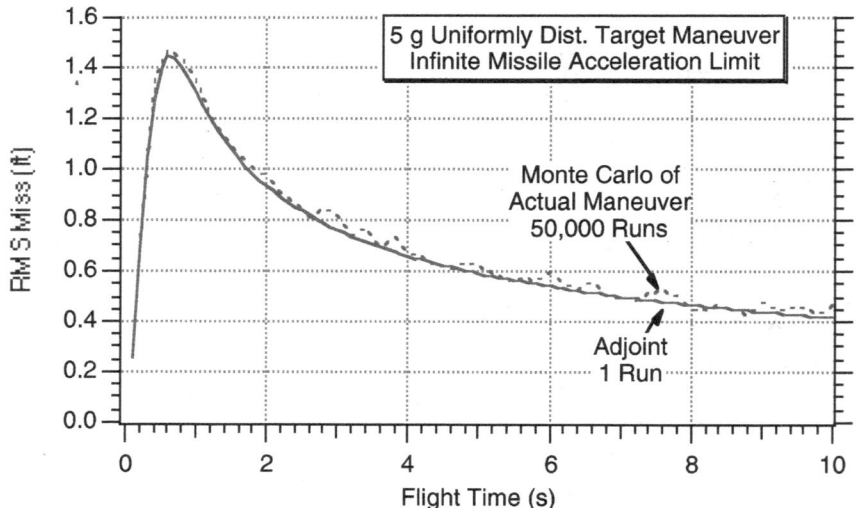

Fig. 5.5 Actual target maneuver and adjoint using shaping filter equivalent yield identical RMS miss distances for uniformly distributed target maneuver.

Fig. 5.5 that the RMS miss distance results of the adjoint are virtually identical to the Monte Carlo RMS miss distance results of the actual random target maneuver. Figure 5.5 certainly justifies the use of adjoints and shaping filters for RMS miss distance evaluation when the missile does not acceleration saturate.

Now let's consider a case in which there is a missile acceleration limit of 30 g, which yields a 6-to-1 missile-to-target-acceleration advantage. Figure 5.6 displays the RMS miss distance as a function of flight time for both target maneuver modeling approaches. We can see that the shaping filter approach significantly overestimates the RMS miss distance due to minor acceleration saturation. By comparing the actual RMS miss due to target maneuver results of Fig. 5.6 with the actual target maneuver results of Fig. 5.5 (where there are no missile acceleration saturation effects) we can see that there is only a slight increase in the RMS miss when there is a 6-to-1 missile-to-target-acceleration advantage. A slight increase in miss due to a large missile-to-target-acceleration advantage is more consistent with the miss distance results of Chapter 6 of Volume 1. *It appears that the shaping filter approach overestimates the RMS miss when the single-lag homing loop is nonlinear due to saturation effects when there is a 6-to-1 missile-to-target-acceleration advantage.*

Now let's consider a case in which there is a missile acceleration limit of 15 g, which yields a 3-to-1 missile-to-target-acceleration advantage. Figure 5.7 displays the RMS miss distance as a function of flight time for both target maneuver modeling approaches. We can see that the shaping filter approach for representing a random target maneuver drastically overestimates the RMS miss distance due

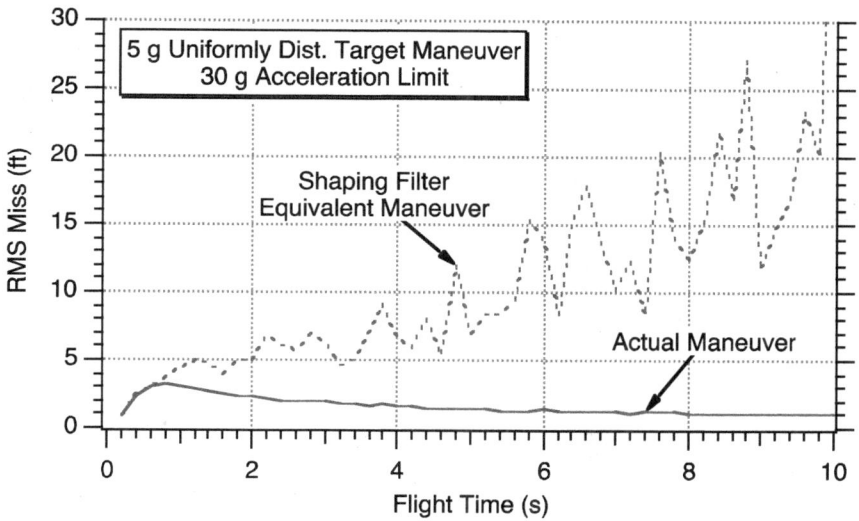

Fig. 5.6   Shaping filter approach overestimates RMS miss when single-lag homing loop is nonlinear and there is a 6-to-1 missile-to-target-acceleration advantage.

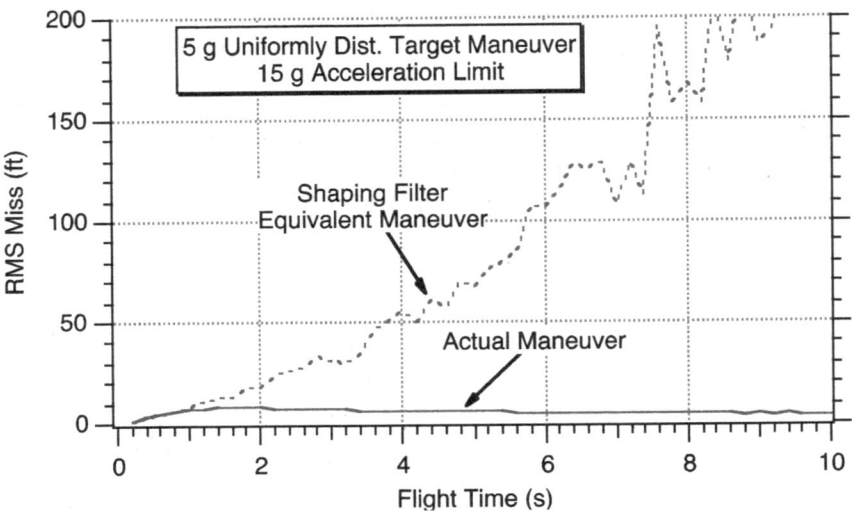

Fig. 5.7 Shaping filter approach dramatically overestimates RMS miss when there is a 3-to-1 missile-to-target-acceleration advantage for uniformly distributed target maneuver.

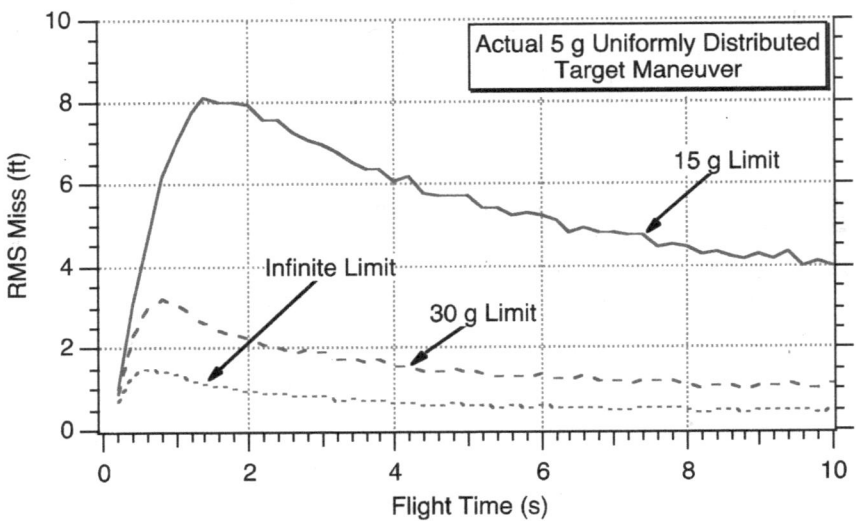

Fig. 5.8 Miss increases with decreasing acceleration limit for actual uniformly distributed target maneuver.

to missile acceleration saturation. In fact, the shaping filter results would have one believe that a missile-to-target-acceleration advantage of 3 to 1 is not sufficient for hitting a maneuvering target. By comparing the actual target maneuver results of Fig. 5.7 with the actual target maneuver results of Fig. 5.5, where there is no missile acceleration saturation, we can see that there should be only a slight increase in the RMS miss when there is a 3-to-1 missile-to-target-acceleration advantage. A slight increase in miss due to a large missile-to-target-acceleration advantage is more consistent with the miss distance results of Chapter 6 of the previous volume.

In the case of a single time constant guidance system where the time constant is small, we have seen that when the actual target maneuver is used in evaluating system performance in the presence of missile acceleration saturation effects, the missile appears to have significant capability against a randomly maneuvering target when proportional navigation is used and the missile-to-target-acceleration advantage is greater than three. Figure 5.8 displays the RMS miss for the actual uniformly distributed target maneuver as a function of flight time for different missile acceleration limits. Here we can see that the degradation in performance due to missile acceleration saturation is important but not necessarily cata-strophic, as would be suggested by using the shaping filter approach.

SUMMARY

In this chapter we have shown that the shaping filter approach for representing random target maneuvers can be very useful and cost effective when missile sat-uration does not occur. Shaping filters are also extremely useful for deriving Kalman filters and various types of guidance laws. However, in evaluating system performance when missile saturation occurs, the use of shaping filters can lead to erroneously pessimistic predictions of the RMS miss distance. For miss distance studies involving missile acceleration saturation, it is recommended that one model the actual target maneuver so that realistic performance projec-tions can be obtained.

CHAPTER 6

Weaving Targets

INTRODUCTION AND BACKGROUND

We saw in Chapter 6 of the previous volume that large miss distances could be induced by the target if a maximum acceleration maneuver was initiated at the proper time to go before intercept. It was also shown that the barrel roll or weave maneuver could also generate large miss distances. Because it is well known that tactical ballistic missiles (TBMs) can spiral or weave into resonance (TBM roll rate equals vehicles' natural pitch frequency) as they re-enter the atmosphere due to either mass or configurational asymmetries, the weave maneuver is of particular interest to the guidance system designer [1, 2].

In this chapter we will first study the influence of the target weave maneuver on a single time constant proportional navigation guidance system. Closed-form solutions for the peak steady-state miss distance as a function of the effective navigation ratio, guidance system time constant, weave maneuver amplitude, and frequency will be derived [3, 4]. Because we have already shown in Volume 1 that the single time constant guidance system seriously underestimates the miss distance, a more realistic, higher-order guidance system will be used to develop normalized miss distance design curves using the normalization factors from the single time constant target maneuver miss distance solutions. The finite acceleration capability of the interceptor also plays an important role in determining system performance. The normalized design curves, which assumed infinite missile acceleration capability, are updated to show how the missile acceleration advantage over the target plays a key role in determining system performance. Finally, methods for improving missile system performance against weaving targets will be explored.

WEAVE MANEUVER IN SINGLE TIME CONSTANT GUIDANCE SYSTEM

Periodic maneuver sequences such as a sinusoidal or weaving target present a challenge for a missile guidance system designer. A planar representation of a weaving target is given by

$$\text{Target Maneuver} = n_T \sin \omega_T t$$

where n_T is the maneuver amplitude, ω_T is the target weave frequency, and t is time. The miss due to a weaving target as a function of flight time can be found using the method of brute force. A nonlinear two-dimensional engagement simulation, based on Listing 2.1 of Volume 1, of a missile guiding on a weaving target appears in Listing 6.1. We can see that the listing is based on a single time constant proportional navigation guidance system and that for the nominal case the missile time constant is 1 s, the effective navigation ratio is 3, the target weave frequency is 3 rad/s, and the target maneuver amplitude is 193.2 ft/s^2 or 6 g. The target always initiates its maneuver at the beginning of flight in this simulation. The program is set up to run in the brute force mode so that the miss distance results for many flight times can be evaluated. In this nonlinear engagement simulation, the initial target downrange position, which is equivalent to the initial missile-target separation, is varied from 500 ft to 40,000 ft in steps of 500 ft, which in the linear world is equivalent to varying the flight time from 0.05 s to 10 s in steps of 0.05 s because the closing velocity is approximately 4000 ft/s. After each run the homing time and miss distance are tabulated. We can also see from the simulation listing that if the target is above the missile at intercept the miss is considered to be positive, whereas if the target is below the missile at intercept the miss is considered to be negative.

LISTING 6.1 NONLINEAR ENGAGEMENT SIMULATION FOR A SINGLE TIME CONSTANT GUIDANCE SYSTEM WITH WEAVING TARGET

```
clear
XNP=3.;
TAU=1.;
XNT=193.2;
W=3.;
n=0;
for RT1IC=500:500:40000
        VM=3000.;
        VT=1000.;
        RM1=0.;
        RM2=0.;
        RT1=RT1IC;
        RT2=0.;
        BETA=0.;
        VT1=-VT*cos(BETA);
        VT2=VT*sin(BETA);
        T=0.;
        S=0.;
```

```
RTM1=RT1-RM1;
RTM2=RT2-RM2;
RTM=sqrt(RTM1^2+RTM2^2);
XLAM=atan2(RTM2,RTM1);
VM1=VM;
VM2=0.;
VTM1=VT1-VM1;
VTM2=VT2-VM2;
VC=-(RTM1*VTM1+RTM2*VTM2)/RTM;
TGO=RTM/VC;
XLAMH=0.;
H=.01;
while VC>0.
if(RTM<1000.)
            H=.0005;
end
            BETAOLD=BETA;
            RT1OLD=RT1;
            RT2OLD=RT2;
            RM1OLD=RM1;
            RM2OLD=RM2;
            VM1OLD=VM1;
            VM2OLD=VM2;
            XLAMHOLD=XLAMH;
            STEP=1;
            FLAG=0;
            while STEP<=1
            if FLAG==1
            STEP=2;
            BETA=BETA+H*BETAD;
            RT1=RT1+H*VT1;
            RT2=RT2+H*VT2;
            RM1=RM1+H*VM1;
            RM2=RM2+H*VM2;
            VM1=VM1+H*AM1;
            VM2=VM2+H*AM2;
            XLAMH=XLAMH+H*XLAMHD;
            T=T+H;
            end
            VT1=-VT*cos(BETA);
            VT2=VT*sin(BETA);
            BETAD=XNT*sin(W*T)/VT;
            RTM1=RT1-RM1;
            RTM2=RT2-RM2;
            RTM=sqrt(RTM1^2+RTM2^2);
            VTM1=VT1-VM1;
```

```
                    VTM2=VT2-VM2;
                    VC=-(RTM1*VTM1+RTM2*VTM2)/RTM;
                    XLAM=atan2(RTM2,RTM1);
                    XLAMHD=(XLAM-XLAMH)/TAU;
                    XNC=XNP*VC*XLAMHD;
                    AM1=-XNC*sin(XLAM);
                    AM2=XNC*cos(XLAM);
                    FLAG=1;
            end
            FLAG=0;
            BETA=.5*(BETAOLD+BETA+H*BETAD);
            RT1=.5*(RT1OLD+RT1+H*VT1);
            RT2=.5*(RT2OLD+RT2+H*VT2);
            RM1=.5*(RM1OLD+RM1+H*VM1);
            RM2=.5*(RM2OLD+RM2+H*VM2);
            VM1=.5*(VM1OLD+VM1+H*AM1);
            VM2=.5*(VM2OLD+VM2+H*AM2);
            XLAMH=.5*(XLAMHOLD+XLAMH+H*XLAMHD);
        end
        if RTM2>0.
            RTMP=RTM;
        else
            RTMP=-RTM;
        end
        n=n+1;
        ArrayT(n)=T;
        ArrayRTMP(n)=RTMP;
    end
    figure
    plot(ArrayT,ArrayRTMP),grid
    xlabel('Flight Time (Sec)')
    ylabel('Miss (Ft)')
    clc
    output=[ArrayT',ArrayRTMP'];
    save datfil.txt output -ascii
    disp 'simulation finished'
```

The nominal case of Listing 6.1 was run, and the miss distance results as a function of flight time appear in Fig. 6.1. We can see that unlike the step target maneuver results of Chapters 3 and 6 of the previous volume, *the miss distance due to a weaving target does not approach zero as the homing time increases!* Depending on the flight time, the miss distance for this example can be as large as 28 ft or as small as zero when the effective navigation ratio is 3. Also note that after an initial transient period, the miss is sinusoidal in nature with a frequency of 3 rad/s, which is identical to the target weave frequency.

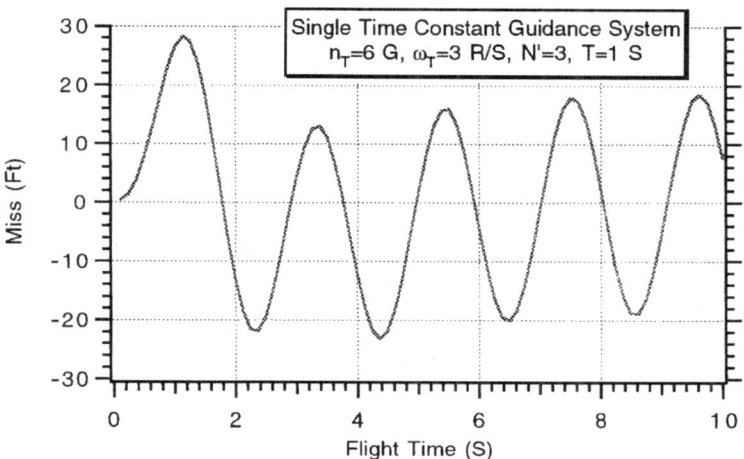

Fig. 6.1 Nonlinear results indicate that weaving target causes miss to oscillate at target weave frequency.

Could these multiple-run nonlinear results be generated with one adjoint run? To find out, we first have to assume that the geometry of the engagement can be linearized in a manner similar to that of Chapters 2, 3, and 20 of the previous volume. In addition, because adjoint theory requires that all inputs to the original system appear as impulses, we also have to find some way of making the sinusoidal maneuver look like an impulse through a linear shaping network. Fortunately, in Table 1.1 of Chapter 1 of the previous volume we showed that a sinusoidal maneuver could be represented as an impulse through a second-order shaping network because

$$\mathcal{L}[n_T \sin \omega_T t] = \frac{\omega_T n_T}{s^2 + \omega_T^2}$$

Figure 6.2 shows a linearized representation of a single time constant linear proportional navigation guidance system driven by a weave target maneuver (that is, impulsive input through second-order shaping network).

Figure 6.2 is drawn in such a way that the adjoint can be taken by inspection. The new adjoint diagram was constructed from Fig. 6.2 using the adjoint rules of Chapter 3 of Volume 1 and appears in Fig. 6.3. In this particular adjoint diagram we are evaluating one disturbance only, the miss due to a weaving target.

An adjoint simulation, based on Fig. 6.3, was constructed and appears in Listing 6.2. Here we can see that the nominal case is set up to be identical to that of the brute force nonlinear engagement simulation of Listing 6.1. However, with the adjoint simulation only one run has to be made to find out how the miss distance varies with flight time.

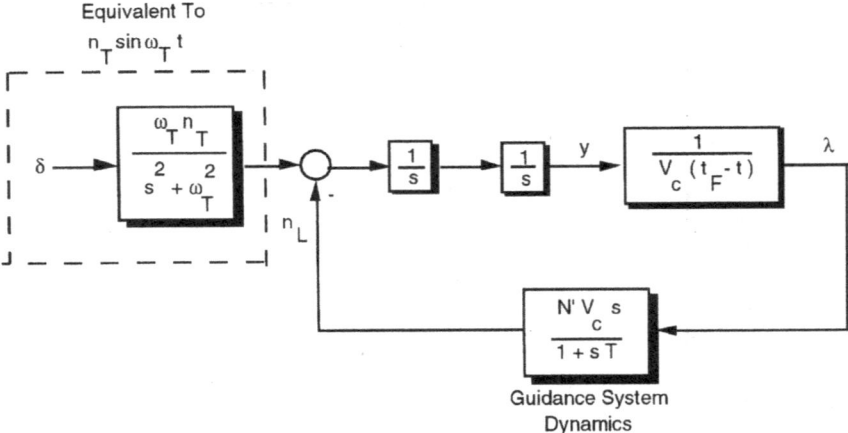

Fig. 6.2   Linearized single time constant guidance system with weave maneuver.

The nominal case of Listing 6.2 was run, and the adjoint miss distance results were plotted alongside the nonlinear, multiple-run results of Fig. 6.1. We can see from Fig. 6.4 that the adjoint results are virtually identical to the nonlinear miss distance results, thus validating the simplified linear model. In the rest of this chapter, we will use the linearized model of the guidance system to generate performance evaluations and to suggest ways of improving performance.

Another case was run with the adjoint simulation in which the target weave frequency was decreased from 3 rad/s to 1.5 rad/s. We can see from Fig. 6.5 that the miss distance increases significantly with the lower weave frequency,

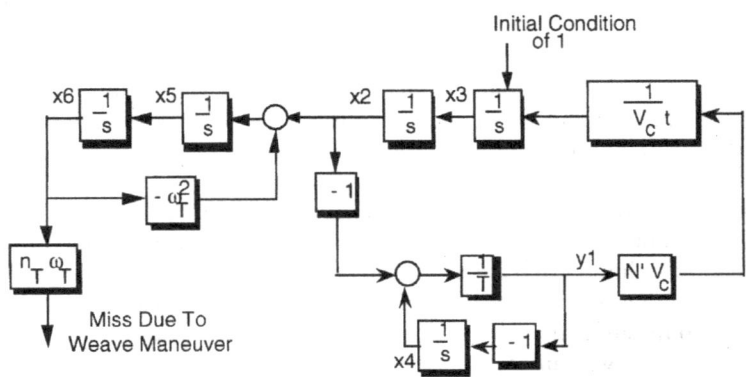

Fig. 6.3   Adjoint of single time constant guidance system with weaving target.

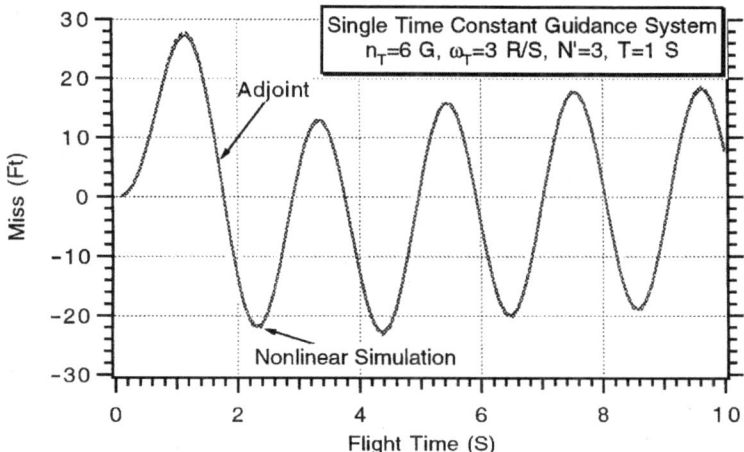

Fig. 6.4   Adjoint simulation results agree with the nonlinear results for the weaving target disturbance.

and the oscillation of the miss with flight time changes to match the new target weave frequency. We now would like to get a deeper understanding of how performance is related to the target weave frequency, guidance system time constant, and effective navigation ratio.

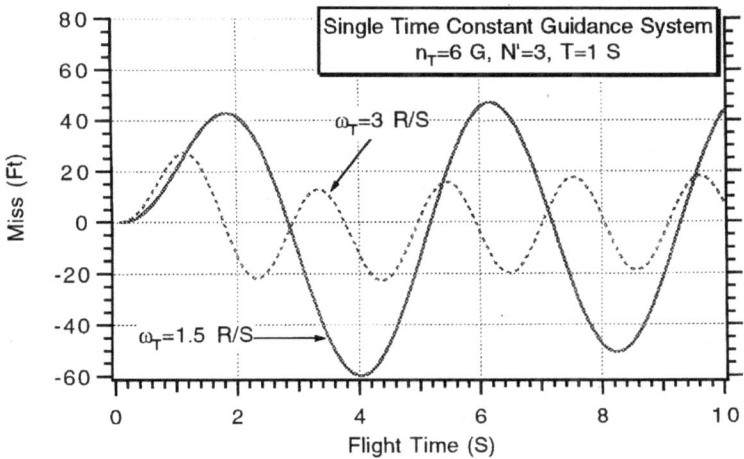

Fig. 6.5   Decreasing weave frequency increases miss distance for this example.

LISTING 6.2   ADJOINT SIMULATION OF SINGLE TIME CONSTANT GUIDANCE SYSTEM AND WEAVING TARGET

```
clear
n=0;
XNT=193.2;
XNP=3.;
TAU=1.;
TF=10.;
VC=4000.;
W=3.;
T=0.;
S=0.;
TP=T+.00001;
X2=0;
X3=1;
X4=0.;
X5=0.;
X6=0.;
H=.01;
while TP<=(TF-1e-5)
        X2OLD=X2;
        X3OLD=X3;
        X4OLD=X4;
        X5OLD=X5;
        X6OLD=X6;
        STEP=1;
        FLAG=0;
        while STEP<=1
           if FLAG==1
           STEP=2;
                   X2=X2+H*X2D;
                   X3=X3+H*X3D;
                   X4=X4+H*X4D;
                   X5=X5+H*X5D;
                   X6=X6+H*X6D;
                   TP=TP+H;
           end
           X2D=X3;
           Y1=(-X2+X4)/TAU;
           TGO=TP+.00001;
           X3D=Y1*XNP/TGO;
           X4D=-Y1;
           X5D=X2-W*W*X6;
           X6D=X5;
           FLAG=1;
        end
```

```
        FLAG=0;
        X2=(X2OLD+X2)/2+.5*H*X2D;
        X3=(X3OLD+X3)/2+.5*H*X3D;
        X4=(X4OLD+X4)/2+.5*H*X4D;
        X5=(X5OLD+X5)/2+.5*H*X5D;
        X6=(X6OLD+X6)/2+.5*H*X6D;
        S=S+H;
        if S>=.09999
          S=0.;
          n=n+1;
          XMWEAVE=XNT*W*X6;
              ArrayTP(n)=TP;
          ArrayXMWEAVE(n)=XMWEAVE;
        end
end
figure
plot(ArrayTP,ArrayXMWEAVE),grid
xlabel('Flight Time (S)')
ylabel('Miss (Ft) ')
clc
output=[ArrayTP',ArrayXMWEAVE'];
save datfil.txt output/ascii
disp 'simulation finished'
```

CLOSED-FORM SOLUTIONS FOR MISS DISTANCE

Because we have already shown that the miss due to a weaving target is a sinusoidal function of the flight time, it only makes sense to look at steady-state miss distances in order to quantify system performance. Closed-form solutions for the miss due to a weaving target can be obtained in the steady state (that is, at large flight times when transients die out). Recall that all of the miss distance formulas for a single time constant guidance system were derived in Chapter 3 of Volume 1 from the generalized adjoint diagram of Fig. 3.16. Figure 3.16 is redrawn and updated to include the miss due to a weaving target MWEAVE, as shown in Fig. 6.6.

In Chapter 3 we found that the miss due to a step target maneuver MNT, expressed in the Laplace transform domain, was given by

$$\text{MNT}(s) = \frac{1 - H(s)}{s^3} * n_T$$

Therefore, the miss due to a weaving target can be found by inspection from Fig. 6.6 and can be expressed as

$$\text{MWEAVE}(s) = \frac{1 - H(s)}{s^2} * \frac{n_T \omega_T}{s^2 + \omega_T^2}$$

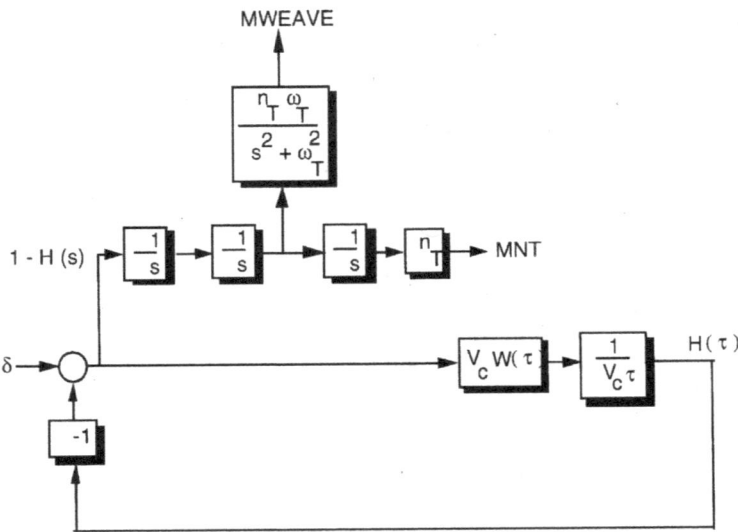

Fig. 6.6 Generalized adjoint diagram showing miss due to weaving target.

where $1 - H(s)$ was shown in Chapter 3 of the previous volume for the single time constant guidance system to be given by

$$1 - H(s) = \left[\frac{s}{s + \dfrac{1}{T}} \right]^{N'}$$

Therefore, the miss due to a weaving target can be expressed as

$$\text{MWEAVE}(s) = \frac{1}{s^2} * \left[\frac{s}{s + \dfrac{1}{T}} \right]^{N'} * \frac{n_T \omega_T}{s^2 + \omega_T^2}$$

The miss distance in the Laplace transform domain can be evaluated directly by first doing a partial fraction expansion of the terms on the right side of the preceding equation and then taking the inverse Laplace transform to find the miss in the time domain. Some of the terms in the resultant complex expression would be transient in nature whereas other terms would be sinusoidal. In the steady state, the transient terms would go to zero and only the sinusoidal terms would be left.

If we are only interested in the steady-state solution, much work can be saved by using a simple technique from electrical engineering. We can rewrite the preceding expression as

$$\frac{\text{Miss}}{\text{weave } n_T}(s) = \frac{32.2}{s^2} \left[\frac{s}{s + \dfrac{1}{T}} \right]^{N'}$$

where n_T is now the maneuver magnitude in units of g and "weave n_T" reminds us that the target maneuver is sinusoidal. If a linear system has a sine wave input with frequency ω_T in units of rad/s, the output in the steady state will also be a sinusoid of the same frequency but of different magnitude and phase. From basic steady-state electrical engineering circuit analysis techniques, it can be shown that the magnitude and phase of the sinusoidal output can be found by replacing s with $j\omega_T$ in the preceding transfer function and then finding the magnitude and phase of the resultant complex transfer function [5]. For example, if the effective navigation ratio is 3, the preceding transfer function becomes

$$\left.\frac{\text{Miss}}{\text{weave } n_T}\right|_{N'=3}(s) = \frac{32.2s}{(s + 1/T)^3}$$

Therefore, the complex weave miss distance transfer function can be derived from the preceding equation by substitution (that is, $s = j\omega_T$) as

$$\left.\frac{\text{Miss}}{\text{weave } n_T}\right|_{N'=3}(j\omega_T) = \frac{32.2j\omega_T}{(j\omega_T + 1/T)^3}$$

The magnitude and phase of this complex transfer function can be written by inspection as

$$\text{Magnitude}\big|_{N'=3} = \frac{32.2\omega_T}{(\omega_T^2 + 1/T^2)^{1.5}}$$

$$\text{Phase}\big|_{N'=3} = \frac{\pi}{2} - 3\tan^{-1}\omega_T T$$

Therefore, the steady-state miss distance due to a weaving target can be written in the time domain as

$$\left.\frac{\text{Miss}}{\text{weave } n_T}\right|_{\substack{N'=3 \\ \text{Steady-State}}} = \text{Magnitude}\big|_{N'=3}\sin\left(\omega_T t_F + \text{Phase}\big|_{N'=3}\right)$$

or

$$\left.\frac{\text{Miss}}{\text{weave } n_T}\right|_{\substack{N'=3 \\ \text{Steady-State}}} = \frac{32.2\omega_T}{\left(\omega_T^2 + \dfrac{1}{T^2}\right)^{1.5}}\sin\left(\omega_T t_F + \frac{\pi}{2} - 3\tan^{-1}\omega_T T\right)$$

Figure 6.7 presents again the adjoint miss distance results as a function of flight time for the case in which the target weave frequency is 3 rad/s while the

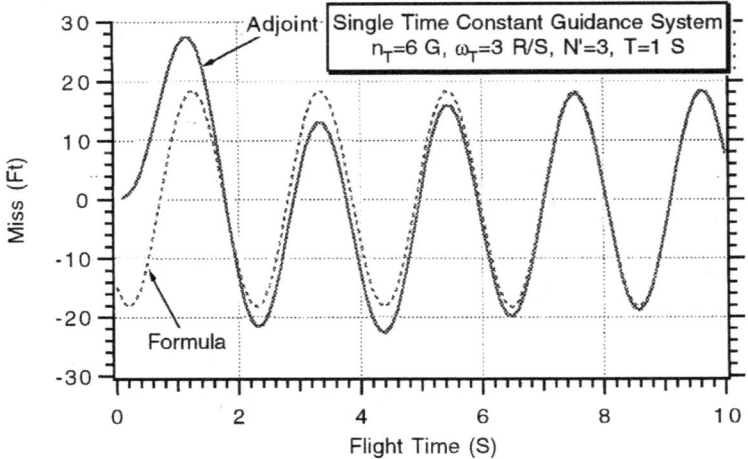

Fig. 6.7   Closed-form miss distance solution agrees with adjoint results.

missile guidance system time constant is 1 s. Superimposed on the figure is the preceding closed-form solution for the miss distance. We can see that after an initial transient period, the closed-form steady-state miss distance solution and computer-generated adjoint results are in excellent agreement, thus confirming the validity of the steady-state analysis.

We have shown mathematically and by simulation that the miss distance due to a weaving target is a sinusoidal function of the flight time. Therefore, it is really a matter of luck on how large or small the miss distance will be. *However, we will show later in Chapter 17 that in three dimensions the peak miss is the real miss.* The peak value of the miss due to a weave maneuver is simply the magnitude of the steady-state miss distance sinusoid. Therefore, the peak miss due to a weave maneuver in a single time constant proportional navigation guidance system with an effective navigation ratio of 3 is given by

$$\frac{\text{Peak Miss}}{\text{weave } n_T}\bigg|_{N'=3} = \frac{32.2\omega_T}{(\omega_T^2 + 1/T^2)^{1.5}} = \frac{32.2\omega_T T^3}{(1 + \omega_T^2 T^2)^{1.5}}$$

Dividing both sides of the equation by T^2 yields

$$\frac{\text{Peak Miss}}{\text{weave } n_T T^2}\bigg|_{N'=3} = \frac{32.2\omega_T T}{(1 + \omega_T^2 T^2)^{1.5}}$$

If we let x be the normalized target weave frequency where

$$x = \omega_T T$$

the peak miss distance formula simplifies further to

$$\frac{\text{Peak Miss}}{\text{weave } n_T T^2}\bigg|_{N'=3} = \frac{32.2x}{(1 + x^2)^{1.5}}$$

Similar expressions can be found for the peak miss distance due to a weave maneuver when the effective navigation ratios are 4 and 5 and can be shown to be

$$\frac{\text{Peak Miss}}{\text{weave } n_T T^2}\bigg|_{N'=4} = \frac{32.2x^2}{(1 + x^2)^{2}}$$

$$\frac{\text{Peak Miss}}{\text{weave } n_T T^2}\bigg|_{N'=5} = \frac{32.2x^3}{(1 + x^2)^{2.5}}$$

Figure 6.8 graphically displays the preceding formulas and shows how the steady-state normalized peak miss distance varies with the normalized target maneuver frequency (that is, product of the target weave frequency and the missile guidance system time constant). We can see from Fig. 6.8 that the peak miss distance is close to a maximum when the normalized target maneuver frequency is near unity. Large weave frequencies do not cause much miss distance because very little target displacement is created. On the other hand, small weave frequencies look like step target maneuvers and thus in the steady state (large flight times) cause very little miss distance. If we were on a collision triangle with the target (with no heading error) and we coasted to the target by turning off the guidance (where $N' = 0$), the peak miss distance would simply be the peak displacement n_T/ω_T^2 caused by the weaving target. Superimposed on Fig. 6.8 is

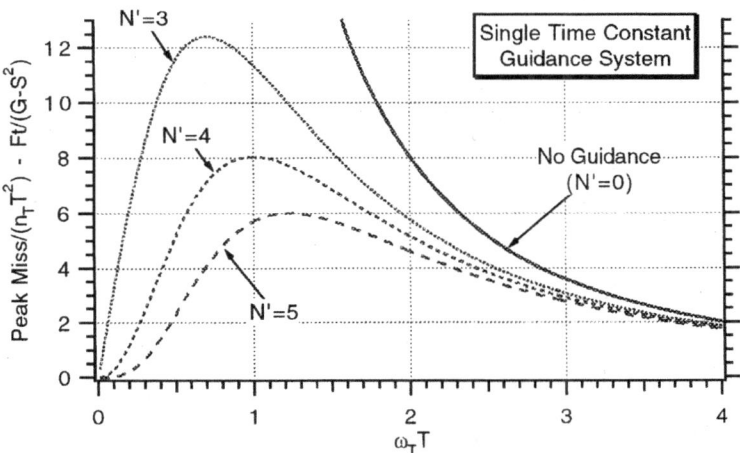

Fig. 6.8   Peak miss distance is maximum when normalized weave frequency is near unity.

the peak displacement or induced miss distance with no missile guidance (where $N' = 0$) caused by the weaving target. We can see that for the single time constant guidance system, guiding with proportional navigation always yields a smaller miss against a weaving target than coasting without guidance. *However, for large values of normalized weave frequency, the miss distance with and without guidance is approximately the same!*

To illustrate the use of the normalized miss distance curves of Fig. 6.8, let us consider a numerical example in which there is a 6-g weaving target with a weave frequency of 2 rad/s. Assuming that the missile guidance system time constant is 1 s and effective navigation ratio is 3, we first compute the normalized weave frequency as

$$\omega_T T = 2 * 1 = 2$$

which results in a normalized miss of approximately 5.5. Therefore, from the ordinate of Fig. 6.8 we can compute the peak steady-state miss distance to be

$$\text{Peak Miss} \approx 5.5 n_T T^2 = 5.5 * 6 * 1^2 = 33 \text{ ft}$$

Reducing the guidance system time constant to 0.5 s changes both the normalized weave frequency and the normalized miss. The new normalized weave frequency is

$$\omega_T T = 2 * 0.5 = 1$$

which results in an increased normalized miss of approximately 11.5. However, the new peak steady-state miss distance is reduced because the guidance system time constant has been halved, or

$$\text{Peak Miss} \approx 11.5 n_T T^2 = 11.5 * 6 * 0.5^2 \approx 17 \text{ ft}$$

Keeping the guidance system constant fixed to 0.5 s but increasing the target weave frequency to 4 rad/s increases the normalized weave frequency back to 2, or

$$\omega_T T = 4 * 0.5 = 2$$

which again results in a normalized miss of approximately 5.5. The new peak steady-state miss distance becomes

$$\text{Peak Miss} \approx 5.5 n_T T^2 = 5.5 * 6 * 0.5^2 \approx 8 \text{ ft}$$

Thus we can see that both the guidance system time constant and target weave frequency are important factors in determining the peak steady-state miss distance.

HIGHER-ORDER GUIDANCE SYSTEM DYNAMICS

The single time constant guidance system model, used in the previous section, was useful because it could be used to derive closed-form solutions for the miss distance due to a weave maneuver. The single time constant guidance system miss distance formulas also suggest normalization factors for the miss distance. We already showed in Chapters 6 and 20 of the previous volume that the disadvantage of the single time constant representation of a missile guidance system is that the miss distance can be seriously underestimated. We saw in Chapter 6 of Volume 1 that a much better and equally convenient representation of a proportional navigation missile guidance system transfer function is a canonic fifth-order binomial given by

$$\frac{n_L}{\lambda} = \frac{N' V_c s}{(1 + sT/5)^5}$$

where T is the total guidance system time constant, n_L is the achieved missile acceleration, and λ is the line-of-sight angle. As was mentioned in Chapters 6 and 20 of the previous volume for this generic interceptor guidance system model, one time constant represents the seeker, another represents the noise filter, and the three other time constants represent the flight-control system dynamics (aerodynamics plus autopilot). It is easy to show that with this canonic guidance system model, the overall guidance system time constant is simply the sum of the five individual time constants or T. The peak steady-state miss distance due to a weaving target for the fifth-order binomial missile homing loop can be evaluated using either the method of adjoints or the method of brute force. Because the adjoint simulation would have to be extensively modified to figure out when steady state was reached and special logic would then have to be developed to capture the maximum miss distance, it was considered easier to use the brute force approach. In addition, the brute force approach can easily be extended to the case where there are significant nonlinearities, whereas the adjoint method would no longer be valid. Listing 6.3 presents the brute force simulation based on linearized geometry, which we have already shown to be valid for the weaving target case. We can see from Listing 6.3 that acceleration saturation effects can be included by simply reducing the value of the acceleration limit XNCLIM from its near-infinite value. The listing shows how the time constant is reduced when the flight times are short to ensure that we are in steady state. The simulation is set up to generate normalized miss distance curves as a function of the normalized target weave frequency.

Figure 6.9 shows how the steady-state normalized peak miss distance due to a weave maneuver varies with the normalized target weave frequency for the fifth-order binomial guidance system. The curves in this figure are similar in shape to the ones of Fig. 6.8, but as expected, the normalized miss distances are much larger. It is interesting to note the steady-state peak miss distance is still

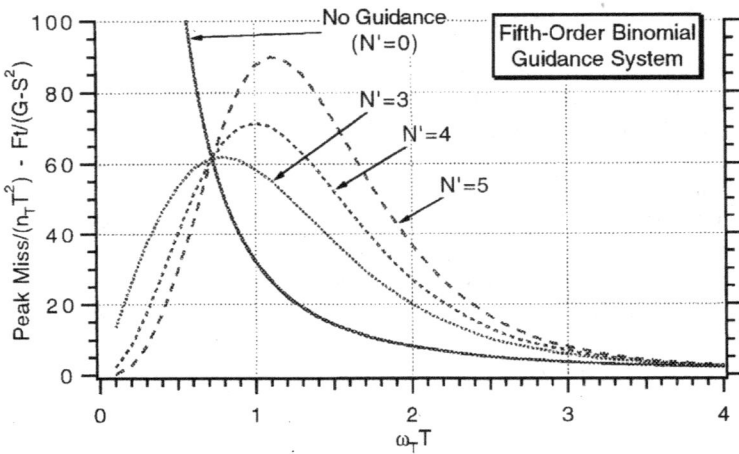

Fig. 6.9   Steady-state peak miss due to weave maneuver is much larger with fifth-order binomial guidance system.

maximum when the normalized target weave frequency is approximately unity. Superimposed on Fig. 6.9 is the zero guidance miss distance or peak displacement n_T/ω_T^2 caused by the weaving target. Surprisingly, we can see that for the fifth-order guidance system, proportional navigation yields only a smaller miss than coasting (where $N' = 0$) when the normalized weave frequency is less than 0.7 (that is, $\omega_T T < 0.7$). *In other words, for normalized weave frequencies greater than 0.7, the weaving target nullifies the effectiveness of a proportional navigation guidance system!*

LISTING 6.3   BRUTE FORCE SIMULATION FOR GENERATING NORMALIZED DESIGN CURVES AGAINST WEAVING TARGET

```
clear
n=0;
VC=4000.;
XNT=32.2;
XNP=3.;
XNCLIM=99999.;
for X=.1:.1:4
        if X<.5
          W=1.;
          TAU=X/W;
        else
          W=X;
          TAU=1.;
```

```
end
XMWEAVEOLD=0.;
XMWEAVEMAX=0.;
for TF=.2:.2:20
    PHASE=0.;
    Y=0.;
    YD=0.;
    XNL=0.;
    D=0.;
    ELAMDH=0.;
    X4=0.;
    X5=0.;
    T=0.;
    H=.01;
    S=0.;
    while T<=(TF-1e-5)
                YOLD=Y;
                YDOLD=YD;
                XNLOLD=XNL;
                DOLD=D;
                ELAMDHOLD=ELAMDH;
                X4OLD=X4;
                X5OLD=X5;
                STEP=1;
                FLAG=0;
    while STEP<=1
                            if FLAG==1
                                STEP=2;
                                Y=Y+H*YD;
                                YD=YD+H*YDD;
                                XNL=XNL+H*XNLD;
                                ELAMDH=ELAMDH+H*ELAMDHD;
                                D=D+H*DD;
                                X4=X4+H*X4D;
                                X5=X5+H*X5D;
                                T=T+H;
                        end
                        YTDD=XNT*sin(W*T);
                        TGO=TF-T+.00001;
                        XLAM=Y/(VC*TGO);
                        DD=5.*(XLAM-D)/TAU;
                        ELAMDHD=5.*(DD-ELAMDH)/TAU;
                        XNC=XNP*VC*ELAMDH;
                        if XNC>XNCLIM
                                XNC=XNCLIM;
                        end
```

```
                            if XNC<-XNCLIM
                                    XNC=-XNCLIM;
                            end
                            X4D=5.*(XNC-X4)/TAU;
                            X5D=5.*(X4-X5)/TAU;
                            XNLD=5.*(X5-XNL)/TAU;
                            YDD=YTDD-XNL;
                            FLAG=1;
                    end
                    FLAG=0;
                    Y=.5*(YOLD+Y+H*YD);
                    YD=.5*(YDOLD+YD+H*YDD);
                    XNL=.5*(XNLOLD+XNL+H*XNLD);
                    D=.5*(DOLD+D+H*DD);
                    ELAMDH=.5*(ELAMDHOLD+ELAMDH+H*ELAMDHD);
                    X4=.5*(X4OLD+X4+H*X4D);
                    X5=.5*(X5OLD+X5+H*X5D);
                end
                XMWEAVE=Y;
                if (XMWEAVE>XMWEAVEOLD & XMWEAVE>XMWEAVEMAX & TF>10.)
                        XMWEAVEMAX=XMWEAVE;
                end
                XMWEAVEOLD=XMWEAVE;
            end
            if X<.5
                XMWEAVEMAX=XMWEAVEMAX/TAU^2;
            end
            n=n+1;
            ArrayX(n)=X;
            ArrayXMWEAVEMAX(n)=XMWEAVEMAX;
end
figure
plot(ArrayX,ArrayXMWEAVEMAX),grid
xlabel('X')
ylabel('Normalized Miss')
clc
output=[ArrayX',ArrayXMWEAVEMAX'];
save datfil.txt output  -ascii
disp 'simulation finished'
```

To illustrate the use of the normalized miss distance curves of Fig. 6.9, let us reconsider the numerical example of the previous section in which there is a 6-g weaving target with a weave frequency of 2 rad/s. Assuming that the missile guidance system time constant is 1 s and effective navigation ratio is 3, we first compute the normalized weave frequency as

$$\omega_T T = 2 * 1 = 2$$

which results in a normalized miss of approximately 20. Therefore, we can compute the peak steady-state miss distance to be

$$\text{Peak Miss} \approx 20 n_T T^2 = 20 * 6 * 1^2 = 120 \text{ ft}$$

which is four times larger than the miss in a single time constant guidance system (that is, 120 ft vs 33 ft). Reducing the guidance system time constant to 0.5 s changes both the normalized weave frequency and the normalized miss. The new normalized weave frequency is

$$\omega_T T = 2 * 0.5 = 1$$

which results in an increased normalized miss of approximately 60. The new peak steady-state miss distance becomes

$$\text{Peak Miss} \approx 60 n_T T^2 = 60 * 6 * 0.5^2 \approx 90 \text{ ft}$$

which is five times larger than the miss in a single time constant guidance system (that is, 90 ft vs 17 ft). Keeping the guidance system constant fixed to 0.5 s but increasing the weave frequency to 4 rad/s increases the normalized weave frequency back to 2, or

$$\omega_T T = 4 * 0.5 = 2$$

which again results in a normalized miss of approximately 20. The new peak steady-state miss distance becomes

$$\text{Peak Miss} \approx 20 n_T T^2 = 20 * 6 * 0.5^2 \approx 30 \text{ ft}$$

which is approximately four times larger than the miss induced with a single time constant guidance system (that is, 30 ft vs 8 ft). Thus we can see that the higher-order guidance system dynamics of the fifth-order binomial guidance system yield much larger miss distances due to a weaving target than does the single time constant representation of the guidance system.

ACCELERATION SATURATION

We have observed in the preceding two sections that both the guidance system dynamics and effective navigation ratio play an important role in determining the miss distance due to a weaving target. The finite acceleration capability of the interceptor is also important in determining the miss distance. Normalized miss distance curves can also be developed when missile acceleration saturation effects are considered. In this case it is hypothesized that miss distance normalization factors remain unchanged but new curves have to be developed for the nondimensional ratio of the missile-to-target acceleration advantage, or

$$\text{Ratio} = n_{\text{LIM}}/n_T$$

where n_{LIM} is the interceptor acceleration limit.

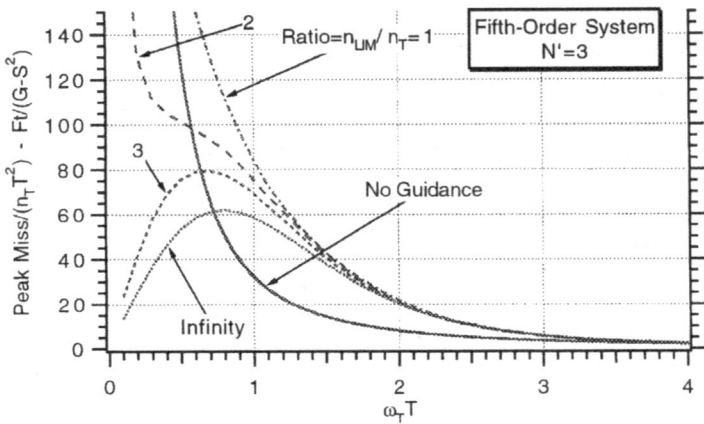

Fig. 6.10 Normalized steady-state peak miss due to weaving target and saturation effects for an effective navigation ratio of 3.

Using the preceding ratio and the normalization factors for the steady-state peak miss due to a weaving target, we can derive normalized miss distance curves by the method of brute force with Listing 6.3. In other words, we can generate normalized miss distance curves by simulating all of the possibilities. We can then infer performance by making extrapolations from the normalized miss distance curves. Of course, detailed checks have to be made to ensure that the normalization factors are correct. Figures 6.10–6.12 present the normalized

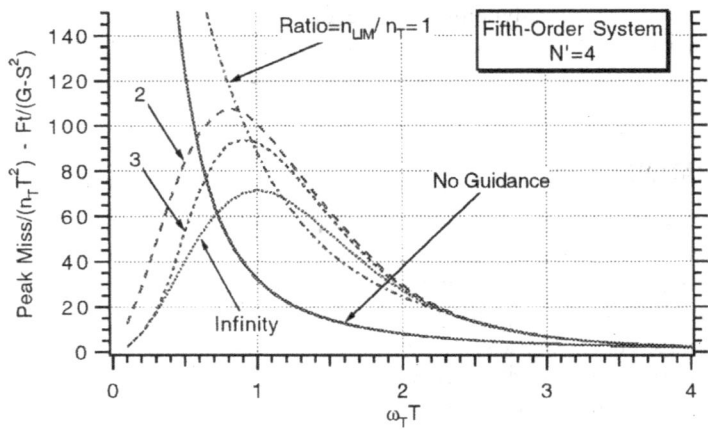

Fig. 6.11 Normalized steady-state peak miss due to weaving target and saturation effects for an effective navigation ratio of 4.

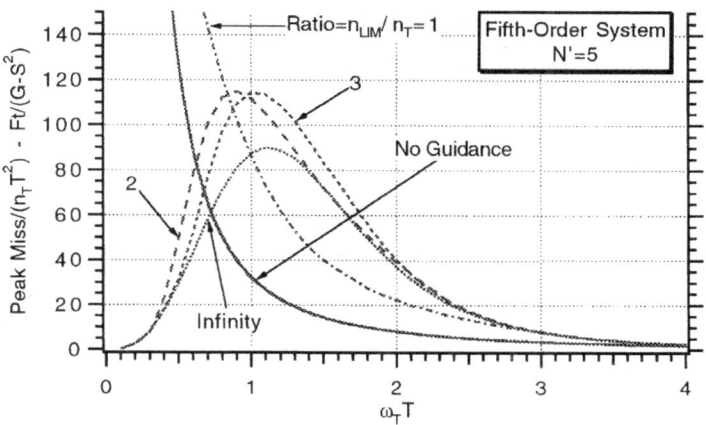

Fig. 6.12 Normalized steady-state peak miss due to weaving target and saturation effects for an effective navigation ratio of 5.

steady-state peak miss distances due to a weaving target for effective navigation ratios ranging from 3 to 5. As expected, we can see that less missile acceleration capability (smaller ratio) means larger miss distances. We can see that at the larger effective navigation ratios (where $N' = 5$), increasing the missile acceleration capability may not always reduce the miss ($\omega_T T = 2$). Under these circumstances, the weaving target causes proportional navigation to be ineffective. This should not be surprising because we know that when the normalized weave frequency is greater than 0.7, doing nothing, or $n_{\mathrm{LIM}}/n_T = 0$, is optimal.

To demonstrate the use of the normalized curves of Figs. 6.10–6.12, let us again consider the same example of the previous section in which there was a 6-g target weave maneuver with weave frequency of 2 rad/s and a proportional navigation missile guidance system with overall time constant of 0.5 s and effective navigation ratio of 3 ($n_T = 6$, $\omega_T = 2$, $T = 0.5$, $N' = 3$). In this case, the normalized weave frequency is 1 ($\omega_T T = 2 * 0.5 = 1$). If the missile acceleration limit is infinite, then the ratio is infinite and we can read from Fig. 6.8 that the steady-state peak miss is 90 ft, or

$$\text{Peak Miss}_\infty = 60 n_T T^2 = 60 * 6 * 0.5^2 = 90 \text{ ft}$$

Reducing the acceleration limit to 18 g reduces the ratio to 3, or

$$\text{Ratio} = n_{\mathrm{LIM}}/n_T = 18/6 = 3$$

For a normalized weave frequency of 1, the new steady-state peak miss increases to 105 ft, or

$$\text{Peak Miss}_{18g} = 70 n_T T^2 = 70 * 6 * 0.5^2 = 105 \text{ ft}$$

Reducing the acceleration limit further to 12 g reduces the ratio to 2, or

$$\text{Ratio} = n_{\text{LIM}}/n_T = 12/6 = 2$$

For a normalized weave frequency of 1, the new steady-state peak miss increases to 113 ft, or

$$\text{Peak Miss}_{12g} = 75n_T T^2 = 75 * 6 * 0.5^2 = 113 \text{ ft}$$

Finally, reducing the acceleration limit even further to 6 g reduces the ratio to 1, or

$$\text{Ratio} = n_{\text{LIM}}/n_T = 6/6 = 1$$

For a normalized weave frequency of 1, the new steady-state peak miss increases to 128 ft, or

$$\text{Peak Miss}_{6g} = 85n_T T^2 = 75 * 6 * 0.5^2 = 128 \text{ ft}$$

In this example, if the missile had no acceleration capability or if the guidance system was turned off, the peak miss would be the maximum value of the weave displacement n_T/ω_T^2, or only 48.4 ft.

If the target weave frequency were increased to 4 rad/s and everything else remained the same, the new normalized weave frequency would be doubled to 2 ($\omega_T T = 4 * 0.5 = 2$). In this case, we can see from Fig. 6.10 that the miss is independent of the missile-to-target acceleration advantage and that the miss would reduce to 30 ft, or

$$\text{Peak Miss}_{\infty,18g,12g,6g} = 20n_T T^2 = 20 * 6 * 0.5^2 = 30 \text{ ft}$$

Again, turning the guidance system off would make the peak miss equivalent to the maximum value of the weave displacement n_T/ω_T^2, or only 12.1 ft.

REDUCING THE TIME CONSTANT TO IMPROVE PERFORMANCE

In general, the safest and most effective method for improving the performance of a proportional navigation guidance system against the weaving target is to reduce the overall guidance system time constant and to increase the missile-to-target acceleration advantage. In aerodynamically controlled missiles, the major contributor to the guidance system time constant is usually the flight-control system time constant, and the limitation on missile acceleration capability is a function of the maximum angle of attack in which a missile can operate without causing flight catastrophe. The ability to speed up the missile flight-control system and the challenge in increasing the missile's maneuverability depend on advances in flight-control system technology. Radome effects will set a lower limit on how small the missile flight-control system time constant can be made without causing stability problems [6], and flight-control system pitch-yaw-roll cross-coupling will place an upper limit on maximum permissible angle of attack.

Although a thorough discussion of the challenges in speeding up a flight-control system and safely achieving high angles of attack are beyond the scope of this text, two numerical examples will be presented in this section showing the benefits to system performance if these goals can be met.

To illustrate the importance of reducing the guidance system time constant, a non-normalized, non-steady-state example was chosen in which there was a 6-g weaving target with a weave frequency of 2 rad/s. Figure 6.13 shows that the miss distance induced by a weaving target on a fifth-order binomial proportional navigation guidance system dramatically decreases with decreasing guidance system time constant. In fact, when the guidance system time constant is 0.1 s there is virtually no miss due to the weaving target!

If we fix the guidance system time constant at 0.1 s, we can see from Fig. 6.14 that although increasing the target weave frequency increases the miss, the miss is still small. We also could have calculated the maximum peak steady-state miss in this example from the normalized curves of Fig. 6.9. For an effective navigation ratio of 3, the curve of Fig. 6.9 is a maximum when the normalized weave frequency is 0.7. That means for this example the actual target weave frequency is 7 rad/s ($\omega_T T = 7 * 0.1 = 0.7$). From Fig. 6.9 we can see that the actual maximum peak miss is approximately 4 ft, or

$$\text{Peak Miss}|_{N'=3} = 63 n_T T^2 = 63 * 6 * 0.1 \approx 4 \text{ ft}$$

Thus we can see that a very small miss distance can be achieved against this difficult maneuver if the guidance system time constant can be reduced to 0.1 s. Of course, we can also see from Fig. 6.9 that turning the guidance system off would also yield the same miss.

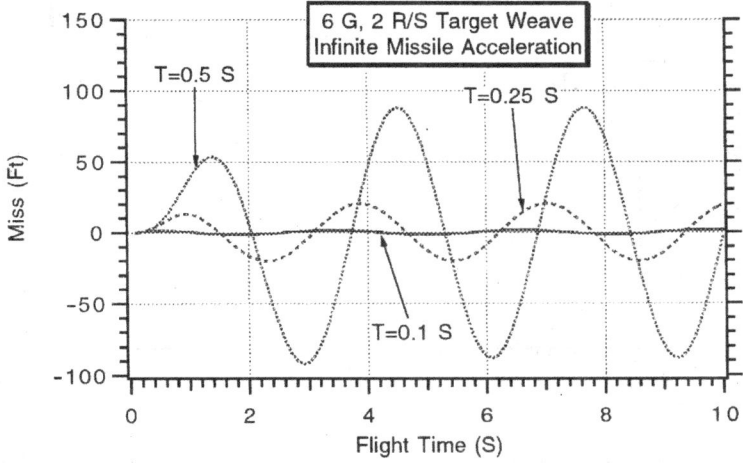

Fig. 6.13   Reducing guidance system time constant dramatically reduces miss.

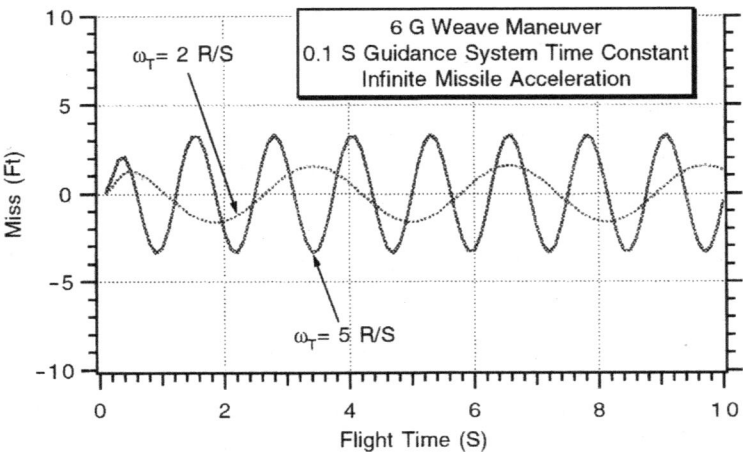

Fig. 6.14   Small guidance time constant yields good performance even when weave frequency increases.

Both previous examples assumed that the missile had infinite acceleration capability. Figure 6.15 shows that when the missile-to-target acceleration advantage decreases from infinity to only two, the miss increases. However, because the guidance system time constant is small, the maximum miss distance is not large. Of course, if the missile guidance system were turned off, the miss would only be approximately 4 ft, regardless of acceleration limit.

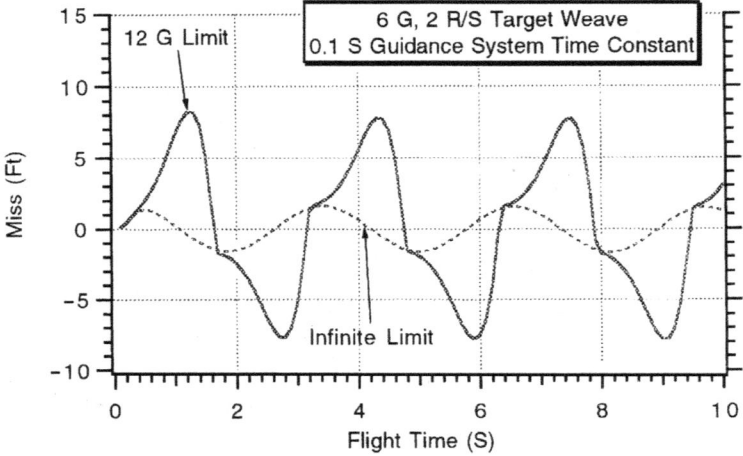

Fig. 6.15   Small miss distances can be achieved even when there is only a two-to-one acceleration advantage.

ADVANCED GUIDANCE TECHNIQUES TO IMPROVE PERFORMANCE

Traditional guidance laws are a form of proportional navigation (PN) in which the acceleration command is proportional to the measured line-of-sight rate. As we saw in Chapter 2 of the previous volume, proportional navigation can also be thought of as a guidance law in which the acceleration command is proportional to the zero effort miss and inversely proportional to the square of the time to go until intercept, or

$$n_c = \frac{N'}{t_{go}^2}[y + \dot{y}t_{go}] = N'V_c\dot{\lambda}$$

The zero effort miss can be thought of as a prediction of how much the missile would miss the target by if the target continued to perform as it had done in the past and the missile issued no further acceleration commands (zero effort). We can see from the preceding equation that the zero effort miss term (bracketed quantity) in proportional navigation assumes that the target is not maneuvering. This does not mean that proportional navigation cannot hit a maneuvering target; it just means that this guidance law is not optimal in the sense that it requires the least acceleration when the target is maneuvering.

If it is known that the target is maneuvering in a step-wise fashion, we saw in Chapter 8 of the previous volume that the zero effort miss could be calculated exactly and a new guidance law result, known as augmented proportional navigation (APN). Mathematically this means that the zero effort miss has an acceleration term based on a constant maneuver, or

$$n_c = \frac{N'}{t_{go}^2}[y + \dot{y}t_{go} + 0.5t_{go}^2\ddot{y}_T] = N'V_c\dot{\lambda} + 0.5N'\ddot{y}_T$$

Although augmented proportional navigation can hit targets maneuvering in different ways (that is, not step maneuvers), it is only optimal for the step target maneuver in the sense that it requires the least acceleration. From an implementation point of view, augmented proportional navigation has one term proportional to the line-of-sight rate and another term proportional to the target acceleration. Therefore, when augmented proportional navigation is implemented, a special filter is required to provide an estimate of both the line-of-sight rate and the instantaneous value of the target acceleration.

We can also derive a special guidance law if it is known in advance that the target is weaving [7, 8]. In this case a simple model, similar to the ones of Chapter 8 of the previous volume for guidance law development, is shown in Fig. 6.16. The second-order shaping network shown in Fig. 6.16 represents the weaving or sinusoidal target maneuver (see Chapter 1 of Volume 1). As was the case for other guidance laws, we are still trying to derive a guidance law that will yield zero miss distance and at the same time minimize the integral of the

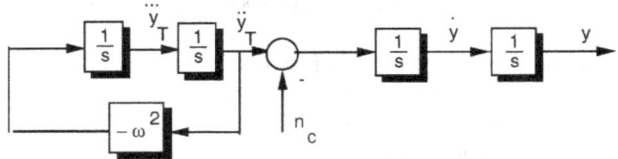

Fig. 6.16 Model for weave guidance law derivation.

acceleration squared, or

$$y(t_F) = 0 \quad \text{subject to minimizing} \int_0^{t_F} n_c^2(t)\mathrm{d}t$$

Figure 6.16 can be expressed in state-space form as

$$\begin{bmatrix} \dot{y} \\ \ddot{y} \\ \dddot{y}_T \\ \ddddot{y}_T \end{bmatrix} = \underbrace{\begin{bmatrix} 0 & 1 & 0 & 0 \\ 0 & 0 & 1 & 0 \\ 0 & 0 & 0 & 1 \\ 0 & 0 & -\omega^2 & 0 \end{bmatrix}}_{F} \begin{bmatrix} y \\ \dot{y} \\ \ddot{y}_T \\ \dddot{y}_T \end{bmatrix} + \underbrace{\begin{bmatrix} 0 \\ -1 \\ 0 \\ 0 \end{bmatrix}}_{G} n_c$$

As was the case in Chapter 8 of the previous volume, if a system is expressed in state-space form, we can also express the final state of the system at any time according to

$$x(t_F) = \Phi(t_F - t)x(t) + \int_t^{t_F} \Phi(t_F - \lambda)G(\lambda)u(\lambda)\mathrm{d}\lambda$$

where $x(t)$ is the system state vector and $\Phi(t)$ is the fundamental matrix. Because the systems dynamics matrix F in this example is time-invariant, the fundamental matrix can be found directly from F according to

$$\Phi(t) = \mathcal{L}^{-1}[(sI - F)^{-1}]$$

yielding

$$\Phi(t) = \begin{bmatrix} 1 & t & \dfrac{(1 - \cos \omega t)}{\omega^2} & \dfrac{(\omega t - \sin \omega t)}{\omega^3} \\[2ex] 0 & 1 & \dfrac{\sin \omega t}{\omega} & \dfrac{(1 - \cos \omega t)}{\omega^2} \\[2ex] 0 & 0 & \cos \omega t & \dfrac{\sin \omega t}{\omega} \\[2ex] 0 & 0 & -\omega \sin \omega t & \cos \omega t \end{bmatrix}$$

Substitution of the Φ and G matrices into the matrix expression for the final state yields four scalar equations in this example. The first of these scalar equations is given by

$$y(t_F) = y(t) + (t_F - t)\dot{y}(t) + \frac{[1 - \cos \omega(t_F - t)]}{\omega^2}\ddot{y}_T(t)$$

$$+ \frac{[\omega(t_F - t) - \sin \omega(t_F - t)]}{\omega^3}\dddot{y}_T(t) - \int_t^{t_T}(t_F - \lambda)n_c(\lambda)d\lambda$$

We can use the same shorthand notation of Chapter 8 of the previous volume and define f_1 and h_1 as

$$f_1(t_F - t) = y(t) + (t_F - t)y(t) + \frac{[1 - \cos \omega(t_F - t)]}{\omega^2}\ddot{y}_T(t)$$

$$+ \frac{[\omega(t_F - t) - \sin \omega(t_F - t)]}{\omega^3}\dddot{y}_T(t)$$

and

$$h_1(t_F - \lambda) = t_F - \lambda$$

so that we can say that

$$y(t_F) = f_1 - \int_t^{t_F}h_1(t_F - \lambda)n_c(\lambda)d\lambda$$

In Chapter 8 of Volume 1 we showed via the Schwartz inequality that the general form of the resultant optimal guidance law based on the preceding formulation is given by

$$n_c(\lambda) = kh_1(t_F - \lambda)$$

where

$$k = f_1(t_F - t)\Big/\int_t^{t_F}h_1^2(t_F - \lambda)d\lambda$$

After some algebra we find that the optimal weave guidance law is given by

$$n_c = \frac{3}{t_{go}^2}\left[y + \dot{y}t_{go} + \frac{1 - \cos \omega t_{go}}{\omega^2}\ddot{y}_T + \frac{\omega t_{go} - \sin \omega t_{go}}{\omega^3}\dddot{y}_T\right]$$

$$= 3V_c\dot{\lambda} + \frac{3}{t_{go}^2}\left[\frac{1 - \cos \omega t_{go}}{\omega^2}\right]\ddot{y}_T + \frac{3}{t_{go}^2}\left[\frac{\omega t_{go} - \sin \omega t_{go}}{\omega^3}\right]\dddot{y}_T$$

We can see that the weave guidance law is similar to other optimal laws we have derived in Chapter 8 of the previous volume in that guidance commands are still proportional to the zero effort miss and inversely proportional to the square of time to go until intercept. From an implementation point of view, assuming that the target weave frequency can be estimated offline and the time to go until intercept is measured, the weave guidance law consists of three terms: one term proportional to the line-of-sight rate, another term proportional to the target acceleration, and a third term proportional to target jerk.

To better understand the relationship between the new guidance law and its predecessors, let us consider the case in which the target weave frequency approaches zero. One can show using Taylor series approximations that the weave guidance law at zero frequency simplifies to

$$\lim_{\omega \to 0} n_{c_{\text{Weave}}} = \frac{3}{t_{go}^2} \left[y + \dot{y} t_{go} + \frac{t_{go}^2}{2} \ddot{y}_T + \frac{t_{go}^3}{6} \dddot{y}_T \right]$$

which is simply augmented proportional navigation with an effective navigation ratio of 3 plus an extra term to account for target jerk. The bracketed term can be recognized as the Taylor series expansion for the zero effort miss for constant target jerk.

It is important to note that the new guidance law requires additional information—an estimate of the target weave frequency, target jerk, and the time to go until intercept. A four-state Kalman filter similar to the three-state filter of Chapter 9 of Volume 1 can be used to provide estimates of the target acceleration and jerk, assuming that the target weave frequency estimate can be derived either from the homing sensors measurements using an extended Kalman filter or from an external sensor (such as ground radar).

Because we have already demonstrated that dynamics within the guidance system will cause miss distance, the preceding guidance law must be modified to account for guidance system lags. With endoatmospheric interceptors, the flight-control system dynamics constitute the bulk of the overall guidance system time constant. If it is known that the target maneuver is sinusoidal in nature, the weave guidance law can be modified to compensate for the known dynamics of the interceptor flight-control system. The compensated weave guidance law [7, 8] is very similar to the optimal guidance law derived in Chapter 8 of the previous volume for a single time constant guidance system, which can be expressed as

$$n_{c_{\text{Weave} \atop \text{Lag}}} = \frac{N'}{t_{go}^2} \left[y + \dot{y} t_{go} + \frac{1 - \cos \omega t_{go}}{\omega^2} \ddot{y}_T \right.$$

$$\left. + \frac{\omega t_{go} - \sin \omega t_{go}}{\omega^3} \dddot{y}_T - n_L T^2 (e^{-x} + x - 1) \right]$$

where x is given by

$$x = \frac{t_{go}}{T}$$

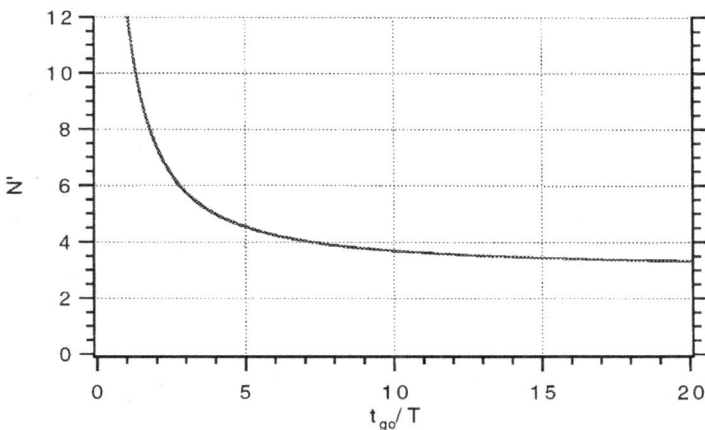

Fig. 6.17   Normalized effective navigation ratio for compensated weave guidance law.

with t_{go} being the time to go until intercept and T being defined as the approximate time constant of the flight-control system. The effective navigation ratio in the compensated weave guidance law is now time-varying and is given by

$$N' = \frac{6x^2(e^{-x} - 1 + x)}{2x^3 + 3 + 6x - 6x^2 - 12xe^{-x} - 3e^{-2x}}$$

The effective navigation ratio for the compensated weave guidance law is identical to the effective navigation ratio of the optimal guidance law of Chapter 8 of the previous volume and is displayed in normalized form in Fig. 6.17. We can see that at the beginning of the flight (long time to go before intercept), the effective navigation ratio is approximately constant and is approaching 3. As we get closer to intercept (small time to go), the effective navigation ratio grows considerably.

To test the effectiveness of the weave and compensated weave guidance laws, the brute force simulation of Listing 6.2 was modified to include various guidance law options (that is, before the FLAG=1 statement). We can see from Listing 6.4 that the guidance system under consideration is a fifth-order binomial even though the compensated weave guidance law assumes a single time constant guidance system. In other words, the compensated weave guidance is actually suboptimal in Listing 6.4.

Figure 6.18 shows that proportional navigation can have substantial miss distances against a 6-g, 2-rad/s weaving target in a fifth-order binomial guidance system with a time constant of 0.25 s. We can also see that uncompensated weave guidance (guidance lags are not accounted for) can substantially reduce the miss.

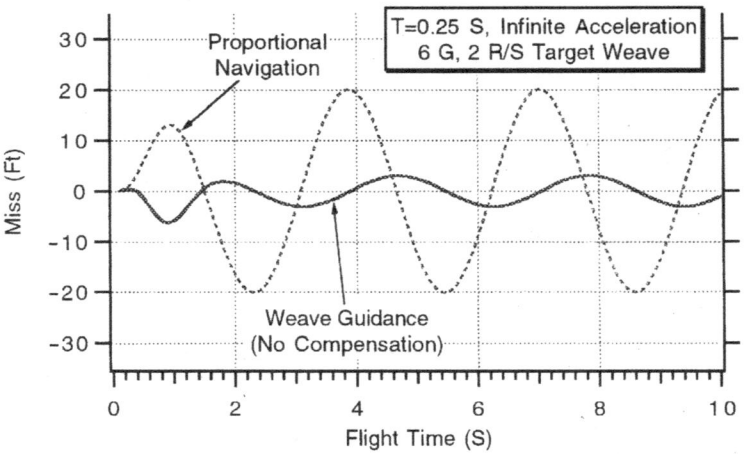

Fig. 6.18   Time lags cause miss—even for uncompensated weave guidance.

LISTING 6.4   BRUTE FORCE SIMULATION FOR GUIDANCE LAW EVALUATION AGAINST WEAVING TARGET

```
clear
n=0;
VC=4000.;
XNT=193.2;
XNP=3.;
XNCLIM=99999999.;
TAU=.25;
W=2.;
WH=2.;
APN=1;
for TF=.1:.1:10
        Y=0.;
        YD=0.;
        XNL=0.;
        D=0.;
        ELAMDH=0.;
        X4=0.;
        X5=0.;
        T=0.;
        H=.01;
        while T<=(TF-.0001)
          YOLD=Y;
          YDOLD=YD;
          XNLOLD=XNL;
          DOLD=D;
```

```
            ELAMDHOLD=ELAMDH;
            X4OLD=X4;
            X5OLD=X5;
            STEP=1;
            FLAG=0;
       while STEP<=1
       if FLAG==1
                                   STEP=2;
                         Y=Y+H*YD;
                         YD=YD+H*YDD;
                         XNL=XNL+H*XNLD;
                         ELAMDH=ELAMDH+H*ELAMDHD;
                         D=D+H*DD;
                         X4=X4+H*X4D;
                         X5=X5+H*X5D;
                         T=T+H;
            end
            YTDD=XNT*sin(W*T);
            YTDDD=W*XNT*cos(W*T);
            TGO=TF-T+.00001;
            XLAM=Y/(VC*TGO);
            DD=5.*(XLAM-D)/TAU;
            ELAMDHD=5.*(DD-ELAMDH)/TAU;
            if APN==1
                         XNC=XNP*VC*ELAMDH;
            elseif APN==2
                         XP=WH*TGO;
                                   XNC=XNP*VC*ELAMDH+XNP*YTDD*(1.-cos(XP))
                                   /XP^2+...
                                   XNP*YTDDD*(XP-sin(XP))/(XP*XP*WH);
            else
                         X=TGO/TAU;
                         XP=WH*TGO;
                         TOP=6.*X*X*(exp(-X)-1.+X);
                         BOT1=2*X*X*X+3.+6.*X-6.*X*X;
                         BOT2=-12.*X*exp(-X)-3.*exp(-2.*X);
                         XNPP=TOP/(.0001+BOT1+BOT2);
                                   XNC=XNPP*VC*ELAMDH+XNPP*YTDD*
                                   (1.cos(XP))/XP^2....
                                            +XNPP*YTDDD*(XP-sin(XP))/(XP*XP*WH)-...
                                   XNPP*XNL*TAU*TAU*(exp(-X)+X-1.)/TGO^2;
            end
            if XNC>XNCLIM
                  XNC=XNCLIM;
            end
            if XNC<-XNCLIM
                         XNC=-XNCLIM;
```

```
                        end
                        X4D=5.*(XNC-X4)/TAU;
                        X5D=5.*(X4-X5)/TAU;
                        XNLD=5.*(X5-XNL)/TAU;
                        YDD=YTDD-XNL;
                        FLAG=1;
                end
                FLAG=0;
                Y=.5*(YOLD+Y+H*YD);
                YD=.5*(YDOLD+YD+H*YDD);
                XNL=.5*(XNLOLD+XNL+H*XNLD);
                D=.5*(DOLD+D+H*DD);
                ELAMDH=.5*(ELAMDHOLD+ELAMDH+H*ELAMDHD);
                X4=.5*(X4OLD+X4+H*X4D);
                X5=.5*(X5OLD+X5+H*X5D);
        end
        n=n+1;
        ArrayTF(n)=TF;
        ArrayY(n)=Y;
end
figure
plot(ArrayTF,ArrayY),grid
xlabel('Flight Time (Sec)')
ylabel('Miss (Ft)')
clc
output=[ArrayTF',ArrayY'];
save datfil.txt output  -ascii
disp 'simulation finished'
```

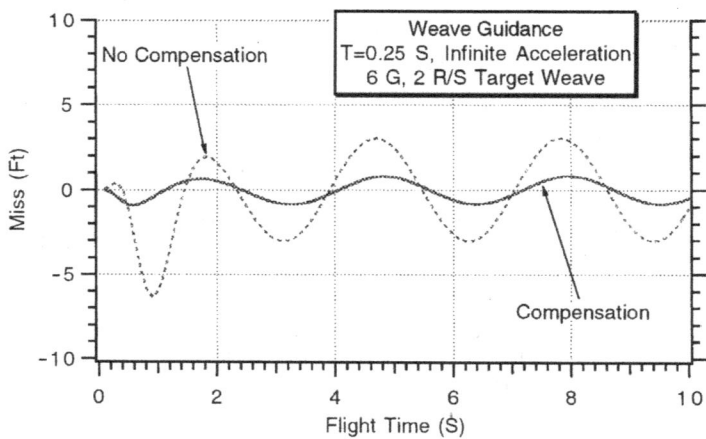

Fig. 6.19   Compensating for guidance system dynamics reduces the miss distance.

Although weave guidance has small miss distances compared to proportional navigation, there is still room for improvement when there are significant guidance system lags. Figure 6.19 shows that when there is imperfect compensation (guidance law optimal for single time constant and there are five time constants in the guidance system being tested), the new guidance law reduces the miss distance even more.

USING THE DISCRETE FOURIER TRANSFORM TO FIND MISS CAUSED BY WEAVING TARGET

Miss-distance curves as a result of a weaving target were first generated in this chapter using a combination of the method of adjoints and brute force. Brute force was required because one adjoint run corresponded to the peak steady-state miss of one target maneuver weave frequency. Thus, even using the method of adjoints required many computer runs to find the peak miss distance for various target weave frequencies. In this section we shall show how using the discrete Fourier transform (DFT) combined with the method of adjoints can evaluate the miss caused by a weaving target for all target weave frequencies in one computer run. This technique was made known to the author by Joel Alpert of Massachusetts Institute of Technology (MIT) Lincoln Laboratory.

To see how the combined method works, let us revisit the single time constant guidance system whose only error source is a weaving target. It was shown in Fig. 6.3 of this chapter that a sinusoidal maneuver could be represented by a shaping network. In this diagram, x2 is the input to the shaping network. In this section we shall show that taking the discrete Fourier transform of x2 yields the steady-state peak miss caused by a weaving target for all target weave frequencies.

Listing 6.5 presents the adjoint of a single time constant guidance system (taken from Listing 3.1 of Volume 1, where x2 is the output of interest and is the impulse response of the guidance system). Note that in this example there is a 3-g target maneuver, the guidance system time constant is 1 s, and the effective navigation ratio is 4. We can see from Listing 6.5 that the x2 output is saved as an array. When the adjoint run is completed, a discrete Fourier transform is taken of x2. The code for the DFT is taken directly from [9]. Statements in Listing 6.5 that differ from the original single time constant adjoint of Listing 3.1 of the previous volume and that pertain to the DFT are highlighted in bold.

LISTING 6.5   COMBINED ADJOINT/DFT APPROACH TO WEAVING TARGET PROBLEM FOR SINGLE TIME CONSTANT GUIDANCE SYSTEM

```
clear all
close all
XNT=96.6;
XNP=4;
```

```
TAU=1.;
VM=3000;
HEDEG=-20;
PI=3.1416;
TF=25.6;
TS=.1;
XNPOINT=TF/TS;
NPOINT=round(XNPOINT);
FS=1./TS;
T=0.;
S=0.;
TP=T+.00001;
X1=0;
X2=0;
X3=1;
X4=0;
H=.01;
HE=HEDEG/57.3;
J=0;
while TP<=(TF-.0001)
      X1OLD=X1;
      X2OLD=X2;
      X3OLD=X3;
      X4OLD=X4;
      STEP=1;
      FLAG=0;
      while STEP <=1
        if FLAG==1
                STEP=2;
                X1=X1+H*X1D;
                X2=X2+H*X2D;
                X3=X3+H*X3D;
                X4=X4+H*X4D;
                TP=TP+H;
        end
        X1D=X2;
        X2D=X3;
        Y1=(X4-X2)/TAU;
        TGO=TP+.00001;
        X3D=XNP*Y1/TGO;
        X4D=-Y1;
        FLAG=1;
      end
      FLAG=0;
      X1=(X1OLD+X1)/2+.5*H*X1D;
      X2=(X2OLD+X2)/2+.5*H*X2D;
```

```
      X3=(X3OLD+X3)/2+.5*H*X3D;
      X4=(X4OLD+X4)/2+.5*H*X4D;

      S=S+H;
      if S>=(TS-.00001)
        S=0;
      J=J+1;
        XMNT=XNT*X1;
        XMHE=-VM*HE*X2;
        X(J)=X2;
    end
end

for K=1:NPOINT
  XR(K)=0;
  XI(K)=0;
  for N=1:NPOINT
    AG = 2*PI*K*N/NPOINT;
    XR(K) = XR(K)+X(N)*cos(AG)/NPOINT;
    XI(K) = XI(K)-X(N)*sin(AG)/NPOINT;
  end
end

IMAX = NPOINT/2;
for I=1:IMAX+1
  F=FS*I/NPOINT;
  XMAG(I)=sqrt(XR(I)^2+XI(I)^2);
  W=2.*PI*F;
  PZ=XMAG(I)*XNT*XNPOINT/FS;
  ArrayW(I)=W;
  ArrayPZ(I)=PZ;
end

figure
plot(ArrayW,ArrayPZ),grid
xlabel('W (r/s)')
ylabel('PZ ')
clc
output=[ArrayW',ArrayPZ'];
save datfil.txt output -ascii
disp 'simulation finished'
```

In this chapter it was shown that the steady-state peak miss in feet caused by a weaving target when the effective navigation ratio was 4 could be derived

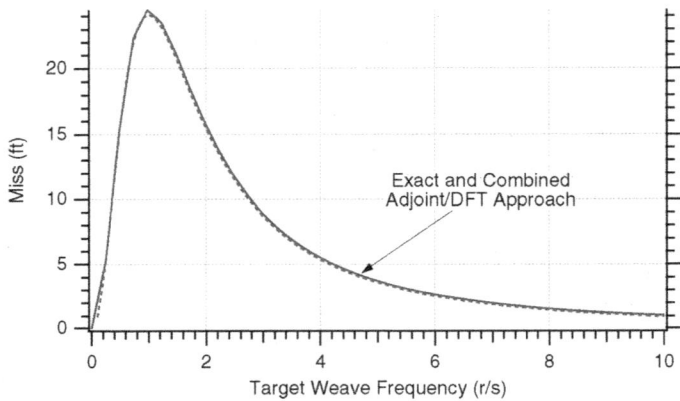

Fig. 6.20   Combined adjoint/DFT approach yields exact miss distances for single time constant guidance system.

analytically. The exact peak miss-distance solution for this example is given by

$$\text{PeakMiss} = \frac{32.2\omega^2 n_T T^4}{(1+\omega^2 T^2)^2} = \frac{32.2 * 3 * 1^4 \omega^2}{(1+\omega^2 1^2)^2} = \frac{96.6\omega^2}{(1+\omega^2)^2}$$

where ω is the target weave frequency in units of rad/s, n_T is the target maneuver amplitude in units of g, and T is the missile guidance system time constant in units of s.

The nominal case of Listing 6.5 was run, and the preceding exact closed-form peak miss-distance solution was superimposed over the adjoint/DFT results for target weave frequencies ranging from 0.1 to 10 rad/s. We can see from Fig. 6.20 that the single-run combined adjoint/DFT approach results are indistinguishable from the exact answers.

The reason this method works should not be surprising. We know that the output x2 represents the impulse response of the adjoint system and that the Fourier transform is simply the convolution of a sine wave with a time-series signal. Therefore, the resulting magnitude of the Fourier transform at each frequency tells us the content of the time-series signal at that frequency. In other words, the magnitude of the Fourier transform of the adjoint impulse response is the same as the magnitude of the output of the adjoint impulse response through an oscillator.

SUMMARY

Normalized design curves have been presented showing how a weaving target influences the miss distance of a generic proportional navigation guidance system. This chapter demonstrated how the target weave frequency and amplitude, missile guidance system time constant, effective navigation ratio, and acceleration capability all play an important role in determining system performance. It was demonstrated that, in general, speeding up a missile guidance system and increasing the missile-to-target acceleration advantage will help reduce the miss distance due to a weaving target. It was also shown how special guidance laws that require more information than proportional navigation can be used to improve system performance.

REFERENCES

[1] Platus, D. H., "Ballistic Re-entry Vehicle Flight Dynamics," *Journal of Guidance, Control, and Dynamics*, Vol. 5, Jan.–Feb. 1982, pp. 4–16.

[2] Chadwick, W. R., and Zarchan, P., "Interception of Spiraling Ballistic Missiles," *Proceedings of American Control Conference*, Seattle, WA, June 1995.

[3] Zarchan, P., "Proportional Navigation and Weaving Targets," *Journal of Guidance, Control, and Dynamics*, Vol. 18, No. 5, 1995, pp. 969–974.

[4] Ohlmeyer, E. J., "Root-Mean-Square Miss Distance of Proportional Navigation Missile Against Sinusoidal Target," *Journal of Guidance, Control, and Dynamics*, Vol. 19, No. 3, 1996, pp. 563–568.

[5] Javid, M., and Brenner, E., *Analysis, Transmission and Filtering of Signals*, McGraw-Hill, New York, 1963.

[6] Nesline, F. W., and Zarchan, P., "Radome Induced Miss Distance in Aerodynamically Controlled Homing Missiles," *Proceedings of AIAA Guidance and Control Conference*, AIAA, New York, Aug. 1984.

[7] Forte, I., and Shinar, J., "Can a Mixed Guidance Strategy Improve Missile Performance," *Journal of Guidance, Control, and Dynamics*, Vol. 11, Jan.–Feb. 1988, pp. 53–59.

[8] Zarchan, P., "The Challenge of Intercepting Spiraling Tactical Ballistic Missiles," *Proceedings of ION Conference*, Cambridge, MA, June 1996.

[9] Ramirez, R. W., *The FFT Fundamentals and Concepts*, Prentice Hall, Upper Saddle River, NJ, 1985, pp. 63–75.

CHAPTER 7

Filtering and Weaving Targets

INTRODUCTION

So far we have shown that we could considerably improve our performance against weaving targets if we could either use a special-purpose guidance law or somehow achieve a smaller guidance system time constant. If we choose to use the guidance law that is optimal against weaving targets, we then have to estimate the target acceleration, target jerk, and target weave frequency.

In this chapter we will explore the various filtering options that can be used against a weaving target by using a step-by-step approach. First we shall see how our original linear three-state Kalman filter from Chapter 9 of the previous volume is able to function in the presence of a weaving target. Although with this filter we cannot use the weave guidance law that was derived in Chapter 6, we can use either proportional navigation, augmented proportional navigation, or optimal guidance. Next we will assume that the target weave frequency is known (that is, estimated or derived using other sensors or phenomenology) and proceed to derive an optimal linear four-state weave Kalman filter that estimates both target acceleration and jerk. This filter can be used with either the weave guidance law or the compensated weave guidance law that were both derived in Chapter 6. Finally, we will assume that the target weave frequency is not known in advance but must also be estimated. In this case an extended five-state Kalman filter that estimates the relative position, relative velocity, target acceleration, jerk, and weave frequency will be derived. All three Kalman filters and appropriate guidance laws will be compared in terms of both performance and robustness.

REVIEW OF ORIGINAL THREE-STATE LINEAR KALMAN FILTER

The original three-state linear Kalman filter from Chapter 9 of the previous volume was derived based on the homing loop model of Fig. 7.1. Recall that in this guidance system model we measured noisy relative position y^* and were attempting to estimate relative position, relative velocity, and target acceleration.

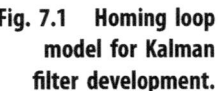

Fig. 7.1 Homing loop model for Kalman filter development.

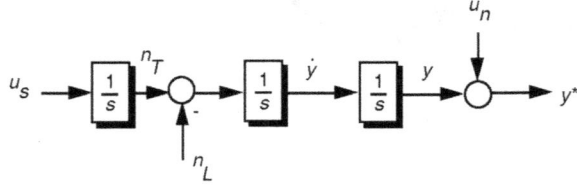

As was the case in Chapter 9 of Volume 1, the achieved missile acceleration n_L was assumed to be known, and the target acceleration was considered to be modeled as white noise through an integrator. It is important to note that we already showed in Chapter 4 of the previous volume that the shaping filter equivalent of a target maneuver with constant amplitude and random starting time is mathematically equivalent (that is, in terms of second-order statistics) to white noise through an integrator.

According to the results of Chapter 4 of Volume 1, the spectral density of the white noise source u_s, depicted in Fig. 7.1, was shown to be

$$\Phi_s = \frac{n_{TMAX}^2}{t_F}$$

where n_{TMAX} is the assumed maximum target maneuver–level magnitude and t_F is the flight time. The model of Fig. 7.1 can be expressed in state space form as

$$\begin{bmatrix} \dot{y} \\ \ddot{y} \\ \dot{n}_T \end{bmatrix} = \begin{bmatrix} 0 & 1 & 0 \\ 0 & 0 & 1 \\ 0 & 0 & 0 \end{bmatrix} \begin{bmatrix} y \\ \dot{y} \\ n_T \end{bmatrix} + \begin{bmatrix} 0 \\ -1 \\ 0 \end{bmatrix} n_L + \begin{bmatrix} 0 \\ 0 \\ u_s \end{bmatrix}$$

Because the systems dynamics matrix of the preceding equation is given by

$$\mathbf{F} = \begin{bmatrix} 0 & 1 & 0 \\ 0 & 0 & 1 \\ 0 & 0 & 0 \end{bmatrix}$$

the continuous fundamental matrix can easily be derived (as was the case in Chapter 9 of the previous volume) as

$$\Phi(t) = \begin{bmatrix} 1 & t & .5t^2 \\ 0 & 1 & t \\ 0 & 0 & 1 \end{bmatrix}$$

By replacing t with the sampling time T_s we can obtain the discrete form of the fundamental matrix as

$$\Phi_k = \begin{bmatrix} 1 & T_s & .5T_s^2 \\ 0 & 1 & T_s \\ 0 & 0 & 1 \end{bmatrix}$$

The discrete measurement equation can be written by inspection of Fig. 7.1 as

$$y_k^* = \begin{bmatrix} 1 & 0 & 0 \end{bmatrix} \begin{bmatrix} y_k \\ \dot{y}_k \\ n_{T_k} \end{bmatrix} + v_k$$

Therefore, the discrete measurement matrix can be written by inspection of the preceding equation as

$$\mathbf{H_k} = \begin{bmatrix} 1 & 0 & 0 \end{bmatrix}$$

The continuous control matrix $\mathbf{G}$ can also be written by inspection of the original state space equation as

$$\mathbf{G} = \begin{bmatrix} 0 \\ -1 \\ 0 \end{bmatrix}$$

Therefore, the discrete control matrix $\mathbf{G_k}$ becomes

$$\mathbf{G_k} = \int_0^{T_s} \Phi(\tau)\mathbf{G}(\tau)d\tau = \int_0^{T_s} \begin{bmatrix} 1 & \tau & .5\tau^2 \\ 0 & 1 & \tau \\ 0 & 0 & 1 \end{bmatrix} \begin{bmatrix} 0 \\ -1 \\ 0 \end{bmatrix} d\tau = \begin{bmatrix} -.5T_s^2 \\ -T_s \\ 0 \end{bmatrix}$$

Recall the discrete Kalman filtering equation is given by

$$\hat{\mathbf{x}}_k = \Phi_k\hat{\mathbf{x}}_{k-1} + \mathbf{G_k}\mathbf{u}_{k-1} + \mathbf{K_k}(\mathbf{z}_k - \mathbf{H}\Phi_k\hat{\mathbf{x}}_{k-1} - \mathbf{H}\mathbf{G_k}\mathbf{u}_{k-1})$$

Substitution of the appropriate matrices into the preceding matrix difference equation yields

$$\begin{bmatrix} \hat{y}_k \\ \hat{\dot{y}}_k \\ \hat{n}_{T_k} \end{bmatrix} = \begin{bmatrix} 1 & T_s & .5T_s^2 \\ 0 & 1 & T_s \\ 0 & 0 & 1 \end{bmatrix} \begin{bmatrix} \hat{y}_{k-1} \\ \hat{\dot{y}}_{k-1} \\ \hat{n}_{T_{k-1}} \end{bmatrix} + \begin{bmatrix} -.5T_s^2 \\ -T_s \\ 0 \end{bmatrix} n_{L_{k-1}} + \begin{bmatrix} K_{1_k} \\ K_{2_k} \\ K_{3_k} \end{bmatrix}$$

$$\times \left(y_k^* - \begin{bmatrix} 1 & 0 & 0 \end{bmatrix} \begin{bmatrix} 1 & T_s & .5T_s^2 \\ 0 & 1 & T_s \\ 0 & 0 & 1 \end{bmatrix} \begin{bmatrix} \hat{y}_{k-1} \\ \hat{\dot{y}}_{k-1} \\ \hat{n}_{T_{k-1}} \end{bmatrix} - \begin{bmatrix} 1 & 0 & 0 \end{bmatrix} \begin{bmatrix} -.5T_s^2 \\ -T_s \\ 0 \end{bmatrix} n_{L_{k-1}} \right)$$

We can multiply out the terms of the preceding matrix equation to yield the linear three-state Kalman filter scalar equations as

$$\text{RES}_k = y_k^* - \hat{y}_{k-1} - T_s\dot{\hat{y}}_{k-1} - .5T_s^2(n_{T_{k-1}} - n_{L_{k-1}})$$

$$\hat{y}_k = \hat{y}_{k-1} + T_s\dot{\hat{y}}_{k-1} + .5T_s^2(\hat{n}_{T_{k-1}} - n_{L_{k-1}}) + K_{1_k}\,\text{RES}_k$$

$$\dot{\hat{y}}_k = \dot{\hat{y}}_{k-1} + T_s(\hat{n}_{T_{k-1}} - n_{L_{k-1}}) + K_{2_k}\,\text{RES}_k$$

$$\hat{n}_{T_k} = \hat{n}_{T_{k-1}} + K_{3_k}\,\text{RES}_k$$

A simulation of the linear three-state linear Kalman filter as part of a missile guidance system is a modified form of Listing 9.2 of the previous volume and appears in Listing 7.1. The simulation now has a single time constant representation of the flight-control system plus a weaving target rather than a constant target maneuver. We can see from the listing that there is a 3-g weaving target with a weave frequency of 2 rad/s. Nominally there is still 1 mrad of measurement noise, but the closing velocity has been increased to 9000 ft/s to reflect a ballistic target engagement. We can see from the listing that the guidance law options for this filter are either proportional navigation, augmented proportional navigation, or optimal guidance (namely, APN=0, 1, or 2, respectively). Because there is a single time constant representation of the flight-control system, the achieved missile acceleration, rather than the commanded acceleration, enters the filtering equations.

LISTING 7.1   ORIGINAL THREE-STATE LINEAR KALMAN FILTER AND WEAVING TARGETCLEAR

```
clear
n=0;
TAU=.5;
APN=0;
VC=9000.;
XNT=96.6;
XNTREAL=96.6;
XNTMAX=96.6;
W=2.;
YIC=0.;
VM=3000.;
HEDEG=0.;
HEDEGFIL=20.;
XNP=3.;
SIGRIN=.001;
TS=.01;
TF=10.;
```

```
Y=YIC;
YD=-VM*HEDEG/57.3;
YDIC=YD;
TS2=TS*TS;
TS3=TS2*TS;
TS4=TS3*TS;
TS5=TS4*TS;
PHIN=XNTMAX*XNTMAX/TF;
RTM=VC*TF;
SIGNOISE=SIGRIN;
SIGPOS=RTM*SIGNOISE;
SIGN2=SIGPOS^2;
P11=SIGN2;
P12=0.;
P13=0.;
P22=(VM*HEDEGFIL/57.3)^2;
P23=0.;
P33=XNTMAX*XNTMAX;
T=0.;
H=.001;
S=0.;
YH=0.;
YDH=0.;
XNTH=0.;
XNC=0.;
XNL=0.;
while T<=TF
        YOLD=Y;
        YDOLD=YD;
        XNLOLD=XNL;
        STEP=1;
        FLAG=0;
        while STEP<=1
                if FLAG==1
                STEP=2;
                        Y=Y+H*YD;
                        YD=YD+H*YDD;
                        XNL=XNL+H*XNLD;
                        T=T+H;
                end
                XNT=XNTREAL*sin(W*T);
                TGO=TF-T+.00001;
                RTM=VC*TGO;
                XLAM=Y/(VC*TGO);
                XLAMD=(RTM*YD+Y*VC)/(RTM^2);
                XNLD=(XNC-XNL)/TAU;
```

```
            YDD=XNT-XNL;
            FLAG=1;
end
FLAG=0;
Y=.5*(YOLD+Y+H*YD);
YD=.5*(YDOLD+YD+H*YDD);
XNL=.5*(XNLOLD+XNL+H*XNLD);
S=S+H;
if S>=(TS-.0001)
            S=0.;
            TGO=TF-T+.000001;
            RTM=VC*TGO;
            SIGNOISE=SIGRIN;
            SIGPOS=RTM*SIGNOISE;
            SIGN2=SIGPOS^2;
            M11=P11+TS*P12+.5*TS2*P13+TS*(P12+TS*P22+.5*TS2*P23);
            M11=M11+.5*TS2*(P13+TS*P23+.5*TS2*P33)+TS5*PHIN/20.;
            M12=P12+TS*P22+.5*TS2*P23+TS*(P13+TS*P23+.5*TS2*P33)...
                    +TS4*PHIN/8.;
            M13=P13+TS*P23+.5*TS2*P33+PHIN*TS3/6.;
            M22=P22+TS*P23+TS*(P23+TS*P33)+PHIN*TS3/3.;
            M23=P23+TS*P33+.5*TS2*PHIN;
            M33=P33+PHIN*TS;
            K1=M11/(M11+SIGN2);
            K2=M12/(M11+SIGN2);
            K3=M13/(M11+SIGN2);
            P11=(1.-K1)*M11;
            P12=(1.-K1)*M12;
            P13=(1.-K1)*M13;
            P22=-K2*M12+M22;
            P23=-K2*M13+M23;
            P33=-K3*M13+M33;
            XLAMNOISE=SIGNOISE*randn;
            YSTAR=RTM*(XLAM+XLAMNOISE);
            RES=YSTAR-YH-TS*YDH-.5*TS*TS*(XNTH-XNL);
            YH=K1*RES+YH+TS*YDH+.5*TS*TS*(XNTH-XNL);
            YDH=K2*RES+YDH+TS*(XNTH-XNL);
            XNTH=K3*RES+XNTH;
            XLAMDH=(YH+YDH*TGO)/(VC*TGO*TGO);
            if APN==0
                    XNC=XNP*(YH+YDH*TGO)/(TGO*TGO);
            elseif APN==1
                    XNC=XNP*(YH+YDH*TGO+.5*XNTH*TGO*TGO)/(TGO*TGO);
            else
                    XS=TGO/TAU;
                    TOP=6.*XS*XS*(exp(-XS)-1.+XS);
```

```
                    BOT1=2*XS*XS*XS+3.+6.*XS-6.*XS*XS;
                    BOT2=-12.*XS*exp(-XS)-3.*exp(-2.*XS);
                    XNPP=TOP/(.0001+BOT1+BOT2);
                    C1=XNPP/(TGO*TGO);
                    C2=XNPP/TGO;
                    C3=.5*XNPP;
                    C4=-XNPP*(exp(-XS)+XS-1.)/(XS*XS);
                    XNC=C1*YH+C2*YDH+C3*XNTH+C4*XNL;
              end
              n=n+1;
              XNTG=XNT/32.2;
              XNTHG=XNTH/32.2;
              ArrayT(n)=T;
              ArrayXNTG(n)=XNTG;
              ArrayXNTHG(n)=XNTHG;
              ArrayY(n)=Y;
              ArrayYSTAR(n)=YSTAR;
       end
end
figure
plot(ArrayT,ArrayXNTG,ArrayT,ArrayXNTHG),grid
title('Acceleration Estimate')
xlabel('Time (Sec) ')
ylabel('Acceleration (G)')
figure
plot(ArrayT,ArrayY,ArrayT,ArrayYSTAR),grid
title('Measurement and Signal')
xlabel('Time (Sec) ')
ylabel('Y (Ft)')
clc
output=[ArrayT',ArrayXNTG',ArrayXNTHG',ArrayY',ArrayYSTAR'];
save datfil.txt output /ascii
disp '*** Simulation Complete'
```

Before we see how well the filter can estimate the weaving target maneuver, it is important to see how much effective noise there is on the filter measurement in this high closing velocity engagement. Figure 7.2 shows that although there is only 1 mrad of measurement noise on the line-of-sight angle, the high closing velocity causes significant noise on the effective measured relative position y^*. Essentially, the filter will have to take two derivatives of this noisy measurement to estimate target acceleration.

The nominal case of Listing 7.1 was run, and Fig. 7.3 shows that the three-state linear Kalman filter's estimate of target acceleration is not very good. The effective high-noise environment prevents the filter from accurately estimating the sinusoidal motion of the target maneuver. Near the end of the flight, where the effective

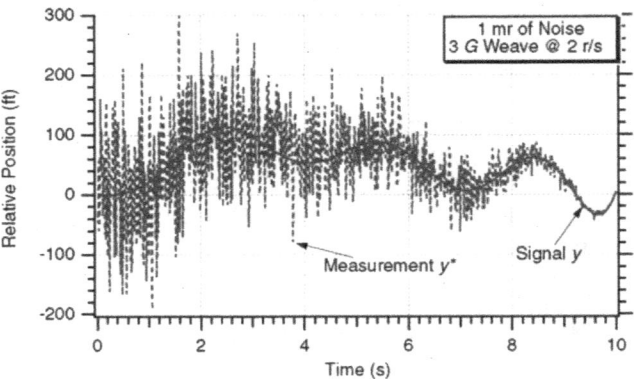

Fig. 7.2   With high closing velocity, 1 mrad of measurement noise translates into a great deal of noise on relative position.

measurement noise on relative position is diminished, the filter estimate improves but lags the actual target maneuver. It is important to note that the Kalman filter is really optimized for a constant target maneuver and is therefore suboptimal in this example because it is mismatched to the real world. However, the presence of process noise in the filter enables the Kalman filter to track all types of target maneuvers. Process noise lets the filter know that its model of the real world may be in error.

Listing 7.1 was modified so that the measurement noise was decreased by an order of magnitude to 0.1 mrad (namely, SIGRIN=.0001). We can see from Fig. 7.4

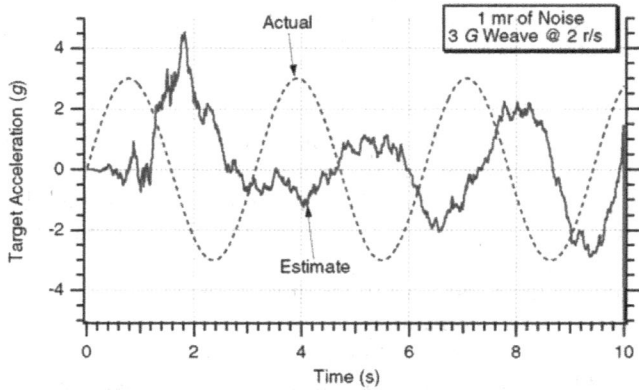

Fig. 7.3   With 1 mrad of measurement noise, the three-state Kalman filter has difficulty in estimating the sinusoidal nature of the weave maneuver.

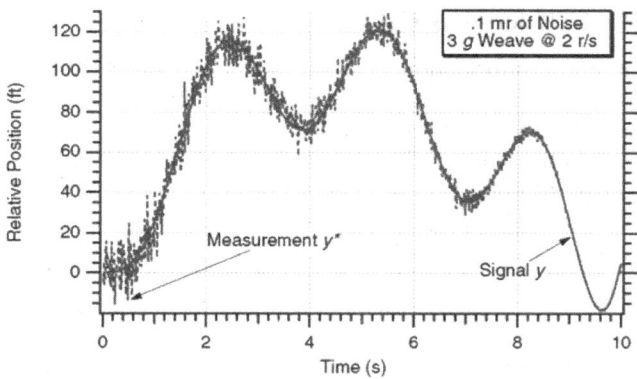

Fig. 7.4   Reducing measurement noise by order of magnitude is beneficial.

that the effective measurement of relative position now more closely resembles the actual relative position. As a consequence of the reduced measurement noise, we can see from Fig. 7.5 that the filter's estimate of the target maneuver now better approximates the sinusoidal nature of the maneuver. The estimate of the target maneuver is nearly perfect, except there is approximately a half-second lag between the actual maneuver and the estimate.

Thus, we can conclude that the original three-state linear Kalman filter of Chapter 9 of the previous volume can track a weaving target quite effectively if the measurement noise can be made small.

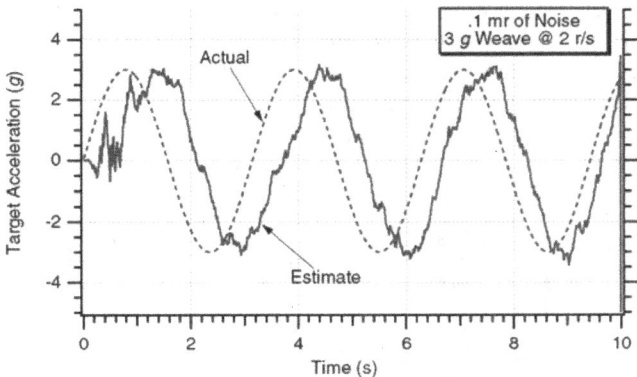

Fig. 7.5   Reducing measurement noise by order of magnitude improves target acceleration estimate.

FOUR-STATE WEAVE KALMAN FILTER

If we had *a priori* information that the target maneuver was sinusoidal in nature, one would think that a better Kalman filter could be designed. To design a Kalman filter optimized to estimate the states of a weaving target, we must first express the sinusoidal target motion in some statistical fashion. First recall from Chapter 1 of the previous volume that the Laplace transform of a sinusoidal signal is given by

$$\pounds(\sin\ \omega t) = \frac{\omega}{s^2 + \omega^2}$$

Therefore, if we assume that the target maneuver is sinusoidal in shape and that the starting time is still uniformly distributed over the flight time, we get the model of Fig. 7.6. Here the input to the sinusoidal transfer function is white noise u_s with spectral density

$$\Phi_s = \frac{n^2_{TMAX}}{t_F}$$

where again n_{TMAX} is the peak of the sinusoidal maneuver and t_F the flight time. It was shown in Chapter 4 of Volume 1 that, mathematically, this is the shaping filter equivalent of a target maneuver with sinusoidal amplitude but random starting time (where the starting time is uniformly distributed over the flight time).

In this homing system model we also effectively measure noisy relative position y^*. If the range from the interceptor to the target is known, it is easy to show that measuring relative position is equivalent to measuring the line-of-sight angle. The linear four-state weave Kalman filter will estimate relative position, relative velocity, target acceleration, and target jerk. The homing loop model of Fig. 6.2 assumes that the achieved missile acceleration n_L and the target weave frequency ω are both known and do not have to be estimated.

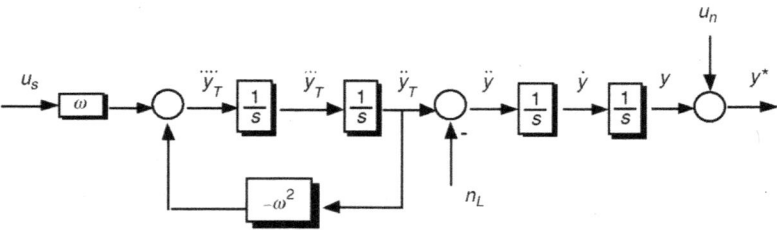

Fig. 7.6   Homing loop model for Kalman filter to be designed for sinusoidal target maneuver.

The model of Fig. 7.6 can be expressed in state space form as

$$\begin{bmatrix} \dot{y} \\ \ddot{y} \\ \dddot{y}_T \\ \ddddot{y}_T \end{bmatrix} = \begin{bmatrix} 0 & 1 & 0 & 0 \\ 0 & 0 & 1 & 0 \\ 0 & 0 & 0 & 1 \\ 0 & 0 & -\omega^2 & 0 \end{bmatrix} \begin{bmatrix} y \\ \dot{y} \\ \ddot{y}_T \\ \dddot{y}_T \end{bmatrix} + \begin{bmatrix} 0 \\ -1 \\ 0 \\ 0 \end{bmatrix} n_L + \begin{bmatrix} 0 \\ 0 \\ 0 \\ \omega u_s \end{bmatrix}$$

The systems dynamics matrix of the preceding equation can be written by inspection and is given by

$$\mathbf{F} = \begin{bmatrix} 0 & 1 & 0 & 0 \\ 0 & 0 & 1 & 0 \\ 0 & 0 & 0 & 1 \\ 0 & 0 & -\omega^2 & 0 \end{bmatrix}$$

The fundamental matrix can be derived from the systems dynamics matrix according to

$$\Phi(\mathbf{t}) = \pounds^{-1}\left[(\mathbf{sI} - \mathbf{F})^{-1}\right]$$

Therefore, the fundamental matrix in the Laplace transform domain can be expressed as

$$\Phi(s) = (\mathbf{sI} - \mathbf{F})^{-1} = \begin{bmatrix} s & -1 & 0 & 0 \\ 0 & s & -1 & 0 \\ 0 & 0 & s & -1 \\ 0 & 0 & s+\omega^2 & 0 \end{bmatrix}^{-1}$$

From the preceding equation we can see that first we must take the inverse of a four-by-four matrix and then take its inverse Laplace transform to find the fundamental matrix in the time domain. After considerable algebra, the continuous fundamental matrix turns out to be

$$\Phi(t) = \begin{bmatrix} 1 & t & \dfrac{1 - \cos \omega t}{\omega^2} & \dfrac{\omega t - \sin \omega t}{\omega^3} \\ 0 & 1 & \dfrac{\sin \omega t}{\omega} & \dfrac{1 - \cos \omega t}{\omega^2} \\ 0 & 0 & \cos \omega t & \dfrac{\sin \omega t}{\omega} \\ 0 & 0 & -\omega \sin \omega t & \cos \omega t \end{bmatrix}$$

By replacing time t with the sampling time T_s we obtain the discrete form of the fundamental matrix as

$$
\Phi_k =
\begin{bmatrix}
1 & T_s & \dfrac{1-\cos x}{\omega^2} & \dfrac{x-\sin x}{\omega^3} \\[2mm]
0 & 1 & \dfrac{\sin x}{\omega} & \dfrac{1-\cos x}{\omega^2} \\[2mm]
0 & 0 & \cos x & \dfrac{\sin x}{\omega} \\[2mm]
0 & 0 & -\omega\sin x & \cos x
\end{bmatrix}
$$

where

$$
x = \omega T_s
$$

The discrete measurement equation can be written by inspection of Fig. 7.6 as

$$
y_k^* = [1 \quad 0 \quad 0 \quad 0]
\begin{bmatrix}
y_k \\
\dot{y}_k \\
\ddot{y}_{T_k} \\
\dddot{y}_{T_k}
\end{bmatrix} + v_k
$$

which means that the discrete measurement matrix is given by

$$
\mathbf{H_k} = [1 \quad 0 \quad 0 \quad 0]
$$

The continuous control matrix $\mathbf{G}$ can be written by inspection of the original state space equation as

$$
\mathbf{G} =
\begin{bmatrix}
0 \\
-1 \\
0 \\
0
\end{bmatrix}
$$

After some algebra, the discrete control matrix $\mathbf{G_k}$ becomes

$$
\mathbf{G_k} = \int_0^{T_s} \Phi(\tau)\mathbf{G}(\tau)d\tau =
\begin{bmatrix}
-.5T_s^2 \\
-T_s \\
0 \\
0
\end{bmatrix}
$$

Finally, the continuous process noise matrix can be written from the system state space equation by inspection as

$$
\mathbf{Q} = \mathbf{E}\left\{ [0 \quad 0 \quad 0 \quad \omega u_s]
\begin{bmatrix}
0 \\
0 \\
0 \\
\omega u_s
\end{bmatrix}
\right\} =
\begin{bmatrix}
0 & 0 & 0 & 0 \\
0 & 0 & 0 & 0 \\
0 & 0 & 0 & 0 \\
0 & 0 & 0 & \omega^2\Phi_s
\end{bmatrix}
$$

After some algebra, the discrete process noise matrix can be derived from the continuous process noise matrix according to

$$\mathbf{Q_k} = \int_0^{T_s} \Phi(\tau)Q\Phi^T(\tau)d\tau = \begin{bmatrix} Q_{11} & Q_{12} & Q_{13} & Q_{14} \\ Q_{12} & Q_{22} & Q_{23} & Q_{24} \\ Q_{13} & Q_{23} & Q_{33} & Q_{34} \\ Q_{14} & Q_{24} & Q_{34} & Q_{44} \end{bmatrix}$$

where

$$Q_{11} = \frac{\Phi_s}{\omega^5}\left[.333x^3 - 2\sin x + 2x\cos x + .5x - .25\sin 2x\right]$$

$$Q_{12} = \frac{\Phi_s}{\omega^4}\left[.5x^2 - x\sin x + .5\sin^2 x\right]$$

$$Q_{13} = \frac{\Phi_s}{\omega^3}\left[\sin x - x\cos x - .5x + .25\sin 2x\right]$$

$$Q_{14} = \frac{\Phi_s}{\omega^2}\left[\cos x + x\sin x - .5\sin^2 x - 1\right]$$

$$Q_{22} = \frac{\Phi_s}{\omega^3}\left[1.5x - 2\sin x + .25\sin 2x\right]$$

$$Q_{23} = \frac{\Phi_s}{\omega^2}\left[1 - \cos x - .5\sin^2 x\right]$$

$$Q_{24} = \frac{\Phi_s}{\omega}\left[\sin x - .5x - .25\sin 2x\right]$$

$$Q_{33} = \frac{\Phi_s}{\omega}\left[.5x - .25\sin 2x\right]$$

$$Q_{34} = .5\Phi_s\sin^2 x$$

$$Q_{44} = \omega\Phi_s[.5x + .25\sin 2x]$$

Recall that in the preceding set of expressions, the process noise and normalized weave frequency have been defined as

$$\Phi_s = \frac{\omega^2 n_{TMAX}^2}{t_F}$$

$$x = \omega T_s$$

Recall that the discrete Kalman filtering equation is given by

$$\hat{x}_k = \Phi_k\hat{x}_{k-1} + G_k u_{k-1} + K_k(z_k - H\Phi_k\hat{x}_{k-1} - HG_k u_{k-1})$$

Substitution of the appropriate matrices into the preceding matrix difference equation yields

$$
\begin{bmatrix} \hat{y}_k \\ \hat{\dot{y}}_k \\ \hat{y}_{T_k} \\ \hat{\dot{y}}_{T_k} \end{bmatrix} =
\begin{bmatrix}
1 & T_s & \dfrac{1-\cos x}{\omega^2} & \dfrac{x-\sin x}{\omega^3} \\
0 & 1 & \dfrac{\sin x}{\omega} & \dfrac{1-\cos x}{\omega^2} \\
0 & 0 & \cos x & \dfrac{\sin x}{\omega} \\
0 & 0 & -\omega\sin x & \cos x
\end{bmatrix}
\begin{bmatrix} \hat{y}_{k-1} \\ \hat{\dot{y}}_{k-1} \\ \hat{y}_{T_{k-1}} \\ \hat{\dot{y}}_{T_{k-1}} \end{bmatrix}
+ \begin{bmatrix} -.5T_s^2 \\ -T_s \\ 0 \\ 0 \end{bmatrix} n_{L_{k-1}}
+ \begin{bmatrix} K_1 \\ K_2 \\ K_3 \\ K_4 \end{bmatrix}
$$

$$
\times \left[y_k^* - \begin{bmatrix} 1 & 0 & 0 & 0 \end{bmatrix}
\begin{bmatrix}
1 & T_s & \dfrac{1-\cos x}{\omega^2} & \dfrac{1-\sin x}{\omega^3} \\
0 & 1 & \dfrac{\sin x}{\omega} & \dfrac{1-\cos x}{\omega^2} \\
0 & 0 & \cos x & \dfrac{\sin x}{\omega} \\
0 & 0 & -\omega\sin x & \cos x
\end{bmatrix}
\begin{bmatrix} \hat{y}_{k-1} \\ \hat{\dot{y}}_{k-1} \\ \hat{y}_{T_{k-1}} \\ \hat{\dot{y}}_{T_{k-1}} \end{bmatrix}
\right.
$$

$$
\left.
- \begin{bmatrix} 1 & 0 & 0 & 0 \end{bmatrix}
\begin{bmatrix} -.5T_s^2 \\ -T_s \\ 0 \\ 0 \end{bmatrix} n_{L_{k-1}}
\right]
$$

We can multiply out the terms of the preceding matrix equation to yield the Kalman filter scalar equations

$$
\mathrm{RES}_k = y_k^* - \hat{y}_{k-1} - T_s\hat{\dot{y}}_{k-1} - \frac{(1-\cos x)}{\omega^2}\hat{y}_{T_{k-1}} - \frac{(x-\sin x)}{\omega^3}\hat{\dot{y}}_{T_{k-1}} + .5T_s^2 n_{L_{k-1}}
$$

$$
\hat{y}_k = \hat{y}_{k-1} + T_s\hat{\dot{y}}_{k-1} + \frac{(1-\cos x)}{\omega^2}\hat{y}_{T_{k-1}} - \frac{(x-\sin x)}{\omega^3}\hat{\dot{y}}_{T_{k-1}}
$$
$$
- .5T_s^2 n_{L_{k-1}} + K_{1_k}\mathrm{RES}_k
$$

$$
\hat{\dot{y}}_k = \hat{\dot{y}}_{k-1} + \frac{(\sin x)}{\omega}\hat{y}_{T_{k-1}} + \frac{(1-\cos x)}{\omega^2}\hat{\dot{y}}_{T_{k-1}} - T_s n_{L_{k-1}} + K_{2_k}\mathrm{RES}_k
$$

$$
\hat{y}_{T_k} = \cos x\,\hat{y}_{T_{k-1}} + \frac{\sin x}{\omega}\hat{\dot{y}}_{T_{k-1}} + K_{3_k}\mathrm{RES}_k
$$

$$
\hat{\dot{y}}_{T_k} = -\omega\sin x\,\hat{y}_{T_{k-1}} + \cos x\,\hat{\dot{y}}_{T_{k-1}} + K_{4_k}\mathrm{RES}_k
$$

Both the Riccati equations and Kalman filtering equations for the linear four-state weave Kalman filter were programmed as part of the homing loop, and the resultant linearized missile-target engagement simulation appears in Listing 7.2. We can see that the simulation has a single time constant representation of the flight-control system plus a 3-g weaving target with a weave frequency of 2 rad/s. Nominally there is 1 mrad of measurement noise on the line-of-sight angle, and the closing velocity is 9000 ft/s to reflect a ballistic target engagement. We can see from Listing 7.2 that the guidance law options for this filter are either proportional navigation, augmented proportional navigation, optimal guidance, weave guidance, or compensated weave guidance (namely, APN=0, 1, 2, 3, or 4, respectively).

The nominal case of Listing 7.2 was run, and Fig. 7.7 shows that the linear four-state weave Kalman filter's estimate of target acceleration is much better than the general purpose linear three-state Kalman filter when the measurement noise is 1 mrad (see Fig. 7.3). Compared with the previous section, when the amount of measurement noise is large, it is now easier to see from the state estimates of the weave Kalman filter that the target maneuver is indeed sinusoidal. In addition, the weave Kalman filter yields estimates of the target jerk. We can see from Fig. 7.8 that the weave Kalman filter provides fairly good estimates of the target jerk.

To see if the four-state weave Kalman filter is truly working properly, it is necessary to examine the errors in the state estimates. Figures 7.9 and 7.10 show that the errors in the estimates of target acceleration and jerk appear to lie within the theoretical bounds (that is, $\pm$ square root of P_{33} and P_{44}, respectively) approximately 68% of the time, indicating that the filter is working properly.

Figures 7.11 and 7.12 indicate that when the measurement noise on the line-of-sight angle is reduced by an order of magnitude to 0.1 mrad, the four-state

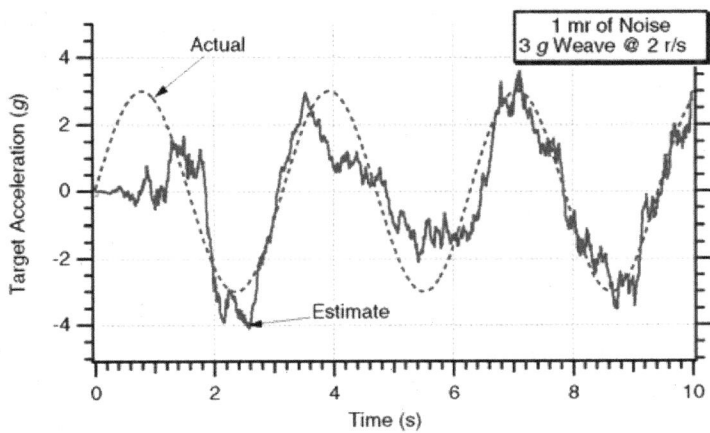

Fig. 7.7   Four-state weave Kalman filter yields better estimates than three-state Kalman filter when measurement noise is large.

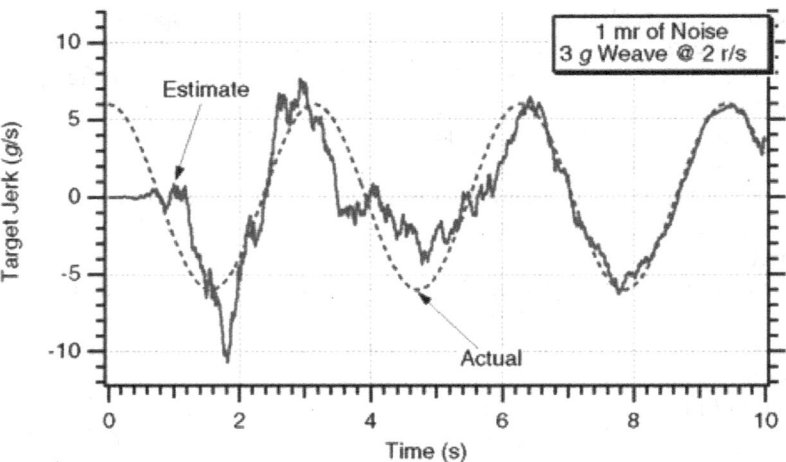

Fig. 7.8   Weave Kalman filter also provides a fairly good estimate of target jerk when measurement noise is large.

weave Kalman filter estimates of target acceleration and jerk improve significantly. In fact, we can see that the estimates of these states are nearly perfect in the low-noise environment.

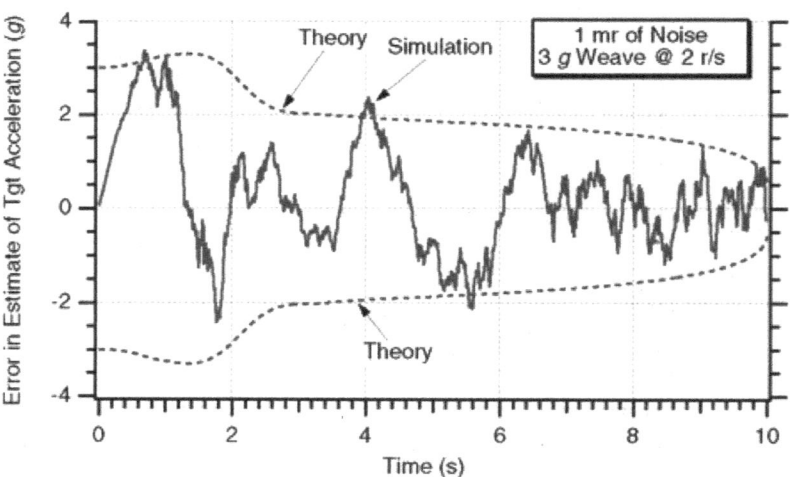

Fig. 7.9   Weave Kalman filter estimation errors for target acceleration are within theoretical bounds.

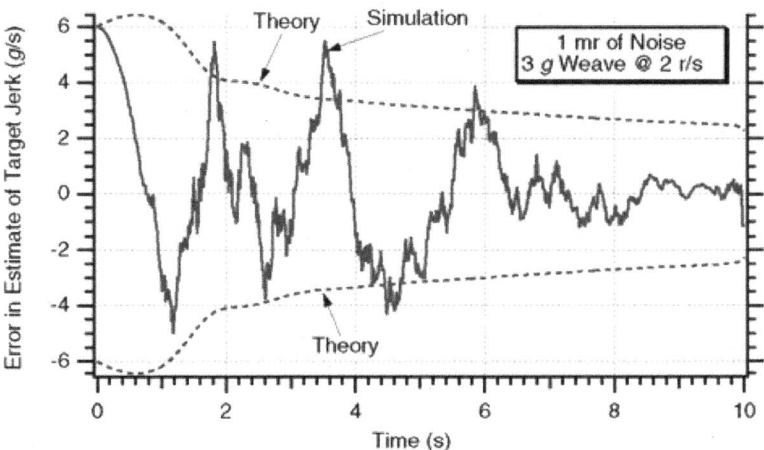

Fig. 7.10   Weave Kalman filter estimation errors for target jerk are within theoretical bounds.

Thus, we can conclude that if the target is maneuvering in a sinusoidal fashion, and we have knowledge of the target wave frequency, superior estimates of the target acceleration and jerk can be obtained with the linear four-state weave Kalman filter.

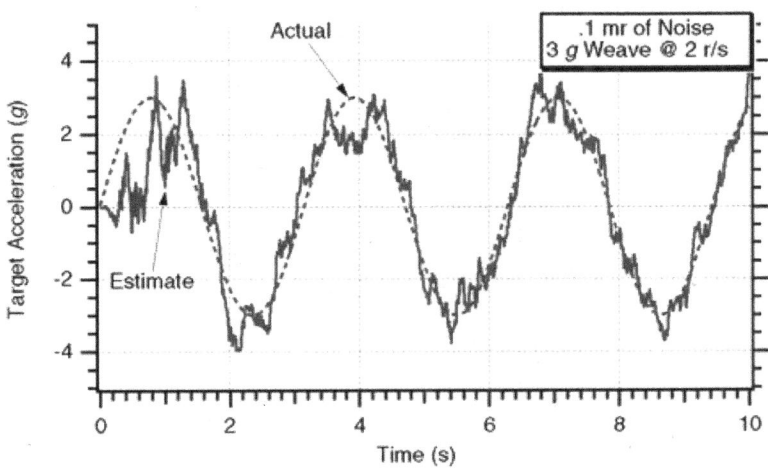

Fig. 7.11   Reducing measurement noise improves four-state weave Kalman filter's target acceleration estimate.

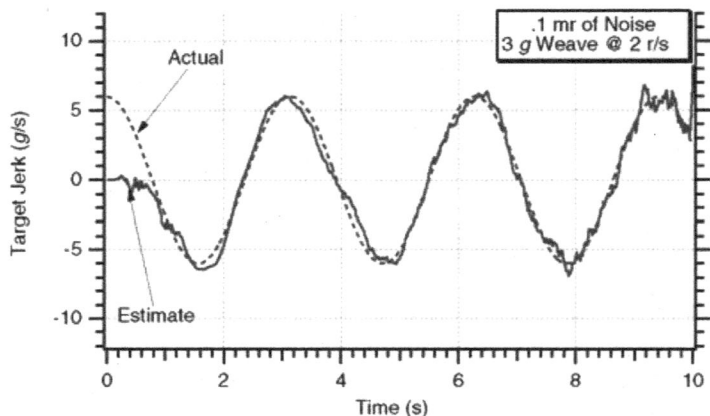

Fig. 7.12   Reducing measurement noise improves weave Kalman filter's target jerk estimate.

LISTING 7.2   WEAVE KALMAN FILTER AND WEAVING TARGET

```
clear
n=0;
TAU=.5;
APN=0;
ORDER=4;
MVR=1;
VC=9000.;
W=2.;
WREAL=2.;
WH=W;
XNT=96.6;
XNTREAL=96.6;
TS=.01;
YIC=0.;
VM=3000.;
HEDEG=0.;
HEDEGFIL=20.;
XNP=3.;
SIGRIN=.001;
SIGGL=0.;
RA=21000.;
SRN=0.;
TF=10.;
QPERFECT=0;
```

```
PHASE=0./57.3;
X=WH*TS;
Y=YIC;
YD=-VM*HEDEG/57.3;
PHIS=WH*WH*XNT*XNT/TF;
RTM=VC*TF;
SIGNOISE=sqrt(SIGRIN^2+(SIGGL/RTM)^2+(SRN*RTM*RTM/(RA*RA))^2);
SIGPOS=RTM*SIGNOISE;
SIGN2=SIGPOS^2;
PHI=zeros(ORDER);
P=zeros(ORDER);
Q=zeros(ORDER);
IDNP=eye(ORDER);
PHI(1,1)=1;
PHI(1,2)=TS;
PHI(1,3)=(1-cos(X))/(WH*WH);
PHI(1,4)=(X-sin(X))/(WH*WH*WH);
PHI(2,2)=1;
PHI(2,3)=sin(X)/WH;
PHI(2,4)=(1-cos(X))/(WH*WH);
PHI(3,3)=cos(X);
PHI(3,4)=sin(X)/WH;
PHI(4,3)=-WH*sin(X);
PHI(4,4)=cos(X);
Q(1,1)=PHIS*(.333*X^3-2*sin(X)+2*X*cos(X)+.5*X-.25*sin(2*X))/(WH^5);
Q(1,2)=PHIS*(.5*X*X-X*sin(X)+.5*sin(X)*sin(X))/(WH^4);
Q(2,1)=Q(1,2);
Q(1,3)=PHIS*(sin(X)-X*cos(X)-.5*X+.25*sin(2*X))/(WH^3);
Q(3,1)=Q(1,3);
Q(1,4)=PHIS*(cos(X)+X*sin(X)-.5*sin(X)*sin(X)-1)/(WH*WH);
Q(4,1)=Q(1,4);
Q(2,2)=PHIS*(1.5*X-2*sin(X)+.25*sin(2*X))/(WH^3);
Q(2,3)=PHIS*(1-cos(X)-.5*sin(X)*sin(X))/(WH*WH);
Q(3,2)=Q(2,3);
Q(2,4)=PHIS*(sin(X)-.5*X-.25*sin(2*X))/WH;
Q(4,2)=Q(2,4);
Q(3,3)=PHIS*(.5*X-.25*sin(2*X))/WH;
Q(3,4)=.5*PHIS*sin(X)*sin(X);
Q(4,3)=Q(3,4);
Q(4,4)=WH*PHIS*(.5*X+.25*sin(2*X));
P(1,1)=SIGN2;
P(2,2)=(VM*HEDEGFIL/57.3)^2;
P(3,3)=XNT*XNT;
P(4,4)=WH*WH*XNT*XNT;
HMAT=[1 0 0 0];
HT=HMAT';
```

```
PHIT=PHI';
T=0.;
H=.001;
S=0.;
XNC=0.;
XNL=0.;
XLAM=Y/RTM;
if MVR==0
      YTDD=XNTREAL;
      YTDDD=0.;
else
      YTDD=XNTREAL*sin(WREAL*T);
      YTDDD=XNTREAL*WREAL*cos(WREAL*T);
end
if QPERFECT==1
      YH=Y;
      YDH=YD;
      YTDDH=YTDD;
      YTDDDH=YTDDD;
else
      YH=0.;
      YDH=0.;
      YTDDH=0.;
      YTDDDH=0.;
end
while T<=(TF-.0001)
      YOLD=Y;
      YDOLD=YD;
      XNLOLD=XNL;
      STEP=1;
      FLAG=0;
      while STEP <=1
            if FLAG==1
            STEP=2;
                  Y=Y+H*YD;
                  YD=YD+H*YDD;
                  XNL=XNL+H*XNLD;
                  T=T+H;
            end
            TGO=TF-T+.000001;
            RTM=VC*TGO;
            XLAM=Y/(VC*TGO);
            if MVR==0
                  YTDD=XNTREAL;
            else
                  YTDD=XNTREAL*sin(WREAL*T);
```

```
            end
        XNLD=(XNC-XNL)/TAU;
        YDD=YTDD-XNL;
        FLAG=1;
end
FLAG=0;
Y=.5*(YOLD+Y+H*YD);
YD=.5*(YDOLD+YD+H*YDD);
XNL=.5*(XNLOLD+XNL+H*XNLD);
S=S+H;
if S>=(TS-.00001)
        S=0.;
        TGO=TF-T+.000001;
        RTM=VC*TGO;
        SIGNOISE=sqrt(SIGRIN^2+(SIGGL/RTM)^2+(SRN*RTM*RTM/(RA*RA))^2);
        SIGPOS=RTM*SIGNOISE;
        SIGN2=SIGPOS^2;
        RMAT=[SIGN2];
        PHIP=PHI*P;
        PHIPPHIT=PHIP*PHIT;
        M=PHIPPHIT+Q;
        HM=HMAT*M;
        HMHT=HM*HT;
        HMHTR=HMHT+RMAT;
        HMHTRINV=inv(HMHTR);
        MHT=M*HT;
        GAIN=MHT*HMHTRINV;
        KH=GAIN*HMAT;
        IKH=IDNP-KH;
        P=IKH*M;
        if MVR==0
                YTDD=XNTREAL;
                YTDDD=0.;
        else
                YTDD=XNTREAL*sin(WREAL*T);
                YTDDD=XNTREAL*WREAL*cos(WREAL*T);
        end
        XLAMNOISE=SIGNOISE*randn;
        YSTAR=RTM*(XLAM+XLAMNOISE);
        RES=YSTAR-YH-TS*YDH-(1-cos(X))*YTDDH/(WH*WH)-(X-sin(X))...
                *YTDDDH/(WH*WH*WH)+.5*TS*TS*XNL;
        YH=YH+TS*YDH+(1-cos(X))*YTDDH/(WH*WH)+(X-sin(X))...
                *YTDDDH/(WH*WH*WH)+GAIN(1,1)*RES-.5*TS*TS*XNL;
        YDH=YDH+sin(X)*YTDDH/WH+(1-cos(X))*YTDDDH/(WH*WH)...
                +GAIN(2,1)*RES-TS*XNL;
        YTDDHNEW=cos(X)*YTDDH+sin(X)*YTDDDH/WH+GAIN(3,1)*RES;
```

```
YTDDDH=-WH*sin(X)*YTDDH+cos(X)*YTDDDH+GAIN(4,1)*RES;
YTDDH=YTDDHNEW;
if APN==0
        XNC=XNP*(YH+YDH*TGO)/(TGO*TGO);
elseif APN==1
        XNC=XNP*(YH+YDH*TGO+.5*YTDDH*TGO*TGO)/(TGO*TGO);
elseif APN==2
        XS=TGO/TAU;
        TOP=6.*XS*XS*(exp(-XS)-1.+XS);
        BOT1=2*XS*XS*XS+3.+6.*XS-6.*XS*XS;
        BOT2=-12.*XS*exp(-XS)-3.*exp(-2.*XS);
        XNPP=TOP/(.0001+BOT1+BOT2);
        C1=XNPP/(TGO*TGO);
        C2=XNPP/TGO;
        C3=.5*XNPP;
        C4=-XNPP*(exp(-XS)+XS-1.)/(XS*XS);
        XNC=C1*YH+C2*YDH+C3*YTDDH+C4*XNL;
elseif APN==3
        XP=WH*TGO;
        XNC=XNP*(YH+YDH*TGO)/(TGO*TGO)+XNP*YTDDH*...
        (1.-cos(XP))/XP^2+XNP*YTDDDH*(XP-sin(XP))/(XP*XP*WH);
else
        XS=TGO/TAU;
        TOP=6.*XS*XS*(exp(-XS)-1.+XS);
        BOT1=2*XS*XS*XS+3.+6.*XS-6.*XS*XS;
        BOT2=-12.*XS*exp(-XS)-3.*exp(-2.*XS);
        XNPP=TOP/(.0001+BOT1+BOT2);
        C1=XNPP/(TGO*TGO);
        C2=XNPP/TGO;
        C3=XNPP*(1.-cos(WH*TGO))/(WH*WH*TGO*TGO);
        C4=-XNPP*(exp(-XS)+XS-1.)/(XS*XS);
        C5=XNPP*(WH*TGO-sin(WH*TGO))/(WH*WH*WH*TGO*TGO);
        XNC=C1*YH+C2*YDH+C3*YTDDH+C4*XNL+C5*YTDDDH;
end
YTDDG=YTDD/32.2;
YTDDHG=YTDDH/32.2;
ERRY=Y-YH;
SP11=sqrt(P(1,1));
SP11P=-SP11;
ERRYD=YD-YDH;
SP22=sqrt(P(2,2));
SP22P=-SP22;
ERRYTDDG=(YTDD-YTDDH)/32.2;
SP33G=sqrt(P(3,3))/32.2;
SP33GN=-SP33G;
ERRYTDDDG=(YTDDD-YTDDDH)/32.2;
```

```
                    SP44G=sqrt(P(4,4))/32.2;
                    SP44GN=-SP44G;
                    YTDDG=YTDD/32.2;
                    YTDDHG=YTDDH/32.2;
                    YTDDDG=YTDDD/32.2;
                    YTDDDHG=YTDDDH/32.2;
                    n=n+1;
                    ArrayT(n)=T;
                    ArrayYTDDG(n)=YTDDG;
                    ArrayYTDDHG(n)=YTDDHG;
                    ArrayYTDDDG(n)=YTDDDG;
                    ArrayYTDDDHG(n)=YTDDDHG;
                    ArrayERRYTDDG(n)=ERRYTDDG;
                    ArraySP33G(n)=SP33G;
                    ArraySP33GN(n)=SP33GN;
                    ArrayERRYTDDDG(n)=ERRYTDDDG;
                    ArraySP44G(n)=SP44G;
                    ArraySP44GN(n)=SP44GN;
            end
end
figure
plot(ArrayT,ArrayYTDDG,ArrayT,ArrayYTDDHG),grid
xlabel('Time (Sec)')
ylabel('Acceleration and Estimate (G)')
figure
plot(ArrayT,ArrayYTDDDG,ArrayT,ArrayYTDDDHG),grid
xlabel('Time (Sec)')
ylabel('Jerk and Estimate (G/S)')
figure
plot(ArrayT,ArrayERRYTDDG,ArrayT,ArraySP33G,ArrayT,ArraySP33GN)...
            ,grid
xlabel('Time (Sec)')
ylabel('Error in Estimate of Acceleration (G)')
figure
plot(ArrayT,ArrayERRYTDDDG,ArrayT,ArraySP44G,ArrayT,ArraySP44GN)...
            ,grid
xlabel('Time (Sec)')
ylabel('Error in Estimate of Jerk (G/S)')
clc
output=[ArrayT',ArrayYTDDG',ArrayYTDDHG',ArrayYTDDDG',...
            ArrayYTDDDHG'];
save datfil.txt output  -ascii
output=[ArrayT',ArrayERRYTDDG',ArraySP33G',ArraySP33GN'...
            ,ArrayERRYTDDDG',ArraySP44G',ArraySP44GN'];
save covfil.txt output  -ascii
disp 'simulation finished'
```

MISS DISTANCE ANALYSIS

To see how the various filtering and guidance law options perform in terms of the RMS miss distance in the presence of a weaving target, the homing loop model of Fig. 7.13 is considered. With this guidance system model, the Kalman filter can be either the general-purpose, linear, three-state filter or the special-purpose, four-state weave filter. When the linear, three-state Kalman filter is used, the possible guidance laws that can be used are either proportional navigation or optimal guidance. Recall that these two guidance laws can be expressed as

$$n_{c_{PN}} = \frac{N'}{t_{go}^2}\left(y + \dot{y}t_{go}\right)$$

$$n_{c_{Optimal}} = \frac{N'}{t_{go}^2}\left[y + \dot{y}t_{go} + \frac{t_{go}^2}{2}\ddot{y}_T - n_L T^2(e^{-x} + x - 1)\right]$$

With proportional navigation, the effective navigation ratio is usually chosen to be a constant in the range of 3 to 5. With optimal guidance, the effective navigation ratio is not constant but can be computed from

$$N' = \frac{6x^2(e^{-x} - 1 + x)}{2x^3 + 3 + 6x - 6x^2 - 12xe^{-x} - 3e^{-2x}}$$

We showed in Chapter 8 of the previous volume that when the missile is very far away from the target (that is, t_{go} is large), the effective navigation ratio

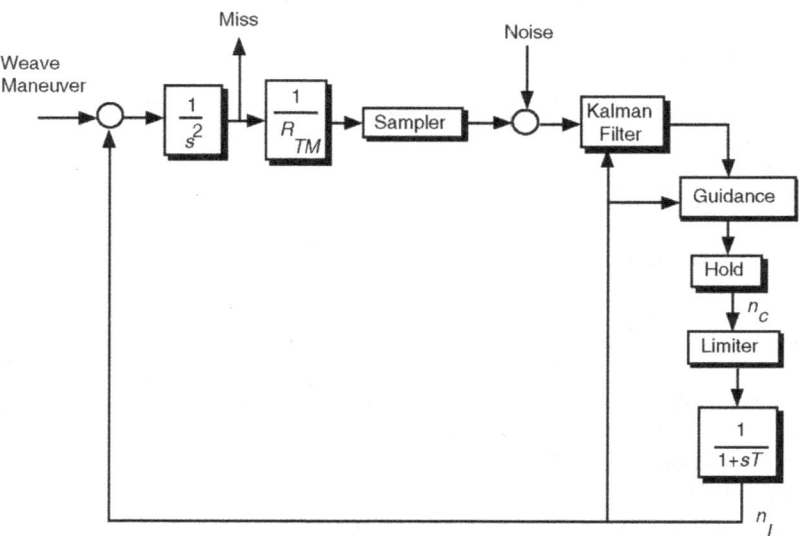

Fig. 7.13   Guidance system model for miss distance analysis.

TABLE 7.1 NOMINAL VALUES FOR EXPERIMENT

Parameter	Value
Autopilot time constant	0.5 s
Missile velocity	3000 ft/s
Closing velocity	9000 ft/s
Target acceleration level	3 g
Target weave frequency	2 rad/s

approaches 3. As the missile gets closer to the target, the navigation ratio increases to a larger number. Both guidance laws make use of the state estimates from the linear three-state Kalman filter (that is, estimates of relative position, relative velocity, and target acceleration). The compensated weave guidance law can only be used with the four-state weave Kalman filter given by

$$n_{C_{\text{Weave Lag}}} = \frac{N'}{t_{go}^2}\left[y + \dot{y}t_{go} + \frac{1 - \cos \omega t_{go}}{\omega^2}\ddot{y}_T + \frac{\omega t_{go} - \sin \omega t_{go}}{\omega^3}\dddot{y}_T \right.$$

$$\left. - n_L T^2(e^{-x} + x - 1)\right]$$

where the effective navigation ratio is the same as it was with the optimal guidance law. Again, this guidance law makes use of the state estimates from the linear four-state weave Kalman filter (that is, estimates of relative position, relative velocity, target acceleration, and target jerk). It is important to note that when this guidance law is used in conjunction with the weave Kalman filter it is assumed that the target weave frequency is known.

The parameters used for the guidance system analysis appear in Table 7.1. We can see from the table that the weaving target has a 3-g maneuver amplitude and a weave frequency of 2 rad/s. The flight-control system time constant is set at 0.5 s. Notice the high closing velocity in Table 7.1 is representative of a ballistic target engagement.

Experiments were run with the different guidance system configurations. Twenty-five Monte Carlo sets were run for flight times ranging from 0.5 s to 10 s in steps of 0.5 s. We can see from Fig. 7.14 that for the case in which there is an infinite missile acceleration capability and 1 mrad of measurement noise, both the three-state Kalman filter using optimal guidance and the four-state weave Kalman filter using the compensated weave guidance law yield approximately the same results. Both guidance laws yield significantly smaller RMS miss distances than proportional navigation.

The previous case assumed that the missile had infinite acceleration capability. If the acceleration limit is set to 10 g (more than a 3-to-1 advantage over the target), we can see from Fig. 7.15 that the RMS miss distance performance of

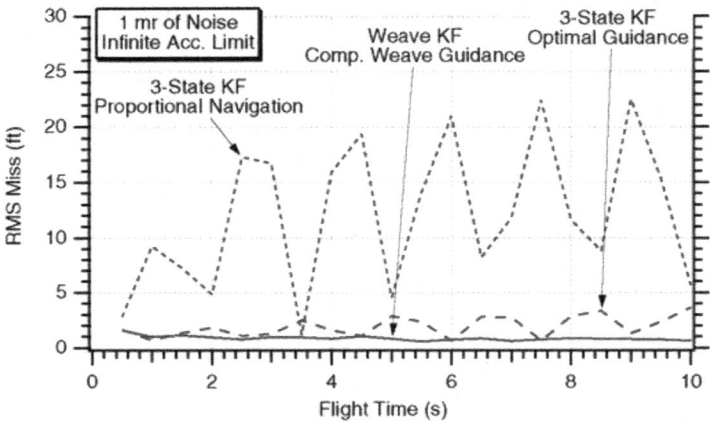

Fig. 7.14   Both optimal guidance and compensated weave guidance have similar performance.

optimal guidance deteriorates significantly. In this particular case we can see that the compensated weave guidance law in combination with the four-state weave Kalman filter yields dramatic performance advantages. Thus, we can conclude that it is only advantageous to use compensated weave guidance rather than optimal guidance if the missile has a small missile-to-target acceleration advantage and knows the target weave frequency.

If we compare Figs. 7.16 and 7.15, we can see that reducing the measurement noise by an order of magnitude to 0.1 mrad improves the performance of

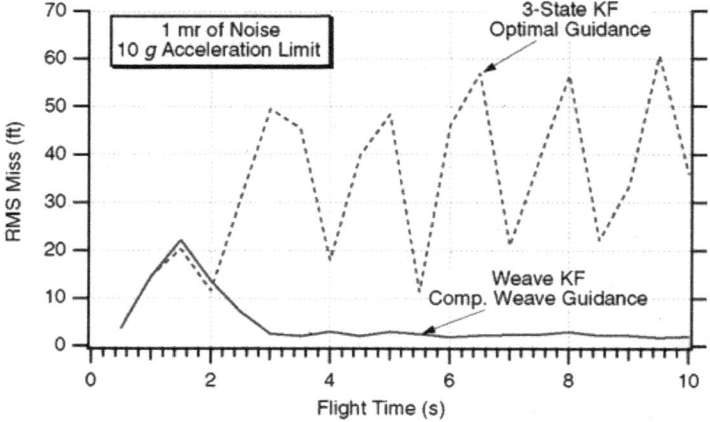

Fig. 7.15   Weave guidance can be superior to optimal guidance when acceleration saturation effects are considered.

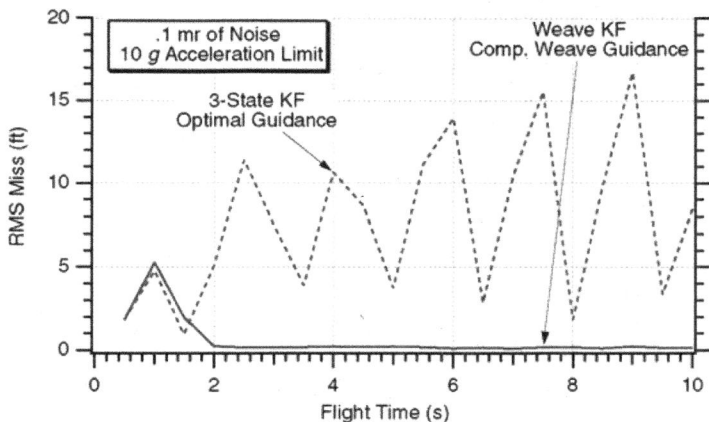

Fig. 7.16   Reducing measurement noise improves performance of both systems, but compensated weave guidance is still better.

both optimal guidance and compensated weave guidance. However, because in this example there is still a small missile-to-target acceleration advantage, it is still better to use compensated weave guidance if the target weave frequency is known.

Recall that the four-state weave Kalman filter and compensated weave guidance law both required knowledge of the target weave frequency. So far it has been assumed that the target weave frequency has been known perfectly. Errors in the knowledge of the target weave frequency will degrade the performance of both the four-state weave Kalman filter and the compensated weave guidance law. Figure 7.17 shows that when the estimated weave frequency is either twice as large or half as small as the actual target weave frequency, significant performance degradation may occur. By comparing Figs. 7.17 and 7.16 we can see that when the target weave frequency is in error we can do just as well and sometimes better by using optimal guidance and the three-state Kalman filter, because this combination does not require knowledge of the target weave frequency. Thus, we can see that if the target weave frequency is not known in advance it must somehow be estimated in real time if we wish to derive the benefits of the compensated weave guidance law.

EXTENDED KALMAN FILTER [1]

To build a Kalman filter that can estimate the target weave frequency, it is first necessary to write the state equations representing our model of the real world. From Fig. 7.6 we can say that the equations for the homing loop with a weave

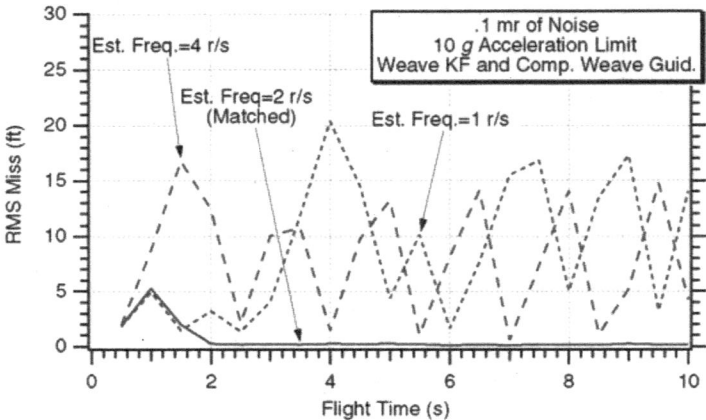

Fig. 7.17 Compensated weave guidance performance can be worse than optimal guidance if estimated weave frequency is in error.

target maneuver are still given by

$$\dot{y} = \dot{y}$$

$$\ddot{y} = \ddot{y}_T - n_L$$

$$\ddot{y}_T = \ddot{y}_T$$

$$\dddot{y}_T = -\omega^2 \ddot{y}_T + \omega u_{s1}$$

We now need an additional equation that says something about the target weave frequency. If the target weave frequency is constant, its derivative must be zero. However, for protection we can say that the derivative of the frequency is simply white noise, or

$$\dot{\omega} = u_{s2}$$

In the preceding five scalar differential equations, u_{s1} and u_{s2} are white process noise sources. Uncertainty in when the target starts to maneuver is reflected in u_{s1}, whereas uncertainty in the fact that the weave frequency may not be a constant is reflected in u_{s2}. In the previous section, for the spectral density of the first white process noise we used

$$\Phi_{s1} = \frac{\omega_{EXP}^2 n_{TMAX}^2}{t_F}$$

where ω_{EXP} can be interpreted as the maximum expected target weave frequency, n_{TMAX} is the maximum target maneuver level, and t_F is the amount of homing time. For now we will simply treat the spectral density of the second white process noise Φ_{s2} as a fudge factor whose value will be determined by experiment.

Because the target weave frequency is a state, the preceding differential equations describing our model of the real world are nonlinear, and the resultant filter will be an extended Kalman filter rather than a linear Kalman filter. However, the measurement equation for this model is still linear and turns out to be

$$y^* = \begin{bmatrix} 1 & 0 & 0 & 0 & 0 \end{bmatrix} \begin{bmatrix} y \\ \dot{y} \\ \ddot{y}_T \\ \dddot{y}_T \\ \omega \end{bmatrix} + v$$

The systems dynamics matrix can be determined from the system state equations as a matrix of partial derivatives given by

$$\mathbf{F} = \begin{bmatrix} \dfrac{\partial \dot{y}}{\partial y} & \dfrac{\partial \dot{y}}{\partial \dot{y}} & \dfrac{\partial \dot{y}}{\partial \ddot{y}_T} & \dfrac{\partial \dot{y}}{\partial \dddot{y}_T} & \dfrac{\partial \dot{y}}{\partial \omega} \\[2ex] \dfrac{\partial \ddot{y}}{\partial y} & \dfrac{\partial \ddot{y}}{\partial \dot{y}} & \dfrac{\partial \ddot{y}}{\partial \ddot{y}_T} & \dfrac{\partial \ddot{y}}{\partial \dddot{y}_T} & \dfrac{\partial \ddot{y}}{\partial \omega} \\[2ex] \dfrac{\partial \ddot{y}_T}{\partial y} & \dfrac{\partial \ddot{y}_T}{\partial \dot{y}} & \dfrac{\partial \ddot{y}_T}{\partial \ddot{y}_T} & \dfrac{\partial \ddot{y}_T}{\partial \dddot{y}_T} & \dfrac{\partial \ddot{y}_T}{\partial \omega} \\[2ex] \dfrac{\partial \dddot{y}_T}{\partial y} & \dfrac{\partial \dddot{y}_T}{\partial \dot{y}} & \dfrac{\partial \dddot{y}_T}{\partial \ddot{y}_T} & \dfrac{\partial \dddot{y}_T}{\partial \dddot{y}_T} & \dfrac{\partial \dddot{y}_T}{\partial \omega} \\[2ex] \dfrac{\partial \dot{\omega}}{\partial y} & \dfrac{\partial \dot{\omega}}{\partial \dot{y}} & \dfrac{\partial \dot{\omega}}{\partial \ddot{y}_T} & \dfrac{\partial \dot{\omega}}{\partial \dddot{y}_T} & \dfrac{\partial \dot{\omega}}{\partial \omega} \end{bmatrix}$$

where the partial derivatives are evaluated at the current state estimates. After taking the appropriate partial derivatives, the systems dynamics matrix turns out to be

$$\mathbf{F} = \frac{\partial f(\mathbf{x})}{\partial \mathbf{x}} = \begin{bmatrix} 0 & 1 & 0 & 0 & 0 \\ 0 & 0 & 1 & 0 & 0 \\ 0 & 0 & 0 & 1 & 0 \\ 0 & 0 & -\hat{\omega}^2 & 0 & -2\hat{\omega}\hat{\ddot{y}}_T \\ 0 & 0 & 0 & 0 & 0 \end{bmatrix}$$

We will use a two-term Taylor series expansion to obtain the fundamental matrix. The number of terms used in the series approach is not critical because the fundamental matrix will only be used in the Riccati equations [2]. The approximate fundamental matrix turns out to be

$$\Phi_k \approx I + FT_s = \begin{bmatrix} 1 & T_s & 0 & 0 & 0 \\ 0 & 1 & T_s & 0 & 0 \\ 0 & 0 & 1 & T_s & 0 \\ 0 & 0 & -\hat{\omega}^2 T_s & 1 & -2\hat{\omega}\hat{\dot{y}}_T T_s \\ 0 & 0 & 0 & 0 & 1 \end{bmatrix}$$

From our model of the real world, the continuous process noise matrix can be found from

$$\mathbf{Q} = E(ww^T) = E\left[\begin{bmatrix} 0 \\ 0 \\ 0 \\ u_{s1} \\ u_{s2} \end{bmatrix} \begin{bmatrix} 0 & 0 & 0 & u_{s1} & u_{s2} \end{bmatrix} \right] = \begin{bmatrix} 0 & 0 & 0 & 0 & 0 \\ 0 & 0 & 0 & 0 & 0 \\ 0 & 0 & 0 & 0 & 0 \\ 0 & 0 & 0 & \Phi_{s1} & 0 \\ 0 & 0 & 0 & 0 & \Phi_{s2} \end{bmatrix}$$

where the spectral densities Φ_{s1} and Φ_{s2} have been previously defined. The discrete process noise matrix can be obtained from the continuous process noise matrix according to

$$Q_k = \int_0^{T_s} \Phi(\tau)\mathbf{Q}\Phi^T(\tau)dt$$

After some algebra we obtain

$$Q_k = \begin{bmatrix} 0 & 0 & 0 & 0 & 0 \\ 0 & 0 & 0 & 0 & 0 \\ 0 & 0 & \Phi_{s1}\dfrac{T_s^3}{3} & \Phi_{s1}\dfrac{T_s^2}{2} & 0 \\ 0 & 0 & \Phi_{s1}\dfrac{T_s^2}{2} & \Phi_{s1}T_s + 4\dfrac{T_s^3}{3}\hat{\omega}^2\hat{\dot{y}}_T\Phi_{s2} & -\hat{\omega}T_s^2\hat{\dot{y}}_T\Phi_{s2} \\ 0 & 0 & 0 & -\hat{\omega}T_s^2\hat{\dot{y}}_T\Phi_{s2} & T_s\Phi_{s2} \end{bmatrix}$$

Note that the elements of the discrete process noise matrix are also evaluated at the current state estimates. Finally, the equations for the extended Kalman filter are simply

$$\hat{y}_k = \bar{y}_k + K_{1_k}(y_k^* - \bar{y}_k)$$
$$\hat{\dot{y}}_k = \bar{\dot{y}}_k + K_{2_k}(y_k^* - \bar{y}_k)$$
$$\hat{y}_{T_k} = \bar{\bar{y}}_{T_k} + K_{3_k}(y_k^* - \bar{y}_k)$$
$$\hat{\ddot{y}}_{T_k} = \bar{\bar{\dot{y}}}_{T_k} + K_{4_k}(y_k^* - \bar{y}_k)$$
$$\hat{\omega}_k = \hat{\omega}_{k-1} + K_{5_k}(x_k^* - \bar{x}_k)$$

where the barred quantities represent projections of the previous state estimates to the current time. Normally, the barred quantities would be obtained by multiplying the previous state estimates by the fundamental matrix to project the states ahead one sampling interval. However, because the fundamental matrix in this example is not exact (because it was obtained by using a two-term Taylor series approximation to a linearized systems dynamics matrix), it is better to use brute force to obtain the necessary projections of all the states. In this case we actually numerically integrate the nonlinear equations of motion forward one sampling interval. Euler integration is used with an integration step size that is much smaller than the sampling interval to accurately numerically integrate the state equations forward.

Listing 7.3 presents a simulation of the extended Kalman filter as part of the homing loop. We can see that routine PROJECT is used to propagate ahead the state estimates one sampling interval. It is important to note that the extended Kalman filter's estimate of the target weave frequency is intentionally initialized incorrectly to -1 rad/s (that is, WHIC$=-1$) rather than to 2 rad/s (W$=$2). The incorrect initialization is used to ensure that the extended Kalman filter is robust to initialization errors.

The nominal case of Listing 7.3 was run in which there was 1 mrad of measurement noise and the second process noise spectral density Φ_{s2} was set to zero. At first, it appears from Fig. 7.18 that the extended Kalman filter is unable to estimate the target weave frequency. However, a closer examination of Fig. 7.18 reveals that the estimated magnitude of the target weave frequency in the steady state is nearly correct (that is, 2.5 rad/s rather than 2 rad/s), but the sign is wrong. If the filter was initialized with a positive frequency, the sign

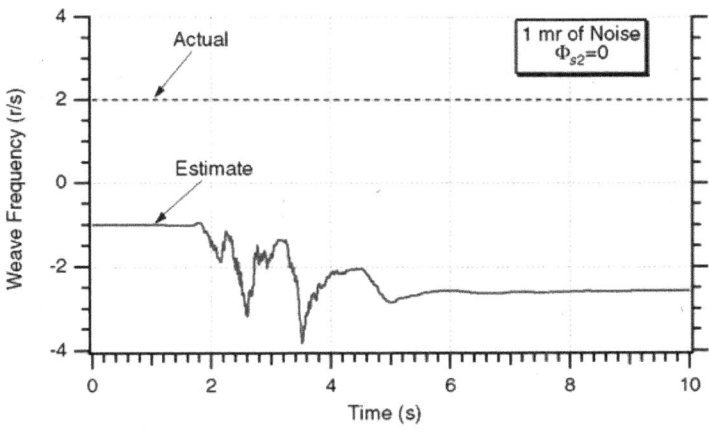

Fig. 7.18   Extended Kalman filter is able to estimate target weave frequency magnitude but not the sign.

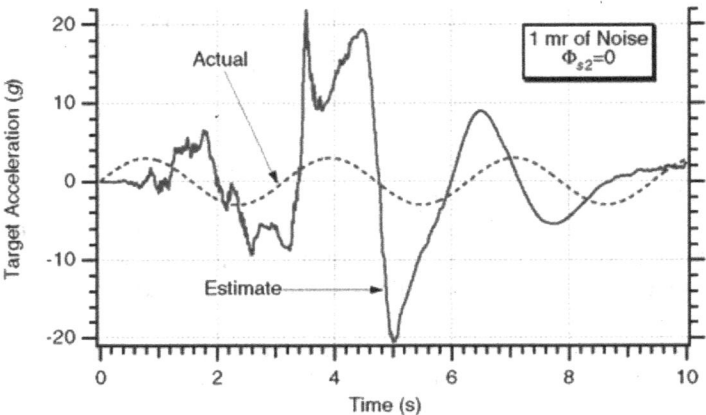

Fig. 7.19   Except for the end of flight, large amount of measurement noise causes bad estimates of target acceleration.

would have been correct. Figures 7.19 and 7.20 reveal that the large amount of measurement noise causes the estimate of the target acceleration and jerk to be fairly bad, except at the end of the flight.

We can see from Fig. 7.21 that reducing the measurement noise by an order of magnitude to 0.1 mrad improves the estimate of the magnitude of the target weave frequency. However, we are still unable to estimate the sign of the target weave frequency. Figures 7.22 and 7.23 now show that the estimates of

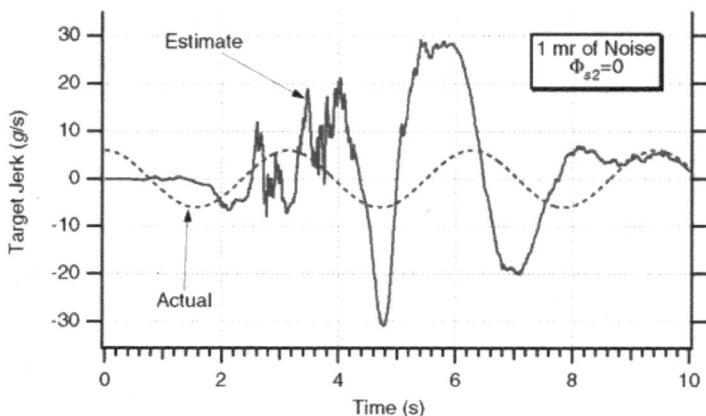

Fig. 7.20   Except for the end of flight, large amount of measurement noise causes bad estimate of target jerk.

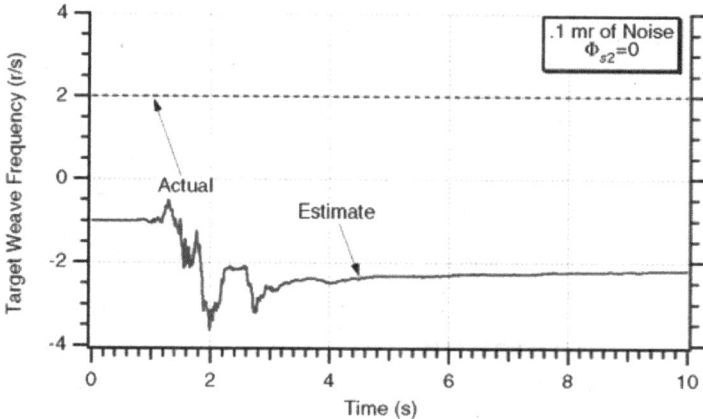

Fig. 7.21   Reducing the measurement noise improves the extended Kalman filter's estimate of the magnitude of the target weave frequency.

target acceleration and jerk are very good for most of the flight with the reduced measurement noise, even though the sign of the target weave frequency is in error. However, these estimates are not quite as good as those estimates that were obtained with the four-state linear weave Kalman filter when the target weave frequency was known precisely (see Figs. 7.11 and 7.12 for comparison).

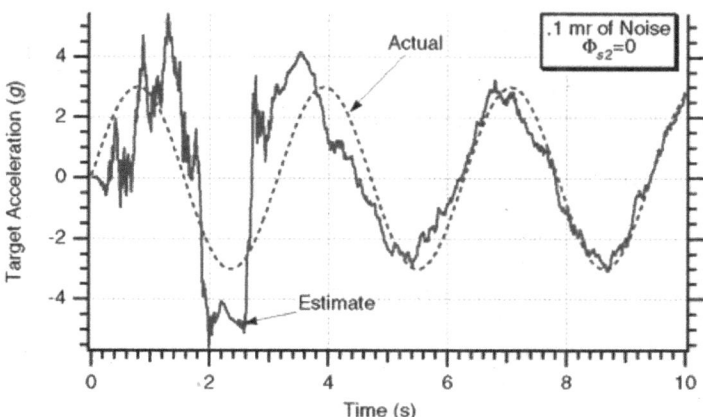

Fig. 7.22   Extended Kalman filter's estimate of target acceleration is fairly good after a transient period when measurement noise is reduced by an order of magnitude.

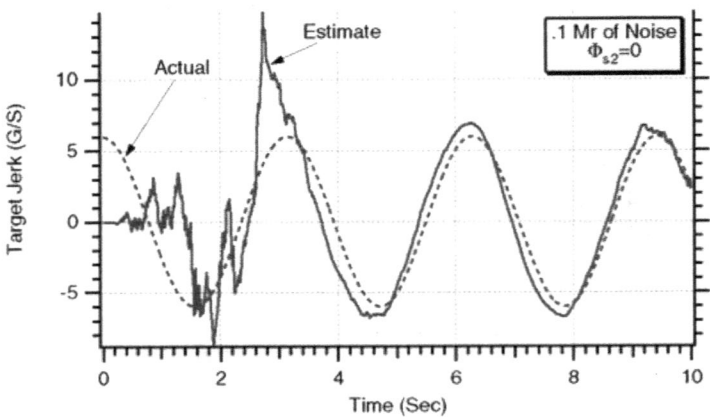

Fig. 7.23   Extended Kalman filter's estimate of target jerk is also fairly good after a transient period when measurement noise is reduced an order of magnitude.

Figure 7.24 shows that if we add a small amount of process noise (namely, $\Phi_{s2}= 0.1$) to the frequency state, our estimate of the target weave frequency magnitude improves. However, if we add too much process noise (namely, $\Phi_{s2}= 1$) to the frequency state, Fig. 7.25 shows that our estimate of the target weave frequency can actually diverge. Therefore, for safety reasons we will simply keep the second process noise source at zero (namely, $\Phi_{s2}= 0$) for future experiments.

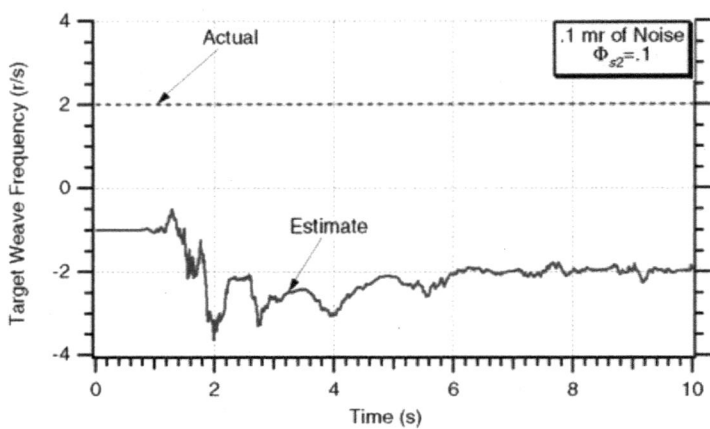

Fig. 7.24   Adding a small amount of process noise to frequency state slightly improves frequency estimate.

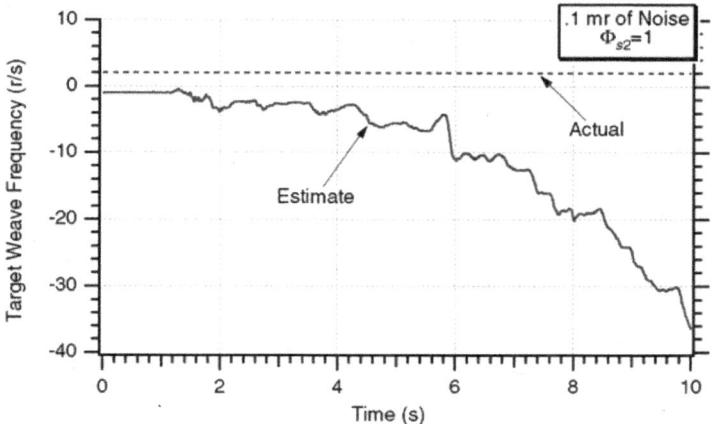

Fig. 7.25 Adding too much process noise to frequency state causes frequency estimate to diverge.

LISTING 7.3 EXTENDED KALMAN FILTER

```
clear
n=0;
PHIS2=0.;
XNT=96.6;
W=2.;
PHASEDEG=0.;
SIGRIN=.0001;
SIGGL=0.;
SRN=0.;
RA=21000.;
WHIC=-1.;
TS=.01;
TF=10.;
PHIS1=W*W*XNT*XNT/TF;
QPERFECT=0;
VC=9000.;
XNP=3.;
XNCLIM=9999999.;
APN=4;
TAU=.5;
HEDEG=0.;
VM=3000.;
QEKF=0;
PHASE=PHASEDEG/57.3;
ORDER=5;
```

```
TGO=TF;
T=0.;
X=W*T;
S=0.;
Y=0.;
YD=-XNT/W-VM*HEDEG/57.3;
YTDD=XNT*sin(W*T);
YTDDD=XNT*W*cos(W*T);
XNC=0.;
XNL=0.;
H=.001;
HP=.001;
TS2=TS*TS;
TS3=TS2*TS;
TS4=TS3*TS;
TS5=TS4*TS;
TS6=TS5*TS;
TS7=TS6*TS;
WH=WHIC;
if QPERFECT==1
        YH=Y;
        YDH=YD;
        YTDDH=YTDD;
        YTDDDH=YTDDD;
        WH=W;
else
        YH=0.;
        YDH=0.;
        YTDDH=0.;
        YTDDDH=0.;
end
PHI=zeros(ORDER);
P=zeros(ORDER);
Q=zeros(ORDER);
IDNP=eye(ORDER);
RTM=VC*TF;
SIGNOISE=sqrt(SIGRIN^2+(SIGGL/RTM)^2+(SRN*RTM*RTM/(RA*RA))^2);
YNOISE=SIGNOISE*RTM;
P(1,1)=YNOISE*YNOISE;
P(2,2)=(VM*20./57.3)^2;
P(3,3)=XNT*XNT;
P(4,4)=(W*XNT)^2;
P(5,5)=W^2;
HMAT=[1 0 0 0 0];
HT=HMAT';
while T<=(TF-.0001)
```

```
YOLD=Y;
YDOLD=YD;
XNLOLD=XNL;
STEP=1;
FLAG=0;
while STEP <=1
        if FLAG==1
                STEP=2;
                Y=Y+H*YD;
                YD=YD+H*YDD;
                XNL=XNL+H*XNLD;
                T=T+H;
        end
        YTDD=XNT*sin(W*T);
        TGO=TF-T+.00001;
        XNLD=(XNC-XNL)/TAU;
        YDD=YTDD-XNL;
        FLAG=1;
end
FLAG=0;
Y=.5*(YOLD+Y+H*YD);
YD=.5*(YDOLD+YD+H*YDD);
XNL=.5*(XNLOLD+XNL+H*XNLD);
S=S+H;
if S>=(TS-.00001)
        S=0.;
        YTDD=XNT*sin(W*T);
        YTDDD=XNT*W*cos(W*T);
        PHI(1,1)=1.;
        PHI(1,2)=TS;
        PHI(2,2)=1.;
        PHI(2,3)=TS;
        PHI(3,3)=1.;
        PHI(3,4)=TS;
        PHI(4,3)=-WH*WH*TS;
        PHI(4,4)=1.;
        PHI(4,5)=-2.*WH*YTDDH*TS;
        PHI(5,5)=1.;
        Q(3,3)=PHIS1*TS*TS*TS/3.;
        Q(3,4)=PHIS1*TS*TS/2.;
        Q(4,3)=Q(3,4);
        Q(4,4)=4.*WH*WH*YTDDH*YTDDH*PHIS2*TS*TS*TS/3.+PHIS1*TS;
        Q(4,5)=-WH*YTDDH*TS*TS*PHIS2;
        Q(5,4)=Q(4,5);
        Q(5,5)=PHIS2*TS;
        PHIT=PHI';
```

```
PHIP=PHI*P;
PHIPPHIT=PHIP*PHIT;
M=PHIPPHIT+Q;
HM=HMAT*M;
HMHT=HM*HT;
RTM=VC*TGO;
SIGNOISE=sqrt(SIGRIN^2+(SIGGL/RTM)^2+(SRN*RTM*RTM/...
      (RA*RA))^2);
YNOISE=SIGNOISE*RTM;
RMAT=[YNOISE^2];
HMHTR=HMHT+RMAT;
HMHTRINV=inv(HMHTR);
MHT=M*HT;
GAIN=MHT*HMHTRINV;
KH=GAIN*HMAT;
IKH=IDNP-KH;
P=IKH*M;
RTM=VC*TGO;
XLAM=Y/RTM;
XNOISE=SIGNOISE*randn;
XLAMS=XLAM+XNOISE;
[YB,YDB,YTDDB,YTDDDB]=PROJECT(T,TS,YH,YDH,YTDDH,YTDDDH,...
      HP,XNL,WH);
RES=RTM*XLAMS-YB;
YH=YB+GAIN(1,1)*RES;
YDH=YDB+GAIN(2,1)*RES;
YTDDH=YTDDB+GAIN(3,1)*RES;
YTDDDH=YTDDDB+GAIN(4,1)*RES;
WH=WH+GAIN(5,1)*RES;
if APN==0
      XNC=XNP*(YH+YDH*TGO)/(TGO*TGO);
elseif APN==1
      XNC=XNP*(YH+YDH*TGO+.5*YTDDH*TGO*TGO)/(TGO*TGO);
elseif APN==2
      XS=TGO/TAU;
      TOP=6.*XS*XS*(exp(-XS)-1.+XS);
      BOT1=2*XS*XS*XS+3.+6.*XS-6.*XS*XS;
      BOT2=-12.*XS*exp(-XS)-3.*exp(-2.*XS);
      XNPP=TOP/(.0001+BOT1+BOT2);
      C1=XNPP/(TGO*TGO);
      C2=XNPP/TGO;
      C3=.5*XNPP;
      C4=-XNPP*(exp(-XS)+XS-1.)/(XS*XS);
      XNC=C1*YH+C2*YDH+C3*YTDDH+C4*XNL;
elseif APN==3
      XP=WH*TGO;
```

```
                XNC=XNP*(YH+YDH*TGO)/(TGO*TGO)+XNP*YTDDH*...
                (1.-cos(XP))/XP^2+XNP*YTDDDH*(XP-sin(XP))/(XP*XP*WH);
else
                XS=TGO/TAU;
                TOP=6.*XS*XS*(exp(-XS)-1.+XS);
                BOT1=2*XS*XS*XS+3.+6.*XS-6.*XS*XS;
                BOT2=-12.*XS*exp(-XS)-3.*exp(-2.*XS);
                XNPP=TOP/(.0001+BOT1+BOT2);
                C1=XNPP/(TGO*TGO);
                C2=XNPP/TGO;
                C3=XNPP*(1.-cos(WH*TGO))/(WH*WH*TGO*TGO);
                C4=-XNPP*(exp(-XS)+XS-1.)/(XS*XS);
                C5=XNPP*(WH*TGO-sin(WH*TGO))/(WH*WH*WH*TGO*TGO);
                XNC=C1*YH+C2*YDH+C3*YTDDH+C4*XNL+C5*YTDDDH;
end
if XNC>XNCLIM
                XNC=XNCLIM;
end
if XNC<-XNCLIM
                XNC=-XNCLIM;
end
ERRYTDD=YTDD-YTDDH;
ERRYTDDG=ERRYTDD/32.2;
ERRYTDDD=YTDDD-YTDDDH;
ERRYTDDDG=ERRYTDDD/32.2;
ERRW=W-WH;
SP44=sqrt(P(4,4));
SP44P=-SP44;
SP33=sqrt(P(3,3));
SP33P=-SP33;
SP33G=SP33/32.2;
SP33PG=SP33P/32.2;
SP44G=SP44/32.2;
SP44PG=SP44P/32.2;
SP55=sqrt(P(5,5));
SP55P=-SP55;
YTDDG=YTDD/32.2;
YTDDHG=YTDDH/32.2;
YTDDDG=YTDDD/32.2;
YTDDDHG=YTDDDH/32.2;
XNCG=XNC/32.2;
n=n+1;
ArrayT(n)=T;
ArrayYTDDG(n)=YTDDG;
ArrayYTDDHG(n)=YTDDHG;
ArrayYTDDDG(n)=YTDDDG;
```

```
                    ArrayYTDDDHG(n)=YTDDDHG;
                    ArrayW(n)=W;
                    ArrayWH(n)=WH;
                    ArrayERRYTDDG(n)=ERRYTDDG;
                    ArraySP33G(n)=SP33G;
                    ArraySP33PG(n)=SP33PG;
                    ArrayERRYTDDDG(n)=ERRYTDDDG;
                    ArraySP44G(n)=SP44G;
                    ArraySP44PG(n)=SP44PG;
                    ArrayERRW(n)=ERRW;
                    ArraySP55(n)=SP55;
                    ArraySP55P(n)=SP55P;
            end
end
figure
plot(ArrayT,ArrayYTDDG,ArrayT,ArrayYTDDHG),grid
xlabel('Time (Sec)')
ylabel('Acceleration and Estimate (G)')
figure
plot(ArrayT,ArrayYTDDDG,ArrayT,ArrayYTDDDHG),grid
xlabel('Time (Sec)')
ylabel('Jerk and Estimate (G/S)')
figure
plot(ArrayT,ArrayW,ArrayT,ArrayWH),grid
xlabel('Time (Sec)')
ylabel('Frequency and Estimate (G/S)')
figure
plot(ArrayT,ArrayERRYTDDG,ArrayT,ArraySP33G,ArrayT,...
      ArraySP33PG),grid
xlabel('Time (Sec)')
ylabel('Error in Estimate of Acceleration (G)')
figure
plot(ArrayT,ArrayERRYTDDDG,ArrayT,ArraySP44G,ArrayT,...
      ArraySP44PG),grid
xlabel('Time (Sec)')
ylabel('Error in Estimate of Jerk (G/S)')
clc
output=[ArrayT',ArrayYTDDG',ArrayYTDDHG',ArrayYTDDDG',...
              ArrayYTDDDHG',ArrayW',ArrayWH'];
save datfil.txt output  -ascii
output=[ArrayT',ArrayERRYTDDG',ArraySP33G',ArraySP33PG',...
          ArrayERRYTDDDG',ArraySP44G',ArraySP44PG'];
save covfil.txt output  -ascii
disp 'simulation finished'

function[YB,YDB,YTDDB,YTDDDB]=PROJECT(TP,TS,YPH,YDPH,...
          YTDDPH,YTDDDPH,HP,XNLP,WPH)
```

```
T=0.;
Y=YPH;
YD=YDPH;
YTDD=YTDDPH;
YTDDD=YTDDDPH;
W=WPH;
XNL=XNLP;
H=HP;
while T<=(TS-.0001)
      YTDDDD=-W*W*YTDD;
      YTDDD=YTDDD+H*YTDDDD;
      YTDD=YTDD+H*YTDDD;
      YDD=YTDD-XNL;
      YD=YD+H*YDD;
      Y=Y+H*YD;
      T=T+H;
end
YB=Y;
YDB=YD;
YTDDB=YTDD;
YTDDDB=YTDDD;
```

To demonstrate the robustness of the five-state extended Kalman filter, another experiment was conducted. This time a case was considered in which the target maneuvered, but not sinusoidally. Instead, the target performed a constant 3-*g* maneuver. Figure 7.26 shows that in this case the extended Kalman filter correctly estimated, after an initial transient period, that the target weave frequency was

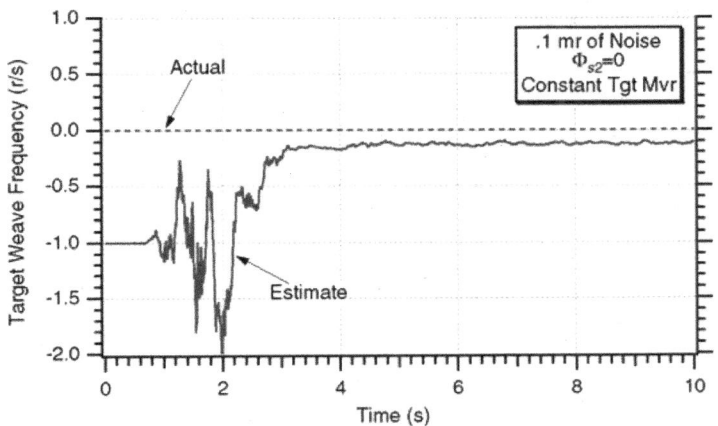

Fig. 7.26   Filter correctly estimates weave frequency of zero in presence of constant target maneuver.

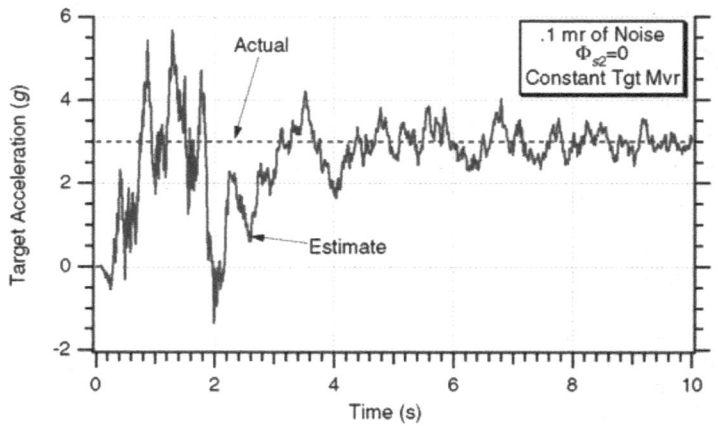

Fig. 7.27   After initial transient period, extended Kalman filter is able to estimate constant target maneuver.

zero. Figure 7.27 shows that after 4 s the filter is able to estimate that the level of the constant target maneuver is 3 g. Finally, Fig. 7.28 shows that the estimate of the target jerk is zero. This should be the case when the target maneuver is constant. Thus, we can conclude that the five-state extended Kalman filter is indeed robust.

Miss distance experiments were conducted with the three-state linear Kalman filter and the five-state extended Kalman filter when both were part of the homing

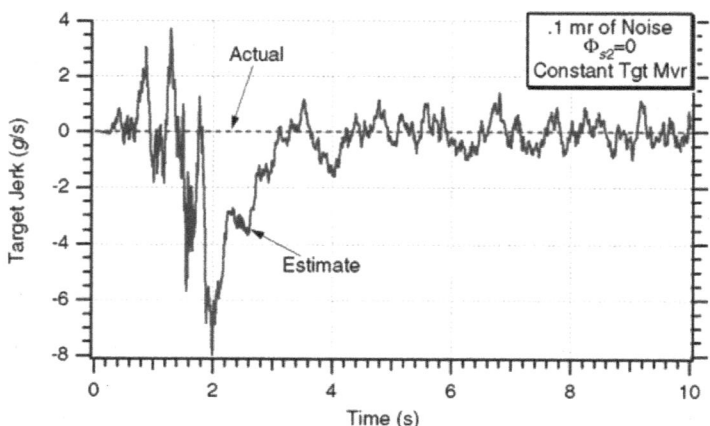

Fig. 7.28   Extended Kalman filter is able to correctly estimate that constant target maneuver has no jerk term.

loop. The guidance law options for the linear Kalman filter were proportional navigation and optimal guidance. The extended Kalman filter was always used in conjunction with the compensated weave guidance law. A close examination of this guidance law reveals that the sign of the target weave frequency does not influence the guidance command. Therefore, the inability of the five-state extended Kalman filter to correctly determine the sign of the target weave frequency should not be important for guidance purposes.

Figure 7.29 shows that when there are no constraints on the missile acceleration and there is 1 mrad of measurement noise, both the three-state linear Kalman filter with optimal guidance and five-state extended Kalman filter with compensated weave guidance yield superior miss distance performance to that of a proportional navigation guidance system. Both optimal guidance and compensated weave guidance yield similar performance as measured by the RMS miss distance. However, we can see from Fig. 7.30 that if there is a 10-*g* missile acceleration limit, optimal guidance degrades severely and the best performance is obtained by the five-state extended Kalman filter with compensated weave guidance. In this case, the performance of optimal guidance and proportional navigation are similar, and both yield much larger RMS miss distances when compared with the compensated weave guidance law.

Reducing the measurement noise by an order of magnitude to 0.1 mrad improves the performance of all the guidance systems. When there is no constraint on the available missile acceleration, Fig. 7.31 again shows that optimal guidance and compensated weave guidance yield much better performance than a proportional navigation guidance system. We can also see from Fig. 7.31 that the performances of both optimal guidance and compensated weave guidance

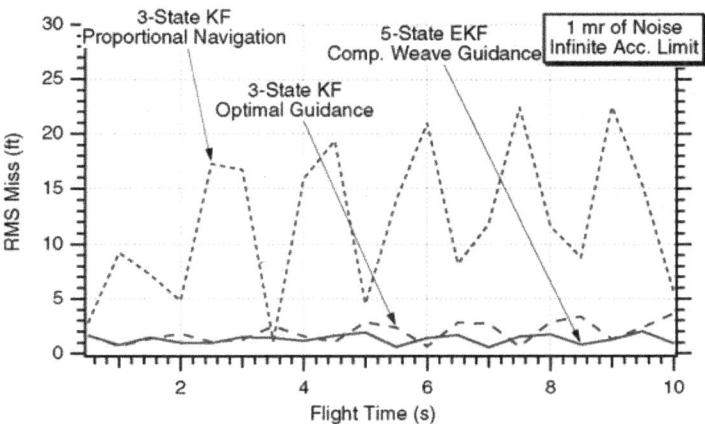

Fig. 7.29   Both compensated weave guidance and optimal guidance yield similar performance when there are no missile acceleration constraints.

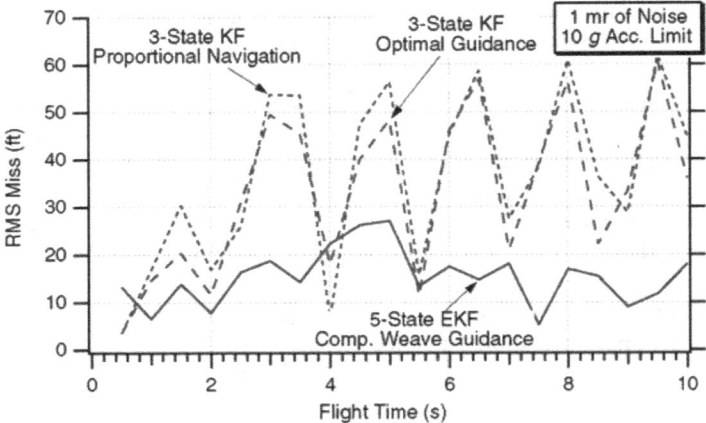

Fig. 7.30   With missile acceleration constraints, five-state extended Kalman filter with compensated weave guidance yields the best results.

are virtually identical. However, we can see from Fig. 7.32 that the performance of the optimal guidance system degrades when there is a 10-*g* missile acceleration limit, and the performance of the extended Kalman filter with the compensated weave guidance law is much better. The trends are identical to the case in which the measurement noise was an order of magnitude larger.

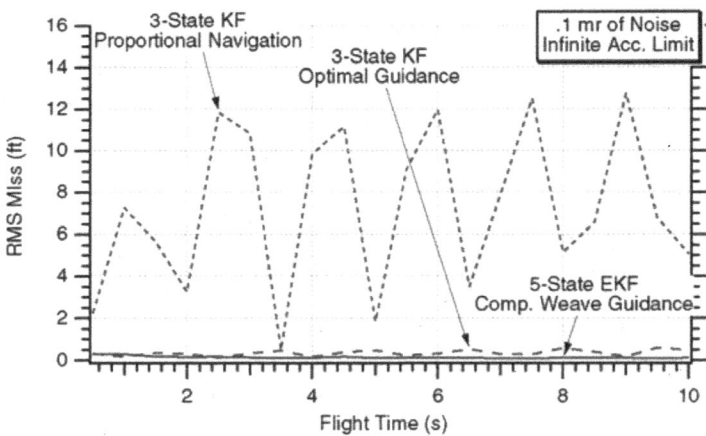

Fig. 7.31   With reduced measurement noise, both compensated weave guidance and optimal guidance still yield similar performance when there are no missile acceleration constraints.

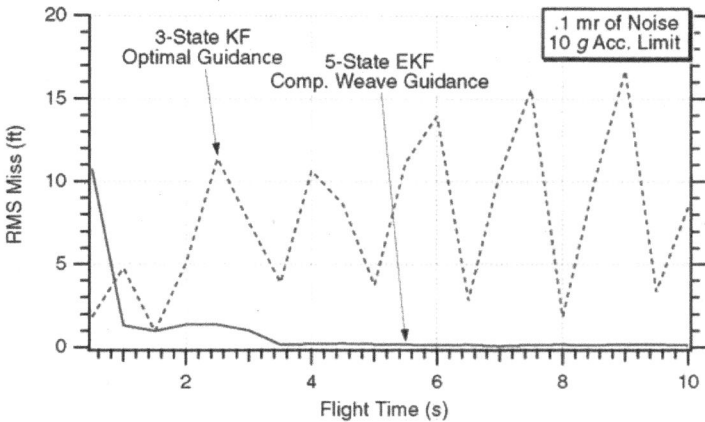

Fig. 7.32   With missile acceleration constraints, five-state extended Kalman filter with compensated weave guidance yields the best results when measurement noise is reduced.

SUMMARY

This chapter shows various filtering options that can be used in conjunction with advance guidance laws to improve the performance of a missile guidance system against weaving targets. A conventional linear three-state Kalman filter can be used in conjunction with the optimal guidance law to yield significant performance improvements compared with proportional navigation. Similar performance improvements can also be obtained with a four-state weave Kalman filter in conjunction with the compensated weave guidance law if the target weave frequency is known. This filtering and guidance approach will perform better than optimal guidance when there is a low missile-to-target acceleration advantage. If the target weave frequency is not known, it can be estimated with a five-state extended Kalman filter. Similar performance improvements can be obtained with this nonlinear filter and guidance law approach.

REFERENCES

[1]   Zarchan, P., "Tracking and Intercepting Spiraling Ballistic Missiles," *Proceedings of IEEE PLANS Conference*, San Diego, CA, March 2000.

[2]   Zarchan, P., and Musoff, H., *Fundamentals of Kalman Filtering: A Practical Approach*, Progress in Astronautics and Aeronautics, AIAA, Reston, VA, 2000.

CHAPTER 8

Filter Bank Approach to Weaving Target Problem

INTRODUCTION

We showed in Chapter 6 that interceptor performance against a weaving target could be improved considerably by use of a special-purpose compensated weave guidance law. Essentially the compensated weave guidance law makes an internal prediction of where the target will be at intercept and also compensates for the missile's own flight-control-system dynamics. This information enables the missile to guide directly to the predicted intercept point using a minimum of energy. If the weave frequency of the target is known in advance, it was shown in Chapter 7 that when the measurement noise was low, a four-state linear weave Kalman filter could be designed to yield excellent estimates of the necessary states to implement the compensated weave guidance law. If, on the other hand, the target weave frequency was not known, we also showed in Chapter 7 that the target weave frequency could be estimated under certain circumstances, with a five-state extended Kalman filter. Because all extended Kalman filters are sensitive to large initialization errors, this type of filter will work only if it is initialized with a target frequency estimate that is close to the actual target weave frequency. In this chapter we shall show how a bank of linear four-state weave Kalman filters, each tuned to a different target weave frequency, can be used to estimate the target weave frequency and improve system performance when the measurement noise is low. It will be demonstrated that the filter bank approach is more robust than an extended Kalman filter when there is a large uncertainty in the target weave frequency.

REVIEW OF FIVE-STATE EXTENDED-KALMAN-FILTER PERFORMANCE

The guidance system used in this chapter for the weaving target problem is identical to the one used in Chapter 7 and is repeated in Fig. 8.1 for convenience. Note

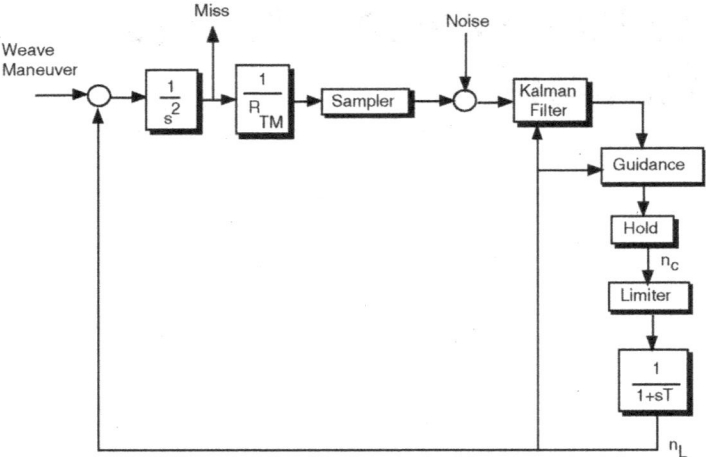

Fig. 8.1   Guidance system model for miss distance analysis.

that the two sources of error in the guidance system are measurement noise and a sinusoidal target maneuver. The Kalman filter, shown in the homing loop of Fig. 8.1, can be either a four-state linear weave Kalman filter, a five-state extended Kalman filter (both of which were fully described in Chapter 7), or a bank of linear weave four-state Kalman filters (which will be fully described later in this chapter). The parameters to be used in the analysis of the Fig. 8.1 guidance system appear in Table 8.1.

It is well known that, in general, the extended Kalman filter is sensitive to initialization errors [1]. This means that if the five-state extended Kalman filter of Chapter 7 is not initialized with a target weave frequency estimate that is fairly

TABLE 8.1   NOMINAL SYSTEM INPUTS FOR VARIOUS STUDIES

Symbol	Definition	Value
n_T	Target acceleration	$3\,g$
n_{LIM}	Missile acceleration limit	$10\,g$
σ_{Noise}	Seeker measurement noise	0.1 mrad
V_c	Closing velocity	9000 ft/s
T	Flight-control-system time constant	0.5 s
ω	Target weave frequency	Varies from 1 to 10 rad/s
t_F	Flight time	10 s
T_s	Sampling time	0.01 s

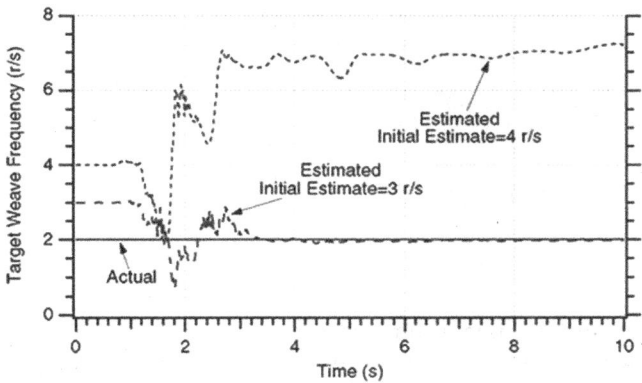

Fig. 8.2   Extended Kalman filter is unable to estimate target weave frequency if initial frequency estimate is in error by 2 rad/s.

close to the actual target weave frequency, the performance of the extended Kalman filter can degrade considerably. For example, Listing 7.3 was run for the case in which the actual target weave frequency was 2 rad/s and the initial estimate of the target weave frequency WHIC was varied. In Fig. 8.2 we can see that if the initial estimate of the target weave frequency is 3 rad/s (in error by 1 rad/s), the extended Kalman filter is able to estimate the target weave frequency after 3 s. However, if the initial estimate of the target weave frequency is 4 rad/s (in error by 2 rad/s), the filter's estimate of the target weave frequency diverges from the truth and is in considerable error.

Thus it appears that we need some other way of estimating the target weave frequency when we have no idea of what the target weave frequency might be.

REVIEW OF FOUR-STATE LINEAR WEAVE KALMAN-FILTER PERFORMANCE

The four-state linear weave Kalman filter was derived in Chapter 7. The states of that filter are relative position y, relative velocity $\dot{y}$, target acceleration $\ddot{y}_T$, and target jerk $\dddot{y}_T$. With this linear Kalman filter it is assumed that the weave frequency of the target ω is known and does not have to be estimated. Chapter 7 showed that the scalar equations for the four-state linear weave Kalman filter are given by

$$\text{RES}_k = y_k^* - \hat{y}_{k-1} - T_s\hat{\dot{y}}_{k-1} - \left(\frac{1-\cos x}{\omega^2}\right)\hat{\ddot{y}}_{T_{k-1}} - \left(\frac{x-\sin x}{\omega^3}\right)\hat{\dddot{y}}_{T_{k-1}} + 0.5T_s^2 n_{L_{k-1}}$$

$$\hat{y}_k = \hat{y}_{k-1} + T_s\hat{\dot{y}}_{k-1} + \left(\frac{1-\cos x}{\omega^2}\right)\hat{\ddot{y}}_{T_{k-1}} + \left(\frac{x-\sin x}{\omega^3}\right)\hat{\dddot{y}}_{T_{k-1}}$$

$$- 0.5T_s^2 n_{L_{k-1}} + K_{1_k}\text{RES}_k$$

$$\hat{y}_k = \hat{y}_{k-1} + \left(\frac{\sin x}{\omega}\right)\hat{\dot{y}}_{T_{k-1}} + \left(\frac{1-\cos x}{\omega^2}\right)\hat{\ddot{y}}_{T_{k-1}} - T_s n_{L_{k-1}} + K_{2_k} \text{RES}_k$$

$$\hat{\dot{y}}_{T_k} = \cos x \hat{\dot{y}}_{T_{k-1}} + \left(\frac{\sin x}{\omega}\right)\hat{\ddot{y}}_{T_{k-1}} + K_{3_k} \text{RES}_k$$

$$\hat{\ddot{y}}_{T_k} = -\omega \sin x \hat{\dot{y}}_{T_{k-1}} + \cos x \hat{\ddot{y}}_{T_{k-1}} + K_{4_k} \text{RES}_k$$

where

$$x = \omega T_s$$

T_s is the sampling time or time between measurements, and ω is the target weave frequency, which is assumed to be known. The Kalman gains (K_1, K_2, K_3, and K_4) are obtained from the matrix Riccati equations. Details concerning the fundamental, measurement, process noise, and measurement noise matrices can also be found in Chapter 7. The filter is optimal if the real target maneuver is a sinusoid, the target weave frequency is known, and the compensated weave guidance law is used.

The compensated weave guidance law, which is optimal in the sense that it requires the least acceleration in the integral squared sense against weave maneuvers, issues guidance commands proportional to the zero effort miss and inversely proportional to the square of time to go until intercept. Chapter 6 showed that the compensated weave guidance law is given by

$$n_c = \frac{N'}{t_{go}^2}\left[y + \dot{y}t_{go} + \left(\frac{1-\cos \omega t_{go}}{\omega^2}\right)\ddot{y}_T\right.$$

$$\left. + \left(\frac{\omega t_{go} - \sin \omega t_{go}}{\omega^3}\right)\dddot{y}_T - n_L T^2(e^{-x} + x - 1)\right]$$

where in this case x (not to be confused with x in Kalman-filter equations) is given by

$$x = \frac{t_{go}}{T}$$

In the preceding equation, t_{go} is the time to go until intercept, and T is defined as the approximate time constant of the flight-control system. Again ω is the target weave frequency, which is assumed to be known. The effective navigation ratio in the compensated weave guidance law is time varying and is given by

$$N' = \frac{6x^2(e^{-x} - 1 + x)}{2x^3 + 3 + 6x - 6x^2 - 12xe^{-x} - 3e^{-2x}}$$

The compensated weave guidance law consists of five terms: the first two terms are related to the line-of-sight rate, the third term is proportional to the target acceleration, a fourth term is proportional to target jerk, and a fifth term is proportional to the achieved missile acceleration. The first two terms in the compensated weave guidance law are proportional navigation. The third and fourth terms in the guidance law compensate for the sinusoidal motion of the target, and the fifth term compensates for the dynamics in the missile flight-control system. Details of the derivation of the compensated weave guidance law can also be found in Chapter 6.

It was also demonstrated in Chapter 7 that the four-state linear weave Kalman filter in conjunction with the compensated weave guidance law worked very well if the target weave frequency was known, as shown in Fig. 8.3. Here we can see that we have virtually perfect estimates of the target acceleration after a very brief period of time.

However, if knowledge of the target weave frequency is in error, the filter's estimate of the target acceleration will deteriorate. For example, consider the case where the actual target weave frequency is 2 rad/s but the filter thinks the target weave frequency is 1 rad/s. In this case we are underestimating the target weave frequency. Figure 8.4 shows that the filter estimate of the target acceleration does not track the actual target acceleration very well when the target weave frequency is underestimated. Figure 8.5 also shows that when the target weave frequency is overestimated at 4 rad/s, the Kalman filter's estimate of target acceleration also deteriorates.

Lack of knowledge of the target weave frequency not only will yield poorer state estimates, but also will influence system performance as measured by the RMS miss distance. A 50-run Monte Carlo miss distance experiment was repeated from Chapter 7 in which the flight times ranged from 0.5 to 10 s in steps of 0.5 s using a modified version of Listing 6.2. Figure 8.6, which is identical to Fig. 7.17

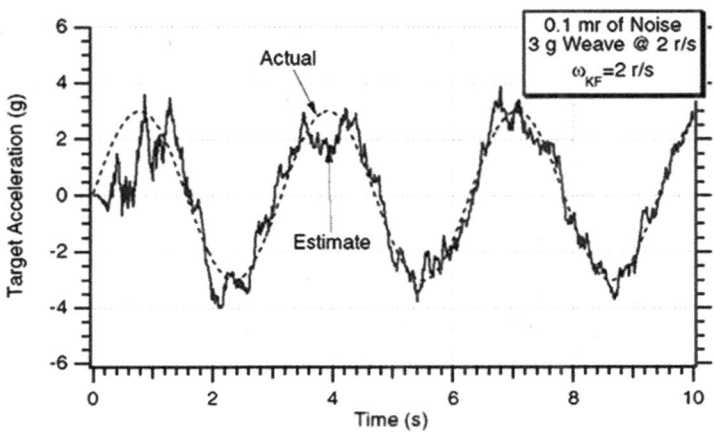

Fig. 8.3   Linear weave four-state Kalman filter works well if target weave frequency is known.

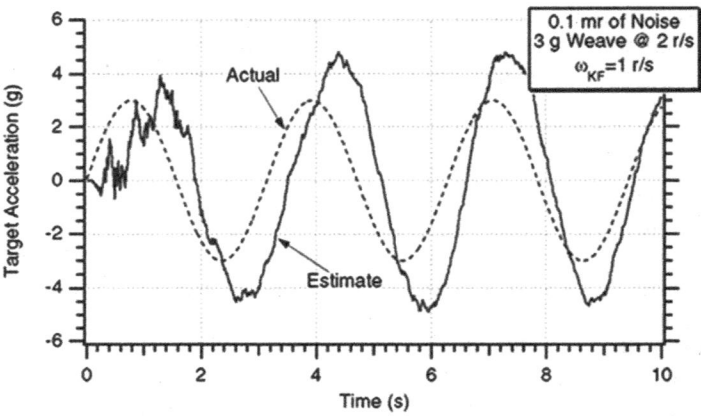

Fig. 8.4   Four-state linear weave Kalman filter's estimate of target acceleration deteriorates when we underestimate target weave frequency.

(except scales have changed), shows that the price paid for the lack of know-ledge of the target weave frequency is increased RMS miss distance. When the target weave frequency is known, the maximum RMS miss distance is 5 ft; however, when knowledge of the target weave frequency is in error on the low side by 1 rad/s, the maximum RMS miss distance is 20 ft. Similarly, when knowledge of the target weave frequency is in error on the high side by 2 rad/s,

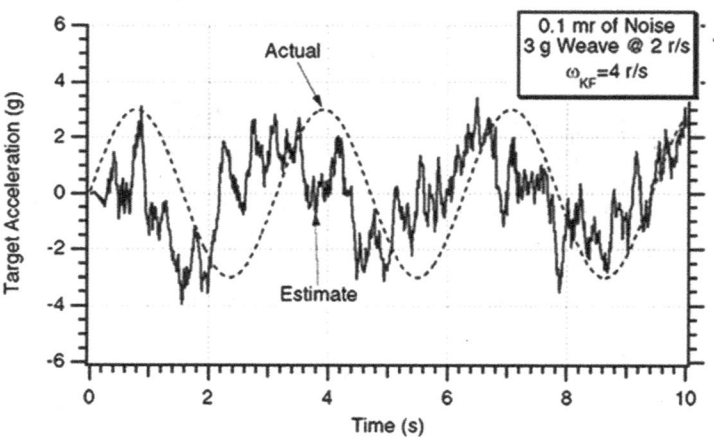

Fig. 8.5   Four-state linear weave Kalman filter's estimate of target acceleration deteriorates when we overestimate target weave frequency.

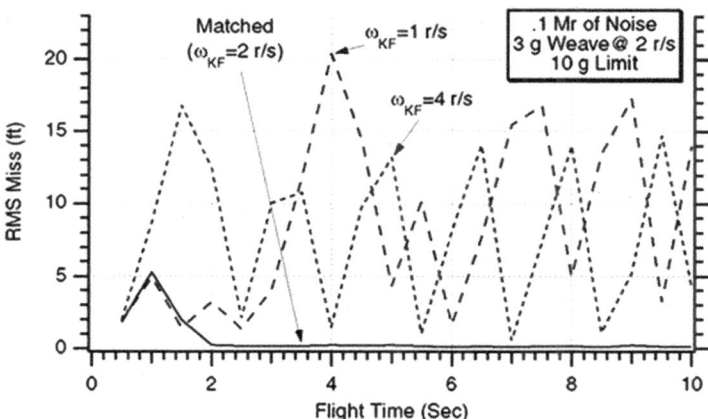

Fig. 8.6 Miss distance performance of four-state linear weave Kalman filter deteriorates when target weave frequency estimate is in error.

the maximum RMS miss distance is 15 ft. Clearly we need a better way of estimating the acceleration of a weaving target when the target weave frequency is unknown.

FILTER BANK METHODOLOGY

The fixed multiple model adaptive estimator (MMAE) technique for determining an unknown parameter within a bank of Kalman filters was first developed by D. T. Magill in 1965 [2]. With this technique, several Kalman filters are run in parallel with each being tuned to a different parameter value. The likelihood function and Bayes' rule are used to determine the probability that a filter is the correct one. Estimates from each of the filters are weighted by the probability that it is the correct one and combined to form a resultant estimate [3–5].

The basic idea of this chapter is to use a bank of linear four-state weave Kalman filters, each assuming a different target weave frequency, operating in parallel (that is, all filters receive the same measurements). Each filter is totally independent of every other filter in the filter bank. It is postulated that by using the fixed MMAE approach at every guidance update, we can determine the probability that a given filter is the correct one (that is, tuned to the correct target weave frequency). The resultant estimates are then obtained by adding the state estimates of each filter multiplied by the probability that the filter is the one tuned to the correct frequency.

With the fixed MMAE approach, each filter is totally independent of every other filter in the filter bank. The fixed MMAE approach makes use of the residual RES and covariance of the residual C of each filter in the filter bank. For the problem concerning the linear four-state weave Kalman filter, the residual and residual covariance are scalars. The ith filter residual and covariance at the kth instant are given by

$$\mathrm{RES}_k(i) = z_k - \mathbf{H}\boldsymbol{\Phi}_k(i)\,\hat{\mathbf{x}}_{k-1}(i) - \mathbf{H}\,\mathbf{G}_k\,\mathbf{u}_{k-1}$$

$$\sigma^2_{\mathrm{RES}_k}(i) = C_k(i) = \mathbf{H}\,\mathbf{M}_k(i)\,\mathbf{H}^{\mathrm{T}} + \mathbf{R}_k$$

It can also be shown that there is something called the likelihood function of the residual f, which is computed using the filter residual and covariance. The likelihood function is also a scalar for the four-state linear weave Kalman filter. The likelihood function for the ith filter at the kth instant is given by the scalar

$$f_k(i) = \frac{1}{\sqrt{2\pi C_k(i)}}\, e^{-0.5\mathrm{RES}_k^2(i)/C_k(i)}$$

The probability $p_k(i)$ that the ith filter is the correct one at the kth instant can be computed according to Bayes' rule as

$$p_k(i) = \frac{f_k(i)p_{k-1}(i)}{\sum_{i=1}^{r} f_k(i)p_{k-1}(i)}$$

$$= \frac{f_k(i)p_{k-1}(i)}{f_k(1)p_{k-1}(1) + f_k(2)p_{k-1}(2) + f_k(3)p_{k-1}(3) + \cdots + f_k(r)p_{k-1}(r)}$$

Note that the preceding equation is recursive and therefore needs initial estimates for the probability that each filter is the correct one. If we assume that it is equally likely that any one of r filters in the filter bank is the correct one, then we can say that [6]

$$p_0(i) = \frac{1}{r}$$

Thus at each instant of time we have a set of r probabilities (numbers ranging from zero to unity) telling us the likelihood that any filter is the correct one. The r probabilities add up to unity. The estimated target weave frequency to be used by the compensated weave guidance law can be found by weighting the frequency assigned to each filter by the probability that the filter is the correct one, or

$$\hat{\omega}_k = p_k(1)\omega(1) + p_k(2)\omega(2) + p_k(3)\omega(3) + \cdots + p_k(r)\omega(r)$$

where $\omega(1)\ldots\omega(r)$ are the tuned frequencies assumed by each of the r filters. The resultant state estimates to also be used by the compensated weave guidance law are obtained by using the state estimates of each individual filter weighted by the probability that the filter is correct, or

$$\hat{\mathbf{x}}_k = p_k(1)\hat{\mathbf{x}}_k(1) + p_k(2)\hat{\mathbf{x}}_k(2) + p_k(3)\hat{\mathbf{x}}_k(3) + \cdots + p_k(r)\hat{\mathbf{x}}_k(r)$$

where $\hat{\mathbf{x}}_k(1)\ldots\hat{\mathbf{x}}_k(r)$ are the state estimates of each of the r filters.

THREE-FILTER BANK EXAMPLE

To better understand the mechanics of the fixed MMAE approach, it is best to first consider a numerical example in which there are only three filters in the filter bank. In this example it is assumed that one of the three filters is actually the correct one (meaning tuned to the correct target weave frequency). If three four-state linear weave Kalman filters are run in parallel, each tuned to a different target weave frequency, the equations of the preceding section simplify. The likelihood function for each of the three filters at the kth instant is given by

$$f_k(1) = \frac{1}{\sqrt{2\pi C_k(1)}} e^{-0.5 \text{RES}_k^2(1)/C_k(1)}$$

$$f_k(2) = \frac{1}{\sqrt{2\pi C_k(2)}} e^{-0.5 \text{RES}_k^2(2)/C_k(2)}$$

$$f_k(3) = \frac{1}{\sqrt{2\pi C_k(3)}} e^{-0.5 \text{RES}_k^2(3)/C_k(3)}$$

where each of the covariances $C_k(i)$ can be found from each filter's set of Riccati equations according to

$$C_k(1) = \mathbf{H}\mathbf{M_k}(1)\mathbf{H}^T + \mathbf{R_k}$$

$$C_k(2) = \mathbf{H}\mathbf{M_k}(2)\mathbf{H}^T + \mathbf{R_k}$$

$$C_k(3) = \mathbf{H}\mathbf{M_k}(3)\mathbf{H}^T + \mathbf{R_k}$$

and the residuals can be found from the error signal within the filter as

$$\text{RES}_k(1) = z_k - \mathbf{H}\mathbf{\Phi_k}(1)\hat{\mathbf{x}}_{\mathbf{k-1}}(1) - \mathbf{H}\mathbf{G_k}\mathbf{u_{k-1}}$$
$$\text{RES}_k(2) = z_k - \mathbf{H}\mathbf{\Phi_k}(2)\hat{\mathbf{x}}_{\mathbf{k-1}}(2) - \mathbf{H}\mathbf{G_k}\mathbf{u_{k-1}}$$
$$\text{RES}_k(3) = z_k - \mathbf{H}\mathbf{\Phi_k}(3)\hat{\mathbf{x}}_{\mathbf{k-1}}(3) - \mathbf{H}\mathbf{G_k}\mathbf{u_{k-1}}$$

The probability that any given filter of the three filters is the correct one can be found from Bayes' rule as

$$p_k(1) = \frac{f_k(1)p_{k-1}(1)}{f_k(1)p_{k-1}(1) + f_k(2)p_{k-1}(2) + f_k(3)p_{k-1}(3)}$$

$$p_k(2) = \frac{f_k(2)p_{k-1}(2)}{f_k(1)p_{k-1}(1) + f_k(2)p_{k-1}(2) + f_k(3)p_{k-1}(3)}$$

$$p_k(3) = \frac{f_k(3)p_{k-1}(3)}{f_k(1)p_{k-1}(1) + f_k(2)p_{k-1}(2) + f_k(3)p_{k-1}(3)}$$

We can see from the preceding set of equations that at any instant of time

$$p_k(1) + p_k(2) + p_k(3) = 1$$

Again, $p_k(1)$ represents the probability that the first filter is correct at the kth instant. To start the preceding recursion for calculating the probabilities that a filter is correct, we assume initially each filter is equally likely to be the correct filter, or

$$p_0(1) = \frac{1}{3}$$

$$p_0(2) = \frac{1}{3}$$

$$p_0(3) = \frac{1}{3}$$

At each instant of time, we calculate the probability that each filter is correct and then weigh the assumed frequency and estimates of each of the three filters according to their probabilities as

$$\hat{\omega}_k = p_k(1)\omega(1) + p_k(2)\omega(2) + p_k(3)\omega(3)$$
$$\hat{\mathbf{x}}_k = p_k(1)\hat{\mathbf{x}}_k(1) + p_k(2)\hat{\mathbf{x}}_k(2) + p_k(3)\hat{\mathbf{x}}_k(3)$$

Listing 8.1 implements the fixed MMAE approach for three four-state linear weave Kalman filters along with the inputs of Table 8.1. This listing is based upon the single four-state linear weave Kalman filter approach of Listing 7.2; however, in Listing 8.1 we only consider the compensated weave guidance law. In addition, in Listing 8.1 the actual target weave frequency is 2 rad/s. One filter in the filter bank is tuned to 1 rad/s, the other is tuned to 2 rad/s, and the third is tuned to 4 rad/s. Therefore, the correct filter in this example is the one tuned to 2 rad/s. The additional code in going from the single-filter approach of Listing 7.2 to the fixed MMAE approach of Listing 8.1 is highlighted in bold.

LISTING 8.1   THREE-FILTER FIXED MMAE APPROACH TO WEAVING TARGET PROBLEM

```
clear
count=0;
TAU=.5;
ORDER=4;
VC=9000.;
XLIM=322.;
W1=1.;
W2=2.;
```

```
W3=4.;
WREAL=2.;
XNT=96.6;
XNTREAL=96.6;
TS=.01;
YIC=0.;
VM=3000.;
HEDEG=0.;
HEDEGFIL=20.;
XNP=3.;
SIGRIN=.0001;
TF=10.;;
PHASE=0./57.3;
X1=W1*TS;
X2=W2*TS;
X3=W3*TS;
Y=YIC;
YD=-VM*HEDEG/57.3;
PHIS1=W1*W1*XNT*XNT/TF;
PHIS2=W2*W2*XNT*XNT/TF;
PHIS3=W3*W3*XNT*XNT/TF;
RTM=VC*TF;
SIGNOISE=SIGRIN;
SIGPOS=RTM*SIGNOISE;
SIGN2=SIGPOS^2;
PHI1=zeros([4,4]);
P1=zeros([4,4]);
Q1=zeros([4,4]);
IDNP=eye(4);
PHI2=zeros([4,4]);
P2=zeros([4,4]);
Q2=zeros([4,4]);
PHI3=zeros([4,4]);
P3=zeros([4,4]);
Q3=zeros([4,4]);

PHI1(1,1)=1;
PHI1(1,2)=TS;
PHI1(1,3)=(1-cos(X1))/(W1*W1);
PHI1(1,4)=(X1-sin(X1))/(W1*W1*W1);
PHI1(2,2)=1;
PHI1(2,3)=sin(X1)/W1;
PHI1(2,4)=(1-cos(X1))/(W1*W1);
PHI1(3,3)=cos(X1);
PHI1(3,4)=sin(X1)/W1;
PHI1(4,3)=-W1*sin(X1);
```

```
PHI1(4,4)=cos(X1);
PHI2(1,1)=1;
PHI2(1,2)=TS;
PHI2(1,3)=(1-cos(X2))/(W2*W2);
PHI2(1,4)=(X2-sin(X2))/(W2*W2*W2);
PHI2(2,2)=1;
PHI2(2,3)=sin(X2)/W2;
PHI2(2,4)=(1-cos(X2))/(W2*W2);
PHI2(3,3)=cos(X2);
PHI2(3,4)=sin(X2)/W2;
PHI2(4,3)=-W2*sin(X2);
PHI2(4,4)=cos(X2);

PHI3(1,1)=1;
PHI3(1,2)=TS;
PHI3(1,3)=(1-cos(X3))/(W3*W3);
PHI3(1,4)=(X3-sin(X3))/(W3*W3*W3);
PHI3(2,2)=1;
PHI3(2,3)=sin(X3)/W3;
PHI3(2,4)=(1-cos(X3))/(W3*W3);
PHI3(3,3)=cos(X3);
PHI3(3,4)=sin(X3)/W3;
PHI3(4,3)=-W3*sin(X3);
PHI3(4,4)=cos(X3);

Q1(1,1)=PHIS1*(.333*X1^3-2*sin(X1)+2*X1*cos(X1)+.5*X1-...
                   .25*sin(2*X1))/(W1^5);
Q1(1,2)=PHIS1*(.5*X1*X1-X1*sin(X1)+.5*sin(X1)*...
                   sin(X1))/(W1^4);
Q1(2,1)=Q1(1,2);
Q1(1,3)=PHIS1*(sin(X1)-X1*cos(X1)-.5*X1+...
                   .25*sin(2*X1))/(W1^3);
Q1(3,1)=Q1(1,3);
Q1(1,4)=PHIS1*(cos(X1)+X1*sin(X1)-.5*sin(X1)*...
                   sin(X1)-1)/(W1*W1);
Q1(4,1)=Q1(1,4);
Q1(2,2)=PHIS1*(1.5*X1-2*sin(X1)+.25*sin(2*X1))/(W1^3);
Q1(2,3)=PHIS1*(1-cos(X1)-.5*sin(X1)*sin(X1))/(W1*W1);
Q1(3,2)=Q1(2,3);
Q1(2,4)=PHIS1*(sin(X1)-.5*X1-.25*sin(2*X1))/W1;
Q1(4,2)=Q1(2,4);
Q1(3,3)=PHIS1*(.5*X1-.25*sin(2*X1))/W1;
Q1(3,4)=.5*PHIS1*sin(X1)*sin(X1);
Q1(4,3)=Q1(3,4);
Q1(4,4)=W1*PHIS1*(.5*X1+.25*sin(2*X1));
```

```
Q2(1,1)=PHIS2*(.333*X2^3-2*sin(X2)+2*X2*cos(X2)...
                    +.5*X2-.25*sin(2*X2))/(W2^5);
Q2(1,2)=PHIS2*(.5*X2*X2-X2*sin(X2)+.5*sin(X2)...
                    *sin(X2))/(W2^4);
Q2(2,1)=Q2(1,2);
Q2(1,3)=PHIS2*(sin(X2)-X2*cos(X2)-.5*X2+.25*...
                    sin(2*X2))/(W2^3);
Q2(3,1)=Q2(1,3);
Q2(1,4)=PHIS2*(cos(X2)+X2*sin(X2)-.5*sin(X2)*...
                    sin(X2)-1)/(W2*W2);
Q2(4,1)=Q2(1,4);
Q2(2,2)=PHIS2*(1.5*X2-2*sin(X2)+.25*sin(2*X2))/(W2^3);
Q2(2,3)=PHIS2*(1-cos(X2)-.5*sin(X2)*sin(X2))/(W2*W2);
Q2(3,2)=Q2(2,3);
Q2(2,4)=PHIS2*(sin(X2)-.5*X2-.25*sin(2*X2))/W2;
Q2(4,2)=Q2(2,4);
Q2(3,3)=PHIS2*(.5*X2-.25*sin(2*X2))/W2;
Q2(3,4)=.5*PHIS2*sin(X2)*sin(X2);
Q2(4,3)=Q2(3,4);
Q2(4,4)=W2*PHIS2*(.5*X2+.25*sin(2*X2));


Q3(1,1)=PHIS3*(.333*X3^3-2*sin(X3)+2*X3*cos(X3)+...
                    .5*X3-.25*sin(2*X3))/(W3^5);
Q3(1,2)=PHIS3*(.5*X3*X3-X3*sin(X3)+.5*sin(X3)*...
                    sin(X3))/(W3^4);
Q3(2,1)=Q3(1,2);
Q3(1,3)=PHIS3*(sin(X3)-X3*cos(X3)-.5*X3+.25*...
                    sin(2*X3))/(W3^3);
Q3(3,1)=Q3(1,3);
Q3(1,4)=PHIS3*(cos(X3)+X3*sin(X3)-.5*sin(X3)*...
                    sin(X3)-1)/(W3*W3);
Q3(4,1)=Q3(1,4);
Q3(2,2)=PHIS3*(1.5*X3-2*sin(X3)+.25*sin(2*X3))/(W3^3);
Q3(2,3)=PHIS3*(1-cos(X3)-.5*sin(X3)*sin(X3))/(W3*W3);
Q3(3,2)=Q3(2,3);
Q3(2,4)=PHIS3*(sin(X3)-.5*X3-.25*sin(2*X3))/W3;
Q3(4,2)=Q3(2,4);
Q3(3,3)=PHIS3*(.5*X3-.25*sin(2*X3))/W3;
Q3(3,4)=.5*PHIS3*sin(X3)*sin(X3);
Q3(4,3)=Q3(3,4);
Q3(4,4)=W3*PHIS3*(.5*X3+.25*sin(2*X3));


P1(1,1)=SIGN2;
P1(2,2)=(VM*HEDEGFIL/57.3)^2;
P1(3,3)=XNT*XNT;
```

```
P1(4,4)=W1*W1*XNT*XNT;

P2(1,1)=SIGN2;
P2(2,2)=(VM*HEDEGFIL/57.3)^2;
P2(3,3)=XNT*XNT;
P2(4,4)=W2*W2*XNT*XNT;

P3(1,1)=SIGN2;
P3(2,2)=(VM*HEDEGFIL/57.3)^2;
P3(3,3)=XNT*XNT;
P3(4,4)=W3*W3*XNT*XNT;

HMAT=[1 0 0 0];
HT=HMAT';
PHIT1=PHI1';
PHIT2=PHI2';
PHIT3=PHI3';
T=0.;
H=.001;
S=0.;
XNC=0.;
XNL=0.;
XLAM=Y/RTM;
YTDD=XNTREAL*sin(WREAL*T);
YTDDD=XNTREAL*WREAL*cos(WREAL*T);

YH1=0.;
YDH1=0.;
YTDDH1=0.;
YTDDDH1=0.;

YH2=0.;
YDH2=0.;
YTDDH2=0.;
YTDDDH2=0.;

YH3=0.;
YDH3=0.;
YTDDH3=0.;
YTDDDH3=0.;

PROB1=.333;
PROB2=.333;
PROB3=.333;
while T<=(TF-.0001)
```

```
S=S+H;
YOLD=Y;
YDOLD=YD;
XNLOLD=XNL;
STEP=1;
FLAG=0;
while STEP <=1
        if FLAG==1
                STEP=2;
                Y=Y+H*YD;
                YD=YD+H*YDD;
                XNL=XNL+H*XNLD;
                T=T+H;
        end
        TGO=TF-T+.000001;
        RTM=VC*TGO;
        XLAM=Y/(VC*TGO);
        YTDD=XNTREAL*sin(WREAL*T);
        XNLD=(XNC-XNL)/TAU;
        YDD=YTDD-XNL;
        FLAG=1;
end
FLAG=0;
Y=.5*(YOLD+Y+H*YD);
YD=.5*(YDOLD+YD+H*YDD);
XNL=.5*(XNLOLD+XNL+H*XNLD);
if S>=(TS-.00001)
        S=0.;
        TGO=TF-T+.000001;
        RTM=VC*TGO;
        SIGPOS=RTM*SIGNOISE;
        SIGN2=SIGPOS^2;
        PHIP1=PHI1*P1;
        PHIPPHIT1=PHIP1*PHIT1;
        M1=PHIPPHIT1+Q1;
        HM1=HMAT*M1;
        HMHT1=HM1*HT;
        HMHTR1=HMHT1(1,1)+SIGN2;
        HMHTRINV1(1,1)=1./HMHTR1;
        MHT1=M1*HT;
        GAIN1=MHT1*HMHTRINV1;
        KH1=GAIN1*HMAT;
        IKH1=IDNP-KH1;
        P1=IKH1*M1;

        PHIP2=PHI2*P2;
```

```
PHIPPHIT2=PHIP2*PHIT2;
M2=PHIPPHIT2+Q2;
HM2=HMAT*M2;
HMHT2=HM2*HT;
HMHTR2=HMHT2(1,1)+SIGN2;
HMHTRINV2(1,1)=1./HMHTR2;
MHT2=M2*HT;
GAIN2=MHT2*HMHTRINV2;
KH2=GAIN2*HMAT;
IKH2=IDNP-KH2;
P2=IKH2*M2;

PHIP3=PHI3*P3;
PHIPPHIT3=PHIP3*PHIT3;
M3=PHIPPHIT3+Q3;
HM3=HMAT*M3;
HMHT3=HM3*HT;
HMHTR3=HMHT3(1,1)+SIGN2;
HMHTRINV3(1,1)=1./HMHTR3;
MHT3=M3*HT;
GAIN3=MHT3*HMHTRINV3;
KH3=GAIN3*HMAT;
IKH3=IDNP-KH3;
P3=IKH3*M3;

CPZ1=HMHTR1;
CPZ2=HMHTR2;
CPZ3=HMHTR3;

YTDD=XNTREAL*sin(WREAL*T);
YTDDD=XNTREAL*WREAL*cos(WREAL*T);
XLAMNOISE=SIGNOISE*randn;
YSTAR=RTM*(XLAM+XLAMNOISE);

RES1=YSTAR-YH1-TS*YDH1-(1-cos(X1))*YTDDH1/...
        (W1*W1)-(X1-sin(X1))*YTDDDH1/(W1*W1*W1)...
        +.5*TS*TS*XNL;
YH1=YH1+TS*YDH1+(1-cos(X1))*YTDDH1/(W1*W1)+...
        (X1-sin(X1))*YTDDDH1/(W1*W1*W1)+GAIN1(1,1)*RES1-...
        .5*TS*TS*XNL;
YDH1=YDH1+sin(X1)*YTDDH1/W1+(1-cos(X1))*YTDDDH1/(W1*W1)...
        +GAIN1(2,1)*RES1-TS*XNL;
    YTDDHNEW1=cos(X1)*YTDDH1+sin(X1)*YTDDDH1/W1+...
        GAIN1(3,1)*RES1;
YTDDDH1=-W1*sin(X1)*YTDDH1+cos(X1)*YTDDDH1+GAIN1(4,1)*RES1;
    YTDDH1=YTDDHNEW1;
```

RES2=YSTAR-YH2-TS*YDH2-(1-cos(X2))*YTDDH2/(W2*W2)-...
    (X2-sin(X2))*YTDDDH2/(W2*W2*W2)+.5*TS*TS*XNL;
YH2=YH2+TS*YDH2+(1-cos(X2))*YTDDH2/(W2*W2)+...
    (X2-sin(X2))*YTDDDH2/(W2*W2*W2)+...
    GAIN2(1,1)*RES2-.5*TS*TS*XNL;
YDH2=YDH2+sin(X2)*YTDDH2/W2+(1-cos(X2))*...
    YTDDDH2/(W2*W2)+GAIN2(2,1)*RES2-TS*XNL;
YTDDHNEW2=cos(X2)*YTDDH2+sin(X2)*YTDDDH2/W2+...
    GAIN2(3,1)*RES2;
YTDDDH2=-W2*sin(X2)*YTDDH2+cos(X2)*YTDDDH2...
    +GAIN2(4,1)*RES2;
YTDDH2=YTDDHNEW2;

RES3=YSTAR-YH3-TS*YDH3-(1-cos(X3))*YTDDH3/...
    (W3*W3)-(X3-sin(X3))*YTDDDH3/(W3*W3*W3)...
    +.5*TS*TS*XNL;
YH3=YH3+TS*YDH3+(1-cos(X3))*YTDDH3/(W3*W3)+...
    (X3-sin(X3))*YTDDDH3/(W3*W3*W3)+...
    GAIN3(1,1)*RES3-.5*TS*TS*XNL;
YDH3=YDH3+sin(X3)*YTDDH3/W3+(1-cos(X3))*...
    $YTDDDH3/(W3*W3)+GAIN3(2,1)*RES3-TS*XNL;$
YTDDHNEW3=cos(X3)*YTDDH3+sin(X3)*YTDDDH3/W3+...
    GAIN3(3,1)*RES3;
YTDDDH3=-W3*sin(X3)*YTDDH3+cos(X3)*YTDDDH3...
    +GAIN3(4,1)*RES3;
YTDDH3=YTDDHNEW3;

F1=exp(-.5*RES1*RES1/CPZ1)/sqrt(6.28*CPZ1);
F2=exp(-.5*RES2*RES2/CPZ2)/sqrt(6.28*CPZ2);
F3=exp(-.5*RES3*RES3/CPZ3)/sqrt(6.28*CPZ3);

PROB1=PROB1*F1/(PROB1*F1+PROB2*F2+PROB3*F3);
PROB2=PROB2*F2/(PROB1*F1+PROB2*F2+PROB3*F3);
PROB3=PROB3*F3/(PROB1*F1+PROB2*F2+PROB3*F3);

WHPZ=W1*PROB1+W2*PROB2+W3*PROB3;
YHPZ=YH1*PROB1+YH2*PROB2+YH3*PROB3;
YDHPZ=YDH1*PROB1+YDH2*PROB2+YDH3*PROB3;
YTDDHPZ=YTDDH1*PROB1+YTDDH2*PROB2+YTDDH3*PROB3;
YTDDDHPZ=YTDDDH1*PROB1+YTDDDH2*PROB2+...
    YTDDDH3*PROB3;

XS=TGO/TAU;
TOP=6.*XS*XS*(exp(-XS)-1.+XS);
BOT1=2*XS*XS*XS+3.+6.*XS-6.*XS*XS;
BOT2=-12.*XS*exp(-XS)-3.*exp(-2.*XS);

```
XNPP=TOP/(.0001+BOT1+BOT2);
C1=XNPP/(TGO*TGO);
C2=XNPP/TGO;
C3=XNPP*(1.-cos(WHPZ*TGO))/(WHPZ*WHPZ*TGO*TGO);
C4=-XNPP*(exp(-XS)+XS-1.)/(XS*XS);
C5=XNPP*(WHPZ*TGO-sin(WHPZ*TGO))/(WHPZ*...
          WHPZ*WHPZ*TGO*TGO);

XNC=C1*YHPZ+C2*YDHPZ+C3*YTDDHPZ+C4*XNL+...
      C5*YTDDDHPZ;
if XNC>XLIM
      XNC=XLIM;
end
if XNC<-XLIM
      XNC=-XLIM;
end
YTDDG=YTDD/32.2;
YTDDHPZG=YTDDHPZ/32.2;
count=count+1;
ArrayT(count)=T;
ArrayPROB1(count)=PROB1;
ArrayPROB2(count)=PROB2;
ArrayPROB3(count)=PROB3;
ArrayWREAL(count)=WREAL;
ArrayWHPZ(count)=WHPZ;
ArrayYTDDG(count)=YTDDG;
ArrayYTDDHPZG(count)=YTDDHPZG;
      end
end
output=[ArrayT',ArrayPROB1',ArrayPROB2',ArrayPROB3',...
      ArrayWREAL',ArrayWHPZ',ArrayYTDDG',...
      ArrayYTDDHPZG'];
save datfil.txt output /ascii
disp 'simulation finished'
clc
figure
plot(ArrayT,ArrayPROB1,ArrayT,ArrayPROB2,ArrayT,...
      ArrayPROB3),grid
xlabel('Time (s) ')
ylabel('Probability')
axis([0 10 0 1.2])
figure
plot(ArrayT,ArrayWREAL,ArrayT,ArrayWHPZ),grid
xlabel('Time (s) ')
ylabel('Frequency (r/s)')
axis([0 10 0 4])
```

```
figure
plot(ArrayT,ArrayYTDDG,ArrayT,ArrayYTDDHPZG),grid
xlabel('Time (s) ')
ylabel('Acceleration (g)')
axis([0 10 -6 6])
```

The nominal case of Listing 8.1 was run, and we can see from Fig. 8.7 that after 3 s the algorithm figures out that the correct filter is most likely the filter tuned to 2 rad/s. After 3.5 s the algorithm is almost certain that the 2 rad/s filter is correct. Figure 8.8 shows that although it takes about 3.5 s to be absolutely certain which filter is correct, the interim estimates of the target weave frequency are quite close to the actual target weave frequency for the entire flight. Figure 8.9 shows that combining the state estimates of the three filters in a weighted manner yields excellent estimates of the target acceleration after only 2 s.

Listing 8.1 was modified to operate in the Monte Carlo mode. In addition, the flight time was varied from 0.5 to 10 s in steps of 0.5 s. For each flight time, 50 runs were made to calculate the RMS miss distance. We can see from Fig. 8.10 that the three-filter bank, which uses filter frequencies of 1, 2, and 4 rad/s, yields RMS miss distances that are usually less than 5 ft. Superimposed on Fig. 8.10 are the RMS miss distance results for the mismatched four-state linear weave Kalman filters. The RMS miss distances for the mismatched filters are usually several times larger than the filter bank results. Clearly the fixed multiple model or MMAE approach can yield significantly improved performance.

Miss-distance results can be further improved if there is less uncertainty in the target weave frequency (that is, more filters in the filter bank). Figure 8.11 shows that if the filter bank frequencies are 1, 2, and 3 rad/s, respectively, the

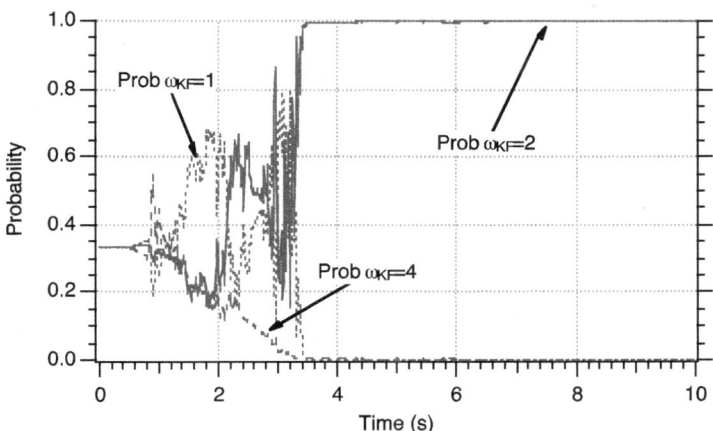

Fig. 8.7   Fixed MMAE approach identifies correct filter after 3 s.

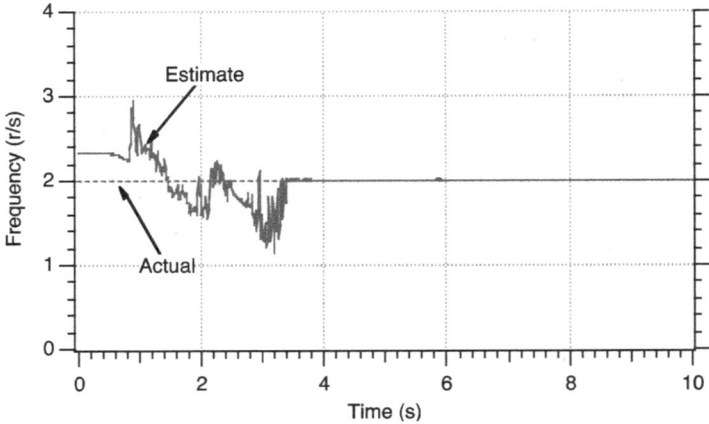

Fig. 8.8   Estimated target weave frequency is always quite close to actual target weave frequency.

RMS miss-distance performance improves slightly because at most we have a 1-rad/s error in knowledge of the target weave frequency rather than 2 rad/s as was the case in Fig. 8.10.

Figure 8.12 compares the optimal performance of the four-state linear weave Kalman filter (that is, when the filter knows the actual target weave frequency is 2 rad/s) and the fixed multiple model results. Although there is room for improvement, we can see that the three-filter bank results are approaching the

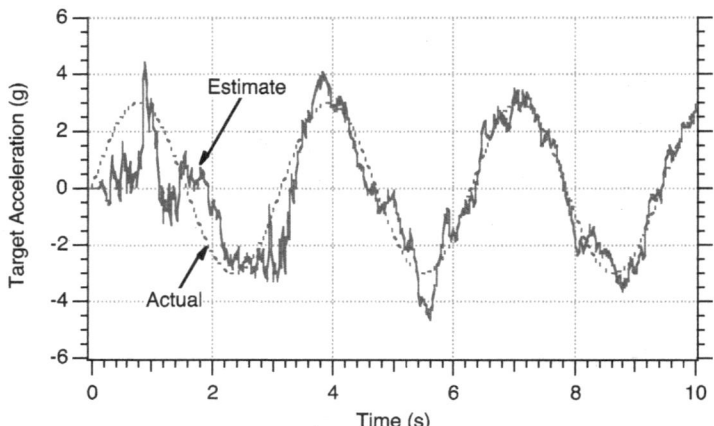

Fig. 8.9   Target acceleration estimate excellent after 3 s.

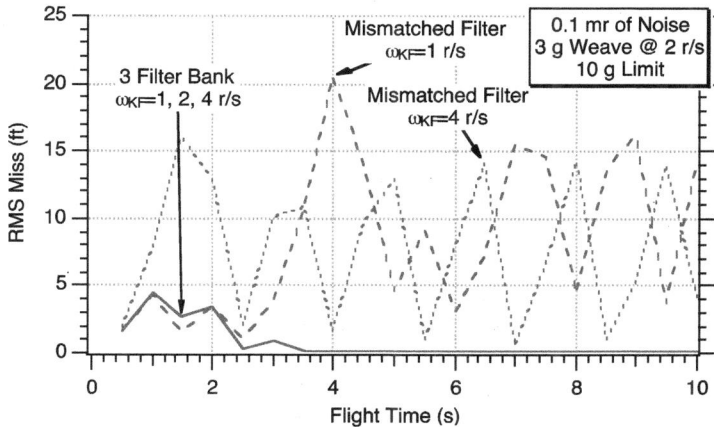

Fig. 8.10   Fixed MMAE approach yields improved performance when target weave frequency is unknown.

near-optimal results when at most there is a 1-rad/s error in knowledge of the target weave frequency.

In the preceding cases it was assumed that the missile acceleration limit was 10 g. Another case was run in which the missile acceleration limit was removed. The performance of the single tuned Kalman filter should improve for flight times of less than 2 s. Figure 8.13 presents the comparison between the optimal

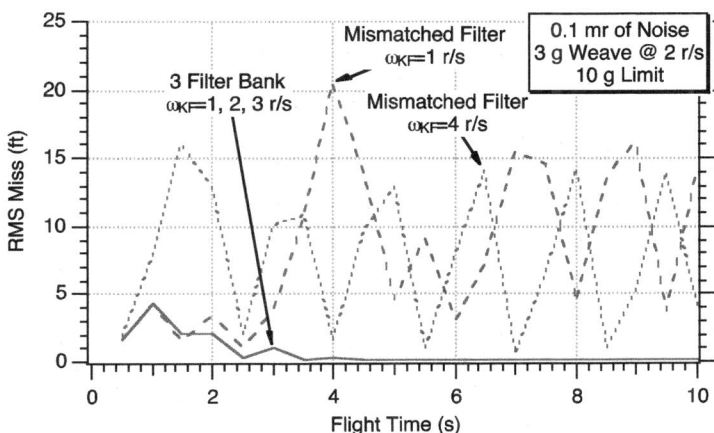

Fig. 8.11   Fixed MMAE performance is improved slightly when there is less uncertainty in actual target weave frequency.

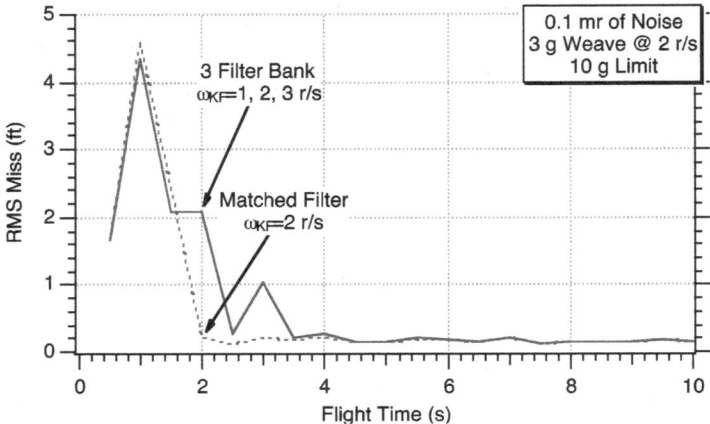

Fig. 8.12   Three-filter bank fixed MMAE approach results yield near-optimal performance.

performance of the single tuned Kalman filter and the filter bank approach. We can see that RMS miss-distance results of the filter bank get even closer to the optimal single tuned filter results when the missile acceleration limit is removed. This means that the filter bank approach is yielding near-optimal performance.

So far we have assumed that one of the filters in the filter bank is tuned to the actual target weave frequency. Actually, this is a requirement for the fixed

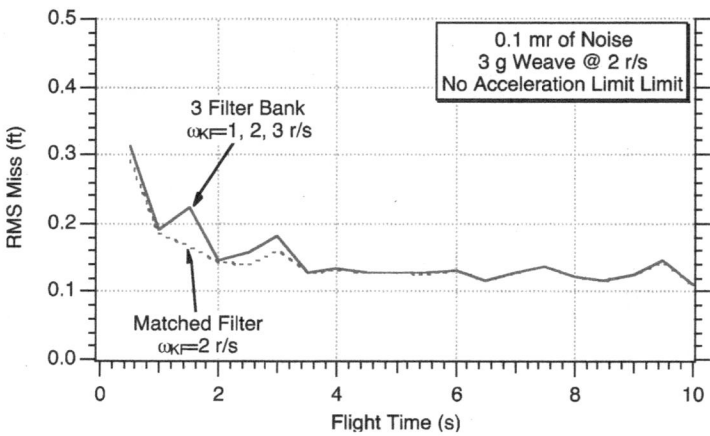

Fig. 8.13   Comparison of optimal weave Kalman filter and three-filter fixed MMAE approach when acceleration limit is removed.

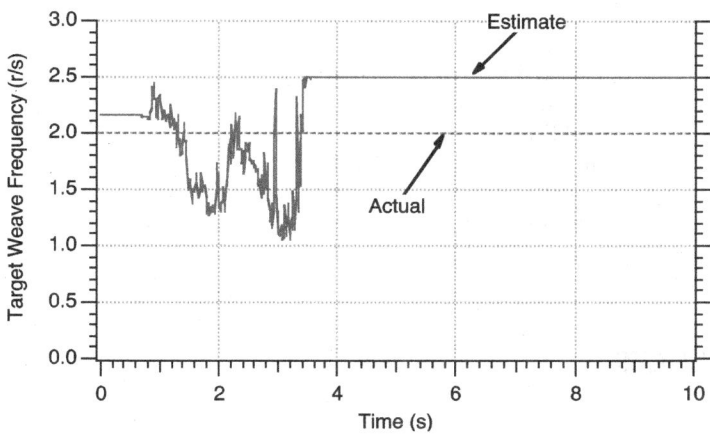

Fig. 8.14   Target weave frequency of filter closest to truth is selected.

MMAE approach to work best. Let us see what happens when all three filters are tuned to the wrong frequency. A case was run where the actual target weave frequency was 2 rad/s, but the three filters in the filter bank were tuned to 1, 2.5, and 3 rad/s, respectively. We can see from Fig. 8.14 that the filter estimate closest to the truth is selected with the fixed MMAE approach. Figure 8.15 indicates that although we do not know the truth, the estimate of target acceleration is excellent after 2 s.

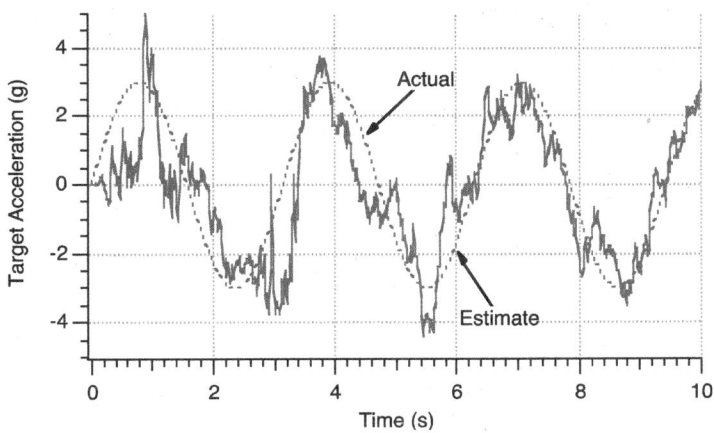

Fig. 8.15   Estimates of target acceleration are excellent even though none of the filters in filter bank are matched to truth.

SUMMARY

It has been demonstrated in this chapter that a bank of linear four-state weave Kalman filters can be used for purposes of estimating the target weave frequency using the fixed MMAE technique. This filter bank approach makes use of each filter's likelihood function and Bayes' rule. It is shown that combining filter outputs in a probabilistic sense yields excellent estimates of the target weave frequency and yields small RMS miss if the seeker measurement noise can be kept to the 0.1-mrad level when the closing velocity is 9000 ft/s.

REFERENCES

[1] Zarchan, P., *Fundamentals of Kalman Filtering: A Practical Approach*, 2nd ed., Progress in Astronautics and Aeronautics, AIAA, Reston, VA, 2005, pp. 606–608, 647–675.

[2] Magill, D. T., "Optimal Adaptive Estimation of Sampled Stochastic Processes," *IEEE Transactions on Automatic Control*, Vol. 10, No. 4, 1965, pp. 434–439.

[3] Bar-Shalom, Y., Li, X. R., and Kirubarajan, T., *Estimation with Applications to Tracking and Navigation*, Wiley, New York, 2001, pp. 440–465.

[4] Shima, T., Oshman, Y., and Shinar, J., "Efficient Multiple Model Adaptive Estimation in Ballistic Missile Interception Scenarios," *Journal of Guidance, Control, and Dynamics*, Vol. 25, No. 4, 2002, pp. 667–675.

[5] Stengel, R., *Optimal Control and Estimation*, Dover, New York, 1994, pp. 402–407.

[6] Welch, G., and Bishop, G., "An Introduction to the Kalman Filter," SIGRAPH 2001, Aug. 2001, pp. 36–39.

[7] Marks, M., "Multiple Models Adaptive Estimation (MMAE) for Improving Guidance Performance Against Weaving Targets," AIAA Paper 2006-6697, Aug. 2006.

[8] Zarchan, P., and Alpert, J., "Using Filter Banks to Improve Performance Against Weaving Targets," AIAA Paper 2006-6700, Aug. 2006.

CHAPTER 9

Predictor – Corrector Guidance

INTRODUCTION

So far, most of the guidance methods we have studied in this text and the previous volume are variants of proportional navigation in which the guidance commands are proportional to the zero effort miss and inversely proportional to the square of time to go. We have seen that guidance laws of this type are very effective and fairly robust. The space community often uses predictor – corrector guidance to achieve intercepts or near-intercepts for their applications. In this chapter we will first see how the predictor – corrector method can be applied to a problem in which other, more conventional guidance laws are also applicable. We will compare all of the guidance laws in terms of performance and robustness. Next, we will select a problem in which the direct application of a proportional naviga-tion – type guidance law is not possible. In this application we will see how predic-tor – corrector guidance can be applied and how it performs. We will also see how robust the predictor – corrector guidance law is to errors in knowledge of acceleration.

SURFACE-TO-SURFACE MISSILE PROBLEM

Let us return to the problem of an impulsively launched surface-to-surface missile that is acted on by only drag and gravity, as was originally modeled in Listing 10.1 of the previous volume. In that simulation, the missile was not guided and we simply studied where the missile would impact the Earth for different launch or flight path angles. In this application we will try to intercept a stationary target located on the surface of the Earth. The only accelerations acting on the missile are drag and gravity; these are displayed in Fig. 9.1.

In the model of Fig. 9.1, the "1" axis is downrange, and the "2" axis is altitude. The downrange and altitude components of the accelerations acting on the missile can be obtained by inspection of Fig. 9.1 and are given by

$$a_{M1} = -\text{Drag} \cos \gamma$$
$$a_{M2} = -\text{Drag} \sin \gamma - g$$

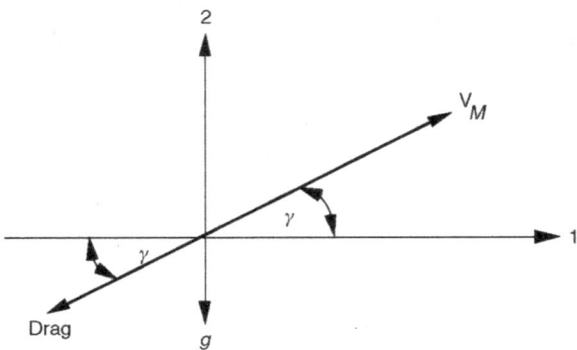

Fig. 9.1 Accelerations acting on impulsively launched missile.

where the drag acceleration can be expressed in terms of gravity g, the dynamic pressure Q, and the ballistic coefficient β as

$$\text{Drag} = \frac{gQ}{\beta}$$

The dynamic pressure can be expressed in terms of the air density ρ and missile velocity V_M as

$$Q = 0.5\rho V_M^2$$

Recall that in Chapter 10 of the previous volume we used an exponential approximation for the air density. Below 30,000 ft we used

$$\rho = 0.002378\ e^{-R_{M2}/30000}\ (R_{M2} < 30,000\,\text{ft})$$

and above 30,000 ft we used

$$\rho = 0.0034\ e^{-R_{M2}/22000}\ (R_{M2} \geq 30,000\,\text{ft})$$

where the altitude R_{M2} was measured in units of ft and the air density had units of slug/ft^3. Because the altitude and downrange missile acceleration components are in an inertial coordinate system, they can be integrated twice to yield the location of the missile. For a proportional navigation guided missile, the acceleration command obtained from the guidance law is assumed to be perpendicular to the line of sight, as shown in Fig. 9.2.

With proportional navigation guidance, the guidance command is simply

$$n_c = N'V_c\dot{\lambda}$$

The component of missile drag perpendicular to the line of sight will appear to the missile as a target maneuver. For a missile with small acceleration capability, excessive acceleration requirements may result from the apparent target

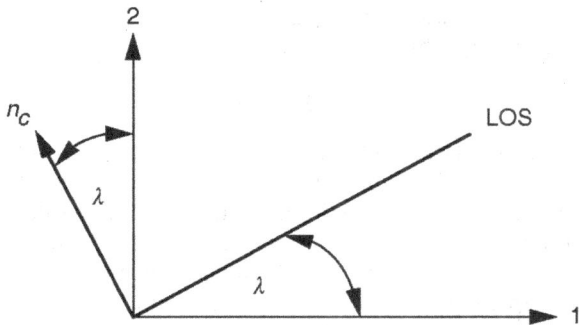

Fig. 9.2 Guidance command is perpendicular to line of sight.

maneuver because of the missile's own drag. We can compensate for missile drag in the guidance law by using a form of augmented proportional navigation, or

$$n_c = N'V_c\dot{\lambda} + 0.5N'(a_{\text{TPLOS}} - \text{Drag}_{\text{PLOS}})$$

where a_{TPLOS} is the target acceleration perpendicular to the line of sight and $\text{Drag}_{\text{PLOS}}$ is the component of the missile drag acceleration that is perpendicular to the line of sight. In this application the target is stationary, so a_{TPLOS} is zero. To calculate the component of drag and gravity perpendicular to the line of sight, we first observe from Fig. 9.1 that the components of drag and gravity in the down-range and altitude directions are given by

$$\text{Drag}_1 = -\text{Drag}\cos\gamma$$
$$\text{Drag}_2 = -\text{Drag}\sin\gamma - g$$

From Fig. 9.3 we can see how the acceleration of drag perpendicular to the line of sight is related to the downrange–altitude coordinate system.

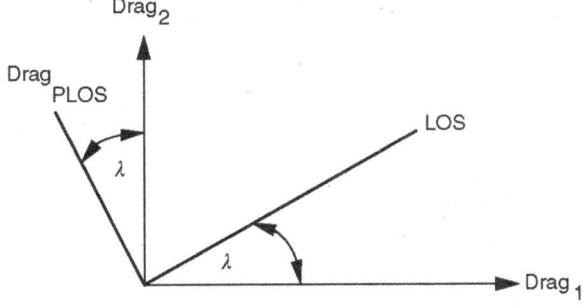

Fig. 9.3 Calculating the component of drag perpendicular to the line of sight.

Therefore, from Fig. 9.3 we can see that the component of drag perpendicular to the line of sight is given by

$$\text{Drag}_{\text{PLOS}} = -\text{Drag}_1 \sin \lambda + \text{Drag}_2 \cos \lambda$$

An engagement simulation of the impulsively launched missile and stationary target appears in Listing 9.1. Initially there is no guidance, and the missile is launched at 3000 ft/s with an initial flight-path angle of 45 deg. For the first 30 s of flight, only drag and gravity act on the missile; however, after 30 s of flight, guidance is turned on and the missile can use either proportional navigation (APN=0) or augmented proportional navigation (APN=1) to guide toward the stationary target located at downrange and altitude coordinates of RT1 and RT2. Because the target is located on the surface of the Earth, RT2 is zero. The simulation has a 5-g acceleration limit on the acceleration commands (XLIMG=5). Notice that the simulation stops when the missile hits the ground rather than when the closing velocity goes negative (missile and target are at point of closest approach). This minor adjustment to our normal stopping condition (that is, normally we stop engagement simulations when the closing velocity goes negative because this is the point of closest approach) will prevent the missile from traveling through the Earth to reach the target.

Cases were run with the engagement simulation of Listing 9.1 in which the target downrange location RT1 was made a parameter. We can see from Fig. 9.4 that when proportional navigation guidance is used, the missile can hit the stationary target when it is located from 40,000 to 60,000 ft downrange of the missile launch point; however, if the target is 70,000 ft downrange of the missile launch point, the missile will hit the ground too soon and will miss the

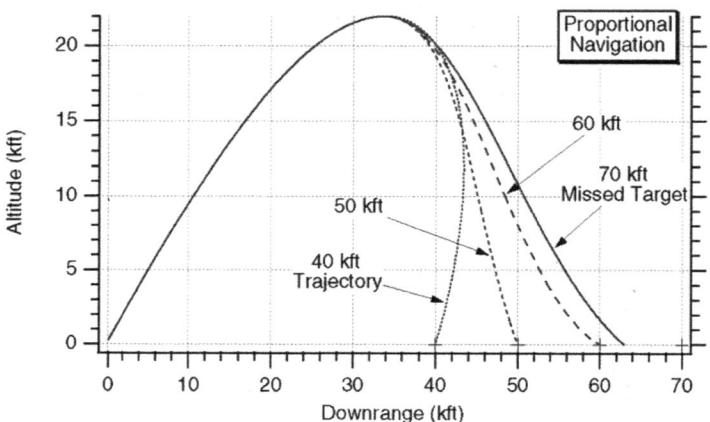

Fig. 9.4   With proportional navigation we can hit stationary targets out to 60,000 ft downrange.

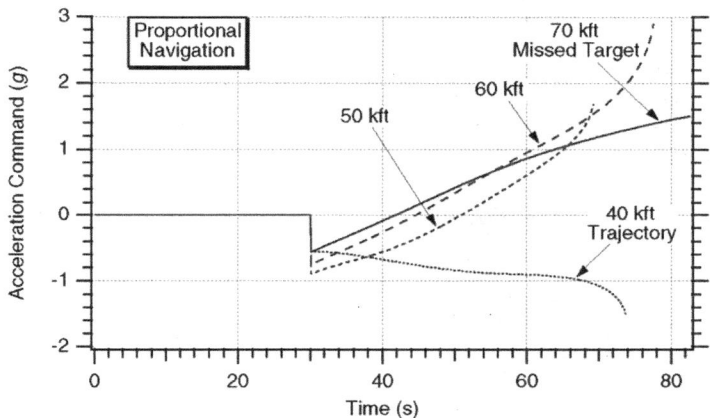

Fig. 9.5 We missed at 70,000 ft because we hit the ground first and not because missile acceleration saturated.

target when proportional navigation guidance is used. Figure 9.5 shows that the lack of success in the 70,000-ft case was not because of acceleration saturation because the acceleration level never reached 5 g. In this case, proportional navigation does not realize that the missile cannot travel through the ground to reach the target. Cases were run in which the drag was reduced, and then the missile was able to hit the target at the 70,000-ft range and beyond. Clearly the drag is causing problems for proportional navigation.

LISTING 9.1 MISSILE–TARGET ENGAGEMENT SIMULATION WITH CONVENTIONAL GUIDANCE

```
clear
n=0;
TS=.1;
APN=0;
XLIMG=5.;
VM=3000.;
BETA=1000.;
T=0.;
S=0.;
GAMDEG=45.;
VM1=VM*cos(GAMDEG/57.3);
VM2=VM*sin(GAMDEG/57.3);
RM1=0.;
RM2=0.;
XNC=0.;
```

```
RT1=60000.;
RT2=0.;
RTM1=RT1-RM1;
RTM2=RT2-RM2;
RTM=sqrt(RTM1^2+RTM2^2);
VTM1=0.-VM1;
VTM2=0.-VM2;
VC=-(RTM1*VTM1+RTM2*VTM2)/RTM;
XLIM=XLIMG*32.2;
XNC=0.;
while ~(T>0.&RM2<=0.)
      if RTM<1000.
        H=.0001;
      else
        H=.01;
      end
      RM1OLD=RM1;
      RM2OLD=RM2;
      VM1OLD=VM1;
      VM2OLD=VM2;
      STEP=1;
      FLAG=0;
      while STEP <=1
        if FLAG==1
        STEP=2;
                RM1=RM1+H*VM1;
                RM2=RM2+H*VM2;
                VM1=VM1+H*AM1;
                VM2=VM2+H*AM2;
                T=T+H;
        end
        RTM1=RT1-RM1;
        RTM2=RT2-RM2;
        RTM=sqrt(RTM1^2+RTM2^2);
        VTM1=0.-VM1;
        VTM2=0.-VM2;
        VC=-(RTM1*VTM1+RTM2*VTM2)/RTM;
        XLAM=atan2(RTM2,RTM1);
        XLAMD=(RTM1*VTM2-RTM2*VTM1)/(RTM*RTM);
        if RM2<30000.
                RHO=.002378*exp(-RM2/30000);
        else
                RHO=.0034*exp(-RM2/22000);
        end
        VM=sqrt(VM1^2+VM2^2);
        Q=.5*RHO*VM*VM;
```

```
  GAM=atan2(VM2,VM1);
  DRAG=Q*32.2/BETA;
  XNE1=-XNC*sin(XLAM);
  XNE2=XNC*cos(XLAM);
  AM1=-DRAG*cos(GAM)+XNE1;
  AM2=-32.2-DRAG*sin(GAM)+XNE2;
  FLAG=1;
end
FLAG=0;
RM1=.5*(RM1OLD+RM1+H*VM1);
RM2=.5*(RM2OLD+RM2+H*VM2);
VM1=.5*(VM1OLD+VM1+H*AM1);
VM2=.5*(VM2OLD+VM2+H*AM2);
S=S+H;
if S>=(TS-.00001)
  S=0.;
  RM1K=RM1/1000.;
  RM2K=RM2/1000.;
  RT1K=RT1/1000.;
  RT2K=RT2/1000.;
  DRAG1=-DRAG*cos(GAM);
  DRAG2=-DRAG*sin(GAM)-32.2;
  DRAGPLOS=-DRAG1*sin(XLAM)+DRAG2*cos(XLAM);
  ATPLOS=0.;
  if T>30.
          if APN==0
                  XNC=3.*VC*XLAMD;
          else
                  XNC=3.*VC*XLAMD+1.5*(ATPLOS-DRAGPLOS);
          end
  else
          XNC=0.;
  end
  if XNC>XLIM
          XNC=XLIM;
  end
  if XNC<-XLIM
          XNC=-XLIM;
  end
  XNCG=XNC/32.2;
  DRAGPLOSG=DRAGPLOS/32.2;
  n=n+1;
  ArrayT(n)=T;
  ArrayRT1K(n)=RT1K;
  ArrayRT2K(n)=RT2K;
  ArrayRM1K(n)=RM1K;
```

```
            ArrayRM2K(n)=RM2K;
            ArrayXNCG(n)=XNC/32.2;
        end
end
RTM
figure
plot(ArrayRT1K,ArrayRT2K,ArrayRM1K,ArrayRM2K),grid
xlabel('Downrange (Ft) ')
ylabel('Altitude (Ft)')
figure
plot(ArrayT,ArrayXNCG),grid
xlabel('Time (sec)')
ylabel('Acceleration of missile (G)')
clc
output=[ArrayT',ArrayRT1K',ArrayRT2K',ArrayRM1K',ArrayRM2K',ArrayXNCG'];
save datfil output -ascii
disp '*** Simulation Complete'
```

As was mentioned previously, the drag component that is perpendicular to the line of sight appears as a target maneuver to the missile and causes potential miss distance problems. Because we should know our own drag, it seems reasonable that we can use augmented proportional navigation (APN=1), as was previously described in this section. When augmented proportional navigation was used, we can see from Fig. 9.6 that now we can hit targets at 70,000 ft and beyond. Thus, we can see that in this application more advanced guidance laws can expand the zone of effectiveness of a missile against stationary targets. Figure 9.7 shows that the missile acceleration profiles for all trajectories

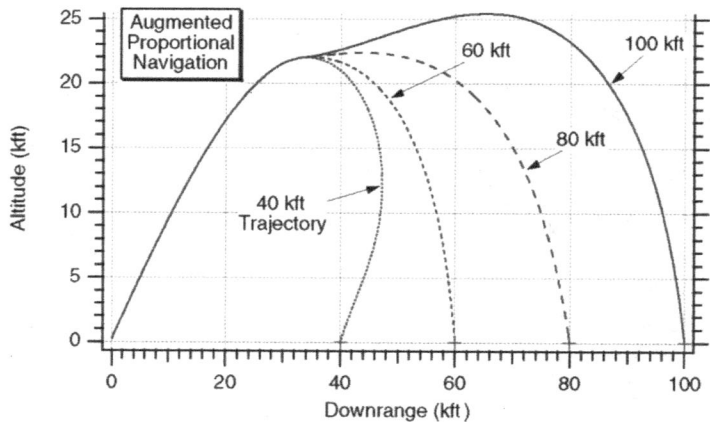

Fig. 9.6 Augmented proportional navigation extends zone of effectiveness of missile.

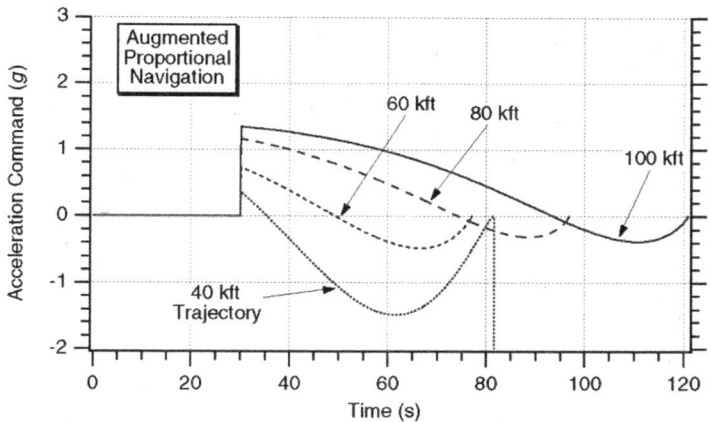

Fig. 9.7 All acceleration profiles are well behaved and do not saturate with augmented proportional navigation.

considered are very reasonable. The acceleration requirements are always well below the simulation limit of 5 g.

PREDICTOR-CORRECTOR GUIDANCE WITH SURFACE-TO-SURFACE INTERCEPT PROBLEM

Predictor-corrector guidance is based on having a model of the interceptor and target and trying to determine at each guidance update how certain changes in important parameters influence the missile trajectory. With predictor-corrector guidance we must first assume a form of the acceleration command in this application. For simplicity, the form of the acceleration command we will assume is a constant. Using the example of the previous section in conjunction with predictor-corrector guidance, we first integrate the missile equations of motion forward assuming no acceleration commands are issued (that is, our nominal constant is zero) to see where the missile impacts the ground. We then measure the difference between where the target really is, RT_1, and where the missile actually hits the ground, x_1, and call the resultant quantity ?x, or

$$\Delta x = R_{T1} - x_1$$

We now add a little bit to the previous constant acceleration command n_c (which is zero the first time we initiate this procedure) and call the new constant acceleration command n'_c. We again integrate the missile equations of motion forward, assuming the new constant acceleration command, and calling the place where the missile impacts the ground x_2. Thus, we can manually construct a partial derivative for the change in impact point with respect to the change in acceleration command by differencing the two impact points and dividing by the difference in the two constant acceleration commands, or

$$\frac{\partial x}{\partial n_c} = \frac{x_2 - x_1}{n'_c - n_c}$$

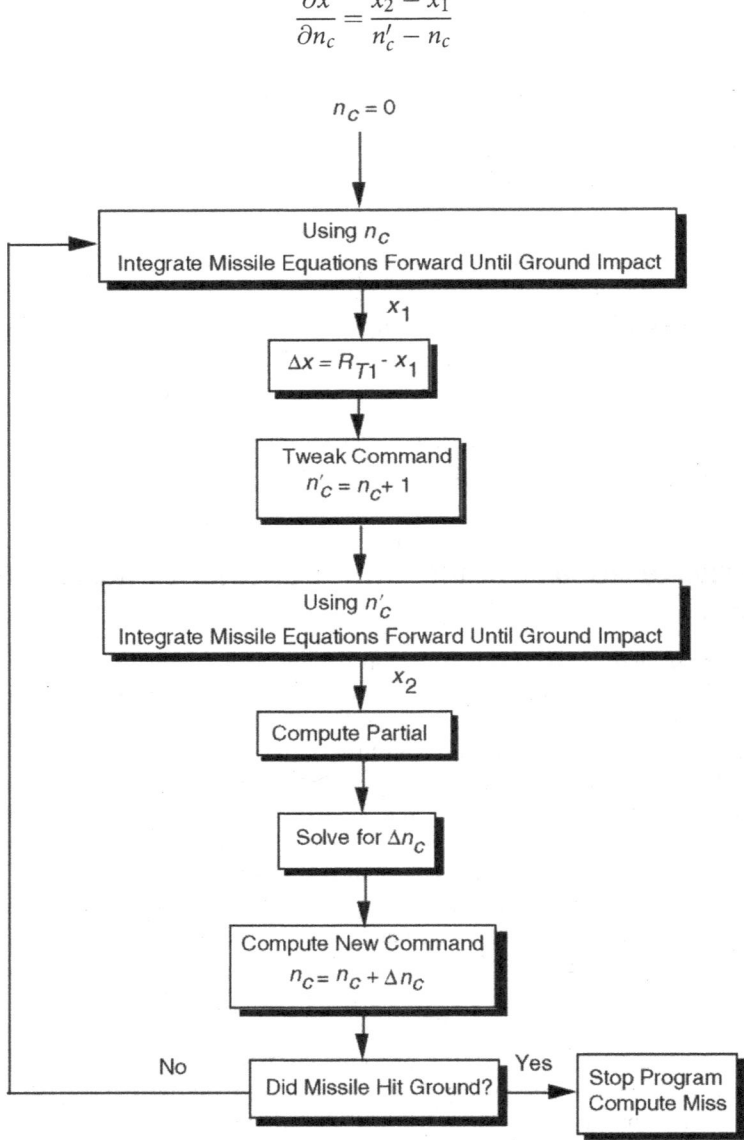

Fig. 9.8   Predictor–corrector guidance logic for first problem.

Using the notation of the preceding two equations, we can compute, using calculus, the total differential as

$$\Delta x = \frac{\partial x}{\partial n_c} \Delta n_c$$

Because we have already calculated Δx and the partial derivative, we can invert the preceding equation and solve for the change in missile acceleration as

$$\Delta n_c = \frac{\Delta x}{\dfrac{\partial x}{\partial n_c}}$$

The new acceleration command, assuming the acceleration command is a constant, in the predictor-corrector scheme is simply the previous command plus the suggested change from the previous equation, or

$$n_{c_k} = n_{c_{k-1}} + \Delta n_c$$

We now repeat the cycle at each guidance update (solving for Δx and partial derivative by integrating equations of motion forward) until the missile hits the ground. The calculations performed at each guidance update are also depicted in Fig. 9.8.

The predictor-corrector guidance scheme is similar to predictive guidance (see Chapter 19 of the previous volume), except that in this application we integrate the equations of motion forward twice rather than once to obtain the desired acceleration command at each guidance update. An engagement simulation using the predictor-corrector method appears in Listing 9.2. As can be seen from the listing, we integrate the equations of motion forward until ground impact twice during each guidance update by calling subroutine PREDICTP. Each time we call subroutine PREDICTP we must supply the current missile and target location, missile speed, ballistic coefficient, and acceleration command. The output of the subroutine is the missile location at ground impact. In Listing 9.2 we have the ability to supply the estimated missile ballistic coefficient BETAH, which may be different than the actual missile ballistic coefficient BETA.

LISTING 9.2 ENGAGEMENT SIMULATION WITH PREDICTOR-CORRECTOR GUIDANCE

```
clear
n=0;
H=.01;
VM=3000.;
BETA=1000.;
```

```
BETAH=1000.;
XLIMG=5.;
T=0.;
S=0.;
GAMDEG=45.;
VM1=VM*cos(GAMDEG/57.3);
VM2=VM*sin(GAMDEG/57.3);
RM1=0.;
RM2=0.;
XNC=0.;
RT1=60000.;
RT2=0.;
RTM1=RT1-RM1;
RTM2=RT2-RM2;
RTM=sqrt(RTM1^2+RTM2^2);
XLIM=XLIMG*32.2;
while ~(T>0.&RM2<=0.)
    if RTM<1000.
%Integration interval increased by factor of 10 to increase speed
        %H=.0001;
        H=.001;
    else
        %H=.01
        H=.1;
    end
    RM1OLD=RM1;
    RM2OLD=RM2;
    VM1OLD=VM1;
    VM2OLD=VM2;
    STEP=1;
    FLAG=0;
    while STEP <=1
        if FLAG==1
        STEP=2;
                RM1=RM1+H*VM1;
                RM2=RM2+H*VM2;
                VM1=VM1+H*AM1;
                VM2=VM2+H*AM2;
            T=T+H;
        end
        RTM1=RT1-RM1;
        RTM2=RT2-RM2;
        RTM=sqrt(RTM1^2+RTM2^2);
        XLAM=atan2(RTM2,RTM1);
        XNE1=-XNC*sin(XLAM);
        XNE2=XNC*cos(XLAM);
```

```
if RM2<30000.
        RHO=.002378*exp(-RM2/30000);
else
        RHO=.0034*exp(-RM2/22000);
end
VM=sqrt(VM1^2+VM2^2);
Q=.5*RHO*VM*VM;
GAM=atan2(VM2,VM1);
DRAG=Q*32.2/BETA;
AM1=-DRAG*cos(GAM)+XNE1;
AM2=-32.2-DRAG*sin(GAM)+XNE2;
FLAG=1;
end
FLAG=0;
RM1=.5*(RM1OLD+RM1+H*VM1);
RM2=.5*(RM2OLD+RM2+H*VM2);
VM1=.5*(VM1OLD+VM1+H*AM1);
VM2=.5*(VM2OLD+VM2+H*AM2);
S=S+H;
if S>=.99999
  S=0.;
  if T>30.
        [X1]=PREDICTP(T,RM1,RM2,VM1,VM2,XNC,RT1,RT2,BETAH);
        DELX=RT1-X1;
        XNCP=XNC+1.;
        [X2]=PREDICTP(T,RM1,RM2,VM1,VM2,XNCP,RT1,RT2,BETAH);
        DXDNC=(X2-X1)/(XNCP-XNC);
        DELXNC=DELX/DXDNC;
        XNC=XNC+DELXNC;
        if XNC>XLIM
                XNC=XLIM;
        end
        if XNC<-XLIM
                XNC=-XLIM;
        end
  end
  RM1K=RM1/1000.;
  RM2K=RM2/1000.;
  RT1K=RT1/1000.;
  RT2K=RT2/1000.;
  XNCG=XNC/32.2;
  n=n+1;
  ArrayT(n)=T;
  ArrayRM1K(n)=RM1K;
  ArrayRM2K(n)=RM2K;
  ArrayRT1K(n)=RT1K;
```

```
                ArrayRT2K(n)=RT2K;
                ArrayXNCG(n)=XNCG;
        end
end
RTM
figure
plot(ArrayRM1K,ArrayRM2K,ArrayRT1K,ArrayRT2K),grid
xlabel('Downrange (Kft)')
ylabel('Altitude (KFt)')
figure
plot(ArrayT,ArrayXNCG),grid
xlabel('Time (Sec)')
ylabel('Acceleration (G)')
clc
output=[ArrayT',ArrayRM1K',ArrayRM2K',ArrayRT1K',ArrayRT2K',ArrayXNCG'];
save datfil output -ascii
disp 'simulation finished'


function[RM1F]= PREDICTP(TP,RM1P,RM2P,VM1P,VM2P,XNC1F,RT1P,RT2P,BETAH)
H=.01;
RM1=RM1P;
RM2=RM2P;
VM1=VM1P;
VM2=VM2P;
XNC=XNC1F;
RT1=RT1P;
RT2=RT2P;
BETA=BETAH;
T=TP;
RTM1=RT1-RM1;
RTM2=RT2-RM2;
RTM=sqrt(RTM1^2+RTM2^2);
while RM2>=0.
        if RTM<1000.
%Integration interval increased by factor of 10 to increase speed
        %H=.0001
        H=.001;
        else
        %H=.01
        H=.1;
        end
        RM1OLD=RM1;
        RM2OLD=RM2;
        VM1OLD=VM1;
        VM2OLD=VM2;
```

```
STEP=1;
FLAG=0;
while STEP <=1
  if FLAG==1
  STEP=2;
        RM1=RM1+H*VM1;
        RM2=RM2+H*VM2;
        VM1=VM1+H*AM1;
        VM2=VM2+H*AM2;
        T=T+H;
  end
  RTM1=RT1-RM1;
  RTM2=RT2-RM2;
  RTM=sqrt(RTM1^2+RTM2^2);
  XLAM=atan2(RTM2,RTM1);
  XNE1=-XNC*sin(XLAM);
  XNE2=XNC*cos(XLAM);
  if RM2<30000.
    RHO=.002378*exp(-RM2/30000);
  else
    RHO=.0034*exp(-RM2/22000);
  end
  VM=sqrt(VM1^2+VM2^2);
  Q=.5*RHO*VM*VM;
  GAM=atan2(VM2,VM1);
  DRAG=Q*32.2/BETA;
  AM1=-DRAG*cos(GAM)+XNE1;
  AM2=-32.2-DRAG*sin(GAM)+XNE2;
  FLAG=1;
  end
  FLAG=0;
  RM1=.5*(RM1OLD+RM1+H*VM1);
  RM2=.5*(RM2OLD+RM2+H*VM2);
  VM1=.5*(VM1OLD+VM1+H*AM1);
  VM2=.5*(VM2OLD+VM2+H*AM2);
end
RM1F=RM1;
```

The predictor-corrector engagement simulation of Listing 9.2 was run for cases in which the target location RT1 was made a parameter. We can see from Fig. 9.9 that, as with augmented proportional navigation, we can hit targets from 40,000 to 100,000 ft downrange of the missile launch point. The acceleration profiles for the various trajectories appear in Fig. 9.10. We can see that when guidance is turned on, after a brief transient period the acceleration commands for the

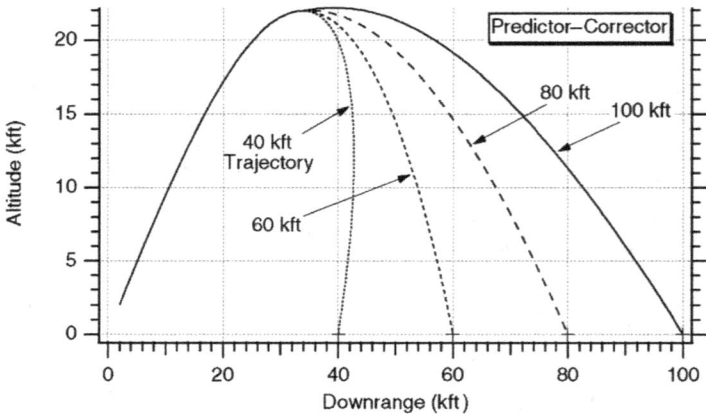

Fig. 9.9   Predictor–corrector guidance is as effective as augmented proportional navigation in this application.

predictor-corrector guidance approach are approximately a constant, thus matching our assumed nominal acceleration profile. The most demanding part of the acceleration command is initially during the transient portion of flight. The acceleration commands in the steady state with predictor-corrector guidance are actually slightly smaller than those obtained with augmented proportional navigation.

Recall from Chapter 19 of Volume 1 that predictive guidance was demonstrated to be very sensitive to modeling errors. To test the robustness of

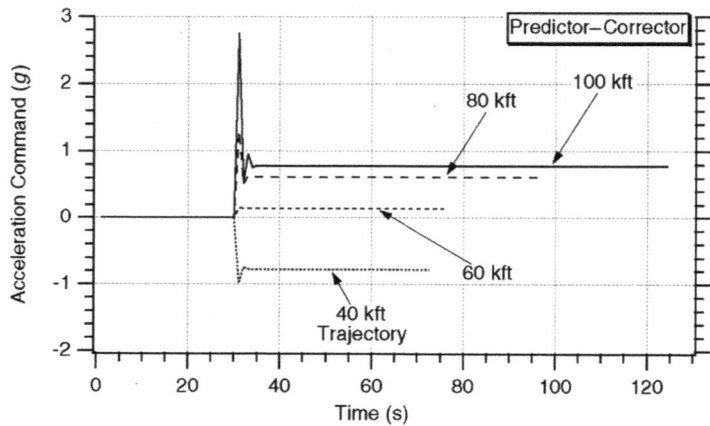

Fig. 9.10   Predictor–corrector guidance commands are near constant in this application.

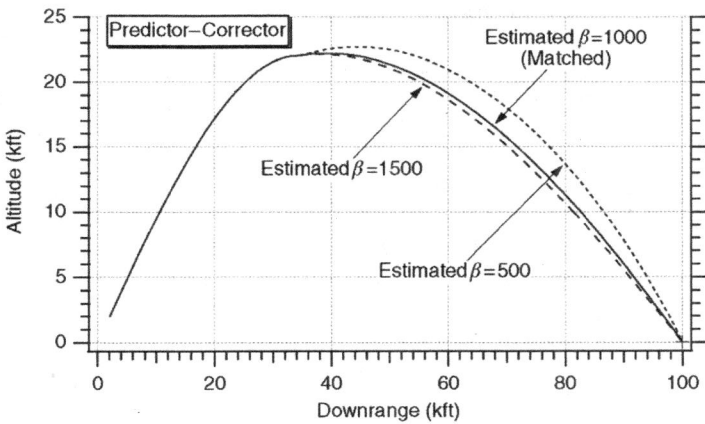

Fig. 9.11   Predictor–corrector guidance still hits target if there is 50% error in knowledge of ballistic coefficient.

predictor-corrector guidance in this application, the estimated ballistic coefficient was varied from the correct value of $1000 \, \text{lb/ft}^2$ to as low as $500 \, \text{lb/ft}^2$ to as high as $1500 \, \text{lb/ft}^2$. We can see from Fig. 9.11 that even with these large ballistic coefficient errors, we are still able to hit the target. Figure 9.12 shows that the acceleration profiles hardly change when there are significant errors in the estimated ballistic coefficient. Therefore, we can conclude that in this application, predictor-corrector guidance does not appear to be sensitive to modeling errors.

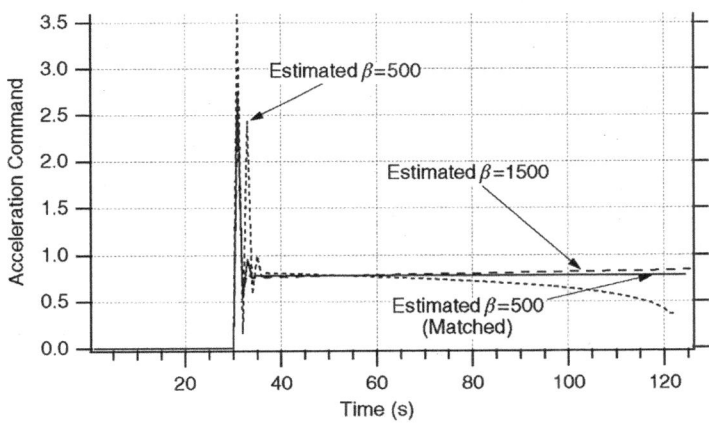

Fig. 9.12   Acceleration commands only change slightly if there is a 50% error in estimate of ballistic coefficient.

ROLL ANGLE OR RATE GUIDANCE PROBLEM

As a further test of predictor-corrector guidance we will consider an unusual problem in which the missile is severely constrained in its ability to issue guidance commands. Consider the case of Fig. 9.13, in which we have a missile that is attempting to hit a stationary target. In this case, the acceleration command of the missile n_c is fixed in amplitude, and we only have control over the orientation of the acceleration magnitude by controlling either the roll angle ϕ or the roll rate. This type of guidance might be representative of a rolling missile that is at a fixed trim angle. From Fig. 9.13 we can see that the components of the missile acceleration command in the x and y directions are given by

$$\ddot{x}_M = n_c \cos \phi$$
$$\ddot{y}_M = n_c \sin \phi$$

In this guidance problem we would like to hit a stationary target at time t_F. Mathematically, this means that

$$x_M(t_F) = x_T$$
$$y_M(t_F) = y_T$$

Therefore, the guidance problem is to figure out what to do with the roll angle or roll rate so that the missile hits the target at the final time.

Before we attempt to use predictor-corrector guidance, let us first see if it is possible to use some form of proportional navigation on this problem. Recall

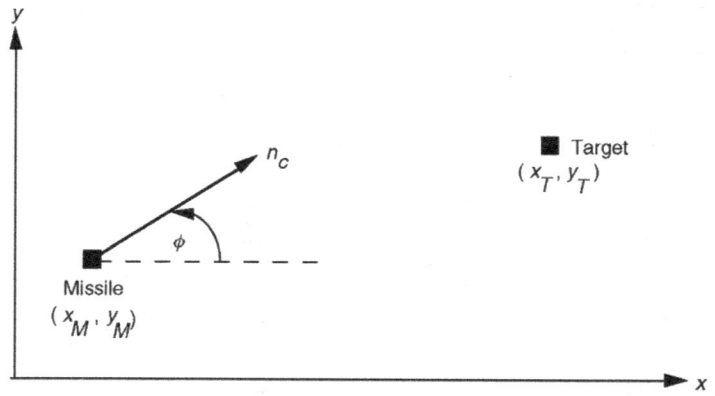

Fig. 9.13   Unusual guidance problem in which we can only control roll angle.

that the relative ranges between the missile and target are given by

$$R_{TM1} = x_T - x_M$$
$$R_{TM2} = y_T - y_M$$

In this problem the target is stationary, and therefore its velocity must be zero. The components of the relative velocity are given by

$$V_{TM1} = \dot{x}_T - \dot{x}_M = -\dot{x}_M$$
$$V_{TM2} = \dot{y}_T - \dot{y}_M = -\dot{y}_M$$

We can think of proportional navigation as a guidance law whose acceleration commands are proportional to the line-of-sight rate or proportional to the zero effort miss and inversely proportional to the square of time to go until intercept (see Chapter 2 of the previous volume). Therefore, if we want to use proportional navigation in this application, the commanded acceleration components in the x and y directions are given by

$$n_{cx} = \frac{N'}{t_{go}^2}(R_{TM1} + V_{TM1}t_{go})$$

$$n_{cy} = \frac{N'}{t_{go}^2}(R_{TM2} + V_{TM2}t_{go})$$

Therefore, from Fig. 9.13, the appropriate roll-angle command given a fixed acceleration level for the implementation of proportional navigation should be

$$\phi = \tan^{-1}\frac{n_{cy}}{n_{cx}} = \tan^{-1}\left[\frac{\dfrac{N'}{t_{go}^2}(R_{TM2} + V_{TM2}t_{go})}{\dfrac{N'}{t_{go}^2}(R_{TM1} + V_{TM1}t_{go})}\right] = \tan^{-1}\left[\frac{R_{TM2} + V_{TM2}t_{go}}{R_{TM1} + V_{TM1}t_{go}}\right]$$

Listing 9.3 is an engagement simulation of the missile attempting to hit the stationary target after 100 s using the roll-angle command of the preceding equation as the guidance law. In this application we are not attempting to control the roll rate. In a practical application we would have to ensure that the resultant roll rates are not too high. The missile is initially at the origin (namely, XM=0, YM=0) and the target has coordinates XT=−4000 ft and YT=5000 ft. In the simulation, the magnitude of the acceleration command is 12 ft/s^2 and the roll-angle command is computed every 0.1 s. Notice that the simulation does not stop at the point of closest approach, but instead stops after 100 s of flight (that is, the time that we would like to hit the target).

LISTING 9.3   AN ENGAGEMENT SIMULATION USING AN UNUSUAL FORM OF PROPORTIONAL NAVIGATION

```
clear
n=0;
XT=-4000.;
YT=5000.;
TF=100.;
H=.01;
TS=.1;
XNC=12.;
XM=0.;
YM=0.;
XMD=0.;
YMD=0.;
S=0.;
T=0.;
PHI=0.;
while T<=TF
      XMOLD=XM;
      YMOLD=YM;
      XMDOLD=XMD;
      YMDOLD=YMD;
      STEP=1;
      FLAG=0;
      while STEP <=1
        if FLAG==1
        STEP=2;
                XM=XM+H*XMD;
                YM=YM+H*YMD;
                XMD=XMD+H*XMDD;
                YMD=YMD+H*YMDD;
                T=T+H;
        end
        XMDD=XNC*cos(PHI);
        YMDD=XNC*sin(PHI);
        FLAG=1;
      end
      FLAG=0;
      XM=(XMOLD+XM)/2+.5*H*XMD;
      YM=(YMOLD+YM)/2+.5*H*YMD;
      XMD=(XMDOLD+XMD)/2+.5*H*XMDD;
      YMD=(YMDOLD+YMD)/2+.5*H*YMDD;
      S=S+H;
      if S>=(TS-.0001)
        S=0.;
```

```
      TGO=TF-T+.0001;
      RTM1=XT-XM;
      RTM2=YT-YM;
      RTM=sqrt(RTM1^2+RTM2^2);
      VTM1=-XMD;
      VTM2=-YMD;
      PHI=atan2(RTM2+VTM2*TGO,RTM1+VTM1*TGO);
      PHIDEG=PHI*57.3;
      n=n+1;
      ArrayT(n)=T;
      ArrayXM(n)=XM;
      ArrayYM(n)=YM;
      ArrayXT(n)=XT;
      ArrayYT(n)=YT;
      ArrayPHIDEG(n)=PHIDEG;
   end
end
RTM
figure
plot(ArrayXM,ArrayYM,ArrayXT,ArrayYT),grid
xlabel('X (Ft)')
ylabel('Y (Ft)')
figure
plot(ArrayT,ArrayPHIDEG),grid
xlabel('Time (Sec)')
ylabel('Roll Rate (Deg/Sec)')
clc
output=[ArrayT',ArrayXM',ArrayYM',ArrayXT',ArrayYT',ArrayPHIDEG'];
save datfil output -ascii
disp 'simulation finished'
```

The nominal case of Listing 9.3 was run, and we can see from Fig. 9.14 that the missile flies directly to the stationary target and scores a direct hit. The roll-angle command required to achieve this successful intercept is displayed in Fig. 9.15. We can see that the roll angle continuously oscillates between −50 and 130 deg. This means that the accelerations are 180 deg apart. In other words, the missile first accelerates in one direction and then, after a very brief period of time, accelerates in the opposite direction. In practice it might be very difficult to achieve the high roll rates needed to switch nearly instantaneously between the two roll angles. From Fig. 9.14 we can see that the angle required for the missile to fly directly to the target is simply

$$\phi = \tan^{-1}\frac{5000}{-4000} = 129 \, \text{deg}$$

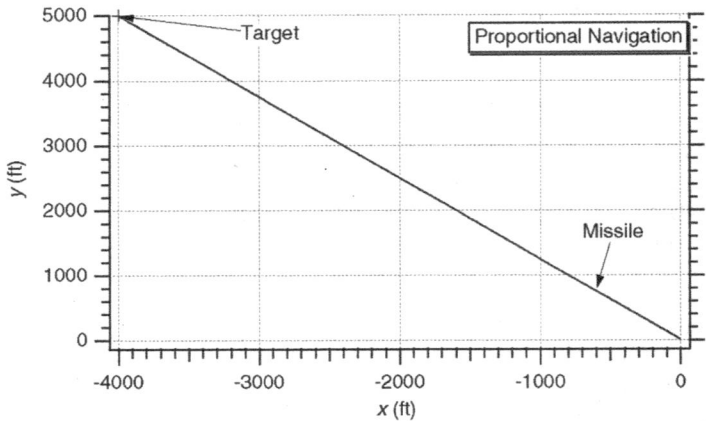

Fig. 9.14   Missile flies directly to target with proportional navigation.

which is approximately the same roll angle chosen by the guidance law. If the guidance law senses that there is too much acceleration and that the missile will arrive at the target too soon, a command is issued in the opposite direction.

Similarly, when the command is in the opposite direction and the guidance law believes that the missile will be late in arriving at the target, a command is issued in the opposite direction.

The target was then moved to all four quadrants [that is, $(-4000, -5000)$, $(-4000, 5000)$, $(4000, -5000)$, and $(4000, 5000)$], and successful intercepts

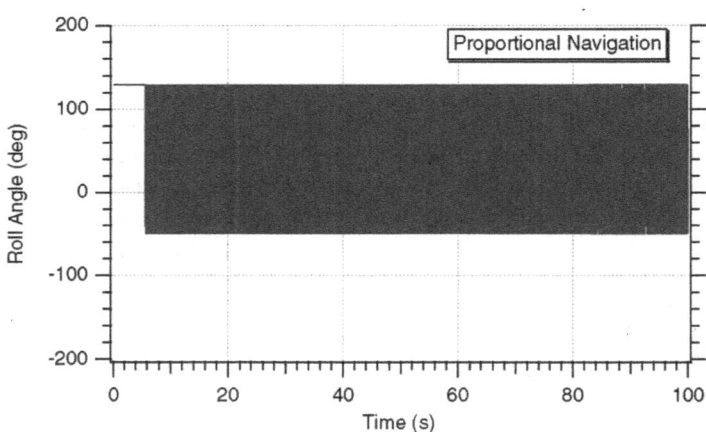

Fig. 9.15   With proportional navigation, missile roll angle oscillates between −50 and 130 deg.

resulted in all cases. Thus, we can conclude that if we ignore the high roll-rate question, at least mathematically proportional navigation–type guidance is effective in this application.

PREDICTOR–CORRECTOR GUIDANCE WITH ROLL-RATE PROBLEM [1–3]

As was mentioned previously in this chapter, predictor–corrector guidance is based on having a model of the interceptor and target and then trying to determine how certain changes in important parameters influence the missile trajectory. In the previous application of predictor–corrector guidance, we assumed a nominal acceleration command that was a constant. In this problem we will assume a linear form for the nominal roll-rate command, as shown in Fig. 9.16. In this model we have specified the nominal maximum and minimum roll rates along with the time of flight. Two parameters are required in our nominal model because we will be attempting to control two parameters.

Using the example of the previous section in conjunction with predictor–corrector guidance, we first integrate the missile equations of motion forward assuming the nominal roll-rate profile of Fig. 9.16 (with initial specified values for the two roll-rate parameters and the flight time). The fixed acceleration command at the appropriate roll-angle profile (that is, obtained by integrating the roll rate) is issued to see where the missile is at the final time t_F. We then measure the difference between where the target really is (x_T, y_T) and where the missile actually ends up (x_1, y_1) and call the resultant quantities Δx and Δy, or

$$\Delta x = x_T - x_1$$
$$\Delta y = y_T - y_1$$

We now add a little bit or tweak one of the roll-rate parameters. For example, let us assume that the new minimum roll-rate parameter is given by

$$\dot{\phi}_{B2} = \dot{\phi}_B + 0.001$$

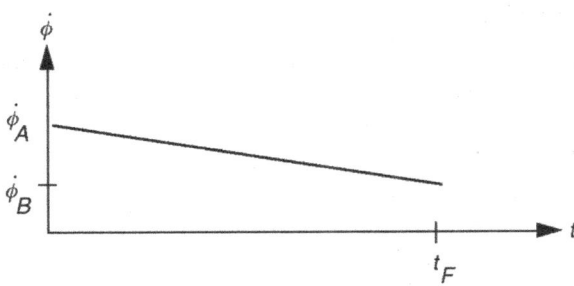

Fig. 9.16   Nominal model for roll-rate profile to be used with predictor–corrector guidance.

We again integrate the missile equations of motion forward using the new minimum roll-rate parameter to see where the missile is at the final time t_F. We then note where the missile actually ends up (x_2, y_2) and manually compute partial derivatives. In this example, we can say that

$$\frac{\partial x}{\partial \dot{\phi}_B} = \frac{x_2 - x_1}{\dot{\phi}_{B2} - \dot{\phi}_B} = \frac{x_2 - x_1}{0.001}$$

$$\frac{\partial y}{\partial \dot{\phi}_B} = \frac{y_2 - y_1}{\dot{\phi}_{B2} - \dot{\phi}_B} = \frac{y_2 - y_1}{0.001}$$

We now add a little bit or tweak the maximum roll-rate parameter. The new maximum roll-rate parameter is given by

$$\dot{\phi}_{A2} = \dot{\phi}_A + 0.001$$

We again integrate the missile equations of motion forward using the new maximum roll-rate parameter to see where the missile is at the final time t_F We then note where the missile actually ends up (x_3, y_3) and manually compute another set of partial derivatives. In this example, we can say that

$$\frac{\partial x}{\partial \dot{\phi}_A} = \frac{x_3 - x_1}{\dot{\phi}_{A3} - \dot{\phi}_A} = \frac{x_3 - x_1}{0.001}$$

$$\frac{\partial y}{\partial \dot{\phi}_A} = \frac{y_3 - y_1}{\dot{\phi}_{A3} - \dot{\phi}_A} = \frac{y_3 - y_1}{0.001}$$

From calculus we know that the total differential can be written as

$$\Delta x = \frac{\partial x}{\partial \dot{\phi}_A} \Delta \dot{\phi}_A + \frac{\partial x}{\partial \dot{\phi}_B} \Delta \dot{\phi}_B$$

$$\Delta y = \frac{\partial y}{\partial \dot{\phi}_A} \Delta \dot{\phi}_A + \frac{\partial y}{\partial \dot{\phi}_B} \Delta \dot{\phi}_B$$

We now have two equations with two unknowns (because we already know the partial derivatives and Δx and Δy). After solving for the new change in the roll-rate parameters, we can say that the new maximum and minimum roll-rate parameters are given by the previous values plus the changes, or

$$\dot{\phi}_{A_k} = \dot{\phi}_{A_{k-1}} + \Delta \dot{\phi}_A$$

$$\dot{\phi}_{B_k} = \dot{\phi}_{B_{k-1}} + \Delta \dot{\phi}_B$$

We now repeat the cycle (that is, solving for Δx, Δy, and partial derivative by integrating equations of motion forward until the final time using the new nominal roll-rate profile). The cycle is also depicted in Fig. 9.17.

Use Nominal Roll-Rate Profile

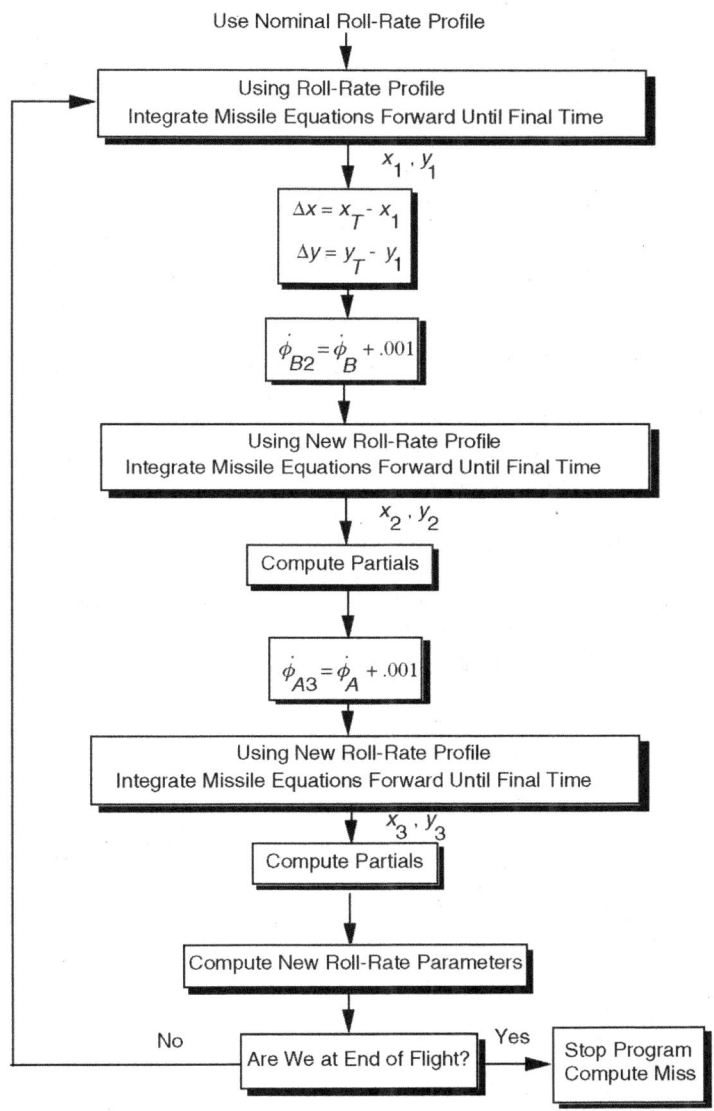

Using Roll-Rate Profile
Integrate Missile Equations Forward Until Final Time

$x_1 \cdot y_1$

$\Delta x = x_T - x_1$
$\Delta y = y_T - y_1$

$\dot{\phi}_{B2} = \dot{\phi}_B + .001$

Using New Roll-Rate Profile
Integrate Missile Equations Forward Until Final Time

$x_2 \cdot y_2$

Compute Partials

$\dot{\phi}_{A3} = \dot{\phi}_A + .001$

Using New Roll-Rate Profile
Integrate Missile Equations Forward Until Final Time

$x_3 \cdot y_3$

Compute Partials

Compute New Roll-Rate Parameters

No ← Are We at End of Flight? → Yes

Stop Program
Compute Miss

Fig. 9.17   Predictor–corrector guidance logic for second problem.

Listing 9.4 is an engagement simulation of the missile attempting to hit the stationary target after 100 s using the previously described predictor–corrector guidance scheme. The missile is initially at the origin (namely, XM=0, YM=0), and the target has coordinates XT=−4000 ft and YT=5000 ft. In the simulation, the magnitude of the acceleration command is 12 ft/s^2 and the roll-angle command is computed every 0.1 s. The nominal maximum and minimum roll rates are 10 deg/s and 5 deg/s, respectively. Again, notice that the simulation does not stop at the point of closest approach but stops after 100 s of flight. In addition, the simulation slightly modifies the formulas for the updated roll rates to be

$$\dot{\phi}_{A_k} = \dot{\phi}_{A_{k-1}} + \text{GAIN}\,\Delta\dot{\phi}_A$$

$$\dot{\phi}_{B_k} = \dot{\phi}_{B_{k-1}} + \text{GAIN}\,\Delta\dot{\phi}_B$$

The addition of the new parameter GAIN will give us an extra degree of flexibility in modifying the iterated roll rates if we run into problems. Nominally, GAIN=1, so we have simulated the scheme that was derived in this section. Unlike the previous application of predictor–corrector guidance in which we called subroutine PREDICTP twice (we only needed to calculate one partial derivative), we now have to call subroutine PREDICT1 three times.

The nominal case of Listing 9.4 was run, and we can see from Fig. 9.18 that a direct hit was scored. The trajectory is totally different from the straight-line approach of proportional navigation. The missile loops around before hitting the target. The required roll-rate profile for the successful intercept is displayed in Fig. 9.19. Here we can see that the roll-rate profile is approximately linear with a maximum roll rate of only 12 deg/s. The required roll rate is much smaller here than it would have been using proportional navigation.

Another case was run in which the target was moved to another location. Figure 9.20 shows that in this case predictor–corrector guidance failed, and the

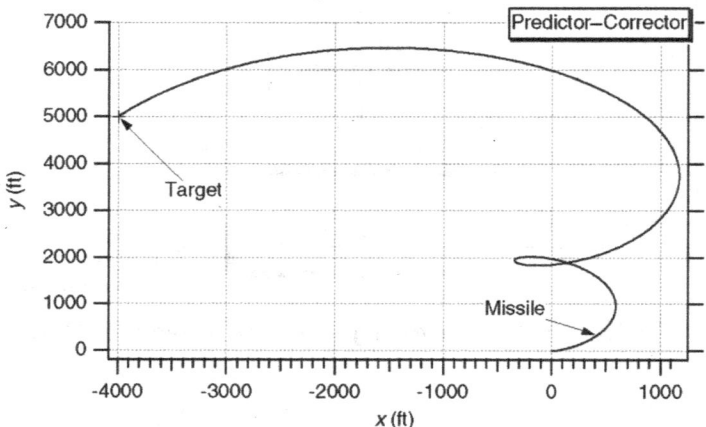

Fig. 9.18 **With predictor–corrector guidance, missile hits target by flying unusual trajectory.**

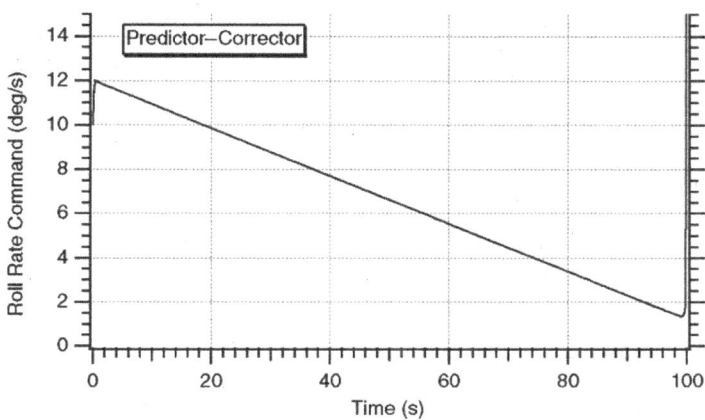

Fig. 9.19   Predictor–corrector guidance yields much smaller roll rates than proportional navigation.

missile did not even get close to the target. Normally predictor–corrector guidance acts as a feedback scheme in which the Δs should be driven to zero. However, we can see from Fig. 9.21 that Δx and Δy do not approach zero, which is another way of letting us know something is wrong.

Numerous experiments were conducted in which various parameters were varied in the simulation to see if predictor–corrector guidance could be made to work in this application. It was discovered that if GAIN could be reduced from 1 to 0.1, the predictor–corrector guidance scheme could be made very

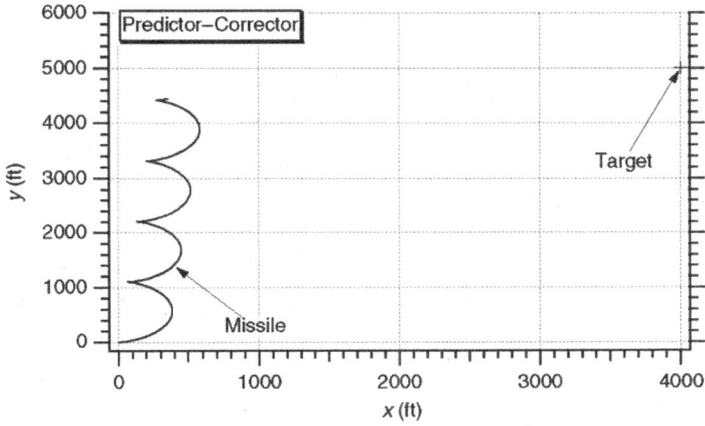

Fig. 9.20   Predictor–corrector guidance has a problem when we move the target.

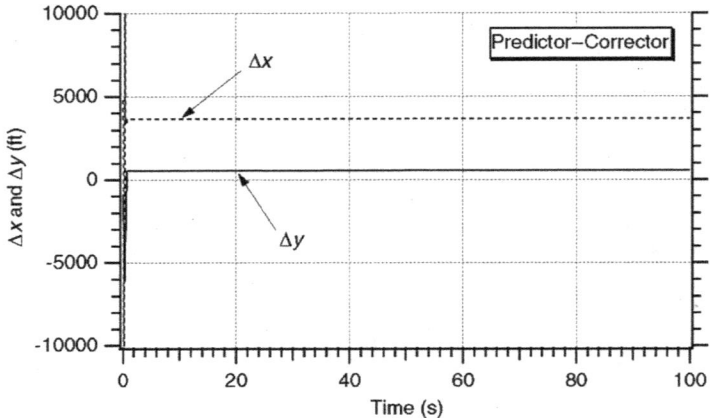

Fig. 9.21   Something is now wrong because errors do not go to zero when target is moved.

effective. Cases were then run in which the stationary target was moved to all four quadrants. We can see from Fig. 9.22 that now the missile can hit the target in any quadrant when GAIN is reduced to the new value of 0.1. The resultant roll-rate profiles are displayed in Fig. 9.23. We can see that no matter where the target is placed, the roll-rate profiles are all well behaved.

To further test the robustness of predictor–corrector guidance, a modeling error was introduced. The actual missile acceleration is 12 ft/s^2, and a 10% error was introduced (namely, ACCERR=1.2) so that subroutine PREDICT1 would

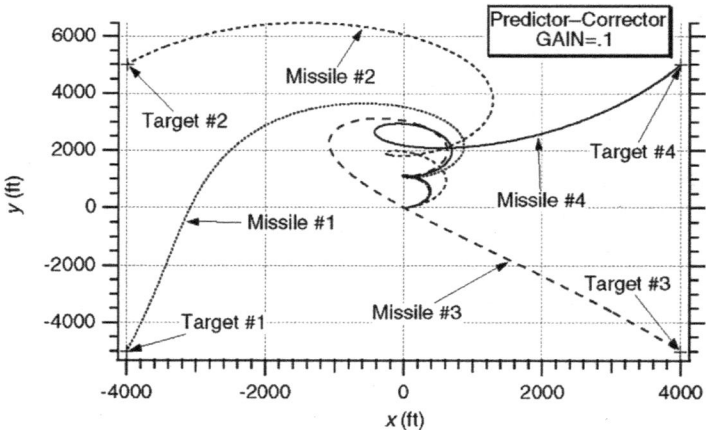

Fig. 9.22   Missile can hit target in any quadrant when gain is reduced from 1 to 0.1.

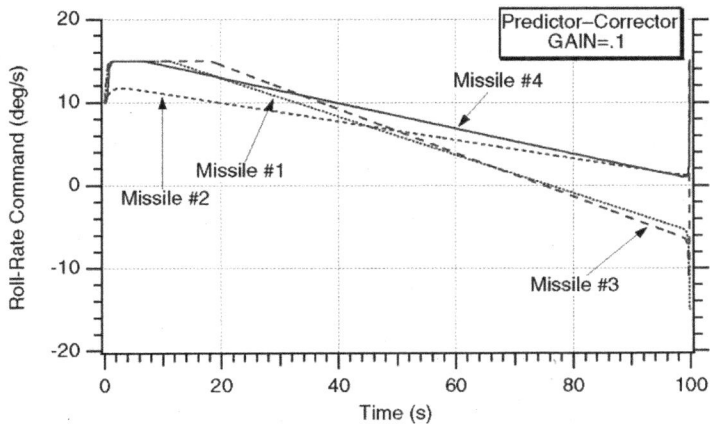

Fig. 9.23   Roll-rate profiles are well behaved and vary with location of target.

be in error. We can see from Fig. 9.24 that the small acceleration error causes the missile to miss the target when predictor–corrector guidance is used. The roll rate command no longer approaches zero. Figure 9.25 shows that the roll-rate profiles start to oscillate when the acceleration error is introduced.

Thus, we can conclude that in this application, predictor–corrector guidance is very sensitive to modeling errors. Its lack of robustness is similar to that of predictive guidance, which is also a brute-force, model-based approach to guidance.

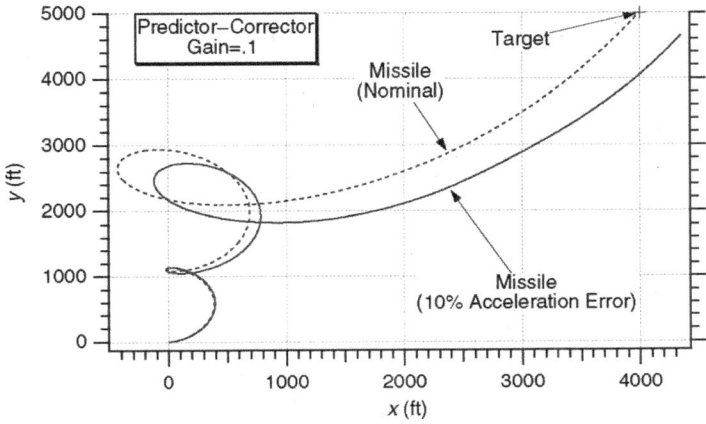

Fig. 9.24   Small error in knowledge of missile acceleration causes predictor guidance to miss target.

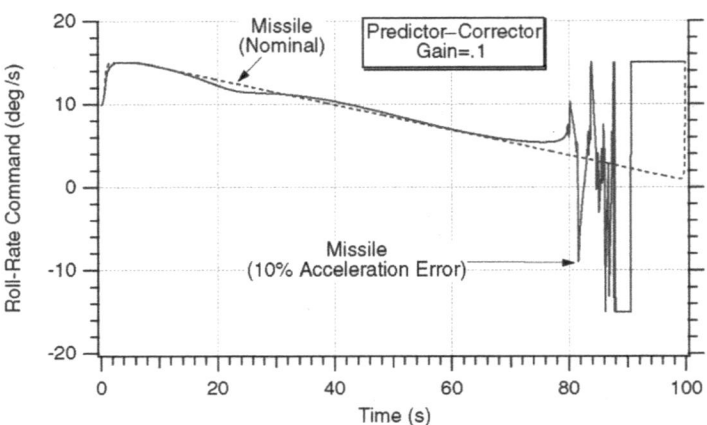

Fig. 9.25   Small error in knowledge of missile acceleration causes roll-rate command to oscillate.

LISTING 9.4   AN UNUSUAL PROBLEM USING PREDICTOR – CORRECTOR GUIDANCE

```
clear
n=0;
GAIN=1.;
XT=-4000.;
YT=5000.;
PHIADDEG=10.;
PHIBDDEG=5.;
PHIDMAXDEG=15.;
TF=100.;
ACCERR=0.;
%This takes about 5 minutes on a 400 MHz G3. Larger integration intervals are not
as accurate
H=.01;
TS=.1;
XNC=12.;
PHIDMAX=PHIDMAXDEG/57.3;
PHI=0.;
XM=0.;
YM=0.;
XMD=0.;
YMD=0.;
PHIAD=PHIADDEG/57.3;
PHIBD=PHIBDDEG/57.3;
S=0.;
T=0.;
```

```
while T<=(TF-.00001)
     XMOLD=XM;
     YMOLD=YM;
     XMDOLD=XMD;
     YMDOLD=YMD;
     PHIOLD=PHI;
     STEP=1;
     FLAG=0;
     while STEP <=1
       if FLAG==1
       STEP=2;
               XM=XM+H*XMD;
               YM=YM+H*YMD;
               XMD=XMD+H*XMDD;
               YMD=YMD+H*YMDD;
               PHI=PHI+H*PHID;
               T=T+H;
       end
       SLOPE=(PHIBD-PHIAD)/TF;
       BINT=PHIBD-SLOPE*TF;
       PHID=SLOPE*T+BINT;
       if PHID>PHIDMAX
               PHID=PHIDMAX;
       end
       if PHID<-PHIDMAX
               PHID=-PHIDMAX;
       end
       XMDD=XNC*cos(PHI);
       YMDD=XNC*sin(PHI);
       FLAG=1;
     end
     FLAG=0;
     XM=(XMOLD+XM)/2+.5*H*XMD;
     YM=(YMOLD+YM)/2+.5*H*YMD;
     XMD=(XMDOLD+XMD)/2+.5*H*XMDD;
     YMD=(YMDOLD+YMD)/2+.5*H*YMDD;
     PHI=(PHIOLD+PHI)/2.+.5*H*PHID;
     S=S+H;
     if S>=(TS-.0001)
       S=0.;
     [X1,Y1]=PREDICT1(T,XM,YM,XMD,YMD,PHI,TF,PHIAD,PHIBD,XNC,H,PHIDMAX,
             ACCERR);
       DELX=XT-X1;
       DELY=YT-Y1;
       PHIBD2=PHIBD+.001;
```

```
    [X2,Y2]=PREDICT1(T,XM,YM,XMD,YMD,PHI,TF,PHIAD,PHIBD2,XNC,H,PHIDMAX,
            ACCERR);
      DXDPB=(X2-X1)/(PHIBD2-PHIBD);
      DYDPB=(Y2-Y1)/(PHIBD2-PHIBD);
      PHIAD3=PHIAD+.001;
    [X3,Y3]=PREDICT1(T,XM,YM,XMD,YMD,PHI,TF,PHIAD3,PHIBD,XNC,H,PHIDMAX,
            ACCERR);
      DXDPA=(X3-X1)/(PHIAD3-PHIAD);
      DYDPA=(Y3-Y1)/(PHIAD3-PHIAD);
      if (DXDPB*DYDPA-DXDPA*DYDPB)==0
            DELPHIAD=0.;
            DELPHIBD=0.;
      else
            DELPHIAD=(DXDPB*DELY-DELX*DYDPB)/(DXDPB*DYDPA-DXDPA*DYDFB);
            DELPHIBD=(DELX*DYDPA-DXDPA*DELY)/(DXDPB*DYDPA-DXDPA*DYDFB);
      end
      PHIAD=PHIAD+GAIN*DELPHIAD;
      PHIBD=PHIBD+GAIN*DELPHIBD;
      PHIDDEG=PHID*57.3;
      n=n+1;
      ArrayT(n)=T;
      ArrayXM(n)=XM;
      ArrayYM(n)=YM;
      ArrayXT(n)=XT;
      ArrayYT(n)=YT;
      ArrayPHIDDEG(n)=PHIDDEG;
    end
end
RTM=sqrt((XT-XM)^2+(YT-YM)^2);
RTM
figure
plot(ArrayXM,ArrayYM,ArrayXT,ArrayYT),grid
xlabel('X (Ft)')
ylabel('Y (Ft)')
figure
plot(ArrayT,ArrayPHIDDEG),grid
xlabel('Time (Sec)')
ylabel('Roll Rate (Deg/Sec)')
clc
output=[ArrayT',ArrayXM',ArrayYM',ArrayXT',ArrayYT',ArrayPHIDDEG'];
save datfil output -ascii
disp 'simulation finished'

function[XF,YF]=PREDICT1(TP,XP,YP,XDP,YDP,THETP,TF,THADP,THBDP,ACC,HP,
THDMAX,ACCERR)
H=HP*10.;
```

```
T=TP;
X=XP;
Y=YP;
XD=XDP;
YD=YDP;
THET=THETP;
THAD=THADP;
THBD=THBDP;
S=0.;
while T<=(TF-.00001)
      XOLD=X;
      YOLD=Y;
      XDOLD=XD;
      YDOLD=YD;
      THETOLD=THET;
      STEP=1;
      FLAG=0;
      while STEP <=1
        if FLAG==1
        STEP=2;
                X=X+H*XD;
                Y=Y+H*YD;
                XD=XD+H*XDD;
                YD=YD+H*YDD;
                THET=THET+H*THETD;
                T=T+H;
           end
           SLOPE=(THBD-THAD)/TF;
           BINT=THBD-SLOPE*TF;
           THETD=SLOPE*T+BINT;
           if THETD>THDMAX
                  THETD=THDMAX;
           end
           if THETD<-THDMAX
                  THETD=-THDMAX;
           end
           XDD=(ACC+ACCERR)*cos(THET);
           YDD=(ACC+ACCERR)*sin(THET);
           FLAG=1;
      end
      FLAG=0;
      X=(XOLD+X)/2+.5*H*XD;
      Y=(YOLD+Y)/2+.5*H*YD;
      XD=(XDOLD+XD)/2+.5*H*XDD;
      YD=(YDOLD+YD)/2+.5*H*YDD;
      THET=(THETOLD+THET)/2.+.5*H*THETD;
```

```
end
XF=X;
YF=Y;
```

We also saw that the application of predictor–corrector guidance is very ad hoc because we must guess a nominal acceleration profile and perform lots of experimentation on various system parameters to achieve success. Thus, the application of predictor–corrector guidance is not as straightforward as the application of proportional navigation types of guidance laws, but under certain circumstances it might be the only guidance choice available.

SUMMARY

Two problems were selected in which a predictor–corrector guidance method was introduced and compared with proportional navigation. We showed that predictor–corrector guidance was similar to predictive guidance in the sense that equations representing our best guess at a model of the real world had to be integrated forward several times at each guidance update. The successful application of this guidance approach often requires significant experimentation to pick appropriate nominal models and to choose appropriate gains.

REFERENCES

[1]   Haro, C., "A Predictor–Corrector Guidance Algorithm Design for a Low L/D Autonomous Re-Entry Vehicle," master's thesis, MIT, Dec. 1998.

[2]   Brau, R., and Powel, R., "Predictor–Corrector Guidance Algorithm for Use in High-Energy Aerobraking System Studies," *Journal of Guidance, Control, and Dynamics*, Vol. 15, No. 3, May–June 1992, pp. 672–678.

[3]   Miel, A., and Wan, T., "Robust Predictor–Corrector Guidance for Aeroassisted Orbital Transfer," *Journal of Guidance, Control, and Dynamics*, Vol. 19, No. 5, Sept.–Oct. 1996, pp. 1134–1141.

CHAPTER 10

Optimal Guidance Law to Minimize Interceptor Jerk

INTRODUCTION

In Chapter 8 of the previous volume we derived guidance laws to minimize the integral of the acceleration squared to yield zero miss distance in sthe absence of flight-control system dynamics. The resultant guidance law was called augmented proportional navigation. Essentially this guidance law was proportional navigation plus an extra term whose magnitude depended on the target maneuver level and could be expressed as

$$n_c = N'V_c\dot{\lambda} + 0.5N'n_T$$

where V_c was the closing velocity, $\dot{\lambda}$ was the line-of-sight rate, n_T was the target acceleration level, and N' was the effective navigation ratio, which was shown to be equal to 3. The issue raised in this chapter is to find out what the guidance law would be if we minimized the integral squared of the target jerk rather than the integral squared of the target acceleration. For example, we may want to minimize the jerk of a vehicle due to various possible design constraints such as structural limitations, vehicle aeroelasticity, or possibly sensor limitations. In addition, we may physically have constraints on the rate with which we can change the missile acceleration, such as in thrust vector–controlled missiles, and we may want to introduce these practical limitations to the guidance law. We shall call the new guidance law that minimizes the integral squared of the target jerk "jerk guidance." In this chapter we shall derive the jerk guidance law and compare it to proportional and augmented proportional navigation in terms of its performance against target maneuvers.

DERIVATION OF THE JERK GUIDANCE LAW [1, 2]

The model of the real world used for the development of jerk guidance is slightly different than the model that was used for the development of augmented proportional navigation. It appears in Fig. 10.1.

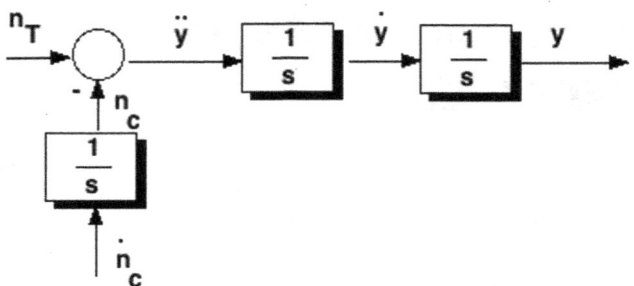

Fig. 10.1   Model for jerk guidance law development.

In the model of Fig. 10.1, we can see that the input guidance command is no longer missile acceleration n_c, but instead missile jerk $\dot{n}_c$. It must be emphasized that using missile jerk as the control may not be realistic; however, we want to see whether there would be any benefit if missile jerk were an input. From Fig. 10.1 we can see that missile jerk is integrated once to yield the commanded missile acceleration n_c. Then the model is the same as the models we used in the past in the sense that relative acceleration $\ddot{y}$ is still target acceleration minus missile acceleration. In addition, the relative acceleration is integrated twice to yield relative position y. At the end of the flight, the miss distance is defined as

$$\text{Miss} = y(t_F)$$

We are trying to develop a guidance law that yields zero miss distance subject to minimizing the integral of the square of missile jerk, or

$$\text{Minimize} \int_0^{t_F} \dot{n}_c^2 \, dt \text{ subject to } y(t_F) = 0$$

We have seen that a system of linear differential equations can always be represented in the following state space form:

$$\dot{x} = Fx + Gu$$

We can express the model of Fig. 10.1 in state space form as

$$\begin{bmatrix} \dot{y} \\ \ddot{y} \\ \dot{n}_T \\ \dot{n}_c \end{bmatrix} = \begin{bmatrix} 0 & 1 & 0 & 0 \\ 0 & 0 & 1 & -1 \\ 0 & 0 & 0 & 0 \\ 0 & 0 & 0 & 0 \end{bmatrix} \begin{bmatrix} y \\ \dot{y} \\ n_T \\ n_c \end{bmatrix} + \begin{bmatrix} 0 \\ 0 \\ 0 \\ 1 \end{bmatrix} \dot{n}_c$$

where F is the 4×4 system dynamics matrix given by

$$F = \begin{bmatrix} 0 & 1 & 0 & 0 \\ 0 & 0 & 1 & -1 \\ 0 & 0 & 0 & 0 \\ 0 & 0 & 0 & 0 \end{bmatrix}$$

and G is the 4×1 vector given by

$$G = \begin{bmatrix} 0 \\ 0 \\ 0 \\ 1 \end{bmatrix}$$

Note that in the preceding formulation n_c *is a state, whereas* $\dot{n}_c$ *is the control in this model of the real world.* The solution to the state space vector differential equation is given at the final time t_F by the vector relationship that was shown in Chapter 8 of the previous volume to be

$$x(t_F) = \Phi(t_F - t)x(t) + \int_t^{t_F} \Phi(t_F - \lambda)G(\lambda)u(\lambda)d\lambda$$

where Φ is the fundamental matrix. In Chapter 1 of the preceding volume we showed that one way of finding the fundamental matrix is by Taylor series expansion, or

$$\Phi(t) = I + Ft + \frac{F^2 t^2}{2} + \frac{F^3 t^3}{6}$$

After some work, we find the fundamental matrix for this problem to be

$$\Phi(t) = \begin{bmatrix} 1 & t & \dfrac{t^2}{2} & -\dfrac{t^2}{2} \\ 0 & 1 & t & -t \\ 0 & 0 & 1 & 0 \\ 0 & 0 & 0 & 1 \end{bmatrix}$$

From the state space equation we can see that

$$u = \dot{n}_c$$

Substitution of the various matrices into the solution to the state space equation yields

$$
\begin{bmatrix} y(t_F) \\ \dot{y}(t_F) \\ n_T(t_F) \\ n_c(t_F) \end{bmatrix} = \begin{bmatrix} 1 & (t_F - t) & \dfrac{(t_F - t)^2}{2} & -\dfrac{(t_F - t)^2}{2} \\ 0 & 1 & (t_F - t) & -(t_F - t) \\ 0 & 0 & 1 & 0 \\ 0 & 0 & 0 & 1 \end{bmatrix} \begin{bmatrix} y(t) \\ \dot{y}(t) \\ n_T(t) \\ n_c(t) \end{bmatrix}
$$

$$
+ \int_t^{t_F} \begin{bmatrix} 1 & (t_F - \lambda) & \dfrac{(t_F - \lambda)^2}{2} & -\dfrac{(t_F - \lambda)^2}{2} \\ 0 & 1 & (t_F - \lambda) & -(t_F - \lambda) \\ 0 & 0 & 1 & 0 \\ 0 & 0 & 0 & 1 \end{bmatrix} \begin{bmatrix} 0 \\ 0 \\ 0 \\ 1 \end{bmatrix} \dot{n}_c(\lambda) d\lambda
$$

Multiplying out the preceding equation yields four scalar equations. The first of the four scalar equations is

$$
y(t_F) = y(t) + (t_F - t)\dot{y}(t) + \frac{(t_F - t)^2}{2} n_T(t) - \frac{(t_F - t)^2}{2} n_c(t) - \int_t^{t_F} \frac{(t_F - \lambda)^2}{2} \dot{n}_c d\lambda
$$

In order to simplify the preceding equation, we can define f_1 and h_1 to be

$$
f_1(t_F - t) = y(t) + (t_F - t)\dot{y}(t) + \frac{(t_F - t)^2}{2} n_T(t) - \frac{(t_F - t)^2}{2} n_c(t)
$$

$$
h_1(t_F - \lambda) = \frac{(t_F - \lambda)^2}{2}
$$

Therefore, the scalar equation becomes

$$
y(t_F) = f_1(t_F - t) - \int_t^{t_F} h_1(t_F - \lambda)\dot{n}_c(\lambda) d\lambda
$$

For zero miss distance $y(t_F) = 0$, the preceding equation becomes

$$
f_1(t_F - t) = \int_t^{t_F} h_1(t_F - \lambda)\dot{n}_c(\lambda) d\lambda
$$

Applying the Schwartz inequality to the preceding expression yields

$$f_1^2(t_F - t) \leq \int_t^{t_F} h_1^2(t_F - \lambda) \int_t^{t_F} \dot{n}_c^2(\lambda) d\lambda$$

Inverting the previous equation and expressing the inequality in terms of the integral of the jerk command squared yields

$$\int_t^{t_F} \dot{n}_c^2(\lambda) d\lambda \geq \frac{f_1^2(t_F - t)}{\int_t^{t_F} h_1^2(t_F - \lambda)}$$

In the preceding equation, the integral of the square of the missile jerk is minimized when the equality sign holds. According to the Schwartz inequality, the equality sign holds when

$$\dot{n}_c(\lambda) = k h_1(t_F - \lambda)$$

where k is a constant. Therefore, substitution yields

$$\int_t^{t_F} k^2 h_1^2(t_F - \lambda) d\lambda = \frac{f_1^2(t_F - t)}{\int_t^{t_F} h_1^2(t_F - \lambda)}$$

Because k is a constant, it can be brought outside the integral. Solving for the constant k in the preceding equation yields

$$k^2 = \frac{f_1^2(t_F - t)}{\int_t^{t_F} h_1^2(t_F - \lambda) \int_t^{t_F} h_1^2(t_F - \lambda)}$$

After taking the square root of both sides of the preceding equation, we find that the constant k can be found from

$$k = \frac{f_1(t_F - t)}{\int_t^{t_F} h_1^2(t_F - \lambda)}$$

and the commanded jerk that minimizes the integral of the jerk squared is given by

$$\dot{n}_c = \frac{f_1(t_F - t)}{\int_t^{t_F} h_1^2(t_F - \lambda)} h_1(t_F - t)$$

Substitution of the definitions of f_1 and h_1 into the preceding equation yields the jerk guidance law

$$\dot{n}_c = \frac{10}{t_{go}^3}(y + \dot{y}t_{go} + 0.5\ddot{y}t_{go}^2)$$

PRELIMINARY EVALUATION OF JERK GUIDANCE LAW

In all the previous guidance laws, we worked with a control that was the commanded acceleration. In the jerk guidance law, the command is the missile jerk. For academic purposes we first have to find a way of implementing proportional navigation in terms of a jerk command so that proportional navigation can be compared to jerk guidance. First we recall that proportional navigation is given by

$$n_c = N'V_c\dot{\lambda}$$

where N' is the effective navigation ratio, which is a designer-chosen constant; V_c is the closing velocity; and $\dot{\lambda}$ is the line-of-sight rate. If N' and V_c are constants, then we can take the derivative of both sides of the preceding equation, yielding

$$\dot{n}_c = N'V_c\ddot{\lambda}$$

where $\ddot{\lambda}$ is the line-of-sight acceleration. In order to find an expression for the line-of-sight acceleration, we first start with the definition of the line-of-sight angle as

$$\lambda = \frac{y}{R_{TM}} = \frac{y}{V_c t_{go}} = \frac{y}{V_c(t_F - t)}$$

where t_F is the intercept time or the final time and R_{TM} is the range from the missile to the target. Using the quotient rule from calculus, we can take the derivative of the preceding expression and obtain an expression for the line-of-sight rate as

$$\dot{\lambda} = \frac{R_{TM}\dot{y} - y\dot{R}_{TM}}{R_{TM}^2} = \frac{R_{TM}\dot{y} + yV_c}{R_{TM}^2} = \frac{V_c t_{go}\dot{y} + yV_c}{V_c^2 t_{go}^2} = \frac{y + \dot{y}t_{go}}{V_c t_{go}^2}$$

because

$$V_c = -\dot{R}_{TM}$$

Taking the derivative again of the line-of-sight rate using the quotient rule from calculus (and after much algebra), we obtain the expression for the line-of-sight acceleration as

$$\ddot{\lambda} = \frac{2(y + \dot{y}t_{go} + 0.5\ddot{y}t_{go}^2)}{V_c t_{go}^3}$$

Therefore, we can express proportional navigation as a jerk command given by

$$\dot{n}_c = N'V_c\ddot{\lambda} = \frac{2N'V_c(y + \dot{y}t_{go} + 0.5\ddot{y}t_{go}^2)}{V_c t_{go}^3} = \frac{2N'(y + \dot{y}t_{go} + 0.5\ddot{y}t_{go}^2)}{t_{go}^3}$$

Let us call the preceding guidance law "jerk proportional navigation." In other words, jerk proportional navigation is proportional navigation issued with a jerk command rather than an acceleration command.

In a simulation of jerk proportional navigation, we have to integrate the preceding equation. Recall that integration requires an initial condition on n_c. Assuming that there is no step in target displacement [i.e., $y(0) = 0$], the initial condition on the relative velocity is given by

$$\dot{y}(0) = -V_M HE$$

Therefore, the initial condition on the acceleration command is given by

$$n_c(0) = N'V_c\dot{\lambda}(0) = N'V_c\left[\frac{V_c t_F \dot{y}(0)}{V_c^2 t_F^2}\right] = \frac{N'\dot{y}(0)}{t_F}$$

Listing 10.1 is an implementation of both jerk proportional navigation and the new jerk guidance law. The only difference between the two guidance laws is the value of the effective navigation ratio. In the new guidance law the effective navigation ratio is 10; we will assume a value of 3 for jerk proportional navigation. From Listing 10.1 we can see that we can have a constant 3-g maneuver (XNTIC=96.6, IMVR=1) or a 3-g vertical-S maneuver at 3 rad/s (IMVR=2, XNTIC=96.6, W=3). In addition, a jerk limit XNCDLIM is available. (XNCDLIM=99999 is default, which means no jerk limit.) The two guidance laws in Listing 10.1 are jerk proportional navigation (QPN=1) and jerk guidance (QPN=0). A calculation is also made of the performance index J, which is the integral of the missile jerk squared.

LISTING 10.1 SIMULATION TO COMPARE JERK GUIDANCE TO JERK PROPORTIONAL NAVIGATION

```
clear
count=0;
QPN=1;
IMVR=1;
VC=4000.;
Y=0.;
VM=3000.;
XNTIC=96.6;
HEDEG=0.;
```

```
W=3.;
TF=10.;
XNP=3.;
XNPP=10.;
XNCDLIM=999999.;
YD=-VM*HEDEG/57.3;
T=0.;
H=.001;
S=0.;
XNC=XNP*YD/TF;
J=0.;
while T<=(TF - 1e-5)
   YOLD=Y;
   YDOLD=YD;
   XNCOLD=XNC;
   JOLD=J;
   STEP=1;
   FLAG=0;
   while STEP <=1
      if FLAG==1
         Y=Y+H*YD;
         YD=YD+H*YDD;
         XNC=XNC+H*XNCD;
         J=J+H*JD;
         T=T+H;
         STEP=2;
      end
      TGO=TF-T+.00001;
      if IMVR==1
         XNT=XNTIC;
      else
         if sin(W*T)>0.
            XNT=XNTIC;
         end
         if sin(W*T)<=0
            XNT=-XNTIC;
         end
      end
      YDD=XNT-XNC;
      if QPN==1
         XNCD=2.*XNP*(Y+YD*TGO+.5*YDD*TGO*TGO)/TGO^3;
      else
         XNCD=XNPP*(Y+YD*TGO+.5*YDD*TGO*TGO)/TGO^3;
      end
      if XNCD>XNCDLIM
```

```
                XNCD=XNCDLIM;
            end
              if XNCD<-XNCDLIM
                  XNCD=-XNCDLIM;
            end
            JD=XNCD^2;
            FLAG=1;
         end
         FLAG=0;
         Y=.5*(YOLD+Y+H*YD);
         YD=.5*(YDOLD+YD+H*YDD);
         XNC=.5*(XNCOLD+XNC+H*XNCD);
         J=.5*(JOLD+J+H*JD);
         S=S+H;
            if S>=.09999
            S=0.;
            count=count+1;
            ArrayT(count)=T;
            ArrayY(count)=Y;
            ArrayXNCG(count)=XNC/32.2;
            ArrayXNCDG(count)=XNCD/32.2;
            ArrayXNTG(count)=XNT/32.2;
            end
end
figure
plot(ArrayT,ArrayY),grid
xlabel('Time (s)')
ylabel('Y (ft) ')
figure
plot(ArrayT,ArrayXNCG),grid
xlabel('Time (s)')
ylabel('XNC (g) ')
figure
plot(ArrayT,ArrayXNCDG),grid
xlabel('Time (s)')
ylabel('XNCD (g/s) ')
figure
plot(ArrayT,ArrayXNTG),grid
xlabel('Time (s)')
ylabel('XNT (g) ')
clc
output=[ArrayT',ArrayY',ArrayXNCG',ArrayXNCDG',ArrayXNTG'];
save datfil.txt output -ascii
disp 'simulation finished'
Y
```

The nominal case of Listing 10.1 was run in which there was a 3-g constant target maneuver (IMVR=1, XNTIC=96.6) and both jerk proportional navigation with a navigation ratio of 3 (QPN=1, XNP=3) and jerk guidance (QPN=0, XNPP=10) were run. We can see the differences in relative position y in Fig. 10.2 for both guidance laws. Both guidance laws yield zero miss because there are no limits on the jerk guidance command (XNCDLIM=999999.). Jerk guidance has a tighter relative trajectory because of its higher effective navigation ratio.

Figure 10.3 shows that both guidance laws have monotonically increasing acceleration profiles against the 3-g constant target maneuver; however, jerk guidance has a smaller maximum acceleration at the end of the flight. Because jerk proportional navigation has an effective navigation ratio of 3, it requires 9 g (or three times the acceleration capability of the target) whereas jerk guidance only requires 5 g at the end of the flight to hit the target. However, we know from Chapter 2 of Volume 1 that the maximum acceleration required by a missile using proportional navigation with an effective navigation ratio of 3 is given by

$$n_c(t_F) = \frac{N'}{N'-2}n_T = \frac{3}{3-2}*96.6 = 290ft/s^2 = 9\,g$$

However, if the effective navigation ratio is increased to 5, we get

$$n_c(t_F) = \frac{N'}{N'-2}n_T = \frac{5}{5-2}*96.6 = 161ft/s^2 = 5\,g$$

which is what Fig. 10.3 is showing for jerk guidance. *In other words, jerk guidance may be exhibiting the same behavior as proportional navigation with an effective navigation ratio of 5!*

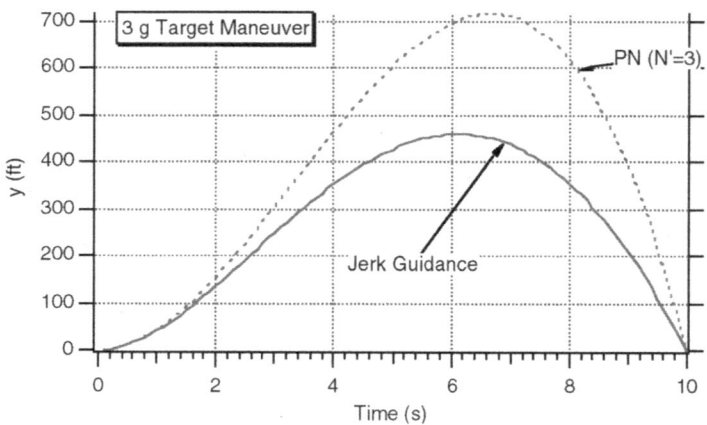

Fig. 10.2   Jerk guidance has a tighter relative trajectory than jerk proportional navigation.

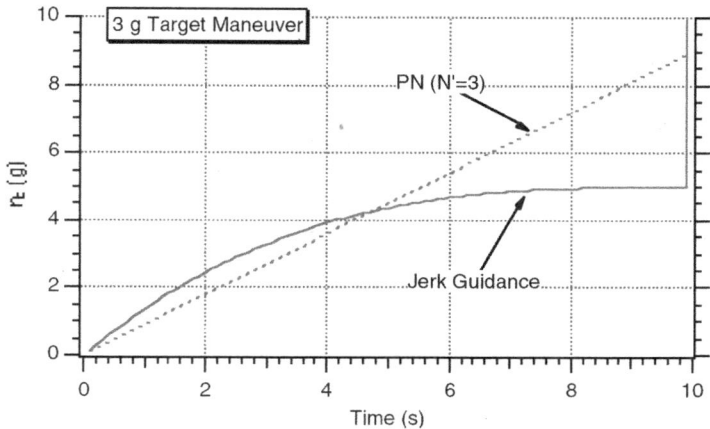

Fig. 10.3   Jerk guidance requires less maximum acceleration than jerk proportional navigation.

Figure 10.4 shows that the jerk command for jerk guidance is monotonically decreasing, whereas the jerk command for jerk proportional navigation due to a 3-g target maneuver is a constant. It is difficult to tell from Fig. 10.4 which jerk command is more desirable. Recall that the purpose of the jerk guidance law is to minimize the integral of the jerk command squared over the time of flight. In Listing 10.1 we integrate the differential equation

$$ j = \dot{n}_c^2 $$

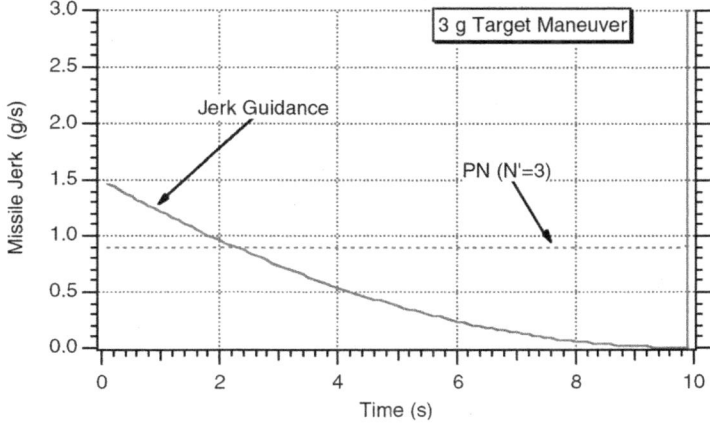

Fig. 10.4   Jerk guidance command for jerk guidance goes to zero at the end of the flight.

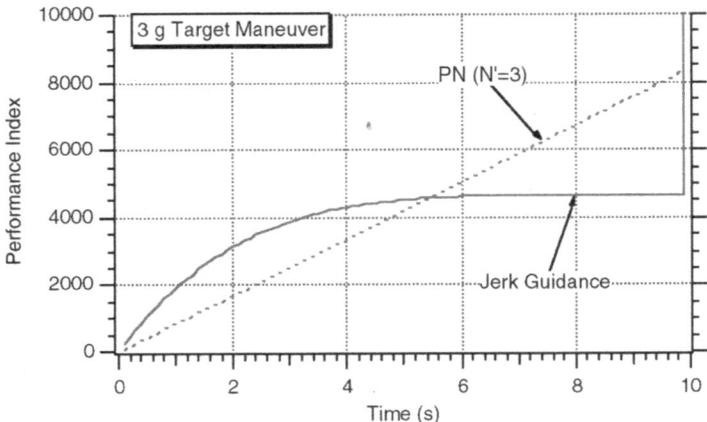

Fig. 10.5   Jerk guidance has a smaller performance index than jerk proportional navigation at the end of the flight.

to find which guidance law yields a smaller performance index J. Figure 10.5 shows that the performance index at the end of the flight for jerk guidance is smaller than that of jerk proportional navigation.

Another set of runs comparing both guidance laws was performed with Listing 10.1 when the only error source was a −20-deg heading error (HEDEG=-20). We can see the differences in relative position y in Fig. 10.6 for both guidance laws. Both guidance laws yield zero miss because there are no limits on the jerk guidance command (XNCDLIM=999999.). In this example, unlike the previous

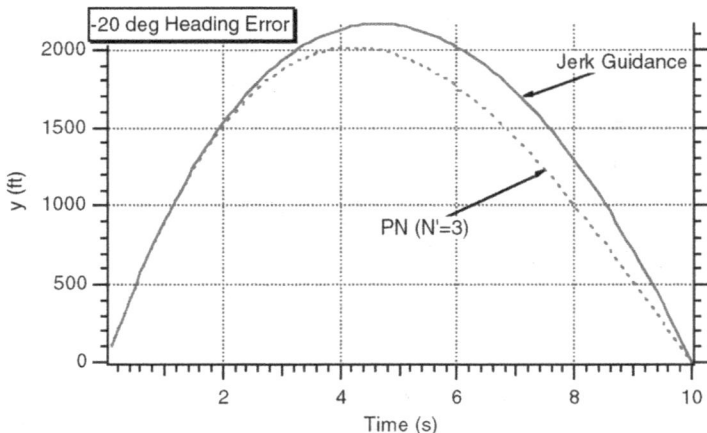

Fig. 10.6   Jerk proportional navigation has a slightly tighter relative trajectory than jerk guidance due to a heading error disturbance.

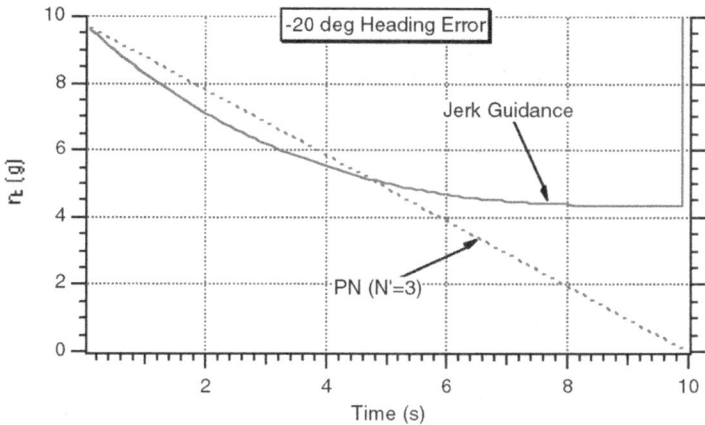

Fig. 10.7   Both guidance laws yield the same maximum acceleration due to heading error.

example where target maneuver was the error source, jerk proportional navigation has a slightly tighter relative trajectory than jerk guidance.

We can see from Fig. 10.7 that both guidance laws have the same maximum acceleration due to heading error at the beginning of the flight. However, the commanded acceleration for jerk proportional navigation goes to zero at the end of the flight, whereas the commanded acceleration due to jerk guidance approaches a nonzero steady-state value.

Figure 10.8 shows that the jerk command for jerk guidance is monotonically decreasing, whereas the jerk command for proportional navigation due to a

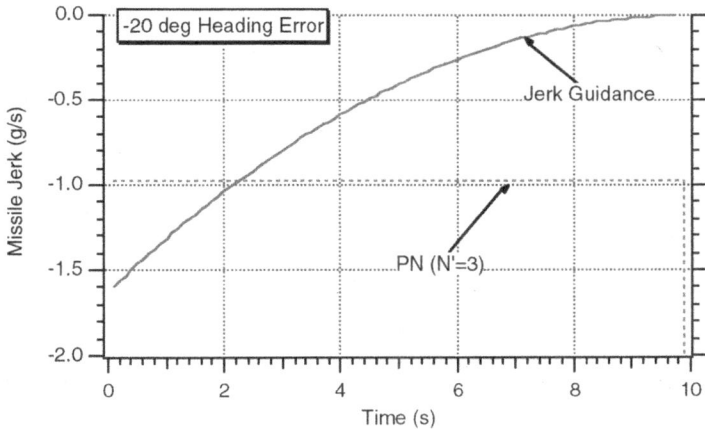

Fig. 10.8   Jerk guidance command for jerk guidance goes to zero at the end of the flight due to heading error disturbance.

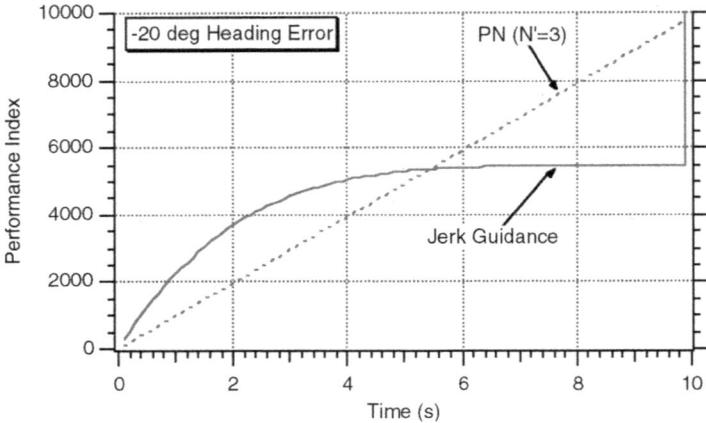

Fig. 10.9 **Jerk guidance has a smaller performance index than jerk proportional navigation at the end of the flight due to heading error disturbance.**

–20-deg heading error is a constant. It is difficult to tell from Fig. 10.8 which jerk command is more desirable. Figure 10.9 shows that the performance index for jerk guidance is smaller than that of jerk proportional navigation at the end of the flight.

Next, a more stressing vertical-S maneuver case was run with Listing 10.1 (IMVR=2); the resultant 3-g, 3-rad/s target maneuver is shown in Fig. 10.10. The

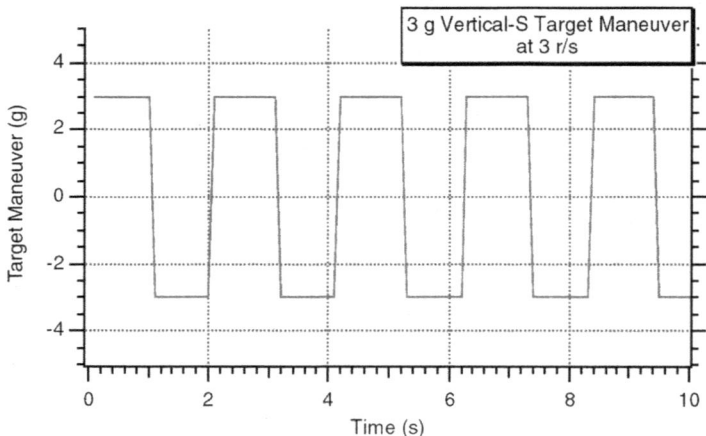

Fig. 10.10 **Vertical-S target maneuver used to test both jerk proportional navigation and jerk guidance.**

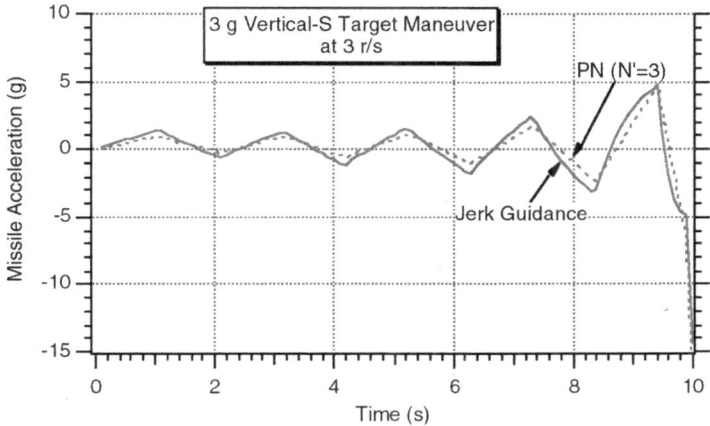

Fig. 10.11   Missile acceleration response for both jerk proportional navigation and jerk guidance are similar.

missile acceleration response is shown in Fig. 10.11. Here we can see that the acceleration response of both jerk proportional navigation and jerk guidance is virtually identical; however, the missile jerk response for both guidance laws is different, as shown in Fig. 10.12. The missile jerk commands appear to follow the target maneuver for jerk proportional navigation (flat response, which gets larger as flight time increases). The response of jerk guidance is similar except

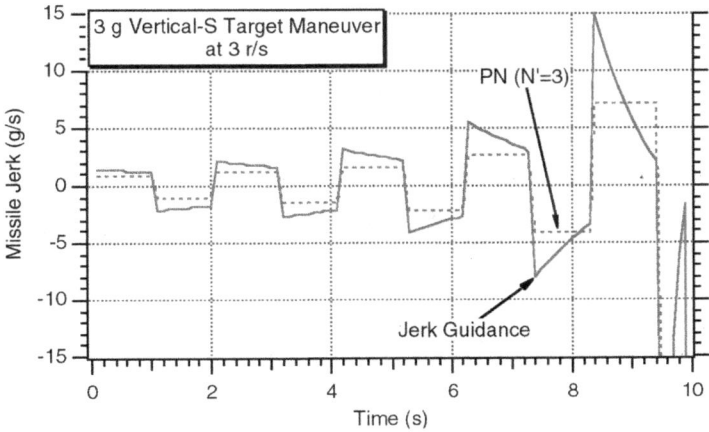

Fig. 10.12   Jerk guidance appears to require larger jerk commands than jerk proportional navigation.

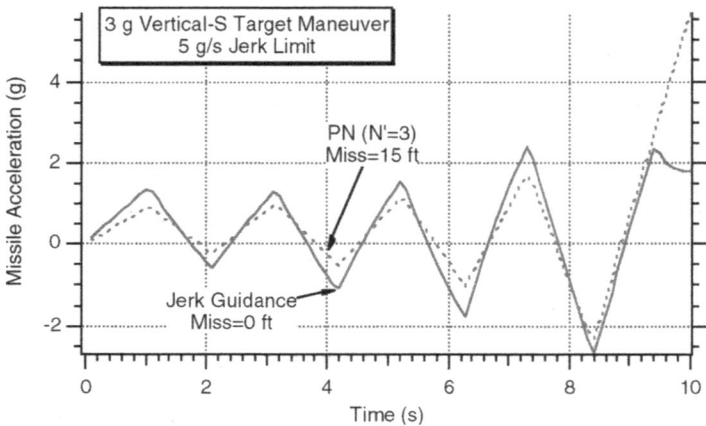

Fig. 10.13   Missile acceleration is larger with jerk proportional navigation than with jerk guidance when there is a 5-*g* jerk limit.

the jerk commands are not flat but peak and then decay. It appears that the maximum value of the jerk commands for jerk guidance exceeds that of proportional navigation. However, when a jerk limit of 5-g/s is placed on the jerk command (XNCDLIM=161), we can see from Figs. 10.13 and 10.14 that the miss due to jerk guidance is zero whereas the miss due to jerk proportional navigation is 15 ft. Figure 10.13 shows that the resultant missile acceleration is larger with proportional navigation than with jerk guidance. Figure 10.14 shows that the cause of the miss for jerk proportional navigation is jerk saturation.

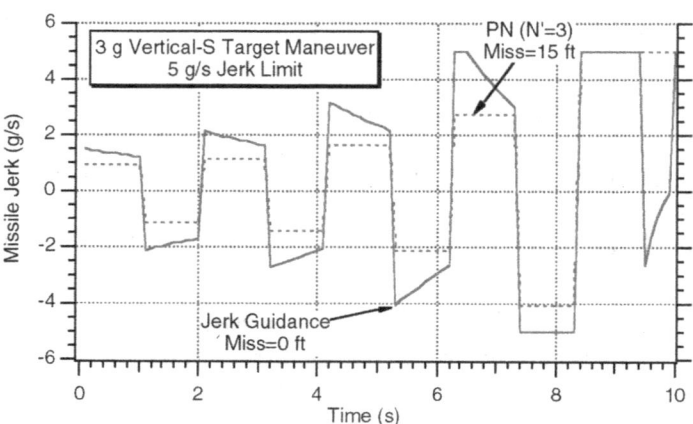

Fig. 10.14   Missile jerk saturates with jerk proportional navigation when there is a 5-g/s jerk limit.

ALTERNATE FORM OF JERK GUIDANCE

Recall that the expression for the jerk guidance command was given by

$$\dot{n}_c = \frac{10}{t_{go}^3}(y + \dot{y}t_{go} + 0.5\ddot{y}t_{go}^2)$$

However, in the previous section we also showed that the line-of-sight acceleration could be expressed by

$$\ddot{\lambda} = \frac{2(y + \dot{y}t_{go} + 0.5\ddot{y}t_{go}^2)}{V_c t_{go}^3}$$

Therefore, by comparing the two preceding equations we can say that

$$y + \dot{y}t_{go} + 0.5\ddot{y}t_{go}^2 = \frac{V_c t_{go}^3 \ddot{\lambda}}{2}$$

Therefore, we can rewrite the jerk guidance law as

$$\dot{n}_c = \frac{10}{t_{go}^3}\left(\frac{V_c t_{go}^3 \ddot{\lambda}}{2}\right) = 5V_c\ddot{\lambda}$$

However, if we integrate both sides of the preceding equation, assuming constant closing velocity, we obtain the jerk guidance law in terms of the acceleration command

$$n_c = 5V_c\dot{\lambda}$$

which is our old friend proportional navigation with an effective navigation ratio of 5. *Can it be that proportional navigation with an effective navigation ratio of 5 also minimizes the jerk?* Let's find out.

Recall from Chapter 2 of the preceding volume that the missile acceleration due to target maneuver for a proportional navigation guidance system is given by

$$n_c = \frac{N'}{N'-2}\left[1 - \left(1 - \frac{t}{t_F}\right)^{N'-2}\right]n_T = \frac{N'}{N'-2}n_T - \frac{N'}{N'-2}n_T\left(1 - \frac{t}{t_F}\right)^{N'-2}$$

Taking derivatives of both sides of the equation yields an expression for the missile jerk as

$$\dot{n}_c = -N'n_T\left(1 - \frac{t}{t_F}\right)^{N'-3}\left(\frac{-1}{t_F}\right) = \frac{N'n_T}{t_F}\left(1 - \frac{t}{t_F}\right)^{N'-3}$$

Finding the integral of the square of the jerk yields

$$z = \int_0^{t_F} \dot{n}_c^2 dt = \frac{N'^2 n_T^2}{t_F^2} \int_0^{t_F} \left(1 - \frac{t}{t_F}\right)^{2N'-6} dt = \frac{N'^2 n_T^2}{t_F^2} \left(\frac{1}{2N' - 5}\right)$$

To find the best N' to minimize z we take the derivative of z and set it equal to zero, yielding

$$\frac{dz}{dN'} = 0 = N'(2N' - 10)$$

The preceding equation has two solutions, which are $N' = 0$ and $N' = 5$. Certainly, setting N' to zero will minimize the missile jerk, but you will also miss the target. *The preceding equation tells us that proportional navigation with an effective navigation ratio of 5 is identical to the jerk guidance law.*

SUMMARY

This chapter introduced a new guidance law known as jerk guidance. It was shown that for the constant target maneuver and heading error cases, the new guidance law minimized the integral of the missile jerk squared. It was also shown that for a very stressing vertical-S target maneuver, the jerk guidance law could yield smaller miss distances than proportional navigation implemented as a missile jerk command when the missile jerk was limited. It was shown that the new guidance law was equivalent to proportional navigation with an effective navigation ratio of 5.

REFERENCES

[1]   Grinfeld, N., and Ben-Asher, J., "Minimal Jerk Guidance Law," *Journal of Guidance, Control and Dynamics*, Vol. 38, 2015, pp. 1520–1525.

[2]   Grinfeld, N., "An Optimal Missile Guidance Law with a Jerk Constraint," master s thesis in Aerospace Engineering, Technion, Israel Institute of Technology, March 2014.

CHAPTER 11

Use of Poisson/Singer Target Maneuver Model for Kalman Filter Design

INTRODUCTION

So far we have considered random step and weave maneuvers for the development of Kalman filters for the homing loop. There are other popular target maneuver models that we have not yet considered, such as the Singer and Poisson models shown in Figs. 11.1 and 11.2. Basically these maneuvers are at the maximum positive or negative target acceleration levels; the switching between levels is random but does not follow a uniform distribution.

We can see from the preceding two figures that both maneuver models appear to vary between maximum positive and negative target accelerations. The Poisson model is also known as the random telegraph signal model [1] and is quite popular in signal processing work. Basically the Poisson model says that the maneuver amplitude is either $+n_T$ or $-n_T$. The length of time for which the maneuver

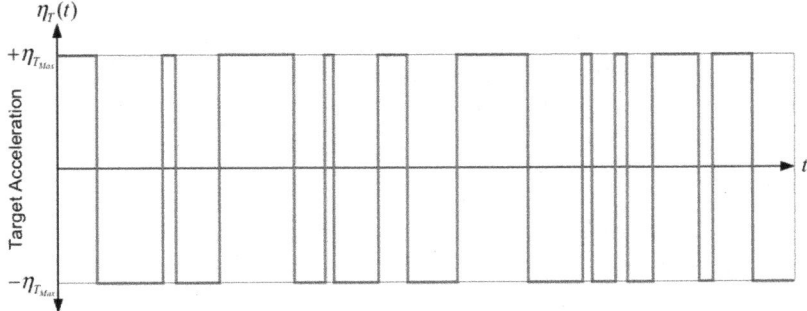

Fig. 11.1 Poisson target maneuver model.

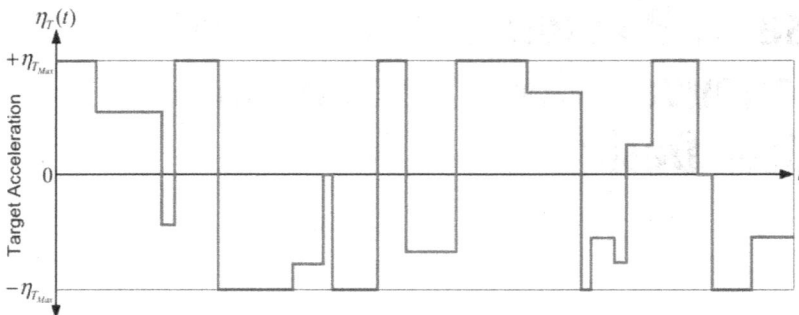

Fig. 11.2   Singer target maneuver model.

remains in either position is random. The number of sign changes (zero crossings) follows a Poisson distribution. If $P(k)$ represents the probability of k sign changes in t_F seconds, then we can say that

$$P(k) = \frac{(vt_F)^k e^{-t/t_F}}{k!}$$

where k is the number of sign changes and v is the average number of zero crossings per second. The Poisson maneuver is probably the most stressing type of target maneuver for a missile guidance system and is often used for worst case scenarios in missile design.

The Singer model [2] is also quite popular in Kalman filter design, and we can see from Fig. 11.2 that it looks quite similar to the Poisson model.

SHAPING NETWORKS FOR POISSON AND SINGER TARGET MANEUVER MODELS

For a step target maneuver whose starting time is equally likely to occur between 0 and t_F seconds, we showed in Chapter 4 of the previous volume that the shaping network for such a random target maneuver is white noise through an integrator when part of the missile homing loop, as shown in Fig. 11.3.

In Fig. 11.3, u_s is a white input with power spectral density

$$\Phi_s = \frac{\beta^2}{t_F}$$

where β is the maximum acceleration level of the target and t_F is the time of flight.

The Poisson target maneuver model can be represented by white noise through a low-pass filter and is sometimes called the random telegraph signal [1]. The Poisson target maneuver model when part of the homing loop appears in Fig. 11.4.

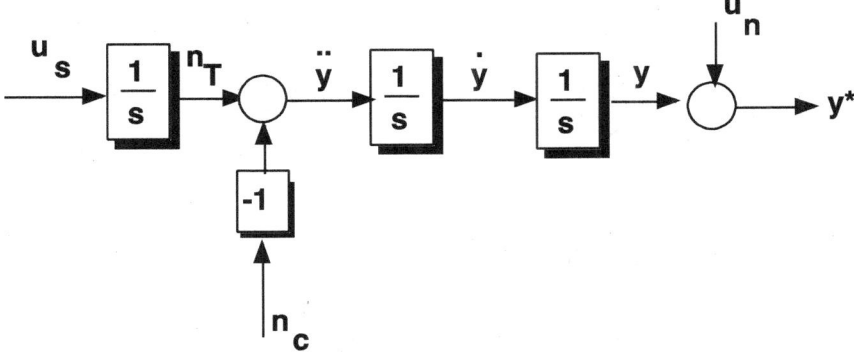

Fig. 11.3   Uniformly distributed constant target maneuver model as part of the homing loop.

In the Poisson target maneuver model, u_s is white noise with power spectral density

$$\Phi_s = \frac{\beta^2}{\nu}$$

where β is the maximum acceleration level of the target and ν is the number of zero crossings per second.

The Singer model was also shown in [2] to be white noise through a low-pass filter. The Singer target maneuver model when part of the homing loop appears in Fig. 11.5.

In the Singer target maneuver model, u_s is white noise with power spectral density

$$\Phi_s = \frac{2\beta^2}{\tau}$$

where τ is a time constant and β is again the maximum acceleration level of the target. All three models discussed so far are very similar. If ν is small, then the

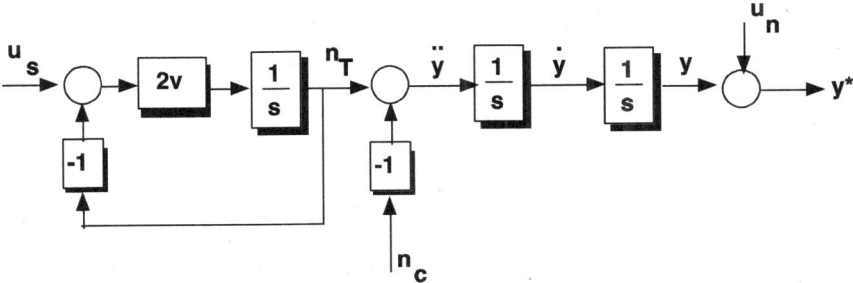

Fig. 11.4   Poisson target maneuver model.

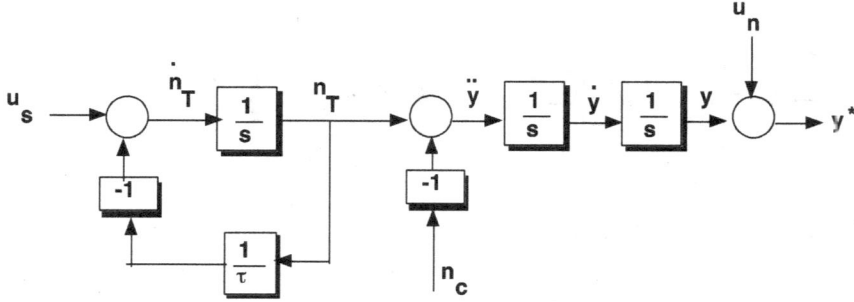

Fig. 11.5   Singer target maneuver model.

systems dynamics matrices of all three models are nearly identical. Both the Singer and Poisson models appear to be very similar because they both have low-pass filter representations of the target maneuver. Eventually in this chapter we will be building Kalman filters based on the Poisson or Singer models, so it is important to look at both models from a state space point of view when they are part of the homing loop.

It was shown in Chapter 9 of Volume 1 that the state space representation of the normal Kalman filter of Fig. 11.3 and the continuous process noise matrix were given by

$$
\begin{bmatrix} \dot{y} \\ \ddot{y} \\ \dot{n}_T \end{bmatrix} = \begin{bmatrix} 0 & 1 & 0 \\ 0 & 0 & 1 \\ 0 & 0 & 0 \end{bmatrix} \begin{bmatrix} y \\ \dot{y} \\ n_T \end{bmatrix} + \begin{bmatrix} 0 \\ -1 \\ 0 \end{bmatrix} n_c + \begin{bmatrix} 0 \\ 0 \\ 1 \end{bmatrix} u_s
$$

$$
Q_{\text{Normal}} = E(ww^T) = \begin{bmatrix} 0 \\ 0 \\ 1 \end{bmatrix} \begin{bmatrix} 0 & 0 & 1 \end{bmatrix} \Phi_s = \begin{bmatrix} 0 & 0 & 0 \\ 0 & 0 & 0 \\ 0 & 0 & 1 \end{bmatrix} \frac{\beta^2}{t_F} = \frac{\beta^2}{t_F} \begin{bmatrix} 0 & 0 & 0 \\ 0 & 0 & 0 \\ 0 & 0 & 1 \end{bmatrix}
$$

The Poisson target maneuver model of Fig. 11.4 can be represented in state space form along with the continuous process noise matrix as

$$
\begin{bmatrix} \dot{y} \\ \ddot{y} \\ \dot{n}_T \end{bmatrix} = \begin{bmatrix} 0 & 1 & 0 \\ 0 & 0 & 1 \\ 0 & 0 & -2v \end{bmatrix} \begin{bmatrix} y \\ \dot{y} \\ n_T \end{bmatrix} + \begin{bmatrix} 0 \\ -1 \\ 0 \end{bmatrix} n_c + \begin{bmatrix} 0 \\ 0 \\ 2v \end{bmatrix} u_s
$$

$$
Q_{\text{Poisson}} = E(ww^T) = \begin{bmatrix} 0 \\ 0 \\ 2v \end{bmatrix} \begin{bmatrix} 0 & 0 & 2v \end{bmatrix} \Phi_s = \begin{bmatrix} 0 & 0 & 0 \\ 0 & 0 & 0 \\ 0 & 0 & 4v^2 \end{bmatrix} \frac{\beta^2}{v} = 4v\beta^2 \begin{bmatrix} 0 & 0 & 0 \\ 0 & 0 & 0 \\ 0 & 0 & 1 \end{bmatrix}
$$

The Singer model of Fig. 11.5 can be expressed in state space form along with the continuous process noise matrix as

$$\begin{bmatrix} \dot{y} \\ \ddot{y} \\ \dot{n}_T \end{bmatrix} = \begin{bmatrix} 0 & 1 & 0 \\ 0 & 0 & 1 \\ 0 & 0 & -\dfrac{1}{\tau} \end{bmatrix} \begin{bmatrix} y \\ \dot{y} \\ n_T \end{bmatrix} + \begin{bmatrix} 0 \\ -1 \\ 0 \end{bmatrix} n_c + \begin{bmatrix} 0 \\ 0 \\ 1 \end{bmatrix} u_s$$

$$Q_{\text{Singer}} = E(ww^T) = \begin{bmatrix} 0 \\ 0 \\ 1 \end{bmatrix} \begin{bmatrix} 0 & 0 & 1 \end{bmatrix} \Phi_s = \frac{2\beta^2}{\tau} \begin{bmatrix} 0 & 0 & 0 \\ 0 & 0 & 0 \\ 0 & 0 & 1 \end{bmatrix}$$

If v of the Poisson model is small and if τ of the Singer model is large, all three systems dynamics matrices are nearly identical. By comparing the process noise values for the Poisson and Singer target maneuver models we can see that the target maneuver representations are equivalent if

$$4v\beta^2 = \frac{2\beta^2}{\tau}$$

which yields

$$\tau = \frac{1}{2v}$$

By comparing the systems dynamics matrix of both Poisson and Singer models we can see that if

$$-\frac{1}{\tau} = -2v$$

then

$$\tau = \frac{1}{2v}$$

The preceding equations mean that both the Poisson and Singer models have identical state space equations and process noise models, provided the time constant of the Singer model matches the number of zero crossings in the Poisson model of the preceding relationship.

In order to build a discrete Kalman filter using either the Poisson or Singer target maneuver model, we need to find the fundamental matrix. We can use the following three-term Taylor series approximation (see Chapter 1 of the previous volume) to get the fundamental matrix for the systems dynamics matrix as

$$\Phi = I + FT_s + 0.5F^2T_s^2$$

It is important to note that the fundamental matrix will only be used in the Riccati equations. Propagation of the state estimates over one sampling interval, required by a Kalman filter, will be obtained by numerical integration of the state space equations. Theoretically we can compute the discrete process noise matrix from the continuous process noise matrix according to

$$Q_k = \int_0^{T_s} \Phi(\tau) Q \Phi^T(\tau) d\tau$$

For simplicity, if we assume that the continuous process noise matrix is constant over the sampling time, we can approximate the preceding equation as

$$Q_k \approx \Phi(\tau) Q \Phi^T(\tau) * T_s$$

MODELING THE POISSON TARGET MANEUVER

Soon we will want to model the Poisson square wave target maneuver as an error disturbance in a Monte Carlo simulation. In this section we will first present the steps in modeling the Poisson target maneuver, and then demonstrate via the method of adjoints that the shaping filter approach in representing the Poisson square wave as white noise through a simple shaping network is correct when no saturation effects occur.

The two steps in simulating an actual Poisson square wave (i.e., not the shaping filter equivalent) in a forward simulation are as follows:

1. Determine the initial sign of the square wave from a Gaussian random number generator with zero mean and unity standard deviation.

2. The length of time between sign changes $\Delta\tau$ is determined by squaring and adding two Gaussian distributions with zero mean and standard deviation given by

$$\sigma = \frac{1}{\sqrt{2v}}$$

The second step for making the Poisson target maneuver is displayed in Fig. 11.6.

In Chapter 4 of the previous volume, Fig. 4.15 modeled a single-time-constant proportional navigation guidance system whose only error source was a uniformly distributed target maneuver. A Monte Carlo simulation of that guidance system appeared in Listing 4.5 of the previous volume. Listing 11.1 modifies Listing 4.5 by replacing the uniformly distributed step target maneuver with the Poisson target maneuver. Statements that pertain to the modeling of the Poisson target maneuver have been highlighted in boldface. In addition, we can see from

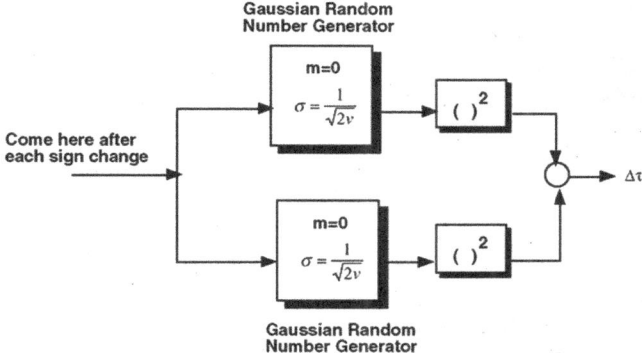

Fig. 11.6   Simulating Poisson target maneuver.

Listing 11.1 that 1000 runs are made (rather than the 50 runs in Listing 4.5 of the previous volume) for each flight time and that the flight time is incremented every 0.2 s (rather than every second in Listing 4.5) in order to get better accuracy.

LISTING 11.1   MONTE CARLO SIMULATION OF SINGLE-TIME-CONSTANT GUIDANCE SYSTEM DRIVEN BY POISSON TARGET MANEUVER

```
clear
count=0;
VC=4000;
XNT=96.6;
VM=3000;
XNP=3;
TAU=1;
RUN=1000;
BETA=96.6;
XNU=.5;
for TF=.2:.2:10,
    Z1=0;
    for I=1:RUN
        QFIRST=1;
        SIG=1./sqrt(2.*XNU);
        PZ=rand(1);
        PZ=PZ-.5;
        if PZ>0
            COEF=1;
        else
            COEF=-1;
```

```
       end;
       XNT=COEF*BETA;
       DELT=9999.;
       TNOW=0;
  Y=0;
  YD=0;
  T=0;
  H=.01;
  S=0;
  XNC=0;
  XNL=0;
  while T<=(TF - 1e-5)
  if QFIRST==1
                      XNOISE1=SIG*randn;
                      XNOISE2=SIG*randn;
                      DELT=XNOISE1^2+XNOISE2^2;
                      QFIRST=0;
                      TNOW=T;
              end;
              if T>=(DELT+TNOW)
                XNT=-XNT;
                QFIRST=1;
              end
  YOLD=Y;
  YDOLD=YD;
  XNLOLD=XNL;
  STEP=1;
  FLAG=0;
  while STEP <=1
    if FLAG==1
          Y=Y+H*YD;
          YD=YD+H*YDD;
          XNL=XNL+H*XNLD;
          T=T+H;
          STEP=2;
  end
  TGO=TF-T+.00001;
  RTM=VC*TGO;
  XLAMD=(RTM*YD+Y*VC)/(RTM^2);
  XNC=XNP*VC*XLAMD;
  XNLD=(XNC-XNL)/TAU;
  YDD=XNT-XNL;
  FLAG=1;
  end;
  FLAG=0;
  Y=.5*(YOLD+Y+H*YD);
```

```
      YD=.5*(YDOLD+YD+H*YDD);
      XNL=.5*(XNLOLD+XNL+H*XNLD);
      S=S+H;
    end;
    Z(I)=Y;
    Z1=Z(I)+Z1;
    XMEAN=Z1/I;
      end;
      SIGMA=0;
      Z1=0;
    Z2=0.;
      for I=1:RUN
      Z1=(Z(I)-XMEAN)^2+Z1;
      Z2=Z(I)^2+Z2;
      if I==1
      SIGMA=0;
        RMS=0.;
      else
      SIGMA=sqrt(Z1/(I-1));
        RMS=sqrt(Z2/(I-1));
      end
      end;
      count=count+1;
      ArrayTF(count)=TF;
      ArrayRMS(count)=RMS;
end;
figure
plot(ArrayTF,ArrayRMS)
title('Shaping filter Monte Carlo results')
xlabel('Time')
ylabel('RMS Miss (ft)')
clc
output=[ArrayTF',ArrayRMS'];
save datfil.txt output -ascii
disp 'simulation finished'
```

As was mentioned at the beginning of this chapter, the shaping filter equivalent of a Poisson target maneuver can be represented by white noise u_s through a low-pass filter with time constant $1/2v$, as shown in Fig. 11.7. In this figure, the white-noise input has spectral density Φ_s given by

$$\Phi_s = \frac{\beta^2}{v}$$

and the initial condition on the integrator has value β (to ensure that the standard deviation of the shaping filter output is β at all times).

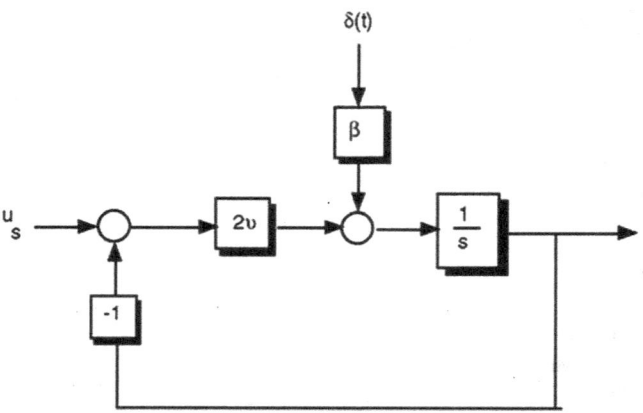

Fig. 11.7   Shaping filter equivalent of Poisson target maneuver.

Because the network of Fig. 11.7 is driven by both an impulse and white noise, we can take its adjoint. The resultant adjoint block diagram of the Poisson target maneuver appears in Fig. 11.8. In this figure we are using the same notation as the adjoint of the single-time-constant homing loop of Fig. 4.17 of the previous volume. The outputs MBT and MIC of Fig. 11.8 represent the adjoint outputs of the miss caused by noise and miss caused by random initial condition on the shaping filter, respectively. Therefore, the total RMS miss caused by the Poisson target maneuver is given by

$$\text{RMS} = \sqrt{\text{MBT}^2 + \text{MIC}^2}$$

The adjoint of the single-time-constant proportional navigation guidance system was taken, and the resultant adjoint simulation appears in Listing 11.2. Only a few modifications to the original adjoint simulation of Listing 4.6 of the previous volume were required, and these statements are highlighted in boldface.

The nominal cases of Listing 11.1 (Monte Carlo code with actual Poisson target maneuver) and Listing 11.2 (adjoint code with shaping filter equivalent of Poisson target maneuver) were run, and the RMS miss distance vs flight results as a result of the Poisson target maneuver are displayed in Fig. 11.9. We can see that the Monte Carlo results, which required 50,000 runs (50 flight times multiplied by 1000 runs per flight time), are identical to the single-run adjoint results.

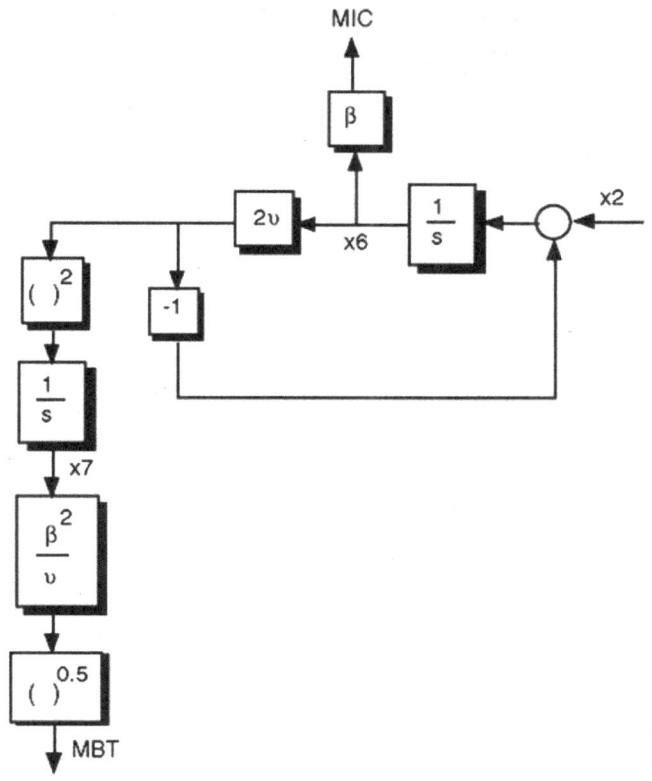

Fig. 11.8   Adjoint of Poisson target maneuver.

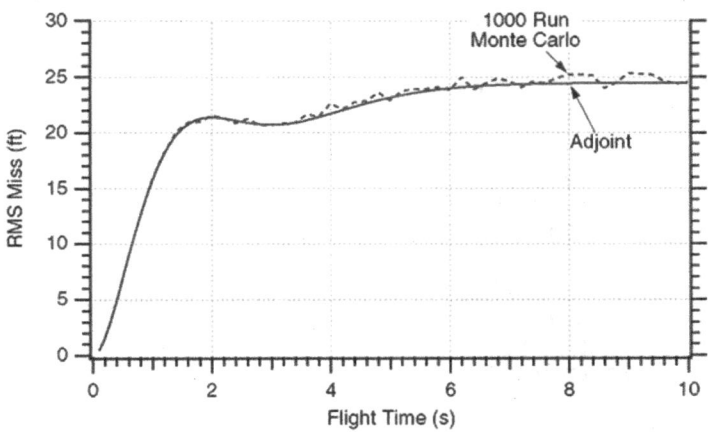

Fig. 11.9   Adjoint and Monte Carlo models agree for Poisson target maneuver disturbance.

LISTING 11.2   ADJOINT OF SINGLE-TIME-CONSTANT GUIDANCE SYSTEM DRIVEN BY POISSON TARGET MANEUVER

```
clear
count=0;
XNT=96.6;
XNP=3;
TAU=1;
TF=10;
T=0;
S=0;
TP=T+.00001;
X1=0;
X2=0;
X3=1;
X4=0;
X5=0.;
X6=0;
X7=0;
H=.01;
XNU=.5;
BETA=96.6
while TP <=(TF - 1e-5)
      STEP=1;
      FLAG=0;
      S=S+H;
      X1OLD=X1;
      X2OLD=X2;
      X3OLD=X3;
      X4OLD=X4;
      X5OLD=X5;
      X6OLD=X6;
      X7OLD=X7;
      while STEP <=1
        if FLAG==1
              STEP=2;
              X1=X1+H*X1D;
              X2=X2+H*X2D;
              X3=X3+H*X3D;
              X4=X4+H*X4D;
              X5=X5+H*X5D;
              X6=X6+H*X6D;
              X7=X7+H*X7D;
              TP=TP+H;
        end;
        X1D=X2;
```

```
        X2D=X3;
        Y1=(X4-X2)/TAU;
        TGO=TP+.00001;
        X3D=XNP*Y1/TGO;
        X4D=-Y1;
        X5D=X1*X1;
        X6D=X2-2.*XNU*X6;
        X7D=(2.*XNU*X6)^2;
        FLAG=1;
    end;
    FLAG=0;
    X1=(X1OLD+X1)/2+.5*H*X1D;
    X2=(X2OLD+X2)/2+.5*H*X2D;
    X3=(X3OLD+X3)/2+.5*H*X3D;
    X4=(X4OLD+X4)/2+.5*H*X4D;
    X5=(X5OLD+X5)/2+.5*H*X5D;
    X6=(X6OLD+X6)/2+.5*H*X6D;
    X7=(X7OLD+X7)/2+.5*H*X7D;
    S=S+H;
    if S>=.000999
        S=0.;
                    XMBT=BETA*sqrt(X7/XNU);
                    XIC=BETA*X6;
                    RMS=sqrt(XMBT^2+XIC^2);
        count=count+1;
        ArrayTP(count)=TP;
        ArrayRMS(count)=RMS;
    end;
end
figure
plot(ArrayTP, ArrayRMS),grid
title('Adjoint model using shaping filter approach')
xlabel('Flight Time (S)')
ylabel('RMS Miss (Ft)')
%axis([00,10,00,30])
clc
output=[ArrayTP',ArrayRMS'];
save datfil.txt output -ascii
disp('Simulation Complete')
```

KALMAN FILTER COMPARISON

In this section we shall compare two different Kalman filter implementations. The first filter will be called "Normal" and is the three-state Kalman filter based on a uniformly distributed constant target maneuver model (i.e., white noise through an integrator representing the maneuver) that has been used often in both the previous and current volumes of the text. The other filter is called "Poisson" and is based on a model that is sometimes called the random telegraph signal (i.e., white noise through a low-pass filter). Listing 11.3 is a Monte Carlo simulation that has options for using either filter in the homing loop with a single-time-constant flight-control system. The simulation is nominally set up to run with a high data rate and low measurement noise (TS=0.01 s and SIGNOISE=0.0001 rad). The magnitude of the target maneuver is 5 g (XNTIC=161 ft/s^2) and the acceleration limit is 10 g (AMAXG=10). Therefore, nominally, we are running with a missile-to-target acceleration advantage of only two. The target maneuver can be either a Poisson maneuver (ICONSTANT=1), a uniformly distributed constant maneuver (ICONSTANT=2), or a uniformly distributed weave maneuver (ICONSTANT=3). The flight control system time constant is 0.2 s. The optimal guidance law, which is described in Chapter 8 of the previous volume, is used for guidance. One hundred runs are made for each of the different flight times, ranging from 0.1 s to 6 s in steps of 0.1 s. Listing 11.3 presents the Monte Carlo simulation; slight modifications enable it to be run in the single flight mode where we can compare the Kalman filter target acceleration estimates to the actual target acceleration. *It is important to note that shaping filters are not used to represent the actual target maneuver. Shaping filters are only used to derive the resultant Kalman filter.*

LISTING 11.3   MONTE CARLO SIMULATION OF TWO DIFFERENT KALMAN FILTER IMPLEMENTATIONS

```
clear
count=0;
RUN=100;
ORDER=3;
IPOISSON=1;
%1=POISSON,2=NORMAL
VC=5.*3280.;
XNTIC=161.;
YIC=0.;
VM=3000.;
HEDEG=20.;
XNP=3.;
SIGNOISE=.0001;
TS=.01;
TAU=.2;
AMAXG=10.;
ICONSTANT=2;
% (1=POISSON,2=UNIF CONST,3=RANDOM SINE)W=2.;
```

```
W=2.;XNU=.5;
TSW=1;
AMAX=AMAXG*32.2;
BETA=XNTIC;
for TF=.1:.1:6.0,
  Z1=0;
  for JJ=1:RUN
      SUM=rand;
      TSTART=TF*SUM;
      SUM=rand;
      COEF=1.;
      PHASE=6.28*SUM;
      Y=0.;
      YD=0.;
      TS2=TS*TS;
      TS3=TS2*TS;
      TS4=TS3*TS;
      TS5=TS4*TS;
      PHI=zeros(ORDER);
      P=zeros(ORDER);
      QC=zeros(ORDER);
      IDNP=zeros(ORDER);
      F=zeros(ORDER);
      RTM=VC*TF;
      SIGPOS=RTM*SIGNOISE;
      SIGN2=SIGPOS^2;
      IDNP(1,1)=1.;
      IDNP(2,2)=1.;
      IDNP(3,3)=1.;
      HMAT(1,1)=1.;
      HMAT(1,2)=0.;
      HMAT(1,3)=0.;
      P(1,1)=SIGN2;
      P(2,2)=(VM*HEDEG/57.3)^2;
      P(3,3)=XNTIC*XNTIC;
      if IPOISSON==1
         F(1,2)=1.;
         F(2,3)=1.;
         F(3,3)=-2.*XNU;
         PHIN=4.*XNU*XNTIC*XNTIC;
      else IPOISSON==2
         F(1,2)=1.;;
         F(2,3)=1.;
         PHIN=XNTIC*XNTIC/10.;
      end
```

```
QC(3,3)=PHIN;
PHI=IDNP+F*TS+.5*F*F*TS*TS;
Q=PHI*QC*PHI'*TS;
T=0.;
H=.001;
S=0.;
YH=0.;
YDH=0.;
XNTH=0.;
XNC=0.;
XNL=0.;
BETA=XNTIC;
QFIRST=1;
SIG=1./sqrt(2.*XNU);
XNOISE=randn;
if XNOISE>0.
  XNTP=BETA;
end
if XNOISE<=0.
  XNTP=-BETA;
end
DELT=9999.;
TNOW=0.;
while T<=(TF-.0001)
  if QFIRST==1
    XNOISE1=SIG*randn;
    XNOISE2=SIG*randn;
    DELT=XNOISE1^2+XNOISE2^2;
    QFIRST=0.;
    TNOW=T;
  end
  if T>=(DELT+TNOW)
    XNTP=-XNTP;
    QFIRST=1;
    end
  YOLD=Y;
  YDOLD=YD;
  XNLOLD=XNL;
  STEP=1;
  FLAG=0;
  while STEP <=1
    if FLAG==1
      STEP=2;
      Y=Y+H*YD;
      YD=YD+H*YDD;
```

```
    XNL=XNL+H*XNLD;
    T=T+H;
  end
  if ICONSTANT==1
    XNTC=XNTP;
  elseif ICONSTANT==2
    if T<TSTART
      XNTC=0.;
    else
      XNTC=COEF*XNTIC;
    end
  else
    if T<TSTART
      XNTC=0.;
    else
      XNTC=XNTIC*sin(W*T+PHASE);
    end
  end
  TGO=TF-T+.00001;
  RTM=VC*TGO;
  XLAM=Y/(VC*TGO);
  XNLD=(XNC-XNL)/TAU;
  YDD=XNTC-XNL;
  FLAG=1;
end
FLAG=0;
Y=.5*(YOLD+Y+H*YD);
YD=.5*(YDOLD+YD+H*YDD);
XNL=.5*(XNLOLD+XNL+H*XNLD);
S=S+H;
if S>=(TS-.00001)
  S=0.;
  TGO=TF-T+.000001;
  RTM=VC*TGO;
  SIGPOS=RTM*SIGNOISE;
  SIGN2=SIGPOS^2;
  RMAT=[SIGN2];
  M=PHI*P*PHI'+Q;
  K = M*HMAT'/(HMAT*M*HMAT' + RMAT);
  P = (IDNP - K*HMAT)*M;
  XLAMNOISE=SIGNOISE*randn;
  YSTAR=RTM*(XLAM+XLAMNOISE);
  [YB,YDB,XNTB]=PROJECT34(T,TS,YH,YDH,XNL,...
          XNTH,XNU,IPOISSON);
  RES(1,1)=YSTAR-YB;
```

```
                    YH=YB+K(1,1)*RES(1,1);
                    YDH=YDB+K(2,1)*RES(1,1);
                    XNTH=XNTB+K(3,1)*RES(1,1);
                    X=TGO/TAU;
                    TAUTGT=1./(2.*XNU);
                    ZEM1H=YH+YDH*TGO-XNL*TAU*TAU*(exp(-X)+X-1.)+.5*XNTH*TGO*TGO;
                    TOP=6.*X*X*(exp(-X)-1.+X);
                    BOT1=2*X*X*X+3.+6.*X-6.*X*X;
                    BOT2=-12.*X*exp(-X)-3.*exp(-2.*X);
                    XNPP=TOP/(.0001+BOT1+BOT2);
                    XNC=XNPP*ZEM1H/TGO^2;
                    if XNC>AMAX
                       XNC=AMAX;
                    end
                    if XNC<-AMAX
                       XNC=-AMAX;
                    end
                 end
              end
              Z(JJ)=Y;
              Z1=Z(JJ)+Z1;
              XMEAN=Z1/JJ;
           end;
           SIGMA=0;
           Z1=0;
           Z2=0.;
           for JJ=1:RUN,
              Z1=(Z(JJ)-XMEAN)^2+Z1;
              Z2=Z(JJ)^2+Z2;
              if JJ==1,
                 SIGMA=0;
                 RMS=0.;
              else
                 SIGMA=sqrt(Z1/(JJ-1));
                 RMS=sqrt(Z2/(JJ-1));
              end;
           end;
           count=count+1;
           ArrayTF(count)=TF;
           ArraySIGMA(count)=SIGMA;
           ArrayXMEAN(count)=XMEAN;
           ArrayRMS(count)=RMS;
        end;
        figure
        plot(ArrayTF',ArrayRMS'),grid
```

```
title('RMS miss for various flight times')
xlabel('Flight Time (S)')
ylabel('RMS MISS (Ft) ')
clc
output=[ArrayTF',ArrayRMS'];
save datfil.txt output -ascii
disp('Simulation Complete')

function [YB,YDB,XNTB]=PROJECT34(TP,TS,YHP,YDHP,XNLP,XNTHP,XNU,...
    IPOISSON)
T=0.;
YH=YHP;
YDH=YDHP;
XNT=XNTHP;
XNL=XNLP;
H=.001;
while T<=(TS-.0001)
  if IPOISSON==1
    XNTD=-2.*XNU*XNT;
    XNT=XNT+H*XNTD;
    YDDH=XNT-XNL;
    YDH=YDH+H*YDDH;
    YH=YH+H*YDH;
  else
    YDDH=XNT-XNL;
    YDH=YDH+H*YDDH;
    YH=YH+H*YDH;
  end
  T=T+H;
end
YB=YH;
YDB=YDH;
XNTB=XNT;
```

Listing 11.3 was run in the single flight mode; target maneuver estimates for the normal and Poisson Kalman filters appear in Figs. 11.10–11.12. We can see from these figures that the Poisson Kalman filter yields noisier estimates for the random step, random weave, and Poisson target maneuvers than the normal Kalman filter.

Listing 11.3 was run in its Monte Carlo mode to generate RMS miss distances; the results appear in Figs. 11.13–11.15. We can see from these figures that the Poisson Kalman filter always yields smaller RMS miss distances for the random step, weave, and Poisson target maneuvers than the normal Kalman filter. The next section will attempt to explain these interesting results.

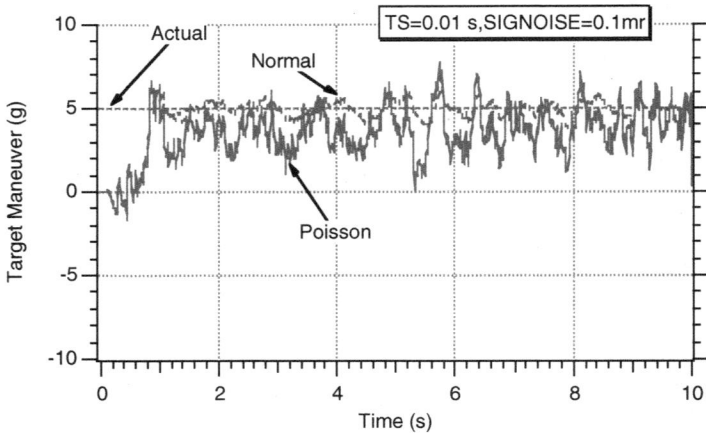

Fig. 11.10   Normal Kalman filter appears to give less noisy estimates of constant target maneuver.

INCREASING THE PROCESS NOISE OF THE NORMAL KALMAN FILTER

It appears from the previous section that the Poisson Kalman filter always yields much smaller RMS miss distances than the normal Kalman filter (i.e., three-state Kalman filter was used extensively in this text and previous volume). However, we also showed earlier in this chapter that basically the structure of both the normal and Poisson three-state Kalman filters were similar but not identical. One major

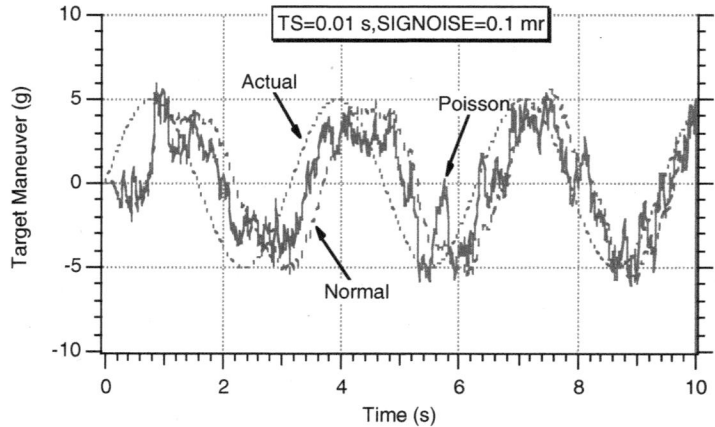

Fig. 11.11   Normal Kalman filter appears to give less noisy estimates of weave target maneuver.

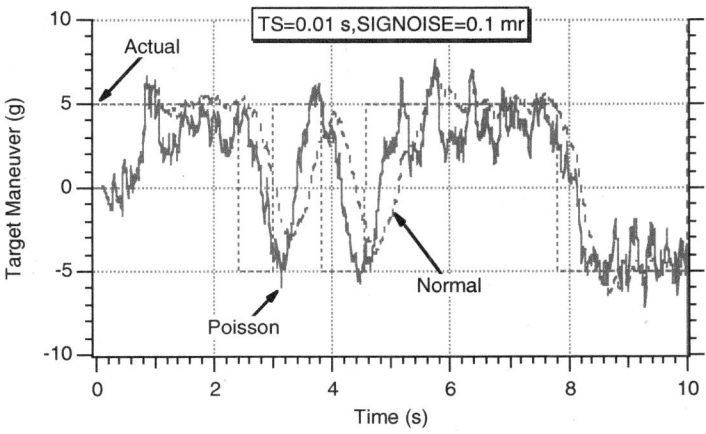

Fig. 11.12 **Normal Kalman filter appears to give less noisy estimates of Poisson target maneuver.**

difference between the two filters was the formula for the process noise matrix. Recall that the continuous process noise matrix for the normal Kalman filter is given by

$$Q_{\text{Normal}} = \frac{\beta^2}{t_F} \begin{bmatrix} 0 & 0 & 0 \\ 0 & 0 & 0 \\ 0 & 0 & 1 \end{bmatrix}$$

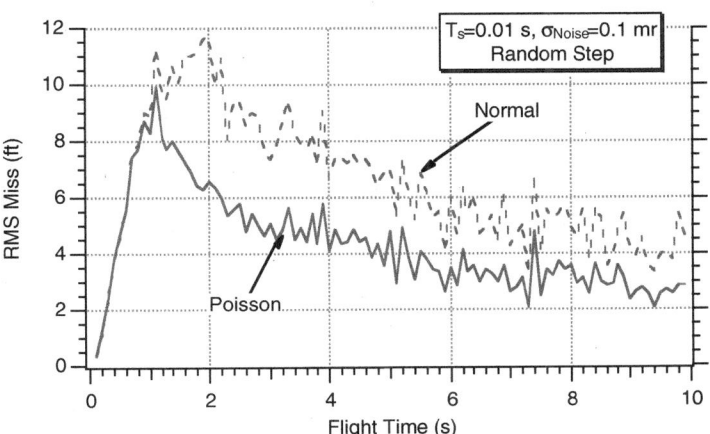

Fig. 11.13 **Poisson Kalman filter yields smaller RMS miss distances against random step target maneuver than normal Kalman filter.**

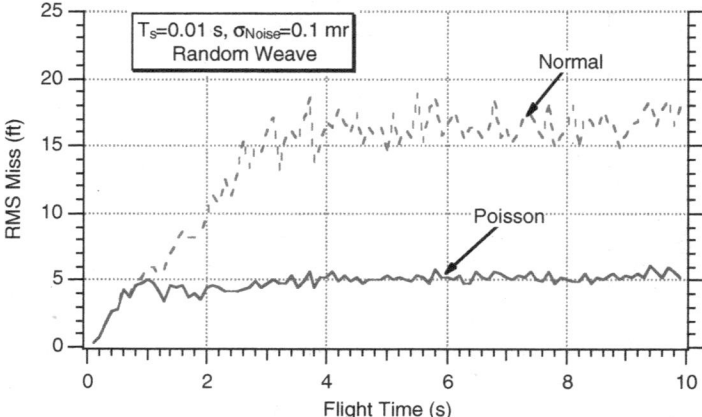

Fig. 11.14   Poisson Kalman filter yields smaller RMS miss distances against random weave target maneuver than normal Kalman filter.

where β is the maximum expected target maneuver level and t_F is the maximum expected homing time. Also recall that the process noise matrix for the Poisson Kalman filter is given by

$$Q_{\text{Poisson}} = 4\nu\beta^2 \begin{bmatrix} 0 & 0 & 0 \\ 0 & 0 & 0 \\ 0 & 0 & 1 \end{bmatrix}$$

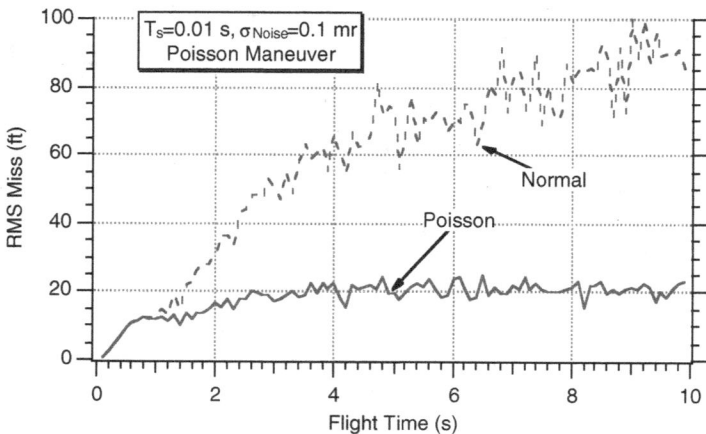

Fig. 11.15   Poisson Kalman filter yields smaller RMS miss distances against Poisson target maneuver than normal Kalman filter.

where v is the number of zero crossings of the Poisson maneuver per second. Therefore, the ratio of the two process noise matrices is given by

$$\frac{Q_{\text{Poisson}}}{Q_{\text{Normal}}} = \frac{4v\beta^2}{\dfrac{\beta^2}{t_F}} = 4vt_F$$

Using the nominal numbers from Listing 11.3 ($v = 0.5$, $t_F = 10$) we obtain

$$\frac{Q_{\text{Poisson}}}{Q_{\text{Normal}}} = 4vt_F = 4 * 0.5 * 10 = 20$$

*In other words, in this example the Poisson Kalman filter has 20 times as much process noise as the normal Kalman filter.*In addition, the systems dynamics matrix for both filters is slightly different. The systems dynamics matrix for the normal Kalman filter is given by

$$F_{\text{Normal}} = \begin{bmatrix} 0 & 1 & 0 \\ 0 & 0 & 1 \\ 0 & 0 & 0 \end{bmatrix}$$

whereas the systems dynamics matrix for the Poisson Kalman filter is given by

$$F_{\text{Poisson}} = \begin{bmatrix} 0 & 1 & 0 \\ 0 & 0 & 1 \\ 0 & 0 & -2v \end{bmatrix}$$

In this example v is a small number, so it appears that the major difference between the two Kalman filter models is in the amount of process noise. In

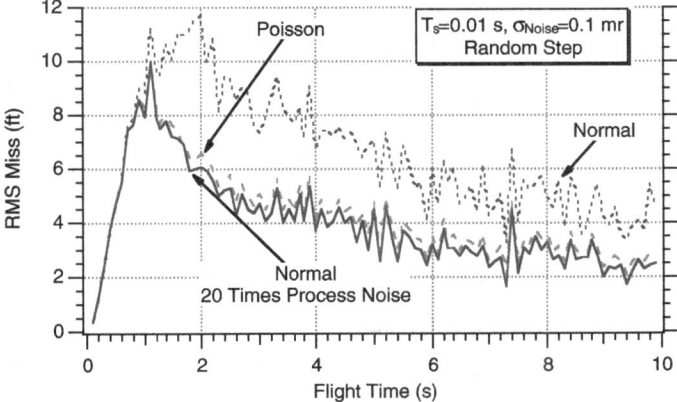

Fig. 11.16 **Normal Kalman filter with 20 times more process noise yields similar RMS miss distance performance to Poisson Kalman filter.**

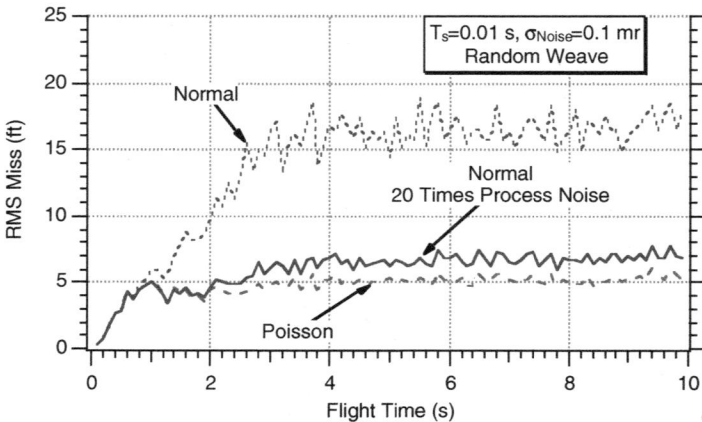

Fig. 11.17   Normal Kalman filter with 20 times more process noise yields similar RMS miss distance performance to Poisson Kalman filter.

other words, in our examples in the preceding section the Poisson Kalman filter had 20 times more process noise than the normal Kalman filter. Let us see if the amount of process noise used is the major reason why the Poisson Kalman filter yielded smaller RMS miss distances than the normal Kalman filter against the various random target maneuvers considered.

Listing 11.1 was run for the different Kalman filter and process noise possibilities; the results are presented in Figs. 11.16–11.18. We can see that for the case of

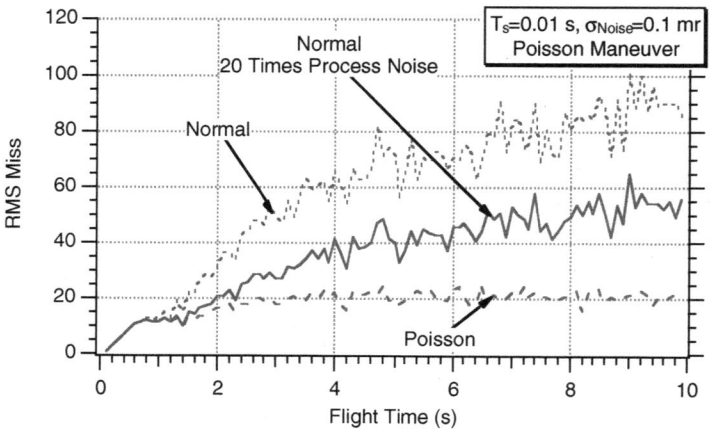

Fig. 11.18   Normal Kalman filter with 20 times more process noise yields larger RMS miss distance performance when compared to a Poisson Kalman filter.

the random step and random weave target maneuvers, both the normal Kalman filter with 20 times more process noise and the Poisson Kalman filter have virtually identical performance in terms of the RMS miss distance. However, if the actual target maneuver is a Poisson maneuver, then the Poisson Kalman filter yields smaller RMS miss distances than the normal Kalman filter with 20 times more process noise. The Poisson target maneuver effect in RMS miss distance performance is probably due to differences in the systems dynamics matrices of the two filters.

SUMMARY

In this chapter we have shown that Kalman filters based on either the Poisson or Singer representation of random target maneuvers are identical. When our normal three-state Kalman filter, based on a random step target maneuver, is compared to a Kalman filter based on the Poisson target maneuver model, the Poisson Kalman filter yields smaller RMS miss distances. However, if the process noise of the normal Kalman filter is increased, it can yield comparable RMS miss distances to the Poisson Kalman filter—except in the case where the actual target maneuver is a Poisson target maneuver.

REFERENCES

[1]   Schwartz, M., *Information, Transmission, Modulation, and Noise*, McGraw-Hill, New York, 1959, pp. 441–448.

[2]   Singer, R., "Estimating Optimal Tracking Performance for Manned Maneuvering Targets," *IEEE Transactions on Aerospace and Electronic Systems*, Vol. AES-6, No. 4, July 1970, pp. 473–483.

CHAPTER 12

Alternative Approaches to Guidance Law Development

INTRODUCTION

So far this text and the previous volume has made use of the Schwartz inequality to derive guidance laws analytically. The Schwartz inequality was used because it was the simplest technique known to the author for the derivation of missile guidance laws. In this chapter we will investigate two alternative ways of developing guidance laws. These techniques can be used either to check analytically derived guidance laws or to derive more advanced guidance laws when the Schwartz inequality technique becomes too cumbersome. The first alternative approach presented is based on optimal control theory and is the most general. The optimal control approach consists of the numerical integration of the nonlinear matrix Riccati differential equation to solve for the guidance law control gains. In addition, a totally different approach for the numerical derivation of guidance laws is also presented in this chapter. This innovative technique is numerically much faster than the optimal control method and is especially convenient when the dynamics of the flight-control system are too complex for any analytical approach. In both approaches, numerical examples are presented showing how the results of the new techniques compare to the analytical expressions for the previously derived guidance laws. Finally, a new guidance law will be developed in this chapter for the case in which there is a significant right-half-plane zero in the flight-control system of a tail-controlled missile. It will be shown that the new guidance law, which is developed numerically, offers improved system performance at very high altitudes.

OPTIMAL CONTROL

Guidance laws do not have to be derived analytically, as has been done so far in this text and the previous volume. For example, optimal control theory [1, 2] can be used to numerically derive guidance laws. With optimal control theory, a performance index is set up, and a nonlinear matrix Riccati differential equation needs to be solved in order to obtain the control gains of the resultant guidance law. At the very least this numerical approach can also be useful as an independent check of the accuracy of the analytically derived guidance laws that were presented earlier in this text and the previous volume. In addition, the numerical approach is useful in situations in which the analytical approach is either impossible or too complex (that is, very high order model of flight-control system dynamics) to apply. To demonstrate the utility of the optimal control approach, we will first state the theoretical equations to be solved and then derive two guidance laws numerically for which we already have closed-form solutions.

Optimal control theory states that guidance laws can be derived either analytically or numerically if our model of the real world can be expressed in state space form as

$$\dot{\mathbf{x}} = \mathbf{F}\mathbf{x} + \mathbf{G}\mathbf{u}$$

where the control $\mathbf{u}$ (guidance command) is linearly related to the states x according to

$$\mathbf{u} = -\mathbf{C}\mathbf{x}$$

In the preceding equation $\mathbf{C}$ is a set of control gains. The expression involving $\mathbf{C}$ is another way of writing a missile guidance law. With optimal control techniques we are trying to minimize a performance index. Previously when we derived guidance laws in this text, we simply tried to make the miss distance zero subject to minimizing the integral of the acceleration squared, or

$$y(t_F) = 0 \quad \text{subject to minimizing} \quad \int_0^{t_F} n_c^2(t)\,dt$$

With standard optimal control techniques we cannot quite minimize the preceding performance index. However, a nearly equivalent performance index J can be expressed in matrix form and is given by

$$J = \mathbf{x}^T(t_F)\mathbf{S}_F\mathbf{x}(t_F) + \int_0^{t_F} \mathbf{u}^T\mathbf{B}\mathbf{u}\,dt$$

where $\mathbf{x}(t_F)$ is the state vector at the final time. The miss $y(t_F)$ is one element of that state vector. By judiciously choosing $\mathbf{S}_F$, we can make the first term of J equal to the miss distance squared $[y^2(t_F)]$. In addition, by making $\mathbf{B}$ a small scalar we can come close to our desire of making the miss zero subject to minimizing the integral of the acceleration squared.

Optimal control theory shows that minimizing J for our state space model of the real world yields a guidance law in which the control gains are linearly related to the states. Obtaining those control gains involves solving the nonlinear matrix

Riccati differential equation

$$\dot{\mathbf{S}} = -\mathbf{F}^T\mathbf{S} - \mathbf{S}\mathbf{F} + \mathbf{C}^T\mathbf{B}\mathbf{C}$$

with boundary value

$$\mathbf{S}(t_F) = \mathbf{S}_F$$

The control gains for the optimal guidance law are related to the solution of the preceding differential equation for S and can be obtained from

$$\mathbf{C} = \mathbf{B}^{-1}\mathbf{G}^T\mathbf{S}$$

There can be significant numerical difficulty in solving the matrix Riccati differential equation numerically because it is a boundary-value problem. In fact, in this text we have not yet solved a boundary-value problem numerically. However, this difficulty can be easily avoided by simply changing variables in order to convert the nonlinear matrix differential equation with a boundary value to one with an initial condition. Recall that from the overdot notation we know that

$$\dot{\mathbf{S}} = \frac{d\mathbf{S}}{dt}$$

Therefore, if we define

$$\tau = t_F - t$$

we can say that

$$\frac{d\mathbf{S}}{dt} = -\frac{d\mathbf{S}}{d\tau}$$

Therefore, by changing variables, the nonlinear matrix differential equation becomes

$$\dot{\mathbf{S}} = \mathbf{F}^T\mathbf{S} + \mathbf{S}\mathbf{F} - \mathbf{C}^T\mathbf{B}\mathbf{C}$$

with initial value

$$\mathbf{S}(0) = \mathbf{S}_F$$

Because we have changed variables, S is now a function of τ or time to go rather than t or time. The control gains for the optimal guidance law are still obtained from

$$\mathbf{C} = \mathbf{B}^{-1}\mathbf{G}^T\mathbf{S}$$

but again, the gains are now functions of time to go rather than a function of time. Double-precision arithmetic plus a very small integration step size are often required for the successful integration of the Riccati equations because of their numerical fragility.

The preceding optimal control equations are valid only for driving the relative position and/or relative velocity or some other quantities to zero at the end of the flight. The technique cannot be used to drive a quantity, such as the relative velocity, to a specified value at the end of the flight, such as was done in using the Schwartz inequality to derive the trajectory-shaping guidance law in the next chapter.

USING OPTIMAL CONTROL TO DERIVE GUIDANCE LAW FOR SINGLE-LAG FLIGHT CONTROL SYSTEM

In Chapter 8 of the previous volume we derived an optimal guidance law when the missile flight-control system was modeled as a single lag. For the single-lag flight-control system, the overall model for guidance law development was displayed in Fig. 8.13 of the previous volume and is repeated here for convenience in Fig. 12.1. Recall that in this model the relative acceleration is simply the difference between target acceleration n_T and the achieved missile acceleration n_L. Integrating the relative acceleration twice yields relative position y. At the end of the flight, the relative position is the miss distance $y(t_F)$.

As was shown in Chapter 8 of Volume 1, the model of Fig. 12.1 can be put in state space form, as shown next. Here we are making the assumption that the target maneuver is constant, and therefore its derivative is zero.

$$
\begin{bmatrix} \dot{y} \\ \ddot{y} \\ \dot{n}_T \\ \dot{n}_L \end{bmatrix} = \begin{bmatrix} 0 & 1 & 0 & 0 \\ 0 & 0 & 1 & -1 \\ 0 & 0 & 0 & 0 \\ 0 & 0 & 0 & -\dfrac{1}{T} \end{bmatrix} \begin{bmatrix} y \\ \dot{y} \\ n_T \\ n_L \end{bmatrix} + \begin{bmatrix} 0 \\ 0 \\ 0 \\ \dfrac{1}{T} \end{bmatrix} n_c
$$

Because the state space equation is given by

$$\dot{\mathbf{x}} = \mathbf{F}\mathbf{x} + \mathbf{G}\mathbf{u}$$

the appropriate state space matrices for the single-lag flight-control system can be written by

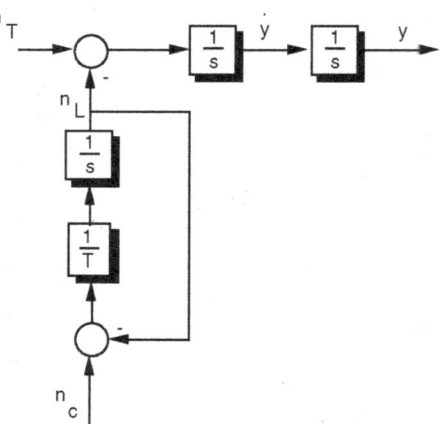

Fig. 12.1   Single-lag flight-control system model for guidance law development.

inspection as

$$\mathbf{F} = \begin{bmatrix} 0 & 1 & 0 & 0 \\ 0 & 0 & 1 & -1 \\ 0 & 0 & 0 & 0 \\ 0 & 0 & 0 & -\dfrac{1}{T} \end{bmatrix}$$

$$\mathbf{G} = \begin{bmatrix} 0 \\ 0 \\ 0 \\ -\dfrac{1}{T} \end{bmatrix}$$

$$\mathbf{u} = n_c$$

As was mentioned at the beginning of this section, the performance index that is usually chosen for guidance problems is one in which we make the miss zero subject to minimizing the integral of the square of acceleration. A convenient way of expressing this desire is shown next for the performance index J. Here the parameter γ is used as a fudge factor in the derivation of the guidance law (meaning that γ numerically prevents the guidance gains from becoming infinite when the time to go approaches zero). Using smaller values of γ will cause us to pay more attention to the miss (that is, make the miss smaller). The performance index for our optimal control problem is given by

$$J = y^2(t_F) + \gamma \int_0^{t_F} n_c^2 \, dt$$

Because the generalized performance index for optimal control problems is given by

$$J = \mathbf{x}^T(t_F)\mathbf{S}_F\mathbf{x}(t_F) + \int_0^{t_F} \mathbf{u}^T \mathbf{B}\mathbf{u} \, dt$$

the matrices in the preceding performance index expression can also be written by inspection as

$$\mathbf{S}_F = \begin{bmatrix} 1 & 0 & 0 & 0 \\ 0 & 0 & 0 & 0 \\ 0 & 0 & 0 & 0 \\ 0 & 0 & 0 & 0 \end{bmatrix}$$

$$\mathbf{B} = \gamma$$

We now have all of the information we need to integrate the matrix Riccati equations and solve for the control gains. In this example the Riccati equations can also be solved in closed form [3, 4]; however, we shall simply use second-order Runge–Kutta numerical integration to obtain the Riccati equation solution.

Analytically, we already know from Chapter 8 of the previous volume that for the single-lag flight-control system, the closed-form solution for the optimal guidance law is given by

$$n_c = \frac{N'}{t_{go}^2}[y + \dot{y}t_{go} + 0.5n_T t_{go}^2 - n_L T^2(e^{-x} + x - 1)]$$

In the preceding equation, the effective navigation ratio can be expressed as

$$N' = \frac{6x^2(e^{-x} - 1 + x)}{2x^3 + 3 + 6x - 6x^2 - 12xe^{-x} - 3e^{-2x}}$$

where T is the flight-control system time constant and x is defined as

$$x = \frac{t_{go}}{T}$$

Alternatively we can also present the optimal guidance law in terms of control gains, or

$$n_c = C_1 y + C_2 \dot{y} + C_3 n_T + C_4 n_L$$

where C_1 through C_4 are the control gains. Equating the preceding expression with the optimal guidance law yields the equations for each of the control gains as

$$C_1 = \frac{N'}{t_{go}^2}$$

$$C_2 = \frac{N'}{t_{go}}$$

$$C_3 = 0.5N'$$

$$C_4 = \frac{-T^2(e^{-x} + x - 1)N'}{t_{go}^2} = \frac{-N'(e^{-x} + x - 1)}{x^2}$$

The preceding equations will be considered the exact solution for the control gains for our guidance problem.

We can numerically integrate the nonlinear matrix Riccati differential equation and solve for the control gains and then compare the numerical results with the preceding closed-form solutions. Listing 12.1 uses the matrices of this section to numerically integrate the nonlinear matrix differential equation with the appropriate initial conditions. The nominal parameters for Listing 12.1 are a flight-control-system time constant of 1 s, a flight time of 10 s, and a performance index weighting of 0.00001. Second-order Runge–Kutta integration is used to numerically integrate the Riccati equation. After some experimentation, an integration step size of 0.0001 s was chosen because it yielded accurate answers and still allowed the program to run quickly. Note that the denominator in the closed-form solution for the effective navigation ratio goes to zero as time to go approaches zero. To avoid that singularity, 0.0001 was added to the denominator, as can be seen from Listing 12.1.

LISTING 12.1   INTEGRATING RICCATI EQUATIONS TO GET OPTIMAL CONTROL GAINS FOR SINGLE-LAG FLIGHT-CONTROL SYSTEM

```
clear
TAU=1;
GAM=.00001;
TF=10;
F=zeros([4,4]);
S=zeros([4,4]);
count=0;
G(1,1)=0;
G(2,1)=0;
G(3,1)=0;
G(4,1)=1./TAU;
F(1,2)=1;
F(2,3)=1;
F(2,4)=-1;
F(4,4)=-1./TAU;
S(1,1)=1;
T=0;
H=.0001;
S1=0;
while ~(T >= (TF-.0001))
        S1=S1+H;
        SOLD=S;
        STEP=1;
        FLAG=0;
        while STEP<=1
           if FLAG==1
                        STEP=2;
                        HSD=H*SD;
                        S=S+HSD;
                        T=T+H;
                end
```

```
            SF=S*F;
            GT=G';
            GTS=GT*S;
            GAMINV=1./GAM;
            C=GAMINV*GTS;
            SFT=SF';
            CT=C';
            CTC=CT*C;
            CTBC=GAM*CTC;
            SFSFT=SF+SFT;
            SD=SFSFT-CTBC;
FLAG=1;
end
FLAG=0;
H2=.5*H;
HSDP=H2*SD;
SS=SOLD+S;
SSP=.5*SS;
S=SSP+HSDP;
if S1>=.009999
            S1=0;
            C1=-C(1,1);
            C2=-C(1,2);
            C3=-C(1,3);
            C4=-C(1,4);
            NP=C2*T;
            XS=T/TAU;
            W1=1./TAU;
            TOP=6.*XS*XS*(exp(-XS)-1.+XS);
            BOT1=2*XS*XS*XS+3.+6.*XS-6.*XS*XS;
            BOT2=-12.*XS*exp(-XS)-3.*exp(-2.*XS);
            XNPP=TOP/(BOT1+BOT2+.0001);
            C1TH=XNPP/(T*T);
            C2TH=XNPP/T;
            C3TH=.5*XNPP;
            C4TH=-XNPP*(exp(-XS)+XS-1.)/(XS*XS);
            count=count+1;
            ArrayT(count)=T;
            ArrayC1(count)=C1;
            ArrayC1TH(count)=C1TH;
            ArrayC2(count)=C2;
            ArrayC2TH(count)=C2TH;
            ArrayC3(count)=C3;
            ArrayC3TH(count)=C3TH;
            ArrayC4(count)=C4;
```

```
                    ArrayC4TH(count)=C4TH;
                    ArrayNP(count)=NP;
                    ArrayXNPP(count)=XNPP;
          end
end
output=[ArrayT',ArrayC1',ArrayC1TH',ArrayC2',ArrayC2TH', . . .
       ArrayC3',ArrayC3TH',ArrayC4',ArrayC4TH',ArrayNP',ArrayXNPP'];
save datfil.txt output-ascii
disp 'simulation finished'
clc
figure
semilogy(ArrayT,ArrayC1,ArrayT,ArrayC1TH),grid
xlabel('Time (s) ')
ylabel('C1')
axis([0 10 .01 1000])
figure
semilogy(ArrayT,ArrayC2,ArrayT,ArrayC2TH),grid
xlabel('Time (s) ')
ylabel('C2')
axis([0 10 .1 200])
figure
semilogy(ArrayT,ArrayC3,ArrayT,ArrayC3TH),grid
xlabel('Time (s) ')
ylabel('C3')
axis([0 10 1 20])
figure
semilogy(ArrayT,-ArrayC4,ArrayT,-ArrayC4TH),grid
xlabel('Time (s) ')
ylabel('-C4')
axis([0 10 .1 20])
figure
semilogy(ArrayT,ArrayNP,ArrayT,ArrayXNPP),grid
xlabel('Time (s) ')
ylabel('Effective Navigation Ratio')
axis([0 10 .1 40])
```

The nominal case of Listing 12.1 was run, and the control gain outputs are displayed along with the analytical solution ("Formula") in Figs. 12.2–12.5. We can see that the numerical and analytical results are in near-perfect agreement, thus verifying the accuracy of the numerical integration in the optimal control approach. Figure 12.6 compares the effective navigation ratios for the optimal guidance law from both the analytical and numerical approaches and shows they are equivalent. Thus we can now feel confident about using Listing 12.1, with minor modifications for the matrices, to generate other guidance laws using the optimal control technique.

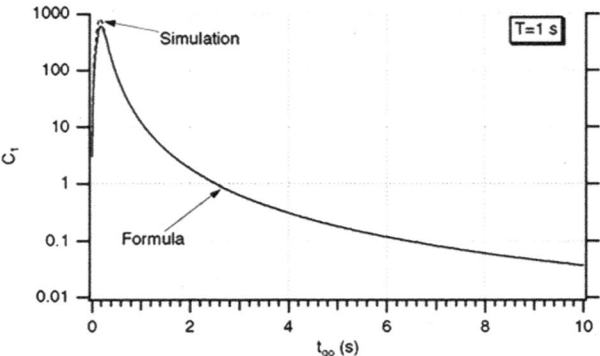

Fig. 12.2   Formula and simulation results agree for first control gain.

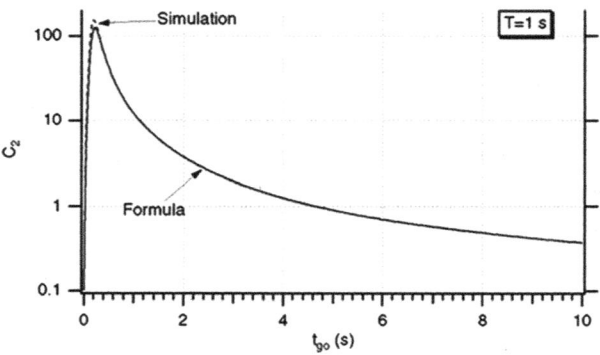

Fig. 12.3   Formula and simulation results agree for second control gain.

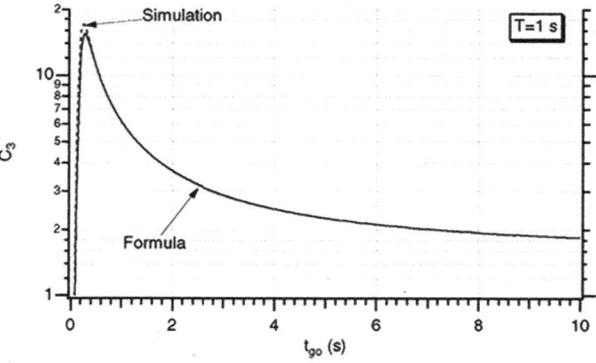

Fig. 12.4   Formula and simulation results agree for third control gain.

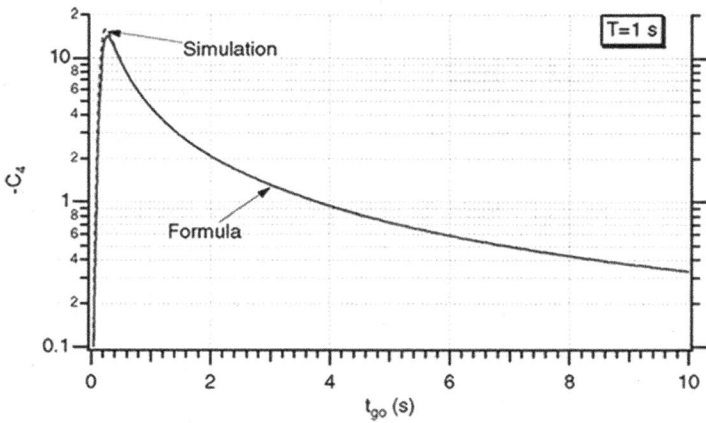

Fig. 12.5   Formula and simulation results agree for fourth control gain.

DERIVING GUIDANCE LAW FOR WEAVING TARGET USING OPTIMAL CONTROL

As another example of the utility of the optimal control approach for deriving guidance laws, let us consider an example involving a weaving target. In this case we shall consider the flight-control system to be perfect (that is, zero time constant). We have already shown in Chapter 6 that a weave maneuver could be represented analytically by a shaping network. The overall model for guidance law development was presented in Fig. 6.16 for the weaving target and is repeated in Fig. 12.7 for convenience.

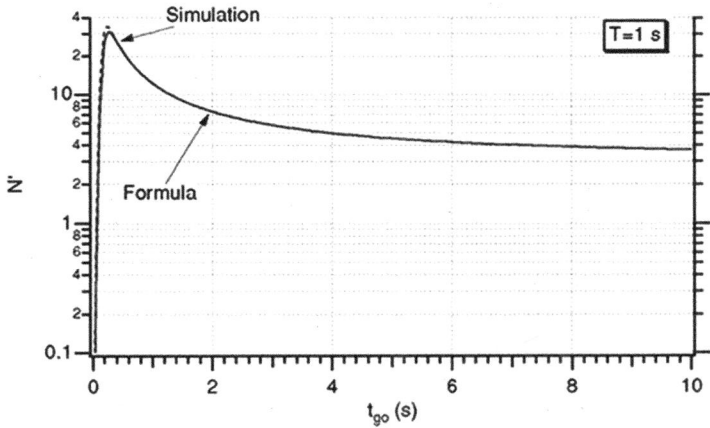

Fig. 12.6   Formula and simulation results agree for effective navigation ratio.

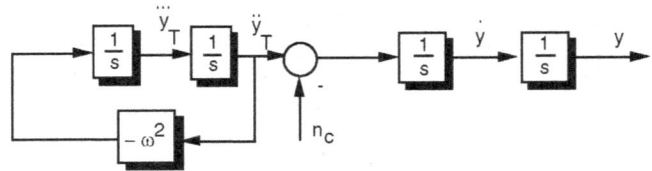

Fig. 12.7   Model for weave guidance law derivation.

As was shown in Chapter 6, the shaping network to the left of the summing junction in Fig. 12.7 represents the sinusoidal maneuver. The model of Fig. 12.7 can be put in state space form as shown here:

$$\begin{bmatrix} \dot{y} \\ \ddot{y} \\ \dot{n}_T \\ \ddot{n}_T \end{bmatrix} = \begin{bmatrix} 0 & 1 & 0 & 0 \\ 0 & 0 & 1 & 0 \\ 0 & 0 & 0 & 1 \\ 0 & 0 & -\omega^2 & 0 \end{bmatrix} \begin{bmatrix} y \\ \dot{y} \\ n_T \\ \dot{n}_T \end{bmatrix} + \begin{bmatrix} 0 \\ -1 \\ 0 \\ 0 \end{bmatrix} n_c$$

As was the case in the preceding section, we will still assume that the performance index to be minimized is given by

$$J = y^2(t_F) + \gamma \int_0^{t_F} n_c^2 \, dt$$

Now that we have formulated the guidance problem, the appropriate matrices required by the matrix Riccati differential equation can be written by inspection as

$$F = \begin{bmatrix} 0 & 1 & 0 & 0 \\ 0 & 0 & 1 & 0 \\ 0 & 0 & 0 & 1 \\ 0 & 0 & -\omega^2 & 0 \end{bmatrix}$$

$$G = \begin{bmatrix} 0 \\ -1 \\ 0 \\ 0 \end{bmatrix}$$

$$S_F = \begin{bmatrix} 1 & 0 & 0 & 0 \\ 0 & 0 & 0 & 0 \\ 0 & 0 & 0 & 0 \\ 0 & 0 & 0 & 0 \end{bmatrix}$$

$$\mathbf{u} = n_c \text{ and } \mathbf{B} = \gamma$$

We have already shown in Chapter 6 that for the weaving target, the closed-form solution for the optimal guidance law, known as weave guidance, is given by

$$
n_c = \frac{3}{t_{go}^2} \left[y + \dot{y}t_{go} + \left(\frac{1 - \cos \omega t_{go}}{\omega^2} \right) n_T + \left(\frac{\omega t_{go} - \sin \omega t_{go}}{\omega^3} \right) \dot{n}_T \right]
$$

Note that in the preceding equation for the weave guidance law, the effective navigation ratio is 3.

As was done in the preceding section, we can also express the guidance law in terms of control gains, or

$$
n_c = C_1 y + C_2 \dot{y} + C_3 n_T + C_4 \dot{n}_T
$$

By comparing the preceding equation with the weave guidance law, we can easily see that the control gains for the weave guidance law are given by

$$
C_1 = \frac{3}{t_{go}^2}
$$

$$
C_2 = \frac{3}{t_{go}}
$$

$$
C_3 = \frac{3}{t_{go}^2} \left(\frac{1 - \cos \omega t_{go}}{\omega^2} \right)
$$

$$
C_4 = \frac{3}{t_{go}^2} \left(\frac{\omega t_{go} - \sin \omega t_{go}}{\omega^3} \right)
$$

The preceding equations can be viewed as the exact solution for the control gains for the weaving guidance law.

We can slightly modify Listing 12.1 to solve for the control gains in the weaving target problem. Listing 12.2 shows that we still numerically integrate the nonlinear matrix Riccati equation with a small integration step size and solve for the control gains. A comparison is made between the results obtained by numerical integration and the preceding closed-form solutions. Statements that have changed from Listing 12.1 in order to find the weave guidance law are highlighted in bold in Listing 12.2. The nominal case considered by Listing 12.2 had a target weave frequency of 1 rad/s, a flight time of 10 s, and a performance index weighting of 0.00001. Again, second-order Runge–Kutta integration was used to numerically integrate the matrix Riccati differential equation, and an integration step size of 0.0001 s was still used to get accurate answers.

LISTING 12.2   INTEGRATING MATRIX RICCATI EQUATION TO GET OPTIMAL CONTROL GAINS FOR WEAVING TARGET

```
clear
W=1;
GAM=.00001;
TF=10;
F=zeros([4,4]);
S=zeros([4,4]);
count=0;
G(1,1)=0;
G(2,1)=-1.;
G(3,1)=0;
G(4,1)=0.;
F(1,2)=1;
F(2,3)=1;
F(3,4)=1.;
F(4,3)=-W*W;
S(1,1)=1;
T=0;
H=.0001;
S1=0;
while ~(T>=(TF-.0001))
        S1=S1+H;
        SOLD=S;
        STEP=1;
        FLAG=0;
        while STEP<=1
          if FLAG==1
                  STEP=2;
                  HSD=H*SD;
                  S=S+HSD;
                  T=T+H;
          end
          SF=S*F;
          GT=G';
          GTS=GT*S;
          GAMINV=1./GAM;
          C=GAMINV*GTS;
          SFT=SF';
          CT=C';
          CTC=CT*C;
          CTBC=GAM*CTC;
          SFSFT=SF+SFT;
          SD=SFSFT-CTBC;
          FLAG=1;
        end
    end
```

```
            FLAG=0;
            H2=.5*H;
            HSDP=H2*SD;
            SS=SOLD+S;
            SSP=.5*SS;
            S=SSP+HSDP;
            if S1>=.009999
                    S1=0;
                    C1=-C(1,1);
                    C2=-C(1,2);
                    C3=-C(1,3);
                    C4=-C(1,4);
                    NP=C2*T;
                    XNPP=3.;
                    C1TH=XNPP/(T*T);
                    C2TH=XNPP/T;
                    C3TH=XNPP*((1.-cos(W*T))/W^2)/T^2;
                    C4TH=XNPP*((W*T-sin(W*T))/W^3)/T^2;
                    count=count+1;
                    ArrayT(count)=T;
                    ArrayC1(count)=C1;
                    ArrayC1TH(count)=C1TH;
                    ArrayC2(count)=C2;
                    ArrayC2TH(count)=C2TH;
                    ArrayC3(count)=C3;
                    ArrayC3TH(count)=C3TH;
                    ArrayC4(count)=C4;
                    ArrayC4TH(count)=C4TH;
                    ArrayNP(count)=NP;
                    ArrayXNPP(count)=XNPP;
            end
    end
output=[ArrayT',ArrayC1',ArrayC1TH',ArrayC2',ArrayC2TH',...
        ArrayC3',ArrayC3TH',ArrayC4',ArrayC4TH',ArrayNP',ArrayXNPP'];
save datfil.txt output-ascii
disp 'simulation finished'
clc
figure
plot(ArrayT,ArrayC3,ArrayT,ArrayC3TH),grid
xlabel('Time (s) ')
ylabel('C3')
axis([0 10 0 1.5])
figure
plot(ArrayT,ArrayC4,ArrayT,ArrayC4TH),grid
xlabel('Time (s) ')
ylabel('C4')
axis([0 10 0 1])
```

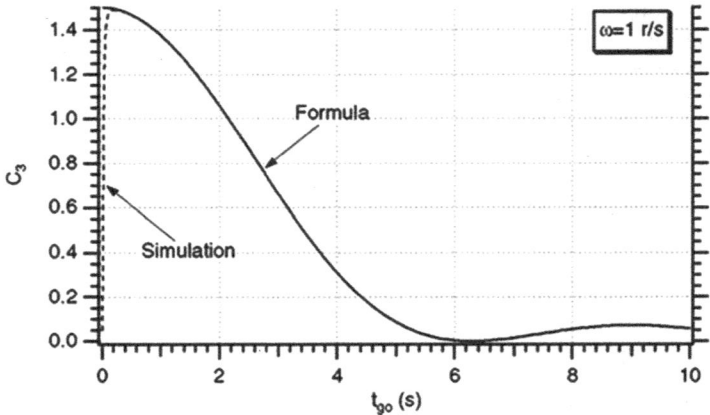

Fig. 12.8   Formula and simulation results agree for third control gain.

The nominal case of Listing 12.2 was run, and the control gain outputs for C_3 and C_4 are displayed along with the analytical solution ("Formula") in Figs. 12.8 and 12.9. We can see that again the numerical and analytical results are in near-perfect agreement, thus again verifying the accuracy of the numerical integration. Therefore, we have again demonstrated that the optimal control technique is a viable alternative approach for numerically deriving missile guidance laws.

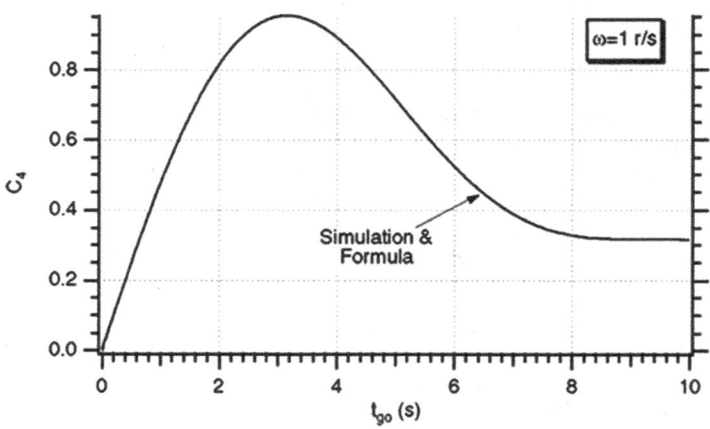

Fig. 12.9   Formula and simulation results agree for fourth control gain.

GUIDANCE PORTION DUE TO MANEUVERING TARGETS

We have seen in this text and the previous volume that all of the guidance laws can be expressed as a navigation ratio times the zero effort miss and divided by the square of time to go until intercept. In addition, we have noted that the first two terms in the guidance law relate to proportional navigation and the other terms relate to target maneuver and the dynamics of the flight-control system. Therefore, we can say that, in general, all guidance laws can be expressed as

$$n_c = \frac{N'}{t_{go}^2} \text{ZEM} = \frac{N'}{t_{go}^2} [y + \dot{y}t_{go} + \text{ZEM}_{TGT} + \text{ZEM}_{FCS}]$$

where ZEM_{TGT} is the zero effort miss due to the target maneuver and ZEM_{FCS} is the zero effort miss due to the flight-control system. In this section we shall present an alternative analytical technique for finding the zero effort miss due to target maneuver and demonstrate that the new technique yields answers that we know to be true.

Reference [5] shows that, in general, the zero effort miss caused by the target maneuver is given by

$$\text{ZEM}_{TGT} = -\int_{t}^{t_F} (\alpha - t_F)n_T(\alpha)d\alpha$$

where α is a dummy variable, n_T is the target maneuver (which does not have to be a constant), and t_F is the time of flight. Let us consider two cases for which we know the answers in order to demonstrate that the preceding relationship is correct.

First let us consider the example of a constant target maneuver. In this case, the closed-form solution for the zero effort miss becomes

$$\text{ZEM}_{\substack{\text{CONST} \\ \text{TGT}}} = -n_T \int_{t}^{t_F} (\alpha - t_F)d\alpha = -n_T \left[\int_{t}^{t_F} \alpha d\alpha - t_F \int_{t}^{t_F} d\alpha \right]$$

Integration of the preceding expression yields

$$\text{ZEM}_{\substack{\text{CONST} \\ \text{TGT}}} = -n_T \left[\frac{\alpha^2}{2} \bigg|_{t}^{t_F} - t_F \alpha \bigg|_{t}^{t_F} \right] = -n_T \left[\frac{t_F^2}{2} - \frac{t^2}{2} - t_F^2 + t_F t \right]$$

which simplifies to

$$\text{ZEM}_{\substack{\text{CONST} \\ \text{TGT}}} = \frac{n_T(t_F - t)^2}{2} = \frac{n_T t_{go}^2}{2}$$

where

$$t_{go} = t_F - t$$

We recognize the zero effort miss caused by constant target maneuver as the term we used in the augmented proportional navigation guidance law, thus verifying the theoretical expression presented at the beginning of this section.

Let us now consider another example in which the target maneuver is not a constant but is a sinusoid with weave frequency ω and arbitrary phase angle ϕ. In this case, we can express the time-varying target acceleration as

$$n_T(t) = a_T \sin(\omega t + \phi)$$

where a_T is the amplitude of the sinusoidal maneuver. By taking the derivative of the preceding expression, we can also find the target jerk as

$$\dot{n}_T(t) = a_T \omega \cos(\omega t + \phi)$$

Substitution of the expression for the sinusoidal target maneuver in the formula for the zero effort miss yields

$$\text{ZEM}_{\text{WEAVE}} = -\int_t^{t_F} (\alpha - t_F) n_T(\alpha) \, d\alpha = -a_T \int_t^{t_F} (\alpha - t_F) \sin(\omega\alpha + \phi) \, d\alpha$$

Let us recall from integral tables that

$$\int \sin(\omega\alpha + \phi) \, d\alpha = -\frac{1}{\omega} \cos(\omega\alpha + \phi)$$

$$\int \alpha \sin(\omega\alpha + \phi) \, d\alpha = -\frac{\alpha}{\omega} \cos(\omega\alpha + \phi) + \frac{1}{\omega^2} \sin(\omega\alpha + \phi)$$

Therefore, the zero effort miss becomes

$$\text{ZEM}_{\text{WEAVE}} = -a_T \left\{ \left[-\frac{\alpha}{\omega} \cos(\omega\alpha + \phi) + \frac{1}{\omega^2} \sin(\omega\alpha + \phi) \right]_t^{t_F} \right.$$
$$\left. + \left[\frac{t_F}{\omega} \cos(\omega\alpha + \phi) \right]_t^{t_F} \right\}$$

After evaluation of the limits and some simplification, we obtain

$$\text{ZEM}_{\text{WEAVE}} = -a_T \left\{ \left(\frac{t}{\omega} - \frac{t_F}{\omega} \right) \cos(\omega t + \phi) \right.$$
$$\left. + \frac{1}{\omega^2} \left[\sin(\omega t_F + \phi) - \sin(\omega t + \phi) \right] \right\}$$

We can expand

$$\sin(\omega t_F + \phi) = \sin[\omega(t_F - t + t) + \phi] = \sin[\omega(t_{\text{go}} + t) + \phi]$$
$$= \sin(\omega t_{\text{go}} + \omega t + \phi)$$

and recognize that

$$\sin(\omega t_{go} + \omega t + \phi) = \sin \omega t_{go} \cos(\omega t_{go} + \phi) + \cos \omega t_{go} \sin(\omega t + \phi)$$

The expression for the zero effort miss caused by the weave maneuver then becomes

$$ZEM_{WEAVE} = -a_T \left\{ \frac{-t_{go}}{\omega} \cos(\omega t + \phi) + \frac{1}{\omega^2} [\sin \omega t_{go} \cos(\omega t + \phi) \right.$$

$$\left. + \cos \omega t_{go} \sin(\omega t + \phi) - \sin(\omega t + \phi)] \right\}$$

Simplification of the preceding expression yields

$$ZEM_{WEAVE} = -a_T \left[\left(\frac{-\omega t_{go} + \sin \omega t_{go}}{\omega^2} \right) \cos(\omega t + \phi) \right.$$

$$\left. + \sin(\omega t + \phi) \left(\frac{\cos \omega t_{go} - 1}{\omega^2} \right) \right]$$

Substitution of the expressions for target acceleration and jerk yields

$$ZEM_{WEAVE} = n_T \left(\frac{1 - \cos \omega t_{go}}{\omega^2} \right) + \dot{n}_T \left(\frac{\omega t_{go} - \sin \omega t_{go}}{\omega^3} \right)$$

The preceding expression is identical to the portion of the zero effort miss related to target acceleration and jerk in the weave guidance law. Thus we have again verified the theoretical formula presented at the beginning of this section.

ALTERNATIVE NUMERICAL APPROACH AS A RESULT OF FLIGHT-CONTROL SYSTEM DYNAMICS

For completeness, an alternative and efficient numerical method for calculating the optimal guidance law as a result of the dynamics of the flight-control system is presented in this section. It is based upon the work of Rusnak and Meir [6]. Consider a flight-control system transfer function $H(s)$, where $H(s) = n_L(s)/n_C(s)$. Reference [6] shows that the optimal control gains can be calculated by first computing the quantity $\Lambda(t_{go})$

$$\Lambda(t_{go}) = \mathscr{L}^{-1} \left[\frac{1}{s^2} \cdot \frac{n_L(s)}{n_C(s)} \right]_{t_{go}} \bigg/ \left[\gamma + \int_0^{t_{go}} \left\{ \mathscr{L}^{-1} \left[\frac{1}{s^2} \cdot \frac{n_L(s)}{n_C(s)} \right]_{t_{go}} \right\}^2 d\tau \right] = \frac{N}{D}$$

where $\mathscr{L}^{-1}$ is the inverse Laplace transform. The symbols N and D have been used as shorthand for numerator and denominator, respectively. Therefore, the inverse Laplace transform of the item in the brackets of the preceding equation represents the time response of the flight-control system to a ramp t acceleration command $n_c(t)$ (the Laplace transform of t is $1/s^2$). The guidance law control gains for the first three states (assuming constant target acceleration) are computed from $\Lambda(t_{go})$ as

$$C_1(t_{go}) = \Lambda(t_{go})$$
$$C_2(t_{go}) = \Lambda(t_{go})t_{go}$$
$$C_3(t_{go}) = 0.5\Lambda(t_{go})t_{go}^2$$

Reference [6] also states that the control gains for the flight-control system states $i = 1$ through N can be expressed as

$$C_{3+i}(t_{go}) = -\Lambda(t_{go})\mathcal{L}^{-1}\left[\frac{1}{s^2}\cdot\frac{n_L(s)}{p_i(0)}\right]_{t_{go}}$$

where p_1 through p_N are the states of the flight-control-system transfer function. Therefore, the quantity in brackets of the preceding equation represents the flight-control system's response to a ramp input t on the derivative of the ith state of the flight-control system. Note that the actual effective navigation ratio is related to $\Lambda(t_{go})$ according to

$$N'(t_{go}) = \Lambda(t_{go})t_{go}^2$$

The best way of illustrating this innovative technique is to work an example for the single time constant flight-control system as was done in [7].

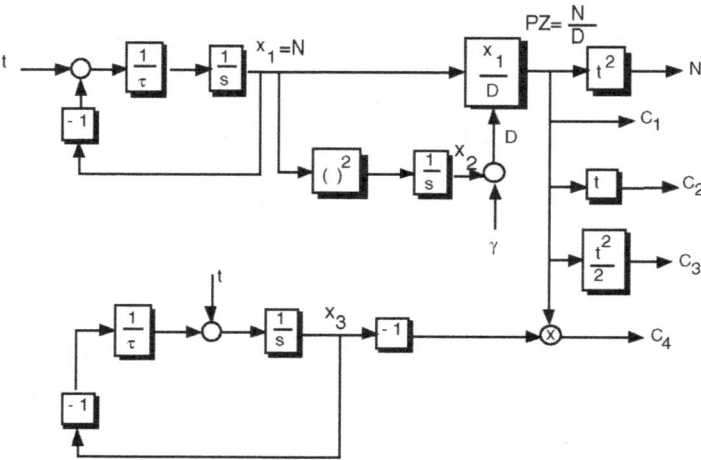

Fig. 12.10   Reference [5] method for finding classic optimal guidance law for single time constant flight-control system.

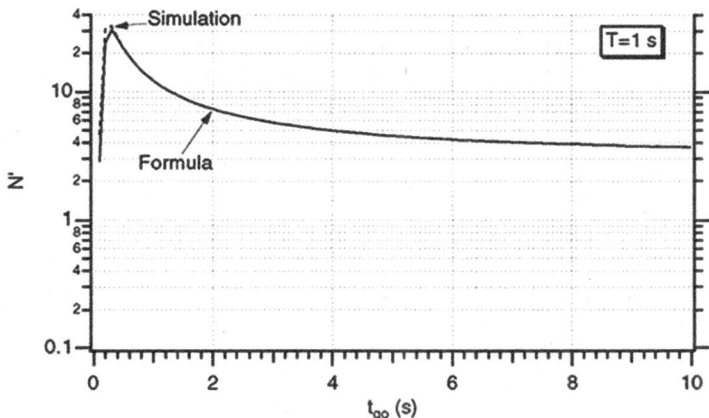

Fig. 12.11 Rusnak and Meir technique is equivalent to Riccati equation method for single-lag flight-control system.

Figure 12.10 is the block diagram equivalent of the preceding equations for the effective navigation ratio and control gains. The resultant differential equations resulting from this block diagram are simple and well behaved numerically.

Recall that for the preceding system, the optimal guidance law can be expressed in terms of the gains as

$$n_c = C_1 y + C_2 \dot{y} + C_3 n_T + C_4 n_L$$

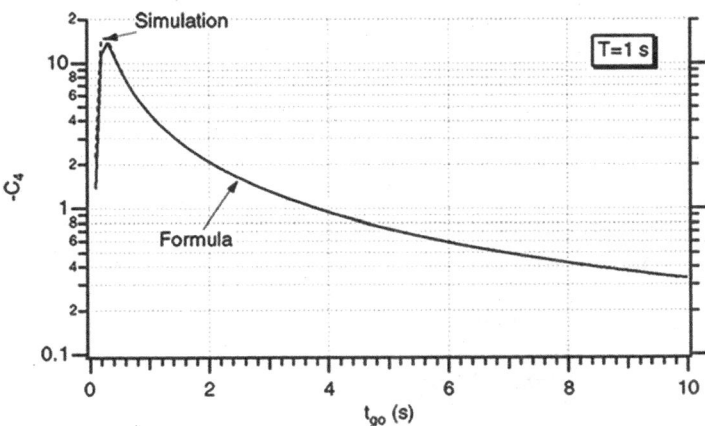

Fig. 12.12 Rusnak and Meir technique yields accurate fourth control gain for single-lag flight-control system.

Figure 12.10 was simulated, and the effective navigation ratio and C_4 gain were compared to the analytical results derived earlier. Listing 12.3 is a simulation of Fig. 12.10 using second-order Runge–Kutta integration with an integration step size of 0.01 s. This technique is not as numerically fragile as the optimal control method. Therefore, this simulation will run approximately 100 times faster than the simulation using the optimal control approach because of the much larger integration step size. Listing 12.3 also presents the formulas for the effective navigation ratio and C_4 gain of the actual optimal guidance law.

The nominal case of Listing 12.3 was run. Figures 12.11 and 12.12 indicate that there is excellent agreement for the effective navigation ratio and fourth control gains using this new technique when compared to the closed-form solutions. Thus we have confidence that no programming errors were made in Listing 12.3 and that a large integration interval yielded the correct answers.

LISTING 12.3   NUMERICAL APPROACH TO GUIDANCE-LAW DEVELOPMENT FOR SINGLE-LAG FLIGHT-CONTROL SYSTEM USING RUSNAK AND MEIR TECHNIQUE

```
clear
count=0;
TAU=1.;
GAM=.00001;
T=0.;
H=.01;
S=0.;
X1=0.;
X2=0.;
X3=0.;
while ~(T >= (10.-.0001))
        S=S+H;
        X1OLD=X1;
        X2OLD=X2;
        X3OLD=X3;
        STEP=1;
        FLAG=0;
        while STEP<=1
                if FLAG==1
                STEP=2;
                        X1=X1+H*X1D;
                        X2=X2+H*X2D;
                        X3=X3+H*X3D;
                        T=T+H;
                end
                X1D=(T-X1)/TAU;
```

```
                    X2D=X1^2;
                    D=X2+GAM;
                    X3D=-X3/TAU+T;
                    FLAG=1;
            end
            FLAG=0;
            X1=.5*(X1OLD+X1+H*X1D);
            X2=.5*(X2OLD+X2+H*X2D);
            X3=.5*(X3OLD+X3+H*X3D);
            if S>=.09999
                    S=0.;
                    PZ=X1/D;
                    XNP=PZ*T*T;
                    C1=PZ;
                    C2=PZ*T;
                    C3=.5*PZ*T*T;
                    C4=-X3*PZ;
                    XS=T/TAU;
                    TOP=6.*XS*XS*(exp(-XS)-1.+XS);
                    BOT1=2*XS*XS*XS+3.+6.*XS-6.*XS*XS;
                    BOT2=-12.*XS*exp(-XS)-3.*exp(-2.*XS);
                    XNPTH=TOP/(BOT1+BOT2+.0001);
                    C4TH=-XNPTH*(exp(-XS)+XS-1.)/(XS*XS);
                    count=count+1;
                    ArrayT(count)=T;
                    ArrayC4(count)=C4;
                    ArrayC4TH(count)=C4TH;
                    ArrayXNP(count)=XNP;
                    ArrayXNPTH(count)=XNPTH;
            end
    end
end
output=[ArrayT',ArrayC4',ArrayC4TH',ArrayXNP',ArrayXNPTH'];
save datfil.txt output-ascii
disp 'simulation finished'
clc
figure
semilogy(ArrayT,ArrayXNP,ArrayT,ArrayXNPTH),grid
xlabel('Time (s) ')
ylabel('NP')
axis([0 10 .1 40])
figure
semilogy(ArrayT,-ArrayC4,ArrayT,-ArrayC4TH),grid
xlabel('Time (s) ')
ylabel('-C4')
axis([0 10 .1 20])
```

DERIVING NEW GUIDANCE LAW FOR CUBIC FLIGHT-CONTROL SYSTEM

So far in this chapter we have used the new techniques to derive guidance laws for which we already had the closed-form solution. This was done to give us confidence that the alternative techniques actually worked. In the next two sections we shall use both new techniques to numerically develop a new guidance law that would not be possible to derive using the Schwartz inequality.

When more accurate representations of the missile flight-control system are used, closed-form solutions for the resultant guidance law become either impossible to derive or unwieldy because of the flight-control system complexity. However, guidance laws do not have to be derived in closed form for them to be useful. The more advanced guidance laws can be developed numerically, and the resultant control gains can be stored in a flight computer for guidance-law implementation [7]. In fact, this approach is similar to the way autopilot gains are stored as a function of flight condition in the missile flight computer [8].

We saw in Chapter 23 of the previous volume that a more realistic model for the flight-control system of a tail-controlled missile (compared to the single-lag representation) is given by the transfer function

$$\frac{n_L}{n_c} = \frac{\left(1 - \frac{s^2}{\omega_z^2}\right)}{\left[(1 + s\tau)\left(1 + \frac{2\zeta}{\omega}s + \frac{s^2}{\omega^2}\right)\right]}$$

In this transfer function we have modeled the "wrong-way" tail effect with left- and right-half-plane zeros. The denominator of the flight-control system consists of a dominant real pole followed by a higher-frequency quadratic. To use optimal control techniques to derive a new guidance law, we must first place the preceding transfer function in state space form. The first step is to multiply out the denominator of the flight-control system transfer function, or

$$\frac{n_L}{n_c} = \left(1 - \frac{s^2}{\omega_z^2}\right) \Big/ \left[1 + \left(\frac{2\zeta}{\omega} + \tau\right)s + \left(\frac{2\zeta\tau}{\omega} + \frac{1}{\omega^2}\right)s^2 + \frac{\tau}{\omega^2}s^3\right]$$

We would like to split the transfer function so that we can start to write differential equations in state space format. This can be accomplished by using the chain rule from calculus, as was done in Chapter 1 of the previous volume, or

$$\frac{n_L}{n_c} = \frac{e}{n_c} * \frac{n_L}{e}$$

Using the preceding equation yields the two transfer functions

$$\frac{e}{n_c} = 1 \Big/ \left[1 + \left(\frac{2\zeta}{\omega} + \tau \right) s + \left(\frac{2\zeta\tau}{\omega} + \frac{1}{\omega^2} \right) s^2 + \frac{\tau}{\omega^2} s^3 \right]$$

$$\frac{n_L}{e} = 1 - \frac{s^2}{\omega_z^2}$$

Cross multiplying and converting to the time domain yields the two differential equations describing the cubic flight-control system as

$$\dddot{e} = \frac{\omega^2}{\tau} \left[n_c - e - \left(\frac{2\zeta}{\omega} + \tau \right) \dot{e} - \left(\frac{2\zeta\tau}{\omega} + \frac{1}{\omega^2} \right) \ddot{e} \right]$$

$$n_L = e - \frac{\ddot{e}}{\omega_z^2}$$

Using the preceding two differential equations, we can express the new model for guidance-law development in block diagram form as shown in Fig. 12.13.

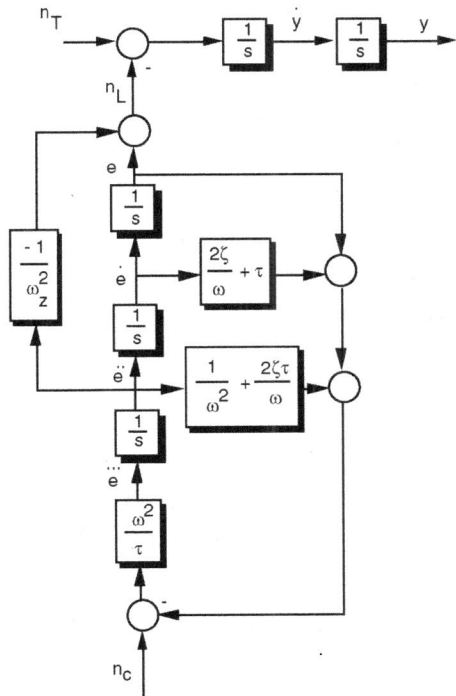

Fig. 12.13   Cubic flight-control system model for guidance-law development.

Figure 12.13 can be converted to state space form by inspection. Thus the state space equation for the homing loop with the cubic flight-control system is given by

$$
\begin{bmatrix} \dot{y} \\ \ddot{y} \\ \dot{n}_T \\ \dot{e} \\ \ddot{e} \\ \dddot{e} \end{bmatrix} =
\begin{bmatrix}
0 & 1 & 0 & 0 & 0 & 0 \\
0 & 0 & 1 & -1 & 0 & \dfrac{1}{\omega_z^2} \\
0 & 0 & 0 & 0 & 0 & 0 \\
0 & 0 & 0 & 0 & 1 & 0 \\
0 & 0 & 0 & 0 & 0 & 1 \\
0 & 0 & 0 & -\dfrac{\omega^2}{\tau} & -\dfrac{\omega^2}{\tau}\left(\dfrac{2\zeta}{\omega}+\tau\right) & -\dfrac{\omega^2}{\tau}\left(\dfrac{1}{\omega^2}+\dfrac{2\zeta\tau}{\omega}\right)
\end{bmatrix}
\begin{bmatrix} y \\ \dot{y} \\ n_T \\ e \\ \dot{e} \\ \ddot{e} \end{bmatrix} +
\begin{bmatrix} 0 \\ 0 \\ 0 \\ 0 \\ 0 \\ \dfrac{\omega^2}{\tau} \end{bmatrix}
$$

Therefore, the systems dynamics and control matrices are given by

$$
F =
\begin{bmatrix}
0 & 1 & 0 & 0 & 0 & 0 \\
0 & 0 & 1 & -1 & 0 & \dfrac{1}{\omega_z^2} \\
0 & 0 & 0 & 0 & 0 & 0 \\
0 & 0 & 0 & 0 & 1 & 0 \\
0 & 0 & 0 & 0 & 0 & 1 \\
0 & 0 & 0 & -\dfrac{\omega^2}{\tau} & -\dfrac{\omega^2}{\tau}\left(\dfrac{2\zeta}{\omega}+\tau\right) & -\dfrac{\omega^2}{\tau}\left(\dfrac{1}{\omega^2}+\dfrac{2\zeta\tau}{\omega}\right)
\end{bmatrix}
$$

$$
G =
\begin{bmatrix} 0 \\ 0 \\ 0 \\ 0 \\ 0 \\ \dfrac{\omega^2}{\tau} \end{bmatrix}
$$

Because we are still trying to minimize the performance index,

$$
J = y^2(t_F) + \gamma \int_0^{t_F} n_c^2 \, dt
$$

we can say that

$$
S_F =
\begin{bmatrix}
1 & 0 & 0 & 0 & 0 & 0 \\
0 & 0 & 0 & 0 & 0 & 0 \\
0 & 0 & 0 & 0 & 0 & 0 \\
0 & 0 & 0 & 0 & 0 & 0 \\
0 & 0 & 0 & 0 & 0 & 0 \\
0 & 0 & 0 & 0 & 0 & 0
\end{bmatrix}
$$

$$
u = n_c \text{ and } B = \gamma
$$

To see what the new guidance law looks like, let us consider a numerical example. The flight-control-system transfer function used to test the new guidance law is considered to be

$$\frac{n_L}{n_c} = \left(1 - \frac{s^2}{5^2}\right) \Big/ \left[(1+s)\left(1 + \frac{2^*0.7}{20}s + \frac{s^2}{20^2}\right)\right]$$

This means that the right-half-plane zero is at the low frequency of 5 rad/s. The dominant time constant is 1 s, and the damping and natural frequency of the flight-control system are 0.7 and 20 rad/s, respectively. This type of transfer function might be representative of a tail-controlled missile, such as the one of Fig. 21.3 of the previous volume, flying at very high altitude and low velocity [7]. Because our model of the real world has six states, the new guidance law will have six control gains. Listing 12.4 is our optimal control program with the new matrices; changes in code (from Listings 12.1 and 12.2) are highlighted in bold.

The nominal case of Listing 12.4 was run. The time constant of the cubic flight-control system is identical to that of the single-lag flight-control system previously considered in Listing 12.1. To compare the new guidance law to the guidance law that was optimal for the single-lag flight-control system, it is easiest to compare the effective navigation ratios. Figure 12.14 shows that both effective navigation ratios approach 3 when the time to go before intercept is very large and that both navigation ratios get large near intercept and go to zero at intercept. However, the new effective navigation ratio also goes negative near intercept in an attempt to compensate for the right-half-plane zero of the flight-control-system transfer function.

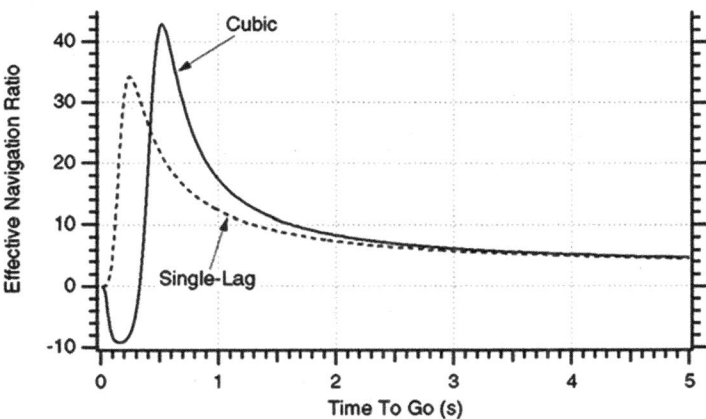

Fig. 12.14 **New guidance-law effective navigation ratio goes negative near intercept.**

The resultant guidance law is given by

$$n_c = C_1 y + C_2 \dot{y} + C_3 n_T + C_4 e + C_5 \dot{e} + C_6 \ddot{e}$$

where the y, $\dot{y}$, and n_T are obtained from a Kalman filter and e, $\dot{e}$, and $\ddot{e}$ can be derived by reconstructing states within the flight-control system. As was already mentioned, there is no need to compute all six control gains in the missile in real time. These gains can be computed as part of the missile design and stored in the onboard computer memory as part of the overall data input. During a missile intercept, the control gains would be determined by table look-up. At each guidance update, the values of time to go, ω_z, ω, τ, and ζ are the inputs to a table look-up algorithm. The outputs are the six control gains C_1 through C_6. The autopilot is usually designed to have a particular response as a function of Mach and altitude, so that ω_z, ω, τ, and ζ can be looked up based on Mach and altitude. Alternatively and more simply, the control gains can be stored and looked up directly as a function of Mach, altitude, and time to go.

LISTING 12.4   INTEGRATING RICCATI EQUATIONS TO GET OPTIMAL CONTROL GAINS FOR CUBIC FLIGHT-CONTROL SYSTEM

```
clear
TAU=1.;
WZ=5.;
W=20.;
Z=.7;
TF=10.;
GAM=.00001;
F=zeros([6,6]);
S=zeros([6,6]);
count=0;
G(1,1)=0.;
G(2,1)=0.;
G(3,1)=0.;
G(4,1)=0.;
G(5,1)=0.;
G(6,1)=W*W/TAU;
F(1,2)=1;
F(2,3)=1;
F(2,4)=-1.;
F(2,6)=1./WZ^2;
F(4,5)=1.;
F(5,6)=1.;
F(6,4)=-W*W/TAU;
F(6,5)=-W*W*(2.*Z/W+TAU)/TAU;
F(6,6)=-W*W*(1./W^2+2.*Z*TAU/W)/TAU;
```

```
S(1,1)=1;
T=0;
H=.0001;
S1=0;
while ~(T >= (TF-.0001))
        S1=S1+H;
        SOLD=S;
        STEP=1;
        FLAG=0;
        while STEP<=1
                if FLAG==1
                        STEP=2;
                        HSD=H*SD;
                        S=S+HSD;
                        T=T+H;
                end
                SF=S*F;
                GT=G';
                GTS=GT*S;
                GAMINV=1./GAM;
                C=GAMINV*GTS;
                SFT=SF';
                CT=C';
                CTC=CT*C;
                CTBC=GAM*CTC;
                SFSFT=SF+SFT;
                SD=SFSFT-CTBC;
                FLAG=1;
        end
        FLAG=0;
        H2=.5*H;
        HSDP=H2*SD;
        SS=SOLD+S;
        SSP=.5*SS;
        S=SSP+HSDP;
        if S1>=.009999
                S1=0;
                C1=-C(1,1);
                C2=-C(1,2);
                C3=-C(1,3);
                C4=-C(1,4);
                C5=-C(1,5);
                C6=-C(1,6);
                NP=C2*T;
                count=count+1;
                ArrayT(count)=T;
```

```
            ArrayC4(count)=C4;
            ArrayC5(count)=C5;
            ArrayC6(count)=C6;
            ArrayNP(count)=NP;
    end
end
output=[ArrayT',ArrayC4',ArrayC5',ArrayC6',ArrayNP'];
save datfil.txt output-ascii
disp 'simulation finished'
clc
figure
plot(ArrayT,ArrayNP),grid
xlabel('Time (s) ')
ylabel('NP')
axis([0 10 -10 50])
```

ALTERNATIVE APPROACH TO CUBIC FLIGHT-CONTROL-SYSTEM GUIDANCE LAW

For completeness, the alternative method was used for the derivation of the guidance law for the cubic flight-control system. Recall that the transfer function of the cubic flight-control system was given by

$$\frac{n_L}{n_c} = \left(1 - \frac{s^2}{\omega_z^2}\right) \Big/ \left[(1 + s\tau)\left(1 + \frac{2\zeta}{\omega}s + \frac{s^2}{\omega^2}\right)\right]$$

Applying the methodology of [6] yields the block diagram of Fig. 12.15. The first three control gains can be found from the solution for the effective navigation ratio as

$$C_1 = \frac{N'}{t_{go}^2}$$

$$C_2 = \frac{N'}{t_{go}}$$

$$C_3 = 0.5N'$$

where t in the block diagram represents t_{go}.

Listing 12.5 simply programs Fig. 12.15. The gains C_1 through C_6 are generated by numerically solving the differential equations of Fig. 12.15, and the program is set up to run a case for the autopilot transfer function

$$\frac{n_L}{n_c} = \left(1 - \frac{s^2}{5^2}\right) \Big/ \left[(1 + s)\left(1 + \frac{2^*0.7}{20}s + \frac{s^2}{20^2}\right)\right]$$

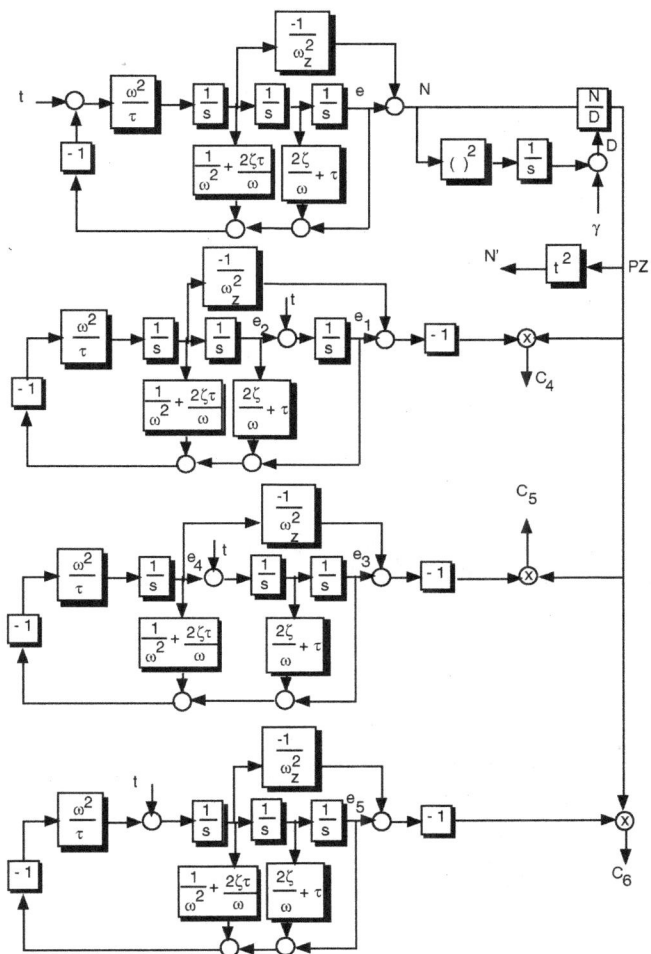

Fig. 12.15   Reference [6] method for finding control gains for cubic autopilot.

Note that the integration step size H is again 100 times larger than the step size required for the optimal control technique.

The nominal case of Listing 12.5 was run, and the resultant effective navigation ratio is displayed in Fig. 12.16. This figure shows that the effective navigation ratio generated by the method of [6] is completely equivalent to that generated by the optimal control method. Thus we have further confirmation that the alternative method for generating guidance gains can be used either as a substitute for the more traditional method or as an independent check in control gain selection.

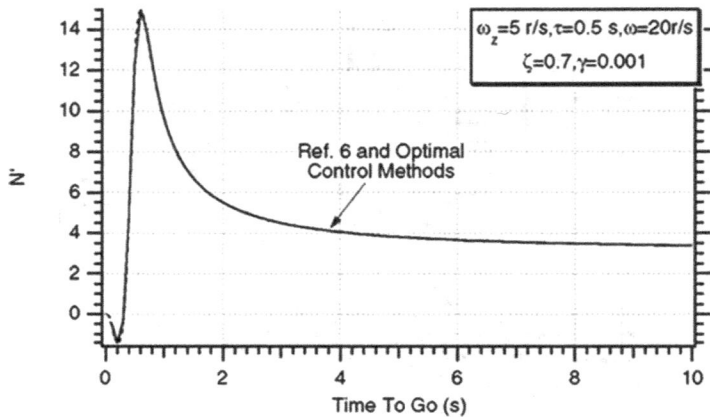

Fig. 12.16   Method of [6] is equivalent to optimal control method for cubic autopilot.

LISTING 12.5   ALTERNATIVE APPROACH FOR OBTAINING CONTROL GAINS FOR CUBIC FLIGHT-CONTROL-SYSTEM GUIDANCE LAW

```
clear
count=0;
TAU=1.;
GAM=.00001;
WZ=5.;
W=20.;
Z=.7;
T=0.;
H=.01;
S=0.;
X=0.;
E=0.;
ED=0.;
EDD=0.;
E1=0.;
E2=0.;
E2D=0.;
E3=0.;
E3D=0.;
E4=0.;
E4D=0.;
E5=0.;
E5D=0.;
E5DD=0.;
while ~(T >= (10.-.0001))
```

```
S=S+H;
XOLD=X;
EOLD=E;
EDOLD=ED;
EDDOLD=EDD;
E1OLD=E1;
E2OLD=E2;
E2DOLD=E2D;
E3OLD=E3;
E3DOLD=E3D;
E4OLD=E4;
E5OLD=E5;
E5DOLD=E5D;
E5DDOLD=E5DD;
STEP=1;
FLAG=0;
while STEP<=1
    if FLAG==1
            STEP=2;
            X=X+H*XD;
            E=E+H*ED;
            ED=ED+H*EDD;
            EDD=EDD+H*EDDD;
            E1=E1+H*E1D;
            E2=E2+H*E2D;
            E2D=E2D+H*E2DD;
            E3=E3+H*E3D;
            E3D=E3D+H*E3DD;
            E4=E4+H*E4D;
            E5=E5+H*E5D;
            E5D=E5D+H*E5DD;
            E5DD=E5DD+H*E5DDD;
            T=T+H;
    end
    EDDD=W*W*(T-(1./W^2+2.*Z*TAU/W)*EDD-(2.*Z/W+TAU)*ED-E)/TAU;
    XN=E-EDD/WZ^2;
    XD=XN^2;
    D=X+GAM;
    PZ=XN/D;
    XNP=T*T*PZ;
    E2DD=W*W*(-(1./W^2+2.*Z*TAU/W)*E2D-(2.*Z/W+TAU)*E2-E1)/TAU;
    E1D=E2+T;
    C4=-(E1-E2D/WZ^2)*PZ;
    E4D=W*W*(-(1./W^2+2.*Z*TAU/W)*E4-(2.*Z/W+TAU)*E3D-E3)/TAU;
    E3DD=E4+T;
    C5=-(E3-E4/WZ^2)*PZ;
```

```
            E5DDD=T-W*W*((1./W^2+2.*Z*TAU/W)*E5DD+(2.*Z/W+TAU)*E5D+E5)/TAU;
            C6=PZ*(-E5+E5DD/WZ^2);
            FLAG=1;
        end
        FLAG=0;
        X=.5*(XOLD+X+H*XD);
        E=.5*(EOLD+E+H*ED);
        ED=.5*(EDOLD+ED+H*EDD);
        EDD=.5*(EDDOLD+EDD+H*EDDD);
        E1=.5*(E1OLD+E1+H*E1D);
        E2=.5*(E2OLD+E2+H*E2D);
        E2D=.5*(E2DOLD+E2D+H*E2DD);
        E3=.5*(E3OLD+E3+H*E3D);
        E3D=.5*(E3DOLD+E3D+H*E3DD);
        E4=.5*(E4OLD+E4+H*E4D);
        E5=.5*(E5OLD+E5+H*E5D);
        E5D=.5*(E5DOLD+E5D+H*E5DD);
        E5DD=.5*(E5DDOLD+E5DD+H*E5DDD);
        if S>=.09999
            S=0.;
            count=count+1;
            ArrayT(count)=T;
            ArrayC4(count)=C4;
            ArrayC5(count)=C5;
            ArrayC6(count)=C6;
            ArrayXNP(count)=XNP;
        end
    end
end
output=[ArrayT',ArrayC4',ArrayC5',ArrayC6',ArrayXNP'];
save datfil.txt output-ascii
disp 'simulation finished'
clc
figure
plot(ArrayT,ArrayXNP),grid
xlabel('Time (s) ')
ylabel('NP')
axis([0 10 -10 50])
```

PERFORMANCE COMPARISON OF GUIDANCE LAWS IN PRESENCE OF CUBIC FLIGHT-CONTROL SYSTEM

In the previous two sections we numerically derived a guidance law that is optimal when the flight-control-system transfer function is given by

$$\frac{n_L}{n_c} = \left(1 - \frac{s^2}{5^2}\right) \Bigg/ \left[(1+s)\left(1 + \frac{2*0.7}{20}s + \frac{s^2}{20^2}\right)\right]$$

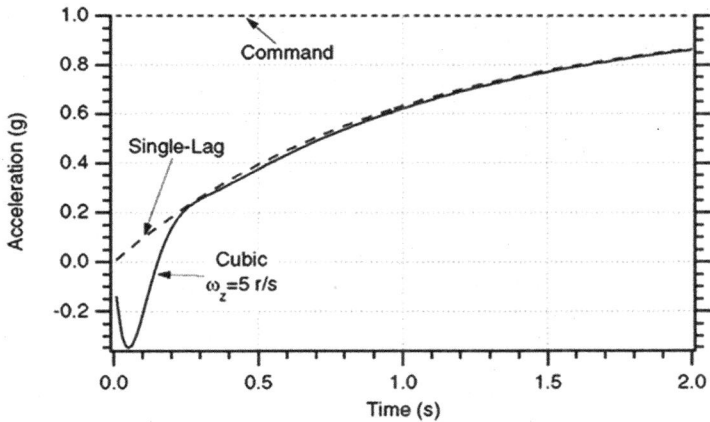

Fig. 12.17 Single-lag flight-control system response does not match cubic response when numerator zero is 5 rad/s.

Figure 12.17 indicates that the step response of the preceding cubic transfer function is totally different than the step response for the single-lag representation when the dominant time constant is 1 s. The single-lag inaccuracy in the step response is mainly caused by the low-frequency numerator zero of 5 rad/s. Figure 12.18 demonstrates that when the numerator zero is increased to 100 rad/s, the single-lag and cubic representations of the flight-control system are virtually identical.

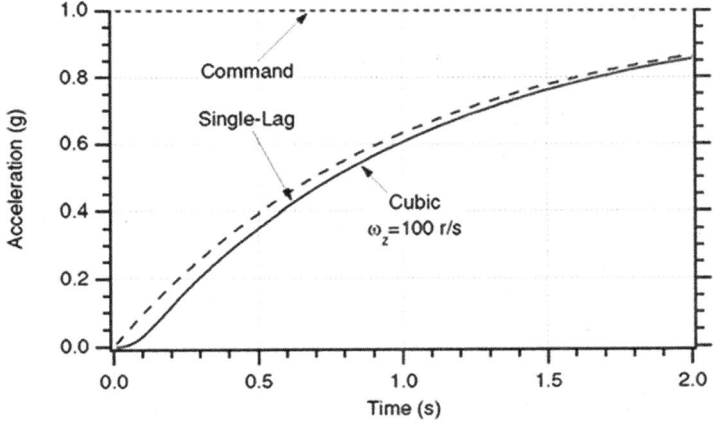

Fig. 12.18 Single-lag flight-control-system response matches cubic response when numerator zero is 100 rad/s.

We are now ready to see if there is a performance benefit in using the numerical solution for the cubic guidance law. As was mentioned earlier, there is no need to compute all six control gains in the missile in real time. These gains can be computed as part of the missile design and stored in the onboard computer memory as part of the overall data input, as was suggested in [7]. To illustrate part of that concept, Listing 12.6 computes the six control gains for the cubic guidance law for a given set of flight-control-system parameters in routine generategains.m using the optimal control approach. The resultant gains, which are a function of the time to go until intercept, are stored in memory and called when needed. For academic purposes, and to get to the heart of the matter, the simulation is set up so that all states are known perfectly (that is, no filtering is required), and the only disturbance is a 0.5-g target maneuver. The program evaluates the miss distances for a variety of flight times. The real flight-control system in the simulation is identical to the cubic transfer function. By setting the parameter APN, we can choose from a variety of guidance laws (0 = proportional navigation, 1 = augmented proportional navigation, 2 = optimal guidance for single-lag flight-control system, and 3 = optimal guidance for cubic flight-control system).

LISTING 12.6   SIMULATION FOR EVALUATING DIFFERENT GUIDANCE LAWS

```
clear
count=0;
TAU=1.;
W=20.;
Z=.7;
WZ=5.;
APN=3;
XNT=16.1;
TS=.01;
GAM=.00001;
XLIM=9999999.;
TFMAX=10.;
XNP=3.;
VC=4000.;
VM=3000.;
if APN==3
        [C1,C2,C3,C4,C5,C6]=GENERATEGAINS(TAU,W,Z,WZ,GAM,TFMAX,TS);
end
for TF=.1:.1:10.
        E=0.;
        ED=0.;
        EDD=0.;
        T=0;
        H=.0001;
        S=0.;
```

```
Y=0.;
YD=0.;
XNC=0.;
RTM=VC*TF;
while T< (TF-.00001)
        S=S+H;
        EOLD=E;
        EDOLD=ED;
        EDDOLD=EDD;
        YOLD=Y;
        YDOLD=YD;
        STEP=1;
        FLAG=0;
        while STEP<=1
                if FLAG==1
                        STEP=2;
                        E=E+H*ED;
                        ED=ED+H*EDD;
                        EDD=EDD+H*EDDD;
                        Y=Y+H*YD;
                        YD=YD+H*YDD;
                        T=T+H;
                end
                TGO=TF-T+.0001;
                RTM=VC*TGO;
                XLAM=Y/RTM;
                XNCG=XNC/32.2;
                EDDD=W*W*(XNC-E-(2.*Z/W+TAU)*ED-(2.*Z*TAU/W+1./W^2)...
                                *EDD)/TAU;
                XNL=E-EDD/WZ^2;
                YDD=XNT-XNL;
                FLAG=1;
        end
        FLAG=0;
        E=.5*(EOLD+E+H*ED);
        ED=.5*(EDOLD+ED+H*EDD);
        EDD=.5*(EDDOLD+EDD+H*EDDD);
        Y=.5*(YOLD+Y+H*YD);
        YD=.5*(YDOLD+YD+H*YDD);
        if S>=(TS-.0001)
                S=0.;
                if APN==0
                        XNC=XNP*(Y+YD*TGO)/(TGO*TGO);
                        XNPP=XNP;
                elseif APN==1
                        XNC=XNP*(Y+YD*TGO+.5*XNT*TGO*TGO)/(TGO*TGO);
```

```
                              XNPP=XNP;
                    elseif APN==2
                              XS=TGO/TAU;
                              TOP=6.*XS*XS*(exp(-XS)-1.+XS);
                              BOT1=2*XS*XS*XS+3.+6.*XS-6.*XS*XS;
                              BOT2=-12.*XS*exp(-XS)-3.*exp(-2.*XS);
                              XNPP=TOP/(.0001+BOT1+BOT2);
                              C1P=XNPP/(TGO*TGO);
                              C2P=XNPP/TGO;
                              C3P=.5*XNPP;
                              C4P=-XNPP*(exp(-XS)+XS-1.)/(XS*XS);
                              XNC=C1P*Y+C2P*YD+C3P*XNT+C4P*XNL;
                    else
                              JJ=fix(TGO/TS)+1;
                              XNC=C1(JJ)*Y+C2(JJ)*YD+C3(JJ)*XNT+C4(JJ)*E...
                                        +C5(JJ)*ED+C6(JJ)*EDD;
                              XNPP=C2(JJ)*TGO;
                    end
                    if XNC>XLIM
                              XNC=XLIM;
                    elseif XNC<-XLIM
                              XNC=-XLIM;
                    end
                    XNCG=XNC/32.2;
              end
       end
       count=count+1;
       ArrayTF(count)=TF;
       ArrayY(count)=Y;
end
figure
plot(ArrayTF,ArrayY),grid
xlabel('Flight Time (s)')
ylabel('Miss (ft) ')
clc
output=[ArrayTF',ArrayY'];
save datfil.txt output-ascii

function[C1,C2,C3,C4,C5,C6]=GENERATEGAINS(TAU,W,Z,WZ,GAM,TF,TS)
F=zeros([6,6]);
S=zeros([6,6]);
G(1,1)=0.;
G(2,1)=0.;
G(3,1)=0.;
G(4,1)=0.;
G(5,1)=0.;
```

```
G(6,1)=W*W/TAU;
F(1,2)=1;
F(2,3)=1;
F(2,4)=-1.;
F(2,6)=1./WZ^2;
F(4,5)=1.;
F(5,6)=1.;
F(6,4)=-W*W/TAU;
F(6,5)=-W*W*(2.*Z/W+TAU)/TAU;
F(6,6)=-W*W*(1./W^2+2.*Z*TAU/W)/TAU;
S(1,1)=1        ;
T=0;
H=.0001;
S1=0;
ICOUNT=1;
while T< (TF-.0001)
        S1=S1+H;
        SOLD=S;
        STEP=1;
        FLAG=0;
        while STEP<=1
                if FLAG==1
                        STEP=2;
                        HSD=H*SD;
                        S=S+HSD;
                        T=T+H;
                end
                SF=S*F;
                GT=G';
                GTS=GT*S;
                GAMINV=1./GAM;
                C=GAMINV*GTS;
                SFT=SF';
                CT=C';
                CTC=CT*C;
                CTBC=GAM*CTC;
                SFSFT=SF+SFT;
                SD=SFSFT-CTBC;
                FLAG=1;
        end
        FLAG=0;
        H2=.5*H;
        HSDP=H2*SD;
        SS=SOLD+S;
        SSP=.5*SS;
        S=SSP+HSDP;
```

```
if S1>=(TS-.0001)
      S1=0;
      C1(ICOUNT)=-C(1,1);
      C2(ICOUNT)=-C(1,2);
      C3(ICOUNT)=-C(1,3);
      C4(ICOUNT)=-C(1,4);
      C5(ICOUNT)=-C(1,5);
      C6(ICOUNT)=-C(1,6);
      ICOUNT=ICOUNT+1;
   end
end
```

A case was first run with Listing 12.6 in which the numerator zero of the flight-control system was set to 100 rad/s (WZ=100). We know from Fig. 12.18 that in this case the single-lag representation of the flight-control system is excellent. Therefore, it comes as no surprise that the performance of the optimal guidance law, as shown in Fig. 12.19, is very good and far superior to the performance of proportional navigation.

Another case was run with Listing 12.6 in which the numerator zero of the flight-control system was reduced to 10 rad/s (WZ=10). We can see from Fig. 12.20 that the performance of the single-lag optimal guidance law deteriorates significantly for flight times less than 25 s. On the other hand, the cubic guidance law yields nearly perfect performance in that the miss distance is always zero. The decrease of the right-half-plane zero does not significantly influence the proportional-navigation results; however, the proportional-navigation results are generally worse than those of both other guidance laws.

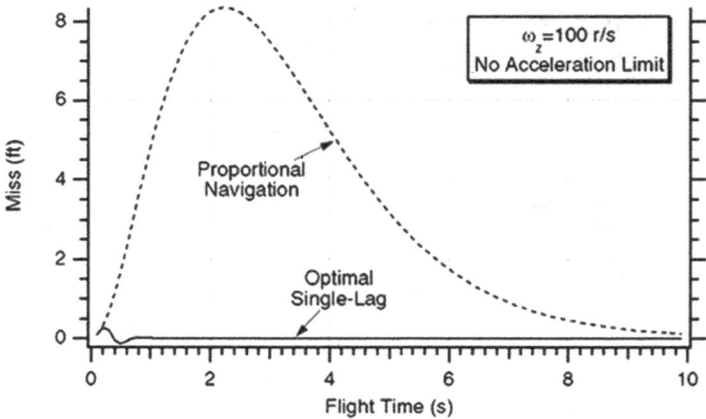

Fig. 12.19   Optimal guidance law for single-lag flight-control system works well when numerator zero is 100 rad/s.

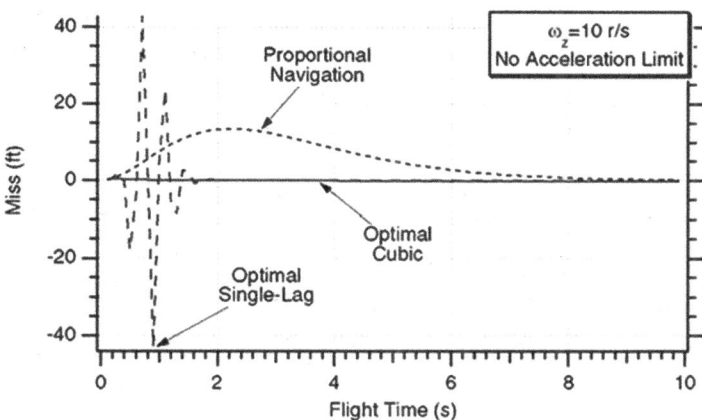

Fig. 12.20   Optimal guidance law for single lag deteriorates when numerator zero is 10 rad/s whereas optimal guidance law for cubic works perfectly.

Finally, a case was run with Listing 12.6 in which the numerator zero of the flight-control system was further reduced to 5 rad/s (WZ=5). We know from Fig. 12.17 that in this case the single-lag representation of the flight-control system is terrible because it totally misses the significant wrong-way tail effect. Therefore, it comes as no surprise that the performance of the optimal single-lag guidance law, as shown in Fig. 12.21, is terrible and in fact is much worse than proportional navigation. The decrease in the right-half-plane zero

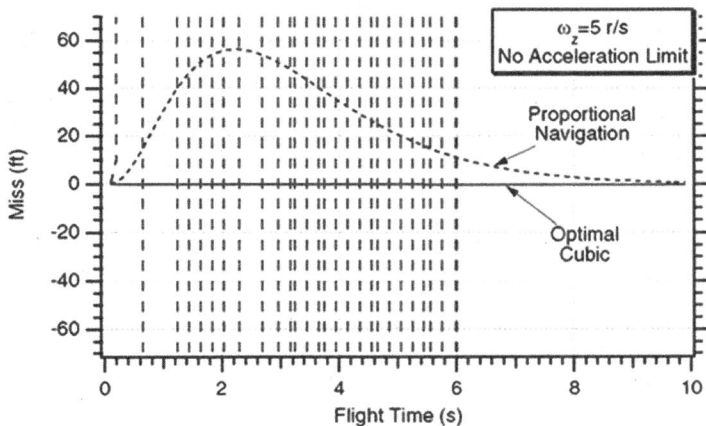

Fig. 12.21   Optimal guidance law for single lag goes unstable when numerator zero is 5 rad/s whereas optimal guidance law for cubic works perfectly.

to 5 rad/s now significantly affects the performance of proportional navigation. Again we can see that the cubic guidance law yields nearly perfect performance.

SUMMARY

In this chapter we have presented two new techniques for deriving missile guidance laws numerically. It was shown that both techniques can accurately numerically derive guidance laws that have previously been derived analytically. In addition, a new guidance law for a complex missile flight-control system was derived using the new techniques. It was shown that the new guidance law can yield significant performance benefits for a tail-controlled missile traveling at low speed and high altitude.

REFERENCES

[1]   Bryson, A. E., and Ho, Y. C., *Applied Optimal Control*, Blaisdell, Waltham, MA, 1969.

[2]   Gelb, A., *Applied Optimal Estimation*, MIT Press, Cambridge, MA, 1974, pp. 356–361.

[3]   Cottrell, R. G., "Optimal Intercept Guidance for Short-Range Tactical Missiles," *AIAA Journal*, Vol. 9, No. 7, 1971, pp. 1414, 1415.

[4]   Nesline, F. W., and Zarchan, P., "A New Look at Classical Versus Modern Homing Guidance," *Journal of Guidance and Control*, Vol. 4, No. 1, 1981, pp. 78–85.

[5]   Asher, R. B., and Matuszewski, J. P., "Optimal Guidance with Maneuvering Targets," *Journal of Spacecraft and Rockets*, Vol. 11, No. 3, 1974, pp. 204–206.

[6]   Rusnak, I., and Meir, L., *Journal of Guidance, Control, and Dynamics*, Vol. 14, No. 5, 1991, pp. 1056–1058.

[7]   Zarchan, P., Greenberg, E., and Alpert, J., "Improving the High Altitude Performance of Tail-Controlled Endoatmospheric Missiles," AIAA Paper 2002–4770, Aug. 2002.

[8]   Wells, B. H., "Tactical Missile Structural Testing and Model Verification for Autopilot Design," *Proceedings of the 1991 AIAA Guidance and Control Conference*, AIAA, Washington, DC, 1991.

CHAPTER 13

Trajectory Shaping Guidance

INTRODUCTION

In all of the guidance work done thus far, the goal has been to hit the target using the least amount of energy. In some applications, in addition to hitting the target, it may also be desirable to shape the missile trajectory near impact. For example, in antitank or antiballistic missile applications we may want to have the missile approach the target at certain strike angles to improve lethality. In this chapter we will show how the guidance problem can be reformulated so that a new guidance law can be developed that both hits the target using minimum energy and travels on the desired trajectory. It will be shown that the new guidance law is actually the same one used to land the Apollo spacecraft on the moon. We will then evaluate the trajectory shaping guidance law and see how it performs in a more realistic nonlinear environment.

PROBLEM SETUP

Before we derive the new guidance law, we must first express mathematically what we desire to do. Let us first revisit our homing loop model for a zero-time constant guidance system, as shown in Fig. 13.1.

As was the case in Chapter 8 from the previous volume for deriving augmented proportional navigation, we are still assuming a constant target maneuver, which means that the derivative of n_T must be zero. Therefore, we can express the model of Fig. 13.1 in matrix form as

$$\begin{bmatrix} \dot{y} \\ \ddot{y} \\ \dot{n}_T \end{bmatrix} = \begin{bmatrix} 0 & 1 & 0 \\ 0 & 0 & 1 \\ 0 & 0 & 0 \end{bmatrix} \begin{bmatrix} y \\ \dot{y} \\ n_T \end{bmatrix} + \begin{bmatrix} \dot{y} \\ -1 \\ 0 \end{bmatrix} n_c$$

This is the same form as the state space matrix differential equation

$$\dot{x} = Fx + Gu$$

Fig. 13.1   Zero-time constant homing loop model for guidance law development.

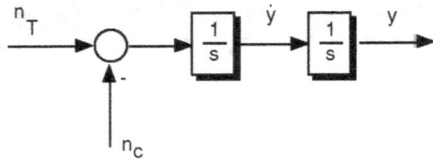

As with our other guidance problems, we still desire to minimize the integral of the commanded acceleration squared. In addition, we want the miss distance to be zero. We will soon see that selecting the relative velocity at the end of flight to be some specified value is the same as shaping the missile trajectory. Mathematically we have just stated that our goals are

$$y(t_F) = 0 \quad \text{and} \quad \dot{y}(t_F) = \dot{y}_F \quad \text{subject to minimizing} \quad \int_0^{t_F} n_c^2(t)\, dt$$

Recall from Chapter 8 of the previous volume that the general solution to the state space differential equation describing the homing loop model is given by the vector relationship

$$x(t_F) = \Phi(t_F - t)\,x(t) + \int_t^{t_F} \Phi(t_F - \lambda)\,G(\lambda)\,u(\lambda)\, d\lambda$$

where Φ is the fundamental matrix. By comparing the state space equation to our matrix equation representing the homing loop model of Fig. 13.1 we can see that F, G, and u are given by

$$F = \begin{bmatrix} 0 & 1 & 0 \\ 0 & 0 & 1 \\ 0 & 0 & 0 \end{bmatrix}$$

$$G = \begin{bmatrix} 0 \\ -1 \\ 0 \end{bmatrix}$$

$$u = n_c$$

We have already shown in Chapter 8 of Volume 1 that for the systems dynamics matrix F under consideration, the continuous fundamental matrix is given by

$$\Phi(t) = \begin{bmatrix} 1 & t & 0.5t^2 \\ 0 & 1 & t \\ 0 & 0 & 1 \end{bmatrix}$$

Substitution of the preceding matrices into the general solution of the state space equation yields

$$
\begin{bmatrix} y(t_F) \\ \dot{y}(t_F) \\ n_T(t_F) \end{bmatrix} = \begin{bmatrix} 1 & t_F - t & 0.5(t_F - t)^2 \\ 0 & 1 & t_F - t \\ 0 & 0 & 1 \end{bmatrix} \begin{bmatrix} y(t) \\ \dot{y}(t) \\ n_T(t) \end{bmatrix}
$$
$$
+ \int_t^{t_F} \begin{bmatrix} 1 & t_F - \lambda & 0.5(t_F - \lambda)^2 \\ 0 & 1 & t_F - \lambda \\ 0 & 0 & 1 \end{bmatrix} \begin{bmatrix} 0 \\ -1 \\ 0 \end{bmatrix} n_c(\lambda) d\lambda
$$

We can multiply out the preceding matrix equation and end up with three scalar equations. Because we are only interested in controlling the relative position and velocity at the end of flight, we can ignore the last scalar equation of the three equations and obtain

$$
y(t_F) = y(t) + (t_F - t)\dot{y}(t) + 0.5(t_F - t)^2 n_T(t) - \int_t^{t_F} (t_F - \lambda)n_c(\lambda)\, d\lambda
$$

$$
\dot{y}(t_F) = \dot{y}(t) + (t_F - t)n_T(t) - \int_t^{t_F} n_c(\lambda)\, d\lambda
$$

We will see in the next section how the Schwartz inequality can be used on the preceding two scalar equations to obtain the trajectory shaping guidance law.

USING THE SCHWARTZ INEQUALITY FOR TRAJECTORY SHAPING GUIDANCE

We will first simplify the guidance problem by using shorthand notation to represent the two scalar equations that were just derived in the preceding section. Let us first define

$$
f_1 = y(t) + (t_F - t)\dot{y}(t) + 0.5(t_F - t)^2 n_T(t)
$$
$$
f_2^* = \dot{y}(t) + (t_F - t)n_T(t)
$$
$$
h_1(\lambda) = t_F - \lambda
$$
$$
h_2(\lambda) = 1
$$

We can now rewrite the two scalar equations at the end of the preceding section in shorthand notation as

$$
y(t_F) = f_1 - \int_t^{t_F} h_1(\lambda)n_c(\lambda)\, d\lambda
$$

$$
\dot{y}(t_F) = f_2^* - \int_t^{t_F} h_2(\lambda)n_c(\lambda)\, d\lambda
$$

Recall that we want the miss distance to be zero $[y(t_F) = 0]$ and we also want to specify the value of the relative velocity at the terminal time. Therefore, the two preceding equations simplify to

$$f_1 = \int_t^{t_F} h_1(\lambda) n_c(\lambda) \, d\lambda$$

$$f_2^* - \dot{y}(t_F) = \int_t^{t_F} h_2(\lambda) n_c(\lambda) \, d\lambda = f_2$$

Let us now combine the two preceding scalar equations into one equation by introducing a new variable δ, or

$$f_1 - \delta f_2 = \int_t^{t_F} [h_1(\lambda) - \delta h_2(\lambda)] n_c(\lambda) \, d\lambda$$

If we apply the Schwartz inequality to the preceding expression we obtain

$$(f_1 - \delta f_2)^2 \le \int_t^{t_F} [h_1(\lambda) - \delta h_2(\lambda)]^2 d\lambda \int_t^{t_F} n_c^2(\lambda) \, d\lambda$$

Inverting the preceding equation and solving for the integral of the square of the commanded acceleration yields

$$\int_t^{t_F} n_c^2(\lambda) \, d\lambda \ge \frac{(f_1 - \delta f_2)^2}{\int_t^{t_F} [h_1(\lambda) - \delta h_2(\lambda)]^2 d\lambda}$$

The integral of the square of the acceleration will be minimized when the equality sign of the preceding inequality holds. According to the Schwartz inequality, the equality sign holds when

$$n_c(\lambda) = K[h_1(\lambda) - \delta h_2(\lambda)]$$

where K is a constant. We will soon use the preceding equation. When the equality sign holds, the integral of the square of the commanded acceleration can be expanded to

$$z = \int_t^{t_F} n_c^2(\lambda) \, d\lambda = \frac{(f_1 - \delta f_2)^2}{\int_t^{t_F} [h_1(\lambda) - \delta h_2(\lambda)]^2 \, d\lambda}$$

$$= \frac{(f_1 - \delta f_2)^2}{\int_t^{t_F} h_1^2(\lambda) \, d\lambda - 2\delta \int_t^{t_F} h_1(\lambda) h_2(\lambda) \, d\lambda + \delta^2 \int_t^{t_F} h_2^2(\lambda) \, d\lambda}$$

In the work that follows we will be performing many algebraic manipulations. To simplify our task, we will again use shorthand notation. We can define

$$\left\|h_1^2\right\| = \int_t^{t_F} h_1^2(\lambda)\,\mathrm{d}\lambda$$

$$\left\|h_2^2\right\| = \int_t^{t_F} h_2^2(\lambda)\,\mathrm{d}\lambda$$

$$\left\|h_1 h_2\right\| = \int_t^{t_F} h_1(\lambda)h_2(\lambda)\,\mathrm{d}\lambda$$

Therefore, our expression for the integral of the square of the commanded acceleration simplifies to

$$z = \frac{(f_1 - \delta f_2)^2}{\left\|h_1^2\right\| - 2\delta\|h_1 h_2\| + \delta^2\left\|h_2^2\right\|}$$

Now we have to choose a value for the variable δ. The best value of δ is one that minimizes the preceding expression. We know from calculus that the value of δ that minimizes the preceding expression can be found by taking the derivative of the preceding expression with respect to δ and setting the result to zero. Using the quotient rule from calculus to take the derivative yields

$$\frac{\mathrm{d}z}{\mathrm{d}\delta} = 0$$

$$= \frac{\left[\left\|h_1^2\right\| - 2\delta\|h_1 h_2\| + \delta^2\left\|h_2^2\right\|\right] 2(f_1 - \delta f_2)(-f_2) - (f_1 - \delta f_2)^2\left[-2\|h_1 h_2\| + 2\delta\left\|h_2^2\right\|\right]}{\left[\left\|h_1^2\right\| - 2\delta\|h_1 h_2\| + \delta^2\left\|h_2^2\right\|\right]^2}$$

We can now solve the preceding expression for δ. After some algebra, we obtain

$$\delta = \frac{f_1\|h_1 h_2\| - f_2\left\|h_1^2\right\|}{f_1\left\|h_2^2\right\| - f_2\|h_1 h_2\|}$$

Because we know that

$$f_1 = \int_t^{t_F} h_1(\lambda)n_c(\lambda)\,\mathrm{d}\lambda$$

and

$$n_c(\lambda) = K[h_1(\lambda) - \delta h_2(\lambda)]$$

Substitution yields

$$f_1 = \int_t^{t_F} h_1(\lambda) K[h_1(\lambda) - \delta h_2(\lambda)] d\lambda$$

If we solve for the constant K, we obtain

$$K = \frac{f_1}{\int_t^{t_F} h_1(\lambda)[h_1(\lambda) - \delta h_2(\lambda)] d\lambda} = \frac{f_1}{\|h_1^2\| - \delta \|h_1 h_2\|}$$

Therefore, the optimal acceleration command can be rewritten as

$$n_c(\lambda) = K[h_1(\lambda) - \delta h_2(\lambda)] = \frac{f_1[h_1(\lambda) - \delta h_2(\lambda)]}{\|h_1^2\| - \delta \|h_1 h_2\|}$$

Substitution of the optimal value of δ in the preceding expression yields

$$n_c(\lambda) = \frac{\left[f_1[h_1(\lambda) - h_2(\lambda)] = \left[\dfrac{f_1\|h_1 h_2\| - f_2\|h_1^2\|}{f_1\|h_2^2\| - f_2\|h_1 h_2\|} \right] \right]}{\|h_1^2\| - \|h_1 h_2\| \left[\dfrac{f_1\|h_1 h_2\| - f_2\|h_1^2\|}{f_1\|h_2^2\| - f_2\|h_1 h_2\|} \right]}$$

After some algebra, we obtain

$$n_c(\lambda) = \frac{f_1 h_1(\lambda)\|h_2^2\| - \|h_1 h_2\|[f_2 h_1(\lambda) + f_1 h_2(\lambda)] + f_2 h_2(\lambda)\|h_1^2\|}{\|h_1^2\|\|h_2^2\| - \|h_1 h_2\|^2}$$

or in the time domain

$$n_c(t) = \frac{f_1 h_1(t)\|h_2^2\| - \|h_1 h_2\|[f_2 h_1(t) + f_1 h_2(t)] + f_2 h_2(t)\|h_1^2\|}{\|h_1^2\|\|h_2^2\| - \|h_1 h_2\|^2}$$

Now we have enough information to evaluate the numerator and denominator of the preceding expression to find the new guidance law. Recall that for this problem

$$f_1 = y(t) + (t_F - t)\dot{y}(t) + 0.5(t_F - t)^2 n_T(t) = y + t_{go}\dot{y} + 0.5 t_{go}^2 n_T$$

$$f_2 = f_2^* - \dot{y}(t_F) = \dot{y}(t) + (t_F - t)n_T(t) - \dot{y}(t_F) = \dot{y} + t_{go}n_T - \dot{y}(t_F)$$

$$h_1(t) = t_F - t = t_{go}$$

$$h_2(t) = 1$$

Therefore, we can easily solve the necessary integrals as

$$\left\|h_1^2\right\| = \int_t^{t_F} h_1^2(\lambda)\, d\lambda = \int_t^{t_F} (t_F - \lambda)^2\, d\lambda = \frac{t_{go}^3}{3}$$

$$\left\|h_2^2\right\| = \int_t^{t_F} h_2^2(\lambda)\, d\lambda = \int_t^{t_F} d\lambda = t_{go}$$

$$\left\|h_1 h_2\right\| = \int_t^{t_F} h_1(\lambda) h_2(\lambda)\, d\lambda = \int_t^{t_F} (t_F - \lambda)\, d\lambda = \frac{t_{go}^2}{2}$$

where the time to go until intercept is given by

$$t_{go} = t_F - t$$

Substituting the preceding expressions into the formula for the acceleration command yields

$$n_c(t) = \left\{ (y + t_{go}\dot{y} + 0.5 t_{go}^2 n_T) t_{go} t_{go} - 0.5 t_{go}^2 \left[(\dot{y} + t_{go} n_T - \dot{y}_F) t_{go} \right. \right.$$

$$\left. \left. + (y + t_{go}\dot{y} + 0.5 t_{go}^2 n_T)(1) \right] + (\dot{y} + t_{go} n_T - \dot{y}_F)(1) \frac{t_{go}^3}{3} \right\} \bigg/ \left[\frac{t_{go}^3}{3} t_{go} - \left(\frac{t_{go}^2}{2} \right)^2 \right]$$

After some algebra, we see that the new trajectory shaping guidance law simplifies to

$$n_c(t) = \frac{6y + 4\dot{y}t_{go} + n_T t_{go}^2 + 2\dot{y}(t_F)t_{go}}{t_{go}^2}$$

The guidance law that landed the Apollo spacecraft on the moon in 1969 used the preceding guidance law. In the Apollo case there was no target acceleration, and the relative velocity at intercept was chosen to be zero (that is, this special case is also known as a rendezvous). Therefore, the Apollo guidance law is simply [1–4]

$$n_c(t)_{Apollo} = \frac{6y + 4\dot{y}t_{go}}{t_{go}^2}$$

ALTERNATE FORM OF TRAJECTORY SHAPING GUIDANCE LAW

We can rewrite the trajectory shaping guidance law of the preceding section as

$$n_c(t) = \frac{4y + 4\dot{y}t_{go} + n_T t_{go}^2 + 2y + 2\dot{y}(t_F)t_{go}}{t_{go}^2} = \frac{4(y + \dot{y}t_{go})}{t_{go}^2}$$

$$+ \frac{2[y + \dot{y}(t_F)t_{go}] + n_T t_{go}^2}{t_{go}^2}$$

Recall that the formula for the line-of-sight angle is given by

$$\lambda = \frac{y}{R_{TM}} = \frac{y}{V_c t_{go}} = \frac{y}{V_c(t_F - t)}$$

Therefore, the line-of-sight rate can be found by differentiating the preceding expression using the quotient rule from calculus. After some algebra, we obtain

$$\dot{\lambda} = \frac{y + \dot{y}t_{go}}{V_c t_{go}^2}$$

Therefore, the trajectory shaping guidance law simplifies to

$$n_c(t) = 4V_c\dot{\lambda} + \frac{2[\lambda V_c + \dot{y}(t_F)]}{t_{go}} + n_T$$

To express the guidance law in terms of a final angle rather than a final relative velocity, we can invert the expression for the line-of-sight rate and solve for the relative velocity, or

$$\dot{y} = \frac{\dot{\lambda}V_c t_{go}^2 - y}{t_{go}} = \frac{\dot{\lambda}V_c t_{go}^2 - \dot{\lambda}V_c t_{go}}{t_{go}} = \dot{\lambda}V_c t_{go}^2 - \lambda V_c$$

We can evaluate the preceding expression at intercept. At the end of the flight, time to go is zero and the final line-of-sight angle is λ_F. Therefore, at the end of the flight we can see that the relative velocity is simply

$$\dot{y}(t_F) = -\lambda(t_F)V_c = -\lambda_F V_c$$

and the trajectory shaping guidance law simplifies to

$$n_c(t) = 4V_c\dot{\lambda} + \frac{2V_c[\lambda - \lambda_F]}{t_{go}} + n_T$$

Thus, we can see that we can think of the trajectory shaping guidance law as one that minimizes the integral of the square of the commanded acceleration, makes the miss zero, and drives the final line-of-sight angle to the designer-chosen value λ_F. The trajectory shaping guidance law appears to be a form of augmented proportional navigation (with an effective navigation ratio of 4 and a different multiplier for the target acceleration term) plus an extra term that is proportional to the difference between the true line-of-sight angle and the desired line-of-sight angle at the end of the flight.

TESTING TRAJECTORY SHAPING GUIDANCE IN THE LINEAR WORLD

Now that the trajectory shaping guidance law has been derived, it is important to first test the new guidance law in the linear world to see if it works as anticipated. In addition, we would like to compare trajectory shaping guidance with proportional navigation in terms of both accuracy and acceleration requirements. Figure 13.2 presents a block diagram of a zero-lag homing loop to be used in evaluating both proportional navigation and trajectory shaping guidance. We can see from the homing loop that the two sources of error considered are target maneuver n_T and heading error HE. We can see from Fig. 13.2 that the two measures of performance will be the miss distance $y(t_F)$ and the final line of sight angle λ_F.

The homing loop model of Fig. 13.2 was programmed, and the resultant engagement simulation appears in Listing 13.1. We can see from the listing that the parameter PN determines the type of guidance law to be used. If PN=1 then proportional navigation is used, whereas if PN=0 the trajectory shaping guidance law is used. When trajectory shaping guidance is used, the final specified line of sight angle is denoted XLAMFDEG and is in units of degrees. The trajectory shaping guidance law assumes that time-to-go information and line-of-sight angle and rate information are available. It is important to note that proportional navigation does not require time-to-go information.

The nominal case of Listing 13.1 was run in which there was a 10-s flight and −20 deg of heading error. When the trajectory shaping guidance law is used it is

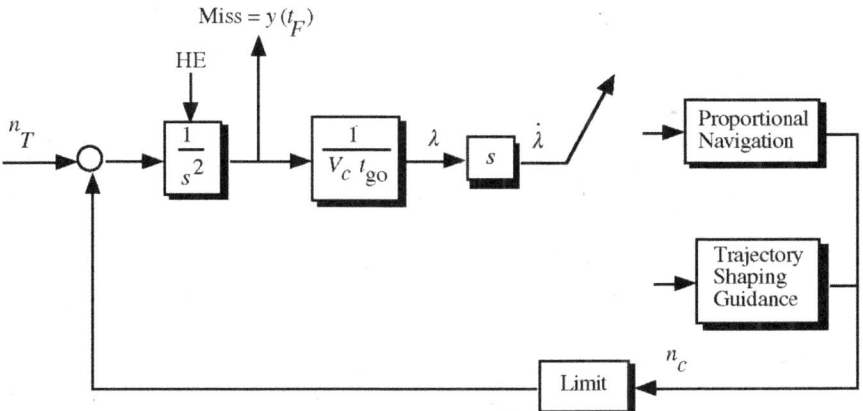

Fig. 13.2   Homing loop model for guidance law comparison.

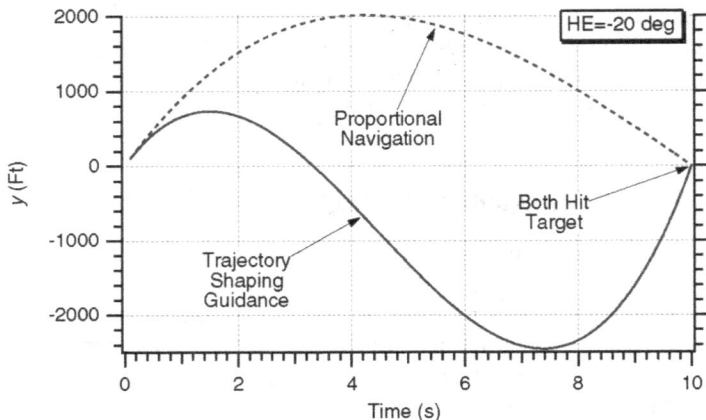

Fig. 13.3   Both guidance laws enable the missile to take out the heading error disturbance and hit the target using different relative trajectories.

specified that the final line-of-sight angle should be −30 deg. We can see from Fig. 13.3 that both guidance laws enable the missile to take out the heading error and hit the target because $y(10) = 0$ in both cases. However, we can see that the relative trajectories are totally different for both guidance laws.

The acceleration requirements for both guidance laws are displayed in Fig. 13.4. We can see that significantly more commanded acceleration is required for trajectory shaping guidance to take out 20 deg of heading error (25 g at the beginning of the flight and approximately −30 g near the end of the flight)

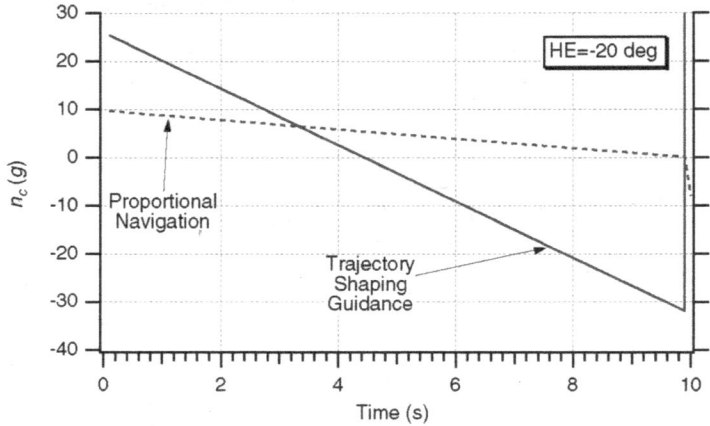

Fig. 13.4   Trajectory shaping guidance requires more acceleration than proportional navigation to take out heading error disturbance.

than is required with proportional navigation (10 g at the beginning of the flight and 0 g near the end of the flight).

LISTING 13.1   LINEAR ENGAGEMENT SIMULATION FOR GUIDANCE LAW COMPARISON

```
clear
n=0;
XNT=0.;
HEDEG=-20.;
XNCLIM=999999.;
PN=0;
XLAMFDEG=-30.;
VC=4000.;
VM=3000.;
TF=10.;
XNP=3.;
XLAMF=XLAMFDEG/57.3;
Y=0.;
YD=-VM*HEDEG/57.3;
T=0.;
H=.001;
S=0.;
while T<=(TF-.0001)
      YOLD=Y;
      YDOLD=YD;
      STEP=1;
      FLAG=0;
      while STEP <=1
        if FLAG==1
      STEP=2;
            Y=Y+H*YD;
            YD=YD+H*YDD;
            T=T+H;
        end
        TGO=TF-T+.00001;
        XLAM=Y/(VC*TGO);
        XLAMD=(Y+YD*TGO)/(VC*TGO*TGO);
        if PN==1
             XNC=XNP*VC*XLAMD;
        else
             XNC=4.*VC*XLAMD+XNT+2.*VC*(XLAM-XLAMF)/TGO;
        end
        if XNC>XNCLIM
             XNC=XNCLIM;
        end
```

```
            if XNC<-XNCLIM
                XNC=-XNCLIM;
            end
            YDD=XNT-XNC;
            FLAG=1;
        end
        FLAG=0;
        Y=.5*(YOLD+Y+H*YD);
        YD=.5*(YDOLD+YD+H*YDD);
        S=S+H;
        if S>=.09999
            S=0.;
            n=n+1;
            XLAMDEG=XLAM*57.3;
            XNCG=XNC/32.2;
            ArrayT(n)=T;
            ArrayY(n)=Y;
            ArrayXNCG(n)=XNCG;
            ArrayXLAMDEG(n)=XLAMDEG;
        end
end
figure
plot(ArrayT,ArrayY),grid
title('Relative Trajectory')
xlabel('Time (Sec) ')
ylabel('Y (Ft)')
figure
plot(ArrayT,ArrayXNCG),grid
title('Commanded Acceleration')
xlabel('Time (Sec) ')
ylabel('XNC (G)')
axis([0 10 -40 30])
figure
plot(ArrayT,ArrayXLAMDEG),grid
title('Line-of-Sight Angle')
xlabel('Time (Sec) ')
ylabel('XLAM (Deg)')
axis([0 10 -30 10])
clc
output=[ArrayT',ArrayY',ArrayXNCG',ArrayXLAMDEG'];
save datfil.txt output -ascii
disp '*** Simulation Complete'
```

Finally, we can see from Fig. 13.5 that with trajectory shaping guidance, the line-of-sight angle matches the design goal of −30 deg at the end of the flight. With proportional navigation, the final line-of-sight angle is not controlled and it is really

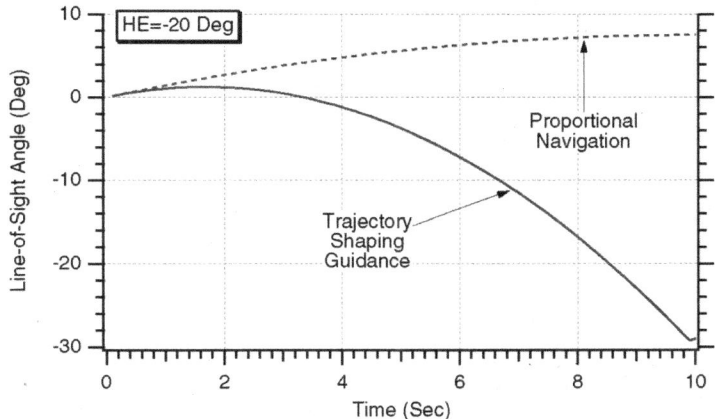

Fig. 13.5   Trajectory shaping guidance law can control final line-of-sight angle.

a matter of luck on what that angle will be (that is, approximately 8 deg in this example). Thus, we can say that simulation results indicate that trajectory shaping guidance appears to be working correctly against the heading error disturbance.

Next, both guidance laws were compared in terms of their response to a 6-g target maneuver. Again, we can see from Fig. 13.6 that both guidance laws enable the missile to hit the maneuvering target because $y(10) = 0$ in both cases. As was the case before, both guidance laws result in relative trajectories that are significantly different.

Again, we can see from Fig. 13.7 that trajectory shaping guidance requires more acceleration than proportional navigation to hit the maneuvering target.

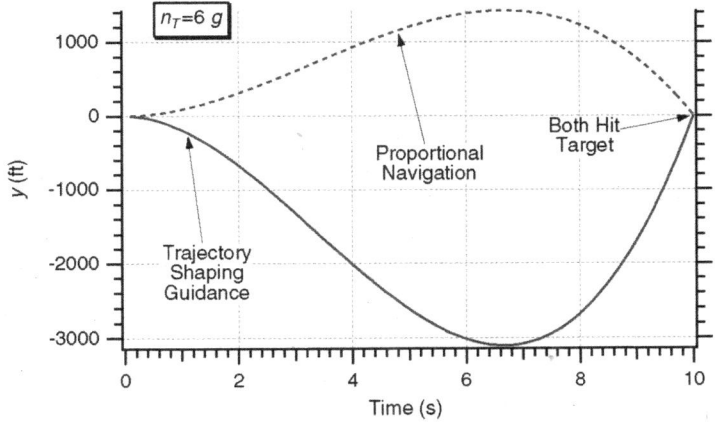

Fig. 13.6   Trajectory shaping guidance law can also hit maneuvering target.

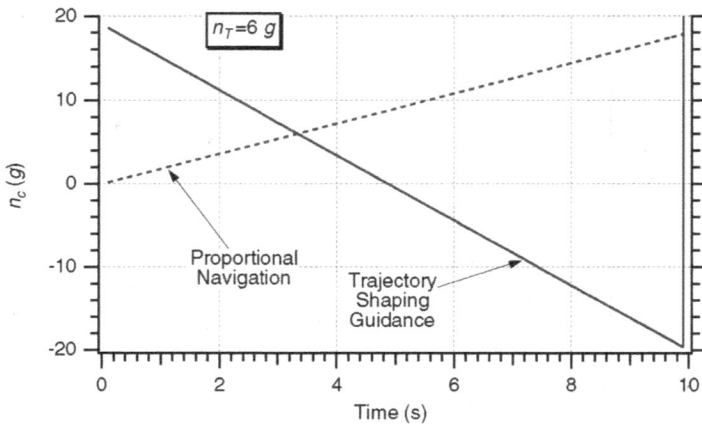

Fig. 13.7   Trajectory shaping guidance law requires more acceleration than proportional navigation against maneuvering target.

For the case of the 6-g maneuvering target and desired final line-of-sight angle of -30 deg, trajectory shaping guidance required 20 g at the beginning of flight and -20 g at the end of flight. On the other hand, proportional navigation, which did not reach the final line-of-sight angle goal, required 0 g at the beginning of flight and nearly 20 g at the end of flight.

From Fig. 13.8 we can see that with trajectory shaping guidance we achieved the goal of the line-of-sight angle becoming -30 deg at the end of the flight. We can also see that, for this example, proportional navigation ended up with a final

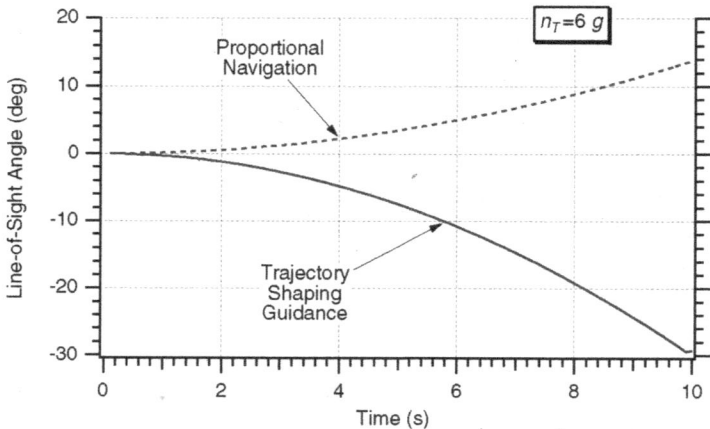

Fig. 13.8   Trajectory shaping guidance law can still control final line-of-sight angle—even in presence of maneuvering target.

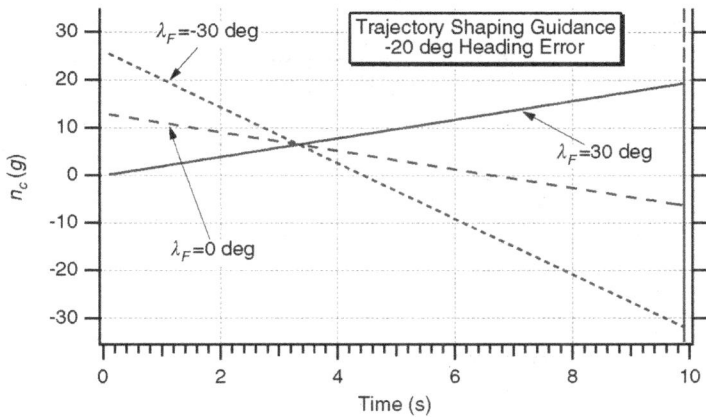

Fig. 13.9   Acceleration requirements depend on final line-of-sight angle specification when disturbance is heading error.

line-of-sight angle of 14 deg. Thus, we can conclude that simulation results indicate that trajectory shaping guidance appears to be working correctly against the target maneuver disturbance.

Other cases were run with the trajectory shaping guidance law for the case in which there was −20 deg of heading error and the final line-of-sight angle is made a parameter. Figure 13.9 shows that the acceleration requirements for the trajectory shaping guidance law are dependent on the final line-of-sight angle. Figure 13.10 shows that the various design goals for the final line-of-sight angle

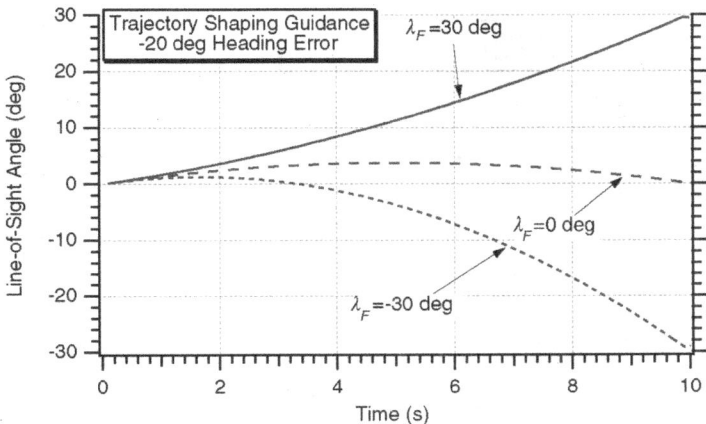

Fig. 13.10   Final line-of-sight angle goals are met in presence of heading error with trajectory shaping guidance.

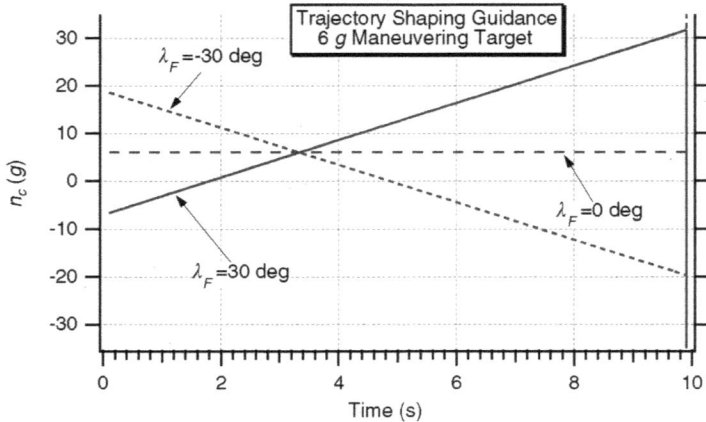

Fig. 13.11   Acceleration requirements still depend on final line-of-sight angle specificaton when disturbance is target maneuver.

are all met with trajectory shaping guidance in the presence of the heading error disturbance, provided adequate acceleration is available.

Finally, even more cases were run with the trajectory shaping guidance law for the situation in which there was a 6-g maneuvering target and the final line-of-sight angle was made a parameter. Figure 13.11 shows that the acceleration requirements for the trajectory shaping guidance law are again dependent on the final line-of-sight angle. Figure 13.12 shows that the various design goals

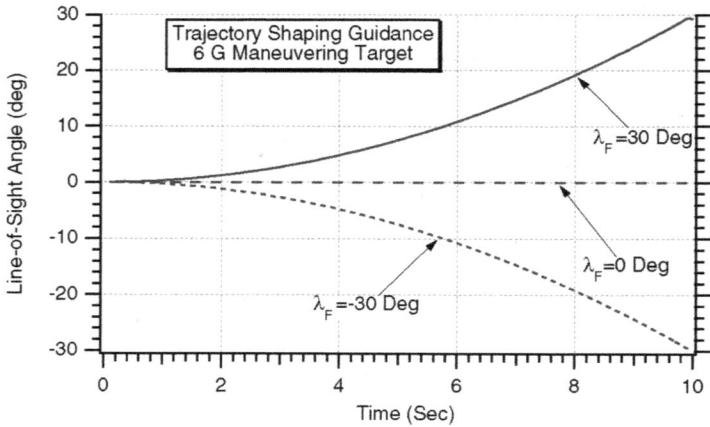

Fig. 13.12   Final line-of-sight angle goals are met in presence of target maneuver with trajectory shaping guidance.

for the final line-of-sight angle are all met with trajectory shaping guidance in the presence of the target maneuver disturbance.

CLOSED-FORM SOLUTIONS

With the proportional and augmented proportional navigation guidance laws, we were able to derive closed-form solutions for the missile acceleration due to a step in target maneuver, heading error, and a step in target displacement for a zero-time constant missile guidance system. The solutions were obtained by solving a linear, first-order, time-varying differential equation as was demonstrated in Chapters 2 and 13 of the previous volume. Let us see if we can use the same techniques to derive acceleration formulas for the trajectory shaping guidance law.

Consider the case in which the only disturbance to the guidance system is target maneuver. From Fig. 13.2 we can see that the relative acceleration is simply target acceleration minus missile acceleration, or

$$\ddot{y} = n_T - n_c$$

Substituting the original expression for the trajectory shaping guidance law into the preceding equation yields

$$\ddot{y} = n_T - n_c(t) = n_T - \left[\frac{6y + 4\dot{y}t_{go} + n_T t_{go}^2 + 2\dot{y}(t_F)t_{go}}{t_{go}^2} \right]$$

$$= \frac{-6y - 4\dot{y}t_{go} - 2\dot{y}(t_F)t_{go}}{t_{go}^2}$$

or more simply

$$\ddot{y} + \frac{4\dot{y}}{t_F - t} + \frac{6y}{(t_F - t)^2} = \frac{-2\dot{y}(t_F)}{t_F - t}$$

with initial conditions

$$y(0) = 0 \quad \text{and} \quad \dot{y}(0) = 0$$

Recall that in Chapters 2 and 8 of the previous volume we obtained closed-form solutions by solving a first-order linear differential equation with time-varying coefficients. However, we now have a second-order linear differential equation with time-varying coefficients. The solution to such an equation is extremely difficult at best. Let us see if we can take another, less conventional, approach to the problem.

Recall that the general solution to the state space equation at the final time

$$\dot{x} = Fx + Gu$$

was given by

$$\mathbf{x}(t_F) = \Phi(t_F - t)\mathbf{x}(t) + \int_t^{t_F} \Phi(t_F - \lambda)\mathbf{G}(\lambda)\mathbf{u}(\lambda)\,d\lambda$$

We can also say that the general solution to the state space equation can be expressed in terms of the initial conditions at time zero as

$$\mathbf{x}(t_F) = \Phi(t_F)\mathbf{x}(0) + \int_0^{t_F} \Phi(t_F - \lambda)\mathbf{G}(\lambda)\mathbf{u}(\lambda)\,d\lambda$$

Therefore, for the homing-loop problem under consideration we have

$$\begin{bmatrix} y(t_F) \\ \dot{y}(t_F) \\ n_T(t_F) \end{bmatrix} = \begin{bmatrix} 1 & t_F & 0.5t_F^2 \\ 0 & 1 & t_F \\ 0 & 0 & 1 \end{bmatrix} \begin{bmatrix} y(0) \\ \dot{y}(0) \\ n_T(0) \end{bmatrix}$$

$$+ \int_0^{t_F} \begin{bmatrix} 1 & t_F - \lambda & 0.5(t_F - \lambda)^2 \\ 0 & 1 & t_F - \lambda \\ 0 & 0 & 1 \end{bmatrix} \begin{bmatrix} 0 \\ -1 \\ 0 \end{bmatrix} n_c(\lambda)\,d\lambda$$

Multiplying out the preceding matrix equation and leaving out the third scalar equation yields

$$y(t_F) = y(0) + t_F\dot{y}(0) + 0.5t_F^2 n_T(0) - \int_0^{t_F} (t_F - \lambda)n_c(\lambda)\,d\lambda$$

$$\dot{y}(t_F) = \dot{y}(0) + t_F n_T(0) - \int_0^{t_F} n_c(\lambda)\,d\lambda$$

We still want to minimize the integral of the square of the commanded acceleration subject to the miss being zero and the relative velocity at the end of flight being specified, or

$$y(t_F) = 0 \quad \text{and} \quad \dot{y}(t_F) = \dot{y}_F \quad \text{subject to minimizing} \int_0^{t_F} n_c^2(t)\,dt$$

We get the same solution for the acceleration command as before, or

$$n_c(t) = \frac{f_1 h_1(t)\|h_2^2\| - \|h_1 h_2\|[f_2 h_1(t) + f_1 h_2(t)] + f_2 h_2(t)\|h_1^2\|}{\|h_1^2\|\|h_2^2\| - \|h_1 h_2\|^2}$$

except that this time the shorthand notation definitions have changed to

$$f_1 = y(0) + t_F\dot{y}(0) + 0.5t_F^2 n_T(0)$$
$$f_2 = \dot{y}(0) + t_F n_T(0) - \dot{y}(t_F)$$
$$h_1(t) = t_F - t = t_{go}$$
$$h_2(t) = 1$$

$$\left\|h_1^2\right\| = \int_0^{t_F} h_1^2(t)\,dt = \int_0^{t_F} (t_F - t)^2 \,d\lambda = \frac{t_F^3}{3}$$

$$\left\|h_2^2\right\| = \int_0^{t_F} h_2^2(t)\,dt = \int_0^{t_F} dt = t_F$$

$$\left\|h_1 h_2\right\| = \int_0^{t_F} h_1(t)h_2(t)\,dt = \int_0^{t_F} (t_F - t)\,dt = \frac{t_F^2}{2}$$

Substitution of the preceding definitions into the formula for the acceleration command yields

$$n_c(t) = \left\{ \left[y(0) + t_F\dot{y}(0) + 0.5t_F^2 n_T(0)\right]t_{go}t_F - 0.5t_F^2\left[\left[\dot{y}(0) + t_F n_T(0) - \dot{y}(t_F)\right]t_{go}.\right.\right.$$

$$+ \frac{\left[y(0) + t_F\dot{y}(0) + 0.5t_F^2 n_T(0)\right](1)\right] + \left[\dot{y}(0) + t_F n_T(0) - \dot{y}(t_F)\right](1)\dfrac{t_F^3}{3}\right\}}{\left[\dfrac{t_F^3}{3}t_F - \left[\dfrac{t_F^2}{2}\right]^2\right]}$$

After much algebra, we obtain

$$n_c(t) = \frac{12y(0)\left(t_{go} - \dfrac{t_F}{2}\right) + t_F\dot{y}(0)(6t_{go} - 2t_F) + t_F^3 n_T(0) + t_F\dot{y}(t_F)(6t_{go} - 4t_F)}{t_F^3}$$

The preceding expression is the closed-form solution for the total missile acceleration due to the various initial conditions or error sources when using the trajectory shaping guidance law. Therefore, the missile acceleration due to an initial condition in relative velocity can be written by inspection from the preceding formula as

$$n_c(t)\big|_{\dot{y}_0} = \frac{t_F\dot{y}(0)(6t_{go} - 2t_F)}{t_F^3} = \frac{2\dot{y}(0)}{t_F}\left[2 - \frac{3t}{t_F}\right]$$

Because the initial relative velocity and heading error are related by

$$\dot{y}(0) = -V_M\text{HE}$$

we can say that the acceleration due to heading error is given by

$$n_c(t)|_{\text{HE}} = \frac{-2V_M\text{HE}}{t_F}\left[2 - \frac{3t}{t_F}\right]$$

Therefore, the acceleration required to take out the heading error is proportional to the amount of heading error and inversely proportional to the amount of homing time. More heading error and less homing time both work in the direction of increasing the missile acceleration requirements.

From the general closed-form acceleration formula, we can see that the acceleration due to a target maneuver is given by

$$n_c(t)|_{n_T} = \frac{t_F^3 n_T(0)}{t_F^3} = n_T(0)$$

We can see from the preceding expression that in this case the missile is simply matching the target acceleration. Therefore, as expected, larger target maneuvers will require more acceleration capability from the missile.

Finally, we can see that the acceleration due to shaping the trajectory to match a desired final relative velocity can also be written by inspection of the total acceleration formula as

$$n_c(t)\Big|_{\dot{y}_F} = \frac{t_F\dot{y}(t_F)(6t_{\text{go}} - 4t_F)}{t_F^3} = \frac{2\dot{y}(t_F)}{t_F}\left[1 - \frac{3t}{t_F}\right]$$

Recall that the final relative velocity can also be expressed in terms of the final line-of-sight angle as

$$\dot{y}(t_F) = -V_c\lambda(t_F)$$

Therefore, the acceleration due to shaping the final line-of-sight angle can be rewritten as

$$n_c(t)|_{\lambda_F} = \frac{-2V_c\lambda(t_F)}{t_F}\left[1 - \frac{3t}{t_F}\right]$$

The acceleration requirements are proportional to the amount of shaping we want to do and inversely proportional to the amount of homing time. Larger desired final line-of-sight angles (that is, more shaping) will require more missile acceleration.

To check the formulas derived in this section, cases were run with the linear engagement simulation of Listing 13.1. First, a case was run with trajectory shaping guidance in which the desired final line-of-sight angle was zero and there was a 6-g target maneuver for a 10-s flight. Recall that the formula for the commanded missile acceleration due to a maneuvering target is given by

$$n_c(t)|_{n_T} = n_T(0)$$

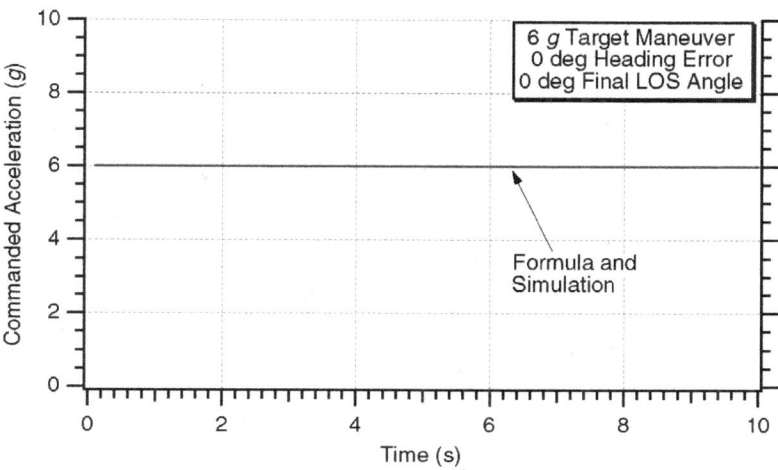

Fig. 13.13   Formula for acceleration due to target maneuver is accurate.

We can see from Fig. 13.13 that the simulation results of Listing 13.1 (namely, XNT=193.2, HEDEG=0, PN=0, XLAMFDEG=0) and the preceding formula are in exact agreement, thus demonstrating that the acceleration due to target maneuver formula is correct.

Listing 13.1 was again run with trajectory shaping guidance in which the desired final line of sight was zero and there was a −20-deg heading error for a 10-s flight. Recall that the formula for the commanded missile acceleration due to heading error is given by

$$n_c(t)|_{HE} = \frac{-2V_M HE}{t_F}\left[2 - \frac{3t}{t_F}\right]$$

We can see from Fig. 13.14 that the simulation results of Listing 13.1 (namely, XNT=0, HEDEG=−20, PN=0, XLAMFDEG=0) and the preceding formula are in exact agreement, thus demonstrating that the acceleration due to heading error formula is also correct.

Finally, Listing 13.1 was run again with trajectory shaping guidance in which the desired final line-of-sight angle is set to −30 deg. In this case there is no heading error or target maneuver; however, the flight time is still 10 s. Recall that the formula for the commanded missile acceleration due to specifying the final line-of-sight angle is given by

$$n_c(t)|_{\lambda_F} = \frac{-2V_c\lambda(t_F)}{t_F}\left[1 - \frac{3t}{t_F}\right]$$

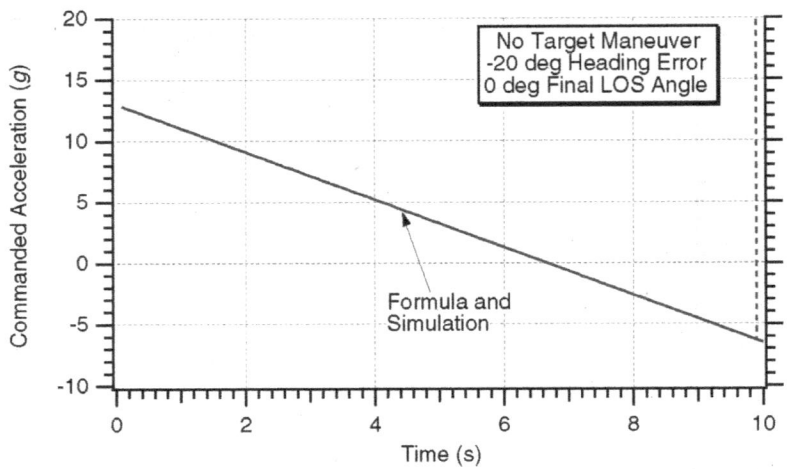

Fig. 13.14   Formula for acceleration heading error is accurate.

We can see from Fig. 13.15 that again the simulation results of Listing 13.1 (namely, XNT=0, HEDEG=0, PN=0, XLAMFDEG=−30) and the preceding formula are in exact agreement, thus demonstrating that the acceleration due to specifying the final line-of-sight angle formula is also correct.

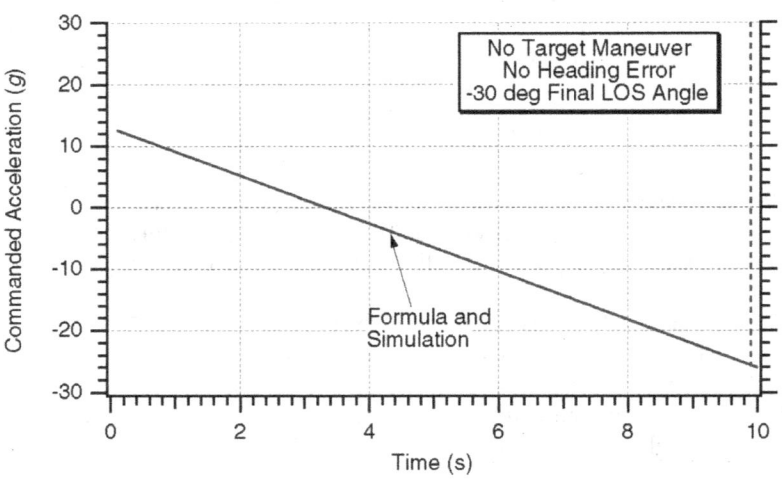

Fig. 13.15   Formula for acceleration of controlling final line-of-sight angle is accurate.

NONLINEAR RESULTS

The trajectory shaping guidance law was really derived for operation in a world in which the geometry was linear. This implies small angle approximations. It is now of interest to see how the new guidance law works in the two-dimensional world in which the equations of motion are nonlinear. Listing 13.2 is a slight modification of the original zero-time constant nonlinear missile-target engagement simulation of Listing 2.1 of the previous volume. An option has been included so that the target can be either stationary (target flight path rate has been modified so there is no division by zero) or moving. It is important to note that the trajectory shaping guidance law requires knowledge of the target acceleration. In the nonlinear engagement simulation, the target acceleration perpendicular to the line of sight is used in the guidance law. The components of the target acceleration in the downrange and altitude direction can be expressed in terms of the target flight-path angle as

$$n_{T1} = n_T \sin \beta$$
$$n_{T2} = n_T \cos \beta$$

Therefore, the target acceleration that appears perpendicular to the line of sight can be obtained from trigonometry and can be expressed as

$$n_{T_{\text{PLOS}}} = -n_{T1} \sin \lambda + n_{T2} \cos \lambda$$

Now the trajectory shaping guidance law for the nonlinear world can be written as

$$n_c(t) = 4V_c\dot{\lambda} + \frac{2V_c[\lambda - \lambda_F]}{t_{\text{go}}} + n_{T_{\text{PLOS}}}$$

where the direction of the commanded acceleration is perpendicular to the line of sight. The new nonlinear engagement simulation appears in Listing 13.2. We can see that the simulation can also be run using proportional navigation by simply setting APN=0.

The nominal case of Listing 13.2 was run for the example in which the target is considered to be stationary (namely, VT=0) and is located 30,000 ft downrange from the missile (namely, RT1IC=30000). The missile is traveling at 3000 ft/s and is initially at 10,000 ft altitude (namely, VM=3000, RM1IC=10000). The geometry is such that the missile is on a collision path with the target (that is, zero heading error). We can see from Fig. 13.16 that when proportional navigation is used, the missile essentially travels in a straight line to the target because it is already on a collision triangle with the target. However, in this application we would like to hit the target vertically for lethality reasons (such as an antitank application). This means that for the trajectory shaping guidance law we would like the final line-of-sight angle to be −90 deg (namely, XLAMFDEG=−90). We can see from Fig. 13.16 that it indeed appears that the trajectory shaping guidance

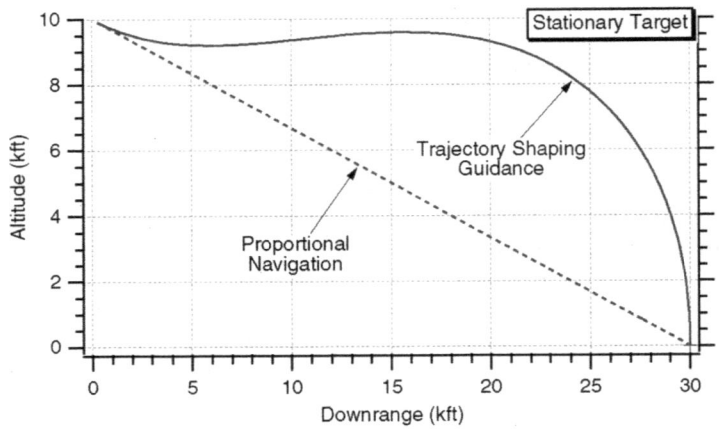

Fig. 13.16   Trajectory shaping also works in the nonlinear world.

law is enabling the missile to hit the target near vertically. In addition, we can see that the trajectory shaping guidance law trajectory is entirely different than the proportional navigation guidance trajectory. Of course, trajectory shaping guidance requires time-to-go information, whereas proportional navigation does not.

LISTING 13.2   NONLINEAR ENGAGEMENT SIMULATION TO TEST TRAJECTORY SHAPING GUIDANCE LAW

```
clear
n=0;
XNTG=0.;
HEDEG=0.;
XNP=3.;
RM1IC=0.;
RM2IC=10000.;
RT1IC=30000.;
RT2IC=0.;
VM=3000.;
VT=0.;
XNCLIMG=9999999.;
APN=1;
XLAMFDEG=-90.;
H=.0001;
XNCLIM=32.2*XNCLIMG;
XLAMF=XLAMFDEG/57.3;
XNT=32.2*XNTG;
RM1=RM1IC;
RM2=RM2IC;
```

```
RT1=RT1IC;
RT2=RT2IC;
BETA=0.;
VT1=-VT*cos(BETA);
VT2=VT*sin(BETA);
HE=HEDEG/57.3;
T=0.;
S=0.;
RTM1=RT1-RM1;
RTM2=RT2-RM2;
RTM=sqrt(RTM1^2+RTM2^2);
XLAM=atan2(RTM2,RTM1);
XLEAD=asin(VT*sin(BETA+XLAM)/VM);
THET=XLAM+XLEAD;
VM1=VM*cos(THET+HE);
VM2=VM*sin(THET+HE);
VTM1=VT1-VM1;
VTM2=VT2-VM2;
VC=-(RTM1*VTM1+RTM2*VTM2)/RTM;
while VC >= 0
        if RTM<1000
          H=.00001;
        else
          H=.0001;
        end
        BETAOLD=BETA;
        RT1OLD=RT1;
        RT2OLD=RT2;
        RM1OLD=RM1;
        RM2OLD=RM2;
        VM1OLD=VM1;
        VM2OLD=VM2;
        STEP=1;
        FLAG=0;
        while STEP <=1
          if FLAG==1
        STEP=2;
                BETA=BETA+H*BETAD;
                RT1=RT1+H*VT1;
                RT2=RT2+H*VT2;
                RM1=RM1+H*VM1;
                RM2=RM2+H*VM2;
                VM1=VM1+H*AM1;
                VM2=VM2+H*AM2;
                T=T+H;
          end
```

```
RTM1=RT1-RM1;
RTM2=RT2-RM2;
RTM=sqrt(RTM1^2+RTM2^2);
VTM1=VT1-VM1;
VTM2=VT2-VM2;
VC=-(RTM1*VTM1+RTM2*VTM2)/RTM;
XLAM=atan2(RTM2,RTM1);
XLAMD=(RTM1*VTM2-RTM2*VTM1)/(RTM*RTM);
TGO=RTM/VC;
if APN==0
        XNC=XNP*VC*XLAMD;
else
        XNT1=XNT*sin(BETA);
        XNT2=XNT*cos(BETA);
        XNTPLOS=-XNT1*sin(XLAM)+XNT2*cos(XLAM);
         XNC=4.*VC*XLAMD+XNTPLOS+2.*VC*(XLAM-XLAMF)/TGO;
end
if XNC>XNCLIM
        XNC=XNCLIM;
end
if XNC<-XNCLIM
        XNC=-XNCLIM;
end
AM1=-XNC*sin(XLAM);
AM2=XNC*cos(XLAM);
VT1=-VT*cos(BETA);
VT2=VT*sin(BETA);
if VT==0.
     BETAD=0.;
else
     BETAD=XNT/VT;
end
 FLAG=1;
end
FLAG=0;
BETA=.5*(BETAOLD+BETA+H*BETAD);
RT1=.5*(RT1OLD+RT1+H*VT1);
RT2=.5*(RT2OLD+RT2+H*VT2);
RM1=.5*(RM1OLD+RM1+H*VM1);
RM2=.5*(RM2OLD+RM2+H*VM2);
VM1=.5*(VM1OLD+VM1+H*AM1);
VM2=.5*(VM2OLD+VM2+H*AM2);
S=S+H;
if S>=.09999
  S=0.;
  n=n+1;
```

```
            RT1K=RT1/1000.;
            RT2K=RT2/1000.;
            RM1K=RM1/1000.;
            RM2K=RM2/1000.;
            XLAMDEG=XLAM*57.3;
            XNCG=XNC/32.2;
            ArrayT(n)=T;
            ArrayRT1K(n)=RT1K;
            ArrayRT2K(n)=RT2K;
            ArrayRM1K(n)=RM1K;
            ArrayRM2K(n)=RM2K;
            ArrayXNCG(n)=XNCG;
            ArrayXLAMDEG(n)=XLAMDEG;
        end
end
RTM
figure
plot(ArrayRT1K,ArrayRT2K,ArrayRM1K,ArrayRM2K),grid
title('Engagement Geometry')
xlabel('Downrange (Kft) ')
ylabel('Altitude (Kft)')
figure
plot(ArrayT,ArrayXNCG),grid
title('Commanded Acceleration')
xlabel('Time (Sec) ')
ylabel('XNC (G)')
axis([0 14 -20 25])
figure
plot(ArrayT,ArrayXLAMDEG),grid
title('Line-of-Sight Angle')
xlabel('Time (Sec) ')
ylabel('XLAM (Deg)')
axis([0 14 -100 0])
clc
output=[ArrayT',ArrayRT1K',ArrayRT2K',ArrayRM1K',...
         ArrayRM2K',ArrayXNCG',ArrayXLAMDEG'];
save datfil.txt output -ascii
disp '*** Simulation Complete'
```

Figure 13.17 shows that the price paid for shaping the trajectory is that considerable acceleration is required by the missile to hit the target. Unlike proportional navigation, which does not require any acceleration to hit the target in this scenario because it is already on a collision triangle, trajectory shaping guidance requires more than 20 g of acceleration at the beginning of the flight and nearly -10 g at the end of the flight.

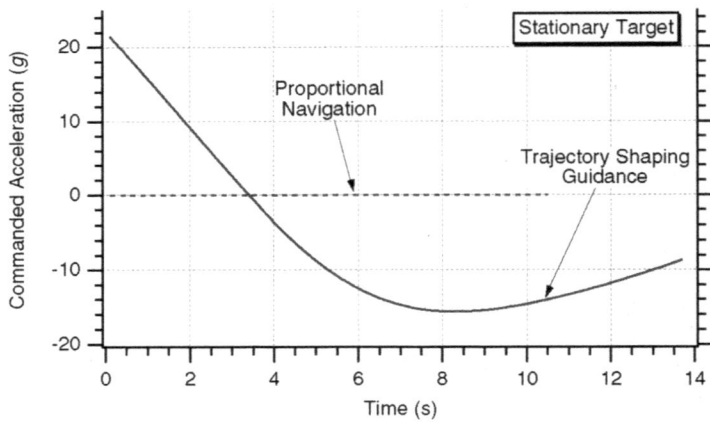

Fig. 13.17   A great deal of acceleration may be required to shape trajectory to get a final line-of-sight angle of −90 deg.

Finally, we can see from Fig. 13.18 that the line-of-sight angle for trajectory shaping guidance achieved the design goal by reaching −90 deg at the end of the flight. Figure 13.18 also shows that proportional navigation, which does not shape the trajectory, ended up with a final line-of-sight angle of −18.4 deg. We can also see from Fig. 13.18 that because proportional navigation enabled the missile to fly directly to the target, the flight time was nearly 4 s shorter than when the trajectory shaping guidance law was used.

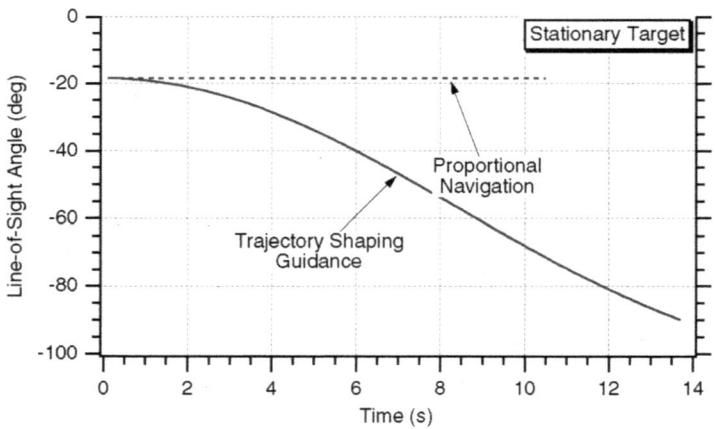

Fig. 13.18   Trajectory shaping enables line-of-sight angle to reach its goal.

It is of considerable interest to see if the formulas we derived in the previous section for the commanded acceleration are useful in predicting the nonlinear results. Recall that in the previous section we showed that the acceleration in units of ft/s^2 required to turn the missile through an angle of λ_F in units of radians is given by

$$n_c(t)_{\lambda_F} = \frac{-2V_c\lambda_F}{t_F}\left[1 - \frac{3t}{t_F}\right]$$

where V_c is the closing velocity in ft/s, t_F is the amount of flight or guidance time in seconds, and t is instantaneous time in seconds. For the problem depicted in Fig. 13.16, the final line-of-sight angle that would have been achieved without trajectory shaping guidance is −18.4 deg (see Fig. 13.18). Therefore, trajectory shaping guidance is attempting to change the angle from −18.4 deg to −90 deg, or a change of −71.6 deg. In addition, we can tell from Fig. 13.18 that the amount of guidance time for trajectory shaping guidance is 13.6 s.

Figure 13.19 indicates that the linear formula for missile acceleration is not very accurate in this example for predicting the nonlinear commanded missile acceleration. It is hypothesized that perhaps the angular change in the line-of-sight angle is too great for linear theory to hold. Another case was run with the nonlinear simulation in which the desired final line-of-sight angle was −30 deg (a change of only 11.6 deg from −18.4 deg that could be obtained with proportional navigation). Because there is less trajectory shaping, the flight time reduces to 10.6 s. We can see from Fig. 13.20 that the formula now matches the nonlinear results quite accurately.

Another, more stressing case was considered in which the target was both moving and maneuvering. In this example, the missile had a −20-deg heading

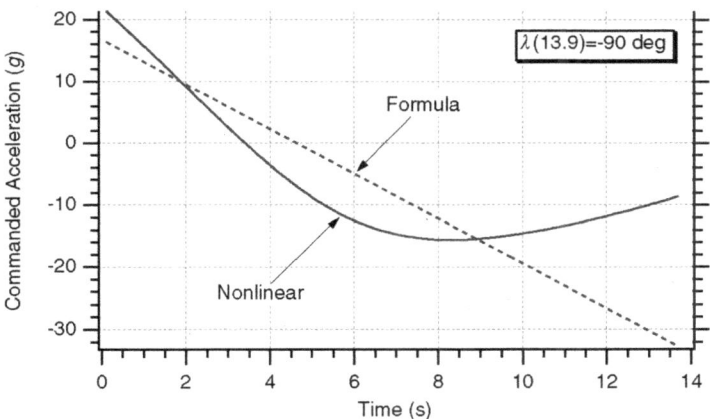

Fig. 13.19   Linear formula is not a great match to nonlinear results.

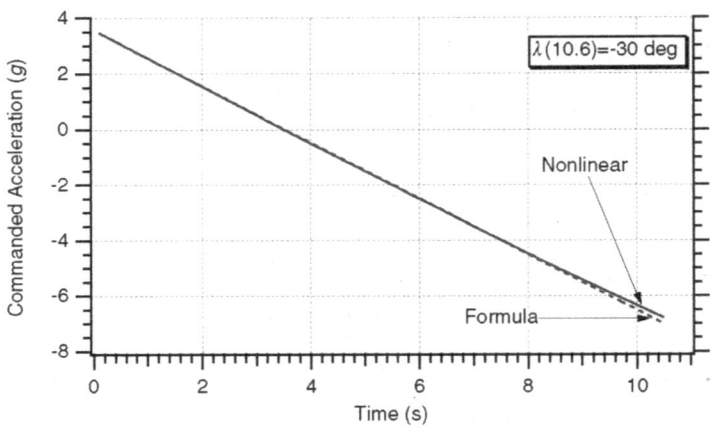

Fig. 13.20   Linear formula is nearly perfect when angular turn is smaller.

error while the target was executing a 6-g maneuver. Two cases were considered—one in which the desired final line-of-sight angle was −30 deg and the other in which the desired final line-of-sight angle was 30 deg. Figure 13.21 shows that when trajectory shaping guidance was used, intercepts were achieved in both cases. It is too difficult to tell from Fig. 13.21 if the final line-of-sight angle design goals have actually been met. Figure 13.22 displays the commanded acceleration profiles that were required in both cases for successful intercepts. We can see that for the −30-deg intercept the maximum positive acceleration was 30 g and

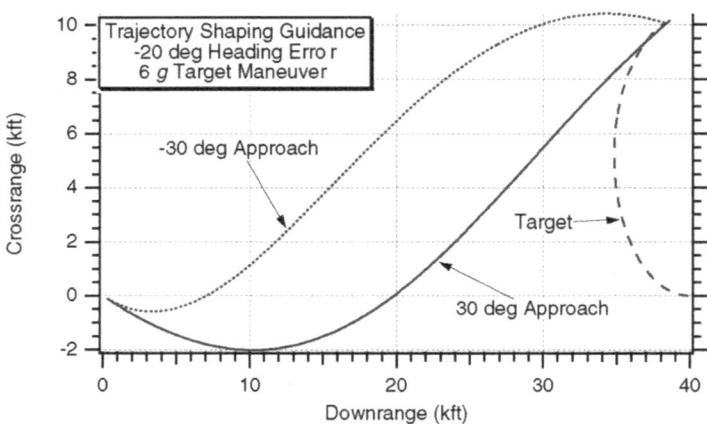

Fig. 13.21   Trajectory shaping guidance works against maneuvering target for different approach angles.

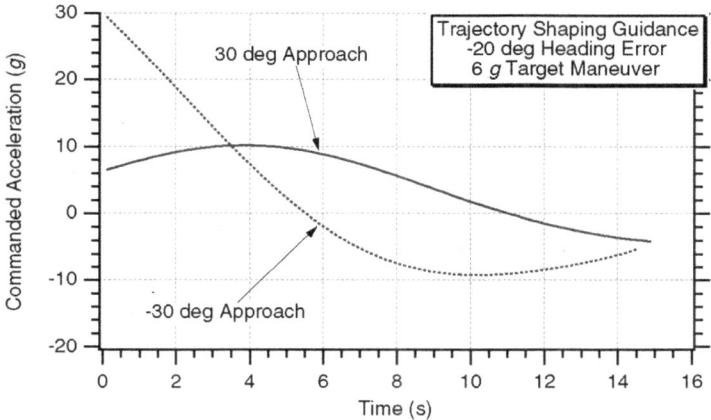

Fig. 13.22 Acceleration requirements are larger with maneuvering target than with stationary target.

the maximum negative acceleration was $-10\,g$. We can also see that for the 30-deg intercept the maximum positive acceleration was only $10\,g$ and the maximum negative acceleration was $-5\,g$. Finally, we can see from Fig. 13.23 that the design goals for the final line-of-sight angles for both cases were met.

Again, it is of considerable interest to see if the formulas we derived in the previous section for the commanded acceleration are useful in predicting the nonlinear results. Recall that the total acceleration in units of ft/s^2 required to turn the missile through an angle of λ_F in units of radians in the presence of target

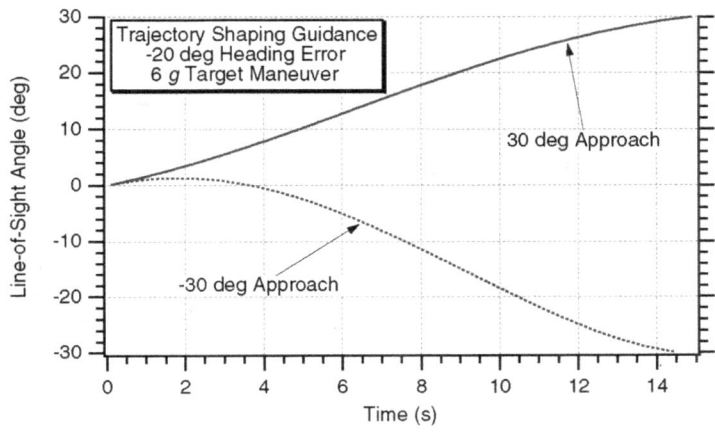

Fig. 13.23 Trajectory shaping guidance meets design goals against maneuvering target.

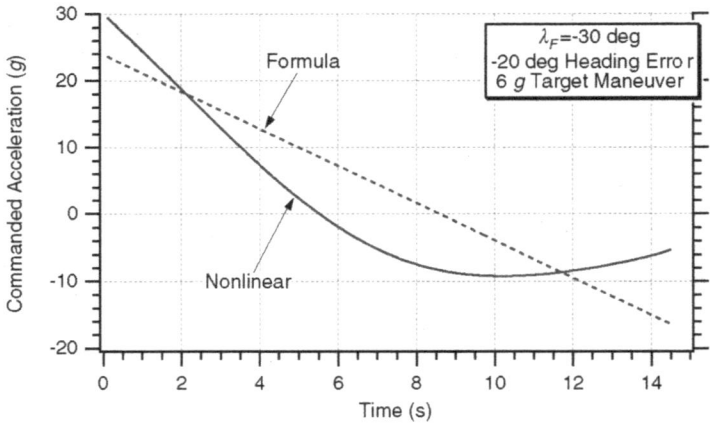

Fig. 13.24   Linear formula is not very accurate for highly maneuvering target.

maneuver n_T in units of ft/s^2 and heading error HE in units of radians is given by

$$n_c(t)\big|_{\text{Total}} = \frac{-2V_c\lambda_F}{t_F}\left[1 - \frac{3t}{t_F}\right] + n_T + \frac{-2V_M\text{HE}}{t_F}\left[2 - \frac{3t}{t_F}\right]$$

For the case of interest, the closing velocity is approximately 4000 ft/s and the flight time turns out to be 14.5 s. We can see from Fig. 13.24 that the match between the formula and simulation results is not very accurate.

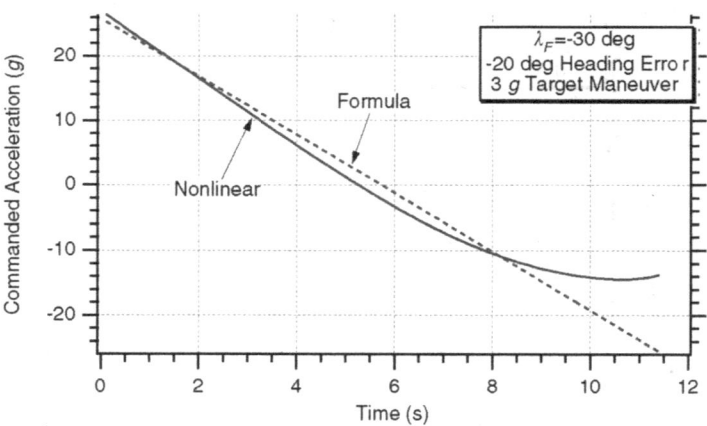

Fig. 13.25   Linear formula is much more accurate when target maneuver level is lower.

It is hypothesized that the reason for the inaccurate comparison of Fig. 13.24 is because of the highly maneuvering target. As the target maneuvers, the portion of the maneuver perpendicular to the line of sight diminishes, making the formula less accurate. To test the hypothesis, another case was run in which the maneuver level decreased to 3 g. The heading error remained at -20 deg, and the desired final line-of-sight angle remained at -30 deg. For this case, the time of flight reduced to 11.4 s. We can see from Fig. 13.25 that now the formula is an excellent approximation to the nonlinear results.

Thus, we can see that the trajectory shaping guidance law also works in the nonlinear world. We have demonstrated that under many circumstances we also have formulas that can be used to predict or explain the resultant acceleration requirements on the missile when trajectory shaping guidance is used.

SUMMARY

In this chapter the trajectory shaping guidance law was derived. It was demonstrated that with this new guidance law we could not only hit the target, but also control the final line-of-sight angle. The price paid for the trajectory shaping was that more acceleration was required to hit the target. Formulas were also derived that could be used to predict the missile acceleration requirements for the new guidance law under a variety of circumstances. It was demonstrated that the formulas were also an accurate indicator of performance in the nonlinear world.

REFERENCES

[1] Battin, R., *An Introduction to the Mathematics and Methods of Astrodynamics*, AIAA, New York, 1987, p. 561.

[2] Lin, C. F., *Modern Navigation, Guidance, and Control Processing*, Prentice Hall, New Jersey, 1991, p. 605.

[3] Ben-Asher, J., and Yaesh, I., *Advances in Missile Guidance Theory*, AIAA, Reston, VA, 1998, pp. 25–88.

[4] Bryson, A. E., and Ho, Y. C., *Applied Optimal Control*, Hemisphere, New York, 1975, Chapter 5.

CHAPTER 14

Guidance Methods Against Stationary Targets

INTRODUCTION

In this chapter we shall discuss two new guidance laws for the special case of shaping the interceptor trajectory against a stationary target without requiring time-to-go information. The first guidance method will control the impact angle and will be compared to the optimal guidance law of Chapter 13. The second method will control the time of arrival of the interceptor at the target and will also drive the seeker look angle to zero at the end of the flight.

BIASED PROPORTIONAL NAVIGATION FOR CONTROLLING IMPACT ANGLE AGAINST STATIONARY TARGETS [1]

In this section we shall discuss the special case of shaping the interceptor trajectory against a stationary target without requiring time-to-go information. This new method will be compared to optimal guidance techniques for shaping the trajectory that require time-to-go information. In Chapter 13 we showed how optimal guidance techniques could be used against maneuvering targets to shape the trajectory of the interceptor so that it could hit the target with a desired final impact angle using minimum acceleration. However, as with other advanced guidance laws, the implementation of optimal guidance requires knowledge of the time to go until intercept. Time-to-go information might be available in radar homing missiles but may not be available for infrared homing missiles. In this section we shall show that for the special case of a stationary target, the missile trajectory can also be shaped by using biased proportional navigation. The form of biased proportional navigation discussed only requires line-of-sight rate information and knowledge of the missile's own velocity for the successful implementation of the guidance law. More complete details on using biased proportional navigation for hitting a stationary target can be found in [1].

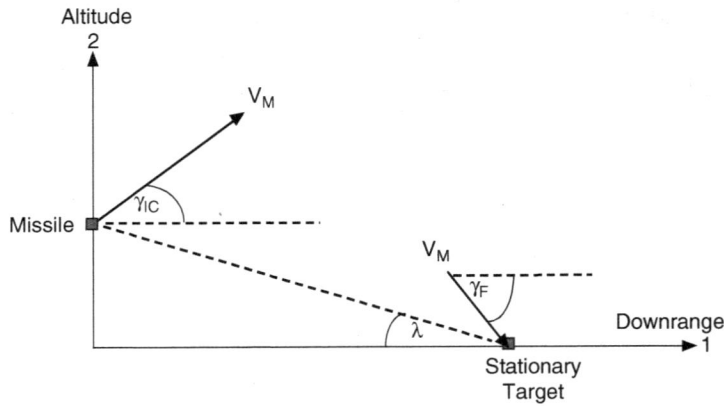

Fig. 14.1   Engagement geometry for stationary target.

The missile-target engagement geometry for a stationary target is depicted in Fig. 14.1. Here we can see that the missile is at an arbitrary initial flight-path angle γ_{IC}. We desire to hit the stationary target at a final flight-path angle γ_F using reasonable amounts of acceleration.

In Chapter 13 we showed that the optimal trajectory guidance law in which we hit the target, minimize the integral of the commanded acceleration squared, and attain a final line-of-sight angle λ_F is given by

$$n_c = 4V_c\dot{\lambda} + n_T + \frac{2V_c(\lambda - \lambda_F)}{t_{go}}$$

where n_c is the commanded missile acceleration, V_c is the closing velocity, n_T is the target maneuver, λ is the line-of-sight angle, $\dot{\lambda}$ is the line-of-sight rate, and t_{gc} is the time to go until intercept. If the target is stationary, there is no target maneuver and the final line-of-sight angle is the final flight-path angle, or

$$n_T = 0$$
$$\gamma_F = \lambda_F$$

Under these conditions, the optimal guidance law against a stationary target simplifies to

$$n_c = 4V_c\dot{\lambda} + \frac{2V_c(\lambda - \gamma_F)}{t_{go}}$$

The optimal guidance law, in addition to hitting the target and achieving a final flight-path angle, is suppose to minimize the integral of the acceleration squared, or

$$\text{Performance Index} = \int_0^{t_F} n_c^2 dt$$

In addition, we shall use the preceding performance index to see how efficient biased proportional navigation is when compared to optimal guidance. Proportional navigation, which does not make use of closing velocity information, can be written in terms of the missile flight-path rate $\dot{\gamma}$ as [2]

$$\dot{\gamma} = N\dot{\lambda}$$

where $\dot{\lambda}$ is the line-of-sight rate and N is the navigation ratio. It is important to note that N in general is not equal to the effective navigation ratio N' previously discussed in Chapter 2 of the previous volume. If the missile acceleration command is perpendicular to the velocity vector, we can say that

$$\dot{\gamma} = \frac{n_c}{V_M}$$

Therefore, we can say that this implementation of proportional navigation, where closing velocity information is not required, can also be written as

$$n_c = V_M \dot{\gamma} = N V_M \dot{\lambda}$$

If we equate this form of proportional navigation, where the acceleration command is perpendicular to the velocity vector, to the traditional form of proportional navigation where the acceleration command is perpendicular to the line of sight we obtain

$$n_c = N V_M \dot{\lambda} = N' V_c \dot{\lambda}$$

which means that the navigation ratio and the effective navigation ratio are related according to

$$N = N' \frac{V_c}{V_M}$$

For a stationary target, the closing velocity is approximately the missile velocity, so in this particular case the two navigation ratios are approximately equal. Biased proportional navigation is proportional navigation plus an extra term and can be expressed in terms of the flight-path rate as

$$\dot{\gamma} = N\dot{\lambda} + \text{bias}$$

where the bias can be zero for different portions of the flight.

Now let us see how we can choose a value for the bias that will help us shape the missile trajectory. If we integrate the preceding equation from 0 to t_F we obtain

$$\gamma_F - \gamma_{IC} = N(\lambda_F - \lambda_{IC}) + \text{bias}\Delta t$$

where Δt is the amount of time the bias is on (that is, has a non-zero value). For a stationary target

$$\lambda_F = \gamma_F$$

we can use substitution to yield

$$\gamma_F - \gamma_{IC} = N(\gamma_F - \lambda_{IC}) + \text{bias}\Delta t$$

Thus we can solve the preceding equation for the bias, yielding

$$\text{bias} = \frac{-\gamma_F(N - 1) + N\lambda_{IC} - \gamma_{IC}}{\Delta t}$$

In other words, if the missile is on an initial flight-path angle γ_{IC} and we wish to hit the target with a final flight-path angle γ_F, we can solve for the bias by selecting a value for the amount of time Δt the bias is on.

Listing 13.2, which contains the optimal trajectory shaping guidance law, has been modified for the case of the stationary target and appears in Listing 14.1. Listing 14.1 provides the option of using either the optimal trajectory shaping guidance law (IGUID=0) or biased proportional navigation (IGUID=1). We can see that the units in Listing 14.1 have been converted to the metric system, and the nominal case is one in which the missile is attempting to hit a stationary target 10 km downrange with a final flight-path angle of −90 deg. The missile is launched with a 30-deg heading error (or initial flight-path angle of 30 deg), and the initial missile velocity is 250 m/s. So far in our work to date the missile acceleration command has always been perpendicular to the line of sight. In this simulation we have the additional capability of seeing what happens if the acceleration command is perpendicular to the missile velocity vector (ICHOICE=1) or to the line of sight (ICHOICE=0). If biased proportional navigation is used (IGUID=1), the nominal case indicates that the bias is turned on at time zero (TBEG=0) and turned off 30 s later (DELT=30). The performance index ($X =$ integral of commanded missile acceleration squared at end of flight), flight-path angle, and missile velocity are written to a file so that results can be compared. Differences between Listing 13.2 and Listing 14.1 are highlighted in bold.

LISTING 14.1   TWO-DIMENSIONAL ENGAGEMENT SIMULATION WITH OPTIMAL TRAJECTORY SHAPING AND BIASED PROPORTIONAL NAVIGATION GUIDANCE OPTIONS

```
n=0 ;
IGUID=0 ;
ICHOICE=0 ;
```

```
GAMDEG=30. ;
GAMIC=GAMDEG/57.3 ;
DELT=30. ;
TBEG=0. ;
TEND=TBEG+DELT ;
XNP=3. ;
RM1IC=0. ;
RM2IC=0. ;
RT1IC=10000.*3.28 ;
RT2IC=0. ;
VM=250.*3.28 ;
XNCLIMG=10. ;
GAMFDEG=-90. ;
H=.0001 ;
XNCLIM=32.2*XNCLIMG ;
RM1=RM1IC ;
RM2=RM2IC ;
RT1=RT1IC ;
RT2=RT2IC ;
VT1=0. ;
VT2=0. ;
T=0. ;
S=0. ;
RTM1=RT1-RM1 ;
RTM2=RT2-RM2 ;
RTM=sqrt(RTM1^2+RTM2^2) ;
XLAM=atan2(RTM2,RTM1) ;
VM1=VM*cos(GAMIC) ;
VM2=VM*sin(GAMIC) ;
VTM1=VT1-VM1 ;
VTM2=VT2-VM2 ;
VC=-(RTM1*VTM1+RTM2*VTM2) / RTM ;
GAMF=GAMFDEG/57.3 ;
BIASDEG=(-GAMFDEG* (XNP-1.) + XNP*XLAM*57.3-GAMDEG)/DELT ;
BIAS=BIASDEG/57.3 ;
X=0. ;
while VC >=0
    if RTM <1000
        H= .00001 ;
    else
        H= .0001 ;
    end
    RM1OLD=RM1 ;
    RM2OLD=RM2 ;
    VM1OLD=VM1 ;
    VM2OLD=VM2 ;
```

```
XOLD=X ;
STEP=1 ;
FLAG=0 ;
while STEP <=1
   if FLAG==1
        STEP=2 ;
        RM1=RM1+H*VM1 ;
        RM2=RM2+H*VM2 ;
        VM1=VM1+H*AM1 ;
        VM2=VM2+H*AM2 ;
        X=X+H*XD ;
        T=T+H ;
   end
   GAM=atan2 (VM2,VM1) ;
   VM=sqrt (VM1^2+VM2^2) ;
   RTM1=RT1-RM1 ;
   RTM2=RT2-RM2 ;
   RTM=sqrt(RTM1^2+RTM2^2) ;
   VTM1=VT1-VM1 ;
   VTM2=VT2-VM2 ;
   VC= -(RTM1*VTM1+RTM2*VTM2) / RTM ;
   XLAM=atan2(RTM2,RTM1) ;
   XLAMD=(RTM1*VTM2-RTM2*VTM1)/(RTM*RTM) ;
   if IGUID==0
      TGO=RTM/VC ;
      XNC=4. *VC*XLAMD+2. *VC*(XLAM-GAMF) / TGO ;
   else
      if T<TBEG
         GAMD=XNP*XLAMD ;
         XNC=VM*GAMD ;
      elseif T<TEND
         GAMD=XNP*XLAMD+BIAS ;
         XNC=VM*GAMD ;
      else
         GAMD=XNP*XLAMD ;
         XNC=VM*GAMD ;
      end
   end
   if XNC>XNCLIM
      XNC=XNCLIM ;
   end
   if XNC<-XNCLIM
      XNC=-XNCLIM ;
   end
   if ICHOICE==0
```

```
        AM1=-XNC*sin (XLAM) ;
        AM2=XNC*cos(XLAM) ;
    else
        AM1=-XNC*sin(GAM) ;
        AM2=XNC*cos(GAM) ;
    end
    XD=XNC*XNC ;
    FLAG=1 ;
  end
  FLAG=0 ;
  RM1= .5* (RM1OLD+RM1+H*VM1) ;
  RM2= .5* (RM2OLD+RM2+H*VM2) ;
  VM1= .5* (VM1OLD+VM1+H*AM1) ;
  VM2= .5* (VM2OLD+VM2+H*AM2) ;
  X= .5*(XOLD+X+H*XD) ;
  S=S+H ;
  if S>= .09999
      S=0. ;
      n=n+1 ;
      RT1K=RT1/3280. ;
      RT2K=RT2/3280. ;
      RM1K=RM1/3280. ;
      RM2K=RM2/3280. ;
      XNCG=XNC/32.2 ;
      GAMDEG=GAM*57.3 ;
      VMM=VM/3.28 ;
      XM=X/(3.28*3.28) ;
      ArrayT (n) =T ;
      ArrayRT1K (n) =RT1K ;
      ArrayRT2K (n) =RT2K ;
      ArrayRM1K (n) =RM1K ;
      ArrayRM2K (n) =RM2K ;
      ArrayXNCG (n) =XNCG ;
      ArrayGAMDEG (n) =GAMDEG ;
      ArrayXM (n) =XM ;
  end
end
figure
plot(ArrayRT1K, ArrayRT2K, ArrayRM1K, ArrayRM2K), grid
title( ' Engagement Geometry ' )
xlabel( ' Downrange (Kft) ' )
ylabel( ' Altitude (Kft) ' )
figure
plot(ArrayT,ArrayXNCG),grid
title( ' Commanded Acceleration ' )
```

```
xlabel( ' Time (Sec) ' )
ylabel( ' XNC (G) ' )
figure
plot(ArrayT, ArrayGAMDEG), grid
title( ' Flight Path Angle ' )
xlabel( ' Time (Sec) ' )
ylabel( ' GAM (Deg) ' )
clc
output=[ArrayT',ArrayRT1K',ArrayRT2K',ArrayRM1K',ArrayRM2K',...
                  ArrayXNCG',ArrayGAMDEG'] ;
save datfil.txt output /ascii
disp ' *** Simulation Complete '
RTM=sqrt(RTM1^2+RTM2^2)
VM=sqrt(VM1^2+VM2^2)
```

For a quick review, the nominal case of Listing 14.1 was run in which the optimal trajectory shaping guidance law was used (IGUID=0) and the missile acceleration command was perpendicular to the line of sight (ICHOICE=0). Another case was run in which the same optimal guidance law was used but with the missile acceleration command perpendicular to the missile velocity vector (ICHOICE=1). We can see from Fig. 14.2 that in both cases the missile trajectories are nearly identical, and the missile hits the stationary target near vertically. However, Fig. 14.3 shows that the commanded acceleration profiles are vastly different. The maximum value of the acceleration command that is perpendicular to the line of sight is much smaller than when the acceleration command is perpendicular to the velocity vector, and the flight time is much longer. Figure 14.4 shows that both guidance command implementations result in the flight-path angle

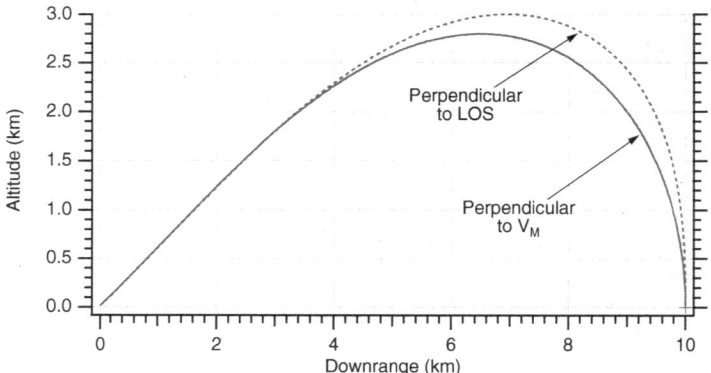

Fig. 14.2   Possible trajectories using optimal guidance for −90-deg impact against stationary target.

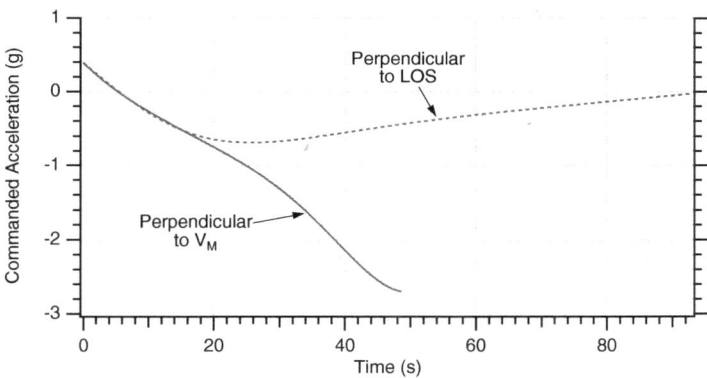

Fig. 14.3 Possible commanded acceleration profiles using optimal guidance for −90-deg impact against stationary target.

approaching −90 deg at intercept. Most importantly, Fig. 14.5 shows that dramatic trajectory shaping causes a severe missile velocity loss when the missile acceleration command is perpendicular to the line of sight, and no velocity loss at all when the missile acceleration command is perpendicular to the velocity vector. From a practical point of view, the velocity loss in this example cannot be tolerated because the actual acceleration capability of the missile diminishes as the velocity of the missile decreases. In future studies of trajectory shaping in this section, we shall assume that the acceleration command is perpendicular to the velocity vector (ICHOICE=1).

Next the optimal trajectory shaping guidance law was compared to biased proportional navigation for the case in which the acceleration commands were

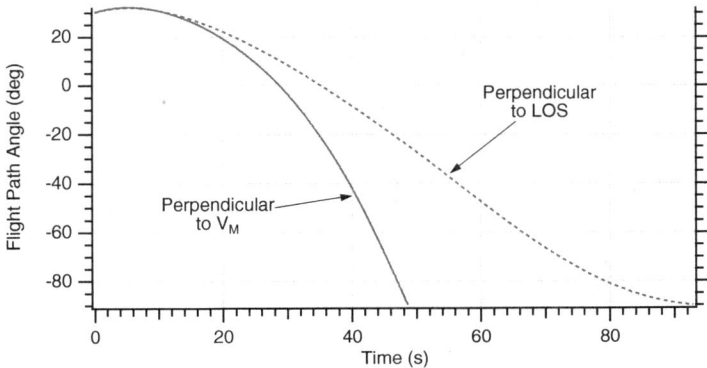

Fig. 14.4 Desired final flight-path angle is achieved with optimal trajectory shaping guidance law.

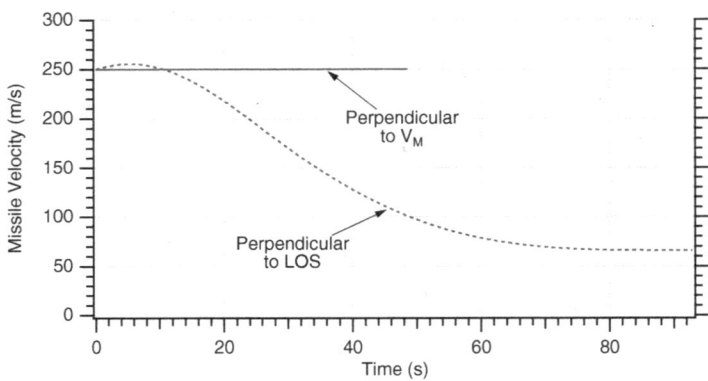

Fig. 14.5   Acceleration command perpendicular to line of sight causes missile velocity to decrease significantly when optimal trajectory shaping guidance law is used.

perpendicular to the velocity vector. Biased proportional navigation was considered when the bias started at time zero and lasted for 30 s. In addition, another case was considered in which the bias started at 10 s and lasted for 30 s. We can see from Fig. 14.6 that both guidance laws result in missile trajectories that hit the stationary target near vertically; however, the optimal guidance law trajectory is much tighter. We can also see from Fig. 14.6 that starting the bias later at 10 s also tightens the trajectory of the missile using biased proportional navigation. Figure 14.7 shows that the commanded acceleration profiles for the various guidance approaches are quite different but the maximum accelerators are similar. As expected, the flight time is much shorter for the tighter trajectories (optimal guidance case and biased proportional navigation when bias starts at

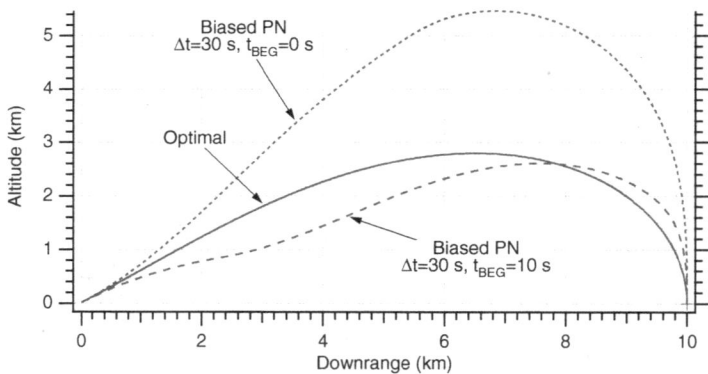

Fig. 14.6   Trajectories for both guidance laws are different.

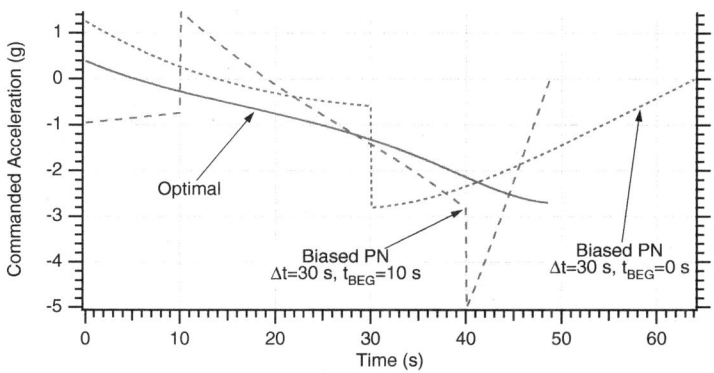

Fig. 14.7 Maximum acceleration command can be similar for both guidance laws.

10 s). Figure 14.8 shows that both guidance laws achieve success in that the flight-path angle approaches −90 deg at intercept. Figure 14.9 shows that although the performance index is smaller for the case in which the optimal guidance law is used, the performance index for biased proportional navigation starting at time zero is only 33% larger. These results indicate that the acceleration requirements for biased proportional navigation against a stationary target should not be significantly higher than those of optimal guidance.

Figure 14.10 demonstrates that when the initial flight-path angle is 30 deg, biased proportional navigation hits the stationary target for a variety of final impact angles. However, Fig. 14.11 shows that more missile acceleration is required as the desired final impact angle increases.

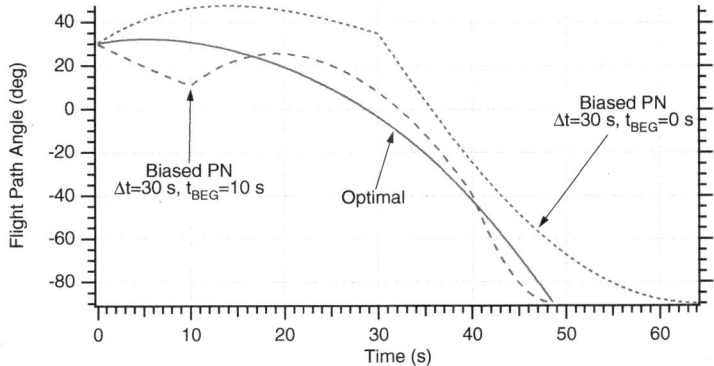

Fig. 14.8 Desired flight-path angle at end is achieved with both guidance laws.

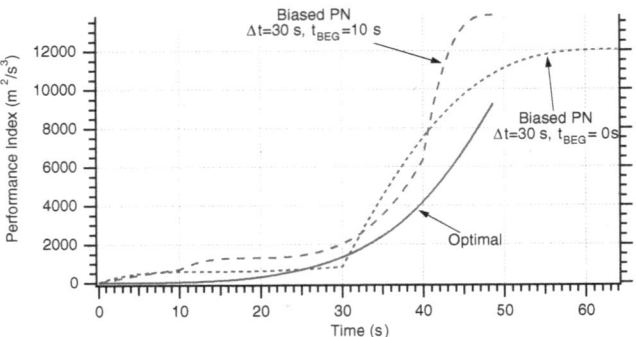

Fig. 14.9   Optimal guidance minimizes performance index.

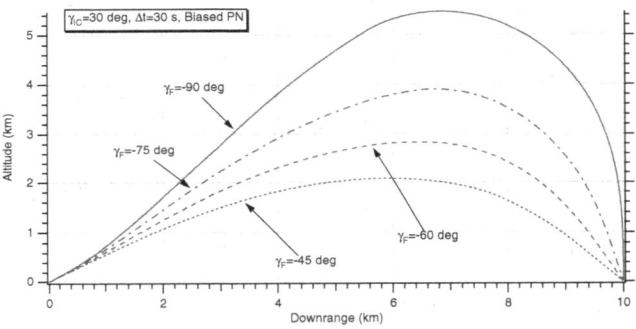

Fig. 14.10   Trajectory changes with changing final flight-path angle when using biased proportional navigation.

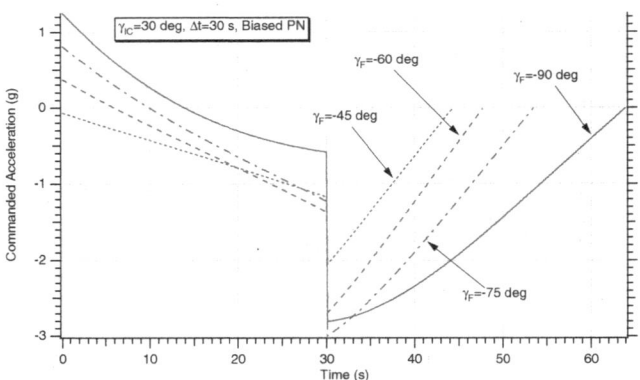

Fig. 14.11   Acceleration requirements increase as final flight-path angle increases for biased proportional navigation.

In this section we have shown that for the problem of hitting a stationary target with a desired impact angle, biased proportional navigation is competitive with the optimal trajectory shaping guidance law. Optimal guidance yields a smaller performance index (smaller integral of acceleration squared) than biased proportional navigation. However, the main advantage of biased proportional navigation over optimal guidance in the case of a stationary target is that time-to-go information is not required.

POLYNOMIAL GUIDANCE [3]

In the first part of this chapter we looked at a missile using biased proportional navigation against a stationary target for purposes of controlling the *missile impact angle*. In this section we shall look at a missile using polynomial guidance against a stationary target for purposes of controlling the *missile time of impact*. The purpose of polynomial guidance would be to enable a number of missiles launched at different times and possibly from different locations to arrive at the same target at the same time. Essentially Lambert guidance, discussed in the previous volume, also enabled a missile to arrive at a specific place at a specific time. The new guidance law discussed in this section makes use of polynomials, which were also used when we investigated Kalman and fading memory filters in the previous volume. Also note that proportional navigation is based on a first-order polynomial, and augmented proportional navigation is based on a second-order polynomial. Therefore, the use of a polynomial reference signal for guidance purposes may not seem so unusual if one really thinks about it.

IMPACT TIME CONTROL PROBLEM FORMULATION

In order to develop a guidance law that will control the impact time, it is first necessary to redraw Fig. 14.1 to include the seeker look angle ε (assuming a strapdown seeker and small angle of attack so that the missile velocity vector and missile body are aligned) and the range from the missile to a stationary target R_{TM}, as shown in Fig. 14.12. In this example we are not trying to control the final impact angle γ, as was done in the previous section of this chapter, but to control the impact time t_{FDES} and drive the seeker look angle ε to zero at the desired impact time.

From Fig. 14.12 we can see that the flight-path angle is the sum of the seeker look angle and the line-of-sight angle, or

$$\gamma = \varepsilon + \lambda$$

In addition, because the target is stationary, we can also say, based on simple trigonometry, that the first derivative of range R_{TM} is related to the missile velocity and seeker look angle according to

$$\dot{R}_{TM} = -V_M \cos \varepsilon$$

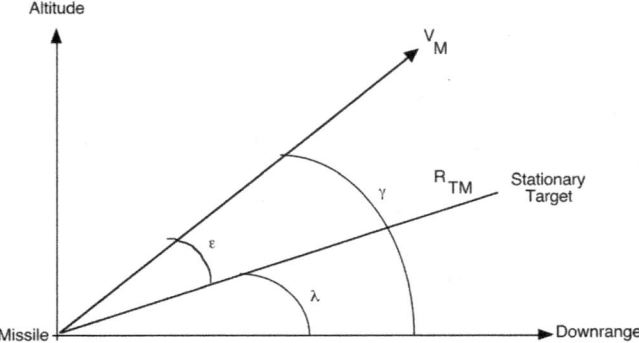

Fig. 14.12   Engagement geometry against a stationary target.

In order to find an alternate expression for the line-of-sight rate, we first expand the sine of the look angle to

$$\sin \varepsilon = \sin (\gamma - \lambda) = \sin \gamma \cos \lambda - \cos \gamma \sin \lambda$$

Using the relationships

$$\sin \gamma = \frac{V_{M2}}{V_M}$$

$$\cos \gamma = \frac{V_{M1}}{V_M}$$

$$\sin \lambda = \frac{-R_{TM2}}{R_{TM}}$$

$$\cos \lambda = \frac{R_{TM1}}{R_{TM}}$$

we can say that

$$\sin \varepsilon = \frac{V_{M2}}{V_M} \frac{R_{TM1}}{R_{TM}} - \frac{V_{M1}}{V_M} \frac{R_{TM2}}{R_{TM}} = \frac{R_{TM1} V_{M2} - R_{TM2} V_{M1}}{R_{TM} V_M}$$

From Chapter 2 of the previous volume we know that the line-of-sight rate can be expressed as

$$\dot{\lambda} = \frac{R_{TM1} V_{TM2} - R_{TM2} V_{TM1}}{R_{TM}^2}$$

Because the target is stationary

$$V_{T1} = V_{T2} = 0$$

the expression for the line-of-sight rate can be simplified to

$$\dot{\lambda} = \frac{-R_{TM1} V_{M2} + R_{TM2} V_{M1}}{R_{TM}^2}$$

Thus we can say that

$$R_{TM}\dot{\lambda} = -V_M \sin \varepsilon$$

Now we would like to make the range from the missile to the target fit an nth order polynomial of the form

$$R_{TM}(t) = x_0 + x_1 t + x_2 t^2 + x_3 t^3 + \ldots + x_{n-1} t^{n-1} + x_n t^n$$

Initially at $t = 0$ we can say that

$$R_{TM}(0) = R_0 = x_0$$

Taking the derivative of range yields

$$\dot{R}_{TM} = x_1 + 2x_2 t + 3x_3 t^2 + \ldots + (n-1)x_{n-1} t^{n-2} + nx_n t^{n-1}$$

Because

$$\dot{R}_{TM} = -V_M \cos \varepsilon$$

at $t = 0$, we can say that

$$\dot{R}_{TM}(0) = x_1 = -V_M \cos \varepsilon_0$$

where ε_0 is the seeker look angle at $t = 0$. We would also like to make the range zero (i.e., zero miss distance) at the desired intercept time t_{FDES}, or

$$R_{TM}(t_{FDES}) = x_0 + x_1 t_{FDES} + x_2 t_{FDES}^2 + x_3 t_{FDES}^3 + \ldots + x_{n-1} t_{FDES}^{n-1} + x_n t_{FDES}^n$$
$$= 0$$

If we desire to make the seeker look angle zero at the desired final time we can say that

$$\dot{R}_{TM}(t_{FDES}) = -V_M \cos \varepsilon_{t_{FDES}} = -V_M \cos (0) = -V_M$$

We can also say that the derivative of range at the desired intercept time is given by

$$\dot{R}_{TM}(t_{FDES}) = x_1 + 2x_2 t_{FDES} + 3x_3 t_{FDES}^2 + \ldots + (n-1)x_{n-1} t_{FDES}^{n-2} + nx_n t_{FDES}^{n-1}$$

or

$$x_1 + 2x_2 t_{FDES} + 3x_3 t_{FDES}^2 + \ldots + (n-1)x_{n-1} t_{FDES}^{n-2} + nx_n t_{FDES}^{n-1} = -V_M$$

In [3] it was shown that n must be at least 4 to have finite terminal acceleration. However, we will start with $n = 3$ because this simplest case is instructive. If
the order of the polynomial is 3, then we have the following four equations with
four unknowns, or

$$x_0 = R_0$$

$$x_1 = -V_M \cos \varepsilon_0$$

$$x_0 + x_1 t_{FDES} + x_2 t_{FDES}^2 + x_3 t_{FDES}^3 = 0$$

$$x_1 + 2x_2 t_{FDES} + 3x_3 t_{FDES}^2 = -V_M$$

The preceding four equations can be written in matrix form as

$$
\begin{bmatrix}
1 & 0 & 0 & 0 \\
0 & 1 & 0 & 0 \\
1 & t_{FDES} & t_{FDES}^2 & t_{FDES}^3 \\
0 & 1 & 2t_{FDES} & 3t_{FDES}^2
\end{bmatrix}
\begin{bmatrix}
x_0 \\
x_1 \\
x_2 \\
x_3
\end{bmatrix}
=
\begin{bmatrix}
R_0 \\
-V_M \cos \varepsilon_0 \\
0 \\
-V_M
\end{bmatrix}
$$

To solve for the coefficients of the polynomial, we multiply both sides of the
equation by the inverse of the square matrix and obtain

$$
\begin{bmatrix}
x_0 \\
x_1 \\
x_2 \\
x_3
\end{bmatrix}
=
\begin{bmatrix}
1 & 0 & 0 & 0 \\
0 & 1 & 0 & 0 \\
1 & t_{FDES} & t_{FDES}^2 & t_{FDES}^3 \\
0 & 1 & 2t_{FDES} & 3t_{FDES}^2
\end{bmatrix}^{-1}
\begin{bmatrix}
R_0 \\
-V_M \cos \varepsilon_0 \\
0 \\
-V_M
\end{bmatrix}
$$

If we had assumed a fourth-order polynomial for the problem we would need
an extra equation because of the additional unknown x_4. The additional equation
chosen is zero final condition on the second derivative of range at the desired
intercept time, or

$$\ddot{R}_{TM}(t_{FDES}) = 0$$

In order to satisfy the preceding relationship, we first write the expression for
range assuming a fourth-order polynomial as

$$R_{TM} = x_0 + x_1 t + x_2 t^2 + x_3 t^3 + x_4 t^4$$

Therefore, the derivative of range is

$$\dot{R}_{TM} = x_1 + 2x_2 t + 3x_3 t^2 + 4x_4 t^3$$

and the second derivative of range is given by

$$\ddot{R}_{TM} = 2x_2 + 6x_3 t + 12x_4 t^2$$

Substituting in the boundary value for the second derivative of range yields

$$\ddot{R}_{TM}(t_{FDES}) = 0 = 2x_2 + 6x_3 t_{FDES} + 12x_4 t_{FDES}^2$$

Therefore, we can solve for the coefficients of the fourth-order polynomial from

$$
\begin{bmatrix} x_0 \\ x_1 \\ x_2 \\ x_3 \\ x_4 \end{bmatrix}
=
\begin{bmatrix}
1 & 0 & 0 & 0 & 0 \\
0 & 1 & 0 & 0 & 0 \\
1 & t_{FDES} & t_{FDES}^2 & t_{FDES}^3 & t_{FDES}^4 \\
0 & 1 & 2t_{FDES} & 3t_{FDES}^2 & 4t_{FDES}^3 \\
0 & 0 & 2 & 6t_{FDES} & 12t_{FDES}^2
\end{bmatrix}^{-1}
\begin{bmatrix} R_0 \\ -V_M \cos \varepsilon_0 \\ 0 \\ -V_M \\ 0 \end{bmatrix}
$$

Similarly we could have assumed a fifth-order polynomial with the additional final condition on the third derivative of range at the desired intercept time to be zero, or

$$\dddot{R}_{TM}(t_{FDES}) = 0$$

As was done before, we first write the expression for range assuming a fifth-order polynomial as

$$R_{TM} = x_0 + x_1 t + x_2 t^2 + x_3 t^3 + x_4 t^4 + x_5 t^5$$

Therefore, the derivative of range for the fifth-order polynomial is

$$\dot{R}_{TM} = x_1 + 2x_2 t + 3x_3 t^2 + 4x_4 t^3 + 5x_5 t^4$$

and the second derivative of range for the fifth-order polynomial is given by

$$\ddot{R}_{TM} = 2x_2 + 6x_3 t + 12x_4 t^2 + 20x_5 t^3$$

while the third derivative of range is

$$\dddot{R}_{TM} = 2 + 6x_3 + 24x_4 t + 60x_5 t^2$$

Substituting in the boundary value for the third derivative of range yields

$$\dddot{R}_{TM}(t_{FDES}) = 0 = 6x_3 + 24x_4 t_{FDES} + 60x_5 t_{FDES}^2$$

The coefficients of the fifth-order polynomial can be obtained from

$$
\begin{bmatrix} x_0 \\ x_1 \\ x_2 \\ x_3 \\ x_4 \\ x_5 \end{bmatrix} =
\begin{bmatrix}
1 & 0 & 0 & 0 & 0 & 0 \\
0 & 1 & 0 & 0 & 0 & 0 \\
1 & t_{FDES} & t_{FDES}^2 & t_{FDES}^3 & t_{FDES}^4 & t_{FDES}^5 \\
0 & 1 & 2t_{FDES} & 3t_{FDES}^2 & 4t_{FDES}^3 & 5t_{FDES}^4 \\
0 & 0 & 2 & 6t_{FDES} & 12t_{FDES}^2 & 20t_{FDES}^3 \\
0 & 0 & 0 & 6 & 24t_{FDES} & 60t_{FDES}^2
\end{bmatrix}^{-1}
\begin{bmatrix} R_0 \\ -V_M \cos \varepsilon_0 \\ 0 \\ -V_M \\ 0 \\ 0 \end{bmatrix}
$$

Reference [3] mentioned that all the coefficients of any nth-order polynomial selected can be found by recursion. Although no formal proof was provided, the thinking made use of symbolic programs, pattern recognition techniques, and intuition that are beyond understanding for most readers of this text, including myself. However, we shall take the practical engineering approach and simply program the results of that amazing intellectual process and compare it to numerical solutions to see if the resulting formula is reasonable. This nonacademic approach was also taken for the method of adjoints, presented in the previous volume. The reader should recall that no proof was offered for the adjoint method, but numerous examples were presented demonstrating the adjoint's validity and utility. It should also be remembered that the derivation of proportional navigation came about 20 years after its first implementation.

Specifically, [3] claimed that the general form of the x_2 coefficient turns out to be

$$
x_2 = \frac{n-1}{2t_{FDES}^2}[(n-2+2\cos\varepsilon_0)V_M t_{FDES} - nR_0]
$$

where n is the order of the polynomial chosen. (We have already looked at $n=3$, 4, and 5.) The purpose of choosing the x_2 coefficient will be revealed in the next section, but first we must see if the preceding equation is correct.

To test the validity of the preceding equation, we will perform some numerical tests. These tests will be to first obtain the numerical value of x_2: first by using matrix inversion to solve for the polynomial coefficients, and then to look at the x_2 coefficient in particular and compare it to the numerical evaluation of the preceding equation. These examples will be performed for the third-, fourth-, and fifth-order polynomials. Let's consider a case in which $R_0 = 10{,}000$ m, $t_{FDES} = 50$ s, $V_M = 250$ m/s, and $\varepsilon_O = 0$ deg. Listing 14.2 calculates x_2 by first performing the matrix inversion, required by the matrix relationships for the polynomial coefficients, and then using the preceding formula looking at the cases for the third-, fourth-, and fifth-order polynomials (i.e., $n = 3, 4, 5$).

LISTING 14.2   CHECKING THE THEORETICAL FORMULA FOR x_2

```
clear
R0=10000;
VM=250;
TFDES=50;
E0=0;
A3=[1 0 0 0;
0 1 0 0;
1 TFDES TFDES^2 TFDES^3;
0 1 2*TFDES 3*TFDES^2];
X0=[R0 -VM*cos(E0) 0 -VM]';
XPZ1=inv(A3)*X0;
X2SIM3=XPZ1(3,1)
N=3;
X2THEORY=(N-1)*((N-2+2*cos(E0))*VM*TFDES-N*R0)/(2*TFDES*TFDES)
A4=[1 0 0 0 0;
0 1 0 0 0;
1 TFDES TFDES^2 TFDES^3 TFDES^4;
0 1 2*TFDES 3*TFDES^2 4*TFDES^3;
0 0 2 6*TFDES 12*TFDES^2];
X0=[R0 -VM*cos(E0) 0 -VM 0]';
XPZ2=inv(A4)*X0;
X2SIM4=XPZ2(3,1)
N=4;
X2THEORY=(N-1)*((N-2+2*cos(E0))*VM*TFDES-N*R0)/(2*TFDES*TFDES)
A5=[1 0 0 0 0 0;
    0 1 0 0 0 0;
    1 TFDES TFDES^2 TFDES^3 TFDES^4 TFDES^5;
    0 1 2*TFDES 3*TFDES^2 4*TFDES^3 5*TFDES^4;
    0 0 2 6*TFDES 12*TFDES^2 20*TFDES^3;
    0 0 0 6 24*TFDES 60*TFDES^2];
X0=[R0 -VM*cos(E0) 0 -VM 0 0]';
XPZ3=inv(A5)*X0;
X2SIM5=XPZ3(3,1)
N=5;
X2THEORY=(N-1)*((N-2+2*cos(E0))*VM*TFDES-N*R0)/(2*TFDES*TFDES)
```

The nominal case of Listing 14.2 was run, and X2SIM3 matched X2THEORY for $N = 3$ (answer = 3), X2SIM4 matched X2THEORY for $N = 4$ (answer = 6), and X2SIM5 matched X2THEORY for $N = 5$ (answer = 10), thus confirming the accuracy of the theoretical equation. From now on we will assume that the formula for $x2$ from [3] is true.

POLYNOMIAL GUIDANCE LAW FOR IMPACT TIME CONTROL

Recall that the fundamental relationship for range rate was obtained by inspection of Fig. 14.12 as

$$\dot{R}_{TM} = -V_M \cos \varepsilon$$

Assuming constant missile velocity, we can take the derivative of the preceding equation and obtain

$$\ddot{R}_{TM} = V_M \dot{\varepsilon} \sin \varepsilon$$

or solving for the seeker look angle rate yields

$$\dot{\varepsilon} = \frac{\ddot{R}_{TM}}{V_M \sin \varepsilon}$$

From Fig. 14.12 we can also see that the flight-path angle can be expressed as

$$\gamma = \varepsilon + \lambda$$

Taking the derivative of the preceding equation yields the flight-path rate

$$\dot{\gamma} = \dot{\varepsilon} + \dot{\lambda}$$

Substituting the seeker look angle rate into the preceding equation yields a new expression for the flight-path rate

$$\dot{\gamma} = \frac{\ddot{R}_{TM}}{V_M \sin \varepsilon} + \dot{\lambda}$$

Recall in the previous section we showed that

$$\dot{R}_{TM} = x_1 + 2x_2 t + 3x_3 t^2 + \ldots + (n-1)x_{n-1}t^{n-2} + nx_n t^{n-1}$$

Taking the derivative again yields

$$\ddot{R}_{TM} = 2x_2 + 6x_3 t + \ldots + (n-1)(n-2)x_{n-1}t^{n-3} + n(n-1)x_n t^{n-2}$$

Therefore, the flight-path rate is given by

$$\dot{\gamma} = \frac{2x_2 + 6x_3 t + \ldots + (n-1)(n-2)x_{n-1}t^{n-3} + n(n-1)x_n t^{n-2}}{V_M \sin \varepsilon} + \dot{\lambda}$$

The preceding guidance law is open loop because it depends on time. Therefore, it will be unable to cope with autopilot lag and acceleration saturation.

However, if we change the reference point by first setting $t = 0$ we get

$$\dot{\gamma} = \frac{2x_2}{V_M \sin \varepsilon} + \dot{\lambda}$$

where x_2 was previously verified from a simulation point of view using Listing 14.2 to be

$$x_2 = \frac{n-1}{2t_{FDES}^2}[(n - 2 + 2\cos \varepsilon_0)V_M t_{FDES} - nR_0]$$

If we now change our reference point and let

$$\varepsilon_0 \rightarrow \varepsilon$$
$$t_{FDES} \rightarrow t_{FDES} - t$$
$$R_0 \rightarrow R_{TM}$$

we find that

$$x_2 = \frac{n-1}{2(t_{FDES} - t)^2}[(n - 2 + 2\cos \varepsilon)V_M(t_{FDES} - t) - nR_{TM}]$$

$$\dot{\gamma} = \frac{2x_2}{V_M \sin \varepsilon} + \dot{\lambda}$$

and the acceleration command is simply

$$n_c = V_M \dot{\gamma}$$

We will now call the preceding set of three equations to be the polynomial guidance law. Listing 14.1 was slightly modified to incorporate the polynomial guidance law and is shown in Listing 14.3 (IGUID=2). The changes from Listing 14.1 to incorporate the polynomial guidance law are highlighted in boldface. Note that the stopping condition for the simulation has been changed. In Listing 14.1 the simulation stopped when the closing velocity went negative, signifying that we were at the point of closest approach to the target. With the polynomial guidance law it is possible to have unusual trajectories in which the range from missile to target might be increasing for a small portion of the flight. Therefore, because the target is on the ground, the simulation is stopped when the missile hits the ground. The nominal case assumes that the order of the polynomial is three (XN=3) and that it is desired to reach the target in 50 s (TFDES=50). The initial missile velocity is 250 m/s (VM=250*3.28), the target location is 10 km downrange (RT1IC=10000*3.28), and the missile acceleration limit is 10 g (XNCLIMG=10).

LISTING 14.3 SIMULATION TO TEST POLYNOMIAL GUIDANCE LAW

```
clear; clc; close all
n=0;
IGUID=2;
XN=3;
TFDES=50.;
GAMDEG=30.;
GAM=GAMDEG/57.3;
DELT=30.;
TBEG=0.;
TEND=TBEG+DELT;
XNP=3.;
RM1IC=0.;
RM2IC=0.;
RT1IC=10000.*3.28;
RT2IC=0.;
VM=250.*3.28;
XNCLIMG=10.;
GAMFDEG=-90.;
H=.0001;
XNCLIM=32.2*XNCLIMG;
RM1=RM1IC;
RM2=RM2IC;
RT1=RT1IC;
RT2=RT2IC;
VT1=0.;
VT2=0.;
T=0.;
S=0.;
RTM1=RT1-RM1;
RTM2=RT2-RM2;
RTM=sqrt(RTM1^2+RTM2^2);
XLAM=atan2(RTM2,RTM1);
VM1=VM*cos(GAM);
VM2=VM*sin(GAM);
VTM1=VT1-VM1;
VTM2=VT2-VM2;
VC=-(RTM1*VTM1+RTM2*VTM2)/RTM;
GAMF=GAMFDEG/57.3;
BIASDEG=(-GAMFDEG*(XNP-1.)+XNP*XLAM*57.3-GAMDEG)/DELT;
BIAS=BIASDEG/57.3;
X=0.;
i=0;
while RM2 >= 0
    i=i+1;
```

```
if RTM < 1000
   H=.000001;
else
   H=.001;
end
GAMOLD=GAM;
XLAMOLD=XLAM;
RM1OLD=RM1;
RM2OLD=RM2;
VM1OLD=VM1;
VM2OLD=VM2;
XOLD=X;
STEP=1;
FLAG=0;
while STEP <=1
   if FLAG==1
      STEP=2;
      GAM=GAM+H*GAMD;
      XLAM=XLAM+H*XLAMD;
      RM1=RM1+H*VM1;
      RM2=RM2+H*VM2;
      VM1=VM1+H*AM1;
      VM2=VM2+H*AM2;
      X=X+H*XD;
      T=T+H;
   end
   VM=sqrt(VM1^2+VM2^2);
   RTM1=RT1-RM1;
   RTM2=RT2-RM2;
   RTM=sqrt(RTM1^2+RTM2^2);
   VTM1=VT1-VM1;
   VTM2=VT2-VM2;
   VC=-(RTM1*VTM1+RTM2*VTM2)/RTM;
   XLAMD=(RTM1*VTM2-RTM2*VTM1)/(RTM*RTM);
   if IGUID==0
      TGO=RTM/VC;
      XNC=4.*VC*XLAMD+2.*VC*(XLAM-GAMF)/TGO;
   elseif IGUID==1
      if T<TBEG
         GAMD=XNP*XLAMD;
         XNC=VM*GAMD;
      elseif T<TEND
         GAMD=XNP*XLAMD+BIAS;
         XNC=VM*GAMD;
      else
         GAMD=XNP*XLAMD;
```

```
    XNC=VM*GAMD;
  end
elseif IGUID==2
  EPS=GAM-XLAM;
  TEMP1=(XN-2+2*cos(EPS))*VM*(TFDES-T)-XN*RTM;
  X2=.5*(XN-1)*TEMP1/(TFDES-T)^2;
  GAMD=2*X2/(VM*sin(EPS))+XLAMD;
  XNC=VM*GAMD;
end
if XNC>XNCLIM
  XNC=XNCLIM;
  GAMD=XNCLIM/VM;
end
if XNC<-XNCLIM
  XNC=-XNCLIM;
  GAMD=-XNCLIM/VM;
end
RHO=.002378*exp(-RM2/22000.);
VM=sqrt(VM1^2+VM2^2);
AM1=-XNC*sin(GAM);
AM2=XNC*cos(GAM);
XD=XNC*XNC;
FLAG=1;
end
FLAG=0;
GAM=.5*(GAMOLD+GAM+H*GAMD);
XLAM=.5*(XLAMOLD+XLAM+H*XLAMD);
RM1=.5*(RM1OLD+RM1+H*VM1);
RM2=.5*(RM2OLD+RM2+H*VM2);
VM1=.5*(VM1OLD+VM1+H*AM1);
VM2=.5*(VM2OLD+VM2+H*AM2);
X=.5*(XOLD+X+H*XD);
S=S+H;
if S>=.09999
  S=0.;
  n=n+1;
  RT1K=RT1/3280.;
  RT2K=RT2/3280.;
  RM1K=RM1/3280.;
  RM2K=RM2/3280.;
  XNCG=XNC/32.2;
  GAMDEG=GAM*57.3;
  VMM=VM/3.28;
  XM=X/(3.28*3.28);
  EPSDEG=EPS*57.3;
```

```
        ArrayT(n)=T;
        ArrayRT1K(n)=RT1K;
        ArrayRT2K(n)=RT2K;
        ArrayRM1K(n)=RM1K;
        ArrayRM2K(n)=RM2K;
        ArrayXNCG(n)=XNCG;
        ArrayGAMDEG(n)=GAMDEG;
        ArrayEPSDEG(n)=EPSDEG;
        ArrayRTM(n)=RTM;
    end
end
figure
plot(ArrayRT1K,ArrayRT2K,ArrayRM1K,ArrayRM2K),grid
title('Engagement Geometry')
xlabel('Downrange (Kft) ')
ylabel('Altitude (Kft)')
figure
plot(ArrayT,ArrayXNCG),grid
title('Commanded Acceleration')
xlabel('Time (Sec) ')
ylabel('XNC (G)')
figure
plot(ArrayT,ArrayEPSDEG),grid
title('Look Angle')
xlabel('Time (Sec) ')
ylabel('EPS (Deg)')
plot(ArrayT,ArrayRTM),grid
title('RANGE')
xlabel('Time (Sec) ')
ylabel('Range (m)')
clc
output=[ArrayT',ArrayRT1K',ArrayRT2K',ArrayRM1K',ArrayRM2K',...
    ArrayXNCG',ArrayEPSDEG',ArrayRTM'];
save datfil.txt output -ascii
disp '*** Simulation Complete'
RTM=sqrt(RTM1^2+RTM2^2)
T
```

Cases were run for third-, fourth-, and fifth-order polynomials. We can see from Fig. 14.13 that all missile trajectories reach the target. It appears that as the order of polynomial increases, the apogee of the missile trajectory increases slightly. However, Fig. 14.14 indicates that the third-order polynomial guidance law resulted in acceleration saturation due to the 10-g acceleration limit, resulting in a 15-ft miss. Increasing the order of the polynomial resulted in zero miss

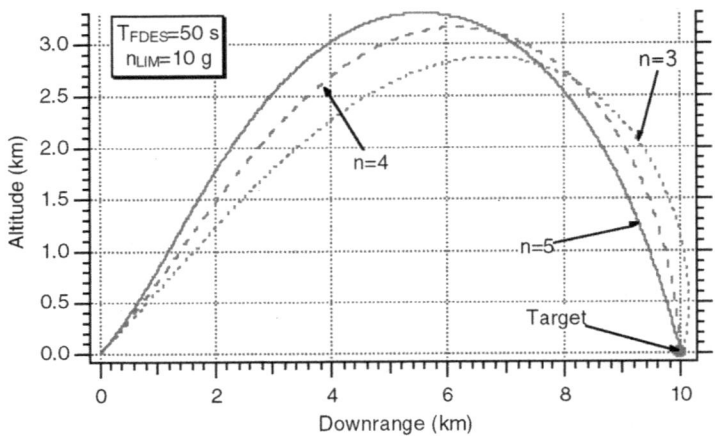

Fig. 14.13   All trajectories reach target in 50 s.

distance, and the time of arrival at the target was exactly 50 s. Figure 14.15 shows that the seeker look angle goes to zero at the end of the flight. The maximum value of the seeker look angle increases with increasing order of the polynomial.

Next, cases were run for the fifth-order polynomial guidance law in which the desired intercept time was varied from 50 s to 100 s. Figure 14.16 shows that the apogee of the missile trajectory increases as the desired intercept time increases. It is important to note that, especially for the 100-s intercept, the missile overshoots the target downrange and then comes back. This type of trajectory means that the

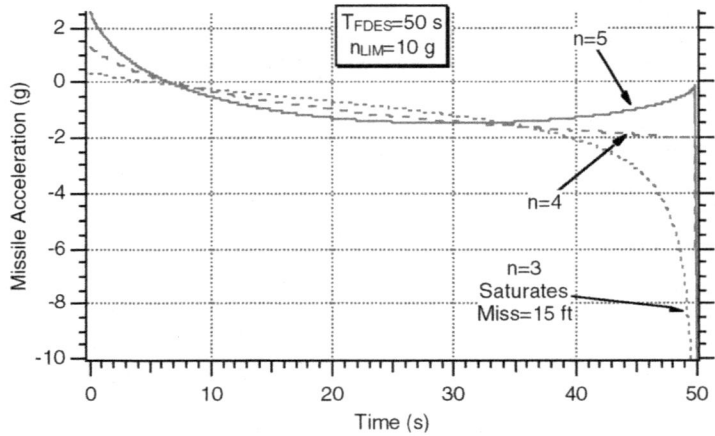

Fig. 14.14   Too low an order polynomial causes acceleration saturation.

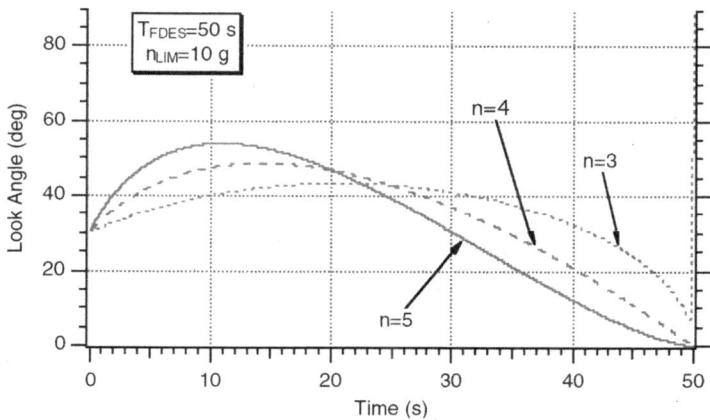

Fig. 14.15   Seeker look angle goes to zero at end of flight.

closing velocity is negative for a portion of the flight—which is why the simulation stopping condition had to be changed. Figure 14.17 shows that the missile maximum acceleration requirements are approximately independent of the missile intercept time. Finally, Fig. 14.18 shows that the maximum seeker look angle increases with increasing intercept time.

CAVEATS

It is important to note that the polynomial guidance law for impact time control will only work perfectly if the missile velocity is constant (i.e., sea-skimming

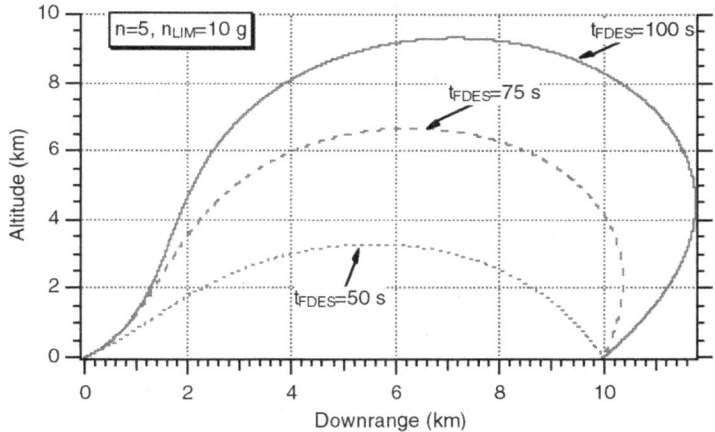

Fig. 14.16   Increasing desired intercept time lofts the trajectory.

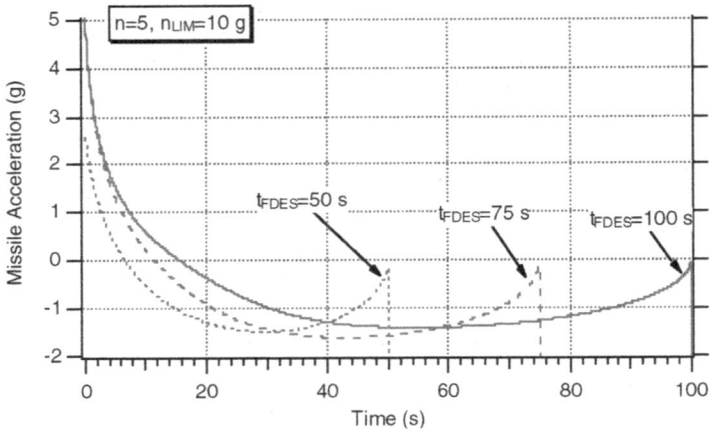

Fig. 14.17   Acceleration requirements are independent of desired flight time.

missiles) and there is sufficient missile acceleration capability. Further work is required to make this guidance law more practical. Some possibilities for future work in this area include using proportional navigation for the very end of the engagement [4], using some sort of predictive scheme to get an average velocity for the engagement, and possibly using some other variable such as the zero effort miss rather than range for the polynomial shaping.

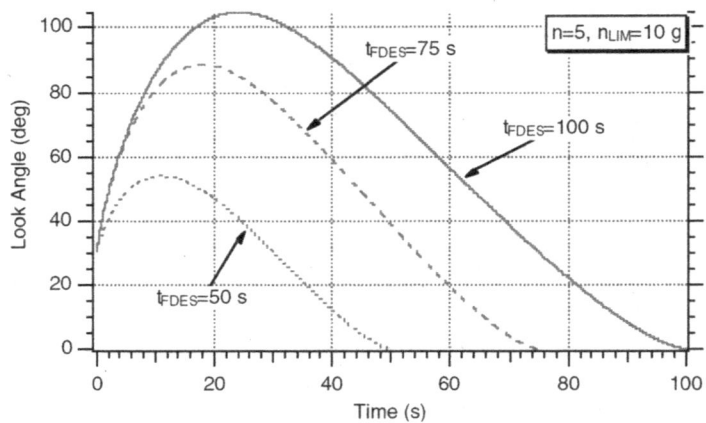

Fig. 14.18   Increasing desired intercept time increases maximum seeker look angle.

SUMMARY

In this chapter we have shown that for a constant velocity missile a guidance law could be developed to hit a stationary target at a specified time while driving the seeker look angle to zero. Cases were investigated for third-, fourth-, and fifth-order polynomials that represented the range from the missile to target. It was shown that lower order polynomial range representations resulted in larger acceleration requirements near the end of the flight but smaller maximum look angles during the flight. The maximum look angle tended to increase as the desired time of flight increased.

REFERENCES

[1] Erer, K. S., and Merttopcuoglu, O., "Indirect Impact-Angle-Control Against Stationary Targets Using Biased Pure Proportional Navigation, " *Journal of Guidance, Control and Dynamics*, Vol. 35, 2012, pp. 700–704.

[2] Jerger, J., "Systems Preliminary Design," D. Van Nostrand Company, Inc., Princeton, N. J., 1960, pp. 182–187.

[3] Tekin, R., "A New Design Framework for Impact Time Control," PhD Thesis, Technical University of Munich, July 2018.

[4] Tekin, R., Erer, K., and Holzapfel, F., "Impact Time Control with Generalized Polynomial Range Formulation," *Journal of Guidance, Control, and Dynamics*, Vol. 41, 2018, pp. 1190–1195.

CHAPTER 15

The Vertical-S Target Maneuver [1]

INTRODUCTION

In this text and the previous volume we have demonstrated how target maneuvers can have a profound effect on missile system performance. Target maneuvers are important from a system designer's point of view because they enable the designer to properly evaluate system performance and to take design actions if the performance against target maneuvers is not acceptable. Target maneuvers are also important for a fighter pilot who is trying to survive in regions of potential missile attack. The fighter pilot is interested in realistic maneuver strategies under conditions in which visual or electronic contact with the enemy missile (or missiles) may not be made. These maneuver strategies are developed from both combat experience and detailed simulation analysis.

Examples of realistic strategies are the barrel roll (weaving or sinusoidal target maneuver) and vertical-S maneuvers. Chapter 6 of the previous volume demonstrated the effectiveness of these evasion strategies. Chapter 6 of this volume went into detail to show how effective the barrel roll target maneuver can be. In addition, methods for improving missile performance against the weaving target were presented in Chapters 7 and 8. In this chapter we shall concentrate on the vertical-S target maneuver, which can be even more effective from a pilot's point of view than the barrel roll maneuver.

SHAPING FILTER APPROACH FOR VERTICAL-S TARGET MANEUVER

In Chapter 4 of the previous volume we showed how white noise through an integrator could be used to represent a step target maneuver whose starting time was equally likely to occur anywhere during the flight. In Chapter 7 of this volume we demonstrated that white noise through a second-order shaping network could be used to represent a weaving target (barrel roll) whose starting time was also equally likely to occur anywhere during the flight. Let's see if we can develop a shaping filter for the vertical-S target maneuver.

The vertical-S target maneuver is depicted in Fig. 15.1 where P is the length or period of the maneuver and t_R is the rise time of the maneuver. With the vertical-S maneuver strategy, the pilot rolls and maneuvers the aircraft so that maximum acceleration n_T is reached. The maximum positive acceleration is held for a while, and the pilot rolls the aircraft 180 deg to reach maximum negative acceleration, as depicted in Fig. 15.1. The amount of time it takes the pilot to roll the aircraft 180 deg is the rise time t_R.

Mathematically the vertical-S maneuver $\ddot{y}_T$ can be expressed by the following set of equations broken into time intervals for one half-period L. According to Fig. 15.1, we can see that

$$0 < t < \frac{t_R}{2}$$

$$\ddot{y}_T = \frac{2n_T}{t_R} t$$

$$\frac{t_R}{2} < t < L - \frac{t_R}{2}$$

$$\ddot{y}_T = n_T$$

$$L - \frac{t_R}{2} < t < L$$

$$\ddot{y}_T = -\frac{2n_T}{t_R} t + \frac{2n_T L}{t_R}$$

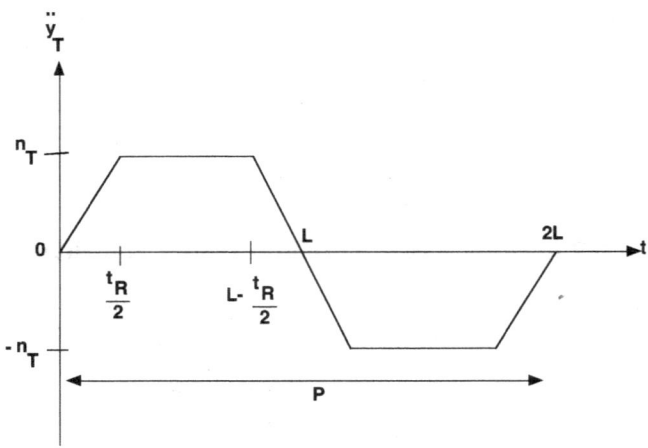

Fig. 15.1   Vertical-S target maneuver.

If we assume that the vertical-S maneuver is periodic, then we can express the vertical-S maneuver as a Fourier series [2]

$$\ddot{y}_T = \sum_{n=1}^{\infty} b_n \sin \frac{n\pi t}{L}$$

where

$$b_n = \frac{2}{L} \int_0^L \ddot{y}_T \sin \frac{n\pi t}{L} dt \quad (n = 1, 2, 3, \ldots)$$

Expanding the preceding integral yields

$$b_n = \frac{2n_T}{L} \left[\int_0^{t_R/2} \frac{2}{t_R} t \sin\frac{n\pi t}{L} dt + \int_{t_R/2}^{L-t_R/2} \sin\frac{n\pi t}{L} dt + \int_{L-t_R/2}^{L} \left(-\frac{2}{t_R} t + \frac{2L}{t_R} \right) \sin\frac{n\pi t}{L} dt \right]$$

$$(n = 1, 2, 3, \ldots)$$

After much algebra we obtain

$$b_n = \frac{8n_T L}{\pi^2 n^2 t_R} \sin\left(\frac{n\pi t_R}{2L}\right) \quad (n = 1, 2, 3, \ldots)$$

If we truncate the Fourier series to only two terms we obtain

$$\ddot{y}_T = \frac{8n_T L}{\pi^2 t_R} \left[\sin\left(\frac{\pi t_R}{2L}\right) \sin\left(\frac{\pi t}{L}\right) + \frac{1}{9}\sin\left(\frac{3\pi t_R}{2L}\right) \sin\left(\frac{3\pi t}{L}\right) \right]$$

Defining

$$L = \frac{P}{2}$$

$$\omega = \frac{2\pi}{P}$$

$$\alpha = \frac{\pi t_R}{2L}$$

enables us to simplify the truncated expression for the vertical-S target maneuver as

$$\ddot{y}_T = \frac{4n_T \left(\sin \alpha \sin \omega t + \dfrac{1}{9}\sin 3\alpha \sin 3\omega t\right)}{\pi \alpha}$$

In order to test the validity and accuracy of the preceding equation for the vertical-S maneuver, the simulation of Listing 15.1 was written to compare the actual vertical-S maneuver with the preceding analytical approximation. In this numerical example it is assumed that the period of the maneuver is 3 s (PZ=3) and the rise time is a parameter that is nominally set to 0.5 s (TR=.5). For zero rise time we set TR=0.01 rather than zero to avoid a division by zero.

LISTING 15.1   CHECKING ANALYTICAL EXPRESSION FOR VERTICAL-S TARGET MANEUVER

```
clear
count=0;
PZ=3.;
TR=.5;
% TR=.01;
XL=PZ/2.;
PI=3.14159;
W=2.*PI/PZ;
XNT=322.;
X=PZ/2.-TR;
ALF=PI*TR/(2.*XL);
T=0.;
for T=0.:.1:6.
     if T<PZ
       if T<TR/2.
       YTDD=2.*XNT*T/TR;
       elseif T<(TR/2.+X)
       YTDD=XNT;
       elseif T<(3.*TR/2.+X)
       YTDD=-2.*XNT*T/TR+2.*XNT+2.*XNT*X/TR;
       elseif T<(3.*TR/2.+2.*X)
       YTDD=-XNT;
       else
       YTDD=2.*XNT*T/TR-4.*XNT-4.*XNT*X/TR;
       end
     elseif T<2.*PZ
     TSTAR=T-PZ;
       if TSTAR<TR/2.
       YTDD=2.*XNT*TSTAR/TR;
       elseif TSTAR<(TR/2.+X)
       YTDD=XNT;
       elseif TSTAR<(3.*TR/2.+X)
       YTDD=-2.*XNT*TSTAR/TR+2.*XNT+2.*XNT*X/TR;
       elseif TSTAR<(3.*TR/2.+2.*X)
       YTDD=-XNT;
       else
       YTDD=2.*XNT*TSTAR/TR-4.*XNT-4.*XNT*X/TR;
       end
```

```
      elseif T<3.*PZ
      TSTAR=T-2.*PZ;
        if TSTAR<TR/2.
        YTDD=2.*XNT*TSTAR/TR;
        elseif TSTAR<(TR/2.+X)
        YTDD=XNT;
        elseif TSTAR<(3.*TR/2.+X)
        YTDD=-2.*XNT*TSTAR/TR+2.*XNT+2.*XNT*X/TR;
        elseif TSTAR<(3.*TR/2.+2.*X)
        YTDD=-XNT;
        else
        YTDD=2.*XNT*TSTAR/TR-4.*XNT-4.*XNT*X/TR;
        end
      else
      TSTAR=T-3.*PZ;
        if TSTAR<TR/2.
        YTDD=2.*XNT*TSTAR/TR;
        elseif TSTAR<(TR/2.+X)
        YTDD=XNT;
        elseif TSTAR<(3.*TR/2.+X)
        YTDD=-2.*XNT*TSTAR/TR+2.*XNT+2.*XNT*X/TR;
        elseif TSTAR<(3.*TR/2.+2.*X)
        YTDD=-XNT;
        else
        YTDD=2.*XNT*TSTAR/TR-4.*XNT-4.*XNT*X/TR;
        end
      end
   XNTH=4.*XNT*(sin(ALF)*sin(W*T)+sin(3.*ALF)*sin(3.*W*T)/9.)/(PI*ALF);
   count=count+1;
   ArrayT(count)=T;
   ArrayYTDD(count)=YTDD;
   ArrayXNTH(count)=XNTH;
end
figure
plot(ArrayT,ArrayYTDD,ArrayT,ArrayXNTH),grid
xlabel('Time (Sec)')
ylabel('Acceleration and Estimate (f/s^2)')
clc
output=[ArrayT',ArrayYTDD',ArrayXNTH'];
save datfil.txt output -ascii
disp 'simulation finished'
```

Listing 15.1 was run for the two cases of rise time (TR=.5 and TR=.01); the results are displayed in Figs. 15.2 and 15.3. We can see from the two rise time cases that the two-term Fourier series is an excellent approximation to the Vertical-S maneuver.

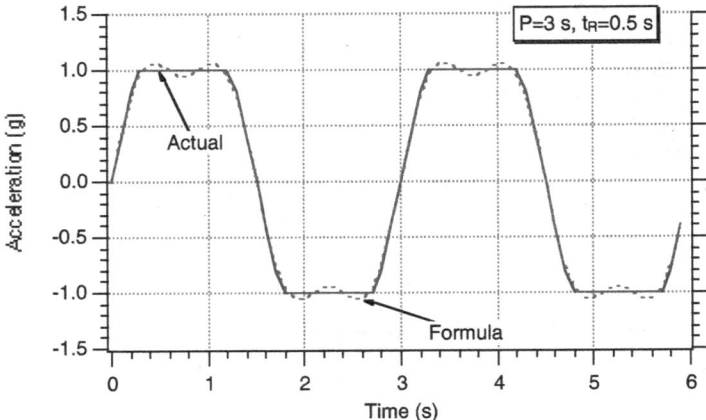

Fig. 15.2   Two-term Fourier series accurately represents vertical-S maneuver with finite rise time.

In order to see if the truncated Fourier series approach to representing a vertical-S target maneuver is accurate from a miss distance point of view, it is necessary to test the approximation in a homing loop environment. Figure 15.4 displays a single time constant guidance system where we program the actual vertical-S maneuver].

A miss distance simulation of the actual vertical-S maneuver whose starting time is equally likely to occur appears in Listing 15.2. In this simulation, the time of the start of the maneuver is obtained from a uniform distribution. One

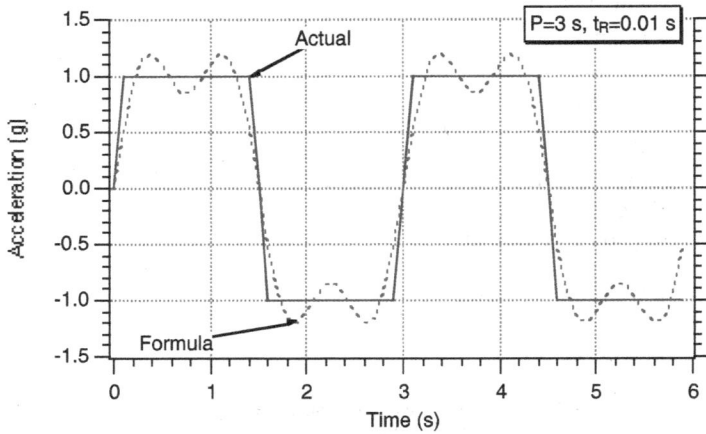

Fig. 15.3   Two-term Fourier series accurately represents vertical-S maneuver with near-zero rise time.

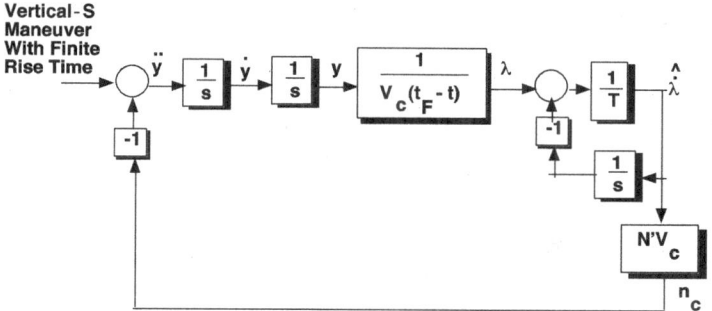

Fig. 15.4   Single time constant guidance system with actual vertical-S target maneuver with finite rise time.

hundred runs are made for homing times ranging from 0.1 s to 10 s for a total of 10,000 runs.

LISTING 15.2   MONTE CARLO MISS DISTANCE SIMULATION FOR ACTUAL VERTICAL-S MANEUVER IN SINGLE TIME CONSTANT GUIDANCE SYSTEM

```
clear
count=0;
RUN=100;
TSTART=0.;
P=3.;
TR=1.;
XL=P/2.;
VC=3000.;
XNT=161.;
TAU=.5;
XNP=3.;
TF=10.;
X=P/2.-TR;
for TF=.1:.1:10,
    Z1=0.;
    for JJ=1:RUN,
        SUM=rand(1);
      TSTART=SUM*TF;
      Y=0.;
      YD=0.;
      XLAMH=0.;
      T=0.;
      H=.01;
      S=0.;
      while T <= (TF - 1e-5)
```

```
YOLD=Y;
YDOLD=YD;
XLAMHOLD=XLAMH;
STEP=1;
FLAG=0;
while STEP <=1
if FLAG==1
  Y=Y+H*YD;
  YD=YD+H*YDD;
  XLAMH=XLAMH+H*XLAMHD;
  T=T+H;
  STEP=2;
end;
if T<TSTART
  YTDD=0.;
elseif (T-TSTART)<P
  if (T-TSTART)<TR/2.
    YTDD=2.*XNT*(T-TSTART)/TR;
  elseif (T-TSTART)<(TR/2.+X)
    YTDD=XNT;
  elseif (T-TSTART)<(3.*TR/2.+X)
    YTDD=-2.*XNT*(T-TSTART)/TR+2.*XNT+2.*XNT*X/TR;
  elseif (T-TSTART)<(3.*TR/2.+2.*X)
    YTDD=-XNT;
  else
    YTDD=2.*XNT*(T-TSTART)/TR-4.*XNT-4.*XNT*X/TR;
  end
elseif (T-TSTART)<2.*P;
  TSTAR=(T-TSTART)-P;
  if TSTAR<TR/2.
    YTDD=2.*XNT*TSTAR/TR;
  elseif TSTAR<(TR/2.+X)
    YTDD=XNT;
  elseif TSTAR<(3.*TR/2.+X)
    YTDD=-2.*XNT*TSTAR/TR+2.*XNT+2.*XNT*X/TR;
  elseif TSTAR<(3.*TR/2.+2.*X)
    YTDD=-XNT;
  else
    YTDD=2.*XNT*TSTAR/TR-4.*XNT-4.*XNT*X/TR;
  end
elseif (T-TSTART)<3.*P
  TSTAR=(T-TSTART)-2.*P;
    if TSTAR<TR/2.
      YTDD=2.*XNT*TSTAR/TR;
  elseif TSTAR<(TR/2.+X)
    YTDD=XNT;
  elseif TSTAR<(3.*TR/2.+X)
```

```
            YTDD=-2.*XNT*TSTAR/TR+2.*XNT+2.*XNT*X/TR;
         elseif TSTAR<(3.*TR/2.+2.*X)
            YTDD=-XNT;
         else
            YTDD=2.*XNT*TSTAR/TR-4.*XNT-4.*XNT*X/TR;
         end
      else
         TSTAR=(T-TSTART)-3.*P;
         if TSTAR<TR/2.
            YTDD=2.*XNT*TSTAR/TR;
         elseif TSTAR<(TR/2.+X)
            YTDD=XNT;
         elseif TSTAR<(3.*TR/2.+X)
            YTDD=-2.*XNT*TSTAR/TR+2.*XNT+2.*XNT*X/TR;
         elseif TSTAR<(3.*TR/2.+2.*X)
            YTDD=-XNT;
         else
            YTDD=2.*XNT*TSTAR/TR-4.*XNT-4.*XNT*X/TR;
         end
      end
      TGO=TF-T+.00001;
      XLAM=Y/(VC*TGO);
      XLAMHD=(XLAM-XLAMH)/TAU;
      XNC=XNP*VC*XLAMHD;
      YDD=YTDD-XNC;
      FLAG=1;
   end;
   FLAG=0;
   Y=.5*(YOLD+Y+H*YD);
   YD=.5*(YDOLD+YD+H*YDD);
   XLAMH=.5*(XLAMHOLD+XLAMH+H*XLAMHD);
   S=S+H;
  end
  Z(JJ)=Y;
  Z1=Z(JJ)+Z1;
  XMEAN=Z1/JJ;
end
SIGMA=0.;
Z1=0.;
Z2=0.;
for I=1:RUN,
  Z1=(Z(I)-XMEAN)^2+Z1;
  Z2=Z(I)^2+Z2;
 if I==1,
    SIGMA=0;
    RMS=0.;
  else
```

```
        SIGMA=sqrt(Z1/(I-1));
        RMS=sqrt(Z2/(I-1));
    end;
end
count=count+1;
ArrayTF(count)=TF;
ArrayRMS(count)=RMS;
end;
figure
plot(ArrayTF',ArrayRMS'),grid
title('RMS miss for various flight times')
xlabel('Flight Time (S)')
ylabel('RMS MISS (Ft) ')
clc
output=[ArrayTF',ArrayRMS'];
save datfil.txt output -ascii
disp('Simulation Complete')
```

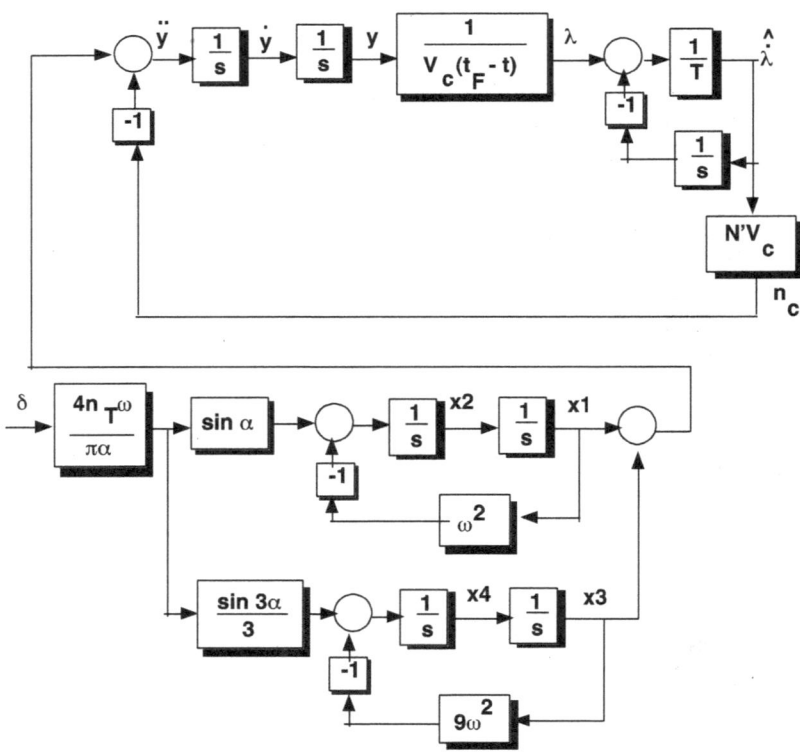

Fig. 15.5   Single time constant guidance system with shaping filter equivalent of vertical-S target maneuver.

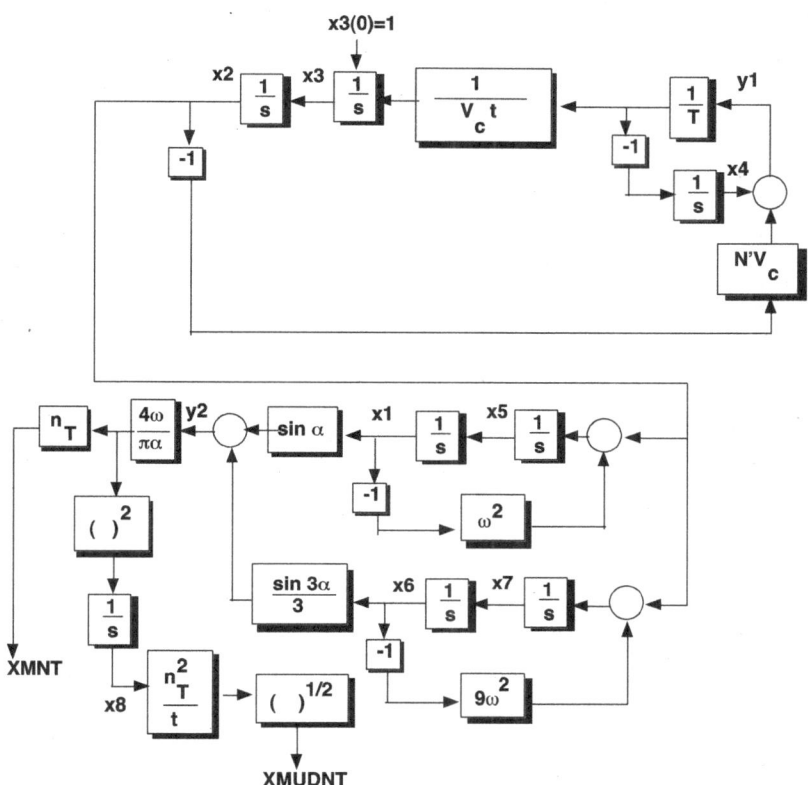

Fig. 15.6 **Adjoint diagram of single time constant guidance system with shaping filter equivalent of uniformly distributed vertical-S maneuver.**

Figure 15.5 represents the vertical-S maneuver as the shaping network derived from the two-term Fourier series previously presented. An adjoint diagram can be constructed from Fig. 15.5, using the rules of adjoints developed in Chapters 3 and 4 of the previous volume. The adjoint of the single time constant guidance system for a uniformly distributed vertical-S maneuver in a single time constant proportional navigation guidance system appears in Fig. 15.6. Listing 15.3 is the resultant adjoint simulation.

LISTING 15.3   ADJOINT SIMULATION OF SINGLE TIME CONSTANT GUIDANCE SYSTEM WITH SHAPING FILTER EQUIVALENT OF A UNIFORMLY DISTRIBUTED VERTICAL-S MANEUVER

```
clear
count=0;
```

```
XNT=161,;
XNP=3.;
TAU=.5;
TF=10.;
VM=2000.;
HEDEG=0.;
P=3.;
XL=P/2.;
TR=1.;
PI=3.1416;
ALF=PI*TR/(2.*XL);
W=2.*PI/P;
VC=3000.;
T=0.;
     S=0.
TP=T+.00001;
X1=0.;
X2=0;
X3=1.;
X4=0;
X5=0.;
X6=0.;
X7=0.;
X8=0.;
H=.01;
HE=HEDEG/57.3;
while TP<=(TF-1e-5)
  X1OLD=X1;
  X2OLD=X2;
  X3OLD=X3;
  X4OLD=X4;
  X5OLD=X5;
  X6OLD=X6;
  X7OLD=X7;
  X8OLD=X8;
  STEP=1;
  FLAG=0;
  while STEP<=1
    if FLAG==1
      STEP=2;
      X1=X1+H*X1D;
      X2=X2+H*X2D;
      X3=X3+H*X3D;
      X4=X4+H*X4D;
      X5=X5+H*X5D;
      X6=X6+H*X6D;
```

```
      X7=X7+H*X7D;
      X8=X8+H*X8D;
      TP=TP+H;
    end
    TGO=TP+.00001;
    X1D=X5;
    X2D=X3;
    Y1=X4-XNP*VC*X2;
    X3D=Y1/(VC*TGO*TAU);
    X4D=-Y1/TAU;
    X5D=X2-W*W*X1;
    X6D=X7;
    X7D=X2-9.*W*W*X6;
    Y2=X1*sin(ALF)+X6*sin(3.*ALF)/9.;
    PZ=4.*W*Y2/(PI*ALF);
    X8D=PZ^2;
    FLAG=1;
  end
  FLAG=0;
  X1=(X1OLD+X1)/2+.5*H*X1D;
  X2=(X2OLD+X2)/2+.5*H*X2D;
  X3=(X3OLD+X3)/2+.5*H*X3D;
  X4=(X4OLD+X4)/2+.5*H*X4D;
  X5=(X5OLD+X5)/2+.5*H*X5D;
  X6=(X6OLD+X6)/2+.5*H*X6D;
  X7=(X7OLD+X7)/2+.5*H*X7D;
  X8=(X8OLD+X8)/2+.5*H*X8D;
  S=S+H;
  if S>=.0999
    S=0.;
    count=count+1;
    XMNT=4.*W*XNT*Y2/(PI*ALF);
    XMUDNT=XNT*sqrt(X8/TGO);
    ArrayTP(count)=TP;
    ArrayXMNT(count)=XMNT;
    ArrayXMUDNT(count)=XMUDNT;
  end
end
figure
plot(ArrayTP,ArrayXMNT),grid
xlabel('Flight Time (Sec)')
ylabel('Target Maneuver Miss (Ft)')
figure
plot(ArrayTP,ArrayXMUDNT),grid
xlabel('Flight Time (Sec)')
ylabel('Random Target Maneuver Miss (Ft)')
```

```
clc
output=[ArrayTP',ArrayXMNT',ArrayXMUDNT'];
save datfil output -ascii.
disp 'simulation finished'
```

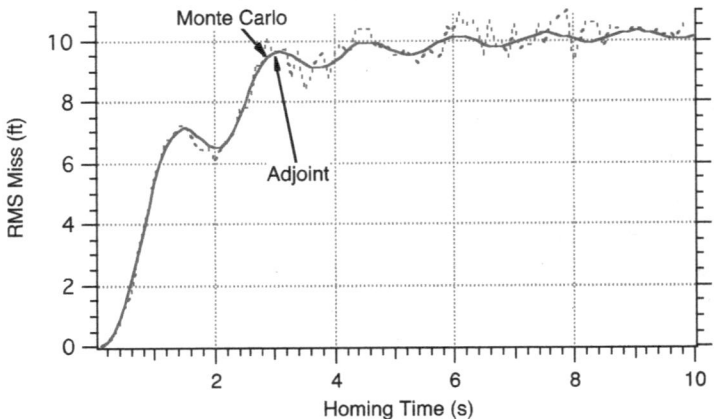

Fig. 15.7   Adjoint and Monte Carlo results agree for vertical-S maneuver with random starting time.

The nominal case of Listings 15.2 and 15.3 were run when the rise time was 1 s (TR=1) and the period was 3 s (PZ=3); the results appear in Fig. 15.7 for homing times ranging from 0.1 s to 10 s. We can see that the RMS miss distance performance of the10,000-run Monte Carlo set of Listing 15.2 is in complete agreement with the single-run adjoint results of Listing 15.3.

KALMAN FILTER FOR VERTICAL-S TARGET MANEUVER

Recall that the three-state Kalman filter, which was optimal against a random step target maneuver, was derived in Chapter 9 of the previous volume. A four-state Kalman filter, which was optimal against a weaving target maneuver, was derived in Chapter 7. In this section we shall derive a six-state Kalman filter that is optimal against a random vertical-S target maneuver.

The homing loop model, upon which the vertical-S target maneuver Kalman filter will be based, is depicted in Fig. 15.8. Here we can see that the vertical-S maneuver with starting time uniformly distributed over the flight time is represented by white noise through the two-term Fourier series shaping network approximation.

From Fig. 15.8 we can begin to write the differential equations describing the system. The four differential equations describing the uniformly distributed vertical-S target maneuver can be written by inspection of Fig. 15.8 and are

$$\dot{x}_1 = x_2$$

$$\dot{x}_2 = -\omega^2 x_1 + \frac{4\omega \sin \alpha}{\pi \alpha} u_S$$

$$\dot{x}_3 = x_4$$

$$\dot{x}_4 = -9\omega^2 x_3 + \frac{4\omega \sin 3\alpha}{3\pi \alpha} u_S$$

where u_s is white noise with power spectral density

$$\Phi_s = \frac{n_T^2}{t_F}$$

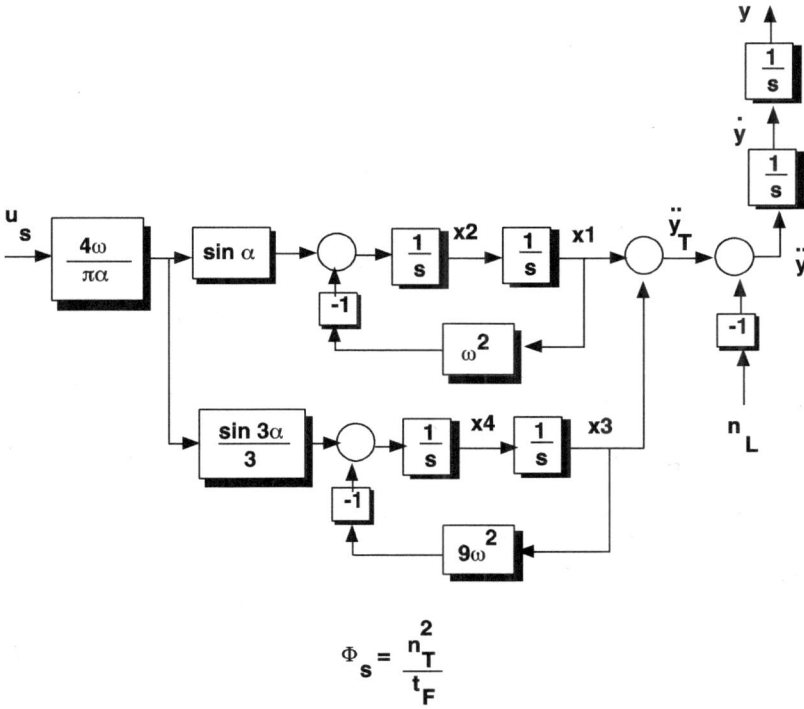

Fig. 15.8 Model for vertical-S maneuver in homing loop.

From the preceding equations and Fig. 15.8 we can write the six state space equations describing the entire homing loop in matrix form as

$$
\begin{bmatrix} \dot{y} \\ \ddot{y} \\ \dot{x}_1 \\ \dot{x}_2 \\ \dot{x}_3 \\ \dot{x}_4 \end{bmatrix} = \begin{bmatrix} 0 & 1 & 0 & 0 & 0 & 0 \\ 0 & 0 & 1 & 0 & 1 & 0 \\ 0 & 0 & 0 & 1 & 0 & 0 \\ 0 & 0 & -\omega^2 & 0 & 0 & 0 \\ 0 & 0 & 0 & 0 & 0 & 1 \\ 0 & 0 & 0 & 0 & -9\omega^2 & 0 \end{bmatrix} \begin{bmatrix} y \\ \dot{y} \\ x_1 \\ x_2 \\ x_3 \\ x_4 \end{bmatrix} + \begin{bmatrix} 0 \\ -1 \\ 0 \\ 0 \\ 0 \\ 0 \end{bmatrix} n_L + \begin{bmatrix} 0 \\ 0 \\ 0 \\ \dfrac{4\omega \sin \alpha}{\pi \alpha} \\ 0 \\ \dfrac{4\omega \sin 3\alpha}{3\pi \alpha} \end{bmatrix} u_s
$$

The continuous process noise matrix can be computed from the last term of the preceding equation as

$$
Q_c = E \left(\begin{bmatrix} 0 \\ 0 \\ 0 \\ \dfrac{4\omega \sin \alpha}{\pi \alpha} \\ 0 \\ \dfrac{4\omega \sin 3\alpha}{3\pi \alpha} \end{bmatrix} u_s \begin{bmatrix} 0 \\ 0 \\ 0 \\ \dfrac{4\omega \sin \alpha}{\pi \alpha} \\ 0 \\ \dfrac{4\omega \sin 3\alpha}{3\pi \alpha} \end{bmatrix}^T u_s^T \right) = \Phi_s \begin{bmatrix} 0 & 0 & 0 & 0 & 0 & 0 \\ 0 & 0 & 0 & 0 & 0 & 0 \\ 0 & 0 & 0 & 0 & 0 & 0 \\ 0 & 0 & 0 & a^2 & 0 & ab \\ 0 & 0 & 0 & 0 & 0 & 0 \\ 0 & 0 & 0 & ab & 0 & b^2 \end{bmatrix}
$$

where

$$
a = \frac{4\omega \sin \alpha}{\pi \alpha}
$$

$$
b = \frac{4\omega \sin 3\alpha}{3\pi \alpha}
$$

$$
\Phi_s = \frac{n_T^2}{t_F}
$$

The measurement equation is linearly related to the states and is simply given by

$$
z = \begin{bmatrix} 1 & 0 & 0 & 0 & 0 & 0 \end{bmatrix} \begin{bmatrix} y \\ \dot{y} \\ x_1 \\ x_2 \\ x_3 \\ x_4 \end{bmatrix} + v
$$

Listing 15.4 is a Monte Carlo simulation of the homing loop with the six-state vertical-S Kalman filter. With this program we can compute the RMS miss distance performance of the six-state Kalman filter in the homing loop. In the program are options for using the random step target maneuver (QMVR=1), the random weave target maneuver (QMVR=2), and the random vertical-S target maneuver (QMVR=3). *It is important to note that in Listing 15.4 the actual target maneuvers are modeled, not their shaping filter equivalents.* From the listing we can see that the single-lag time constant representation of the flight-control system is quite small (TAU=.05), and therefore the augmented proportional guidance law can be used to provide reasonable RMS miss distances. The nominal value of the measurement noise is 0.1 mrad (SIGRIN =.0001), and the sampling time is 0.01 s (TS=.01). Equivalent Monte Carlo simulations with Kalman filters in the single time constant homing loop were set up for the three different random target maneuvers. The simulation based on the three-state Kalman filter of Listing 9.3 of Chapter 9 of the previous volume, which is optimal against a random step maneuver, is used in the comparison that follows in this section. In addition, the linear four-state Kalman filter of Listing 7.2 of Chapter 7, which is optimal against a weaving target if the weave frequency of the target is known, is also used in the comparison that follows in this section.

LISTING 15.4   SIX-STATE VERTICAL-S TARGET MANEUVER KALMAN FILTER IN HOMING LOOP

```
count=0;
RUN=100;
TAU=.05;
ORDER=6;
TSTART=0.;
PZ=3.;
TR=.5;
XL=PZ/2.;
PI=3.14159;
W=2.*PI/PZ;
WH=W;
WREAL=W;
VC=2.5*3280.;
XNT=322.;
XNTREAL=XNT;
TS=.01;
VM=2.*3280.;
HEDEGFIL=20.;
XNP=3.;
SIGRIN=.0001;
SIGGL=0.;
RA=21000.;
SRN=0.;
TF=10.;
QPERFECT=0;
```

```
PHASE=0./57.3;
X=WH*TS;
Y=0;
YD=0.;
XLIM=1288.;

      TSTART=0.;
      XNTIC=XNT;
      XL=PZ/2.;
      X=PZ/2.-TR;
      PI=3.1416;
      ALF=PI*TR/(2.*XL);
      W=2.*PI/PZ;
      PHIS=XNT*XNT/6.;
      RTM=VC*TF;
      SIGNOISE=sqrt(SIGRIN^2+(SIGGL/RTM)^2+(SRN*RTM*RTM/(RA*RA))^2);
      SIGPOS=RTM*SIGNOISE;
      SIGN2=SIGPOS^2;
      PHI=zeros(ORDER);
      P=zeros(ORDER);
      QC=zeros(ORDER);
      Q=zeros(ORDER);
      F=zeros(ORDER);
      IDNP=eye(ORDER);
      F(1,2)=1.;
      F(2,3)=1.;
      F(2,5)=1.;
      F(3,4)=1.;
      F(4,3)=-W*W;
      F(5,6)=1.;
      F(6,5)=-9.*W*W;
      PHI=IDNP+F*TS+.5*F*F*TS*TS;
      TMPA=4.*W*sin(ALF)/(PI*ALF);
      TMPB=4.*W*sin(3*ALF)/(3.*PI*ALF);
      QC(4,4)=PHIS*TMPA*TMPA;
      QC(4,6)=PHIS*TMPA*TMPB;
      QC(6,4)=PHIS*TMPA*TMPB;
      QC(6,6)=PHIS*TMPB*TMPB;
      Q=PHI*QC*PHI'*TS;
      P(1,1)=SIGN2;
      P(2,2)=(VM*HEDEGFIL/57.3)^2;
      P(3,3)=322.^2;
      P(4,4)=(W*322.)^2;
      P(5,5)=(322.)^2;
      P(6,6)=(W*322.)^2;
      HMAT=[1 0 0 0 0 0];
```

```
T=0.;
H=.001;
S=0.;
XNC=0.;
XNL=0.;
XLAM=Y/RTM;
YTDD=0;
YTDDD=0;
YH=0.;
YDH=0.;
X1H=0.;
X2H=0.;
X3H=0.;
X4H=0.;
Y=0.;
YD=0.;
while T<=(TF-.0001)
  YOLD=Y;
  YDOLD=YD;
  XNLOLD=XNL;
  STEP=1;
  FLAG=0;
  while STEP <=1
    if FLAG==1
      STEP=2;
      Y=Y+H*YD;
      YD=YD+H*YDD;
      XNL=XNL+H*XNLD;
      T=T+H;
    end
    TGO=TF-T+.000001;
    RTM=VC*TGO;
    XLAM=Y/(VC*TGO);
    if T<TSTART
      YTDD=0.;
    elseif (T-TSTART)<PZ
      if (T-TSTART)<TR/2.
        YTDD=2.*XNT*(T-TSTART)/TR;
      elseif (T-TSTART)<(TR/2.+X)
        YTDD=XNT;
      elseif (T-TSTART)<(3.*TR/2.+X)
        YTDD=-2.*XNT*(T-TSTART)/TR+2.*XNT+2.*XNT*X/TR;
      elseif (T-TSTART)<(3.*TR/2.+2.*X)
        YTDD=-XNT;
      else
        YTDD=2.*XNT*(T-TSTART)/TR-4.*XNT-4.*XNT*X/TR;
```

```
      end
    elseif (T-TSTART)<2.*PZ
      TSTAR=(T-TSTART)-PZ;
      if TSTAR<TR/2.
         YTDD=2.*XNT*TSTAR/TR;
      elseif TSTAR<(TR/2.+X)
         YTDD=XNT;
      elseif TSTAR<(3.*TR/2.+X)
         YTDD=-2.*XNT*TSTAR/TR+2.*XNT+2.*XNT*X/TR;
      elseif TSTAR<(3.*TR/2.+2.*X)
         YTDD=-XNT;
      else
         YTDD=2.*XNT*TSTAR/TR-4.*XNT-4.*XNT*X/TR;
      end
    elseif (T-TSTART)<3.*PZ
      TSTAR=(T-TSTART)-2.*PZ;
      if TSTAR<TR/2.
         YTDD=2.*XNT*TSTAR/TR;
      elseif TSTAR<(TR/2.+X)
         YTDD=XNT;
      elseif TSTAR<(3.*TR/2.+X)
         YTDD=-2.*XNT*TSTAR/TR+2.*XNT+2.*XNT*X/TR;
      elseif TSTAR<(3.*TR/2.+2.*X)
         YTDD=-XNT;
      else
         YTDD=2.*XNT*TSTAR/TR-4.*XNT-4.*XNT*X/TR;
      end
    else
      TSTAR=(T-TSTART)-3.*PZ;
      if TSTAR<TR/2.
         YTDD=2.*XNT*TSTAR/TR;
      elseif TSTAR<(TR/2.+X)
         YTDD=XNT;
      elseif TSTAR<(3.*TR/2.+X)
         YTDD=-2.*XNT*TSTAR/TR+2.*XNT+2.*XNT*X/TR;
      elseif TSTAR<(3.*TR/2.+2.*X)
         YTDD=-XNT;
      else
         YTDD=2.*XNT*TSTAR/TR-4.*XNT-4.*XNT*X/TR;
      end
    end
    XNLD=(XNC-XNL)/TAU;
    YDD=YTDD-XNL;
    FLAG=1;
  end
```

```
        FLAG=0;
        Y=.5*(YOLD+Y+H*YD);
        YD=.5*(YDOLD+YD+H*YDD);
        XNL=.5*(XNLOLD+XNL+H*XNLD);
        S=S+H;
        if S>=(TS-.00001)
          S=0.;
          TGO=TF-T+.000001;
          RTM=VC*TGO;
          SIGNOISE=sqrt(SIGRIN^2+(SIGGL/RTM)^2+(SRN*RTM*RTM/(RA*RA))^2);
          SIGPOS=RTM*SIGNOISE;
          SIGN2=SIGPOS^2;
          RMAT=[SIGN2];
          M=PHI*P*PHI'+Q;
          K = M*HMAT'/(HMAT*M*HMAT' + RMAT);
          P = (IDNP - K*HMAT)*M;
          [YB,YDB,X1B,X2B,X3B,X4B]=PROJECT6S(T,TS,YH,YDH,XNL,...
             X1H,X2H,X3H,X4H,W,H);
          XLAMNOISE=SIGNOISE*randn;
          YSTAR=RTM*(XLAM+XLAMNOISE);
          RES(1,1)=YSTAR-YB;
          YH=YB+K(1,1)*RES(1,1);
          YDH=YDB+K(2,1)*RES(1,1);
          X1H=X1B+K(3,1)*RES(1,1);
          X2H=X2B+K(4,1)*RES(1,1);
          X3H=X3B+K(5,1)*RES(1,1);
          X4H=X4B+K(6,1)*RES(1,1);
          XNTH=X1H+X3H;
          XNC=XNP*(YH+YDH*TGO+.5*XNTH*TGO*TGO)/(TGO*TGO);
          if XNC>XLIM
             XNC=XLIM;
          end
          if XNC<-XLIM
             XNC=-XLIM;
          end
          count=count+1;
          ArrayT(count)=T;
          ArrayYTDD(count)=YTDD;
          ArrayXNTH(count)=XNTH;
        end
      end
figure
plot(ArrayT',ArrayYTDD',ArrayT',ArrayXNTH'),grid
title('RMS miss for various flight times')
xlabel('Flight Time (S)')
```

```
ylabel('RMS MISS (Ft) ')
clc
output=[ArrayT',ArrayYTDD',ArrayXNTH'];
save datfil.txt output -ascii
disp('Simulation Complete')

function [YH,YDH,X1H,X2H,X3H,X4H]=project6S(TP,TS,YP,YDP,XNLP,X1HP,X2HP,X3HP,
X4HP,W,HP)
T=0.;
Y=YP;
YD=YDP;
X1=X1HP;
X2=X2HP;
X3=X3HP;
X4=X4HP;
XNL=XNLP;
H=HP;
while T<=(TS-.0001)
   XNT=X1+X3;
   YDD=XNT-XNL;
   YD=YD+H*YDD;
   Y=Y+H*YD;
   X1D=X2;
   X1=X1+H*X1D;
   X2D=-W*W*X1;
   X2=X2+H*X2D;
   X3D=X4;
   X3=X3+H*X3D;
   X4D=-9.*W*W*X3;
   X4=X4+H*X4D;
      T=T+H;
end
YH=Y;
YDH=YD;
X1H=X1;
X2H=X2;
X3H=X3;
X4H=X4;
```

All three filters were run against the vertical-S target maneuver in a single flight mode. We can see from Fig. 15.9 that the performance of the three-state Kalman filter yields slightly noisy estimates that significantly lag the actual target maneuver. Figure 15.10 shows that the performance of the four-state weave Kalman filter with the weave frequency matching the period P of the

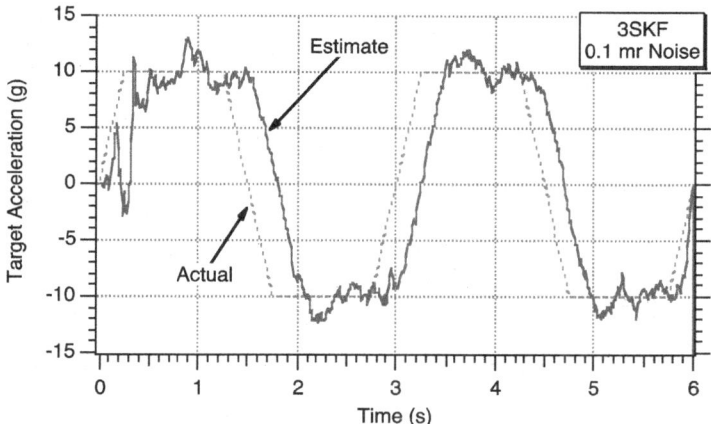

Fig. 15.9   Three-state Kalman filter gives reasonable estimates of vertical-S maneuver but with a large lag.

vertical-S maneuver according to

$$\omega = \frac{2\pi}{P}$$

gives smoother estimates than the three-state filter but is not able to follow the shape of the vertical-S maneuver very well. Finally, from Fig. 15.11 we can see

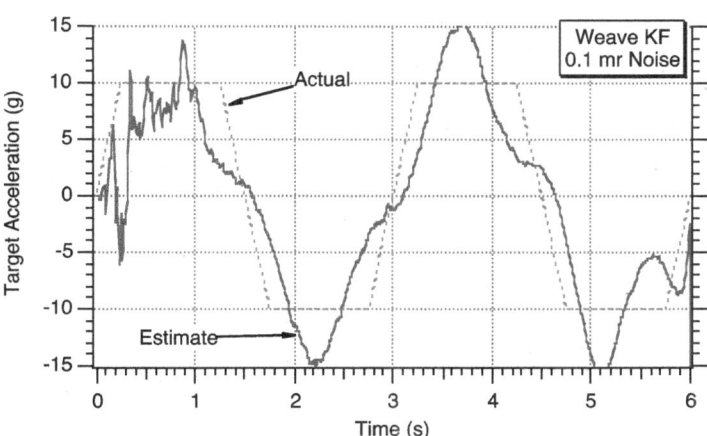

Fig. 15.10   Four-state Kalman filter estimates vertical-S maneuver with less lag but more overshoot.

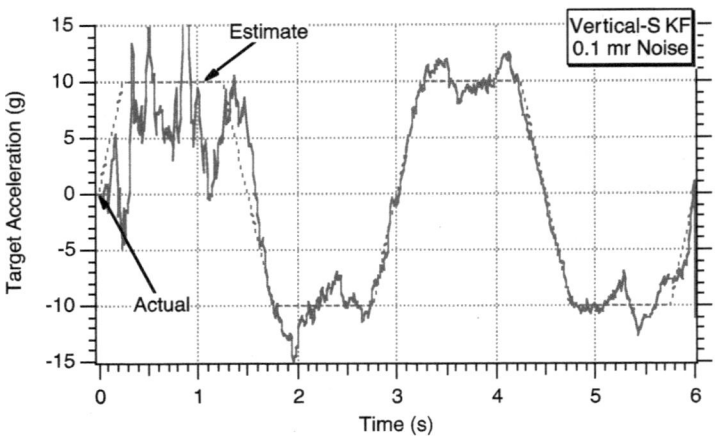

Fig. 15.11   Six-state Kalman filter gives best estimates of vertical-S maneuver.

that, as expected, the six-state vertical-S Kalman filter yields the best target maneuver estimates because the six-state Kalman filter was tuned for the vertical-S maneuver].

All three simulations (Listing 9.2 of the previous volume, Listing 7.2 of this volume, and Listing 15.4) were run in the Monte Carlo mode for the case where the vertical-S maneuver level is 10 g while the missile acceleration limit is 20 g. Figure 15.12 shows that when the rise time of the vertical-S target maneuver is 0.01 s (i.e., nearly a square wave), the six-state filter performs better than either the three-state Kalman filter (i.e., tuned for a random step maneuver) or

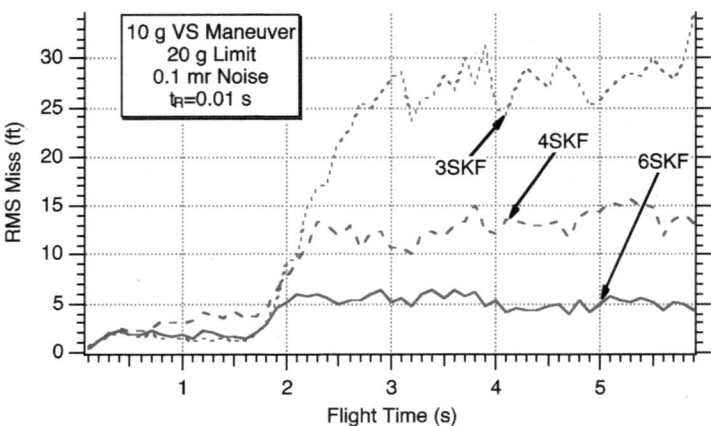

Fig. 15.12   Six-state Kalman filter yields best miss distance performance against vertical-S maneuver with rise time of 0.01 s.

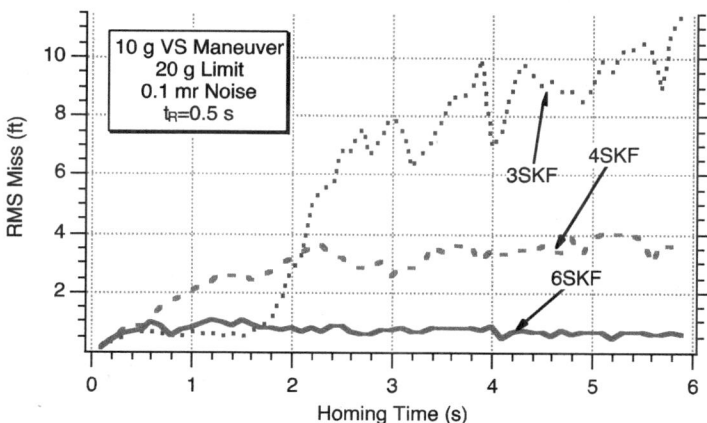

Fig. 15.13   Six-state Kalman filter yields best miss distance performance against vertical-S maneuver with rise time of 0.5 s.

the four-state Kalman filter (i.e., tuned for a sinusoidal maneuver). Figure 15.13 shows that when the vertical-S rise time is increased to 0.5 s, the RMS miss distances from all three filters is significantly reduced, but the six-state Kalman filter results in the smallest RMS miss distances.

Because the six-state Kalman filter is tuned for the vertical-S target maneuver, the results of Figs. 15.11–15.13 are not surprising. An experiment was conducted to see how all three filters performed when the actual target maneuver was a random step or random weave. Figures 15.14 and 15.15 show that the six-state

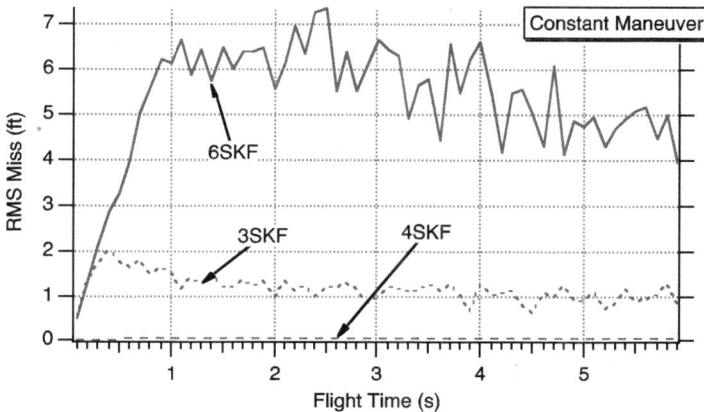

Fig. 15.14   Six-state Kalman filter performs worse than the other Kalman filters when the target maneuver is a random constant.

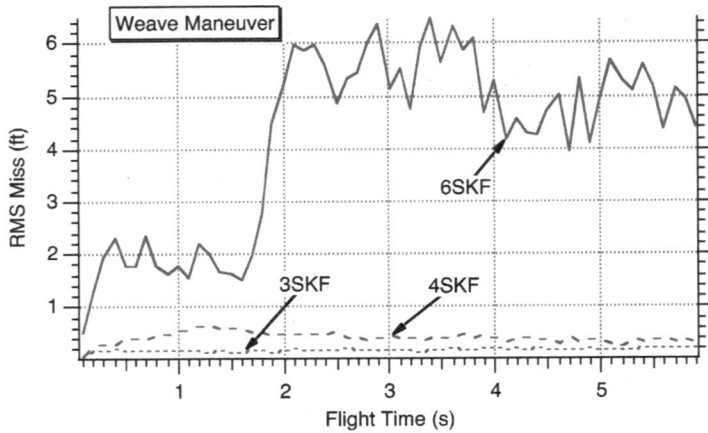

Fig. 15.15   Six-state Kalman filter performs worse than the other Kalman filters when the target maneuver is a random weave.

vertical-S Kalman filter performs worse than the three-state and four-state Kalman filters under these circumstances. These results indicate that some sort of filter bank scheme, where each filter in the bank is tuned to a different type of target maneuver, would be required to make the six-state vertical-S Kalman filter more practical.

SUMMARY

This chapter derived a two-term Fourier series approximation to represent the vertical-S target maneuver. A six-state Kalman filter was derived based on the approximate Fourier series representation of the vertical-S target maneuver. It was shown that when the actual target maneuver was a random vertical-S the six-state Kalman filter outperformed both the three-state Kalman filter (derived assuming a random constant target maneuver) and the four-state weave Kalman filter. However, if the actual target maneuver was either a random constant or random weave, the six-state Kalman filter did not perform as well as the three- and four-state Kalman filters.

REFERENCES

[1]   Zarchan, P., "Representation of Realistic Evasive Maneuvers by the Use of Shaping Filters," *Journal of Guidance and Control*, Vol. 2, 1979, pp. 290–295.

[2]   Selby, S. M., *CRC Standard Mathematical Tables*, 20th ed., Chemical Rubber Company, Cleveland, OH, July 1972, p. 464.

CHAPTER 16

Improving the Performance of a Tail-Controlled Missile at High Altitudes

INTRODUCTION

We saw in Chapter 12 that at high altitudes the right-half plane zero of the airframe transfer function of a tail-controlled missile can get small enough to cause severe miss distance problems. It was demonstrated in Chapter 12 that if we took into account an accurate representation of the flight-control system transfer function in the derivation of the guidance law, the miss distance performance could be improved significantly. The resultant guidance law was not derived in closed form, but instead, the control gains for the guidance law were computed offline and the resultant control gains were to be stored in a lookup table as a function of time to go and flight condition. In this chapter we shall work with a simplified model of the flight-control system that includes the airframe right-half plane zero. A new guidance law will be presented in closed form and evaluated to see if performance can be improved at high altitudes where the importance of the right-half plane zero is critical.

ZERO OVER POLE GUIDANCE LAW

We saw in Chapter 23 of the previous volume that a realistic model for the flight-control system of a tail-controlled missile is given by the transfer function

$$\frac{n_L}{n_c} = \frac{\left(1 - \dfrac{s^2}{\omega_z^2}\right)}{\left[(1 + s\tau)\left(1 + \dfrac{2\zeta}{\omega}s + \dfrac{s^2}{\omega^2}\right)\right]}$$

where τ is the dominant time constant, ζ is the damping, ω is the frequency of the flight-control system, and ω_z is the right-half plane zero of the airframe. The preceding transfer function models the "wrong-way" tail effect with left- and right-half plane zeros of the airframe ω_z. In this chapter we consider a simplified version of the preceding flight-control system transfer function that is given by

$$\frac{n_L}{n_c} = \frac{1 - \dfrac{s}{\omega_z}}{1 + sT}$$

where T is the dominant time constant of the flight-control system, ω_z is the right-half plane zero of the airframe, n_L is the achieved missile acceleration, and n_c is the commanded missile acceleration. Based on the preceding model, a guidance law can be derived that minimizes the integral of the acceleration squared and drives the miss to zero. The resultant guidance law is called the "zero over pole" guidance law [1, 2] and is given by

$$n_c = \frac{N'}{t_{go}^2}\left[y + \dot{y}t_{go} + 0.5n_T t_{go}^2 - e_{PZ}T^2(e^{-x} + x - 1)\right]$$

where

$$e_{PZ} = n_L + \frac{n_c}{T\omega_z}$$

and

$$x = \frac{t_{go}}{T}$$

The expression for the effective navigation ratio is

$$N' = \frac{-\dfrac{t_{go}^3}{T\omega_z} + \left(1 + \dfrac{T}{\omega_z}\right)t_{go}^2(e^{-x} + x - 1)T}{y + \dfrac{t_{go}^3}{3} + \left(1 + \dfrac{1}{T\omega_z}\right)T^3\left[0.5\left(1 - \dfrac{3}{T\omega_z}\right) + x\left(1 + \dfrac{1}{T\omega_z}\right) - x^2 - 2xe^{-x} + \dfrac{2e^{-x}}{T\omega_z} - 0.5e^{-2x}\left(1 + \dfrac{1}{T\omega_z}\right)\right]}$$

The derivation of the guidance law has been skipped because it is too complex and lengthy for this text. However, in the next section we will demonstrate, via computerized optimal control techniques from Chapter 12, that the zero over pole guidance law is correct. However, first, let us see what happens to the guidance law as the airframe zero approaches infinity.

When the airframe zero approaches infinity, the effective navigation of the preceding expression reduces to

$$\lim N'_{\omega_z \to \infty} = \frac{t_{go}^2(e^{-x} + x - 1)T}{y + \dfrac{t_{go}^3}{3} + T^3\left[0.5 + x - x^2 - 2xe^{-x} - 0.5e^{-2x}\right]}$$

If we let γ approach zero in the preceding expression, we obtain

$$\lim_{\substack{\omega_z \to \infty \\ \gamma \to 0}} N' = \frac{t_{go}^2(e^{-x} + x - 1)T}{\dfrac{t_{go}^3}{3} + T^3\left[0.5 + x - x^2 - 2xe^{-x} - 0.5e^{-2x}\right]}$$

which is also the expression for the effective navigation ratio for the optimal guidance law when the flight-control system was represented by a single time constant (see Chapter 8 of the previous volume). In addition, because

$$e_{PZ} = n_L + \frac{n_c}{T\omega_z}$$

we get

$$\lim e_{PZ_{\omega_z \to \infty}} = n_L$$

Therefore, we can begin to have confidence that the zero over pole guidance law appears to be reasonable.

INTEGRATING OPTIMAL CONTROL EQUATION TO CONFIRM ZERO OVER POLE GUIDANCE LAW

The zero over pole representation of the flight-control system can be represented by the block diagram of Fig. 16.1.

In order to see if Fig. 16.1 really does represent the zero over pole flight-control system, we can see from the figure that

$$se_{PZ} = \frac{1}{T}\left[-e_{PZ} + \left(\frac{1}{T\omega_z} + 1\right)n_c\right]$$

$$n_L = e_{PZ} - \frac{n_c}{T\omega_z}$$

Working on the first equation of the preceding set, we obtain

$$e_{PZ}\left(s + \frac{1}{T}\right) = \frac{1}{T}\left(1 + \frac{1}{T\omega_z}\right)n_c$$

which can be expressed as

$$e_{PZ} = \frac{\dfrac{1}{T}\left(1 + \dfrac{1}{T\omega_z}\right)n_c}{s + \dfrac{1}{T}}$$

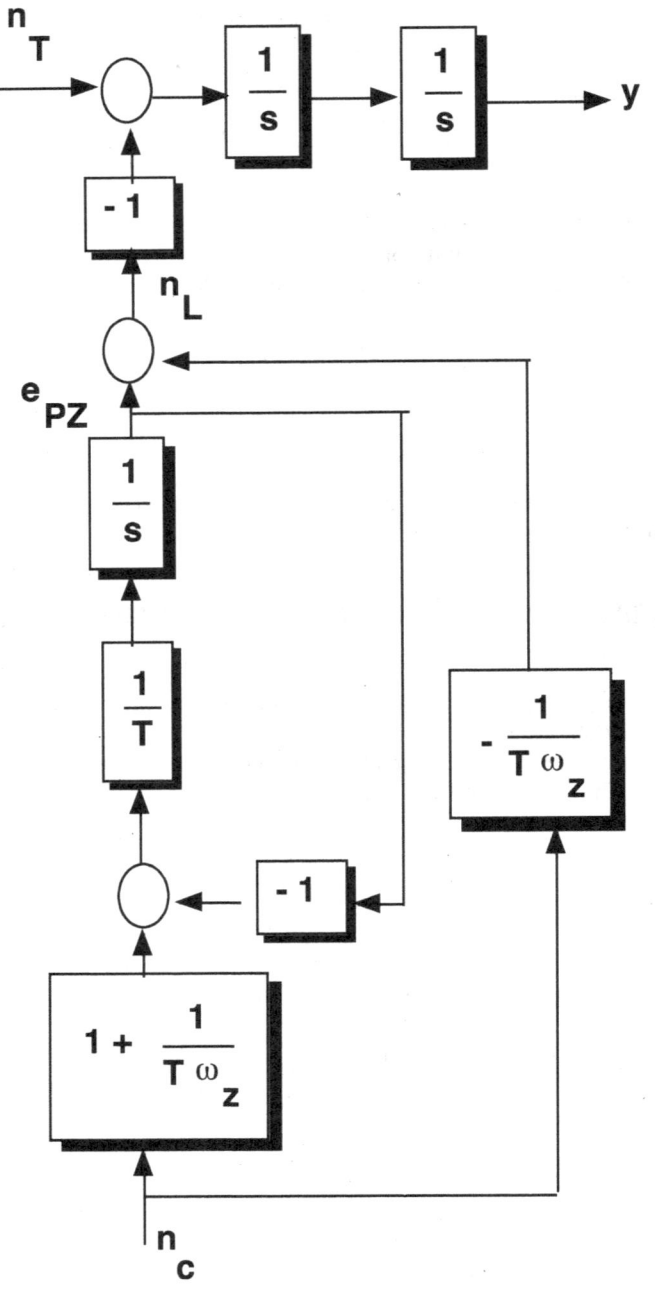

Fig. 16.1　Model of homing loop for derivation of zero over pole guidance law.

Therefore, by substituting the preceding equation into the achieved acceleration equation we obtain

$$n_L = \frac{\frac{1}{T}\left(1 + \frac{1}{T\omega_z}\right)n_c}{s + \frac{1}{T}} - \frac{n_c}{T\omega_z}$$

which yields the desired transfer function

$$\frac{n_L}{n_c} = \frac{1 - s/\omega_z}{1 + sT}$$

From the Laplace transform representation of e_{pz} we can see that the differential equation describing the zero over pole autopilot is given by

$$\dot{e}_{PZ} = \frac{1}{T}\left[-e_{PZ} + \left(\frac{1}{T\omega_z} + 1\right)n_c\right]$$

We can put the preceding differential equation and the remainder of the homing loop of Fig. 16.1 into state space form as

$$\begin{bmatrix} \dot{y} \\ \ddot{y} \\ \dot{n}_T \\ \dot{e}_{PZ} \end{bmatrix} = \begin{bmatrix} 0 & 1 & 0 & 0 \\ 0 & 0 & 1 & -1 \\ 0 & 0 & 0 & 0 \\ 0 & 0 & 0 & \frac{-1}{T} \end{bmatrix} \begin{bmatrix} y \\ \dot{y} \\ n_T \\ e_{PZ} \end{bmatrix} + \begin{bmatrix} 0 \\ 1 \\ \frac{1}{T\omega_z} \\ 0 \\ \frac{1}{T}\left(1 + \frac{1}{T\omega_z}\right) \end{bmatrix} n_c$$

As was previously mentioned in Chapter 12, optimal control theory states that guidance laws can be derived either analytically or numerically if our model of the real world can be expressed in state space form as

$$\dot{\mathbf{x}} = \mathbf{Fx} + \mathbf{Gu}$$

where the control $\mathbf{u}$ (guidance command) is linearly related to the states $\mathbf{x}$ according to

$$\mathbf{u} = -\mathbf{Cx}$$

In the preceding equation, $\mathbf{C}$ is a set of control gains.

With standard optimal control techniques, we minimize the performance index J, which is given by

$$J = \mathbf{x}^T(t_F)\mathbf{S}_F\mathbf{x}(t_F) + \int_0^{t_F} \mathbf{u}^T\mathbf{Bu}\, dt$$

where $\mathbf{x}(t_F)$ is the state vector at the final time. The miss $y(t_F)$ is one element of that state vector. By judiciously choosing $\mathbf{S}_F$, we can make the first term of J equal to the miss distance squared $[y^2(t_F)]$. In addition, by making **B** a small scalar γ, we can come close to our desire of making the miss zero, subject to minimizing the integral of the acceleration squared.

Optimal control theory shows that minimizing J for our state space model of the real world yields a guidance law in which the control gains are linearly related to the states. Obtaining those control gains involves solving the nonlinear matrix Riccati differential equation

$$\dot{\mathbf{S}} = -\mathbf{F}^T\mathbf{S} - \mathbf{S}\mathbf{F} + \mathbf{C}^T\mathbf{B}\mathbf{C}$$

with boundary value

$$\mathbf{S}(t_F) = \mathbf{S}_F$$

The control gains for the optimal guidance law are related to the solution of the preceding differential equation for S and can be obtained from

$$\mathbf{C} = \mathbf{B}^{-1}\mathbf{G}^T\mathbf{S}$$

Note that the preceding equations are based on solving a boundary value problem. However, we can convert the nonlinear matrix differential equation with a boundary value to one with an initial condition. Recall that from the overdot notation we know that

$$\dot{\mathbf{S}} = \frac{d\mathbf{S}}{dt}$$

Therefore, if we define

$$\tau = t_F - t$$

we can say that

$$\frac{d\mathbf{S}}{dt} = -\frac{d\mathbf{S}}{d\tau}$$

Therefore, by changing variables, the nonlinear matrix differential equation becomes

$$\dot{\mathbf{S}} = \mathbf{F}^T\mathbf{S} + \mathbf{S}\mathbf{F} - \mathbf{C}^T\mathbf{B}\mathbf{C}$$

with initial value

$$\mathbf{S}(0) = \mathbf{S}_F$$

Because we have changed variables, **S** is now a function of τ or time to go rather than t or time. The control gains for the optimal guidance law are still obtained from

$$\mathbf{C} = \mathbf{B}^{-1}\mathbf{G}^T\mathbf{S}$$

but again the gains are now functions of time to go rather than functions of time. A very small integration step size is often required for the successful integration of the Riccati equations because of their numerical fragility.

The performance index to be minimized is

$$J = y^2(t_F) + \gamma \int_0^{t_F} n_c^2 dt$$

From our state space representation of Fig. 16.1 and from the preceding perform-ance index, the important matrices for our problem are given by

$$F = \begin{bmatrix} 0 & 1 & 0 & 0 \\ 0 & 0 & 1 & -1 \\ 0 & 0 & 0 & 0 \\ 0 & 0 & 0 & \frac{-1}{T} \end{bmatrix} G = \begin{bmatrix} 0 \\ 1 \\ T\omega_z \\ 0 \\ \frac{1}{T}\left(1 + \frac{1}{T\omega_z}\right) \end{bmatrix}$$

$$S_F = \begin{bmatrix} 1 & 0 & 0 & 0 \\ 0 & 0 & 0 & 0 \\ 0 & 0 & 0 & 0 \\ 0 & 0 & 0 & 0 \end{bmatrix}$$

$$\mathbf{u} = n_c \quad \text{and} \quad \mathbf{B} = \gamma$$

These matrices will be used in the numerical derivation of the zero over pole gui-dance law. In order to compare the numerical solution of the guidance law to the closed-form solution that was presented at the beginning of this chapter, we have to change the form of the closed-form solution. Because the closed-form solution was given by

$$n_c = \frac{N'}{t_{go}^2}\left[y + \dot{y}t_{go} + 0.5n_T t_{go}^2 - e_{PZ}T^2(e^{-x} + x + 1)\right]$$

and we want the guidance law to look like

$$n_c = C_1 y + C_2 \dot{y} + C_3 n_T + C_4 e_{PZ}$$

we can say that

$$C_1 = \frac{N'}{t_{go}^2}$$

$$C_2 = \frac{N'}{t_{go}}$$

$$C_3 = 0.5N'$$

$$C_4 = \frac{-T^2(e^{-x} + x - 1)N'}{t_{go}^2} = \frac{-N'(e^{-x} + x - 1)}{x^2}$$

where the effective navigation ratio is

$$N' = \frac{-\dfrac{t_{go}^3}{T\omega_z} + \left(1 + \dfrac{T}{\omega_z}\right)t_{go}^2(e^{-x} + x - 1)T}{\gamma + \dfrac{t_{go}^3}{3} + \left(1 + \dfrac{1}{T\omega_z}\right)T^3\left[0.5\left(1 - \dfrac{3}{T\omega_z}\right) + x\left(1 + \dfrac{1}{T\omega_z}\right) - x^2 - 2xe^{-x} + \dfrac{2e^{-x}}{T\omega_z} - 0.5e^{-2x}\left(1 + \dfrac{1}{T\omega_z}\right)\right]}$$

Listing 16.1 is modeled after programs in Chapter 12 that solve for the optimal control gains. In this program the computed control gains are compared to the theoretical control gains that were stated without proof to check their veracity.

LISTING 16.1   SOLVING FOR OPTIMAL CONTROL GAINS FOR ZERO OVER POLE AUTOPILOT

```
clear
TAU=.5;
GAM=.00001;
WZ=10;
TF=10;
F=zeros([4,4]);
S=zeros([4,4]);
count=0;
G(2,1)=1./(TAU*WZ);
G(4,1)=(1.+1./(TAU*WZ))/TAU;
F(1,2)=1;
F(2,3)=1;
F(2,4)=-1;
F(4,4)=-1./TAU;
S(1,1)=1;
T=0;
H=.0001;
S1=0;
while ~(T >= (TF-.0001))
       S1=S1+H;
       SOLD=S;
       STEP=1;
       FLAG=0;
       while STEP<=1
         if FLAG==1
         STEP=2;
         HSD=H*SD;
```

```
        S=S+HSD;
                T=T+H;
        end
        SF=S*F;
        GT=G';
        GTS=GT*S;
        GAMINV=1./GAM;
        C=GAMINV*GTS;
        SFT=SF';
        CT=C';
        CTC=CT*C;
        CTBC=GAM*CTC;
        SFSFT=SF+SFT;
        SD=SFSFT-CTBC;
        FLAG=1;
    end
    FLAG=0;
    H2=.5*H;
    HSDP=H2*SD;
    SS=SOLD+S;
    SSP=.5*SS;
    S=SSP+HSDP;
    if S1>=.009999
        S1=0;
        C1=-C(1,1);
        C2=-C(1,2);
        C3=-C(1,3);
        C4=-C(1,4);
        NP=C2*T;
        XS=T/TAU;
    TEMP1=T*T*TAU*(exp(-XS)-1.+XS);
    TOP=-(T^3)/(TAU*WZ)+(1.+1./(TAU*WZ))*TEMP1;
    TEMP2=.5*(1.-3./(TAU*WZ))+XS*(1.+1./(TAU*WZ))-XS*XS;
    TEMP3=-2.*XS*exp(-XS);
    TEMP4=2.*exp(-XS)/(TAU*WZ)-.5*exp(-2.*XS)*(1.+1./(TAU*WZ));
    BOT=GAM+T*T*T/3.+(1.+1./(TAU*WZ))*TAU^3*(TEMP2+TEMP3+TEMP4);
    XNPP=TOP/BOT;
        C1TH=XNPP/(T*T);
        C2TH=XNPP/T;
        C3TH=.5*XNPP;
        C4TH=-XNPP*(exp(-XS)+XS-1.)/(XS*XS);
        count=count+1;
        ArrayT(count)=T;
        ArrayC1(count)=C1;
        ArrayC1TH(count)=C1TH;
        ArrayC2(count)=C2;
```

```
            ArrayC2TH(count)=C2TH;
            ArrayC3(count)=C3;
            ArrayC3TH(count)=C3TH;
            ArrayC4(count)=C4;
            ArrayC4TH(count)=C4TH;
            ArrayNP(count)=NP;
            ArrayXNPP(count)=XNPP;
        end
end
output=[ArrayT',ArrayC1',ArrayC1TH',ArrayC2',ArrayC2TH',...
        ArrayC3',ArrayC3TH',ArrayC4',ArrayC4TH',ArrayNP',ArrayXNPP'];
save datfil.txt output -ascii
disp 'simulation finished'
clc
figure
semilogy(ArrayT,ArrayC1,ArrayT,ArrayC1TH),grid
xlabel('Time (s) ')
ylabel('C1')
axis([0 10 .01 1000])
figure
semilogy(ArrayT,ArrayC2,ArrayT,ArrayC2TH),grid
xlabel('Time (s) ')
ylabel('C2')
axis([0 10 .1 200])
figure
semilogy(ArrayT,ArrayC3,ArrayT,ArrayC3TH),grid
xlabel('Time (s) ')
ylabel('C3')
axis([0 10 1 20])
figure
semilogy(ArrayT,-ArrayC4,ArrayT,-ArrayC4TH),grid
xlabel('Time (s) ')
ylabel('-C4')
axis([0 10 .1 20])
figure
semilogy(ArrayT,ArrayNP,ArrayT,ArrayXNPP),grid
xlabel('Time (s) ')
ylabel('Effective Navigation Ratio')
axis([0 10 .1 40])
```

The nominal case of Listing 16.1 was run, and the results are displayed in Figs. 16.2–16.6. We can see from these figures that the theoretical gains whose formulas were presented in this chapter and the gains obtained by integrating the optimal control equation are identical, thus confirming the accuracy of the zero over pole guidance law.

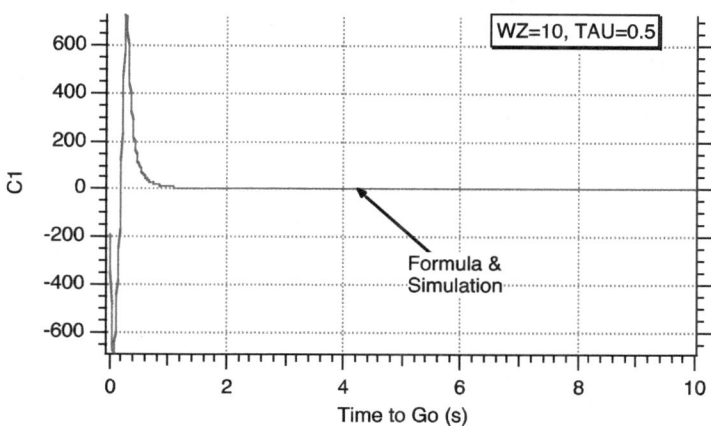

Fig. 16.2   First control gain matches closed-form solution.

PERFORMANCE OF GUIDANCE LAW WHEN REAL AUTOPILOT IS ZERO OVER POLE

In this section we will conduct an academic experiment to see how the zero over pole guidance law performs in the homing loop if the real autopilot could, in fact, be accurately represented by a zero over pole transfer function. If the new guidance law does not do well here it will not be worthwhile performing other experiments. An idealized model of the homing loop appears in Fig. 16.7. In this model it is assumed that all the states required by the guidance law are known perfectly.

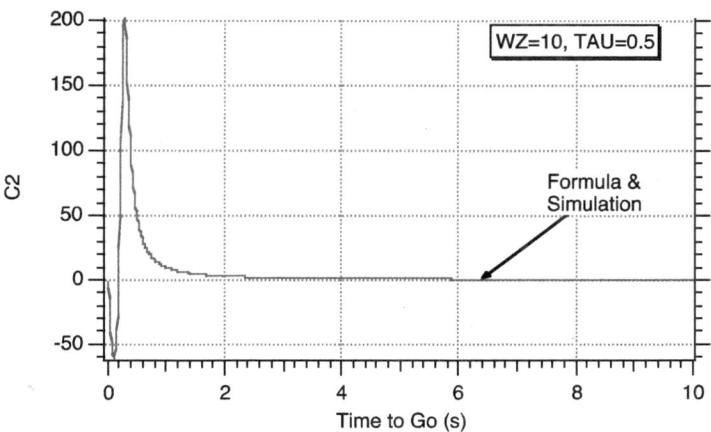

Fig. 16.3   Second control gain matches closed-form solution.

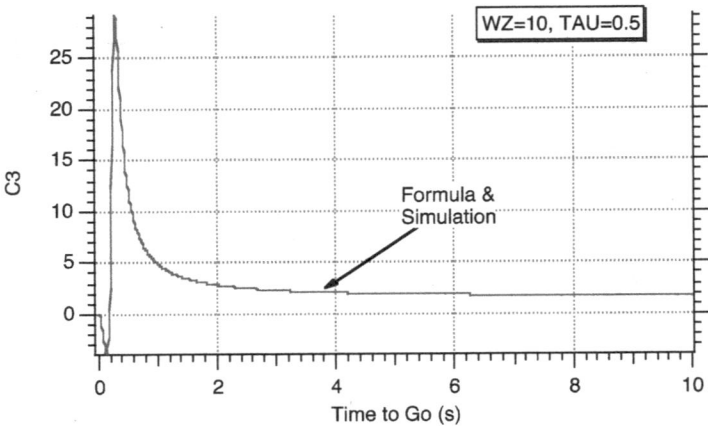

Fig. 16.4   Third control gain matches closed-form solution.

However, the model of Fig. 16.7 can run a variety of guidance laws by changing the control gains.

The homing loop of Fig. 16.2 was programmed and appears in Listing 16.2. In this brute force simulation, the only error disturbance is a constant 5-g step target maneuver starting at the beginning of the flight. The flight time is varied from 0.1 s to 10 s in steps of 0.1 s. The flight-control system time constant is 0.5 s, and the right-half plane zero is nominally at 10 rad/s. The guidance law can be changed through the parameter APN. If APN=0 we get proportional navigation

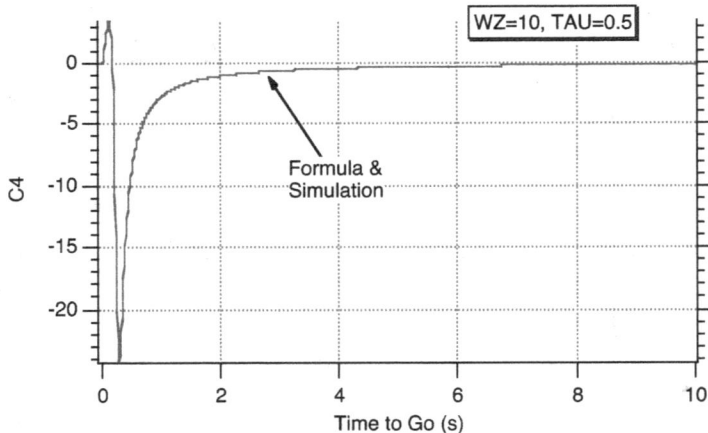

Fig. 16.5   Fourth control gain matches closed-form solution.

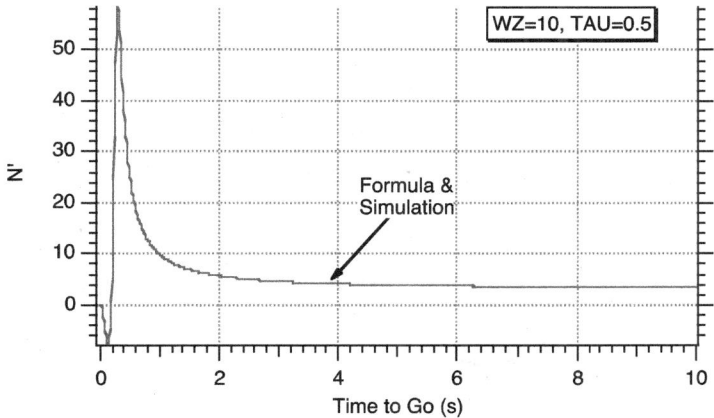

Fig. 16.6   Effective navigation ratio matches closed-form solution.

with a navigation ratio of 3. If APN=1 we get optimal guidance, and if APN=2 we get the zero over pole guidance law. The acceleration limit XNCLIM is set to a large number to avoid acceleration saturation.

LISTING 16.2   SIMULATION OF HOMING LOOP WITH A ZERO OVER POLE AUTOPILOT

```
clear
VC=4000.;
GAM=.0001;
APN=2;
XNT=161;
YIC=0.;
VM=3000.;
HEDEG=0.;
TAU=.5;
WZ=10;
XNP=3.;
n=0.;
for TF=.1:.1:10
    Y=0;
    YD=0;
    E=0.;
    T=0.;
    H=.01;
    XNL=0;
    XNCLIM=9999999;
    while T<=(TF-1e-5)
```

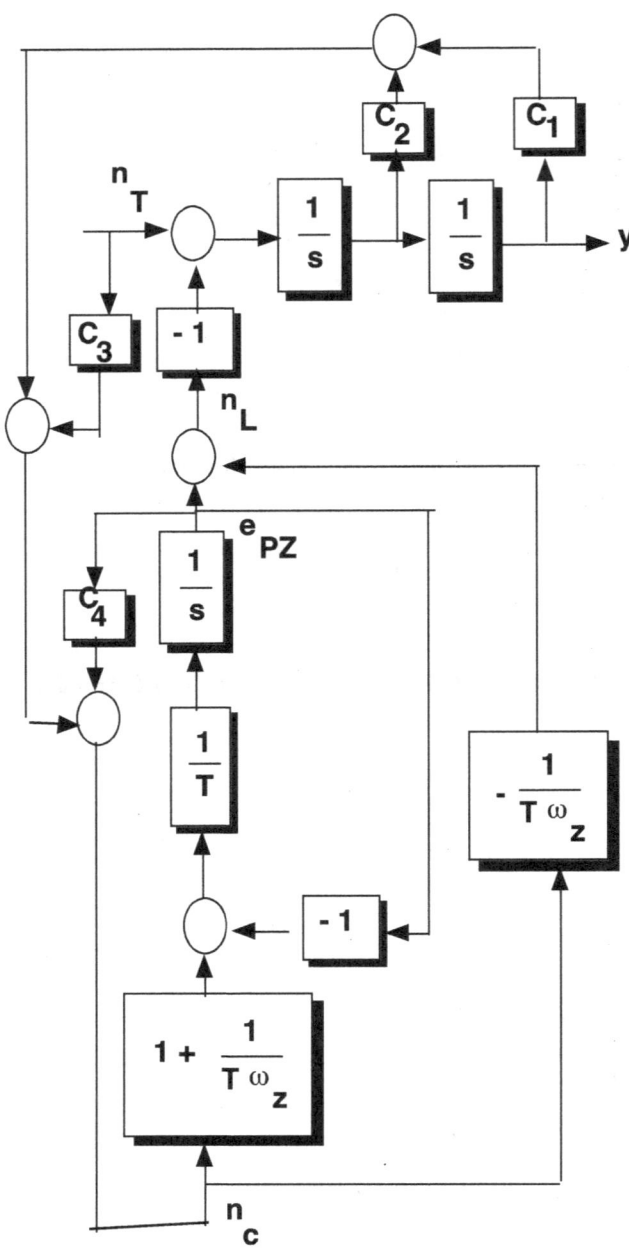

Fig 16.7 Homing loop with zero over pole flight-control system and zero over pole guidance law.

```
YOLD=Y;
YDOLD=YD;
EOLD=E;
STEP=1;
FLAG=0;
while STEP<=1
  if FLAG==1
  STEP=2;
  Y=Y+H*YD;
  YD=YD+H*YDD;
  E=E+H*ED;
  T=T+H;
  end
  TGO=TF-T+.00001;
  XLAMD=(Y+YD*TGO)/(VC*TGO*TGO);
  if APN==0
    XNC=XNP*(Y+YD*TGO)/TGO^2;
  elseif APN==1
    X=TGO/TAU;
    TOP=6.*X*X*(exp(-X)-1.+X);
    BOT1=2*X*X*X+3.+6.*X-6.*X*X;
    BOT2=-12.*X*exp(-X)-3*exp(-2.*X);
    XNPP=TOP/(.0001+BOT1+BOT2);
    XNEW=XNPP*XNL*(exp(-X)+X-1.)/(X*X);
    XNC=XNPP*VC*XLAMD+.5*XNPP*XNT-XNEW;
  else
    XS=TGO/TAU;
    TEMP1=TGO*TGO*TAU*(exp(-XS)-1.+XS);
    TOP=-(TGO^3)/(TAU*WZ)+(1.+1./(TAU*WZ))*TEMP1;
    TEMP2=.5*(1.-3./(TAU*WZ))+XS*(1.+1./(TAU*WZ))-XS*XS;
    TEMP3=-2.*XS*exp(-XS);
    TEMP4=2.*exp(-XS)/(TAU*WZ)-.5*exp(-2.*XS)*(1.+1./(TAU*WZ));
BOT=GAM+TGO*TGO*TGO/3.+(1.+1./(TAU*WZ))*TAU^3*(TEMP2+TEMP3+TEMP4);
    XNPP=TOP/BOT;
    C1TH=XNPP/(TGO*TGO);
    C2TH=XNPP/TGO;
    C3TH=.5*XNPP;
    C4TH=-XNPP*(exp(-XS)+XS-1.)/(XS*XS);
    XNC=C1TH*Y+C2TH*YD+C3TH*XNT+C4TH*E;
  end
  if XNC>XNCLIM
  XNC=XNCLIM;
  end
  if XNC<-XNCLIM
  XNC=-XNCLIM;
  end
```

```
        ED=(XNC-E)/TAU;
        XNL=E-ED/WZ;
        YDD=XNT-XNL;
        FLAG=1;
     end
     FLAG=0;
     Y=.5*(YOLD+Y+H*YD);
     YD=.5*(YDOLD+YD+H*YDD);
     E=.5*(EOLD+E+H*ED);
   end
   n=n+1;
   ArrayTF(n)=TF;
   ArrayY(n)=Y;
end
figure
plot(ArrayTF,ArrayY),grid
xlabel('Flight Time (Sec)')
ylabel('Miss (ft)')
clc
output=[ArrayTF',ArrayY',];
save datfil.txt output -ascii
disp 'simulation finished'
```

The nominal case of Listing 16.2 was run when the right-half plane zero was 50 rad/s (WZ=50) for the three different guidance laws. We can see from Fig. 16.8 that proportional navigation has very large miss distances for flight times less than 4 s, whereas both optimal and zero over pole guidance have

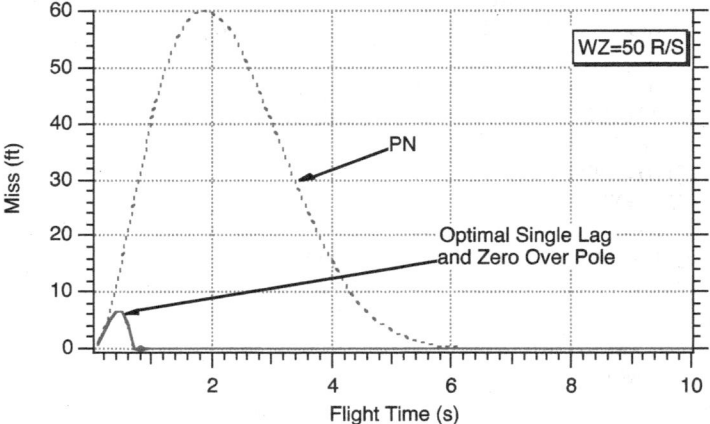

Fig. 16.8   Performance of optimal and zero over pole guidance is superior to proportional navigation when airframe zero is 50 rad/s.

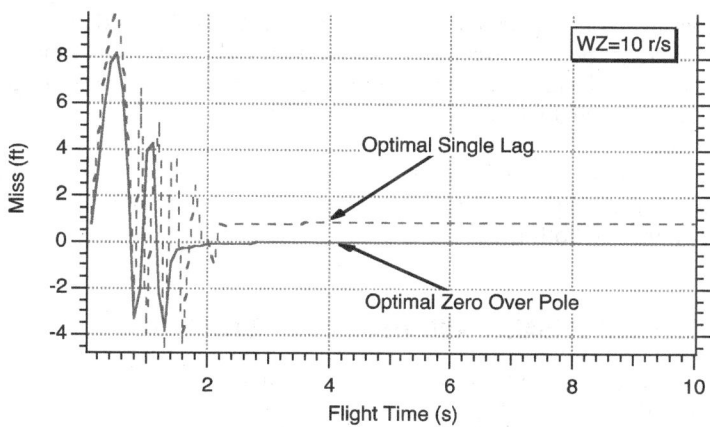

Fig. 16.9 **Performance of zero over pole guidance is slightly better than optimal guidance when airframe zero is 10 rad/s.**

very small miss distances and have identical results. The reason why the results are identical is that the airframe zero is high.

Listing 16.2 was rerun when the right-half plane zero was reduced to 10 rad/s (WZ=10). We can see from Fig. 16.9 that for flight times greater than 2 s, both guidance laws have excellent performance but the zero over pole guidance law is slightly better. The miss distances for both guidance laws oscillate for flight times less than 2 s, but zero over pole guidance has slightly fewer oscillations.

Another case was run when the airframe zero was further reduced to 5 rad/s. We can see from Fig. 16.10 that now the performance of zero over pole guidance is

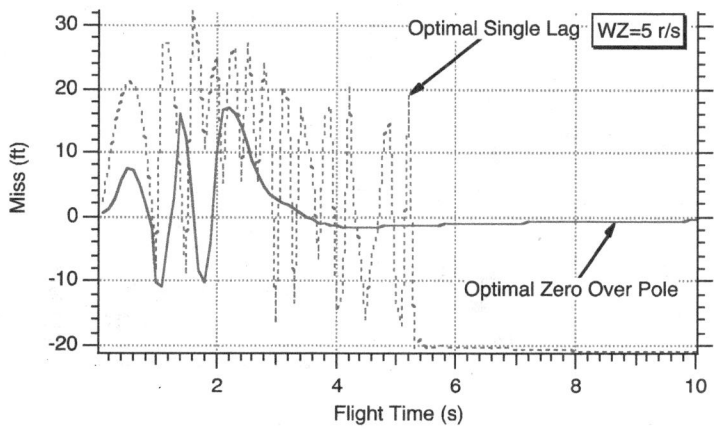

Fig. 16.10 **Zero over pole guidance is superior to optimal guidance when airframe zero is 5 rad/s.**

vastly superior to optimal guidance. When the flight time is less than 5 s, optimal guidance oscillates between large miss distances and when the flight time is greater than 5 s, the miss distance is -20 ft. On the other hand, zero over pole guidance only oscillates for flight times less than 2 s and then settles out to a near-zero miss distance.

Thus, we can see that for very low airframe zero frequencies, zero over pole guidance is superior to optimal guidance. As the right-half plane zero increases, the performance of both guidance laws is identical.

ZERO OVER POLE GUIDANCE AND THREE-LOOP AUTOPILOT

In the preceding section we saw that if the zero over pole guidance law was perfectly matched to the real flight-control system, significant performance benefits over optimal guidance could be achieved if the airframe zero frequency was very low. We would now like to see if this new guidance law performs better than optimal guidance when the flight-control system is more realistic.

Figure 16.11 presents a diagram of the homing loop with the linearized airframe of the missile originally presented in Chapter 21 of the previous volume

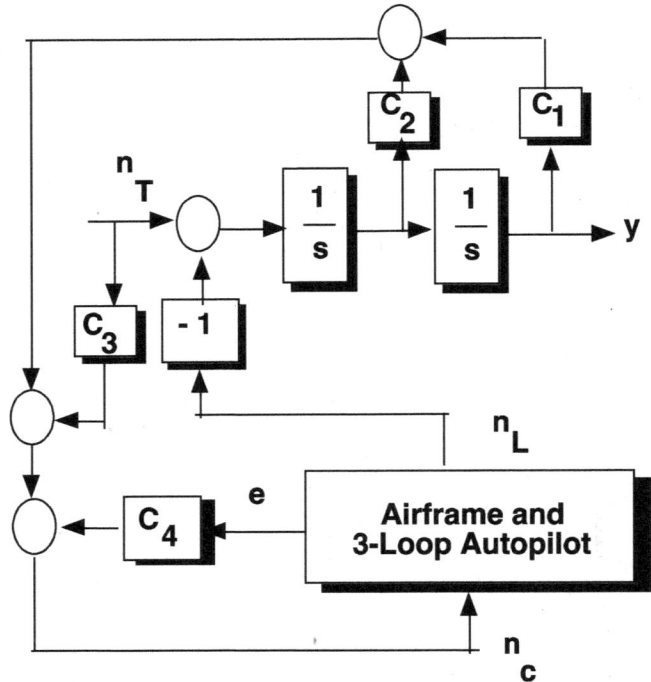

Fig. 16.11   Homing loop with three-loop autopilot and linearized airframe.

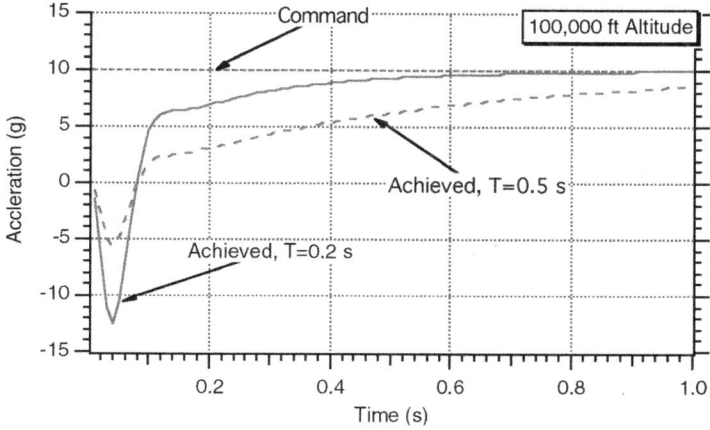

Fig. 16.12   Step response of three-loop autopilot at 100,000 ft altitude.

along with the three-loop autopilot of Chapter 23 of the previous volume. In addition, the choice of control gains determines the guidance that is to be used.

In this section we will be conducting academic experiments at 100,000 ft altitude. (The missile airframe of Chapter 21 of the previous volume was not designed to work at this extreme altitude.) Before any miss distance experiments are conducted, the step response program of Listing 23.1 of the previous volume was run to ensure that the three-loop autopilot still worked if we were to operate at 100,000 ft altitude. We can see from Fig. 16.12 that we have a stable flight-control system.

Listing 16.3 is a simulation of the linearized airframe and three-loop autopilot of Listing 23.1 and homing loop model of Listing 23.3 of Chapter 23 of Volume 1. This time the only disturbance is a 5-g step target maneuver and a 10-g acceleration limit. An option for different guidance laws exists: proportional navigation (APN=0), optimal guidance (APN=1), and zero over pole guidance (APN=2). Multiple cases were run for flight times ranging from 0.1 s to 10 s in steps of 0.1 s to generate adjoint-type curves (without running an adjoint). The missile considered from Chapter 21 of the previous volume cannot really operate at 100,000 ft altitude; however, in order to demonstrate what happens when the airframe zero is low, it was necessary to go to this very high altitude ($\omega_z = 7.6$ rad/s).

LISTING 16.3   SIMULATION OF HOMING LOOP WITH THREE-LOOP AUTOPILOT AND LINEARIZED AIRFRAME

```
clear
ALT=100000.;
TAU=.5;
```

```
XNT=161.;
HEDEG=0.;
TFMAX=10.;
XLIM=322.;
APN=2;
WCR=50.;
H=.001;
TF=3.;
% APN=0 is PN, APN=1 is OGS, APN=2 is Zero Over Pole
XNP=3.;
WACT=150.;
ZACT=.7;
TS=.01;
VC=4000.;
VM=3000.;
FR=3.;
DIAM=1.;
XL=20.;
CTW=0.;
CRW=6.;
HW=2.;
CTT=0.;
CRT=2.;
HT=2.;
XN=4.;
XCG=10.;
XHL=19.5;
VM=3000.;
XNCG=5.;
ZETA=.7;
HEDEGFIL=2.;
A=1000.;
GAM=.001
if ALT<=30000
   RHO=.002378*exp(-ALT/30000.);
else
   RHO=.0034*exp(-ALT/22000.);
end
WGT=1000.;
XNLLIN=0.;
XACC=XCG;
SWING=.5*HW*(CTW+CRW);
STAIL=.5*HT*(CTT+CRT);
SREF=3.1416*DIAM*DIAM/4.;
XLP=FR*DIAM;
SPLAN=(XL-XLP)*DIAM+1.33*XLP*DIAM/2.;
```

```
XCPN=2*XLP/3;
AN=.67*XLP*DIAM;
AB=(XL-XLP)*DIAM;
XCPB=(.67*AN*XLP+AB*(XLP+.5*(XL-XLP)))/(AN+AB);
XCPW=XLP+XN+.7*CRW-.2*CTW;
XMACH=VM/A;
XIYY=WGT*(3*((DIAM/2)^2)+XL*XL)/(12*32.2);
TMP1=(XCG-XCPW)/DIAM;
TMP2=(XCG-XHL)/DIAM;
TMP3=(XCG-XCPB)/DIAM;
TMP4=(XCG-XCPN)/DIAM;
B=sqrt(XMACH^2-1);
Q=.5*RHO*VM*VM;
P1=WGT*XNCG/(Q*SREF);
Y1=2+8*SWING/(B*SREF)+8*STAIL/(B*SREF);
Y2=1.5*SPLAN/SREF;
Y3=8*STAIL/(B*SREF);
Y4=2*TMP4+8*SWING*TMP1/(B*SREF)+8*STAIL*TMP2/(B*SREF);
Y5=1.5*SPLAN*TMP3/SREF;
Y6=8*STAIL*TMP2/(B*SREF);
P2=Y2-Y3*Y5/Y6;
P3=Y1-Y3*Y4/Y6;
ALFTR=(-P3+sqrt(P3*P3+4.*P2*P1))/(2.*P2);
DELTR=-Y4*ALFTR/Y6-Y5*ALFTR*ALFTR/Y6;
CNA=2+1.5*SPLAN*ALFTR/SREF+8*SWING/(B*SREF)+8*STAIL/(B*SREF);
CND=8*STAIL/(B*SREF);
CMAP=2*TMP4+1.5*SPLAN*ALFTR*TMP3/SREF+8*SWING*TMP1/(B*SREF);
CMA=CMAP+8*STAIL*TMP2/(B*SREF);
CMD=8*STAIL*TMP2/(B*SREF);
XMA=Q*SREF*DIAM*CMA/XIYY;
XMD=Q*SREF*DIAM*CMD/XIYY;
ZA=-32.2*Q*SREF*CNA/(WGT*VM);
ZD=-32.2*Q*SREF*CND/(WGT*VM);
WZ=sqrt((XMA*ZD-ZA*XMD)/ZD);
WAF=sqrt(-XMA);
ZAF=.5*WAF*ZA/XMA;
XK1=-VM*(XMA*ZD-XMD*ZA)/(1845*XMA);
XK2=XK1;
TA=XMD/(XMA*ZD-XMD*ZA);
XK3=1845*XK1/VM;
W=(TAU*WCR*(1+2.*ZAF*WAF/WCR)-1)/(2*ZETA*TAU);
W0=W/sqrt(TAU*WCR);
Z0=.5*W0*(2*ZETA/W+TAU-WAF^2/(W0*W0*WCR));
XKC=(-W0^2/WZ^2-1.+2.*Z0*W0*TA)/(1.-2.*Z0*W0*TA+W0*W0*TA*TA);
XKA=XK3/(XK1*XKC);
XK0=-W*W/(TAU*WAF*WAF);
```

```
XK=XK0/(XK1*(1+XKC));
WI=XKC*TA*W0*W0/(1+XKC+W0^2/WZ^2);
XKR=XK/(XKA*WI);
XKDC=1.+1845./(XKA*VM);
n=0;
for TF=.1:.1:10
  X=0.;
  T=0;
  S=0.;
  Y=0.;
  YD=-VM*HEDEG/57.3;
  XNC=0.;
  XNL=0.;
  DELD=0.;
  DEL=0.;
  XNLPZ=0.;
  E=0.;
  ED=0.;
  while T<=(TF-1e-5)
    S=S+H;
    XOLD=X;
    YOLD=Y;
    YDOLD=YD;
    DELOLD=DEL;
    DELDOLD=DELD;
    EOLD=E;
    EDOLD=ED;
    STEP=1;
    FLAG=0;
    while STEP<=1
      if FLAG==1
        STEP=2;
        X=X+H*XD;
        Y=Y+H*YD;
        YD=YD+H*YDD;
        DEL=DEL+H*DELD;
        DELD=DELD+H*DELDD;
        E=E+H*ED;
        ED=ED+H*EDD;
        T=T+H;
      end
      TGO=TF-T+.0001;
      RTM=VC*TGO;
      XLAM=Y/RTM;
      XLAMD=(Y+YD*TGO)/(VC*TGO*TGO);
      if APN==0
```

```
     XNC=XNP*VC*XLAMD;
   elseif APN==1
     XS=TGO/TAU;
     TOP=6.*XS*XS*(exp(-XS)-1.+XS);
     BOT1=2*XS*XS*XS+3.+6.*XS-6.*XS*XS;
     BOT2=-12.*XS*exp(-XS)-3.*exp(-2.*XS);
     XNPP=TOP/(.0001+BOT1+BOT2);
     XNLPZ=XNL*32.2;
     XNEW=XNPP*XNLPZ*(exp(-XS)+XS-1.)/(XS*XS);
     XNC=XNPP*VC*XLAMD+.5*XNPP*XNT-XNEW;
   else
     XS=TGO/TAU;
     TEMP1=TGO*TGO*TAU*(exp(-XS)-1.+XS);
     TOP=-(TGO^3)/(TAU*WZ)+(1.+1./(TAU*WZ))*TEMP1;
     TEMP2=.5*(1.-3./(TAU*WZ))+XS*(1.+1./(TAU*WZ))-XS*XS;
     TEMP3=-2.*XS*exp(-XS);
     TEMP4=2.*exp(-XS)/(TAU*WZ)-.5*exp(-2.*XS)*(1.+1./(TAU*WZ));
     BOT=GAM+TGO*TGO*TGO/3.+(1.+1./(TAU*WZ))*TAU^3*...
        (TEMP2+TEMP3+TEMP4);
     XNPP=TOP/BOT;
     C1TH=XNPP/(TGO*TGO);
     C2TH=XNPP/TGO;
     C3TH=.5*XNPP;
     C4TH=-XNPP*(exp(-XS)+XS-1.)/(XS*XS);
     XNLPZ=XNL*32.2;
     EPZ=XNLPZ+1/(TAU*WZ);
     XNC=C1TH*Y+C2TH*YD+C3TH*XNT+C4TH*EPZ;
   end
   if XNC>XLIM
     XNC=XLIM;
   end
   if XNC<-XLIM
     XNC=-XLIM;
   end
   XNCG=XNC/32.2;
   THD=XK3*(E+TA*ED);
   DELC=XKR*(X+THD);
   DELDD=WACT*WACT*(DELC-DEL-2.*ZACT*DELD/WACT);
   EDD=WAF*WAF*(DEL-E-2.*ZAF*ED/WAF);
   XNL=XK1*(E-EDD/WZ^2);
   XD=WI*(THD+XKA*(XNL-XNCG*XKDC));
   XNLPZ=XNL*32.2;
   YDD=XNT-XNLPZ;
   FLAG=1;
 end
 FLAG=0;
```

```
        X=.5*(XOLD+X+H*XD);
        Y=.5*(YOLD+Y+H*YD);
        YD=.5*(YDOLD+YD+H*YDD);
        DEL=.5*(DELOLD+DEL+H*DELD);
        DELD=.5*(DELDOLD+DELD+H*DELDD);
        E=.5*(EOLD+E+H*ED);
        ED=.5*(EDOLD+ED+H*EDD);
      end
      n=n+1;
      ArrayTF(n)=TF;
      ArrayY(n)=Y;
    end
    figure
    plot(ArrayTF,ArrayY),grid
    xlabel('Flight Time (Sec)')
    ylabel('Miss (ft)')
    clc
    output=[ArrayTF',ArrayY',];
    save datfil.txt output -ascii
    disp 'simulation finished'
```

The nominal case of Listing 16.3 was run for the case where the desired autopilot
time constant was 0.5 s. In this academic example we can see from Figs. 16.13 and
16.14 the zero over pole guidance law yields slightly smaller miss distances than
the optimal guidance law. Reducing the autopilot time constant from 0.5 s to
0.2 s enables the zero over pole guidance law again to yield slightly better
results than the optimal guidance law.

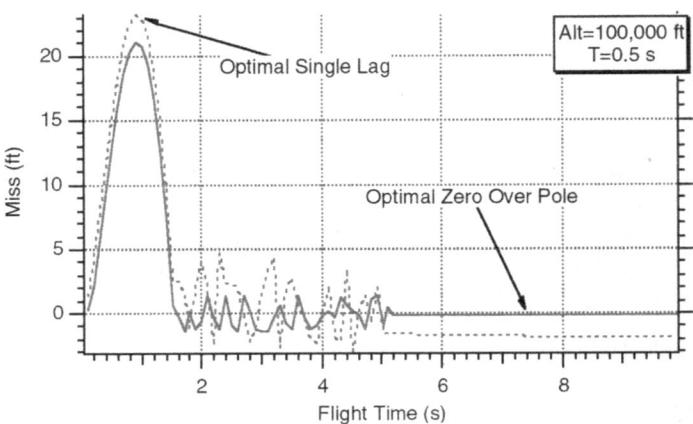

**Fig. 16.13   At 100,000 ft altitude, zero over pole guidance is slightly better than optimal
guidance when autopilot time constant is 0.5 s.**

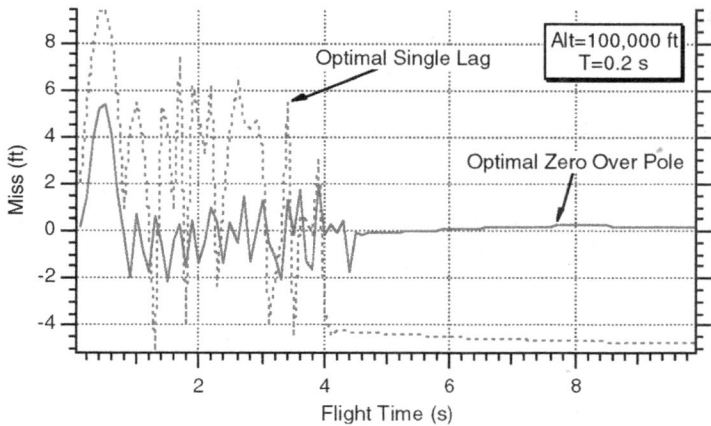

Fig. 16.14   At 100,000 ft altitude, zero over pole guidance is slightly better than optimal guidance when autopilot time constant is 0.2 s.

As a further test of the zero over pole guidance law an academic experiment was conducted in which the missile altitude was increased from 100,000 ft to 150,000 ft where the airframe zero reduces from 7.6 r/s to 3.9 r/s. We can see from Fig. 16.15 that in this case the zero over pole guidance law is superior to the optimal guidance law. It must be emphasized that the experiment is not realistic for an endoatmospheric missile because the altitude is too high. However Fig. 16.15 emphasizes that the new guidance law can be beneficial if the airframe zero of a tail-controlled missile is very low.

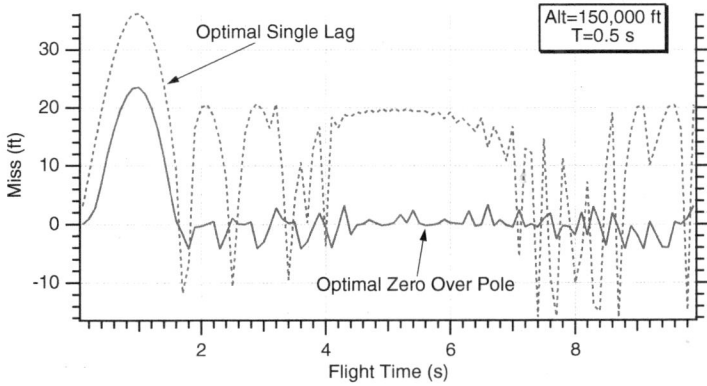

Fig. 16.15   At 150,000 ft altitude the zero over pole guidance law is significantly better than optimal guidance when the autopilot time constant is 0.2 s.

SUMMARY

A new closed-form guidance law was introduced in an attempt to improve the performance of a tail-controlled missile operating at very high altitude. Although the performance of the new guidance law was slightly better than optimal guidance for the example chosen, it did not offer dramatic performance benefits. The computerized guidance law of Chapter 12, which was specifically derived for the complex model of the flight-control system transfer function, still offers the most benefits for a tail-controlled missile operating at very high altitudes.

REFERENCES

[1]   Zarchan, P., "Derivation of Optimal Control Gains for a 'Zero Over Pole' Autopilot," Raytheon Memorandum, Oct. 1969.

[2]   Stallard, D. V., "Near-Optimal Stochastic Terminal Controllers," PhD thesis, Massachusetts Institute of Technology, 1971.

CHAPTER 17

Engagement Simulations in Three Dimensions

INTRODUCTION

So far, all of the engagement simulations presented in this book have been in either one or two dimensions. This was done to make it easier for the reader to readily understand all of the concepts presented in the text. In this chapter we shall provide several examples on how to convert important elements of engagement simulation code to three dimensions in both the tactical and strategic worlds. Complete three-dimensional tactical and strategic engagement simulations will be presented to illustrate all important points discussed.

WEAVING TARGETS IN THREE DIMENSIONS

So far, all of the guidance laws presented in the text have been presented in only one dimension. The extension of the guidance laws to three dimensions is quite straightforward and can most easily be done by making use of the zero effort miss concept. Recall that the zero effort miss is simply the miss distance that would result if the target continued to do what it is currently doing and the interceptor issued no further acceleration commands. We have previously demonstrated that all of the guidance laws presented in this text and the previous volume can be expressed in terms of the zero effort miss. The resultant guidance command is equal to the effective navigation ratio times the zero effort miss perpendicular to the line of sight divided by the square of time to go until intercept. The only difference between all of the guidance laws presented in this text is the way in which the zero effort miss is computed. In three dimensions we can define the relative position and velocity components between the missile and target as

$$R_{TM1} = x_T - x_M$$
$$R_{TM2} = y_T - y_M$$
$$R_{TM3} = z_T - z_M$$

$$V_{TM1} = \dot{x}_T - \dot{x}_M$$
$$V_{TM2} = \dot{y}_T - \dot{y}_M$$
$$V_{TM3} = \dot{z}_T - \dot{z}_M$$

Let us start with the simplest guidance law, proportional navigation. For proportional navigation, the components of the zero effort miss in three dimensions are simply

$$ZEM_1 = R_{TM1} + V_{TM1}t_{go}$$
$$ZEM_2 = R_{TM2} + V_{TM2}t_{go}$$
$$ZEM_3 = R_{TM3} + V_{TM3}t_{go}$$

The weave guidance law, on the other hand, has a more complex expression for the zero effort miss that we shall deal with later in this section. As was mentioned earlier, for guidance purposes we are only interested in the component of the zero effort miss that is perpendicular to the line of sight. To find the component of the zero effort miss that is perpendicular to the line of sight, we must first find the component of the zero effort miss that is parallel to the line of sight. If the zero-effort-miss vector **ZEM** is defined as

$$\mathbf{ZEM} = ZEM_1\mathbf{i} + ZEM_2\mathbf{j} + ZEM_3\mathbf{k}$$

then a unit vector along the line of sight can be expressed as

$$\mathbf{1}_{RTM} = \frac{R_{TM1}\mathbf{i} + R_{TM2}\mathbf{j} + R_{TM3}\mathbf{k}}{R_{TM}}$$

where the relative range between the missile and target is simply

$$R_{TM} = \sqrt{R_{TM1}^2 + R_{TM2}^2 + R_{TM3}^2}$$

The vector parallel to the line of sight has a magnitude equal to the dot product of the two vectors (zero effort miss and line of sight) and is along the same direction as the unit line-of-sight vector. Therefore, the zero-effort-miss vector parallel to the line of sight is given by

$$\mathbf{ZEM_{PAR}} = ZEMDOTRTM\left(\frac{R_{TM1}\mathbf{i} + R_{TM2}\mathbf{j} + R_{TM3}\mathbf{k}}{R_{TM}}\right)$$

where the quantity ZEMDOTRTM can be computed as

$$ZEMDOTRTM = \frac{ZEM_1 R_{TM1} + ZEM_2 R_{TM2} + ZEM_3 R_{TM3}}{R_{TM}}$$

The zero-effort-miss vector perpendicular to the line-of-sight $\mathbf{ZEM_{PER}}$ is simply the vector difference between the zero-effort-miss vector **ZEM** and the zero effort miss parallel to the line-of-sight $\mathbf{ZEM_{PAR}}$, or

$$\mathbf{ZEM_{PER}} = \mathbf{ZEM} - \mathbf{ZEM_{PAR}}$$

Therefore, we can say that

$$\mathbf{ZEM_{PER}} = \text{ZEM}_{PER1}\mathbf{i} + \text{ZEM}_{PER2}\mathbf{j} + \text{ZEM}_{PER3}\mathbf{k}$$

where the components of the zero effort miss perpendicular to the line of sight can be computed as

$$\text{ZEM}_{PER1} = \text{ZEM}_1 - \frac{\text{ZEMDOTRTM} * R_{TM1}}{R_{TM}}$$

$$\text{ZEM}_{PER2} = \text{ZEM}_2 - \frac{\text{ZEMDOTRTM} * R_{TM2}}{R_{TM}}$$

$$\text{ZEM}_{PER3} = \text{ZEM}_3 - \frac{\text{ZEMDOTRTM} * R_{TM3}}{R_{TM}}$$

As was mentioned earlier, the desired interceptor guidance commands are proportional to the zero effort miss perpendicular to the line of sight and inversely proportional to the square of time to go. Therefore, the individual acceleration components of the guidance command are given by

$$a_{M1} = \frac{N'\text{ZEMPER}_1}{t_{go}^2}$$

$$a_{M2} = \frac{N'\text{ZEMPER}_2}{t_{go}^2}$$

$$a_{M3} = \frac{N'\text{ZEMPER}_3}{t_{go}^2}$$

where the effective navigation ratio N' is a constant set equal to three for proportional navigation.

The compensated weave guidance law, which is optimal in the sense that it requires the least acceleration against target weave maneuvers in the presence of a single-time-constant missile flight-control system, also issues guidance commands proportional to the zero effort miss and inversely proportional to the square of time to go until intercept. In one dimension the compensated weave guidance law was shown in Chapter 20 of the previous volume to be

$$n_c = \frac{N'}{t_{go}^2}\left[y + \dot{y}t_{go} + \left(\frac{1 - \cos \omega t_{go}}{\omega^2}\right)\ddot{y}_T \right.$$

$$\left. + \left(\frac{\omega t_{go} - \sin \omega t_{go}}{\omega^3}\right)\dddot{y}_T - n_L T^2(e^{-x} + x - 1) \right]$$

where x is given by

$$x = \frac{t_{go}}{T}$$

In the preceding expressions, t_{go} is the time to go until intercept, and the bracketed quantity is the zero effort miss. The effective navigation ratio for the compensated weave guidance law is not a constant, but is given by

$$N' = \frac{6x^2(e^{-x} - 1 + x)}{2x^3 + 3 + 6x - 6x^2 - 12xe^{-x} - 3e^{-2x}}$$

We can see that the compensated weave guidance law consists of five terms: the first two terms are related to the line-of-sight rate, the third term is proportional to the target acceleration, the fourth term is proportional to target jerk, and the fifth term is proportional to the achieved missile acceleration. The first two terms in the guidance law are proportional navigation, the third and fourth terms in the guidance law compensate for the sinusoidal motion of the target, and the fifth term compensates for the flight-control-system dynamics. Details of the derivation of the compensated weave guidance law can be found in Chapter 6.

Thus the zero effort miss for the compensated weave guidance law in three dimensions can be expressed as

$$ZEM_1 = R_{TM1} + V_{TM1}t_{go} + \left(\frac{1 - \cos \omega t_{go}}{\omega^2}\right)a_{T1}$$

$$+ \left(\frac{\omega t_{go} - \sin \omega t_{go}}{\omega^3}\right)\dot{a}_{T1} - T^2(e^{-x} + x - 1)a_{M1}$$

$$ZEM_2 = R_{TM2} + V_{TM2}t_{go} + \left(\frac{1 - \cos \omega t_{go}}{\omega^2}\right)a_{T2}$$

$$+ \left(\frac{\omega t_{go} - \sin \omega t_{go}}{\omega^3}\right)\dot{a}_{T2} - T^2(e^{-x} + x - 1)a_{M2}$$

$$ZEM_3 = R_{TM3} + V_{TM3}t_{go} + \left(\frac{1 - \cos \omega t_{go}}{\omega^2}\right)a_{T3}$$

$$+ \left(\frac{\omega t_{go} - \sin \omega t_{go}}{\omega^3}\right)\dot{a}_{T3} - T^2(e^{-x} + x - 1)a_{M3}$$

where a_{M1}, a_{M2}, and a_{M3} are the three components of the achieved missile accelerations. The target acceleration components are represented by a_{T1}, a_{T2}, and a_{T3}, whereas the components of target jerk are given by $\dot{a}_{T1}$, $\dot{a}_{T2}$, and $\dot{a}_{T3}$. The implementation of the compensated weave guidance law in three dimensions is the same as the implementation of proportional navigation. Therefore, the only difference between the two guidance laws is in the expressions for the zero effort miss.

Now let us consider a numerical example to test our implementation of guidance laws in three dimensions. A weaving or spiraling target is chosen as the threat. The acceleration equations of a three-dimensional spiraling target are given by

$$a_{T1} = a_T \sin \omega t$$
$$a_{T2} = a_T \cos \omega t$$
$$a_{T3} = 0$$

where the subscript 1 indicates the downrange direction, the subscript 2 denotes the altitude direction, and the subscript 3 denotes the cross-range direction. Thus in this example the target is traveling in the cross-range direction and spiraling in the downrange and altitude directions. Because the compensated weave guidance law requires target jerk, we can calculate it exactly by differentiating the preceding equations. The resultant components of the target jerk are given by

$$\dot{a}_{T1} = a_T \omega \cos \omega t$$
$$\dot{a}_{T2} = -a_T \omega \sin \omega t$$
$$\dot{a}_{T3} = 0$$

A three-dimensional tactical engagement simulation, utilizing the preceding equations, appears in Listing 17.1. From Listing 17.1 we can see that the target is traveling at 1000 ft/s and is spiraling at 3 rad/s with an acceleration level of 6 g. The missile is traveling at 3000 ft/s and has an infinite acceleration limit. There is a single-lag time constant of 1 s representing the dynamics of the flight-control system in each of the three channels of the guidance system. From Listing 17.1 we can see that if QPN is set to 1, the missile uses proportional navigation for guidance; if QPN is set to 0, the missile uses the compensated weave guidance law. The program is set up to run a number of cases, each of which has a different initial missile-target separation. Each separation corresponds to a different flight time. In this way we can generate adjoint-type curves using the brute-force approach as we did in Chapter 3 of the previous volume for the two-dimensional nonlinear engagement simulation. At the end of each flight, the components of the miss, along with the total miss, are printed out as a function of the engagement time. The only source of error in the simulation is the spiraling target maneuver.

LISTING 17.1   THREE-DIMENSIONAL TACTICAL ENGAGEMENT SIMULATION WITH SPIRALING TARGET

```
clear
count=0;
QPN=1;
```

```
TAU=1.;
W=3.;
AT=193.2;
VT=1000.;
VM=3000.;
XNP=3.;
XNCLIM=9999999999.;
for RT3IC=40000:-500:500
        RM1=0.;
        RM2=10000.;
        RM3=0.;
        RT1=0.;
        RT2=10000.;
        RT3=RT3IC;
        VT1=-AT/W;
        VT2=0.;
        VT3=-VT;
        T=0.;
        S=0.;
        RTM1=RT1-RM1;
        RTM2=RT2-RM2;
        RTM3=RT3-RM3;
        RTM=sqrt(RTM1^2+RTM2^2+RTM3^2);
        VM1=0.;
        VM2=0.;
        VM3=VM;
        VTM1=VT1-VM1;
        VTM2=VT2-VM2;
        VTM3=VT3-VM3;
        VC=-(RTM1*VTM1+RTM2*VTM2+RTM3*VTM3)/RTM;
        AM1=0.;
        AM2=0.;
        AM3=0.;
        while VC>=0
                if RTM<1000
                H=.0002;
                else
                H=.01;
                end
                RT1OLD=RT1;
                RT2OLD=RT2;
                RT3OLD=RT3;
                RM1OLD=RM1;
                RM2OLD=RM2;
                RM3OLD=RM3;
                VM1OLD=VM1;
```

```
VM2OLD=VM2;
VM3OLD=VM3;
VT1OLD=VT1;
VT2OLD=VT2;
VT3OLD=VT3;
AM1OLD=AM1;
AM2OLD=AM2;
AM3OLD=AM3;
STEP=1;
FLAG=0;
while STEP<=1
      if FLAG==1
            STEP=2;
            RT1=RT1+H*VT1;
            RT2=RT2+H*VT2;
            RT3=RT3+H*VT3;
            RM1=RM1+H*VM1;
            RM2=RM2+H*VM2;
            RM3=RM3+H*VM3;
            VM1=VM1+H*AM1;
            VM2=VM2+H*AM2;
            VM3=VM3+H*AM3;
            VT1=VT1+H*AT1;
            VT2=VT2+H*AT2;
            VT3=VT3+H*AT3;
            AM1=AM1+H*AM1D;
            AM2=AM2+H*AM2D;
            AM3=AM3+H*AM3D;
            T=T+H;
      end
      RTM1=RT1-RM1;
      RTM2=RT2-RM2;
      RTM3=RT3-RM3;
      RTM=sqrt(RTM1^2+RTM2^2+RTM3^2);
      VTM1=VT1-VM1;
      VTM2=VT2-VM2;
      VTM3=VT3-VM3;
      VC=-(RTM1*VTM1+RTM2*VTM2+RTM3*VTM3)/RTM;
      TGO=RTM/VC;
      AT1=AT*sin(W*T);
      AT2=AT*cos(W*T);
      AT3=0.;
      if QPN==1
            ZEM1=RTM1+VTM1*TGO;
            ZEM2=RTM2+VTM2*TGO;
            ZEM3=RTM3+VTM3*TGO;
```

```
else
        AT1D=AT*W*cos(W*T);
        AT2D=-AT*W*sin(W*T);
        AT3D=0.;
        X=TGO/TAU;
        TOP=6.*X*X*(exp(-X)-1.+X);
        BOT1=2*X*X*X+3.+6.*X-6.*X*X;
        BOT2=-12.*X*exp(-X)-3.*exp(-2.*X);
        XNP=TOP/(.0001+BOT1+BOT2);
        ZEM1=RTM1+VTM1*TGO+AT1*(1.-cos(W*TGO))/W^2+...
                (AT1D*(W*TGO-sin(W*TGO))/W^3)-...
                AM1*TAU*TAU*(exp(-X)+X-1.);
        ZEM2=RTM2+VTM2*TGO+AT2*(1.-cos(W*TGO))/W^2+...
                (AT2D*(W*TGO-sin(W*TGO))/W^3)-...
                AM2*TAU*TAU*(exp(-X)+X-1.);
        ZEM3=RTM3+VTM3*TGO+AT3*(1.-cos(W*TGO))/W^2+...
                (AT3D*(W*TGO-sin(W*TGO))/W^3)-...
                AM3*TAU*TAU*(exp(-X)+X-1.);
end
ZEMDOTRTM=(ZEM1*RTM1+ZEM2*RTM2+ZEM3*RTM3)/RTM;
ZEMPER1=ZEM1-ZEMDOTRTM*RTM1/RTM;
ZEMPER2=ZEM2-ZEMDOTRTM*RTM2/RTM;
ZEMPER3=ZEM3-ZEMDOTRTM*RTM3/RTM;
AM1P=XNP*ZEMPER1/(TGO*TGO);
AM2P=XNP*ZEMPER2/(TGO*TGO);
AM3P=XNP*ZEMPER3/(TGO*TGO);
AM1D=(AM1P-AM1)/TAU;
AM2D=(AM2P-AM2)/TAU;
AM3D=(AM3P-AM3)/TAU;
if AM1>XNCLIM
        AM1=XNCLIM;
end
if AM1<-XNCLIM
        AM1=-XNCLIM;
end
if AM2>XNCLIM
        AM2=XNCLIM;
end
if AM2<-XNCLIM
        AM2=-XNCLIM;
end
if AM3>XNCLIM
        AM3=XNCLIM;
end
if AM3<-XNCLIM
        AM3=-XNCLIM;
```

```
                        end
                        XNCG=sqrt(AM1^2+AM2^2+AM3^2)/32.2;
                        FLAG=1;
                end
                FLAG=0;
                RT1=.5*(RT1OLD+RT1+H*VT1);
                RT2=.5*(RT2OLD+RT2+H*VT2);
                RT3=.5*(RT3OLD+RT3+H*VT3);
                RM1=.5*(RM1OLD+RM1+H*VM1);
                RM2=.5*(RM2OLD+RM2+H*VM2);
                RM3=.5*(RM3OLD+RM3+H*VM3);
                VM1=.5*(VM1OLD+VM1+H*AM1);
                VM2=.5*(VM2OLD+VM2+H*AM2);
                VM3=.5*(VM3OLD+VM3+H*AM3);
                VT1=.5*(VT1OLD+VT1+H*AT1);
                VT2=.5*(VT2OLD+VT2+H*AT2);
                VT3=.5*(VT3OLD+VT3+H*AT3);
                AM1=.5*(AM1OLD+AM1+H*AM1D);
                AM2=.5*(AM2OLD+AM2+H*AM2D);
                AM3=.5*(AM3OLD+AM3+H*AM3D);
                S=S+H;
                if S>=.09999
                        S=0.;
                        RT1K=RT1/1000.;
                        RT2K=RT2/1000.;
                        RT3K=RT3/1000.;
                        RM1K=RM1/1000.;
                        RM2K=RM2/1000.;
                        RM3K=RM3/1000.;
                end
        end
        count=count+1;
        ArrayT(count)=T;
        ArrayRTM1(count)=RTM1;
        ArrayRTM2(count)=RTM2;
        ArrayRTM3(count)=RTM3;
        ArrayRTM(count)=RTM;
end
output=[ArrayT',ArrayRTM1',ArrayRTM2',ArrayRTM3',ArrayRTM'];
save datfil.txt output -ascii
disp 'simulation finished'
clc
figure
plot(ArrayT,ArrayRTM1),grid
xlabel('Time (s) ')
ylabel('RTM1 (ft)')
```

```
axis([0 10 -30 30])
figure
plot(ArrayT,ArrayRTM2),grid
xlabel('Time (s) ')
ylabel('RTM2 (ft)')
axis([0 10 -30 30])
figure
plot(ArrayT,ArrayRTM3),grid
xlabel('Time (s) ')
ylabel('RTM3 (ft)')
axis([0 10 -30 30])
figure
plot(ArrayT,ArrayRTM),grid
xlabel('Time (s) ')
ylabel('RTM (ft)')
axis([0 10 0 30])
```

The nominal case of Listing 17.1 was run using the proportional navigation guidance law. The three components of the miss vs flight time are shown in Figs. 17.1, 17.2, and 17.3. We can see that the miss oscillates with flight time in Figs. 17.1 and 17.2, and the amplitude of the miss agrees with the single plane results of Fig. 6.7. This should not be surprising because identical inputs were used. We can see from Fig. 17.3 that the miss is virtually zero in the cross-range direction because there is no target maneuver in that direction. However, we can see from Fig. 17.4 that the total miss is virtually a constant vs flight time after an initial transient period. This should not be surprising because the target weave was

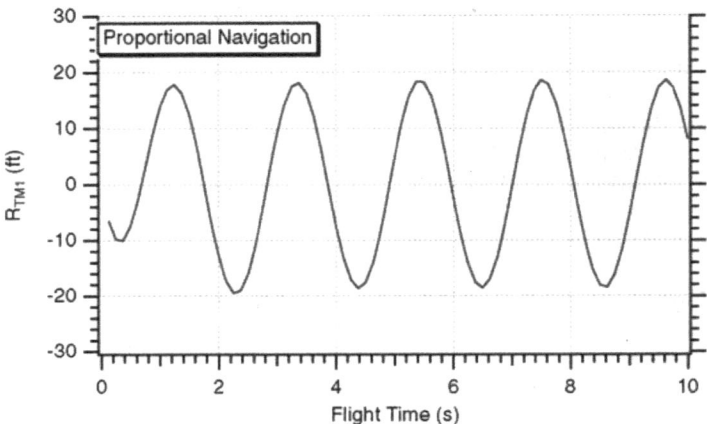

Fig. 17.1   Downrange component of miss oscillates at same frequency as 3-rad/s weaving target maneuver.

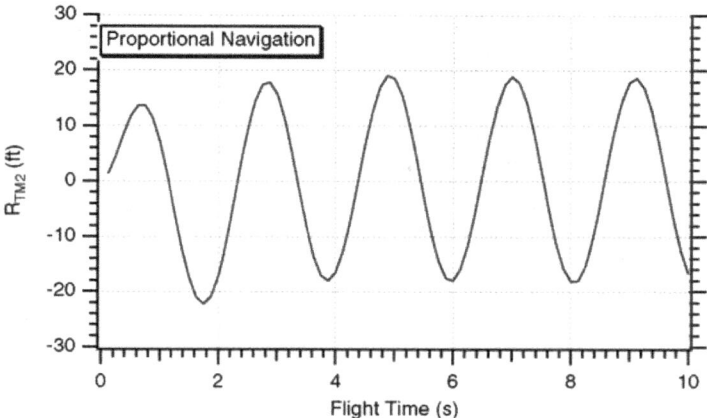

Fig. 17.2   Altitude component of miss oscillates at same frequency as 3-rad/s weaving target maneuver.

a sine wave in one direction and a cosine wave in the other direction, which yields circular motion. Apparently the total miss is also along a circle in the downrange and altitude direction. Therefore, we have to be careful when looking at single plane results for a weaving target. One gets the impression from single plane results that the miss can be large or small depending on the flight time. In actuality, the peak miss distance for the single plane results will define the radius of a circle or the three-dimensional miss distance for the weaving target.

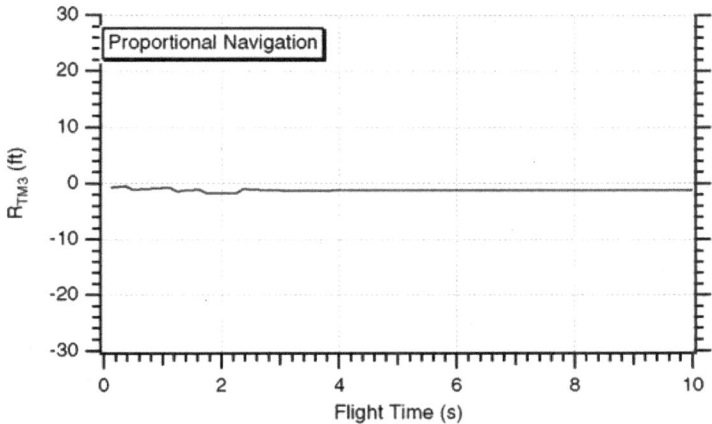

Fig. 17.3   Cross-range component of miss is virtually zero because target does not maneuver in that channel.

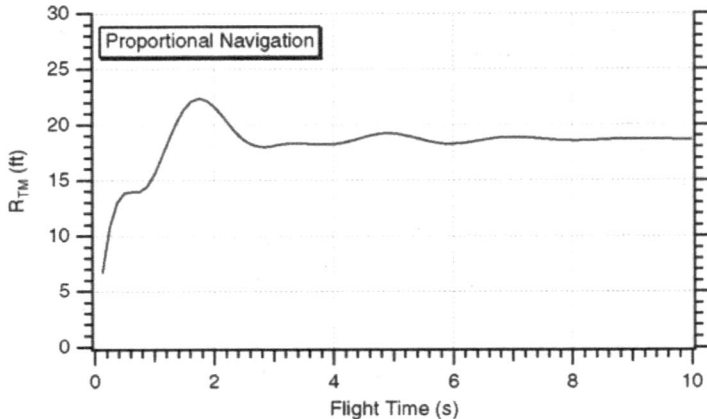

Fig. 17.4   Total three-dimensional miss caused by 3-rad/s weaving target approaches a constant in steady state.

Listing 17.1 was rerun using the compensated weave guidance law (QPN=0). Figure 17.5 displays the total miss vs flight time for the same case of the 3-rad/s spiraling 6-g target and the flight-control-system time constant of 1 s. We can see that with the compensated weave guidance law, the miss is reduced from nearly 20 ft to virtually zero when we move to three dimensions, as predicted by theory and the planar results.

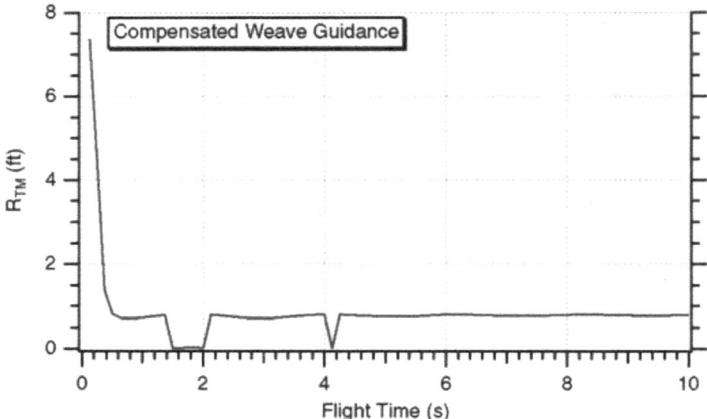

Fig. 17.5   Compensated weave guidance also dramatically reduces the total miss in three dimensions.

BALLISTIC TARGET TRAJECTORY GENERATOR IN THREE DIMENSIONS

Let us now consider a three-dimensional example of a ballistic target trajectory. For simplicity let us consider an example in which a ballistic target is impulsively launched from Monte Carlo to Las Vegas. Because we are neglecting the boost phase of the target, only gravity acts on the ballistic target. As was shown in Chapter 15 from the previous volume, a convenient coordinate system for the simulation of our strategic engagements is an Earth-centered Cartesian coordinate system, as shown in Fig. 17.6. Because this coordinate system is fixed in inertial space (even though the Earth rotates), all missile acceleration differential equations can be integrated directly to yield velocity and position, without having to worry about Coriolis effects.

Because only gravity acts on the ballistic missile of our example, the three differential equations describing the acceleration of a target in a gravity field can be derived from Newton's law of universal gravitation in the Earth-centered Cartesian coordinate system as [2]

$$\ddot{x} = \frac{-gm\,x}{(x^2 + y^2 + z^2)^{1.5}}$$

$$\ddot{y} = \frac{-gm\,y}{(x^2 + y^2 + z^2)^{1.5}}$$

$$\ddot{z} = \frac{-gm\,z}{(x^2 + y^2 + z^2)^{1.5}}$$

where x_T, y_T, and z_T are component distances to the ballistic missile measured from the center of the Earth, and gm is the gravitational parameter with value

$$gm = 1.4077 * 10^{16}\ \text{ft}^3/\text{s}^2$$

in the English system of units.

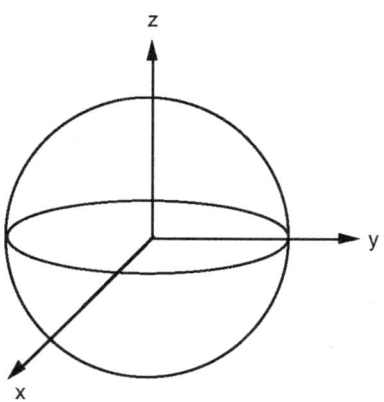

We would like to relate our inertial Earth-centered system (also known as the ECI coordinate system), where we will integrate the target equations of motion, to a system in which we can draw maps (with longitude and latitude) using tools such as the Mapping Toolbox in MATLAB. It is convenient to use an Earth coordinate system for the drawing of maps (also known as the Earth-centered Earth-fixed or ECEF coordinate system). Our Earth coordinate system (x_e, y_e, z_e) is related to

Fig. 17.6   Earth-centered coordinate system.

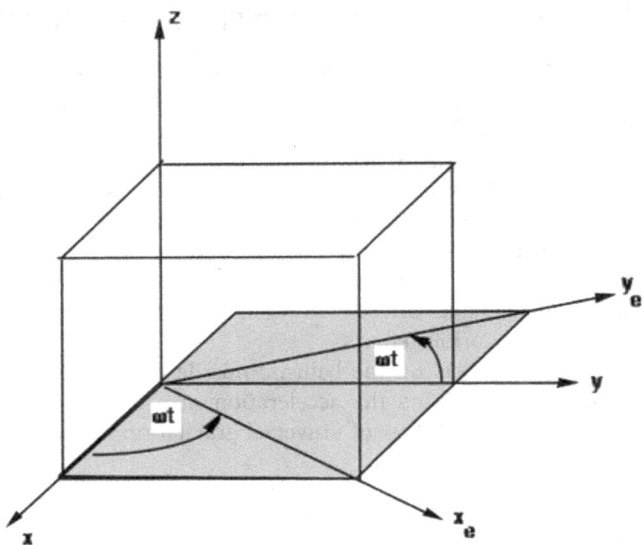

Fig. 17.7   Relationship between ECI and ECEF coordinate system.

the inertial coordinate system (x, y, z) through the rotation of the Earth ω, as shown in Fig. 17.2 [2]. In Fig. 17.7 the Earth rotates at ω rad/s, and t is time.

By inspection of Fig. 17.7, we can see that we can convert ECI coordinates to ECEF coordinates by using the trigonometric relationships

$$x_e = x \cos \omega t + y \sin \omega t$$

$$y_e = -x \sin \omega t + y \cos \omega t$$

$$z_e = z$$

Because the Earth rotates once per day, Earth rotation is given approximately by

$$\omega \approx \frac{360°}{1 \text{ day}} = \frac{6.28 \text{ rad}}{86,400 \text{ s}} = 7.27 * 10^{-5} \text{rad/s}$$

Therefore, an object can be expressed in either ECI or ECEF coordinates. An object that is in ECEF coordinates can also be converted to longitude, latitude, and altitude according to Fig. 17.8 [1].

If an object is in space, the distance from the center of the Earth to the object is simply the radius of the Earth plus the altitude of the object, or

$$r = a + \text{alt} = \sqrt{x_e^2 + y_e^2 + z_e^2}$$

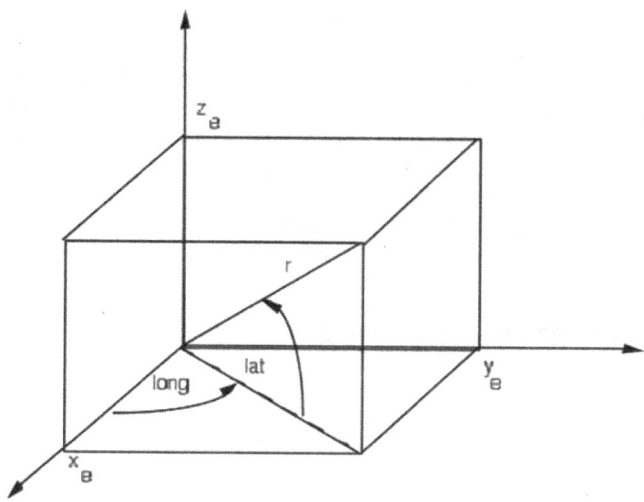

Fig. 17.8   Relationship between ECEF and mapping coordinates.

Longitude and latitude can be derived from Fig. 17.8 as

$$\text{long} = \tan^{-1}\left(\frac{y_e}{x_e}\right)$$

$$\text{lat} = \tan^{-1}\left(\frac{z_e}{\sqrt{x_e^2 + y_e^2}}\right)$$

In addition, if an object is expressed in mapping coordinates, we can convert longitude and latitude to ECEF coordinates according to

$$x_e = r\cos(\text{lat})\cos(\text{long})$$
$$y_e = r\cos(\text{lat})\sin(\text{long})$$
$$z_e = r\sin(\text{lat})$$

We now have enough information to simulate an impulsively launched ballistic missile. All that we are missing is the three-dimensional Lambert routine to provide the initial velocity of the ballistic missile. The new three-dimensional Lambert routine is shown in Listing 17.2. We can see that it is a straightforward extension of the already derived two-dimensional efficient Lambert routine in Chapter 17 of the previous volume. Reference [3] is another good source for understanding the code. In Listing 17.2 we enter the longitude and latitude of the target launch point (Monte Carlo) and its final destination (Las Vegas). It is specified that it will take the ballistic missile 2000 s to reach its destination. The missile starting and ending points are converted to ECI coordinates. We can see from the listing that the missile's destination must be modified in ECI coordinates because the Earth rotates significantly in 2000 s. As was already mentioned, all integrations are carried out in ECI coordinates. The inertial missile trajectory outputs are first converted to ECEF coordinates and then to longitude and latitude. In addition, the downrange (from missile launch point) and altitude of the missile are computed. The longitude and latitude outputs of the missile can be supplied to the MATLAB Mapping Toolbox in order to obtain geographical context.

LISTING 17.2 MATLAB BALLISTIC MISSILE TRAJECTORY GENERATOR

```
clear
count=0;
SWITCH1=0;
XLONGTDEG=7.42;
XLATTDEG=43.75;
XLATFDEG=36.175;
XLONGFDEG=-115.136;
TF=2000.;
A=2.0926E7;
GM=1.4077E16;
W=-6.283185/86400.;
XLONGF=XLONGFDEG/57.3;
XLATF=XLATFDEG/57.3;
XLONGT=XLONGTDEG/57.3;
XLATT=XLATTDEG/57.3;
XLONGF=XLONGF-W*TF;
XF=A*cos(XLATF)*cos(XLONGF);
YF=A*cos(XLATF)*sin(XLONGF);
ZF=A*sin(XLATF);
XT=A*cos(XLATT)*cos(XLONGT);
YT=A*cos(XLATT)*sin(XLONGT);
ZT=A*sin(XLATT);
[VRX,VRY,VRZ]=LAMBERT3D(XT,YT,ZT,TF,XF,YF,ZF,SWITCH1);
XTD=VRX;
YTD=VRY;
ZTD=VRZ;
XTINIT=XT;
YTINIT=YT;
ZTINIT=ZT;
T=0.;
H=.001;
S=0.;
ALTNM=(sqrt(XT^2+YT^2+ZT^2)-A)/6076.;
```

```
while ALTNM>-1
      XTOLD=XT;
      YTOLD=YT;
      ZTOLD=ZT;
      XTDOLD=XTD;
      YTDOLD=YTD;
      ZTDOLD=ZTD;
      STEP=1;
      FLAG=0;
      while STEP<=1
              if FLAG==1
      STEP=2;
                      XT=XT+H*XTD;
                      YT=YT+H*YTD;
                      ZT=ZT+H*ZTD;
                      XTD=XTD+H*XTDD;
                      YTD=YTD+H*YTDD;
                      ZTD=ZTD+H*ZTDD;
      T=T+H;
      end
      TEMPBOTT=(XT^2+YT^2+ZT^2)^1.5;
              XTDD=-GM*XT/TEMPBOTT;
              YTDD=-GM*YT/TEMPBOTT;
              ZTDD=-GM*ZT/TEMPBOTT;
              ALTNM=(sqrt(XT^2+YT^2+ZT^2)-A)/6076.;
              FLAG=1;
      end
      FLAG=0;
      XT=.5*(XTOLD+XT+H*XTD);
      YT=.5*(YTOLD+YT+H*YTD);
      ZT=.5*(ZTOLD+ZT+H*ZTD);
      XTD=.5*(XTDOLD+XTD+H*XTDD);
      YTD=.5*(YTDOLD+YTD+H*YTDD);
      ZTD=.5*(ZTDOLD+ZTD+H*ZTDD);
      S=S+H;
      if S>=9.9999
              S=0.;
              XTE=XT*cos(W*T)-YT*sin(W*T);
              YTE=XT*sin(W*T)+YT*cos(W*T);
              ZTE=ZT;
              XLATT= atan2(ZTE, sqrt(XTE^2+YTE^2));
              XLATTDEG=57.3*XLATT;
              XLONGT= atan2(YTE, XTE);
              XLONGTDEG=57.3*XLONGT;
              DISTRTNM=distance3d(XTE,YTE,ZTE,XTINIT,YTINIT,ZTINIT);
              count=count+1;
```

```
                ArrayT(count)=T;
                ArrayDISTRTNM(count)=DISTRTNM;
                ArrayALTNM(count)=ALTNM;
                ArrayXLONGTDEG(count)=XLONGTDEG;
                ArrayXLATTDEG(count)=XLATTDEG;
        end
end
figure
plot(ArrayDISTRTNM,ArrayALTNM),grid
xlabel('Downrange (Nmi)')
ylabel('Altitude (Nmi) ')
clc
output=[ArrayT',ArrayDISTRTNM',ArrayALTNM'];
save datfil.txt output -ascii
output=[ArrayT',ArrayXLONGTDEG',ArrayXLATTDEG'];
csvwrite('trajfil.txt',output)

disp 'simulation finished'

function DISTKM=distance3d(XT,YT,ZT,XF,YF,ZF)
R=sqrt(XT^2+YT^2+ZT^2);
RF=sqrt(XF^2+YF^2+ZF^2);
A=2.0926E7;
CBETA=(XT*XF+YT*YF+ZT*ZF)/(R*RF);
if CBETA<=1.
        BETA=acos(CBETA);
        DISTKM=A*BETA/3280.;
else
        DISTKM=(XF-XT)/3280.;
end

function [VRX,VRY,VRZ]=LAMBERT3D(XT,YT,ZT,TF,XF,YF,ZF,SWITCH1)
PI=3.1415926535898;
DEG_PER_RAD=57.3;
EARTH_RADIUS=2.0926E7;
GM=1.4077E16;
HALFPI=PI/2.;
FT_PER_KM=3280.;
RAD_PER_SEC=1/57.3;
RF0X=XF-XT;
RF0Y=YF-YT;
RF0Z=ZF-ZT;
RODOTRF=XT*XF+YT*YF+ZT*ZF;
RODOTRF0=XT*RF0X+YT*RF0Y+ZT*RF0Z;
R0MAG=sqrt(XT^2+YT^2+ZT^2);
RFMAG=sqrt(XF^2+YF^2+ZF^2);
```

```
RF0MAG=sqrt(RF0X^2+RF0Y^2+RF0Z^2);
RATIO=R0MAG/RFMAG;
GMDIVR0=GM/R0MAG;
COS_T=R0DOTRF/(R0MAG*RFMAG);
VNUMER=GMDIVR0*(1.-COS_T);
T_MIN=0.;
if  SWITCH1==0
        G_MIN=HALFPI-acos(R0DOTRF0/(R0MAG*RF0MAG));
        G_MAX=HALFPI;
        THETA=acos(COS_T);
else
        G_MIN=-HALFPI;
        G_MAX=-HALFPI+acos(R0DOTRF0/(R0MAG*RF0MAG));
        THETA=2.*PI-acos(COS_T);
end
SIN_T=sin(THETA);
COT_HALFT=1./tan(THETA/2.);
GAMMA=(G_MAX+G_MIN)/2.;
GOLD=G_MIN;
TOLD=0.;
T=0.;
ITERS=1;

S=.5*(R0MAG+RFMAG+RF0MAG);
BL=sqrt(R0MAG*RFMAG)*cos(THETA/2.)/S;
BT=sqrt(8.*GM/(S*S*S))*TF;

while  (abs(TF-T)>(.00000001*TF))
        SIN_G=sin(GAMMA);
        COS_G=cos(GAMMA);
        TAN_G=SIN_G/COS_G;
        COS_TPLUSG=cos(THETA+GAMMA-2.*PI);
        TERM1=(RATIO*COS_G-COS_TPLUSG)*COS_G;
        RV0MAG=sqrt(VNUMER/TERM1);
        LAMBDA=RV0MAG*RV0MAG/GMDIVR0;
        if LAMBDA<1.9999999
          TERM0=sqrt(2./LAMBDA-1.);
          TERM1=(TAN_G*(1.-COS_T)+(1.-LAMBDA)*SIN_T)/...
                    ((2.-LAMBDA)*RATIO);
          TERM2=(COS_G+COS_G)/(LAMBDA*TERM0*TERM0*TERM0);
          TERM3= atan2(TERM0, (COS_G*COT_HALFT-SIN_G));
          T=(R0MAG/(RV0MAG*COS_G))*(TERM1+TERM2*TERM3);
        elseif  LAMBDA>2.0000001
          TERM0=sqrt(1.-2./LAMBDA);
          TERM1=(TAN_G*(1.-COS_T)+(1.-LAMBDA)*SIN_T)/...
                    ((2.-LAMBDA)*RATIO);
```

```
          TERM2=COS_G/(LAMBDA*TERM0*TERM0*TERM0);
          TERM3=SIN_G-COS_G*COT_HALFT;
          TERM3=log((TERM3-TERM0)/(TERM3+TERM0));
          T=(R0MAG/(RV0MAG*COS_G))*(TERM1-TERM2*TERM3);
        else
          TERM0=COS_G*COT_HALFT;
          TERM1=TERM0-SIN_G;
          TERM0=(3*TERM0*TERM1+1.)/(TERM1*TERM1*TERM1);
          T=TERM0*(2.*R0MAG)/(3.*RV0MAG);
        end
        if T>TF & GAMMA<G_MAX
          G_MAX=GAMMA;
        end

        if T<0. & GAMMA<G_MAX
          G_MAX=GAMMA;
        end
        if  T<TF & GAMMA>G_MIN
          G_MIN=GAMMA;
          T_MIN=T;
        end
        if (T<0.)
          NEXT=(G_MIN+G_MAX)/2.;
          GOLD=G_MIN;
          TOLD=T_MIN;
        else
          NEXT=GAMMA+(TF-T)*(GAMMA-GOLD)/(T-TOLD);
          if NEXT>=G_MAX
            NEXT=(GAMMA+G_MAX)/2.;
          elseif NEXT<=G_MIN
            NEXT=(GAMMA+G_MIN)/2.;
          end
          GOLD=GAMMA;
          TOLD=T;
        end
        GAMMA=NEXT;
        ITERS=ITERS+1;
    if ITERS>100
        break
      end
  end
end
if SWITCH1==0
            GAMMA=GAMMA;
        ANGLE=HALFPI-GAMMA;
        SINA=sin(ANGLE);
        COSA=cos(ANGLE);
```

```
        V1X=XT;
        V1Y=YT;
        V1Z=ZT;
        V2X=XF;
        V2Y=YF;
        V2Z=ZF;
        MAG1=sqrt(V1X*V1X+V1Y*V1Y+V1Z*V1Z);
        DOTMAG=V1X*V2X+V1Y*V2Y+V1Z*V2Z;
        CROSSX=V1Y*V2Z-V1Z*V2Y;
        CROSSY=V1Z*V2X-V1X*V2Z;
        CROSSZ=V1X*V2Y-V1Y*V2X;
        CROSSMAG=sqrt(CROSSX*CROSSX+CROSSY*CROSSY+CROSSZ*CROSSZ);
        C2=MAG1*SINA/CROSSMAG;
        C1=COSA/MAG1-DOTMAG*C2/(MAG1*MAG1);
        RTEMPX=C1*V1X;
        RTEMPY=C1*V1Y;
        RTEMPZ=C1*V1Z;
        VUNITX=C2*V2X;
        VUNITY=C2*V2Y;
        VUNITZ=C2*V2Z;
        VUNITX=VUNITX+RTEMPX;
        VUNITY=VUNITY+RTEMPY;
        VUNITZ=VUNITZ+RTEMPZ;
else
        ANGLE=GAMMA-HALFPI;
        SINA=sin(ANGLE);
        COSA=cos(ANGLE);
        V1X=XT;
        V1Y=YT;
        V1Z=ZT;
        V2X=XF;
        V2Y=YF;
        V2Z=ZF;
        MAG1=sqrt(V1X*V1X+V1Y*V1Y+V1Z*V1Z);
        DOTMAG=V1X*V2X+V1Y*V2Y+V1Z*V2Z;
        CROSSX=V1Y*V2Z-V1Z*V2Y
        CROSSY=V1Z*V2X-V1X*V2Z;
        CROSSZ=V1X*V2Y-V1Y*V2X;
        CROSSMAG=sqrt(CROSSX*CROSSX+CROSSY*CROSSY+CROSSZ*CROSSZ);
        C2=MAG1*SINA/CROSSMAG;
        C1=COSA/MAG1-DOTMAG*C2/(MAG1*MAG1);
        RTEMPX=C1*V1X;
        RTEMPY=C1*V1Y;
        RTEMPZ=C1*V1Z;
        VUNITX=C2*V2X;
        VUNITY=C2*V2Y;
```

```
            VUNITZ=C2*V2Z;
            VUNITX=VUNITX+RTEMPX;
            VUNITY=VUNITY+RTEMPY;
            VUNITZ=VUNITZ+RTEMPZ;
end
VRX=RV0MAG*VUNITX;
VRY=RV0MAG*VUNITY;
VRZ=RV0MAG*VUNITZ;
```

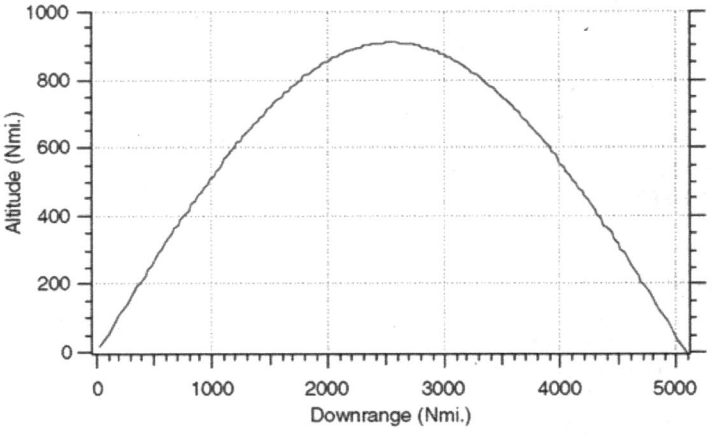

Fig. 17.9   Ballistic target trajectory.

The nominal case of Listing 17.2 was run, and Fig. 17.9 presents a planar view of the actual target trajectory. We can see that the distance from Monte Carlo to Las Vegas is nearly 5000 n miles. The target apogee is approximately 900 n miles.

INTERCEPT POINT PREDICTION FOR BALLISTIC TARGETS

Soon in this chapter we shall attempt to write a three-dimensional engagement simulation for a missile intercepting a ballistic target. To initially aim the missile, we will have to make a prediction of the intercept point. In other words, we want to know the future location of the target at a desired intercept time. One way of finding the intercept point is by integrating the ballistic target equations of motion forward until the desired intercept time. This method certainly works but can be time consuming—especially if a small integration step size is used and the desired intercept time is a large number. A much better method, which is only applicable to ballistic targets (targets that are only influenced by gravity), is to use the numerical solution to Kepler's problem. With

TABLE 17.1   COMPARISON IN PREDICTED INTERCEPT POINT BETWEEN DIRECT INTEGRATION AND KEPLER

Subroutine	x, km	y, km	z, km	$\dot{x}$, km/s	$\dot{y}$, km/s	$\dot{z}$, km/s
Integration	2547.4175	−3583.1082	6758.2423	−3.558504	−3.245617	−0.379447
Kepler	2547.4148	−3583.1074	6758.2384	−3.558509	−3.245615	−0.379455

the Kepler method, the initial target states (position and velocity) are known, and we desire to predict the target states at some given time in the future. Many algorithms exist for solving Kepler's problem; the one used in this text was chosen simply because it appeared in the open literature [4] and not because it is the best. In this section we will not derive the Kepler routine but simply demonstrate that it works.

Let us consider a ballistic target prediction problem in which the initial conditions of the ballistic target are identical to those of the previous ballistic target simulation. We desire to predict the location and velocity of the ballistic target 1000 s after target launch. Listing 17.3 directly integrates the target equations of motion for 1000 s and prints the final states of the target. In addition, at the beginning of Listing 17.3 the target's initial conditions (converted to km and km/s) are fed into a Kepler routine to also predict the final states of the target. Quantities related to the Kepler subroutine are highlighted in bold.

The sample case of Listing 17.3 was run, and comparisons between numerical integration and the use of the Kepler subroutine are presented in Table 17.1. We can see that the difference in answers between the two methods is very small. Thus in the future we shall use the Kepler subroutine for the prediction of the intercept point of a ballistic target.

LISTING 17.3   COMPARING DIRECT INTEGRATION WITH KEPLER PROPAGATION

```
%        Obtained from output of Listing C17L2
clear
XT=14990432.9744621;
YT=1952093.10305573;
ZT=14469752.1663352;
XTD=996.773566434768;
YTD=-14954.8124604715;
ZTD=17528.1931768263;
TF=1000.;
A=2.0926E7;
GM=1.4077E16;
W=-6.283185/86400.;
```

```
T=0.;
H=.001;
S=0.;
T0=0.;
T1=TF;
X0(1)=XT/3280.;
X0(2)=YT/3280.;
X0(3)=ZT/3280.;
X0(4)=XTD/3280.;
X0(5)=YTD/3280.;
X0(6)=ZTD/3280.;
[X1]=KEPLER1(X0,T0,T1);
while T<=TF
        XTOLD=XT;
        YTOLD=YT;
        ZTOLD=ZT;
        XTDOLD=XTD;
        YTDOLD=YTD;
        ZTDOLD=ZTD;
        STEP=1;
        FLAG=0;
        while STEP<=1
                if FLAG==1
        STEP=2;
                        XT=XT+H*XTD;
                        YT=YT+H*YTD;
                        ZT=ZT+H*ZTD;
                        XTD=XTD+H*XTDD;
                        YTD=YTD+H*YTDD;
                        ZTD=ZTD+H*ZTDD;
         T=T+H;
        end
        TEMPBOTT=(XT^2+YT^2+ZT^2)^1.5;
                XTDD=-GM*XT/TEMPBOTT;
                YTDD=-GM*YT/TEMPBOTT;
                ZTDD=-GM*ZT/TEMPBOTT;
                FLAG=1;
        end
        FLAG=0;
        XT=.5*(XTOLD+XT+H*XTD);
        YT=.5*(YTOLD+YT+H*YTD);
        ZT=.5*(ZTOLD+ZT+H*ZTD);
        XTD=.5*(XTDOLD+XTD+H*XTDD);
        YTD=.5*(YTDOLD+YTD+H*YTDD);
        ZTD=.5*(ZTDOLD+ZTD+H*ZTDD);
        S=S+H;
```

```
        if S>=9.9999
            S=0.;
        end
    end
end
XTKM=XT/3280.;
YTKM=YT/3280.;
ZTKM=ZT/3280.;
XTDKM=XTD/3280.;
YTDKM=YTD/3280.;
ZTDKM=ZTD/3280.;
ERRX=XTKM-X1(1);
ERRY=YTKM-X1(2);
ERRZ=ZTKM-X1(3);
ERRXD=XTDKM-X1(4);
ERRYD=YTDKM-X1(5);
ERRZD=ZTDKM-X1(6);
XTKM
X1(1)
YTKM
X1(2)
ZTKM
X1(3)
XTDKM
X1(4)
YTDKM
X1(5)
ZTDKM
X1(6)
ERRX
ERRY
ERRZ
ERRXD
ERRYD
ERRZD

function [X1]=KEPLER1(X0,T0,T1)
GMX=398923.;
REX=6380.;
TLIMIT=1.E-10;
KN=10;
MUQR = 1.;
DT = T1 - T0;
DX=10.;
if (abs(DT) <TLIMIT)
        for I=1:6
        X1(I) = X0(I);
```

```
      end
   end

TIME_FACTOR = sqrt(REX^3/GMX);
VEX = REX/TIME_FACTOR;
DT = DT/TIME_FACTOR;
for I=1:3
      R0(I) = X0(I)/REX;
      V0(I) = X0(I+3)/VEX;
end
R0MAG = sqrt(R0(1)^2 +R0(2)^2 +R0(3)^2);
V0MAG = sqrt(V0(1)^2 +V0(2)^2 +V0(3)^2);
D0 = R0(1)*V0(1) +R0(2)*V0(2) +R0(3)*V0(3);
SIGMA0 = D0/MUQR;
ALP0 = 2./R0MAG - V0MAG*V0MAG;
if  ALP0 == 0.
     A0 = 1.E30;
else
     A0 = 1./ALP0;
end

X = ALP0*DT;
if  ALP0 <= 0.
     X = .1*DT/R0MAG;
end
for K=1:KN
      if ALP0<0.
              Y = ALP0*X*X;
         YQR = sqrt(-Y);
         CY = (1.-cosh(YQR))/Y;
         SY = (sinh(YQR) -YQR)/(YQR^3);
      elseif ALP0==0.
         Y = 0.;
         CY = .5;
         SY = 1./6.;
      else
         Y = ALP0*X*X;
         YQR = sqrt(Y);;
         CY = (1.-cos(YQR))/Y;
         SY = (YQR -sin(YQR))/(YQR^3);
      end
         U1 = X*(1.-Y*SY);
         U2 = X*X*CY;
         U3 = X*X*X*SY;
         FX = R0MAG*U1 +SIGMA0*U2 +U3 -DT*MUQR;
         DFX = SIGMA0*U1 +(1. -ALP0*R0MAG)*U2 +R0MAG;
```

```
        DFX2 = SIGMA0*(1. -Y*CY) +(1. -ALP0*R0MAG)*U1;
        SDFX = DFX/(abs(DFX));
        DX0 = 16.*DFX*DFX;
        DX1 = 20.*FX*DFX2;
        DX2 = 16.*DFX*DFX - 20.*FX*DFX2;
        if  DX2 > 0.
        DX = 5.*FX/(DFX +SDFX*sqrt(DX2));
        else
        DX = .5*X;
        end
        X =X -DX;
end

RMAG =DFX;
F = 1. -U2/R0MAG;
G = DT -U3/MUQR;
DF = -MUQR*U1/(RMAG*R0MAG);
DG = 1. -U2/RMAG;
for I=1:3
        X1(I) = (F*R0(I) +G*V0(I))*REX;
        X1(I+3) = (DF*R0(I) +DG*V0(I))*VEX;
end
```

STRATEGIC MISSILE-TARGET ENGAGEMENT SIMULATION

We now have enough information to write an engagement simulation. All we have to do is to add the missile equations to the existing target simulation of Listing 17.2. We will assume that the missile is also impulsively launched (TLAUNCH seconds after the target is launched) toward the predicted intercept point. The nearly exact predicted intercept point is obtained from the Kepler subroutine of the preceding section. In addition, we will add the ability to introduce deterministic intercept point prediction errors (PREDERR set to desired intercept point prediction error). Because the missile is initially stationary (before it is launched) but is on a rotating Earth, care must be taken on how to integrate the missile differential equations. A careful examination of the code reveals that the missile differential equations are different before and after launch. At some point near the end of the flight (200 s before intercept in this code), the missile guidance system is turned on to take out any remaining errors.

Once the missile guidance system is turned on, we have assumed that proportional navigation is used as the homing guidance law. As shown at the beginning of this chapter, we can express proportional navigation in terms of the zero effort miss perpendicular to the line of sight; an example of its implementation in three dimensions has already been presented in Listing 5.1. Recall that the individual missile guidance acceleration components for proportional navigation are given by

$$a_{\text{XMGUID}} = \frac{N'\text{ZEMPER}_1}{t_{go}^2}$$

$$a_{\text{YMGUID}} = \frac{N'\text{ZEMPER}_2}{t_{go}^2}$$

$$a_{\text{ZMGUID}} = \frac{N'\text{ZEMPER}_3}{t_{go}^2}$$

If the missile is not yet in the homing phase of flight, the individual guidance acceleration components are zero. Therefore, the acceleration differential equations for the impulsive missile after launch become

$$\ddot{x}_M = \frac{-gmx_M}{(x_M^2 + y_M^2 + z_M^2)^{1.5}} + a_{\text{XMGUID}}$$

$$\ddot{y}_M = \frac{-gmy_M}{(x_M^2 + y_M^2 + z_M^2)^{1.5}} + a_{\text{YMGUID}}$$

$$\ddot{z}_M = \frac{-gmz_M}{(x_M^2 + y_M^2 + z_M^2)^{1.5}} + a_{\text{ZMGUID}}$$

Before missile launch, the acceleration differential equations are set to zero so that the missile remains on the ground. The missile-target engagement simulation appears in Listing 17.4. All equations that are related either to the missile or to the relative equations are highlighted in bold. Latitude and longitude information for the missile and target trajectories are written to the comma-delimited file TRAJ-FIL.TXT. The missile attempting to intercept the threat from Monte Carlo (and thereby protecting Las Vegas) is launched from Atlantic City.

LISTING 17.4 THREE-DIMENSIONAL STRATEGIC MISSILE-TARGET ENGAGEMENT SIMULATION (LAMBERT AND KEPLER SUBROUTINES NOT INCLUDED)

```
clear
count=0;
SWITCH1=0;
SWITCHM=0;
TLAUNCH=200.;
XLONGTDEG=7.42;
XLATTDEG=43.75;
XLATFDEG=36.175;
XLONGFDEG=-115.136;
XLONGMDEG=-74.423;
XLATMDEG=39.364;
PREDERR=10.*6076.;
```

```
XNCLIM=161.;
XLATMDEGIC=XLATMDEG;
XLONGMDEGIC=XLONGMDEG;
TFTOT=2000.;
TF=1000.;
A=2.0926E7;
GM=1.4077E16;
W=-6.283185/86400.;
QBOOSTM=1;
XLONGF=XLONGFDEG/57.3;
XLATF=XLATFDEG/57.3;
XLONGT=XLONGTDEG/57.3;
XLATT=XLATTDEG/57.3;
XLONGM=XLONGMDEG/57.3;
XLATM=XLATMDEG/57.3;
XLONGF=XLONGF-W*TF;
XF=A*cos(XLATF)*cos(XLONGF);
YF=A*cos(XLATF)*sin(XLONGF);
ZF=A*sin(XLATF);
XT=A*cos(XLATT)*cos(XLONGT);
YT=A*cos(XLATT)*sin(XLONGT);
ZT=A*sin(XLATT);
[XTD,YTD,ZTD]=LAMBERT3D(XT,YT,ZT,TFTOT,XF,YF,ZF,SWITCH1);
XTINIT=XT;
YTINIT=YT;
ZTINIT=ZT;
XM=A*cos(XLATM)*cos(XLONGM);
YM=A*cos(XLATM)*sin(XLONGM);
ZM=A*sin(XLATM);
XMD=A*W*cos(XLATM)*sin(XLONGM);
YMD=-A*W*cos(XLATM)*cos(XLONGM);
ZMD=0.;
RTM1=XT-XM;
RTM2=YT-YM;
RTM3=ZT-ZM;
RTM=sqrt(RTM1^2+RTM2^2+RTM3^2);
VTM1=XTD-XMD;
VTM2=YTD-YMD;
VTM3=ZTD-ZMD;
VC=-(RTM1*VTM1+RTM2*VTM2+RTM3*VTM3)/RTM;
T=0.;
H=.001;
T0=0.;
T1=TF;
X0(1)=XT/3280.;
X0(2)=YT/3280.;
```

```
X0(3)=ZT/3280.;
X0(4)=XTD/3280.;
X0(5)=YTD/3280.;
X0(6)=ZTD/3280.;
[X1]=KEPLER1(X0,T0,T1);
XTF=X1(1)*3280.;
YTF=X1(2)*3280.;
ZTF=X1(3)*3280.;
XTF=XTF+PREDERR;
S=0.;
DELV=0.;
ALTNM=(sqrt(XT^2+YT^2+ZT^2)-A)/3280.;
while VC>0
        if RTM>5000.
                H=.01;
        else
                H=.00001;
        end
        XTOLD=XT;
        YTOLD=YT;
        ZTOLD=ZT;
        XTDOLD=XTD;
        YTDOLD=YTD;
        ZTDOLD=ZTD;
        XMOLD=XM;
        YMOLD=YM;
        ZMOLD=ZM;
        XMDOLD=XMD;
        YMDOLD=YMD;
        ZMDOLD=ZMD;
        DELVOLD=DELV;
        STEP=1;
        FLAG=0;
        while STEP<=1
                if FLAG==1
        STEP=2;
                        XT=XT+H*XTD;
                        YT=YT+H*YTD;
                        ZT=ZT+H*ZTD;
                        XTD=XTD+H*XTDD;
                        YTD=YTD+H*YTDD;
                        ZTD=ZTD+H*ZTDD;
                        if T<TLAUNCH & QBOOSTM==1
                                XM=XM;
                                YM=YM;
                                ZM=ZM;
```

```
                        XMD=XMD;
                        YMD=YMD;
                        ZMD=ZMD;
                else
                        XM=XM+H*XMD;
                        YM=YM+H*YMD;
                        ZM=ZM+H*ZMD;
                        XMD=XMD+H*XMDD;
                        YMD=YMD+H*YMDD;
                        ZMD=ZMD+H*ZMDD;
                end
                DELV=DELV+H*DELVD;
   T=T+H;
   end
   TEMPBOTT=(XT^2+YT^2+ZT^2)^1.5;
        XTDD=-GM*XT/TEMPBOTT;
        YTDD=-GM*YT/TEMPBOTT;
        ZTDD=-GM*ZT/TEMPBOTT;
        ALTNM=(sqrt(XT^2+YT^2+ZT^2)-A)/3280.;
        RTM1=XT-XM;
        RTM2=YT-YM;
        RTM3=ZT-ZM;
        VTM1=XTD-XMD;
        VTM2=YTD-YMD;
        VTM3=ZTD-ZMD;
        RTM=sqrt(RTM1^2+RTM2^2+RTM3^2);
        VC=-(RTM1*VTM1+RTM2*VTM2+RTM3*VTM3)/RTM;
        TGO=RTM/VC;
        if TGO<200. & T>(TLAUNCH+50.)
                ZEM1=RTM1+VTM1*TGO;
                ZEM2=RTM2+VTM2*TGO;
                ZEM3=RTM3+VTM3*TGO;
                ZEMDOTRTM=(ZEM1*RTM1+ZEM2*RTM2+ZEM3*RTM3)/RTM;
                ZEMPER1=ZEM1-ZEMDOTRTM*RTM1/RTM;
                ZEMPER2=ZEM2-ZEMDOTRTM*RTM2/RTM;
                ZEMPER3=ZEM3-ZEMDOTRTM*RTM3/RTM;
              ZEMPERLOSKM=sqrt(ZEMPER1^2+ZEMPER2^2+ZEMPER3^2)/3280.;
                AXMGUID=3.*ZEMPER1/(TGO^2);
                AYMGUID=3.*ZEMPER2/(TGO^2);
                AZMGUID=3.*ZEMPER3/(TGO^2);
        else
                AXMGUID=0.;
                AYMGUID=0.;
                AZMGUID=0.;
        end
        if AXMGUID>XNCLIM
```

```
                    AXMGUID=XNCLIM;
            elseif AXMGUID<-XNCLIM
                    AXMGUID=-XNCLIM;
            end
            if AYMGUID>XNCLIM
                    AYMGUID=XNCLIM;
            elseif AYMGUID<-XNCLIM
                    AYMGUID=-XNCLIM;
            end
            if AZMGUID>XNCLIM
                    AZMGUID=XNCLIM;
            elseif AZMGUID<-XNCLIM
                    AZMGUID=-XNCLIM;
            end
            if T>TLAUNCH
                    TEMPBOTM=(XM^2+YM^2+ZM^2)^1.5;
                    XMDD=-GM*XM/TEMPBOTM+AXMGUID;
                    YMDD=-GM*YM/TEMPBOTM+AYMGUID;
                    ZMDD=-GM*ZM/TEMPBOTM+AZMGUID;
            else
                    XMDD=0.;
                    YMDD=0.;
                    ZMDD=0.;
            end
            DELVD=sqrt(AXMGUID^2+AYMGUID^2+AZMGUID^2);
            FLAG=1;
    end
    FLAG=0;
    XT=.5*(XTOLD+XT+H*XTD);
    YT=.5*(YTOLD+YT+H*YTD);
    ZT=.5*(ZTOLD+ZT+H*ZTD);
    XTD=.5*(XTDOLD+XTD+H*XTDD);
    YTD=.5*(YTDOLD+YTD+H*YTDD);
    ZTD=.5*(ZTDOLD+ZTD+H*ZTDD);
    if T<TLAUNCH & QBOOSTM==1
            XM=A*cos(XLATMDEGIC/57.3)*cos(XLONGMDEGIC/57.3-W*T);
            YM=A*cos(XLATMDEGIC/57.3)*sin(XLONGMDEGIC/57.3-W*T);
            ZM=A*sin(XLATMDEGIC/57.3);
            XMD=A*W*cos(XLATMDEGIC/57.3)*sin(XLONGMDEGIC/57.3-W*T);
            YMD=-A*W*cos(XLATMDEGIC/57.3)*cos(XLONGMDEGIC/57.3-W*T);
            ZMD=0.;
    else
            XM=.5*(XMOLD+XM+H*XMD);
            YM=.5*(YMOLD+YM+H*YMD);
            ZM=.5*(ZMOLD+ZM+H*ZMD);
            XMD=.5*(XMDOLD+XMD+H*XMDD);
```

```
                YMD=.5*(YMDOLD+YMD+H*YMDD);
                ZMD=.5*(ZMDOLD+ZMD+H*ZMDD);
        end
        DELV=.5*(DELVOLD+DELV+H*DELVD);
        TGOM=TF-T;
        if T>=TLAUNCH & QBOOSTM==1
                [XMD,YMD,ZMD]=LAMBERT3D(XM,YM,ZM,TGOM,XTF,YTF,ZTF,SWITCHM);
                QBOOSTM=0;
                XMDOLD=XMD;
                YMDOLD=YMD;
                ZMDOLD=ZMD;
        end
        S=S+H;
        if S>=9.9999
                S=0.;
                XTE=XT*cos(W*T)-YT*sin(W*T);
                YTE=XT*sin(W*T)+YT*cos(W*T);
                ZTE=ZT;
                XLATT=atan2(ZTE, sqrt(XTE^2+YTE^2));
                XLATTDEG=57.3*XLATT;
                XLONGT=atan2(YTE, XTE);
                XLONGTDEG=57.3*XLONGT;
                DISTRTNM=distance3d(XTE,YTE,ZTE,XTINIT,YTINIT,ZTINIT);
                XME=XM*cos(W*T)-YM*sin(W*T);
                YME=XM*sin(W*T)+YM*cos(W*T);
                ZME=ZM;
                XLATM=atan2(ZME, sqrt(XME^2+YME^2));
                XLATMDEG=57.3*XLATM;
                XLONGM=atan2(YME, XME);
                XLONGMDEG=57.3*XLONGM;
                DISTRMNM=distance3d(XME,YME,ZME,XTINIT,YTINIT,ZTINIT);
                ALTMNM=(sqrt(XM^2+YM^2+ZM^2)-A)/3280.;
                AXMGUIDG=sqrt(AXMGUID^2+AYMGUID^2+AZMGUID^2)/32.2;
                count=count+1;
                ArrayT(count)=T;
                ArrayDISTRTNM(count)=DISTRTNM;
                ArrayALTNM(count)=ALTNM;
                ArrayDISTRMNM(count)=DISTRMNM;
                ArrayALTMNM(count)=ALTMNM;
                ArrayAXMGUIDG(count)=AXMGUIDG;
                ArrayXLONGTDEG(count)=XLONGTDEG;
                ArrayXLATTDEG(count)=XLATTDEG;
                ArrayXLONGMDEG(count)=XLONGMDEG;
                ArrayXLATMDEG(count)=XLATMDEG;
        end
end
```

```
XTE=XT*cos(W*T)-YT*sin(W*T);
YTE=XT*sin(W*T)+YT*cos(W*T);
ZTE=ZT;
XLATT=atan2(ZTE, sqrt(XTE^2+YTE^2));
XLATTDEG=57.3*XLATT;
XLONGT=atan2(YTE, XTE);
XLONGTDEG=57.3*XLONGT;
DISTRTNM=distance3d(XTE,YTE,ZTE,XTINIT,YTINIT,ZTINIT);
XME=XM*cos(W*T)-YM*sin(W*T);
YME=XM*sin(W*T)+YM*cos(W*T);
ZME=ZM;
XLATM=atan2(ZME, sqrt(XME^2+YME^2));
XLATMDEG=57.3*XLATM;
XLONGM=atan2(YME, XME);
XLONGMDEG=57.3*XLONGM;
DISTRMNM=distance3d(XME,YME,ZME,XTINIT,YTINIT,ZTINIT);
ALTMNM=(sqrt(XM^2+YM^2+ZM^2)-A)/3280.;
AXMGUIDG=sqrt(AXMGUID^2+AYMGUID^2+AZMGUID^2)/32.2;
count=count+1;
ArrayT(count)=T;
ArrayDISTRTNM(count)=DISTRTNM;
ArrayALTNM(count)=ALTNM;
ArrayDISTRMNM(count)=DISTRMNM;
ArrayALTMNM(count)=ALTMNM;
ArrayAXMGUIDG(count)=AXMGUIDG;
ArrayXLONGTDEG(count)=XLONGTDEG;
ArrayXLATTDEG(count)=XLATTDEG;
ArrayXLONGMDEG(count)=XLONGMDEG;
ArrayXLATMDEG(count)=XLATMDEG;
RTM
DELV
figure
plot(ArrayDISTRTNM,ArrayALTNM,ArrayDISTRMNM,ArrayALTMNM),grid
xlabel('Downrange (km)')
ylabel('Altitude (km) ')
figure
plot(ArrayT,ArrayAXMGUIDG),grid
xlabel('Time (s))')
ylabel('Acceleration (g) ')
axis([0 1000 0 .5])
clc
output=[ArrayT',ArrayDISTRTNM',ArrayALTNM',ArrayDISTRMNM',...ArrayALTMNM',
          ArrayAXMGUIDG'];
save datfil.txt output -ascii
output=[ArrayT',ArrayXLONGTDEG',ArrayXLATTDEG',ArrayXLONGMDEG',...Array
          XLATMDEG'];
```

```
csvwrite('trajfil.txt',output)
disp 'simulation finished'
% 3d Kepler and Lambert routines previously presented in this chapter
```

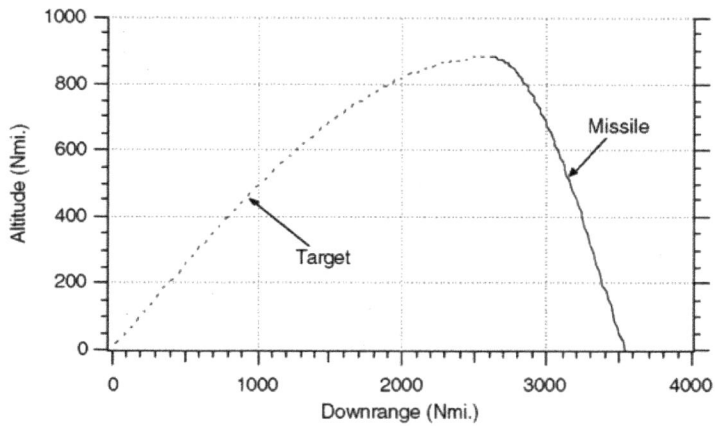

Fig. 17.10   Missile and target trajectories.

The nominal case of Listing 17.4 was run in which there was 10 n miles of intercept point prediction error. Figure 17.10 displays the missile and target trajectories. We can see that intercept takes place about 2500 n miles from the target launch site at approximately 900 n miles altitude. Figure 17.11 plots the total

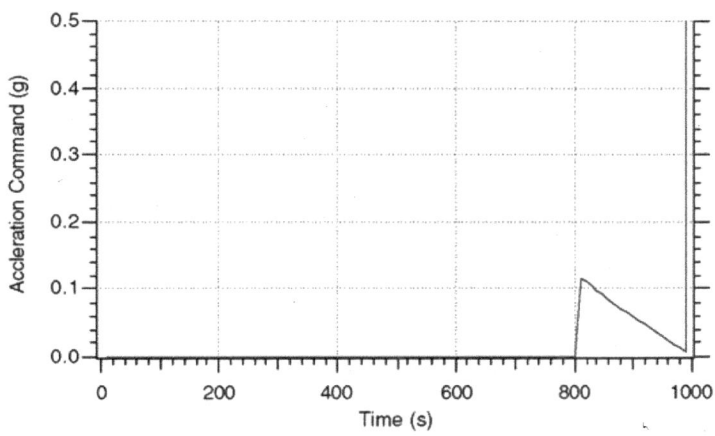

Fig. 17.11   Acceleration required to take out prediction error.

commanded acceleration required for the missile to hit the target. Because homing guidance started at 800 s (200 s before intercept), the acceleration command was zero before that. A maximum acceleration of 0.1 g was required to take out the prediction error in 200 s.

TRAJECTORY SHAPING GUIDANCE IN THREE DIMENSIONS

In Chapter 13 we showed that the trajectory shaping guidance law in one dimension was given by

$$n_c(t) = \frac{6y + 4\dot{y}t_{go} + n_T t_{go}^2 + 2\dot{y}(t_F)t_{go}}{t_{go}^2}$$

where y was the relative position between the target and missile, $\dot{y}$ was the relative velocity between target and missile, n_T was the target acceleration, and $\dot{y}(t_F)$ was the desired relative velocity at intercept between the target and missile. At the beginning of this chapter we showed how several of the guidance laws presented in the text could be programmed in three dimensions. In this section we shall also show how we can convert the trajectory shaping guidance law to three dimensions. Before we can convert the preceding guidance law to three dimensions, it is first necessary to write the trajectory shaping guidance law in a slightly different form than was presented in Chapter 13. Expanding the final relative velocity term of the trajectory shaping guidance law yields

$$n_c = \frac{6y + 4\dot{y}t_{go} + n_T t_{go}^2 + 2[\dot{y}_T(t_F) - \dot{y}_M(t_F)]t_{go}}{t_{go}^2}$$

If we assume that the target velocity does not change very much during the flight, then we can say that

$$\dot{y}_T(t_F) \approx \dot{y}_T$$

and the trajectory shaping guidance law can then be rewritten as

$$n_c = \frac{6y + 4\dot{y}t_{go} + n_T t_{go}^2 + 2[\dot{y}_T - \dot{y}_M(t_F)]t_{go}}{t_{go}^2}$$

Because we know that

$$\dot{y} = \dot{y}_T - \dot{y}_M$$

we can also say that

$$n_c = \frac{6y + 4\dot{y}t_{go} + n_T t_{go}^2 + 2[\dot{y} + \dot{y}_M - \dot{y}_M(t_F)]t_{go}}{t_{go}^2}$$

or

$$n_c = \frac{6y + 6\dot{y}t_{go} + n_T t_{go}^2 + 2[\dot{y}_M - \dot{y}_M(t_F)]t_{go}}{t_{go}^2}$$

LISTING 17.5 TRAJECTORY SHAPING GUIDANCE LAW IN THREE-DIMENSIONAL SIMULATION

```
clear
n=0;
XNTG=4.;
VT=1000.;
VM=3000.;
RM1=0.;
RM2=10000.;
RM3=-1000.;
RT1=30000.;
RT2=10000.;
RT3=0.;
GAMFPDEG= -30.;
GAMFYDEG= 20.;
XNT=32.2*XNTG;
BETA=0.;
VT1=-VT*cos(BETA);
VT2=VT*sin(BETA);
VT3=0;
GAMFP=GAMFPDEG/57.3;
GAMFY=GAMFYDEG/57.3;
[VM1,VM2,VM3,TF]=LAUNCHLOGIC(RM1,RM2,RM3,RT1,RT2,RT3,VT1,VT2,...
                            VT3,VM);
H=.0001;
T=0.;
S=0.;
RTM1=RT1-RM1;
RTM2=RT2-RM2;
RTM3=RT3-RM3;
RTM=sqrt(RTM1^2+RTM2^2+RTM3^2);
VTM1=VT1-VM1;
VTM2=VT2-VM2;
VTM3=VT3-VM3;
VC=-(RTM1*VTM1+RTM2*VTM2+RTM3*VTM3)/RTM;
VM1F=VM*cos(GAMFP)*cos(GAMFY);
VM2F=VM*sin(GAMFP);
VM3F=VM*cos(GAMFP)*sin(GAMFY);
```

```
while ~(VC<0)
      if RTM<1000.
        H=.00001;
      else
        H=.0001;
      end
      BETAOLD=BETA;
      RT1OLD=RT1;
      RT2OLD=RT2;
      RT3OLD=RT3;
      RM1OLD=RM1;
      RM2OLD=RM2;
      RM3OLD=RM3;
      VM1OLD=VM1;
      VM2OLD=VM2;
      VM3OLD=VM3;
      STEP=1;
      FLAG=0;
      while STEP<=1
      if FLAG==1
            STEP=2;
            BETA=BETA+H*BETAD;
            RT1=RT1+H*VT1;
            RT2=RT2+H*VT2;
            RT3=RT3+H*VT3;
            RM1=RM1+H*VM1;
            RM2=RM2+H*VM2;
            RM3=RM3+H*VM3;
            VM1=VM1+H*AM1;
            VM2=VM2+H*AM2;
            VM3=VM3+H*AM3;
            T=T+H;
      end
      RTM1=RT1-RM1;
      RTM2=RT2-RM2;
      RTM3=RT3-RM3;
      RTM=sqrt(RTM1^2+RTM2^2+RTM3^2);
      VTM1=VT1-VM1;
      VTM2=VT2-VM2;
      VTM3=VT3-VM3;
      VC=-(RTM1*VTM1+RTM2*VTM2+RTM3*VTM3)/RTM;
      TGO=RTM/VC;
      BETAD=XNT/VT;
      XNT1=XNT*sin(BETA);
      XNT2=XNT*cos(BETA);
      XNT3=0.;
```

```
        VT1=-VT*cos(BETA);
        VT2= VT*sin(BETA);
        VT3=0.;
        VM=sqrt(VM1*VM1+VM2*VM2+VM3*VM3);
        XNC1=((6.*RTM1+6.*VTM1*TGO)/TGO^2) +2.0*(VM1-VM1F)/(TGO)+XNT1;
        XNC2=((6.*RTM2+6.*VTM2*TGO)/TGO^2) +2.0*(VM2-VM2F)/(TGO)+XNT2;
        XNC3=((6.*RTM3+6.*VTM3*TGO)/TGO^2) +2.0*(VM3-VM3F)/(TGO)+XNT3;
        XNCG=sqrt(XNC1^2+XNC2^2+XNC3^2)/32.2;
        XNCDOTVM=(XNC1*VM1+XNC2*VM2+XNC3*VM3)/VM;
        AM1=XNC1-XNCDOTVM*VM1/VM;
        AM2=XNC2-XNCDOTVM*VM2/VM;
        AM3=XNC3-XNCDOTVM*VM3/VM;
        GAMYDEG=57.3*atan2(VM3,VM1);
        GAMPDEG=57.3*atan2(VM2,sqrt(VM1^2+VM3^2));
        FLAG=1;
end
FLAG=0;
BETA=.5*(BETAOLD+BETA+H*BETAD);
RT1=.5*(RT1OLD+RT1+H*VT1);
RT2=.5*(RT2OLD+RT2+H*VT2);
RT3=.5*(RT3OLD+RT3+H*VT3);
RM1=.5*(RM1OLD+RM1+H*VM1);
RM2=.5*(RM2OLD+RM2+H*VM2);
RM3=.5*(RM3OLD+RM3+H*VM3);
VM1=.5*(VM1OLD+VM1+H*AM1);
VM2=.5*(VM2OLD+VM2+H*AM2);
VM3=.5*(VM3OLD+VM3+H*AM3);
S=S+H;
if S>=.0999
    S=0.;
    n=n+1;
    RT1K=RT1/1000.;
    RT2K=RT2/1000.;
    RT3K=RT3/1000.;
    RM1K=RM1/1000.;
    RM2K=RM2/1000.;
    RM3K=RM3/1000.;
    ArrayT(n)=T;
    ArrayRM1K(n)=RM1K;
    ArrayRM2K(n)=RM2K;
    ArrayRM3K(n)=RM3K;
    ArrayRT1K(n)=RT1K;
    ArrayRT2K(n)=RT2K;
    ArrayRT3K(n)=RT3K;
    ArrayGAMPDEG(n)GAMPDEG;
    ArrayGAMFPDEG(n)=GAMFPDEG;
```

```
      ArrayGAMYDEG(n)=GAMYDEG;
      ArrayGAMFYDEG(n)=GAMFYDEG;
      ArrayXNCG(n)=XNCG;
    end
end
n=n+1;
      RT1K=RT1/1000.;
      RT2K=RT2/1000.;
      RT3K=RT3/1000.;
      RM1K=RM1/1000.;
      RM2K=RM2/1000.;
      RM3K=RM3/1000.;
      ArrayT(n)=T;
      ArrayRM1K(n)=RM1K;
      ArrayRM2K(n)=RM2K;
      ArrayRM3K(n)=RM3K;
      ArrayRT1K(n)=RT1K;
      ArrayRT2K(n)=RT2K;
      ArrayRT3K(n)=RT3K;
      ArrayGAMPDEG(n)=GAMPDEG;
      ArrayGAMFPDEG(n)=GAMFPDEG;
      ArrayGAMYDEG(n)=GAMYDEG;
      ArrayGAMFYDEG(n)=GAMFYDEG;
      ArrayXNCG(n)=XNCG;
      figure
      plot(ArrayRM1K,ArrayRM2K,ArrayRT1K,ArrayRT2K),grid
      xlabel('Downrange (kft)')
      ylabel('Altitude (kft)')
figure
plot(ArrayRM1K,ArrayRM3K,ArrayRT1K,ArrayRT3K),grid
xlabel(apos;Downrange (kft)')
ylabel('Crossrange (kft)')
figure
plot(ArrayT,ArrayGAMPDEG,ArrayT,ArrayGAMFPDEG),grid
xlabel('Time (Sec)')
ylabel('Pitch Reentry Angle (deg)')
figure
plot(ArrayT,ArrayGAMYDEG,ArrayT,ArrayGAMFYDEG),grid
xlabel('Time (Sec)')
ylabel('Yaw Reentry Angle (deg)')
figure
plot(ArrayT,ArrayXNCG),grid
xlabel('Time (Sec)')
ylabel('Commanded Acceleration (g)')
axis([00,10,00,60])
```

```
clc
output=[ArrayT',ArrayRM1K',ArrayRM2K',ArrayRM3K',ArrayRT1K',...
        ArrayRT2K',ArrayRT3K',ArrayGAMPDEG',ArrayGAMFPDEG',...
        ArrayGAMYDEG',ArrayGAMFYDEG'];
save datfil.txt output -ascii
disp 'simulation finished'
RTM

function [VM1,VM2,VM3,TF]=LAUNCHLOGIC(RM1,RM2,RM3,RT1,RT2,RT3,...
                                      VT1,VT2,VT3,VM)
for TF=.1:.1:10,
        RTM1=RT1-RM1;
        RTM2=RT2-RM2;
        RTM3=RT3-RM3;
        RT1F=RT1+VT1*TF;
        RT2F=RT2+VT2*TF;
        RT3F=RT3+VT3*TF;
        THET=asin((RT2F-RM2)/(VM*TF));
        PSI=atan2(RT3F-RM3,RT1F-RM1);
        VM1=VM*cos(THET)*cos(PSI);
        VM2=VM*sin(THET);
        VM3=VM*cos(THET)*sin(PSI);
        RM1F=RM1+VM1*TF;
        RM2F=RM2+VM2*TF;
        RM3F=RM3+VM3*TF;
        RTM1F=RT1F-RM1F;
        RTM2F=RT2F-RM2F;
        RTM3F=RT3F-RM3F;
        RTMF=sqrt(RTM1F^2+RTM2F^2+RTM3F^2);
        VTM1=VT1-VM1;
        VTM2=VT2-VM2;
        VTM3=VT3-VM3;
        VC=-(RTM1F*VTM1+RTM2F*VTM2+RTM3F*VTM3)/RTMF;
        if VC<0
                break
        end
end
```

An examination of the term in brackets of the preceding expression indicates that we are attempting to make the missile velocity $\dot{y}_M$ reach a specified value at the end of the flight. Soon we will show that this is equivalent to controlling the missile flight-path angle. By duplicating the expression for the trajectory shaping guidance law in each of the Earth's inertial coordinates, we can express the guidance law in three dimensions by inspection as

$$n_{c_1} = \frac{6R_{TM1} + 6V_{TM1}t_{go} + n_{T1}t_{go}^2 + 2[V_{M1} - V_{M1}(t_F)]t_{go}}{t_{go}^2}$$

$$n_{c_2} = \frac{6R_{TM2} + 6V_{TM2}t_{go} + n_{T2}t_{go}^2 + 2[V_{M2} - V_{M2}(t_F)]t_{go}}{t_{go}^2}$$

$$n_{c_3} = \frac{6R_{TM3} + 6V_{TM3}t_{go} + n_{T3}t_{go}^2 + 2[V_{M3} - V_{M3}(t_F)]t_{go}}{t_{go}^2}$$

where R_{TM} and V_{TM} are relative position and velocity, respectively. In the preceding expressions, 1, 2, and 3 represent downrange, altitude, and cross range, respectively, in the Earth or inertial coordinate system. Thus the guidance commands are in each of those directions. For the trajectory shaping guidance law, we want to make the total acceleration perpendicular to the missile velocity vector. This will ensure that the missile velocity will remain constant throughout the flight. A similar approach was taken in [5].

If we want to make the missile hit the target at desired flight-path angles γ_{PF} and γ_{YF}, we can say that the desired missile velocity components at the end of the flight are given by

$$V_{M1}(t_F) = V_M \cos \gamma_{PF} \cos \gamma_{YF}$$

$$V_{M2}(t_F) = V_M \sin \gamma_{PF}$$

$$V_{M3}(t_F) = V_M \cos \gamma_{PF} \sin \gamma_{YF}$$

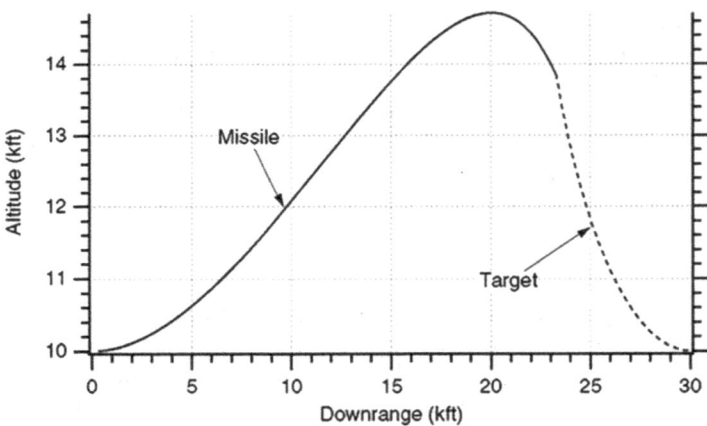

Fig. 17.12   Engagement viewed in altitude-downrange plane.

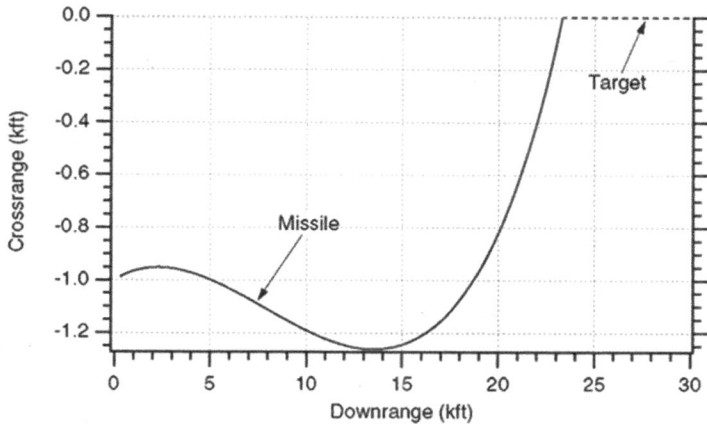

Fig. 17.13 Engagement viewed in crossrange-downrange plane.

where V_M is the total missile velocity. The trajectory shaping guidance law was implemented in the three-dimensional simulation of Listing 17.5. In the nominal case of the simulation, the target is executing a constant 4-g maneuver perpendicular to the target velocity vector in the altitude-downrange plane. In addition, it is desired that the missile hit the target with a pitch flight-path angle of -30 deg and a yaw flight-path angle of 20 deg. The simulation calculates the instantaneous pitch and yaw flight-path angles, as well as the miss distance, to see if the trajectory shaping guidance law meets its objectives. A missile launch logic subroutine, valid for flight times of less than 10 s, is included to place the missile on a collision triangle (assuming no target maneuver) with the target.

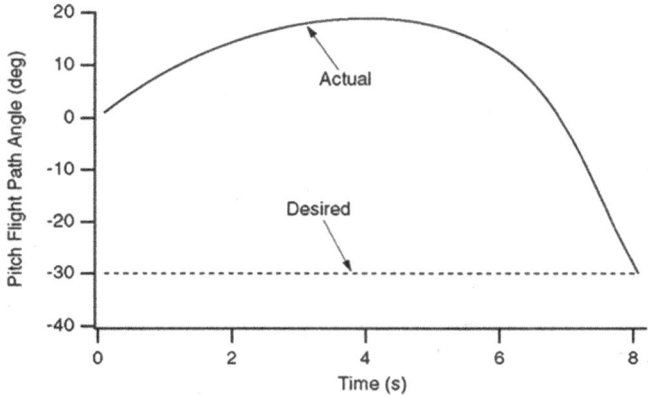

Fig. 17.14 Trajectory shaping guidance law enables missile to achieve pitch flight-path-angle objective.

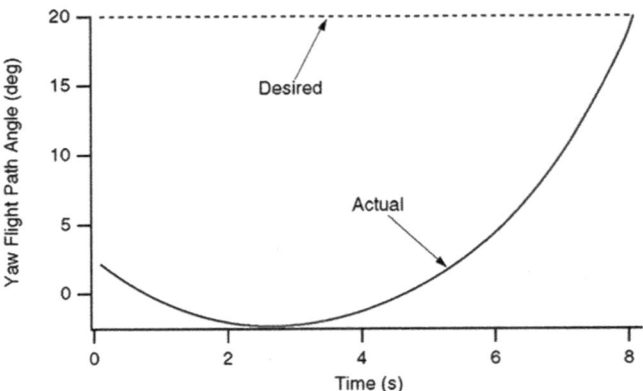

Fig. 17.15   Trajectory shaping guidance law enables missile to achieve yaw flight-path-angle objective.

The nominal case of Listing 17.5 was run. The resultant altitude-downrange and crossrange-downrange trajectories of Figs 17.12 and 17.13 indicate that the missile is hitting the target. In addition, we can see that there is much curvature to the missile trajectory, indicating that a great deal of trajectory shaping has taken place.

Figures 17.14 and 17.15 present the missile pitch and yaw flight-path angle profiles, respectively. We can see from both figures that the missile is meeting the desired pitch and yaw flight-path-angle objectives at the end of the flight; however, we also can see from Fig. 17.16 that the price paid for the trajectory

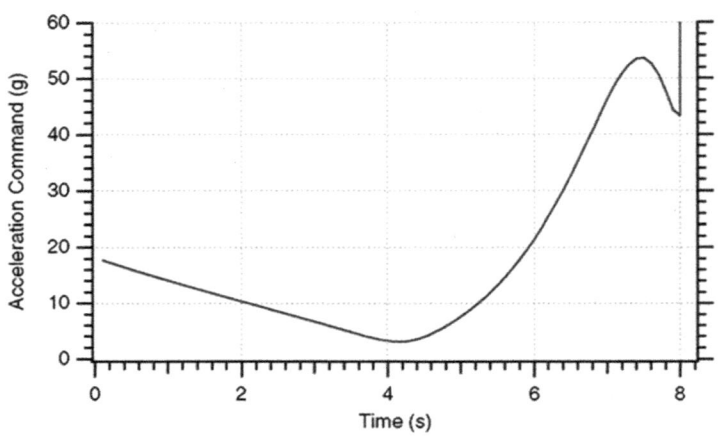

Fig. 17.16   A great deal of acceleration is required to make trajectory shaping guidance law work.

shaping is a very large commanded missile acceleration. Chapter 13 goes into detail in showing how the acceleration requirements of trajectory shaping guidance can be reduced.

SUMMARY

This chapter has shown how three-dimensional simulations can be used to convey important information. In the tactical world we have seen how new insights against spiraling targets can be gained in three dimensions. We saw that the miss caused by a weaving target does not oscillate in three dimensions as it does in one or two dimensions but in fact approaches steady state as the flight time increases. Code has been provided to extend the Lambert routine to three dimensions in our strategic engagement simulations. New code for a Kepler subroutine has been provided so that we could do intercept point prediction for three-dimensional ballistic targets without having to resort to numerical integration.

REFERENCES

[1] Miller, R., and Reddy, F., "Mapping the World in Pascal," *BYTE*, Vol. 12, Dec. 1987, pp. 329–334.

[2] Bate, R. R., Mueller, D. D., and White, J. E., *Fundamentals of Astrodynamics*, Dover, New York, 1971.

[3] Nelson, S. L., and Zarchan, P., "Alternative Approach to the Solution of Lambert's Problem," *Journal of Guidance, Control, and Dynamics*, Vol. 15, No. 4, 1992, pp. 1003–1009.

[4] Vinti, J. P., *Orbital and Celestial Mechanics*, G. J. Der, and N. L. Bonavito, eds., Progress in Astronautics and Aeronautics Series, AIAA, Reston, VA, 1998.

[5] Ohlmeyer, E. J., and Phillips, C. A., "Generalized Vector Explicit Guidance," *Journal of Guidance, Control, and Dynamics*, Vol. 29, No. 2, 2006, pp. 261–268.

CHAPTER 18

Smallest Possible Miss Distance for a Radar Homing Missile

INTRODUCTION

In this chapter we shall find the minimum achievable miss distance with a radar homing missile. Formulas will be developed that depend on error sources beyond the control of the missile designer for establishing the minimum RMS miss distance. Techniques for achieving the minimum possible RMS miss are discussed, and examples demonstrating their utility are presented.

SMALLEST POSSIBLE MISS DISTANCE FOR A RADAR HOMING MISSILE

In preliminary analysis it is sometimes of interest to find out the smallest possible miss distance a radar homing missile can achieve [1]. This calculation can be done by assuming the missile has infinite acceleration capability and that the two main sources of error are glint noise and target maneuver. The missile designer has no control over these error sources because they depend on the target. The standard deviation of the glint noise σ_{GL} is approximately one-fifth of the target length that is perpendicular to the line of sight. This noise is highly correlated and can be modeled as white noise through a low-pass filter where the time constant of the low-pass filter T_{GL} can range from 0.1 s to 0.25 s [2].

However, if a frequency agile radar is available, the glint can simply be approximated as white noise as far as the guidance system is concerned [1]. If frequency agility is applied at the rate f_s, then

$$f_s = \frac{1}{T_s}$$

where T_s is the sampling time of the frequency agile radar. The spectral density of the white glint noise in units squared per Hz is related to the standard deviation of the glint according to

$$\Phi_{GL} = \sigma_{GL}^2 T_s$$

The preceding relationship is identical to the one we used in Chapter 4 of the previous volume with the sampling time T_s being replaced by the integration interval.

Two factors will determine system performance if the missile has infinite acceleration capability. The first factor is the Kalman filter used to estimate the target states. If the Kalman filter process noise model is matched to the expected target maneuver and if the filter measurement noise model is matched to the actual measurement noise, then the filter is optimal. For the three-state Kalman filter discussed throughout this text, and the previous volume the first state is relative position, and the square root of the first diagonal element of the covariance matrix represents the smallest possible RMS value of the error in the estimate of position. Because we cannot control the missile any better than we can estimate the states, this quantity also represents the smallest possible RMS miss distance. The second factor is the guidance law. If we use an optimal guidance law (also matched to the shape of the expected maneuver), we should be able to achieve the minimum possible RMS miss, provided we have adequate missile acceleration capability and that the flight time is long enough so that system transients such as heading error do not contribute to the miss.

If the sampling time of the guidance system is sufficiently small, the performance of a discrete Kalman filter will approach that of a continuous Kalman filter. In this section we shall show that the steady-state, closed-form solutions for the diagonal elements of the covariance matrix of a continuous Kalman filter that depends on the glint noise and target maneuver levels can be found.

The original three-state linear Kalman filter of Chapter 9 of the previous volume was derived based on the homing loop model of Fig. 18.1. Recall that in this guidance system model we measured noisy relative position y^* and were attempting to estimate relative position, relative velocity, and target acceleration. As was the case in Chapter 9 of the previous volume, the achieved missile acceleration n_L was assumed to be known, and the target acceleration was considered to be modeled as a white noise through an integrator. It was shown in Chapter 4 of

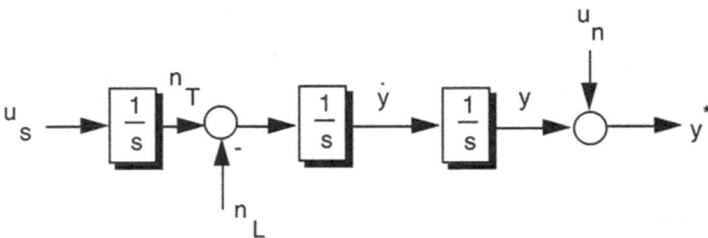

Fig. 18.1   Homing loop model to three-state Kalman filter development.

the previous volume that the shaping filter equivalent of a target maneuver with constant amplitude and random starting time (or uniformly distributed target maneuver) is mathematically equivalent in terms of second-order statistics to white noise through an integrator. It was shown in Chapter 5 of this text that as long as there is no acceleration saturation, the RMS miss resulting from the shaping filter equivalent of the random target maneuver and the actual maneuver are equivalent.

According to the results of Chapter 4 of the previous volume, the spectral density of the white noise source u_s depicted in Fig. 18.1 is shown to be

$$\Phi_S = \frac{n_{T\text{MAX}}^2}{t_F}$$

where $n_{T\text{MAX}}$ is the assumed maximum target maneuver level magnitude and t_F the flight time over which the starting time of the maneuver is equally likely to occur. In this example, u_n is the measurement noise, which is assumed to be white glint noise. The model of Fig. 18.1 can be expressed in state space form as

$$\begin{bmatrix} \dot{y} \\ \ddot{y} \\ \dot{n}_T \end{bmatrix} = \begin{bmatrix} 0 & 1 & 0 \\ 0 & 0 & 1 \\ 0 & 0 & 0 \end{bmatrix} \begin{bmatrix} y \\ \dot{y} \\ n_T \end{bmatrix} + \begin{bmatrix} 0 \\ -1 \\ 0 \end{bmatrix} n_L + \begin{bmatrix} 0 \\ 0 \\ u_s \end{bmatrix}$$

Because the systems dynamics matrix of the preceding equation is given by

$$\mathbf{F} = \begin{bmatrix} 0 & 1 & 0 \\ 0 & 0 & 1 \\ 0 & 0 & 0 \end{bmatrix}$$

and the continuous process noise matrix can be found from

$$Q = E \left\{ \begin{bmatrix} 0 \\ 0 \\ u_s \end{bmatrix} \begin{bmatrix} 0 & 0 & u_s \end{bmatrix} \right\} = \begin{bmatrix} 0 & 0 & 0 \\ 0 & 0 & 0 \\ 0 & 0 & \Phi_s \end{bmatrix} = \frac{n_{T\text{MAX}}^2}{t_F} \begin{bmatrix} 0 & 0 & 0 \\ 0 & 0 & 0 \\ 0 & 0 & 1 \end{bmatrix}$$

The measurement equation can be seen from Fig. 18.1 to be

$$y^* = y + u_n = \begin{bmatrix} 1 & 0 & 0 \end{bmatrix} \begin{bmatrix} y \\ \dot{y} \\ n_T \end{bmatrix} + u_n$$

Therefore, the measurement matrix can be written by inspection of the previous equation as

$$H = \begin{bmatrix} 1 & 0 & 0 \end{bmatrix}$$

and the measurement noise matrix is a scalar given by

$$R = E[u_n u_n^T] = \Phi_n$$

   The differential equation for the covariance matrix of a continuous Kalman filter is given by [3]

$$\dot{P} = -PH^{T}R^{-1}HP + PF^{T} + FP + Q$$

where P is the covariance matrix and has the property of being symmetric. We can solve the preceding equation in the steady state by setting the derivative of the covariance matrix to zero, or

$$0 = -PH^{T}R^{-1}HP + PF^{T} + FP + Q$$

Substitution of the appropriate matrices into the preceding equation yields

$$\begin{bmatrix} 0 & 0 & 0 \\ 0 & 0 & 0 \\ 0 & 0 & 0 \end{bmatrix} = - \begin{bmatrix} P_{11} & P_{12} & P_{13} \\ P_{12} & P_{22} & P_{23} \\ P_{13} & P_{23} & P_{33} \end{bmatrix} \begin{bmatrix} 1 \\ 0 \\ 0 \end{bmatrix} \frac{1}{\Phi_n} \begin{bmatrix} 1 & 0 & 0 \end{bmatrix} \begin{bmatrix} P_{11} & P_{12} & P_{13} \\ P_{12} & P_{22} & P_{23} \\ P_{13} & P_{23} & P_{33} \end{bmatrix}$$

$$+ \begin{bmatrix} P_{11} & P_{12} & P_{13} \\ P_{12} & P_{22} & P_{23} \\ P_{13} & P_{23} & P_{33} \end{bmatrix} \begin{bmatrix} 0 & 0 & 0 \\ 1 & 0 & 0 \\ 0 & 1 & 0 \end{bmatrix} + \begin{bmatrix} 0 & 1 & 0 \\ 0 & 0 & 1 \\ 0 & 0 & 0 \end{bmatrix}$$

$$\times \begin{bmatrix} P_{11} & P_{12} & P_{13} \\ P_{12} & P_{22} & P_{23} \\ P_{13} & P_{23} & P_{33} \end{bmatrix} + \begin{bmatrix} 0 & 0 & 0 \\ 0 & 0 & 0 \\ 0 & 0 & \Phi_s \end{bmatrix}$$

After taking symmetry into account and multiplying out the equations, we get the following six scalar algebraic equations:

$$P_{11}^2 = 2P_{12}\Phi_n$$

$$P_{12}^2 = 2P_{23}\Phi_n$$

$$P_{13}^2 = \Phi_s\Phi_n$$

$$P_{11}P_{12} = \Phi_n(P_{22} + P_{13})$$

$$P_{11}P_{13} = P_{23}\Phi_n$$

$$P_{12}P_{13} = P_{33}\Phi_n$$

After much algebra we find that

$$P_{11} = 2\Phi_s^{1/6}\Phi_n^{5/6}$$

Because the square root of P_{11} represents the error in the estimate of relative position, taking the square root of both sides of the preceding equation yields the best the filter can estimate relative position, or

$$\sqrt{P_{11}} = \sqrt{2\Phi_s^{1/6}\Phi_n^{5/6}}$$

Because we cannot control the missile any better than we can estimate, the preceding equation also represents the smallest possible achievable RMS miss.

In order to illustrate the use of the preceding formula, let us assume we have a guidance system in which there is a 0.01-s sampling time, a standard deviation of 10 ft of glint noise, and a 5-g target maneuver whose starting time is equally likely to occur anywhere (that is, uniformly distributed) during a 10-s flight. Therefore, the power spectral density (in units squared per Hertz) of the target maneuver and glint noise are given by

$$\Phi_s = \frac{n_{TMAX}^2}{t_F} = \frac{161^2}{10} = 2592$$

$$\Phi_N = \sigma_{GL}^2 T_s = 10^2 * 0.01 = 1$$

Therefore, the standard deviation of the first diagonal element of the covariance matrix or RMS value of the smallest possible miss is

$$\sqrt{P_{11}} = \sqrt{2\Phi_s^{1/6}\Phi_n^{5/6}} = \sqrt{2 * 2592^{1/6} * 1^{5/6}} = 2.72 \text{ ft}$$

In the real word we want to build a discrete three-state Kalman filter. As was shown in Chapter 9 of the previous volume, the gains for such a filter are obtained by numerically solving the recursive matrix Riccati difference equations by iteration. As was shown in Chapter 9 of the previous volume, the difference equations to be solved are

$$M_k = \Phi_k P_{k-1} \Phi_k^T + Q_k$$

$$K_k = M_k H^T [H M_k H^T + R_k]^{-1}$$

$$P_k = (I - K_k H) M_k$$

where P_k is the covariance matrix after an update, Φ_k is the fundamental matrix, Q_k is the discrete process noise matrix, M_k is the covariance matrix before an update, R_k is the discrete noise matrix, and K_k is the Kalman gain matrix. In the preceding difference equations we need to find R_k, Φ_k, and Q_k so that we can iterate and solve for the discrete covariance matrix P_k. It was shown in Chapter 9 of the previous volume that the discrete fundamental matrix can easily be derived from the systems dynamics matrix $\mathbf{F}$ as

$$\Phi_k = \begin{bmatrix} 0 & T_s & 0.5T_s^2 \\ 0 & 1 & T_s \\ 0 & 0 & 1 \end{bmatrix}$$

The discrete measurement noise R_k is simply the variance of the glint noise, or

$$R_k = \sigma_{GL}^2$$

Finally, Chapter 9 of the previous volume showed that the discrete process noise matrix can be found from the continuous process noise matrix as

$$
\begin{aligned}
Q_k &= \int_0^{T_s} \Phi(\tau) Q \Phi^T(\tau)\, d\tau \\
&= \int_0^{T_s}
\begin{bmatrix}
1 & \tau & 0.5\tau^2 \\
0 & 1 & \tau \\
0 & 0 & 1
\end{bmatrix}
\begin{bmatrix}
0 & 0 & 0 \\
0 & 0 & 0 \\
0 & 0 & \Phi_s
\end{bmatrix}
\begin{bmatrix}
1 & 0 & 0 \\
\tau & 1 & 0 \\
0.5\tau^2 & \tau & 1
\end{bmatrix} d\tau \\
&= \Phi_s
\begin{bmatrix}
\dfrac{T_s^5}{20} & \dfrac{T_s^4}{8} & \dfrac{T_s^3}{6} \\[2ex]
\dfrac{T_s^4}{8} & \dfrac{T_s^3}{3} & \dfrac{T_s^2}{2} \\[2ex]
\dfrac{T_s^3}{6} & \dfrac{T_s^2}{2} & T_s
\end{bmatrix}
\end{aligned}
$$

Listing 18.1 presents a simulation of the discrete Riccati equations solved for different times of flight for the case of a 5-g uniformly distributed target maneuver and 10 ft of glint noise. The formula, obtained by solving the steady-state continuous Riccati equations for the square root of the first diagonal element, is presented for comparison. In theory, as the sampling time gets smaller, the continuous and discrete answers for the square root of the first diagonal element of the covariance matrix should agree.

LISTING 18.1   SOLVING DISCRETE RICCATI EQUATIONS

```
clear
n=0;
VC=5.*3280.;
XNTIC=161.;
VM=3000.;
HEDEG=20.;
SIGNOISE=10.;
TS=.01;
for TF=.2:.2:10.0,
    TS2=TS*TS;
    TS3=TS2*TS;
    TS4=TS3*TS;
    TS5=TS4*TS;
    PHIS=XNTIC*XNTIC/TF;
```

```
PHIN=SIGNOISE*SIGNOISE*TS;
SIGPOS=SIGNOISE;
SIGN2=SIGPOS^2;
P11=SIGN2;
P12=0.;
P13=0.;
P22=(VM*HEDEG/57.3)^2;
P23=0.;
P33=XNTIC*XNTIC;
T=0.;
for T=TS:TS:TF,
  TGO=TF-T+.000001;
  RTM=VC*TGO;
  SIGPOS=SIGNOISE;
  SIGN2=SIGPOS^2;
  M11=P11+TS*P12+.5*TS2*P13+TS*(P12+TS*P22+.5*TS2*P23);
  M11=M11+.5*TS2*(P13+TS*P23+.5*TS2*P33)+TS5*PHIS/20.;
  M12=P12+TS*P22+.5*TS2*P23+TS*(P13+TS*P23+.5*TS2*P33);
  M12=M12+TS4*PHIS/8.;
  M13=P13+TS*P23+.5*TS2*P33+PHIS*TS3/6.;
  M22=P22+TS*P23+TS*(P23+TS*P33)+PHIS*TS3/3.;
  M23=P23+TS*P33+.5*TS2*PHIS;
  M33=P33+PHIS*TS;
  K1=M11/(M11+SIGN2);
  K2=M12/(M11+SIGN2);
  K3=M13/(M11+SIGN2);
  P11=(1.-K1)*M11;
  P12=(1.-K1)*M12;
  P13=(1.-K1)*M13;
  P22=-K2*M12+M22;
  P23=-K2*M13+M23;
  P33=-K3*M13+M33;
  SP11=sqrt(P11);
end;
FORM=sqrt(2.*(PHIS^.16667)*(PHIN^.83333));
n=n+1;
ArrayT(n)=T;
ArrayFORM(n)=FORM;
ArraySP11(n)=SP11;
end;
figure
plot(ArrayT',ArrayFORM',ArrayT',ArraySP11'),grid
title('RMS miss for various flight times')
xlabel('Flight Time (S)')
ylabel('RMS MISS (Ft) ')
clc
```

```
output=[ArrayT',ArrayFORM',ArraySP11'];
save datfil.txt output -ascii
disp('Simulation Complete')
```

Listing 18.1 was run for the cases in which the sampling time was 0.1 s and 0.01 s. We can see from Fig. 18.2 that the formula and discrete solutions are virtually identical when the sampling time is 0.01 s; however, when the sampling time is 0.1 s there is more of a difference between the formula and the solution obtained by solving the discrete Riccati equations. As expected, we can also see that better performance can be obtained in a homing guidance system with smaller sampling times.

A homing loop, based on the linearized guidance system model used in Chapter 9 of the previous volume, is shown in Fig. 18.3. In this homing loop the optimal guidance law, derived in Chapter 8 of the previous volume, is used against the random uniformly distributed target maneuver. The homing loop includes the three-state Kalman filter that is required to estimate the states required by the optimal guidance law. In addition, there is a single time constant representation of the flight-control system that indicates that there is a delay between the commanded and achieved missile acceleration. The transfer function of the flight-control system is given by

$$\frac{n_L}{n_c} = \frac{1}{1 + sT}$$

where T is the flight-control system time constant. In Fig. 18.3 we can also see that the acceleration command is not limited. As was previously mentioned, the two sources of error in this homing loop are glint noise and a uniformly distributed target maneuver.

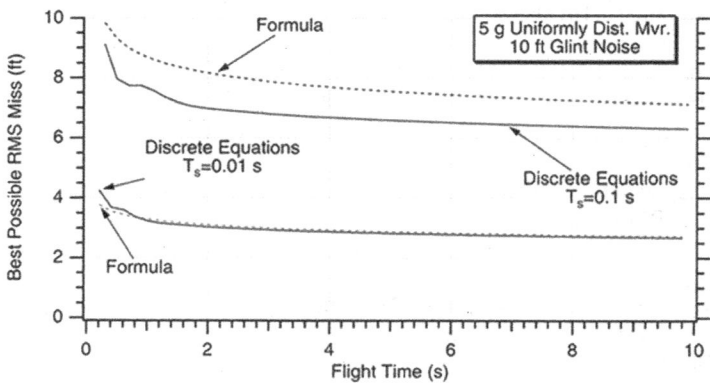

Fig. 18.2   Formula agrees with simulation results when sampling time is small.

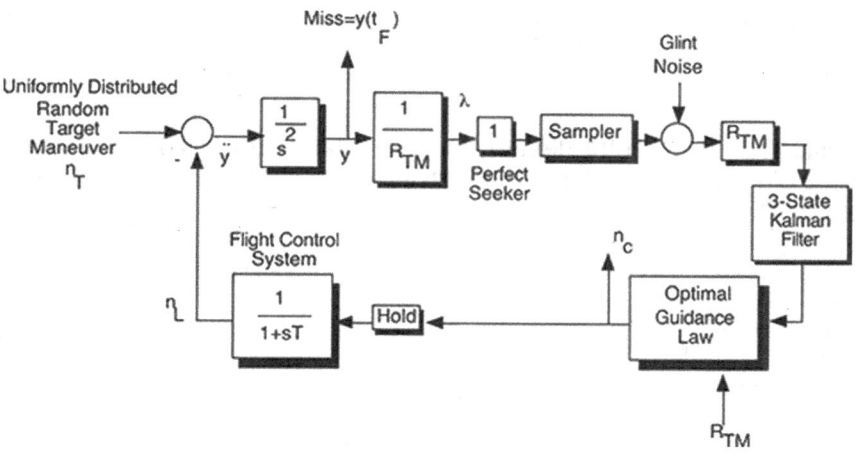

Fig. 18.3   Homing loop with random error sources, a three-state Kalman filter, and optimal guidance law.

In Chapter 8 of the previous volume it was shown the optimal guidance law for a single time constant guidance system where the target is executing a constant maneuver is given by

$$n_{COG} = \frac{N'}{t_{go}^2}[y + \dot{y}t_{go} + 0.5n_T t_{go}^2 - n_L T^2(e^{-x} + x - 1)]$$

where the effective navigation ratio can be expressed as

$$N' = \frac{6x^2(e^{-x} - 1 + x)}{2x^3 + 3 + 6x - 6x^2 - 12xe^{-x} - 3e^{-2x}}$$

and x is defined as

$$x = \frac{t_{go}}{T}$$

Listing 18.2 models the homing loop of Fig. 18.3 with the uniformly distributed target maneuver and the uncorrelated or white glint noise. Because there is measurement noise and a random target maneuver modeled in Listing 18.2, the simulation has to be run in the Monte Carlo mode (repeated simulation trials, each of which has a different maneuver starting time and different sign of the maneuver amplitude) in order to calculate the RMS miss distance. The flight time is also varied in this simulation, and 100-run Monte Carlo sets are made for flight times ranging from 0.2 s to 10 s in steps of 0.2 s. Therefore, each case studied consists of 5000 runs (100 runs times 50 flight times). We can see from Listing 18.2 that the three-state Kalman filter and optimal guidance law are also modeled.

LISTING 18.2   MONTE CARLO SIMULATION OF THREE-STATE KALMAN FILTER, OPTIMAL GUIDANCE LAW, UNCORRELATED GLINT NOISE, AND UNIFORMLY DISTRIBUTED TARGET MANEUVER IN HOMING LOOP

```
%This simulation runs very slowly due to the small integration interval
%and large number of runs
%Preallocation
clear
Z=zeros(size(1:1000));
I=zeros(size(1:50));
TF=zeros(size(1:50));
count=0;
VC=5.*3280.;
XNTIC=161.;
YIC=0.;
VM=3000.;
HEDEG=20.;
XNP=3.;
SIGNOISE=10.;
TS=.01;
TAU=.2;
RUN=100;
AMAXG=99999999.;
PZ1=.0001;
AMAX=AMAXG*32.2;
PHIN=SIGNOISE*SIGNOISE*TS;
for TF=.2:.2:10.0,
    Z1=0.;
    for JJ=1:RUN
      SUM=rand(1);
      TSTART=TF*SUM;
      PZ=rand(1);
      PZ=PZ-.5;
      if PZ > 0
            COEF=1;
      else
            COEF=-1;
      end;
      Y=0.;
      YD=0.;
      TS2=TS*TS;
```

```
TS3=TS2*TS;
TS4=TS3*TS;
TS5=TS4*TS;
PHIS=XNTIC*XNTIC/TF;
RTM=VC*TF;
SIGPOS=SIGNOISE;
SIGN2=SIGPOS^2;
P11=SIGN2;
P12=0.;
P13=0.;
P22=(VM*HEDEG/57.3)^2;
P23=0.;
P33=XNTIC*XNTIC;
T=0.;
H=.001;
S=0.;
YH=0.;
YDH=0.;
XNTH=0.;
XNC=0.;
XNL=0.;
while T <= (TF - 1e-5)
        YOLD=Y;
        YDOLD=YD;
        XNLOLD=XNL;
        STEP=1;
        FLAG=0;
        while STEP <=1
          if FLAG==1
                Y=Y+H*YD;
                YD=YD+H*YDD;
                XNL=XNL+H*XNLD;
                T=T+H;
                STEP=2;
          end;
          if T<TSTART
                XNTC=0.;
          else
                XNTC=COEF*XNTIC;
          end;
          TGO=TF-T+.00001;
          RTM=VC*TGO;
          XLAM=Y/(VC*TGO);
          XNLD=(XNC-XNL)/TAU;
          XNT=XNTC;
          YDD=XNT-XNL;
```

```
        FLAG=1;
end;
FLAG=0;
Y=.5*(YOLD+Y+H*YD);
YD=.5*(YDOLD+YD+H*YDD);
XNL=.5*(XNLOLD+XNL+H*XNLD);
S=S+H;
if S>=(TS - 1e-5)
  S=0.;
  TGO=TF-T+.000001;
  RTM=VC*TGO;
  SIGPOS=SIGNOISE;
  SIGN2=SIGPOS^2;
  M11=P11+TS2*P12+.5*TS2*P13+TS*(P12+TS*P22+.5*TS2*P23);
  M11=M11+.5*TS2*(P13+TS*P23+.5*TS2*P33)+TS5*PHIS/20.;
  M12=P12+TS*P22+.5*TS2*P23+TS*(P13+TS*P23+.5*TS2*P33);
  M12=M12+TS4*PHIS/8.;
  M13=P13+TS*P23+.5*TS2*P33+PHIS*TS3/6.;
  M22=P22+TS*P23+TS*(P23+TS*P33)+PHIS*TS3/3.;
  M23=P23+TS*P33+.5*TS2*PHIS;
  M33=P33+PHIS*TS;
  K1=M11/(M11+SIGN2);
  K2=M12/(M11+SIGN2);
  K3=M13/(M11+SIGN2);
  P11=(1.-K1)*M11;
  P12=(1.-K1)*M12;
  P13=(1.-K1)*M13;
  P22=-K2*M12+M22;
  P23=-K2*M13+M23;
  P33=-K3*M13+M33;
  YNOISE=SIGNOISE*randn;
  YSTAR=Y+YNOISE;
  RES=YSTAR-YH-TS*YDH-.5*TS*TS*(XNTH-XNL);
  YH=K1*RES+YH+TS*YDH+.5*TS*TS*(XNTH-XNL);
  YDH=K2*RES+YDH+TS*(XNTH-XNL);
  XNTH=K3*RES+XNTH;
  X=TGO/TAU;
  ZEM2H=YH+YDH*TGO-XNL*TAU*TAU*(exp(-X)+X-1.). . .
          +.5*XNTH*TGO*TGO;
  TOP=6.*X*X*(exp(-X)-1.+X);
  BOT1=2*X*X*X+3.+6.*X-6.*X*X;
  BOT2=-12.*X*exp(-X)-3.*exp(-2.*X);
  XNPP=TOP/(PZ1+BOT1+BOT2);
```

```
                    XNEW=XNPP*XNL*(exp(-X)+X-1.)/(X*X);
                    XNC=XNPP*ZEM2H/TGO^2;
                    if XNC>AMAX
                            XNC=AMAX;
                    end;
                    if XNC<-AMAX
                            XNC=-AMAX;
                    end;
            end;
        end;
        SP11=sqrt(P11);
        Z(JJ)=Y;
        Z1=Z(JJ)+Z1;
        XMEAN=Z1/JJ;
    end;
    SIGMA=0.;
    Z1=0.;
    Z2=0.;
    for JJ=1:RUN,
        Z1=(Z(JJ)-XMEAN)^2+Z1;
        Z2=Z(JJ)^2+Z2;
        if JJ==1,
                SIGMA=0.;
                RMS=0.;
        else
                SIGMA=sqrt(Z1/(JJ-1));
                RMS=sqrt(Z2/(JJ-1));
        end;
    end;
    FORM=sqrt(2.*(PHIS^.16667)*(PHIN^.8333));
    count=count+1;
    ArrayTF(count)=TF;
    ArrayRMS(count)=RMS;
    ArraySP11(count)=SP11;
    ArrayFORM(count)=FORM;
end;
figure
plot(ArrayTF',ArrayRMS',ArrayTF',ArrayFORM',ArrayTF',ArraySP11'),grid
title('RMS miss for various flight times')
xlabel('Flight Time (S)')
ylabel('RMS MISS (Ft) ')
clc
output=[ArrayTF',ArrayRMS',ArraySP11',ArrayFORM'];
save datfil.txt output -ascii
disp('Simulation Complete')
```

The nominal case of Listing 18.2 was run, and RMS miss vs flight time results from the Monte Carlo simulation are presented in Fig. 18.4. Superimposed on the graph are the predictions from the discrete matrix Riccati equations (square root of the first diagonal element of the covariance matrix). Also superimposed on Fig. 18.4 are the smallest possible RMS miss results obtained from the formula

$$\sqrt{P_{11}} = \sqrt{2\Phi_s^{1/6}\Phi_n^{5/6}}$$

We can see that after the transients have died out, the Monte Carlo RMS miss distance results are in very close agreement with the predictions from both the discrete and continuous Riccati equations. This close RMS miss distance agreement means that we are controlling the missile via the optimal guidance law as well as we can estimate the relative position state via the three-state Kalman filter. *This means that, in the absence of missile acceleration saturation effects, we can do no better with any other guidance law against the uniformly distributed target maneuver.*

Listing 18.2 was rerun with the flight-control system time constant increased from 0.2 s to 1 s. Figure 18.5 shows that the Monte Carlo RMS miss distance results are now larger than the theoretical projections of the Riccati equations. Because the optimal guidance law is suppose to cancel out the single-lag flight-control system dynamics perfectly, something appears to be wrong because the RMS miss should not increase in the absence of missile acceleration saturation effects.

In order to understand what happened when the flight-control system time constant was increased, let us review the expression for the navigation ratio in the optimal guidance law. As was previously mentioned, the expression for the

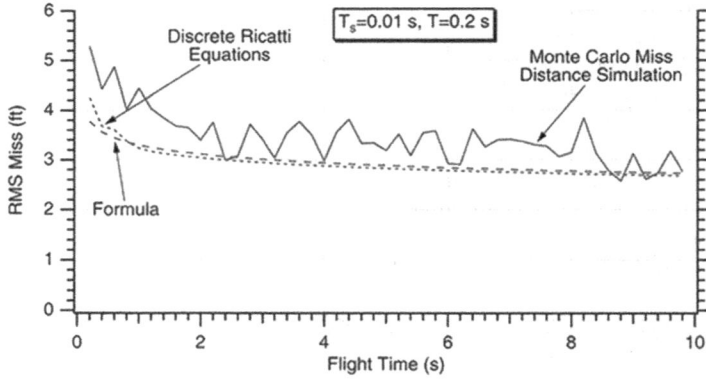

Fig. 18.4   Optimal guidance is nearly able to achieve smallest possible RMS miss distance.

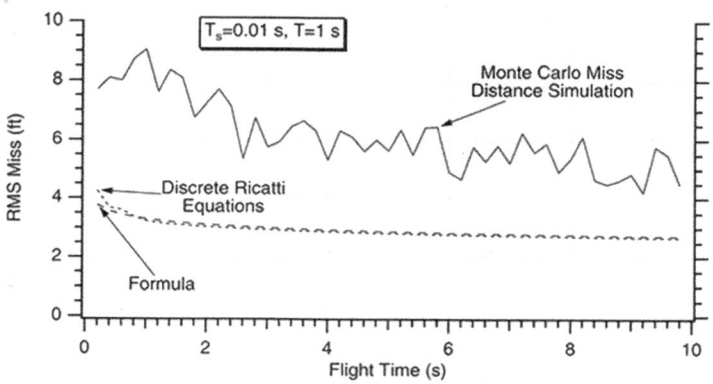

Fig. 18.5   Increasing flight-control system time constant increases RMS miss.

optimal navigation ratio is given by

$$N' = \frac{6x^2(e^{-x} - 1 + x)}{2x^3 + 3 + 6x - 6x^2 - 12xe^{-x} - 3e^{-2x}}$$

where

$$x = \frac{t_{go}}{T}$$

If we examine the denominator of the formula for N' we can see that as t_{go} or x goes to zero, the denominator goes to zero. A careful examination of Listing 18.2 indicates that a division by zero was avoided by adding the term γ to the denominator in order to prevent the expression from blowing up as t_{go} approached zero, or

$$N' = \frac{6x^2(e^{-x} - 1 + x)}{\gamma + 2x^3 + 3 + 6x - 6x^2 - 12xe^{-x} - 3e^{-2x}}$$

The term γ was set to 0.0001 and the optimal guidance law appeared to work rather well in the experiments conducted so far.

The miss distance experiment of Fig. 18.5 was rerun, except this time γ was reduced from 0.0001 to 0.00000001. We can see from Fig. 18.6 that making γ closer to zero enabled the optimal guidance law to yield RMS miss distance results that were much closer to the theoretical predictions of the Riccati equations.

It appears from Fig. 18.6 that miraculous performance improvement can be obtained by reducing γ in the formula for the effective navigation ratio when the flight-control system time constant is large. However, reducing γ has the effect of increasing the effective navigation ratio and missile acceleration near

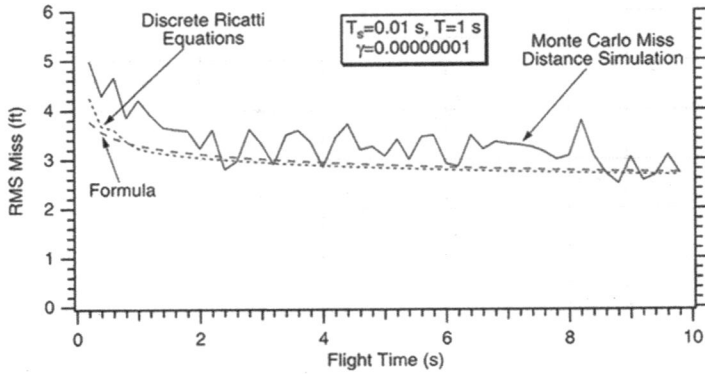

Fig. 18.6   Reducing γ enables optimal guidance to achieve smallest possible RMS miss when flight-control system time constant is large.

the end of the flight. An increase in missile acceleration near the end of the flight can lead to acceleration saturation and increased miss distance. Figure 18.7 shows that for the case of a large flight-control system time constant ($T = 1$ s), we only get very small miss distances when the missile acceleration limit is infinite. For a 3-to-1 (15-g missile acceleration limit) and a 5-to-1 (25-g missile acceleration limit), missile acceleration advantages over the target the miss distances can be quite large.

Let us go back to the original nominal case in which the flight-control system time constant was 0.2 s and the value of γ used in the optimal guidance law was 0.0001. Figure 18.8 first shows that the RMS miss distances are much smaller for a

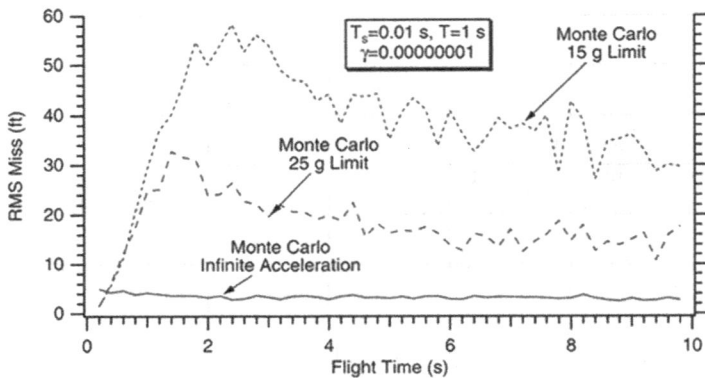

Fig. 18.7   Acceleration saturation effects influence our ability to compensate for large flight-control system time constants.

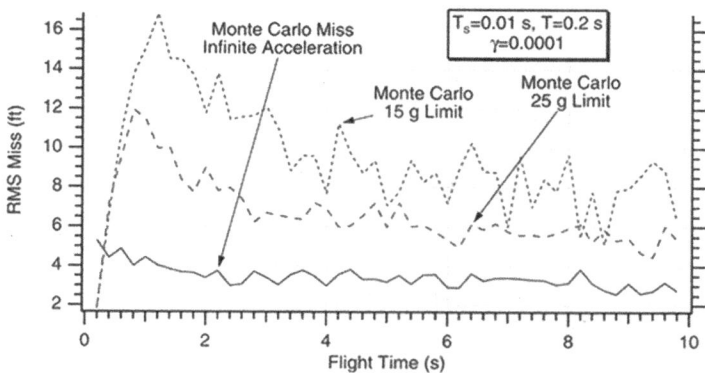

Fig. 18.8 Reducing the flight-control system time constant allows us to get closer to the lowest possible theoretical RMS miss distance.

5-to-1 and a 3-to-1 missile to target acceleration advantage than they were in Fig. 18.7. Next, Fig. 18.8 shows that with a 5-to-1 missile to target acceleration advantage we are not too far from the smallest possible RMS miss.

SUMMARY

In this chapter we have developed formulas for the minimum possible RMS miss distance for the case in where there is white glint noise, a uniformly distributed target maneuver, and infinite missile acceleration capability. These error sources are up to the target rather than the missile designer, so they place a lower limit on the best achievable performance of a radar homing missile that has infinite acceleration capability. Finite missile acceleration capability plus other sources of error will only serve to increase the miss distance.

REFERENCES

[1]   Rusnak, I., "Bounds on the RMS Miss of Radar-Guided Missiles," *Journal of Guidance, Control and Dynamics*, Vol. 3, No. 6, Nov.–Dec. 2010, pp. 1718–1723.

[2]   Garnell, P., and East, D. J., *Guided Weapon Control Systems*, Pergamon Press, Oxford, England, 1977, pp. 15–17.

[3]   Zarchan, P., and Musoff, H., *Fundamentals of Kalman Filtering: A Practical Approach*, 3rd ed, AIAA Progress Series, Reston, VA, 2009, pp. 219–255.

CHAPTER 19

Comparison of Differential Game Guidance with Optimal Guidance

INTRODUCTION

In Chapter 18 we showed that the minimum possible RMS miss distance could be achieved if the Kalman filter was matched to the real world, the missile had infinite acceleration capability, the sampling time was small, and the missile employed an optimal guidance law that was matched to the shape of the target maneuver assuming that it was known. In the example presented in Chapter 18, the three-state discrete Kalman filter knew (via the process noise) and the guidance law knew that the target was performing a random uniformly distributed constant maneuver. It was shown that when there was infinite missile acceleration available, the RMS miss distances that could be achieved approached the estimation capability of the optimal filter. However, when there was only a 3-to-1 missile-to-target acceleration advantage, RMS miss distance performance started to degrade. In this chapter we shall see if other guidance approaches can yield better performance than optimal guidance when the Kalman filter is not matched to the target maneuver and there is a low missile-to-target acceleration advantage.

Pursuit-evasion differential game theory can be used to derive a guidance law that does not depend on the knowledge of the future target maneuver. An important family of pursuit-evasion games assumes linear kinematics and bounded controls [1–8]. However, usually differential game theory guidance laws with bounded controls have not been considered for practical application in endoatmospheric missiles because of their bang-bang nature (meaning the missile guidance command is either at plus or minus the maximum missile acceleration capability), which may lead to excessive induced drag or severe actuator requirements. In this chapter we shall initially disregard the practical objections to differential game guidance with bounded controls in order to see if there might be performance benefits to this unusual missile guidance approach. If there are substantial performance benefits to differential game guidance with bounded controls, we will see if something simple can be done to make this type of guidance law more practical.

In this chapter we shall first review the various types of stressing target maneuvers we shall consider for the missile guidance law comparison. Then, we shall use the three-state Kalman filter of Chapter 9 of the previous volume (its derivation based on a constant amplitude target maneuver and starting time uniformly distributed over the flight time). The three-state Kalman filter will be used for estimating states of a variety of other challenging target maneuvers. Under these circumstances the Kalman filter will no longer be optimal but should still work. The Kalman filter will estimate the states required for the different guidance laws considered, and a performance comparison will be made. The performance comparison will be made under very stressing conditions when there is only a 2-to-1 missile-to-target acceleration advantage.

TRADITIONAL GUIDANCE LAW REVIEW

Proportional navigation is probably the world's most popular guidance law because of its robustness and ease of implementation. The guidance law can be expressed as

$$n_{c_{PN}} = N' V_c \dot{\lambda}$$

where $n_{c_{PN}}$ is the missile acceleration command perpendicular to the line of sight, N' is a designer-chosen gain usually in the range of 3 to 5, V_c is the closing velocity, and $\dot{\lambda}$ is the line-of-sight rate. Here it is assumed that the line-of-sight rate and closing velocity can be measured by the seeker. Proportional navigation can also be written in terms of the zero effort miss ZEM (meaning the miss that would result if the target continued to do what it was doing and the missile did not issue acceleration commands) as

$$\text{ZEM}_{PN} = y + \dot{y} t_{go}$$

$$n_{c_{PN}} = N' V_c \dot{\lambda} = \frac{N' \text{ZEM}_{PN}}{t_{go}^2}$$

It can be shown that proportional navigation is an optimal guidance law if the missile has ideal dynamics (that is, zero-lag guidance system), the target does not maneuver, and $N' = 3$. Here optimal means that under ideal circumstances proportional navigation can yield zero miss distance for the least amount of missile acceleration (that is, the integral of the acceleration squared over the flight time is minimized). If the target is maneuvering, proportional navigation still works, but another guidance law might do better.

The augmented proportional navigation guidance law assumes that the target is executing a constant maneuver of magnitude n_T. The guidance law is proportional navigation plus an extra term to account for the target maneuver and is given by

$$n_{c_{APN}} = N'V_c\dot{\lambda} + 0.5N'n_T$$

where it is assumed that the target acceleration can be estimated. It can be shown that augmented proportional navigation is an optimal guidance law (that is, zero miss and the integral of the acceleration squared is minimized) if the target executes a constant maneuver and $N' = 3$. Augmented proportional navigation can also be written in terms of the zero effort miss as

$$n_{c_{APN}} = \frac{N' \, \text{ZEM}_{APN}}{t_{go}^2}$$

where

$$\text{ZEM}_{APN} = y + \dot{y}t_{go} + 0.5n_T t_{go}^2$$

Again, if the target executes another type of maneuver, augmented proportional navigation still works, but another guidance law might do better.

It was shown in Chapter 8 of the previous volume that missile flight-control system dynamics can cause miss distance if the target maneuvers a short time before intercept. The flight-control system dynamics cause an unwanted delay between the commanded acceleration n_c and the achieved missile acceleration n_L. Let us assume that the dynamics of the missile flight-control system can be represented by the single-lag network

$$\frac{n_L}{n_c} = \frac{1}{1 + sT}$$

where T is the approximate time constant of the missile flight-control system. Large miss distances can result if the time constant is big, so it is desirable to keep the flight-control system time constant as small as possible. If we also assume that the target is executing a constant maneuver and the time constant of the missile flight-control system is known, then it can be shown that the optimal guidance law that compensates for the flight-control system dynamics is given by

$$n_{c_{OG}} = \frac{N'}{t_{go}^2}[y + \dot{y}t_{go} + 0.5n_T t_{go}^2 - n_L T^2(e^{-x} + x - 1)]$$

where the effective navigation ratio is no longer constant and can be expressed as

$$N' = \frac{6x^2(e^{-x} - 1 + x)}{2x^3 + 3 + 6x - 6x^2 - 12xe^{-x} - 3e^{-2x}}$$

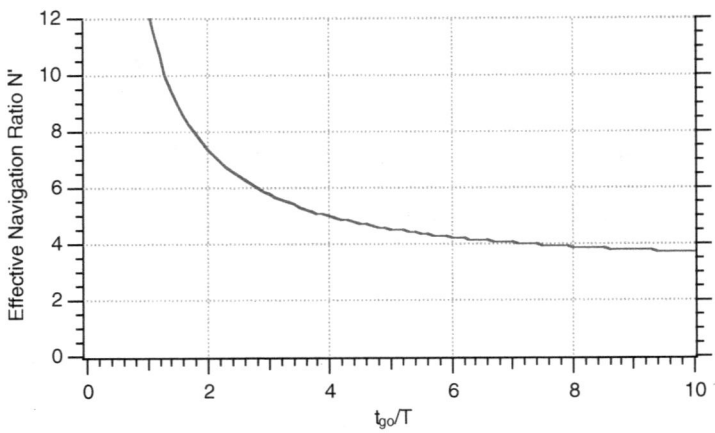

Fig. 19.1   Normalized effective navigation ratio for optimal guidance law.

and x is defined as

$$x = \frac{t_{go}}{T}$$

The effective navigation ratio for the optimal guidance law is displayed in Fig. 19.1. We can see that at the beginning of the flight (long time to go), the effective navigation ratio is approximately constant and is approaching 3. As we get closer to intercept (small time to go), the effective navigation ratio grows considerably.

DIFFERENTIAL GAME GUIDANCE LAW

Much work has been done in the past 30-plus years on guidance laws based on differential game theory with bounded controls [1–8]. The concept of differential pursuit-evasion games is based on assuming that the target maneuvering strategy is the best for maximizing the miss distance. Against such target maneuvering, the optimal strategy of the interceptor missile is the guidance law based on the solution of a differential game. This guidance law is robust with respect to the actual target maneuver by guaranteeing the smallest miss distance against any bounded target maneuver. This guidance law, like the guidance laws of the previous section, is also based on the zero effort miss concept but is bang-bang in nature (missile executes either maximum positive or negative acceleration) because the control effort is not penalized. The bang-bang nature of this guidance law may not make it suitable for endoatmospheric application (possibly too much induced drag or excessive actuator requirements), but for the moment we will put this practical concern aside. The differential game guidance law with bounded controls considered in this chapter is based on the zero effort miss from the optimal guidance law, assuming ideal target dynamics, and is given by

$$n_{c_{DG}} = n_{MAX}\text{sign}(\text{ZEM}_{DG})$$

where the zero effort miss *does not depend on the future target maneuver* and is given by

$$\text{ZEM}_{DG} = y + \dot{y}t_{go} - n_L T^2(e^{-x} + x - 1)$$

where

$$x = \frac{t_{go}}{T}$$

We can see that the differential game guidance zero effort miss calculation *does not depend on any assumptions concerning the target maneuver type or even the magnitude of the current target acceleration.* The fact that this guidance law does not require knowledge of the target maneuver means that this guidance law should perform equally as well (or as poorly) against all types of target maneuvers.

In the preceding guidance law it is assumed that the target dynamics are ideal in the sense that the target can execute its maneuver instantaneously without a lag. If there are target dynamics that can be represented by a first-order lag with time constant T_T and these dynamics are known to the pursuing interceptor, then the zero effort miss becomes

$$\text{ZEM}_{DG} = y + \dot{y}t_{go} - n_L T^2(e^{-x} + x - 1) + n_T T_T^2(e^{-x_T} + x_T - 1)$$

where

$$x_T = \frac{t_{go}}{T_T}$$

If target maneuver dynamics are considered, then the target maneuver level must be estimated with a Kalman filter. If target maneuver dynamics are neglected ($T_T = 0$), then we can see from the zero effort miss calculation that the target maneuver level does not have to be estimated.

TARGET MANEUVERS

It is important to note that because we will consider the missile acceleration limit in the studies that will follow it is important to model the actual target maneuvers rather than their shaping filter equivalent. Chapter 5 showed that if a shaping filter representation of a random target maneuver is used, then the resultant RMS miss distance projection will yield unduly pessimistic results.

For the guidance law comparison it is desirable to pick a set of random target maneuvers of varying degrees of difficulty for the pursuing interceptor. The first target maneuver considered, and one that has been used extensively in this text, and the previous volume is the uniformly distributed target maneuver. This constant maneuver's starting time is random and equally likely to occur anywhere during the flight (uniformly distributed over the flight time). The magnitude of the maneuver is fixed, but for any given flight the maneuver is equally likely to be either positive or negative. An example of a 5-g uniformly distributed target maneuver is shown in Fig. 19.2.

Another maneuver that is often used in missile guidance system analysis is the Poisson (or random telegraph signal) target maneuver. This target maneuver, although not the most realistic, is probably the most challenging to intercept of all possible target maneuvers because it stresses the missile guidance system. The Poisson square wave or random telegraph signal is defined as a maneuver of amplitude $+\beta$ or $-\beta$. The length of time for which the maneuver $n_T(t)$ remains in either the positive or negative direction is random. In particular, the number of times the maneuver changes sign (zero crossings) during 1 s is given by the Poisson distribution. If $P(k)$ represents the probability of k sign changes in t_F seconds, then we can say that

$$P(k) = \frac{(\upsilon t_F)^k e^{-\upsilon t_F}}{k!}$$

where k is the number of sign changes and υ is the average number of zero crossings per second. Details on how to model the Poisson maneuver in both the forward and adjoint time domains can be found in Chapter 11. An example of a 5-g Poisson target maneuver with an average of 0.5 zero crossings per second is displayed in Fig. 19.3. Generally speaking, if good performance can be obtained

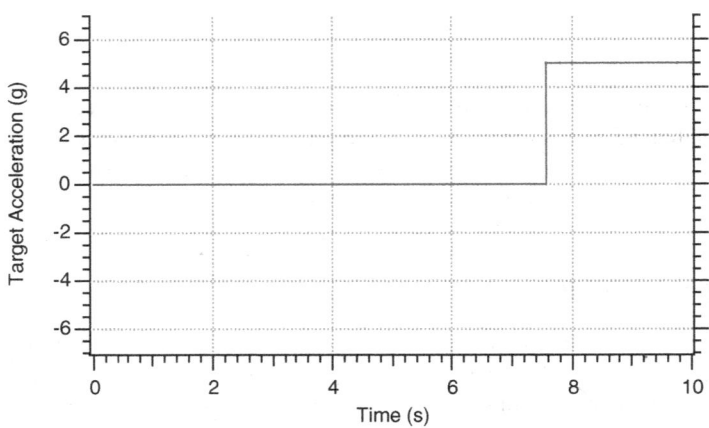

Fig. 19.2   Sample uniformly distributed constant 5-g target maneuver.

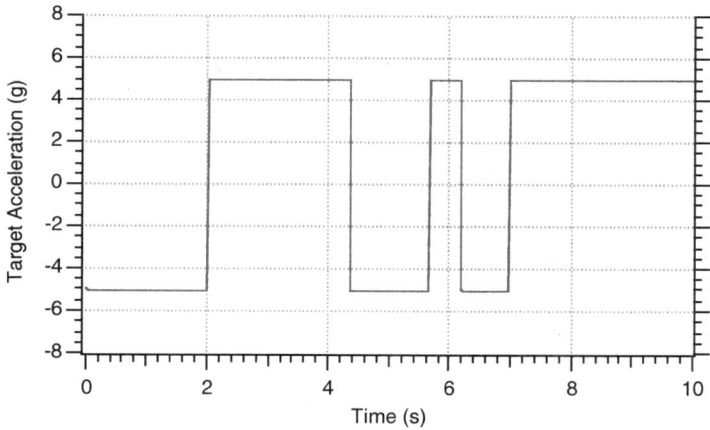

Fig. 19.3   Sample 5-*g* Poisson target maneuver with an average of 0.5 zero crossings per second.

against the Poisson maneuver, even better performance will be obtained against other types of target maneuvers.

A random weave maneuver, originally discussed in Chapter 6, is a sinusoidal maneuver of constant frequency with a starting time that is uniformly distributed over the flight time and whose phase is uniformly distributed between 0 deg and 360 deg. A sample 5-*g*, 2-rad/s random weave maneuver is displayed in Fig. 19.4.

The random vertical-S maneuver is one in which the target is always at maximum positive or negative acceleration but the sign of the acceleration is

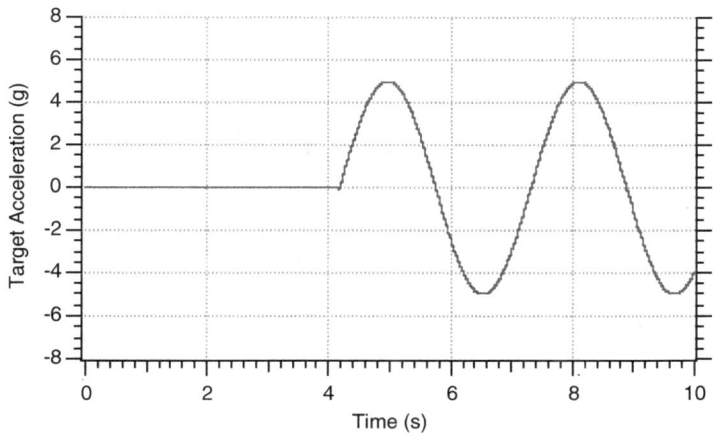

Fig. 19.4   Sample random 5-*g*, 2-rad/s target weave maneuver.

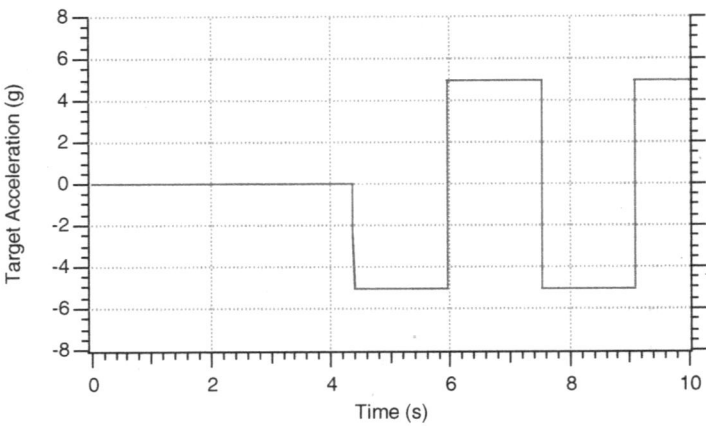

Fig. 19.5   Sample 5-*g*, 2-rad/s random vertical-S target maneuver.

periodically reversed. The starting time of the maneuver is uniformly distributed over the flight time. A sample random 5-*g*, 2-rad/s vertical-S maneuver is displayed in Fig. 19.5.

GUIDANCE LAW COMPARISON

Figure 19.6 shows the homing loop when measurement angle noise and random target maneuver are considered as the two sources of error that can cause miss distance. Here the three-state Kalman filter, which was originally derived in Chapter 9 of the previous volume, is used to estimate the states required for guidance.

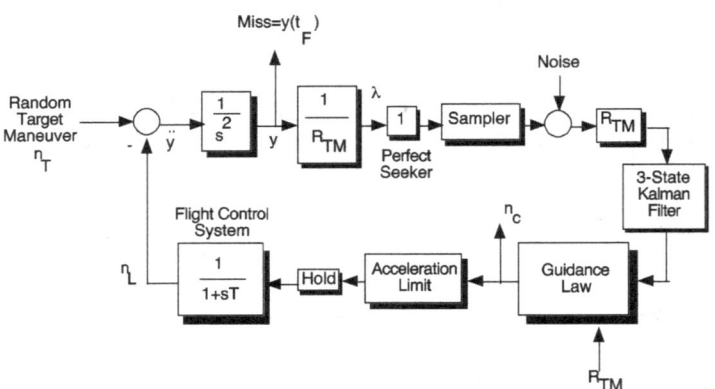

Fig. 19.6   Homing loop with measurement noise and three-state Kalman filter.

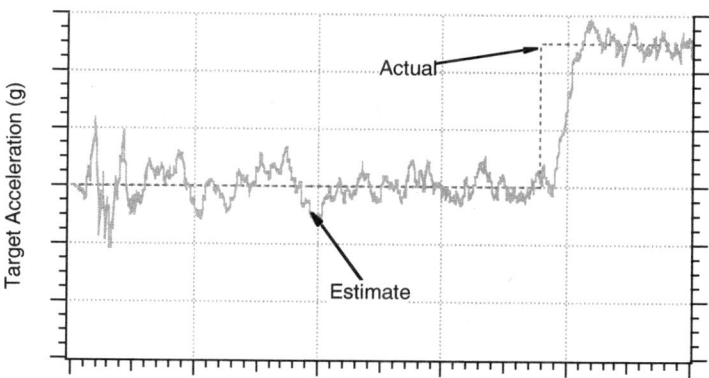

Fig. 19.7   Estimating random constant target maneuver with three-state Kalman filter.

Using the three-state Kalman filter, sample estimates of each of the target maneuver types are presented in Figs. 19.7 to 19.10. In the following examples, the angle measurement noise standard deviation is 0.1 mrad and the sampling time is 0.01 s. The assumed process noise adjusts the bandwidth of the Kalman filter. The value of the continuous process noise matrix (used in deriving the discrete process noise matrix) in the English system of units was assumed to be

$$
Q_c = \begin{bmatrix} 0 & 0 & 0 \\ 0 & 0 & 0 \\ 0 & 0 & \dfrac{n_{TMAX}^2}{t_F} \end{bmatrix} = \begin{bmatrix} 0 & 0 & 0 \\ 0 & 0 & 0 \\ 0 & 0 & \dfrac{161^2}{10} \end{bmatrix}
$$

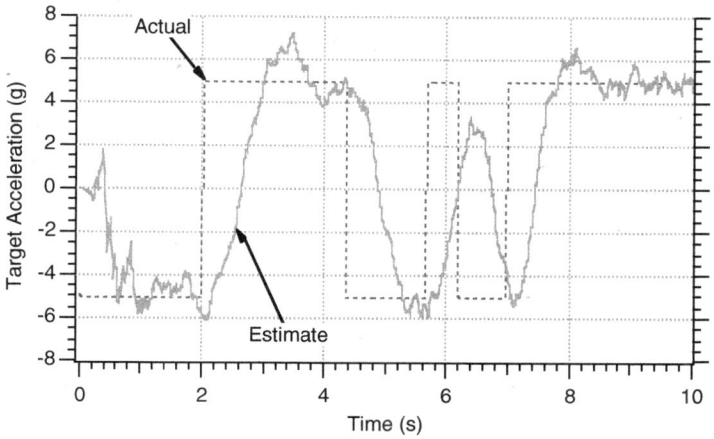

Fig. 19.8   Estimating Poisson target maneuver with three-state Kalman filter.

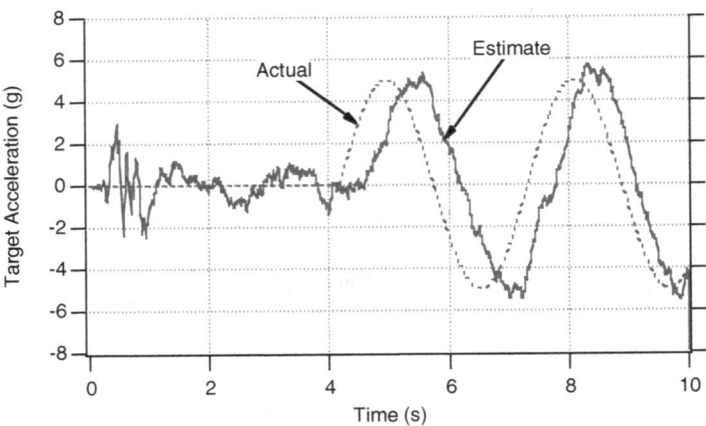

Fig. 19.9   Estimating random weave target maneuver with three-state Kalman filter.

We can see from Fig. 19.7 that the estimate of the random constant target maneuver is excellent. This should not be surprising because the three-state Kalman filter is optimal for this type of target maneuver. The worst estimate of the target maneuver is for the Poisson maneuver case of Fig. 19.8. In this case the filter cannot keep up with the zero crossings of the target. More process noise or a wider bandwidth filter would have allowed the filter to follow the target motion better at the expense of much noisier estimates. Figures 19.9 and 19.10 show that the filter can provide reasonable estimates of both the 2-rad/s

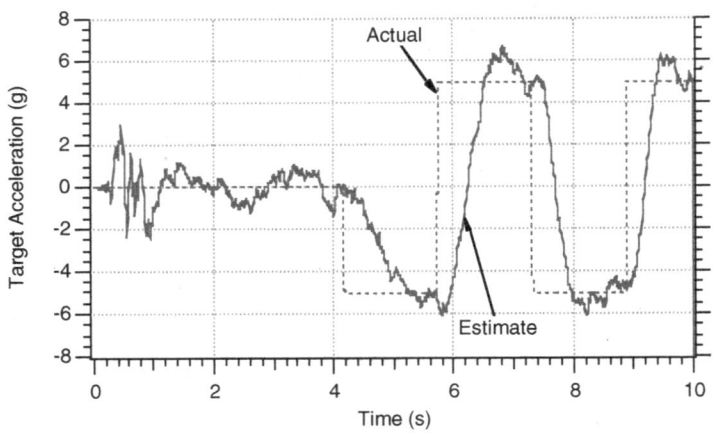

Fig. 19.10   Estimating random vertical-S target maneuver with three-state Kalman filter.

TABLE 19.1   GUIDANCE SYSTEM PARAMETERS

Parameter	Definition	Value
n_T	Target maneuver magnitude	5 g
σ_{Noise}	Measurement noise standard deviation	0.0001 rad
T	Flight-control system time constant	0.2 s
T_s	Kalman filter sampling time	0.01 s
a_{MAX}	Missile acceleration limit	10 g
V_c	Closing velocity	5 km/s

weaving and vertical-S target maneuvers. However, the filter estimates in these two examples lag the actual target maneuver.

The system parameters of our homing loop shown in Fig. 19.6 are defined in Table 19.1. We can immediately see that system performance will be a challenge because there is only a 2-to-1 missile-to-target acceleration advantage.

A Monte Carlo simulation of Fig. 19.6 is used to generate RMS miss distance performance and appears in Listing 19.1. Nominally 50-run Monte Carlo sets are run for each of 100 flight times (ranging from 0.1 s to 10 s in steps of 0.1 s) in order to generate guidance law comparisons.

LISTING 19.1   MONTE CARLO SIMULATION FOR COMPARING GUIDANCE LAW EFFECTIVENESS WHEN NOISE AND FILTERING ARE CONSIDERED

```
% Runs very slowly because of small integration interval
%Preallocation
clear
Z=zeros(size(1:1000));
I=zeros(size(1:50));
TF=zeros(size(1:50));
count=0;
TSW=1;
VC=5.*3280.;
XNTIC=161;
YIC=0.;
VM=3000.;
HEDEG=20.;
XNP=3.;
SIGNOISE=.0001;
TS=.01;
TAU=.2;
NOISE=1;
```

```
RUN=50;
% TYPE OF GUIDANCE (0=PN,1=OG,2=DG,3=HYBRID)
APN=1;
XLIM=322;
XNU=.5;
W=2;
% TYPE OF MANEUVER (1=POISSON,2=UNIF CONST,3=RANDOM SINE,4=RANDOM VS
ICONSTANT=2;
for TF=.1:.1:10.0,
      Z1=0;
      for I=1:RUN
             SUM=rand(1);
             TSTART=TF*SUM;
             PZ=rand(1);
             PZ=PZ-.5;
             if PZ > 0
                    COEF=1;
             else
                    COEF=-1;
             end;
      SUM=rand(1);
      PHASE=6.28*SUM;
             Y=YIC;
             YD=0;
             TS2=TS*TS;
             TS3=TS2*TS;
             TS4=TS3*TS;
             TS5=TS4*TS;
             PHIN=XNTIC*XNTIC/10;
             RTM=VC*TF;
             SIGPOS=RTM*SIGNOISE;
             SIGN2=SIGPOS^2;
             P11=SIGN2;
             P12=0.;
             P13=0.;
             P22=(VM*HEDEG/57.3)^2;
             P23=0.;
             P33=XNTIC*XNTIC;
             T=0.;
             H=.001;
             S=0.;
             YH=0.;
             YDH=0.;
             XNTH=0.;
             XNC=0.;
             XNL=0.;
```

```
BETA=XNTIC;
QFIRST=1;
SIG=1./sqrt(2.*XNU);
XNOISE=randn;
if XNOISE>0
      XNTP=BETA;
else
      XNTP=-BETA;
end;
DELT=9999.;
TNOW=0.;
while T <= (TF - 1e-5)
      if QFIRST==1
            XNOISE1=SIG*randn;
            XNOISE2=SIG*randn;
            DELT=XNOISE1^2+XNOISE2^2;
            QFIRST=0;
            TNOW=T;
      end;
      if T>= (DELT+TNOW)
            XNTP=-XNTP;
            QFIRST=1;
      end;
      YOLD=Y;
      YDOLD=YD;
      XNLOLD=XNL;
      STEP=1;
      FLAG=0;
      while STEP <=1
            if FLAG==1
            Y=Y+H*YD;
            YD=YD+H*YDD;
            XNL=XNL+H*XNLD;
            T=T+H;
      STEP=2;
            end;
            if ICONSTANT==1
                  XNTC=XNTP;
            elseif ICONSTANT==2
                  if T<TSTART
                        XNTC=0.;
                  else
                        XNTC=COEF*XNTIC;
                  end;
            elseif ICONSTANT==3
                  if T<TSTART
```

```
                              XNTC=0.;
               else
                              XNTC=XNTIC*sin(W*T+PHASE);
               end;
       else
               if T<TSTART
                      XNTC=0.;
               else
                      XNTC=COEF*XNTIC*sign(sin(W*(T-TSTART)));
               end;
       end;
       TGO=TF-T+.00001;
       RTM=VC*TGO;
       XLAM=Y/(VC*TGO);
       XLAMD=(RTM*YD+Y*VC)/(RTM^2);
       XNLD=(XNC-XNL)/TAU;
       YDD=XNTC-XNL;
       FLAG=1;
end;
FLAG=0;
Y=.5*(YOLD+Y+H*YD);
YD=.5*(YDOLD+YD+H*YDD);
XNL=.5*(XNLOLD+XNL+H*XNLD);
S=S+H;
if S>=(TS - 1e-5)
       S=0.;
       TGO=TF-T+.000001;
       RTM=VC*TGO;
       SIGPOS=RTM*SIGNOISE;
       SIGN2=SIGPOS^2;
       M11=P11+TS*P12+.5*TS2*P13+TS*(P12+TS*P22
               +.5*TS2*P23);
       M11=M11+.5*TS2*(P13+TS*P23+.5*TS2*P33)
               +TS5*PHIN/20.;
       M12=P12+TS*P22+.5*TS2*P23+TS*
               (P13+TS*P23+.5*TS2*P33)...
               +TS4*PHIN/8.;
       M13=P13+TS*P23+.5*TS2*P33+PHIN*TS3/6.;
       M22=P22+TS*P23+TS*(P23+TS*P33)+PHIN*TS3/3.;
       M23=P23+TS*P33+.5*TS2*PHIN;
       M33=P33+PHIN*TS;
       K1=M11/(M11+SIGN2);
       K2=M12/(M11+SIGN2);
       K3=M13/(M11+SIGN2);
       P11=(1.-K1)*M11;
       P12=(1.-K1)*M12;
```

```
P13=(1.-K1)*M13;
P22=-K2*M12+M22;
P23=-K2*M13+M23;
P33=-K3*M13+M33;
      XLAMNOISE=SIGNOISE*randn;
YSTAR=RTM*(XLAM+XLAMNOISE);
RES=YSTAR-YH-TS*YDH-.5*TS*TS*(XNTH-XNC);
YH=K1*RES+YH+TS*YDH+.5*TS*TS*(XNTH-XNC);
YDH=K2*RES+YDH+TS*(XNTH-XNC);
XNTH=K3*RES+XNTH;
XLAMDH=(YH+YDH*TGO)/(VC*TGO*TGO);
      X=TGO/TAU;
      ZEM1H=YH+YDH*TGO-XNL*TAU*TAU
            *(exp(-X)+X-1.);
      ZEM2H=YH+YDH*TGO-XNL*TAU*TAU
            *(exp(-X)+X-1.)+...
         .5*XNTH*TGO*TGO;
if APN==0
XNC=XNP*(YH+YDH*TGO)/TGO^2;
elseif APN==1
X=TGO/TAU;
TOP=6.*X*X*(exp(-X)-1.+X);
BOT1=2*X*X*X+3.+6.*X-6.*X*X;
BOT2=-12.*X*exp(-X)-3.*exp(-2.*X);
XNPP=TOP/(.0001+BOT1+BOT2);
            XNC=XNPP*ZEM2H/TGO^2;
      elseif APN==2
            XNC=XLIM*sign(ZEM1H);
      else
         if TGO>TSW
               TOP=6.*X*X*(exp(-X)-1.+X);
               BOT1=2*X*X*X+3.+6.*X-6.*X*X;
               BOT2=-12.*X*exp(-X)-3.*exp(-2.*X);
               XNPP=TOP/(.0001+BOT1+BOT2);
               XNEW=XNPP*XNL*(exp(-X)+X-1.)/
                     (X*X);
               XNC=XNPP*ZEM2H/TGO^2;
               if XNC>XLIM
                     XNC=XLIM;
               end
               if XNC<-XLIM
                     XNC=-XLIM;
               end
         else
               XNC=XLIM*sign(ZEM1H);
         end;
```

```
                    end;
                    if XNC>XLIM
                            XNC=XLIM;
                    elseif XNC<-XLIM
                            XNC=-XLIM;
                    end;
                end;
            end;
            Z(I)=Y;
            Z1=Z(I)+Z1;
            XMEAN=Z1/I;
        end;
        SIGMA=0;
        Z1=0;
        Z2=0.;
        for I=1:RUN,
            Z1=(Z(I)-XMEAN)^2+Z1;
                        Z2=Z(I)^2+Z2;
            if I==1,
                SIGMA=0;
                        RMS=0.;
            else
                SIGMA=sqrt(Z1/(I-1));
                        RMS=sqrt(Z2/(I-1));
            end;
        end;
        count=count+1;
        ArrayTF(count)=TF;
        ArraySIGMA(count)=SIGMA;
        ArrayXMEAN(count)=XMEAN;
        ArrayRMS(count)=RMS;
end;
figure
plot(ArrayTF',ArrayRMS'),grid
title('RMS miss for various flight times')
xlabel('Flight Time (S)')
ylabel('RMS MISS (Ft) ')
clc
output=[ArrayTF',ArrayRMS'];
save datfil.txt output -ascii
disp('Simulation Complete')
```

Listing 19.1 was first run in the Monte Carlo mode for the proportional navigation guidance law for the case of the 5-g uniformly distributed constant target maneuver when the missile acceleration limit was 15 g and 10 g. We can see from

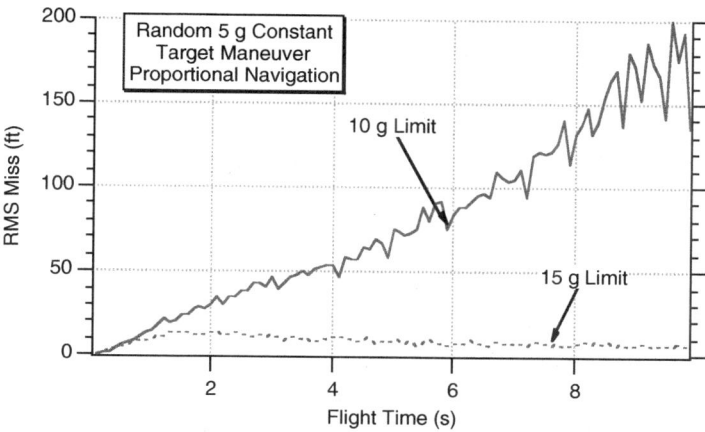

Fig. 19.11   Proportional navigation RMS miss is too large in presence of 2-to-1 missile-to-target acceleration advantage.

Fig. 19.11 that when the missile only has a 2-to-1 acceleration advantage over the target, the miss is unacceptable. Because we are only considering missile-to-target acceleration advantages of 2 to 1 in this chapter, proportional navigation guidance will be eliminated from further consideration.

Next Monte Carlo runs were made for both the optimal guidance (APN=1) and differential game (APN=2) guidance laws with bounded controls in the case of a 5-g random constant ideal target maneuver (ICONSTANT=2) and a 10-g missile

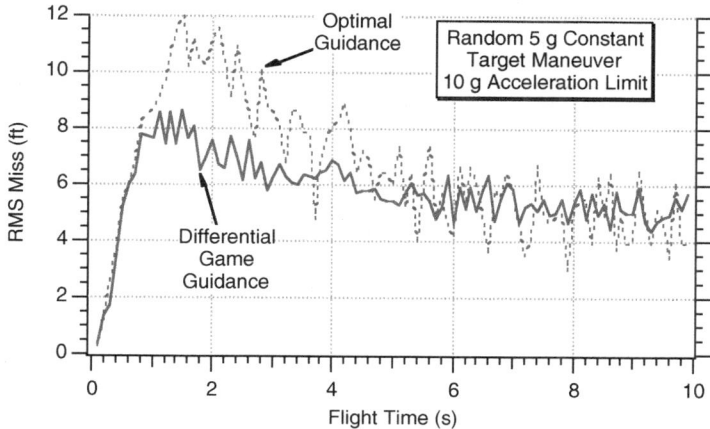

Fig. 19.12   Both optimal guidance and differential game guidance with bounded controls have similar performance against a random constant maneuver if there is only a 2-to-1 missile-to-target acceleration advantage.

acceleration limit. We can see from Fig. 19.12 that against the random constant target maneuver, both guidance laws yield similar performance for flight times greater than 4 s; however, for smaller flight times, differential game guidance with bounded controls yielded slightly smaller RMS miss distances.

Other Monte Carlo sets of runs were made for both optimal guidance and differential game guidance laws with bounded controls for the case of a 5-*g* Poisson target maneuver (ICONSTANT=1) with an average of 0.5 zero crossings per second and a 10-*g* missile acceleration limit. From Fig. 19.13 we can see that against this difficult target maneuver, differential game guidance with bounded controls performed much better than optimal guidance.

Other Monte Carlo sets of runs were made for both guidance laws for the case of a 5-*g*, 2-rad/s random weaving target maneuver (ICONSTANT=3) with a 10-*g* missile acceleration limit. From Fig. 19.14 we can see that against the weaving target maneuver, the differential game guidance again performed much better than optimal guidance.

Other Monte Carlo sets of runs were made for different guidance laws for the case of a 5-*g*, 2-rad/s random vertical-S target maneuver (ICONSTANT=4) with a 10-*g* missile acceleration limit. From Fig. 19.15 we can see that against the vertical-S target maneuver, differential game guidance with bounded controls is again better than optimal guidance.

From the previous set of results we can see that if there was 0.1 mrad of measurement noise, the sampling time was 0.01 s, and there was a 2-to-1 missile-to-target acceleration advantage, the differential game guidance law with bounded controls generally yields better performance than the optimal guidance

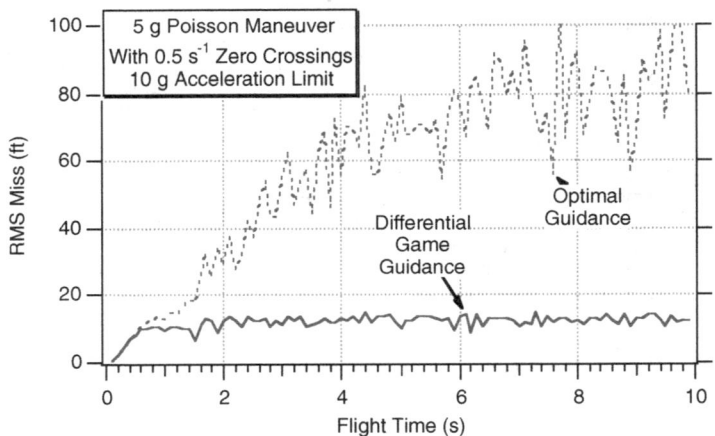

Fig. 19.13   Differential game guidance with bounded controls is more effective than optimal guidance against a Poisson target maneuver if there is only a 2-to-1 missile-to-target acceleration advantage.

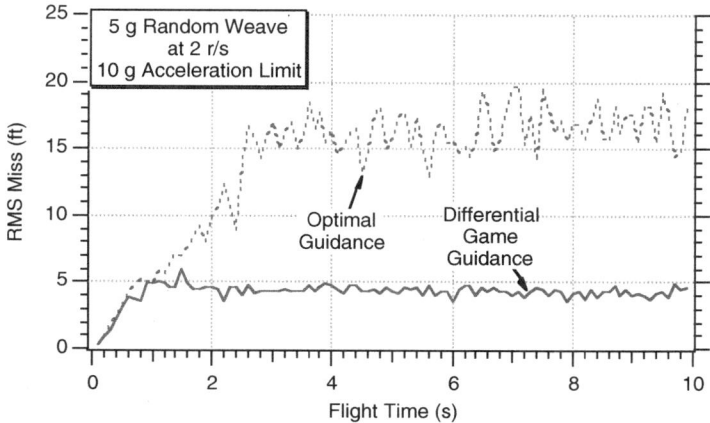

Fig. 19.14   Differential game guidance with bounded controls is more effective than optimal guidance against a random weaving target maneuver if there is only a 2-to-1 missile-to-target acceleration advantage.

law against any of the four random target maneuvers considered. To see if the performance benefit is due to the low measurement noise or the small sampling time, let us now see how optimal guidance and differential game guidance with bounded controls compare when the measurement noise is increased by an order of magnitude against random weave maneuver. The missile-to-target

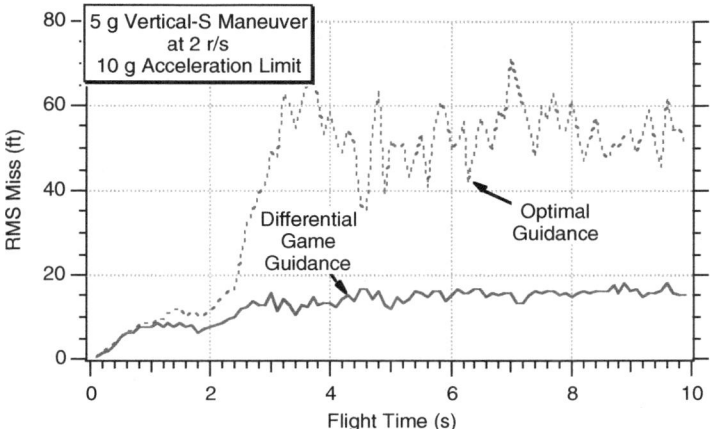

Fig. 19.15   Differential game guidance with bounded controls is more effective than optimal guidance against a random vertical-S target maneuver if there is only a 2-to-1 missile-to-target acceleration advantage.

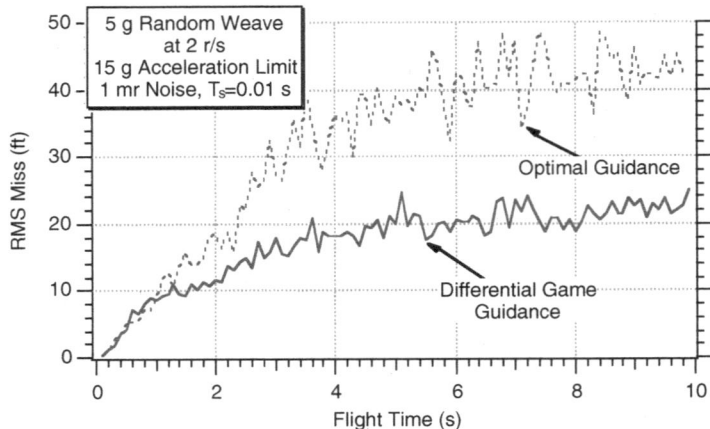

Fig. 19.16   Differential game guidance with bounded controls is more effective than optimal guidance against a random weave target maneuver when the measurement noise is increased.

acceleration limit is increased to 3 to 1. Figure 19.16 shows that when the measurement noise is increased by an order of magnitude, the RMS miss distance increases significantly but differential game guidance with bounded controls still yields better performance than optimal guidance. If the measurement noise is kept at 0.1 mrad but the sampling time is increased to 0.1 s while the missile-to-target

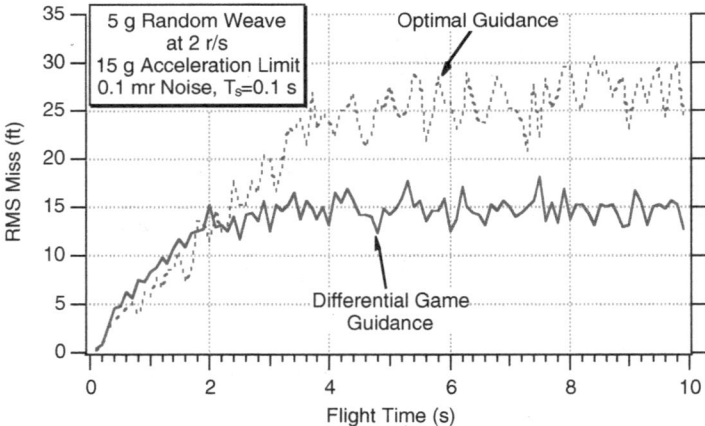

Fig. 19.17   Differential game guidance with bounded controls is more effective than optimal guidance against a random weave target maneuver when the sampling time is increased.

acceleration advantage is 3 to 1, Fig. 19.17 shows that when the sampling time is increased by an order of magnitude, the RMS miss distance increases significantly but differential game guidance with bounded controls still yields better performance than optimal guidance.

MAKING DIFFERENTIAL GAME GUIDANCE MORE PRACTICAL

So far we have demonstrated that differential game guidance with bounded controls appears to be more effective than optimal guidance for a variety of very difficult target maneuvers when the missile-to-target acceleration advantage is low. The advantages of differential game guidance with bounded controls has been well known for more than three decades by the academic community but has not been met with enthusiasm by practicing engineers. As was mentioned previously, one of the main practical objections to the implementation of differential game guidance with bounded controls has been the chattering back and fourth of the acceleration command, as is illustrated in the single-flight results of Fig. 19.18. Here it can be seen that the missile acceleration command due to the vertical-S target maneuver constantly switches back and fourth between maximum positive and negative acceleration. It has been thought that this chattering could lead to excessive induced drag or excessive actuator requirements, which would negate any of the idealized performance benefits illustrated in this chapter.

It is hypothesized that most of the miss distance improvements offered by the differential game guidance law with bounded controls occur in actions taken near the very end of the flight. Therefore, a possible solution to this potential induced drag problem is to limit the application of differential game guidance with

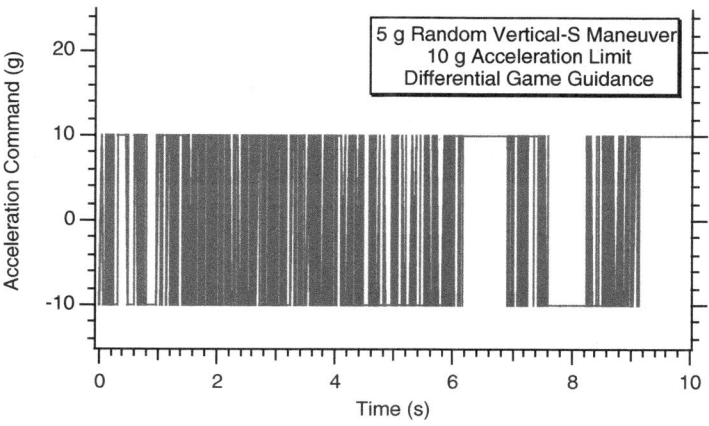

Fig. 19.18   Sample single-run missile acceleration command profile for differential game guidance with bounded controls indicates a great deal of chattering.

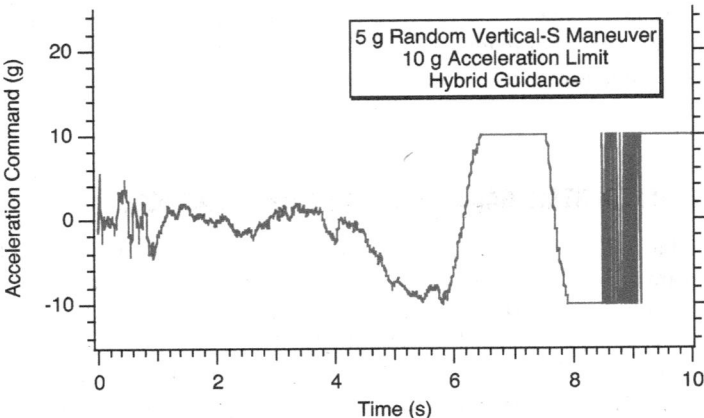

Fig. 19.19 Sample single-run missile acceleration command profile for hybrid guidance indicates most of the chattering has been eliminated.

bounded controls to the last few seconds of flight. In other words, optimal guidance could be used for most of the flight, and then TSW (a parameter in Listing 19.1) seconds before the end of the flight, differential game guidance with bounded controls could be used. This type of simple switching logic could be termed *hybrid guidance* and can be found in Listing 19.1 (APN=3).

Figure 19.19 illustrates the use of hybrid guidance for a single flight against the vertical-S target maneuver with TSW=2. Here we can see a gentler missile

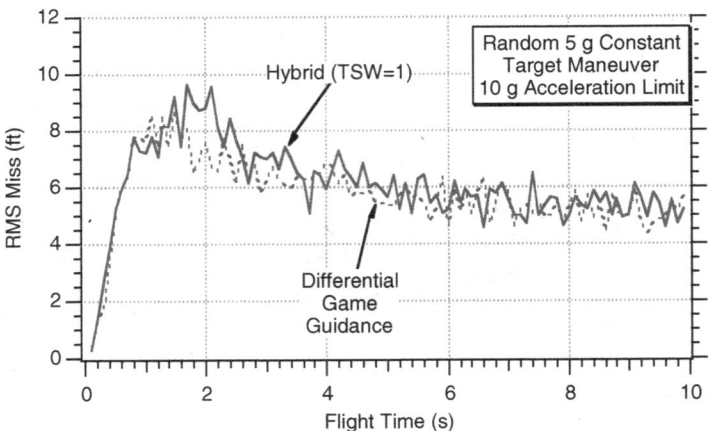

Fig. 19.20 Hybrid (TSW=1 s) and differential game guidance with bounded controls yield similar performance against random constant target maneuver.

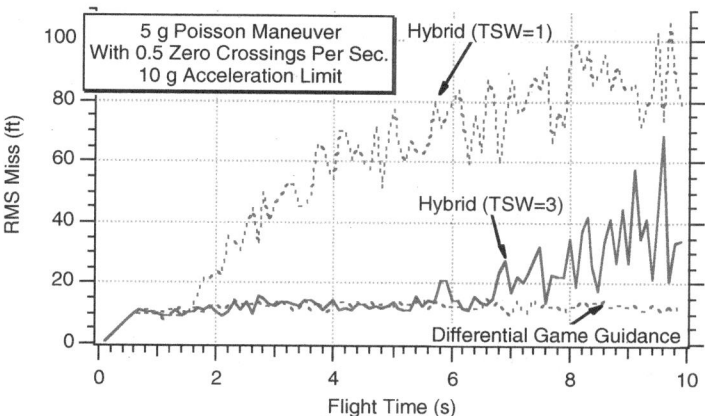

Fig. 19.21   Hybrid (TSW=3 s) and differential game guidance with bounded controls yield similar performance against Poisson target maneuver.

acceleration command profile when compared to Fig. 19.18. Chattering has been removed from 80% of the flight.

In order to see if results comparable to differential game guidance with bounded controls could be obtained with hybrid guidance, a set of Monte Carlo results was generated for each of the four different random target maneuvers previously considered. The goal of the experiment was to see how small TSW could be made so that performance degradation due to chattering either would not result or would be considerably alleviated. Figures 19.20–19.23 show that hybrid guidance

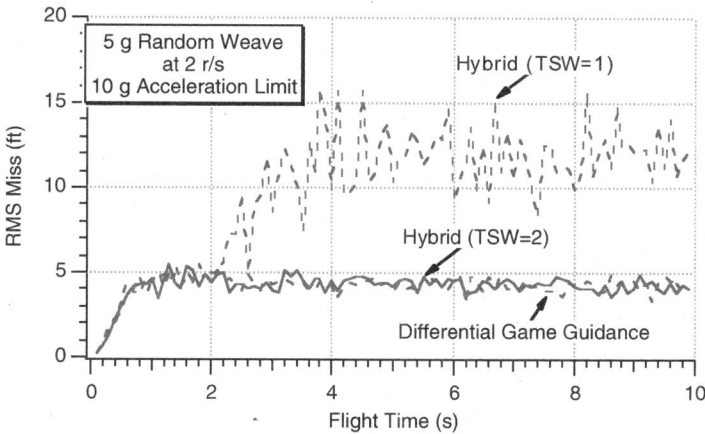

Fig. 19.22   Hybrid (TSW=2 s) and differential game guidance with bounded controls yield similar performance against random weave target maneuver.

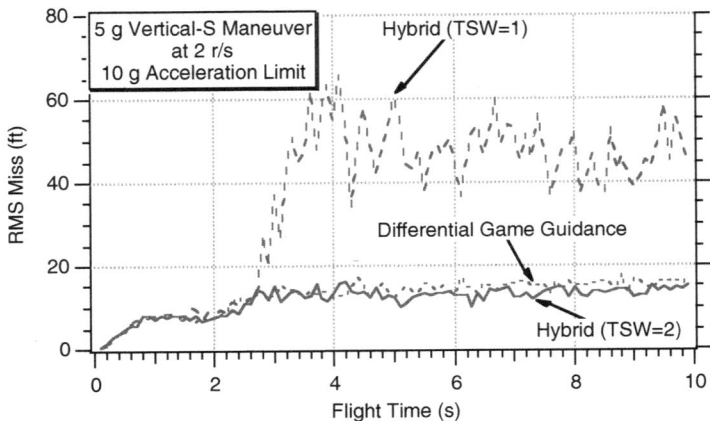

Fig. 19.23   Hybrid (TSW=2 s) and differential game guidance with bounded controls yield similar performance against random vertical-S target maneuver.

can yield performance very similar to that of differential game guidance with bounded controls with values of TSW ranging from 1 s to 3 s. Thus, depending on the application, hybrid guidance may yield significant and achievable performance advantages over optimal guidance against very challenging target maneuvers.

TARGET DYNAMICS

It was previously mentioned in this chapter that the target could also be modeled as having a single time constant representation between the commanded and achieved target acceleration. If the target dynamics were known, the zero effort miss would have an extra term to account for those dynamics, as shown in the section that discussed the differential game guidance law with bounded controls. Figure 19.24 shows what happens to performance in the case of the random weave target maneuver when differential game guidance with bounded controls is used, assuming target time constants of 0 s and 0.2 s. We can see that miss distance performance improves slightly if there is a target time constant of 0.2 s—even though the guidance law is not compensating for the dynamics. However, there is a more substantial improvement in performance when the guidance law compensates for the target dynamics.

In Fig. 19.24 it was assumed that when the target dynamics were known perfectly, the differential game guidance law with bounded controls could yield much smaller miss distances. Figure 19.25 shows that the differential game guidance law with bounded controls is not very sensitive to knowing the exact target time

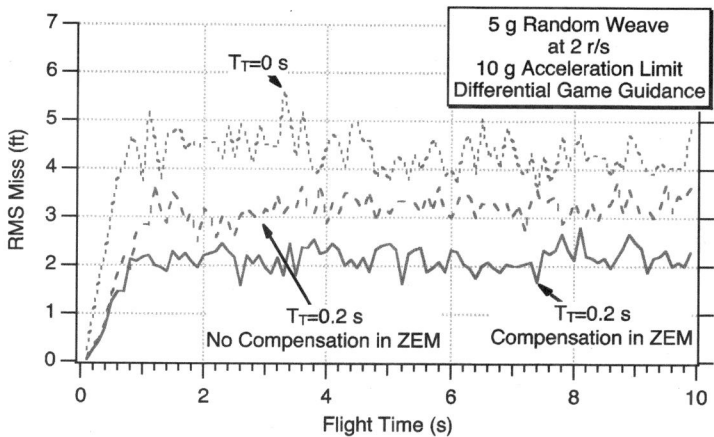

Fig. 19.24   Compensating for target dynamics can yield better system performance when the target time constant is 0.2 s.

constant because nearly identical performance is achieved if we overestimate or underestimate the target time constant.

Finally Fig. 19.26 shows that if the target time constant is increased from 0.2 s to 1 s, the miss distances are much smaller and it is not as important to compensate for the target time constant.

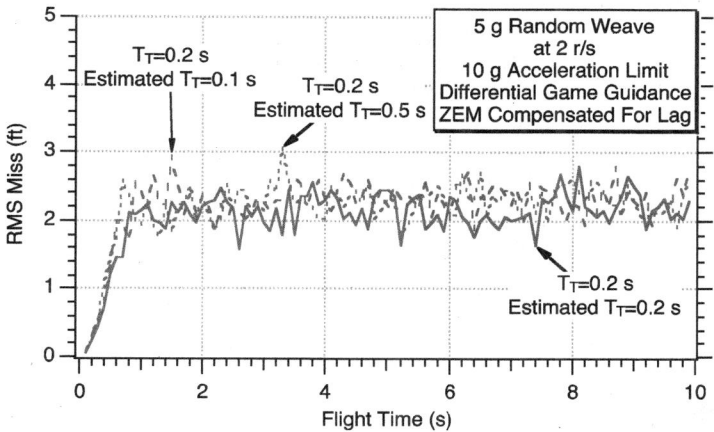

Fig. 19.25   Differential game guidance law with bounded controls does not have to know target time constant exactly.

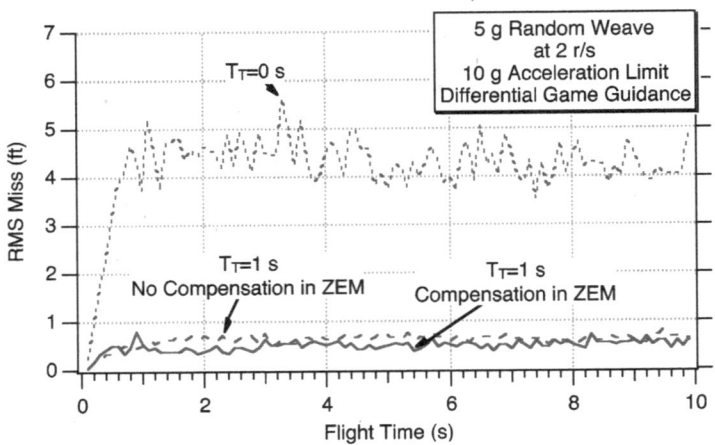

Fig. 19.26   Compensating for target dynamics is not as important when the target time constant is increased from 0.2 s to 1 s.

SUMMARY

In this chapter the differential game guidance law with bounded controls was introduced. In theory (that is, in noise-free scenarios), this guidance law guarantees the smallest miss distance against *any* bounded target maneuver, including the entire family of unknown and/or random target maneuvers. It has been demonstrated that in realistic noisy scenarios with an estimator in the loop, the differential game guidance with bounded controls offers considerable performance improvements over optimal guidance when the missile-to-target acceleration advantage is low. The chapter also demonstrated how the chattering caused by the bang-bang nature of differential game guidance with bounded controls could be reduced dramatically without significant performance degradation by simply applying this guidance law in the last few seconds of flight. Other guidance laws that can potentially eliminate the typical chattering of the guidance laws based on differential game theory with bounded controls can be found in [9]. Finally, it has been shown that if there is a target time constant, performance improvements could be achieved by compensating for the target dynamics in the computation of the zero effort miss.

REFERENCES

[1]   Gutman, S., and Leitmann, G., "Optimal Strategies in the Neighborhood of a Collision Course," *AIAA Journal*, Vol. 14, No. 9, 1976, pp. 1210–1212.

[2]   Gutman, S., "On Optimal Guidance for Homing Missiles," *Journal of Guidance and Control*, Vol. 3, No. 4, 1979, pp. 296–300.

[3]   Shinar, J., and Gutman, S., "Three Dimensional Optimal Pursuit and Evasion with Bounded Control," *IEEE Transactions on Automatic Control*, Vol. AC-25, No. 3, 1980, pp. 492–496.

[4]   Shinar, J., "Solution Techniques for Realistic Pursuit-Evasion Games," *Advances in Control and Dynamic Systems*, C. T. Leondes, ed., Vol. 17, Academic Press, New York, 1981, pp. 63–124.

[5]   Shinar, J., "Interception of Maneuvering Targets in Theater Missile Defense," *Theater Ballistic Missile Defense*, B. Naveh, ed., Progress in Aeronautics and Astronautics, Vol. 192, American Institute of Aeronautics and Astronautics, Washington D.C., 2000.

[6]   Shima, T., and Shinar, J., "Time Varying Linear Pursuit-Evasion Game Models with Bounded Controls," *Journal of Guidance, Control and Dynamics*, Vol. 25, No. 3, 2002, pp. 425–432.

[7]   Shinar, J., and Turetsky, V., "Missile Guidance Laws Based on Pursuit-Evasion Game Formulations," *Automatica*, Vol. 39, No. 4, 2003, pp. 607–618.

[8]   Shinar, J., Turetsky, V., and Oshman, Y., "Integrated Estimation/Guidance Design Approach for Improved Homing Against Randomly Maneuvering Targets," *Journal of Guidance, Control and Dynamics*, Vol. 30, No. 1, Jan.–Feb. 2007, pp. 154–161.

[9]   Shinar, J., and Turetsky, V., "Interceptor Guidance Laws—Robustness and Implementation," 50th Israel Annual Conference on Aviation and Astronautics, Tel Aviv–Haifa, Feb. 2010.

CHAPTER 20

Boost-Phase Filtering Options

INTRODUCTION

Intercepting ballistic missiles during their boost phase is attractive because ballistic missiles are easy to detect and track while they are thrusting and, if the intercept is successful, the raid size for subsequent missile defense layers is thinned. In addition, decoys and other countermeasures, which can be very important for midcourse phase intercepts (that is, when the target is only under the influence of gravity) are usually considered to be more difficult to devise against boosting targets. An important part of the boost-phase intercept problem is the problem of tracking the boosting target and predicting where it will be in the future. Robust filtering options that require a minimum of computation are highly desired for the boost-phase intercept application. This chapter [1] takes the approach of simplifying the boost-phase filtering problem down to its most basic levels so that two different fundamental approaches can be explored.

It is usually thought that the more *a priori* information one gives a Kalman filter, the better the Kalman filter will perform. This is certainly true if the *a priori* information is correct. If the *a priori* information has slight errors, however, considerable filtering errors may result. In fact if the *a priori* information is incorrect, filter divergence may result. Process noise reflects how much confidence we have in the mathematical representation of the model of the real world embedded in the Kalman filter. Using large amounts of process noise in the filter design is an engineer's way of telling the filter that we have very little confidence in our model of the real world and that the filter should always pay attention to the measurements. Using small amounts or zero process noise in the filter design means that we are so confident in our model of the real world that the filter can eventually stop paying attention to the measurements. (Kalman gains eventually go to zero with zero process noise.) In order to make a Kalman filter robust to errors in *a priori* information, sufficient process noise must be added to the filter—sometimes massive amounts of process noise! Although increased process noise can often eliminate filter divergence, it also increases the errors in the state estimates. Sometimes it is possible that when *a priori* information is in error one might have been better off with a simpler Kalman filter, requiring less process noise, that does not require *a priori* information.

Although there is an extensive body of academic papers on high-order filters for tracking applications, many systems that are built use low-order decoupled filtering techniques [2, 3]. For both simplicity and ease of understanding we will only consider the low-order tracking filters in this chapter. Two filtering options for tracking a boosting intercontinental ballistic missile (ICBM) will be investigated in this chapter. First, the performance of a decoupled two-state Kalman filter, sometimes called a position-velocity filter, will be investigated assuming that the current ICBM acceleration magnitude and direction are known perfectly. Next, the degradation in performance of the two-state Kalman filter will be studied assuming that the current acceleration magnitude is known exactly but the current direction is not known and must be guessed. In this case the filter will assume that the ICBM is performing a gravity turn even though it is not. Next a simple linear three-state decoupled polynomial Kalman filter, sometimes called a position-velocity-acceleration filter, for tracking the boosting ICBM will also be considered. This type of filter, which does not require *a priori* information, is often used in the tactical missile world for tracking unpredictable maneuvering aircraft. Performance comparisons will be made between the two filtering options.

ICBM MODEL

To get at the heart of the matter and to avoid unnecessary complexity, the physics of the real world will be kept as simple as possible and an ICBM model from the open literature will be used. One such generic ICBM model appears in the American Physical Society (APS) report [4] and is representative of a two-stage, liquid-fueled ICBM that is capable of traveling 12,000 km. The magnitude of the longitudinal acceleration profile of the APS two-stage liquid ICBM during its boost phase appears in Fig. 20.1.

Figure 20.1 indicates that the ICBM boost phase lasts 240 s, and the discontinuities in the acceleration profile are due to staging events. At the end of the first-stage burn at 120 s, the peak ICBM longitudinal acceleration is approximately 6 g; at the end of the second-stage burn at 240 s, the peak acceleration is approximately 13 g. The job of the Kalman filter will be to estimate the position, velocity, and possibly the acceleration of the boosting ICBM.

ICBM GUIDANCE

Usually an ICBM flies straight up for a while and then performs a gravity turn while it is in the atmosphere to minimize loading and drag effects. Once the ICBM is out of the atmosphere, or when the dynamic pressure falls below a

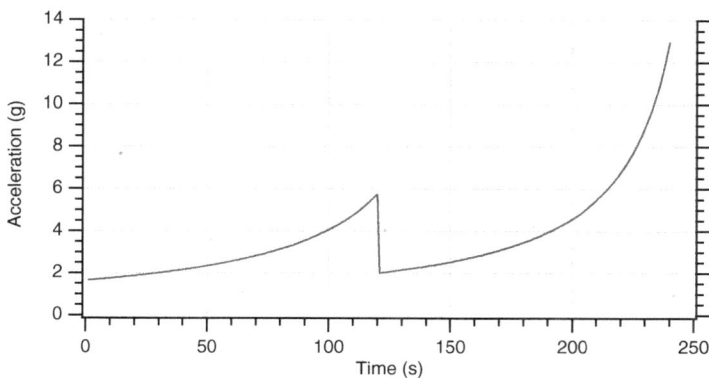

Fig. 20.1   Two-stage APS liquid ICBM acceleration profile.

certain level, the ICBM can perform closed-loop guidance so that it can reach its intended target [5]. Lambert guidance is one possible method of steering a boosting ICBM to its intended target at the desired impact time. Lambert guidance involves the numerical solution to Lambert's problem and was discussed fully in Chapter 17 of the previous volume [6–8]. Essentially, at each instant of time if the ICBM knows where it is and where it wants to go and how long it should take to get to its destination, the solution to Lambert's problem tells the ICBM at each instant of time the magnitude and direction of the required velocity vector.

As was shown in Chapter 16 of the previous volume, with a gravity turn the ICBM's thrust vector is aligned with the velocity vector in an attempt to drive the angle of attack to zero in order to minimize drag. The differential equations for the acceleration components of the ICBM while performing the gravity turn in an earth-centered inertial (ECI) coordinate system was shown to be given by

$$a_x = \frac{-gmx}{(x^2 + y^2)^{1.5}} + a_T \frac{\dot{x}}{(x^2 + y^2)^{.5}}$$

$$a_y = \frac{-gmy}{(x^2 + y^2)^{1.5}} + a_T \frac{\dot{y}}{(\dot{x}^2 + \dot{y}^2)^{.5}}$$

The first term in the preceding differential equation is due to gravity, and the second term is the gravity turn portion of the boosting target's acceleration.

Listing 20.1, which is based on Listing 17.3 of the previous volume, is a two-dimensional trajectory generator for the APS ICBM traveling on a 7000-km trajectory. Here the ICBM goes straight up for 20 s (TUPT=20), then performs an open-loop gravity turn for the next 80 s (TGRAVEND=100), and finally uses closed-loop Lambert guidance to get to its final destination. Note that the trajectory generator also calculates the dynamic pressure Q of the ICBM. The dynamic

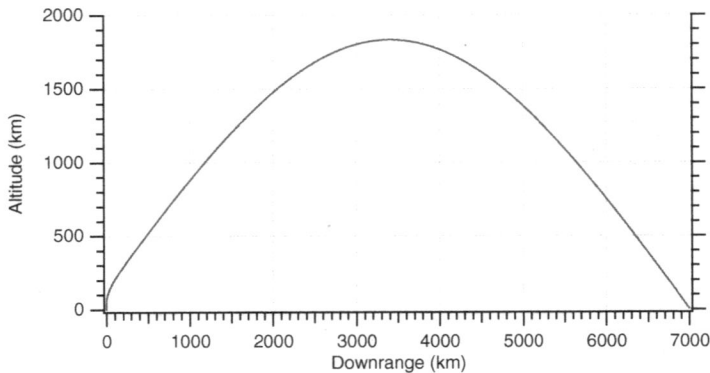

Fig. 20.2   Nominal 7000-km ICBM trajectory.

pressure acting on the ICBM is given by the expression

$$Q = 0.5\rho V^2$$

where V is the booster's velocity in ft/s and ρ is air density in slug/ft^3. In the English system of units, air density ρ was shown in Chapter 10 of the previous volume to be (above 30,000 ft altitude) approximated by

$$\rho = 0.0034 e^{-\text{alt}/22000}$$

where alt is the booster altitude in feet. The nominal case of Listing 20.1 was run, and the resultant lofted (TLOFT=500) 7000-km ICBM trajectory appears in Fig. 20.2. We can see that the apogee of the trajectory is approximately 1800 km.

LISTING 20.1   TRAJECTORY GENERATOR FOR NOMINAL ICBM

```
clear
count=0;
RDESKM=7000.;
TF=2000.;
TFINISH=999999.;
TLOFT=500.;
TGRAVEND=100.;
GAMDEGIC=89.8;
TUPT=20.;
RDESRKM=560.;
LEFT=1;
QBOOST=1;
QOOMPH=1;
QFINISH=1;
```

```
QZERO=0;
CW=0;
SWITCH=0;
QFIRST=1;
GAMDEG=GAMDEGIC;
H=.01;
T=0.;
S=0.;
A=2.0926E7;
GM=1.4077E16;
ALTNM=0.;
ALT=ALTNM*6076.;
ANGDEG=0.;
ANG=ANGDEG/57.3;
XLONGM=ANG;
X=(A+ALT)*cos(ANG);
Y=(A+ALT)*sin(ANG);
ALT=sqrt(X^2+Y^2)-A;
XFIRST=X;
YFIRST=Y;
X1=cos(1.5708-GAMDEG/57.3+ANG);
Y1=sin(1.5708-GAMDEG/57.3+ANG);
AXT=0.;
AYT=0.;
XLONGTDEG=57.3*RDESKM*3280./A;
XLONGRDEG=57.3*RDESRKM*3280./A;
TF=252.+.223*RDESKM-(5.44E-6)*RDESKM*RDESKM;
TF=TF+TLOFT;
XLONGT=XLONGTDEG/57.3;
XLONGR=XLONGRDEG/57.3;
XF=A*cos(XLONGT);
YF=A*sin(XLONGT);
XR=A*cos(XLONGR);
YR=A*sin(XLONGR);
AXT=0.;
AYT=0.;
Z=0;
ZF=0;
ZFIRST=0;
while ~(ALTNM<-1 | T>TFINISH)
        XOLD=X;
        YOLD=Y;
        X1OLD=X1;
        Y1OLD=Y1;
        STEP=1;
        FLAG=0;
```

```
while STEP <=1
      if FLAG==1
             STEP=2;
             X=X+H*XD;
             Y=Y+H*YD;
             X1=X1+H*X1D;
             Y1=Y1+H*Y1D;
             T=T+H;
      end;
      if T<120.
             WGT=-2622*T+440660.;
             TRST=725850.;
      elseif T<240.
             WGT=-642.*T+168120.;
             TRST=182250.;
      else
             WGT=5500.;
             TRST=0.;
      end
      AT=32.2*TRST/WGT;
      XD=X1;
      YD=Y1;
      VEL=sqrt(XD^2+YD^2);
      TEMBOT=(X^2+Y^2)^1.5;
      X1D=-GM*X/TEMBOT+AXT;
      Y1D=-GM*Y/TEMBOT+AYT;
      ALT=sqrt(X^2+Y^2)-A;
      ACCG=sqrt(AXT^2+AYT^2)/32.2;
      FLAG=1;
end
FLAG=0;
X=(XOLD+X)/2+.5*H*XD;
Y=(YOLD+Y)/2+.5*H*YD;
X1=(X1OLD+X1)/2+.5*H*X1D;
Y1=(Y1OLD+Y1)/2+.5*H*Y1D;
S=S+H;
Z=0.;
ZF=0.;
TGOLAM=TF-T;
if (QBOOST==1 & T>TGRAVEND)
      TGOLAM=TF-T;
      [VRX,VRY,VRZ]=LAMBERT3D(X,Y,Z,TGOLAM,XF,YF,ZF,SWITCH);
      DELX=VRX-X1;
      DELY=VRY-Y1;
      DEL=sqrt(DELX^2+DELY^2);
      if (T<240 & DEL>500.)
```

```
                    AXT=AT*DELX/DEL;
                    AYT=AT*DELY/DEL;
            elseif DEL<500.
                    TRST=0.;
                    QBOOST=0;
                    AXT=0.;
                    AYT=0.;
                    X1=VRX;
                    Y1=VRY;
                    X1OLD=X1;
                    Y1OLD=Y1;
            else
                    QBOOST=0;
                    AXT=0.;
                    AYT=0.;
            end
    elseif (T>=TUPT & T<= TGRAVEND & QFIRST==1)
            QFIRST=0;
            VEL=sqrt(XD^2+YD^2);
            X1=VEL*cos(1.5708-GAMDEGIC/57.3+ANG);
            Y1=VEL*sin(1.5708-GAMDEGIC/57.3+ANG);
            X1OLD=X1;
            Y1OLD=Y1;
            AXT=AT*X1/VEL;
            AYT=AT*Y1/VEL;
    elseif (T>=TUPT & T<=TGRAVEND)
            VEL=sqrt(XD^2+YD^2);
            AXT=AT*X1/VEL;
            AYT=AT*Y1/VEL;
    elseif T<=TUPT
            RTMAG=sqrt(X^2+Y^2);
            AXT=AT*X/RTMAG;
            AYT=AT*Y/RTMAG;
    end
    if S>=.9999
            S=0.;
            DISTNM=distance3dkm(X,Y,Z,XFIRST,YFIRST,ZFIRST);
            ALTNM=(sqrt(X^2+Y^2)-A)/3280.;
            RMAG=sqrt(X^2+Y^2);
            VMAG=sqrt(XD^2+YD^2);
            GAMDEG=90-57.3*acos((X*XD+Y*YD)/(RMAG*VMAG));
            RHO=.0034*exp(-ALT/22000.);
            Q=.5*RHO*VEL*VEL;
            RRMAG=sqrt(XR^2+YR^2);
            RRTMAG=sqrt((X-XR)^2+(Y-YR)^2);
            ELDEG=90.-57.3*acos((XR*(X-XR)+YR*(Y-YR))/(RRMAG*RRTMAG));
```

```
                    if (ELDEG>2. & ELDEG<85)
                            ISEE=1;
                    else
                            ISEE=0;
                    end
                    count=count+1;
                    ArrayT(count)=T;
                    ArrayDISTNM(count)=DISTNM;
                    ArrayALTNM(count)=ALTNM;
                    ArrayQ(count)=Q;
                    ArrayISEE(count)=ISEE;
            end
end
figure
plot(ArrayDISTNM,ArrayALTNM),grid
xlabel('Downrange (km)')
ylabel('Altitude (km)')
clc
output=[ArrayT',ArrayDISTNM',ArrayALTNM',ArrayQ',ArrayISEE'];
save datfil.txt output -ascii
disp 'simulation finished'

% LAMBERT3D can be found in Listing 17.2

% distance3dkm can also be found in Listing 17.2
```

For the boost-phase portion of the ICBM trajectory, or the first 240 s of flight, the dynamic pressure, in the English system of units, is depicted in Fig. 20.3. As was previously mentioned, usually a gravity turn is performed while the ICBM is in the atmosphere to reduce drag and loading effects. However, when the dynamic pressure is low enough, closed-loop guidance can begin. If we assume that ICBM guidance cannot begin until the dynamic pressure drops below 600 lb/ft^2 (that is,

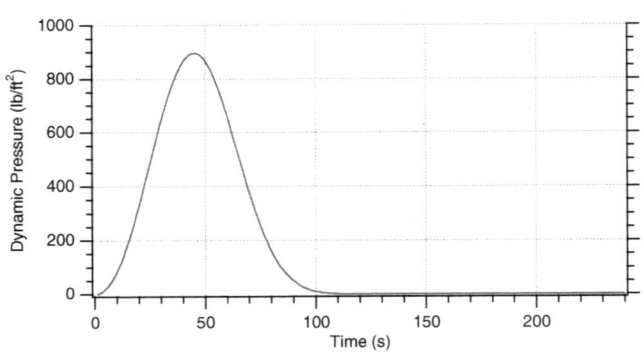

Fig. 20.3   Dynamic pressure for ICBM boost-phase portion of flight.

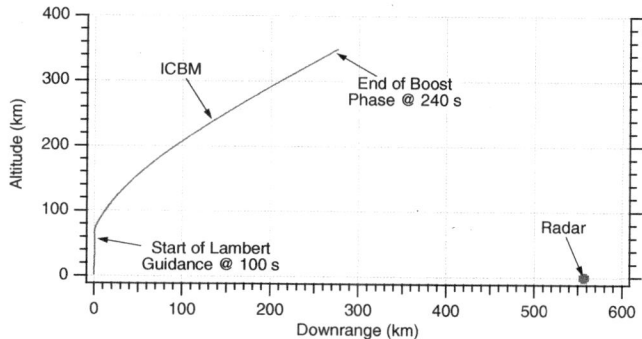

Fig. 20.4   Boost-phase portion of 7000-km ICBM trajectory.

the actual number depends on booster design), then we can say that for this example, closed-loop Lambert guidance cannot begin until 65 s after the ICBM takes off. However, in order to be conservative, it shall be assumed that closed-loop guidance can begin at 100 s where the dynamic pressure is below 50 lb/ft^2.

The boost-phase portion of the ICBM 7000-km trajectory is shown in Fig. 20.4. In this example the boosting ICBM first goes straight up for 20 s, performs a gravity turn for the next 80 s, and then switches to Lambert guidance at 100 s after launch. Also shown in Fig. 20.4 is a tracking radar located approximately 560 km from the ICBM launch site. The radar will not be able to see the target immediately due to the curvature of the Earth and radar elevation angle constraints. If a minimum radar elevation angle of 2 deg is required for acquisition, then running Listing 20.1 also indicates that the radar will not be able to see the ICBM for the first 90 s of flight (that is, when ISEE=1).

Figure 20.5 displays the downrange and altitude acceleration profiles of the boosting ICBM. The figure indicates that for the first 20 s the downrange

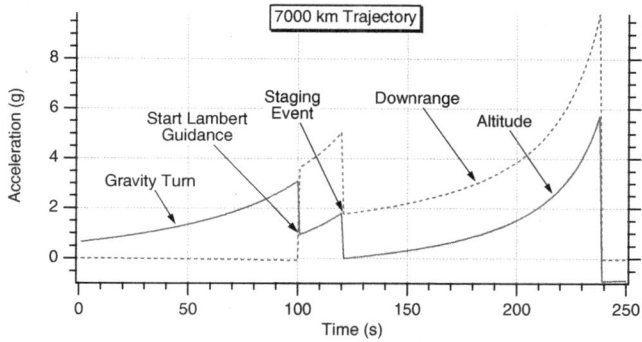

Fig. 20.5   ICBM acceleration components during boost phase for 7000-km trajectory.

acceleration is zero because the APS ICBM is going straight up. As was mentioned before, the ICBM goes straight up for 20 s, a gravity turn is employed from 20 s to 100 s, and Lambert guidance is used for the remainder of the flight. There is a staging event at 120 s.

FILTERING OPTIONS

For radar tracking applications, range and angle measurements are available. A fully coupled extended Kalman filter (EKF) can be designed to estimate the position, velocity, and acceleration of the target. With this approach the filter will require accurate state and covariance initialization and have a certain computational burden. An alternative approach that has been used in older tracking systems [2, 3] is to transform the actual radar measurements to pseudo position measurements and then to build decoupled linear polynomial Kalman filters in each part of the coordinate system under consideration. Such an approach offers a considerable reduction in computation over the EKF approach, which means that many more targets can be tracked for a given flight computer throughput capacity. In addition, linear polynomial Kalman filters are very robust and are insensitive to filter initialization issues, which can be very important in military applications. Therefore, in each of the filtering approaches considered in this chapter, the filters will be considered to be decoupled, and only the equations in one of the channels will be presented.

If *a priori* information concerning the current acceleration profile of the ICBM is available, a two-state Kalman filter can be designed to estimate the ICBM's position and velocity. Because the acceleration of the boosting ICBM is assumed to be known, it does not have to be estimated. If acceleration information, also known as a *template*, is not available, a three-state polynomial Kalman filter can be designed to estimate the position, velocity, and acceleration of the boosting ICBM.

Figure 20.4 presented the geometry for a surface-based radar tracking an ICBM during its boost phase. In this generic model it is assumed that the radar measures range and angle to the ICBM and that the radar tracks the target with an angular accuracy of 1 mrad and a range accuracy of 10 m. Measurements are taken every second. Due to previously mentioned radar horizon and elevation angle constraints, the radar sees the ICBM at 90 s after target launch.

The range and angle from the radar to the target or ICBM is given by

$$R_T = \sqrt{(x_T - x_R)^2 + (y_T - y_R)^2}$$

$$\theta_T = \tan^{-1}\left(\frac{y_T - y_R}{x_R - x_T}\right)$$

where the coordinates of the radar x_R, y_R are assumed to be known and the target location x_T, y_T is unknown. As was previously mentioned, although the radar actually measures range and angle, one can pretend the radar measures down-range and altitude in order to avoid building an EKF. Creating such position pseudo measurements will allow us to build a linear decoupled polynomial Kalman filter.

The target location can be expressed in terms of the radar location, angle to the target, and range to the target as

$$x_T = -R_T \cos \theta_T + x_R$$
$$y_T = R_T \sin \theta_T + y_R$$

Using the chain rule from calculus yields

$$\Delta x_T = \frac{\partial x_T}{\partial R_T} \Delta R_T + \frac{\partial x_T}{\partial \theta_T} \Delta \theta_T$$

$$\Delta y_T = \frac{\partial y_T}{\partial R_T} \Delta R_T + \frac{\partial y_T}{\partial \theta_T} \Delta \theta_T$$

After taking the partial derivatives we obtain

$$\Delta x_T = -\cos \theta_T \Delta R_T + R_T \sin \theta_T \Delta \theta_T$$
$$\Delta y_T = \sin \theta_T \Delta R_T + R_T \cos \theta_T \Delta \theta_T$$

Squaring both sides of the preceding two equations, ignoring cross-coupling effects, and taking expectations yields

$$\sigma_x^2 = \cos^2 \theta_T \sigma_R^2 + R_T^2 \sin^2 \theta_T \sigma_\theta^2$$
$$\sigma_y^2 = \sin^2 \theta_T \sigma_R^2 + R_T^2 \cos^2 \theta_T \sigma_\theta^2$$

where it has been assumed that

$$E(\Delta x_T^2) = \sigma_x^2$$

$$E(\Delta y_T^2) = \sigma_y^2$$

$$E(\Delta R_T^2) = \sigma_R^2$$

$$E(\Delta \theta_T^2) = \sigma_\theta^2$$

TWO-STATE TEMPLATE-BASED FILTER

If cross coupling effects are neglected, two decoupled two-state linear polynomial Kalman filters can be built. One filter is in downrange and the other in altitude so that the position and velocity of the booster can be estimated. A template-based Kalman filter assumes that a perfect acceleration template of the boosting target is available in which the current components of the booster acceleration a_{Tx}, a_{Ty} are known exactly. Under these conditions the equations of the downrange template-based two-state linear polynomial Kalman filter become

$$\text{Res}_k = x^*_{T_k} - \hat{x}_{T_{k-1}} - \hat{\dot{x}}_{T_{k-1}} T_s - 0.5 a_{T_{k-1}} T_s^2$$

$$\hat{x}_{T_k} = \hat{x}_{T_{k-1}} + \hat{\dot{x}}_{T_{k-1}} T_s + 0.5 a_{T_{k-1}} T_s^2 + K_{1_k} \text{Res}_k$$

$$\hat{\dot{x}}_{T_k} = \hat{\dot{x}}_{T_{k-1}} + a_{T_{k-1}} T_s + K_{2_k} \text{Res}_k$$

where Res_k stands for the filter residual. There is an identical set of equations for the Kalman filter in the altitude direction. The Kalman gains for the downrange and altitude filters will be different because the variance of the pseudo measurement noise in each channel is different. The process noise matrix for both decoupled filters is given by

$$\mathbf{Q}_k = \Phi_s \begin{bmatrix} \dfrac{T_s^3}{3} & \dfrac{T_s^2}{2} \\ \dfrac{T_s^2}{2} & T_s \end{bmatrix}$$

where Φ_s is the process noise power spectral density in units squared per Hz. In practice, Φ_s is chosen by extensive computer experiments in which practical errors are introduced to ensure that the filter is robust.

THREE-STATE FILTER

Kalman filters that are not template-based do not have to be invented because they have been used in the tactical missile world for many decades for tracking unpredictable maneuvering aircraft targets. These filters simply take derivatives and are known as *linear polynomial Kalman filters*. For tracking an ICBM during the boost phase in a two-dimensional world, two decoupled three-state linear polynomial Kalman filters can also be built, one in downrange and the other in altitude, to estimate the position, velocity, and acceleration of the boosting ICBM. The equations of the downrange three-state linear polynomial Kalman filter are given by [9, 10]

$$\text{Res}_k = x^*_{T_k} - \hat{x}_{T_{k-1}} - \hat{\dot{x}}_{T_{k-1}} T_s - 0.5 \hat{\ddot{x}}_{T_{k-1}} T_s^2$$

$$\hat{x}_{T_k} = \hat{x}_{T_{k-1}} + \hat{\dot{x}}_{T_{k-1}} T_s + 0.5 \hat{\ddot{x}}_{T_{k-1}} T_s^2 + K_{1_k} \text{Res}_k$$

$$\hat{\dot{x}}_{T_k} = \hat{\dot{x}}_{T_{s_{k-1}}} + \hat{\ddot{x}}_{T_{k-1}} T_s + K_{2_k} \text{Res}_k$$

$$\hat{\ddot{x}}_{T_k} = \hat{\ddot{x}}_{T_{s_{k-1}}} + K_{3_k} \text{Res}_k$$

where Res_k again represents the filter residual. There is an identical set of equations in the altitude direction. The Kalman gains for the downrange and altitude filters will be different because the variance of the pseudo measurement noise is different in each channel. The process noise matrix for both decoupled filters is given by

$$
Q = \Phi_s \begin{bmatrix} \dfrac{T_s^5}{20} & \dfrac{T_s^4}{8} & \dfrac{T_s^3}{6} \\[2ex] \dfrac{T_s^4}{8} & \dfrac{T_s^3}{3} & \dfrac{T_s^2}{2} \\[2ex] \dfrac{T_s^3}{6} & \dfrac{T_s^2}{2} & T_s \end{bmatrix}
$$

where Φ_s is the process noise power spectral density in units squared per Hz. For all studies in this chapter with the three-state linear polynomial Kalman filter, the value of process noise is determined by experiment and is set to 260.

Listing 20.2 is a simulation of the ICBM trajectory generator with the just-discussed two- and three-state linear decoupled polynomial Kalman filters. If IFILTER=1, then the two-state template filter is used where the target acceleration is assumed to be known. In addition, we can see that the filter does not depend on knowledge of the target's initial position and velocity but initializes itself by using a two-state least squares filter for the first few measurements. If IFILTER=2, then the three-state linear polynomial Kalman filter is used. In this case we can see that the filter also does not depend on knowledge of the target's initial position, velocity, and acceleration estimates but initializes itself by using a three-state least squares filter for the first few measurements [10]. The radar elevation angle is calculated, and it is assumed that the radar cannot see the target unless the elevation angle is between 2 deg and 85 deg. Filtering does not start until the radar can see the target (ISEE=1).

LISTING 20.2   SIMULATION OF TWO POSSIBLE LINEAR DECOUPLED POLYNOMIAL KALMAN FILTERS FOR TRACKING AN ICBM DURING BOOST PHASE

```
clear
count=0;
IFILTER=1;
RDESKM=7000.;
TF=2000.;
TFINISH=240.;
TLOFT=500.;
TGRAVEND=100.;
GAMDEGIC=89.8;
TUPT=20.;
RDESRKM=560.;
SWITCH=0;
PHIS=0.;
PHIS1=260.;
```

```
ERR=0.;
TS=1.;
SIGTHET=.001;
SIGR=10.*3.28;
QGRAV=0;
ORDER =2;
LEFT=1;
QBOOST=1;
QFINISH=1;
QFIRST=1;
GAMDEG=GAMDEGIC;
HINT=.01;
XH=0.;
XDH=0.;
XDDH=0.;
YH=0.;
YDH=0.;
YDDH=0.;
PHI=zeros([2,2]);
P=zeros([2,2]);
Q=zeros([2,2]);
IDNPZ=eye(2);
P(1,1)=99999999999.;
P(2,2)=99999999999.;
PP(1,1)=99999999999.;
PP(2,2)=99999999999.;
PHI(1,1)=1;
PHI(1,2)=TS;
PHI(2,2)=1;
HMAT(1,1)=1.;
HMAT(1,2)=0.;
PHIT=PHI';
HT=HMAT';
Q(1,1)=PHIS*TS^3/3.;
Q(1,2)=PHIS*TS^2/2.;
Q(2,1)=Q(1,2);
Q(2,2)=PHIS*TS;
P11=99999999999.;
P12=0.;
P13=0.;
P22=99999999999.;
P23=0.;
P33=99999999999.;
P11P=99999999999.;
P12P=0.;
P13P=0.;
```

```
P22P=99999999999.;
P23P=0.;
P33P=99999999999.;
XN=0.;
T=0.;
S=0.;
A=2.0926E7;
GM=1.4077E16;
ALTNM=0.;
ALT=ALTNM*6076.;
ANGDEG=0.;
ANG=ANGDEG/57.3;
XLONGM=ANG;
X=(A+ALT)*cos(ANG);
Y=(A+ALT)*sin(ANG);
ALT=sqrt(X^2+Y^2)-A;
XFIRST=X;
YFIRST=Y;
X1=cos(1.5708-GAMDEG/57.3+ANG);
Y1=sin(1.5708-GAMDEG/57.3+ANG);
AXT=0.;
AYT=0.;
XLONGTDEG=57.3*RDESKM*3280./A;
XLONGRDEG=57.3*RDESRKM*3280./A;
TF=252.+.223*RDESKM-(5.44E-6)*RDESKM*RDESKM;
TF=TF+TLOFT;
XLONGT=XLONGTDEG/57.3;
XLONGR=XLONGRDEG/57.3;
XF=A*cos(XLONGT);
YF=A*sin(XLONGT);
XR=A*cos(XLONGR);
YR=A*sin(XLONGR);
AXT=0.;
AYT=0.;
Z=0;
ZF=0;
ZFIRST=0;
while ~(ALT< -1 | T>TFINISH)
      XOLD=X;
      YOLD=Y;
      X1OLD=X1;
      Y1OLD=Y1;
      STEP=1;
      FLAG=0;
      while STEP <=1
            if FLAG==1
```

```
                STEP=2;
                X=X+HINT*XD;
                Y=Y+HINT*YD;
                X1=X1+HINT*X1D;
                Y1=Y1+HINT*Y1D;
                T=T+HINT;
        end
        if T<120.
                WGT=-2622*T+440660.;
                TRST=725850.;
        elseif T<240.
                WGT=-642.*T+168120.;
                TRST=182250.;
        else
                WGT=5500.;
                TRST=0.;
        end
        AT=32.2*TRST/WGT;
        XD=X1;
        YD=Y1;
        VEL=sqrt(XD^2+YD^2);
        TEMBOT=(X^2+Y^2)^1.5;
        X1D=-GM*X/TEMBOT+AXT;
        Y1D=-GM*Y/TEMBOT+AYT;
        ALT=sqrt(X^2+Y^2)-A;
        ACCG=sqrt(AXT^2+AYT^2)/32.2;
        FLAG=1;
end
FLAG=0;
X=(XOLD+X)/2+.5*HINT*XD;
Y=(YOLD+Y)/2+.5*HINT*YD;
X1=(X1OLD+X1)/2+.5*HINT*X1D;
Y1=(Y1OLD+Y1)/2+.5*HINT*Y1D;
S=S+HINT;
Z=0.;
ZF=0.;
ZR=0;
TGOLAM=TF-T;
if (QBOOST==1 && T>TGRAVEND)
        TGOLAM=TF-T;
        [VRX,VRY,VRZ]=LAMBERT3D(X,Y,Z,TGOLAM,XF,YF,ZF,SWITCH);
        DELX=VRX-X1;
        DELY=VRY-Y1;
        DEL=sqrt(DELX^2+DELY^2);
        if (T<240 & DEL>500.)
                AXT=AT*DELX/DEL;
```

```
                    AYT=AT*DELY/DEL;
        elseif DEL<500.
                    TRST=0.;
                    QBOOST=0;
                    AXT=0.;
                    AYT=0.;
                    X1=VRX;
                    Y1=VRY;
                    X1OLD=X1;
                    Y1OLD=Y1;
        else
                    QBOOST=0;
                    AXT=0.;
                    AYT=0.;
    end
    elseif (T>=TUPT & T<=TGRAVEND & QFIRST==1)
            QFIRST=0;
            VEL=sqrt(XD^2+YD^2);
            X1=VEL*cos(1.5708-GAMDEGIC/57.3+ANG);
            Y1=VEL*sin(1.5708-GAMDEGIC/57.3+ANG);
            X1OLD=X1;
            Y1OLD=Y1;
            AXT=AT*X1/VEL;
            AYT=AT*Y1/VEL;
    elseif (T>=TUPT & T<=TGRAVEND)
            VEL=sqrt(XD^2+YD^2);
            AXT=AT*X1/VEL;
            AYT=AT*Y1/VEL;
    elseif T<=TUPT
            RTMAG=sqrt(X^2+Y^2);
            AXT=AT*X/RTMAG;
            AYT=AT*Y/RTMAG;
    end
    if S>=(TS-.00001)
            S=0.;
            DISTNM=distance3dkm(X,Y,Z,XFIRST,YFIRST,ZFIRST);
            ALTNM=(sqrt(X^2+Y^2)-A)/3280.;
            RMAG=sqrt(X^2+Y^2);
            VMAG=sqrt(XD^2+YD^2);
            GAMDEG=90-57.3*acos((X*XD+Y*YD)/(RMAG*VMAG));
            RHO=.0034*exp(-ALT/22000.);
            QPRES=.5*RHO*VEL*VEL;
            RRKM=sqrt((X-XR)^2+(Y-YR)^2)/3280.;
            DISTRKM=distance3dkm(XR,YR,ZR,XFIRST,YFIRST,ZFIRST);
            ALTRKM=(sqrt(XR^2+YR^2)-A)/3280.;
            RRMAG=sqrt(XR^2+YR^2);
```

```
RRTMAG=sqrt((X-XR)^2+(Y-YR)^2);
ELDEG=90.-57.3*acos((XR*(X-XR)+YR*(Y-YR))/(RRMAG*RRTMAG));
if (ELDEG>2. & ELDEG<85)
        ISEE=1;
else
        ISEE=0;
end
if ISEE==1
        XN=XN+1;
        if IFILTER==1
                XK1=2.*(2.*XN-1.)/(XN*(XN+1));
                XK2=6./(XN*(XN+1)*TS);
        else
                XK1=3*(3*XN*XN-3*XN+2)/(XN*(XN+1)*(XN+2));
                XK2=18*(2*XN-1)/(XN*(XN+1)*(XN+2)*TS);
                XK3=60/(XN*(XN+1)*(XN+2)*TS*TS);
        end
        TS2=TS*TS;
        TS3=TS2*TS;
        TS4=TS3*TS;
        TS5=TS4*TS;
        THET=atan2(Y-YR,XR-X);
        R=sqrt((XR-X)^2+(Y-YR)^2);
        THETNOISE=SIGTHET*randn;
        RNOISE=SIGR*randn;
        RMEAS=R+RNOISE;
        THETMEAS=THET+THETNOISE;
        XTS=XR-RMEAS*cos(THETMEAS);
        YTS=YR+RMEAS*sin(THETMEAS);
        XTNOISE=X-XTS;
        SIGX=sqrt((cos(THET)*SIGR)^2+(R*sin(THET)*SIGTHET)^2);
        YTNOISE=Y-YTS;
        SIGY=sqrt((sin(THET)*SIGR)^2+(R*cos(THET)*SIGTHET)^2);
        if IFILTER==1
                RMAT(1,1)=SIGX^2;
                PHIP=PHI*P;
                PHIPPHIT=PHIP*PHIT;
                M=PHIPPHIT+Q;
                HM=HMAT*M;
                HMHT=HM*HT;
                HMHTR=HMHT+RMAT;
                HMHTRINV(1,1)=1./HMHTR;
                MHT=M*HT;
                K=MHT*HMHTRINV;
                KH=K*HMAT;
                IKH=IDNPZ-KH;
```

```
P=IKH*M;
if XK1>K(1,1)
        XK1PZ=XK1;
        XK2PZ=XK2;
else
        XK1PZ=K(1,1);
        XK2PZ=K(2,1);
end
if QGRAV==0
        XDDH=X1D*(1.+ERR);
else
        VELH=sqrt(XDH^2+YDH^2);
        XDDH=AT*XDH/(VELH+.0001);
end
RES=XTS-XH-TS*XDH-.5*TS*TS*XDDH;
XH=XH+XDH*TS+.5*TS*TS*XDDH+XK1PZ*RES;

XDH=XDH+XDDH*TS+XK2PZ*RES;
RMATP(1,1)=SIGY^2;
PHIPP=PHI*PP;
PHIPPHITP=PHIPP*PHIT;
MP=PHIPPHITP+Q;
HMP=HMAT*MP;
HMHTP=HMP*HT;
HMHTRP=HMHTP+RMATP;
HMHTRINVP(1,1)=1./HMHTRP;
MHTP=MP*HT;
KP=MHTP*HMHTRINVP;
KHP=KP*HMAT;
IKHP=IDNPZ-KHP;
PP=IKHP*MP;
if XK1>K(1,1)
        XK1PZP=XK1;
        XK2PZP=XK2;
else
        XK1PZP=KP(1,1);
        XK2PZP=KP(2,1);
end
if QGRAV==0
        YDDH=Y1D*(1.+ERR);
else
        YDDH=AT*YDH/(VELH+.0001);
end
RESP=YTS-YH-TS*YDH-.5*TS*TS*YDDH;
YH=YH+YDH*TS+.5*TS*TS*YDDH+XK1PZP*RESP;
YDH=YDH+YDDH*TS+XK2PZP*RESP;
```

```
else
    M11=P11+TS*P12+.5*TS2*P13+TS*(P12+TS*P22+.5*TS2*P23);
    M11=M11+.5*TS2*(P13+TS*P23+.5*TS2*P33)+TS5*PHIS1/20.;
    M12=P12+TS*P22+.5*TS2*P23+TS*(P13+TS*P23+.5*TS2*P33);
    M12=M12+TS4*PHIS1/8.;
    M13=P13+TS*P23+.5*TS2*P33+PHIS1*TS3/6.;
    M22=P22+TS*P23+TS*(P23+TS*P33)+PHIS1*TS3/3.;
    M23=P23+TS*P33+.5*TS2*PHIS1;
    M33=P33+PHIS1*TS;
    BOT=M11+SIGX*SIGX;
    K1=M11/BOT;
    K2=M12/BOT;
    K3=M13/BOT;
    FACT=1.-K1;
    P11=FACT*M11;
    P12=FACT*M12;
    P13=FACT*M13;
    P22=-K2*M12+M22;
    P23=-K2*M13+M23;
    P33=-K3*M13+M33;
    if XK1>K1
            XK1PZ=XK1;
            XK2PZ=XK2;
            XK3PZ=XK3;
    else
            XK1PZ=K1;
            XK2PZ=K2;
            XK3PZ=K3;
    end
    RES=XTS-XH-TS*XDH-.5*TS*TS*XDDH;
    XH=XH+XDH*TS+.5*TS*TS*XDDH+XK1PZ*RES;
    XDH=XDH+XDDH*TS+XK2PZ*RES;
    XDDH=XDDH+XK3PZ*RES;

M11P=P11P+TS*P12P+.5*TS2*P13P+TS*(P12P+TS*P22P+.5*TS2*P23P);

M11P=M11P+.5*TS2*(P13P+TS*P23P+.5*TS2*P33P)+TS5*PHIS1/20.;

M12P=P12P+TS*P22P+.5*TS2*P23P+TS*(P13P+TS*P23P+.5*TS2*P33P);
            M12P=M12P+TS4*PHIS1/8.;
            M13P=P13P+TS*P23P+.5*TS2*P33P+PHIS1*TS3/6.;
            M22P=P22P+TS*P23P+TS*(P23P+TS*P33P)+PHIS1*TS3/3.;
            M23P=P23P+TS*P33P+.5*TS2*PHIS1;
            M33P=P33P+PHIS1*TS;
            BOTP=M11P+SIGY*SIGY;
            K1P=M11P/BOTP;
            K2P=M12P/BOTP;
```

```
                              K3P=M13P/BOTP;
                              FACTP=1.-K1P;
                              P11P=FACTP*M11P;
                              P12P=FACTP*M12P;
                              P13P=FACTP*M13P;
                              P22P=-K2P*M12P+M22P;
                              P23P=-K2P*M13P+M23P;
                              P33P=-K3P*M13P+M33P;
                              if XK1>K1
                                      XK1PZP=XK1;
                                      XK2PZP=XK2;
                                      XK3PZP=XK3;
                              else
                                      XK1PZP=K1P;
                                      XK2PZP=K2P;
                                      XK3PZP=K3P;
                              end
                              RESP=YTS-YH-TS*YDH-.5*TS*TS*YDDH;
                              YH=YH+YDH*TS+.5*TS*TS*YDDH+XK1PZP*RESP;
                              YDH=YDH+YDDH*TS+XK2PZP*RESP;
                              YDDH=YDDH+XK3PZP*RESP;
                    end
          end
          if IFILTER==1
                    ERRXTD=(X1-XDH);
                    SP22=sqrt(P(2,2));
                    ERRYTD=(Y1-YDH);
                    SP22P=-SP22;
                    count=count+1;
                    ArrayT(count)=T;
                    ArrayDISTNM(count)=DISTNM;
                    ArrayALTNM(count)=ALTNM;
                    ArrayERRXTD(count)=ERRXTD;
                    ArraySP22(count)=SP22;
                    ArraySP22P(count)=-SP22;
          else
                    ERRXTD=(X1-XDH);
                    SP22=sqrt(P22);
                    ERRYTD=(Y1-YDH);
                    SP22P=-SP22;
                    count=count+1;
                    ArrayT(count)=T;
                    ArrayDISTNM(count)=DISTNM;
                    ArrayALTNM(count)=ALTNM;
                    ArrayERRXTD(count)=ERRXTD;
                    ArraySP22(count)=SP22;
```

```
                      ArraySP22P(count)=-SP22;
              end
          end
end
figure
plot(ArrayDISTNM,ArrayALTNM),grid
xlabel('Downrange (km)')
ylabel('Altitude (km)')
figure
plot(ArrayT,ArrayERRXTD,ArrayT,ArraySP22,ArrayT,ArraySP22P),grid
xlabel('Time (s)')
ylabel('Velocity Error (f/s)')
axis([90 240 -500 500])
clc
output=[ArrayT',ArrayDISTNM',ArrayALTNM',ArrayERRXTD',ArraySP22',ArraySP22P'];
save datfil.txt output -ascii
disp 'simulation finished'

% LAMBERT3D can be found in Listing 17.2

% distance3dkm can be found in Listing 17.2
```

Listing 20.2 was first run assuming that the two-state template-based filter did not require any process noise (IFILTER=1, PHIS=0). Figure 20.6 shows single flight results that indicate when there is no process noise, the velocity error of the template-based filter starts to diverge from the covariance matrix predictions. Initially the results of Fig. 20.6 might seem strange because we have perfect knowledge of the target's current acceleration states; however, we do not have information on where the target is going and when it will arrive at its destination. From a filtering point of view this means that our predictions to the next

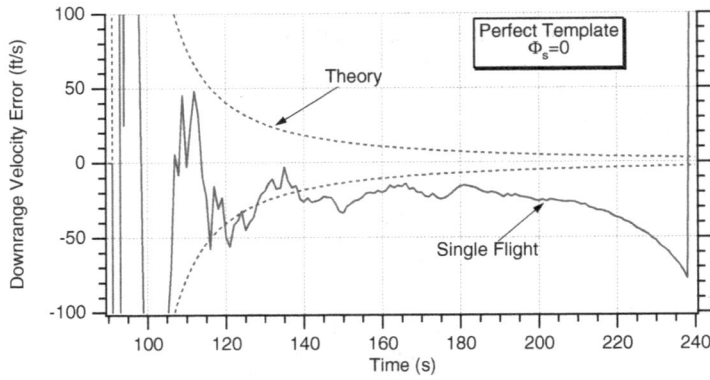

Fig. 20.6   Filter errors in estimates diverge when there is zero process noise.

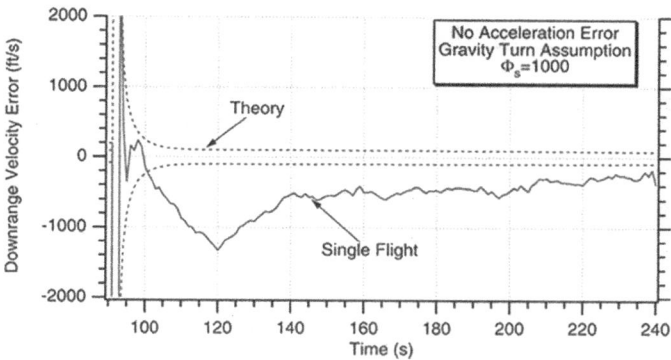

Fig. 20.7   With perfect acceleration template, two-state Kalman filter yields excellent downrange velocity estimates.

measurement, via the fundamental matrix, are not exact. Making the process noise zero means that the filter will eventually stop paying attention to the measurements and filter divergence results, as is indicated in Fig. 20.6. It is important to note that in this chapter we will be making judgments based on single flight results. Normally a Monte Carlo analysis is required for precise estimates in filtering work; however, usually a disaster, such as the filter divergence shown in Fig. 20.6, can be detected in a single run.

The simple engineering fix to filter divergence is to increase the filter process noise. Figure 20.7 indicates that divergence is eliminated and filter consistency is achieved when Φ_s is increased from 0 to 100. The figure indicates that, on a single flight basis, the error in the estimate of downrange velocity is consistent with the theoretical predictions of the covariance matrix obtained from the Riccati equations. Perfect knowledge of the current target's acceleration magnitude and direction in the template-based filter helps keep downrange velocity errors to less than 40 ft/s by the end of the ICBM boost phase.

Next, a 10% acceleration template error was introduced into our knowledge of the ICBM's acceleration magnitude (ERR=0.1). It was still assumed that the current direction of the ICBM longitudinal acceleration was known perfectly. Figure 20.8 indicates that an acceleration template error is hardly visually noticeable.

However, Fig. 20.9 shows that even a small 10% template error significantly influences two-state template-based Kalman filter consistency. The figure demonstrates that the downrange error in the velocity estimate diverges from the theoretical bounds for a 10% template error. Because our model of the real world has additional errors, more process noise is required by the filter to obtain filter consistency.

Figure 20.10 shows that when the filter process noise is increased by an order of magnitude (Φ_s increased from 100 to 1000), the two-state template-based Kalman filter becomes consistent and divergence is no longer an issue when

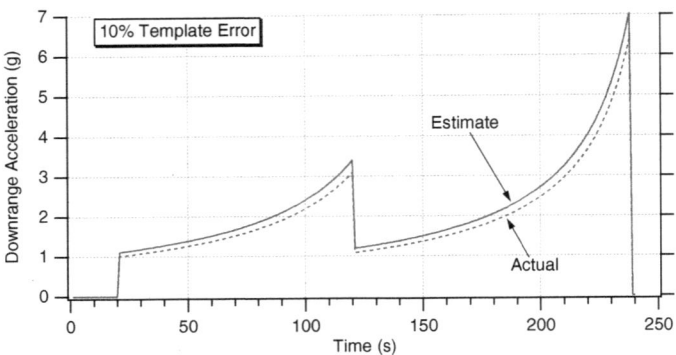

Fig. 20.8   Ten-percent errors slightly degrade downrange acceleration template.

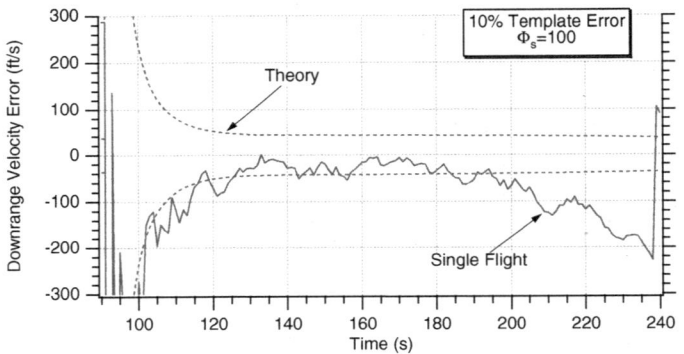

Fig. 20.9   Ten-percent template errors cause two-state Kalman filter to diverge.

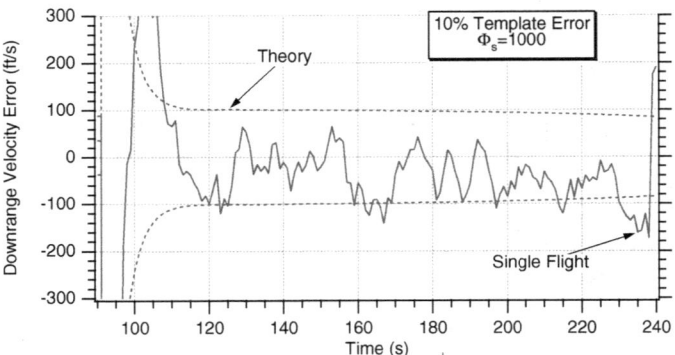

Fig. 20.10   More process noise is required by two-state Kalman filter to eliminate divergence when there is template error.

there is a 10% filter template error. The price paid for increasing the process noise is that the theoretical errors in the velocity estimate at 240 s increase from 40 ft/s ($\Phi_s = 100$) to 90 ft/s ($\Phi_s = 1000$).

So far all of the Kalman filter results have depended on a template in which the current booster acceleration direction was known exactly. Because in reality it is impossible to know where the ICBM is going and how it will get there, an assumption must be made on the direction of the future booster acceleration vector. One such popular assumption is to assume the ICBM acceleration direction is that of a gravity turn [9]. As was shown in the previous section, the gravity turn assumption implies that the booster thrust vector is always aligned with its velocity vector.

Many consider the gravity turn assumption to be reasonable because a gravity turn will minimize loading and drag while the booster is in the atmosphere. However, a gravity turn is not a closed-loop guidance law that can enable an ICBM to reach its intended target in the desired time. An ICBM performs gravity turn–type maneuvers in the atmosphere to minimize drag and loading effects, or to provide gentle maneuvering if the dynamic pressure is too high. When the dynamic pressure is low enough, an ICBM must use a closed-loop guidance law (in our case the ICBM is guiding using Lambert guidance) to reach its intended target. Therefore, there will be a model mismatch with the gravity turn assumption of the Kalman filter and what is happening in the real world when the dynamic pressure is low. In our real world model the target is performing a gravity turn until 100 s, and then the target switches to Lambert guidance when the dynamic pressure for this example is close to zero.

Figure 20.11 shows that when the gravity turn assumption is incorporated into the two-state template-based Kalman filter (QGRAV=1) there is divergence because there is not sufficient process noise ($\Phi_s = 1000$).

Figure 20.12 shows that when the filter process noise is increased by more than an order of magnitude (Φ_s increased from 1000 to 50,000), the two-state template-based Kalman filter becomes consistent and divergence is no longer an issue with

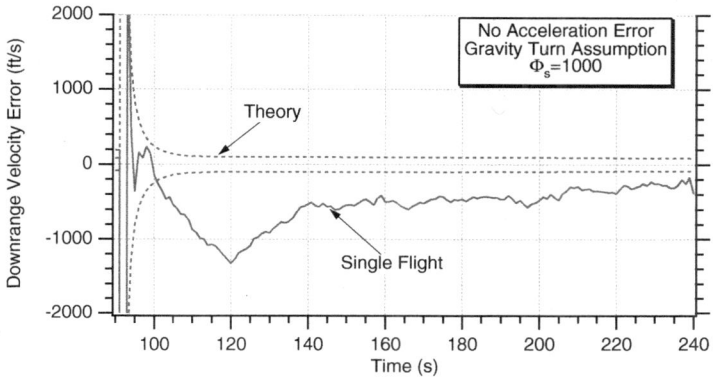

Fig. 20.11   Kalman filter diverges with gravity turn assumption because there is not sufficient process noise.

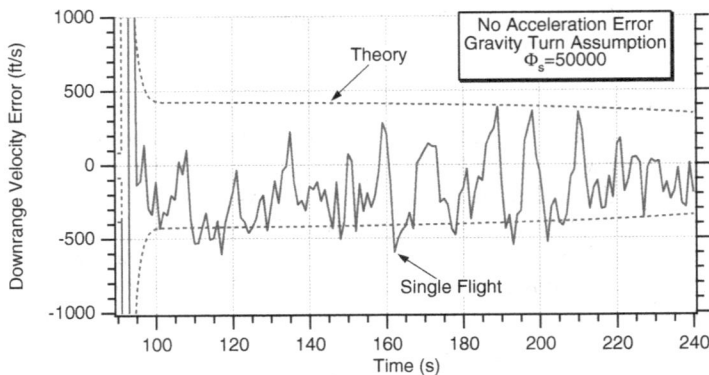

Fig. 20.12   Kalman filter is consistent with gravity turn assumption when process noise is increased by more than an order of magnitude.

the gravity turn assumption. The price paid for increasing the process noise is that the theoretical errors in the velocity estimate at 240 s increases from 40 ft/s ($\Phi_s = 100$) to 350 ft/s ($\Phi_s = 50,000$).

Therefore, it has been demonstrated that our template-based gravity turn two-state Kalman filter is very sensitive to modeling errors—even when knowledge of the current acceleration magnitude and direction is perfect. Large amounts of process noise are required to prevent filter divergence when the acceleration direction is not known. Errors in our downrange velocity estimates can increase by an order of magnitude when minor errors in our knowledge of the real world are introduced. Does a Kalman filter exist that is not template-based and is more robust than the Kalman filter just considered?

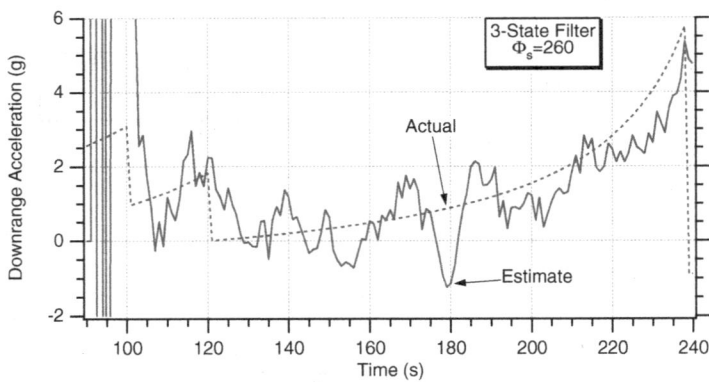

Fig. 20.13   Three-state linear polynomial Kalman filter does not require a template to estimate ICBM acceleration.

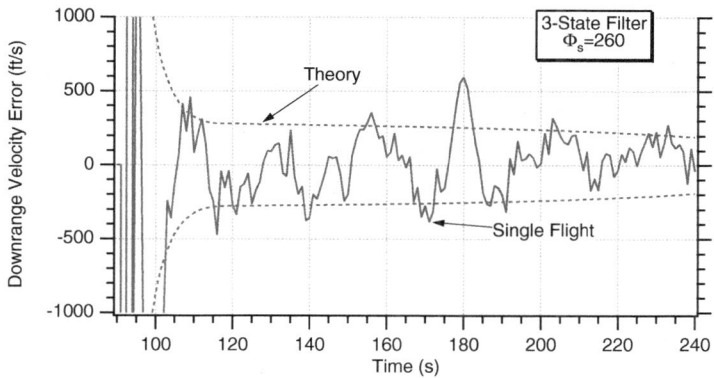

Fig. 20.14   Three-state Kalman filter is consistent.

Listing 20.2 was run for the case of the three-state filter (IFILTER=2) with a value of process noise that was determined by experiment (PHIS1=260). Figure 20.13 shows single flight results for the acceleration estimate. The figure indicates that there is no problem in estimating the target acceleration without a template with a three-state linear polynomial Kalman filter.

Figure 20.14 shows that for the nominal trajectory the filter is consistent and that the error in the estimate of downrange velocity for the three-state polynomial Kalman filter is approximately 200 ft/s. This error in the estimate is not as good as the two-state filter with a perfect template and perfect knowledge of the future intention of the target; however, it is better than the two-state gravity turn template-based Kalman filter (see Fig. 20.12).

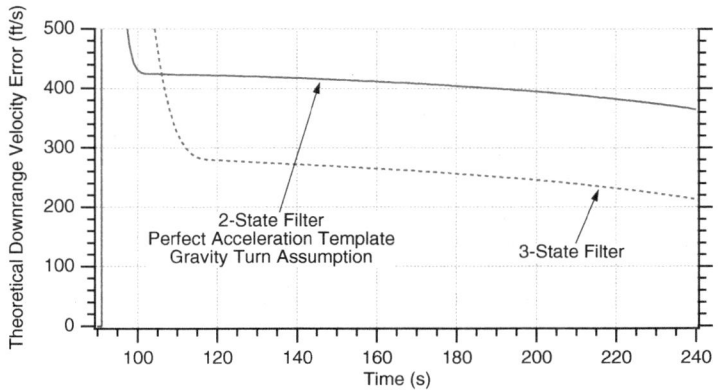

Fig. 20.15   Three-state Kalman filter is superior on a theoretical basis to two-state, perfect acceleration template, gravity turn Kalman filter.

So far sufficient process noise has been added to the two Kalman filters considered in order to obtain filter consistency. With a consistent filter, the covariance matrix of the Riccati equations can be used to obtain accurate performance projections concerning the errors in the estimate. Figure 20.15 presents the theoretical error in the downrange velocity estimate for both the three-state linear polynomial Kalman filter and the two-state, perfect template, gravity turn assumption Kalman filter when they both have sufficient process noise. The figure indicates that the resultant errors in the downrange velocity estimate for the three-state filter are nearly half of those for the two-state, perfect template, gravity turn assumption Kalman filter.

SUMMARY

For the boost-phase intercept problem, it has been demonstrated that when a template-based Kalman filter has perfect *a priori* information it offers significantly better performance than a linear three-state polynomial Kalman filter that does not require such information. However, when the two-state template-based filter has slight errors, this chapter shows that equivalent or sometimes better performance can be obtained with the simpler three-state polynomial Kalman filter.

REFERENCES

[1]   Zarchan, P., "Boost Phase Filtering Options: Is Simpler Better?," *Journal of Guidance, Control, and Dynamics*, Vol. 33, Nov.–Dec. 2010, pp. 1724–1731.

[2]   Blackman, S. S., *Multiple Target Tracking With Radar Applications*, Artech House, Norwood, MA, 1986, pp. 19–44.

[3]   Brookner, E., *Tracking and Kalman Filtering Made Easy*, John Wiley & Sons, New York, 1998, pp. 3–104.

[4]   Kleppner, D., and Lamb, F. K. (eds.), *Boost Phase Intercept Systems for National Missile Defense*, American Physical Society, July 2003, pp. 263–278.

[5]   White, J. E., "Guidance and Targeting for the Strategic Target System," *Journal of Guidance, Control, and Dynamics*, Vol. 15, Nov.–Dec. 1992, pp. 1313–1319.

[6]   Battin, R. H., *An Introduction to the Mathematics and Methods of Astrodynamics*, AIAA Education Series, New York, 1987.

[7]   Brand, T. J., "A New Approach to Lambert Guidance," Charles Stark Draper Laboratory, Rept. R-694, Cambridge, MA, June 1971.

[8]   Nelson, S. L., and Zarchan, P., "Alternative Approach to the Solution of Lambert's Problem," *Journal of Guidance, Control, and Dynamics*, Vol. 15, July–Aug. 1992, pp. 1003–1009.

[9]   Blackman, S., and Popoli, R., *Design and Analysis of Modern Tracking Systems*, Artech House, Boston, MA, 1999, pp. 241–250.

[10] Zarchan, P., *Fundamentals of Kalman Filtering: A Practical Approach*, 3rd ed, Progress in Astronautics and Aeronautics, AIAA, Reston, VA 2009, pp. 159–165.

CHAPTER 21

Kinematics of Intercepting a Ballistic Target

In Chapter 17 of the previous volume we showed how a long-range ballistic target, using the principals of Lambert guidance, could fly to its intended destination. In this chapter we shall investigate the amount of velocity that is required by a pursuing interceptor to hit the target during the target's ballistic or midcourse portion of flight. In addition, standard graphical methods of conveying interceptor performance shall be introduced in this chapter. To simplify matters and to keep the discussion generic, we shall assume that both the ballistic target and pursuing interceptor are impulsively launched, but not at the same time, so that they both get up to speed immediately. In addition, for simplicity we shall initially consider two-dimensional engagements on a round, nonrotating Earth where gravity is determined by Newton's law of universal gravitation and atmospheric effects are neglected. It shall be assumed that the interceptor knows where the ballistic target will be at the desired intercept time [that is, predicted intercept point (PIP) is known perfectly]. From Chapter 15 of the previous volume we know that the impulsive ballistic target differential equations of motion in two dimensions are given by

$$\ddot{x}_T = -gm \frac{x_T}{(x_T^2 + y_T^2)^{1.5}}$$

$$\ddot{y}_T = -gm \frac{y_T}{(x_T^2 + y_T^2)^{1.5}}$$

where x_T and y_T are component distances from the center of the Earth to the target and gm is the gravitational parameter with value in the English system of units of

$$gm = 1.4077 * 10^{16} \text{ft}^3/\text{s}^2$$

In Chapter 17, a general three-dimensional Lambert routine was presented in Listing 17.2 showing how to calculate the initial velocity required by the ballistic target to reach its final destination at the desired time. The call to the Lambert routine was given by

[VRX,VRY,VRZ]=LAMBERT3D (XT,YT,ZT,TGOLAM,XF,YF,ZF,SWITCH);

where XT,YT,ZT was a three-dimensional vector describing the initial target location; TGOLAM was the allotted time to go for the target to reach its final destination; and XF,YF,ZF was a three-dimensional vector describing the target's intended destination. The Lambert routine's output is VRX,VRY,VRZ, which is a three-dimensional vector describing the components of the required initial target velocity.

Listing 21.1, which is based on Listing 17.3, simulates an impulsive intercontinental ballistic missile (ICBM) in two dimensions. The z coordinate in this simulation has been set to zero, and all quantities that depend on latitude have been set to zero as well so that the two-dimensional problem will work with the three-dimensional Lambert routine. As was previously mentioned, the Lambert routine, which is *not* included in Listing 21.1, can be found in Listing 17.3. A careful examination of Listing 21.1 reveals that the time-of-flight formula for the minimum energy trajectory of the target has been programmed. The derivation of the formula for the time of flight for a minimum energy trajectory $t_{F_{ME}}$ is derived in Chapter 15 of the previous volume and is given by

$$t_{F_{ME}} = 252 + 0.223DR_{km} - 5.44 * 10^{-6}DR^2_{km}$$

where DR_{km} is the desired distance of travel in km. The trajectory can be lofted by adding TLOFT in units of seconds to $t_{F_{ME}}$ or depressed by subtracting TLOFT from $t_{F_{ME}}$. From a three-dimensional point of view, the ICBM in Listing 21.1 is flying across the equator (latitude is zero) starting at 0-deg longitude and traveling 10,000 km or approximately a quarter of the way around the world.

LISTING 21.1 TWO-DIMENSIONAL TRAJECTORY SIMULATION OF IMPULSIVE TWO-DIMENSIONAL ICBM

```
clear
count=0;
TS=1.;
RDESKM=10000.;
TLOFT=0.;
TFTOT=252.+.223*RDESKM-(5.44E-6)*RDESKM*RDESKM;
TFTOT=TFTOT+TLOFT;
A=2.0926E7;
GM=1.4077E16;
XLONGFDEG=57.3*RDESKM*3280./A;
XLONGTDEG=0.;
XLATTDEG=0.;
```

```
XLATFDEG=0.;
SWITCH=0;
T=0.;
S=0.;
XLONGF=XLONGFDEG/57.3;
XLATF=XLATFDEG/57.3;
XF=A*cos(XLATF)*cos(XLONGF);
YF=A*cos(XLATF)*sin(XLONGF);
ZF=0.;
XLONGT=XLONGTDEG/57.3;
XLATT=XLATTDEG/57.3;
XT=A*cos(XLATT)*cos(XLONGT);
YT=A*cos(XLATT)*sin(XLONGT);
ZT=0.;
XTINIT=XT;
YTINIT=YT;
ZTINIT=0.;
RTINIT=sqrt(XTINIT^2+YTINIT^2+ZTINIT^2);
H=.01;
ALTTKM=(sqrt(XT^2+YT^2)-A)/3280.;
TGOLAM=TFTOT-T;
[VRX,VRY,VRZ]=LAMBERT3D(XT,YT,ZT,TGOLAM,XF,YF,ZF,SWITCH);
XTD=VRX;
YTD=VRY;
ZTD=VRZ;
VBOT=sqrt(XTD^2+YTD^2)/3280.;
while ALTTKM>-1
      XTOLD=XT;
      YTOLD=YT;
      XTDOLD=XTD;
      YTDOLD=YTD;
      STEP=1;
      FLAG=0;
      while STEP<=1
            if FLAG==1
                  STEP=2;
                  XT=XT+H*XTD;
                  YT=YT+H*YTD;
                  XTD=XTD+H*XTDD;
                  YTD=YTD+H*YTDD;
                  T=T+H;
            end;
            TEMPBOTT=(XT^2+YT^2)^1.5;
            XTDD=-GM*XT/TEMPBOTT;
            YTDD=-GM*YT/TEMPBOTT;
            ALTTKM=(sqrt(XT^2+YT^2)-A)/3280.;
```

```
        FLAG=1;
    end
    FLAG=0;
    XT=.5*(XTOLD+XT+H*XTD);
    YT=.5*(YTOLD+YT+H*YTD);
    XTD=.5*(XTDOLD+XTD+H*XTDD);
    YTD=.5*(YTDOLD+YTD+H*YTDD);
    S=S+H;
    if S>=(TS-.0001)
        S=0.;
        ALTTKM=(sqrt(XT^2+YT^2)-A)/3280.;
        DISTRTKNM=distance3dkm(XT,YT,ZT,XTINIT,YTINIT,ZTINIT);
        count=count+1;
        ArrayT(count)=T;
        ArrayDISTRTKNM(count)=DISTRTKNM;
        ArrayALTTKM(count)=ALTTKM;
    end
end
figure
plot(ArrayDISTRTKNM,ArrayALTTKM),grid
xlabel('Downrange (km)')
ylabel('Altitude (km)')
clc
output=[ArrayT',ArrayDISTRTKNM',ArrayALTTKM'];
save datfil.txt output -ascii
disp 'simulation finished'
% LAMBERT3D can be found in Listing 17.2
% distance3dkm can be found in Listing 17.2
```

Minimum energy (TLOFT=0), lofted (TLOFT=500), and depressed (TLOFT=−200) trajectories for the ballistic target were run using Listing 21.1, and the resultant trajectories are compared in Fig. 21.1. The initial velocity required for the minimum energy trajectory was 7.19 km/s, 7.33 km/s was required for the lofted trajectory, and 7.23 km/s was required for the depressed trajectory. We can see from Fig. 21.1 that all the trajectories have different apogees and arrive at the 10,000-km destination at different times.

Now let us examine how much velocity would be required by a pursuing impulsive interceptor to hit the ballistic target at a desired intercept time t_F. It is assumed that the interceptor is launched some time after the target is launched and that the interceptor launch point is located z km downrange from the target launch point, as shown in Fig. 21.2.

A multiple-run simulation in which the desired intercept time TF is varied for each run appears in Listing 21.2. From Listing 21.2 we can see that the interceptor launch time is nominally 300 s after the target is launched (TLAUNCH=300) and the distance from the target launch point to the missile launch point is nominally

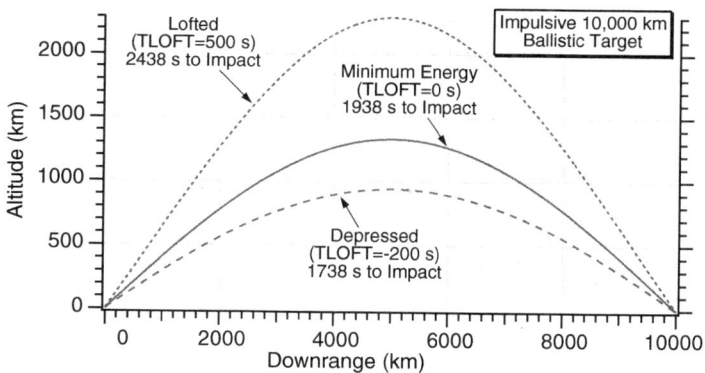

Fig. 21.1   Possible trajectories for 10,000-km impulsive ICBM.

1000 km   (XLONGMDEGICKM=1000).   The   target   is   flying   a   10,000-km
(RDESKM=10000) minimum energy trajectory (TLOFT=0). A perfect prediction
routine predict44.m tells the interceptor where the impulsively launched target
will be at the desired intercept time TF. This prediction routine knows that the
target is impulsive and takes the initial position and velocity of the target and integrates the target differential equations forward to the desired intercept time, thus
yielding the predicted intercept point (XTFACT and YTFACT). At the desired interceptor launch time (TLAUNCH), a three-dimensional Lambert routine (with z component set to zero) calculates the required interceptor velocity vector so that it will
arrive at the predicted intercept point at the desired intercept time. The equations
of motion for the interceptor and target are numerically integrated forward, and
the engagement simulation is stopped at the point of closest approach, which is
the miss distance. After each run, the interceptor flight time (TF-TLAUNCH) and
the required interceptor velocity (VMBO) are tabulated for interceptor burnout velocities less than 8 km/s. In this simulation, the precise miss distance is not calculated because the integration interval is not small enough in either the main
program or the prediction routine to get a very accurate calculation of the miss.
Decreasing the integration interval by an order of magnitude would have
yielded very long running times. Under the condition of a large integration

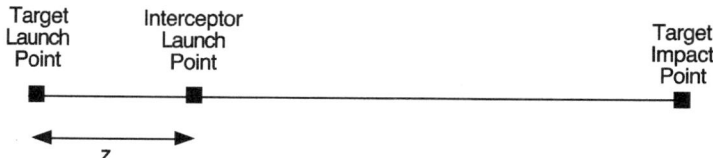

Fig. 21.2   Distance between interceptor and target launch points.

interval, a miss of several hundred feet will be considered a direct hit. If the integration interval was reduced by an order of magnitude, the miss distances would be near zero. Flights are considered a success if the required missile velocity is less than 8 km/s, the miss is less than 1000 ft, and the altitude of intercept is greater than 50 km. We can see from Listing 21.2 that one run involves cases in which the desired intercept time is varied from 60 s after the interceptor is launched to 20 s before the target would impact the ground in steps of 20 s.

LISTING 21.2 TWO-DIMENSIONAL KINEMATIC MULTIPLE-RUN ENGAGEMENT SIMULATION

```
clear
count=0;
TLAUNCH=300.;
TF=1400.;
TS=1.;
XLONGMDEGICKM=1000.;
XLATMDEGICKM=0.;
RDESKM=10000.;
TFTOT=2000.;
ALTMKMIC=0.;
TLOFT=0.;
QEARTH=0;
TFTOT=252.+.223*RDESKM-(5.44E-6)*RDESKM*RDESKM;
TFTOT=TFTOT+TLOFT;
for TF=(TLAUNCH+60):20:(TFTOT-20),
        SWITCH=0;
        SWITCHM=0;
        ALTM=ALTMKMIC*3280.;
        TFTOTP=TFTOT;
        RDESKMP=RDESKM;
        A=2.0926E7;;
        GM=1.4077E16;
        XLONGFDEG=57.3*RDESKM*3280./A;
        XLONGMDEGIC=XLONGMDEGICKM/111.;
        XLATMDEGIC=0.;
        XLONGMDEG=XLONGMDEGIC;
        XLATMDEG=XLATMDEGIC;
        QGUID=0;
        QLAUNCH=0;
        QFIRST=1;
        XLONGTDEG=0.;
        XLATTDEG=0.;
        XLATFDEG=0.;
        TBO=0.;
```

```
QBOOST=1;
QBOOSTM=1;
SWITCH=0;
SWITCHM=0;
T=0.;
S=0.;
AXT=0.;
AYT=0.;
ATP=0.;
XLONGF=XLONGFDEG/57.3;
XLATF=XLATFDEG/57.3;
XF=A*cos(XLATF)*cos(XLONGF);
YF=A*cos(XLATF)*sin(XLONGF);
ZF=0.;
XLONGT=XLONGTDEG/57.3;
XLONGM=XLONGMDEG/57.3;
XLATM=XLATMDEG/57.3;
XLATT=XLATTDEG/57.3;
XT=A*cos(XLATT)*cos(XLONGT);
YT=A*cos(XLATT)*sin(XLONGT);
ZT=0.;
XTINIT=XT;
YTINIT=YT;
ZTINIT=0.;
RTINIT=sqrt(XTINIT^2+YTINIT^2+ZTINIT^2);
XM=(A+ALTM)*cos(XLATM)*cos(XLONGM);
YM=(A+ALTM)*cos(XLATM)*sin(XLONGM);
ZM=0.;
XMINIT=XM;
YMINIT=YM;
ZMINIT=ZM;
RMINIT=sqrt(XMINIT^2+YMINIT^2+ZMINIT^2);
ATP=1.;
AXM=0.;
AYM=0.;
AMP=0.;
H=.01;
ALTTKM=(sqrt(XT^2+YT^2)-A)/3280.;
XMD=0.;
YMD=0.;
ACC=0.;
AXMGUID=0.;
AYMGUID=0.;
PREDERRKM=0.;
ZEM1=0.;
ZEM2=0.;
```

```
ALTMKM=(sqrt(XM^2+YM^2)-A)/3280.;
ALTTKM=(sqrt(XT^2+YT^2)-A)/3280.;
TGOLAM=TFTOT-T;
[VRX,VRY]=LAMBERT3D(XT,YT,ZT,TGOLAM,XF,YF,ZF,SWITCH);
XTD=VRX;
YTD=VRY;
ZTD=0.;
[XTFACT,YTFACT]=predict44(T,XT,YT,XTD,YTD,TF,TFTOT,XF,YF);
VBOT=sqrt(XTD^2+YTD^2)/3280.;
RTM1=XT-XM;
RTM2=YT-YM;
RTM=sqrt(RTM1^2+RTM2^2);
VTM1=XTD-XMD;
VTM2=YTD-YMD;
VC=-(RTM1*VTM1+RTM2*VTM2)/RTM;
while ~((T>(TLAUNCH+50.)) & VC<0. & RTM<10000.)
      if RTM<1000
             H=.0001;
      else
             H=.01;
      end
      XTOLD=XT;
      YTOLD=YT;
      XTDOLD=XTD;
      YTDOLD=YTD;
      XMOLD=XM;
      YMOLD=YM;
      XMDOLD=XMD;
      YMDOLD=YMD;
      STEP=1;
      FLAG=0;
      while STEP <=1
             if FLAG==1
                    STEP=2;
                    XT=XT+H*XTD;
                    YT=YT+H*YTD;
                    XTD=XTD+H*XTDD;
                    YTD=YTD+H*YTDD;
                    XM=XM+H*XMD;
                    YM=YM+H*YMD;
                    XMD=XMD+H*XMDD;
                    YMD=YMD+H*YMDD;
                    T=T+H;
             end
             TEMPBOTT=(XT^2+YT^2)^1.5;
             XTDD=-GM*XT/TEMPBOTT;
```

```
            YTDD=-GM*YT/TEMPBOTT;
            RTM1=XT-XM;
            RTM2=YT-YM;
            VTM1=XTD-XMD;
            VTM2=YTD-YMD;
            RTM=sqrt(RTM1^2+RTM2^2);
            VC=-(RTM1*VTM1+RTM2*VTM2)/RTM;
            TGO=RTM/VC;
            if T>TLAUNCH
                    TEMPBOTM=(XM^2+YM^2)^1.5;
                    XMDD=-GM*XM/TEMPBOTM;
                    YMDD=-GM*YM/TEMPBOTM;
            else
                    XMDD=0.;
                    YMDD=0.;
            end
            ALTMKM=(sqrt(XM^2+YM^2)-A)/3280.;
            ALTTKM=(sqrt(XT^2+YT^2)-A)/3280.;
            ALTT=sqrt(XT^2+YT^2)-A;
            FLAG=1;
    end
    FLAG=0;
    XT=.5*(XTOLD+XT+H*XTD);
    YT=.5*(YTOLD+YT+H*YTD);
    XTD=.5*(XTDOLD+XTD+H*XTDD);
    YTD=.5*(YTDOLD+YTD+H*YTDD);
    XM=.5*(XMOLD+XM+H*XMD);
    YM=.5*(YMOLD+YM+H*YMD);
    XMD=.5*(XMDOLD+XMD+H*XMDD);
    YMD=.5*(YMDOLD+YMD+H*YMDD);
    S=S+H;
    if T>=TLAUNCH
            TGOLAMM=TF-T;
            XTF=XTFACT;
            YTF=YTFACT;
            ZTF=0.;
            QLAUNCH=1;
    end
    TGOLAMM=TF-T;
    if ((T>=TLAUNCH) & QBOOSTM==1)
            QBOOSTM=0;
    [VRXM,VRYM,VRZM]=LAMBERT3D(XM,YM,ZM,TGOLAMM,XTF,YTF,ZTF,
        SWITCHM);
            XMD=VRXM;
            YMD=VRYM;
            XMDOLD=VRXM;
```

```
                        YMDOLD=VRYM;
                        VBOM=sqrt(XMD^2+YMD^2)/3280.;
                        QLAUNCH=1;
                end
                if S>=(TS-.0001)
                        S=0.;
                        ALTTKM=(sqrt(XT^2+YT^2)-A)/3280.;
                        DISTRTKM=distance3dkm(XT,YT,ZT,XTINIT,YTINIT,ZTINIT);
                        ALTMKM=(sqrt(XM^2+YM^2)-A)/3280.;
                        DISTRMKM=distance3dkm(XM,YM,ZM,XTINIT,YTINIT,ZTINIT);
                        VTK=sqrt(XTD^2+YTD^2)/3280.;
                        ATG=sqrt(XTDD^2+YTDD^2)/32.2;
                        VMKM=sqrt(XMD^2+YMD^2)/3280.;
                        VTKM=sqrt(XTD^2+YTD^2)/3280.;
                end
        end
        if (VBOM<8. & RTM<1000. & ALTTKM>50.)
                count=count+1;
                ArrayTFTL(count)=TF-TLAUNCH;
                ArrayVBOM(count)=VBOM;
        end
end
figure
plot(ArrayTFTL,ArrayVBOM),grid
xlabel('Interceptor Flight Time (s)')
ylabel('Interceptor Velocity (km/s) ')
clc
output=[ArrayTFTL',ArrayVBOM'];
save datfil.txt output -ascii
disp 'simulation finished'

% This is m file for PREDICT44.M
function [xtf,ytf]=predict44(tp,xtp,ytp,xtdp,ytdp,tf,tftot,xf,yf)
t=tp;
switch1=0;
xt=xtp;
yt=ytp;
zt=0.;
xtd=xtdp;
ytd=ytdp;
ztd=0.;
zf=0.;
a=2.0926E7;
gm=1.4077E16;
qboost=1;
h=.01;
```

```
s=0.;
axt=0.;
ayt=0.;
tgolam=tftot-t;
[vrx,vry,vrz]=LAMBERT3D(xt,yt,zt,tgolam,xf,yf,zf,switch1);
xtd=vrx;
ytd=vry;
ztd=0;
while t<=(tf-.00001)
        xtold=xt;
        ytold=yt;
        xtdold=xtd;
        ytdold=ytd;
        step=1;
        flag=0;
        while step <=1
                if flag==1
                        xt=xt+h*xtd;
                        yt=yt+h*ytd;
                        xtd=xtd+h*xtdd;
                        ytd=ytd+h*ytdd;
                        t=t+h;
                        step=2;
                end
                tembot=(xt^2+yt^2)^1.5;
                xtdd=-gm*xt/tembot;
                ytdd=-gm*yt/tembot;
                flag=1;
        end;
        flag=0;
        xt=(xtold+xt)/2+.5*h*xtd;
        yt=(ytold+yt)/2+.5*h*ytd;
        xtd=(xtdold+xtd)/2+.5*h*xtdd;
        ytd=(ytdold+ytd)/2+.5*h*ytdd;
end
xtf=xt;
ytf=yt;
% LAMBERT3D can be found in Listing 17.2
% distance3dkm can be found in Listing 17.2
```

The nominal case of Listing 21.2 was run in which the target is on a 10,000-km-minimum energy trajectory and the interceptor is launched 300 s after the target is launched. Additional cases were also run in which the distance from the interceptor launch point to the target launch point was considered to be 1000 km, 5000 km, and 9000 km, respectively. We can see from Fig. 21.3 that for

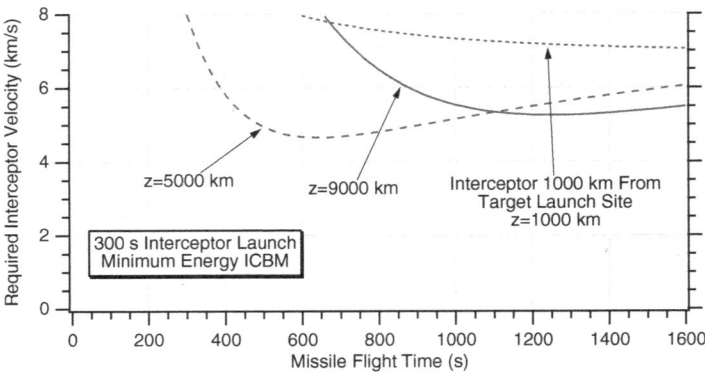

Fig. 21.3   Required initial interceptor velocity against a minimum-energy ICBM depends on interceptor launch point location and flight time.

very short interceptor flight times the required interceptor velocity can be quite large. However, depending on the launch point separation of the interceptor and target, there appears to be a minimum required interceptor velocity. We can see from Fig. 21.3 that a 5.3-km/s interceptor is required if the interceptor is launched 9000 km from the target launch point (or 1000 km from the target impact point) and the flight time is approximately 1400 s.

We can examine specific cases of the preceding example in more detail for the scenario in which the interceptor launch point is 9000 km downrange from the target launch point. Figure 21.4 presents the interceptor–target engagement geometries for an interceptor located 9000 km downrange of the target launch

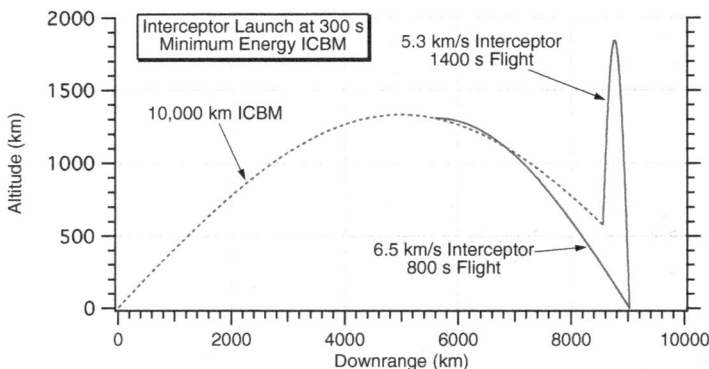

Fig. 21.4   Two sample engagement geometries when the interceptor launch point is 9000 km from the target launch point.

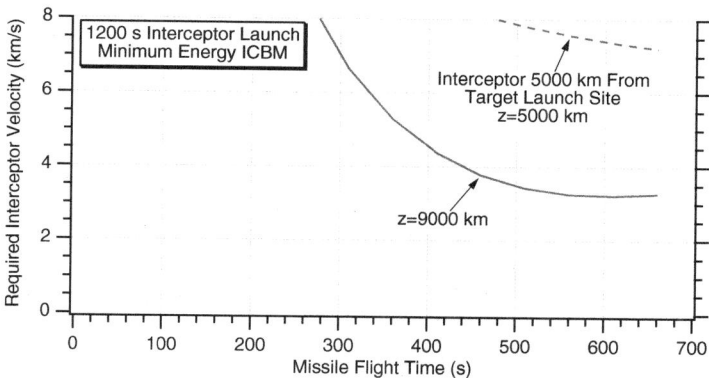

Fig. 21.5   Increasing interceptor launch time requires interceptor launch point to be closer to target impact point.

point—one in which the interceptor flight time is 800 s (6.5 km/s initial interceptor velocity is required according to Fig. 21.3) and the other in which the interceptor flight time is 1400 s (5.3 km/s initial interceptor velocity is required according to Fig. 21.3). We can see that when the flight time is shorter, the interceptor flies directly to the intercept point; however, when the flight time is much longer, the interceptor has to fly a lofted trajectory in order to reach the intercept point under the influence of gravity only. In practice, homing sensor constraints would have to be checked to see if such a lofted trajectory is practical.

Figure 21.5 shows that if the interceptor launch time is increased to 1200 s from 300 s it is not possible for the interceptor launch site to be only 1000 km from the target launch site. In addition, we can see that lower interceptor velocities are only possible if the interceptor is near the target impact point ($z = 9000$ km).

Figure 21.6 examines in more detail two cases taken from Fig. 21.5. The first case shows an interceptor launched at 1200 s and located 5000 km from the target launch site. In this case the interceptor has an initial velocity of 7.4 km/s and travels for 610 s (as indicated from Fig. 21.5). In this case, Fig. 21.6 indicates that the interceptor flies directly to the target. The second case shows an interceptor launched at 1200 s and located 9000 km from the target launch site. In this case, the interceptor has an initial velocity of 3.2 km/s but also travels for 610 s (as indicated from Fig. 21.5). In this case, Fig. 21.6 indicates a lofted trajectory for the interceptor because the interceptor is very close to the impact point of the target and has a great deal of flight time.

OPERATIONAL AREA

Although considerable insight has been gained from the two-dimensional results of the previous section, even more insight can be gained by examining the three-

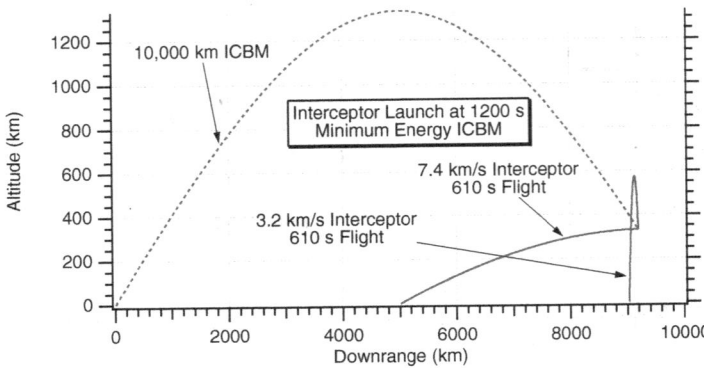

Fig. 21.6   Two sample engagement geometries for two different interceptor launch points with the same flight time.

dimensional case. In addition, we can speed up the simulation of Listing 21.2 significantly by eliminating the numerical integration of the differential equations for the interceptor and target. This can be done because both the interceptor and target are impulsive. The three-dimensional Kepler subroutine that first appeared in Listing 17.3 can be used to accurately predict where the target will be at the desired intercept time, and the three-dimensional Lambert routine that first appeared in Listing 17.2 can be used to calculate the required interceptor velocity and direction so that the interceptor will arrive at the intercept point at the desired intercept time. The avoidance of numerical integration can speed up the simulation of Listing 21.2 by several orders of magnitude. The resultant computational savings allow us to vary more parameters and run many more cases in order to get a more complete picture. We can set up a program in which the target trajectory is fixed so we can figure out where the interceptor should be placed for a successful intercept. Displaying information in this fashion is known as an operational area.

Listing 21.3 develops an operational area for an interceptor with an initial velocity that cannot exceed VBOLIM against targets that can fly different types of trajectories (for example, TLOFT=0 yields minimum energy trajectory). In addition, checks are performed to ensure that intercepts are made above 50-km altitude. In this three-dimensional simulation, the impulsive target is flying a minimum-energy 10,000-km trajectory along the equator. The interceptor is placed at different downrange locations specified by longitude and latitude. The initial interceptor downrange locations are longitude XLONGMDEG expressed in degrees and varied from 0 deg to 150 deg in steps of 5 deg. (0 deg corresponds to the interceptor being at the target launch point, and 150 deg corresponds to the interceptor being more than 16,000 km from the target launch point.) The initial interceptor cross range locations are latitude XLATMDEG expressed in degrees and vary from −40 deg to 40 deg in steps of 2.5 deg. The interceptor launch time is varied from 300 s after target launch to 1600 s after target launch in steps of 100 s. Latitude and longitude have been approximately converted to distances by assuming that 1 deg corresponds to 111 km (which is valid along the equator but overestimates distances as the latitude increases). The desired intercept time is varied from 60 s after interceptor launch to the time the target would impact the ground (TFTOT) in steps of 50 s. If the required interceptor velocity is less than VBOLIM, the intercept is considered a success and the initial interceptor location is noted. In addition, cases are not considered in which the target velocity would be greater than 7.5 km/s.

LISTING 21.3   GENERATING THREE-DIMENSIONAL OPERATIONAL AREAS

```
clear
count=0;
TLAUNCH=300.;
RDESKM=10000.;
ALTMKMIC=0.;
TLOFT=0.;
VBOLIM=5.;
I=1;
for XLONGMDEG=0:5:150,
        for XLATMDEG=-40:2.5:40,
                for TLAUNCH=300:100:1600,
                        ALTM=ALTMKMIC*3280.;
                        A=2.0926E7;
                        GM=1.4077E16;
                        XLONGFDEG=57.3*RDESKM*3280./A;
                        XLATFDEG=0.;
                        PI=3.14159;
                        XLONGTDEG=0.;
                        XLATTDEG=0.;
                        SWITCH=0;
                        SWITCHM=0;
                        T=0.;
                        XLONGF=XLONGFDEG/57.3;
                        XLATF=XLATFDEG/57.3;
                        XF=A*cos(XLATF)*cos(XLONGF);
                        YF=A*cos(XLATF)*sin(XLONGF);
                        ZF=A*sin(XLATF);
                        XLONGT=XLONGTDEG/57.3;
                        XLONGM=XLONGMDEG/57.3;
                        XLATM=XLATMDEG/57.3;
                        XLATT=XLATTDEG/57.3;
                        XT=A*cos(XLATT)*cos(XLONGT);
```

```
                    YT=A*cos(XLATT)*sin(XLONGT);
                    ZT=A*sin(XLATT);
                    XM=(A+ALTM)*cos(XLATM)*cos(XLONGM);
                    YM=(A+ALTM)*cos(XLATM)*sin(XLONGM);
                    ZM=(A+ALTM)*sin(XLATM);
                    DISTFKM=distance3dkm(XF,YF,ZF,XT,YT,ZT);
                    TFTOT=252.+.223*DISTFKM-(5.44E-6)*DISTFKM*DISTFKM;
                    TFTOT=TFTOT+TLOFT;
                    for TF=(TLAUNCH+60.):50:TFTOT,
                    TGOLAM=TFTOT-T;
% CALCULATE TARGET VELOCITY REQUIRED TO REACH ITS DESTINATION
        [VRX,VRY,VRZ]=LAMBERT3D(XT,YT,ZT,TGOLAM,XF,YF,ZF,SWITCH);
                    XTD=VRX;
                    YTD=VRY;
                    ZTD=VRZ;
% CALCULATE TARGET STATES AT DESIRED INTERCEPT TIME
                    T0=0.;
                    T1=TF;
                    X0(1)=XT/3280.;
                    X0(2)=YT/3280.;
                    X0(3)=ZT/3280.;
                    X0(4)=XTD/3280.;
                    X0(5)=YTD/3280.;
                    X0(6)=ZTD/3280.;
                    [X1]=KEPLER1(X0,T0,T1);
                    XTF=X1(1)*3280.;
                    YTF=X1(2)*3280.;
                    ZTF=X1(3)*3280.;
                    ALTFKM=(sqrt(XTF^2+YTF^2+ZTF^2)-A)/3280.;
                    if ALTFKM<50.
                        break
                    end
% CALCULATE MISSILE VELOCITY REQUIRED TO INTERCEPT TARGET AT DESIRED
INTERCEPT TIME
                    TGOLAMM=TF-TLAUNCH;
    [VRXM,VRYM,VRZM]=LAMBERT3D(XM,YM,ZM,TGOLAMM,XTF,YTF,ZTF,SWITCHM);
                    XMD=VRXM;
                    YMD=VRYM;
                    ZMD=VRZM;
% CALCULATE MISSILE STATES AT DESIRED INTERCEPT TIME
                    T1=TF;
                    T0=TLAUNCH;
                    X0(1)=XM/3280.;
                    X0(2)=YM/3280.;
                    X0(3)=ZM/3280.;
                    X0(4)=XMD/3280.;
```

```
                                    X0(5)=YMD/3280.;
                                    X0(6)=ZMD/3280.;
                                    [X1]=KEPLER1(X0,T0,T1);
                                    XMF=X1(1)*3280.;
                                    YMF=X1(2)*3280.;
                                    ZMF=X1(3)*3280.;
                                    XMDF=X1(4)*3280.;
                                    YMDF=X1(5)*3280.;
                                    ZMDF=X1(6)*3280.;
                                    VBOM=sqrt(XMD^2+YMD^2+ZMD^2)/3280.;
                                    VBOT=sqrt(XTD^2+YTD^2+ZTD^2)/3280.;
                                    if (VBOM<VBOLIM & VBOT<7.5)
count=count+1;
                                        ArrayTF(count)=TF;
                                        ArrayTLAUNCH(count)=TLAUNCH;
                                        ArrayXLONGM(count)=XLONGMDEG*111.;
                                        ArrayXLATM(count)=XLATMDEG*111.;
                                        ArrayVBOM(count)=VBOM;
                                    end
                                end
                        end
                end
end
figure
plot(ArrayXLONGM,ArrayXLATM,'r+'),grid
xlabel('Downrange (km)')
ylabel('Crossrange (km)')
clc
output=[ArrayTF',ArrayTLAUNCH',ArrayXLONGM',ArrayXLATM',ArrayVBOM'];
save datfil.txt output -ascii
disp 'simulation finished'
% LAMBERT3D can be found in Listing 17.2
% distance3dkm can be found in Listing 17.2
% KEPLER1 can be found in Listing 17.3
```

The nominal case of Listing 21.3 was run for a 6-km/s interceptor against the 10,000-km target flying a minimum energy trajectory. Figure 21.7 displays the resultant operational area, indicated by plus signs, and the target trajectory, represented by the solid straight line. Thus, the operational area represents places from which an interceptor can be launched and hit the target at some point in its trajectory. Missing from the plot are the interceptor launch and intercept times that resulted in a success. Some plus signs correspond to many possible combinations of launch and intercept times whereas other plus signs may be the result of a unique combination. We can see that the operational area is very large because the interceptor speed can be as much as 6 km/s. We can also see

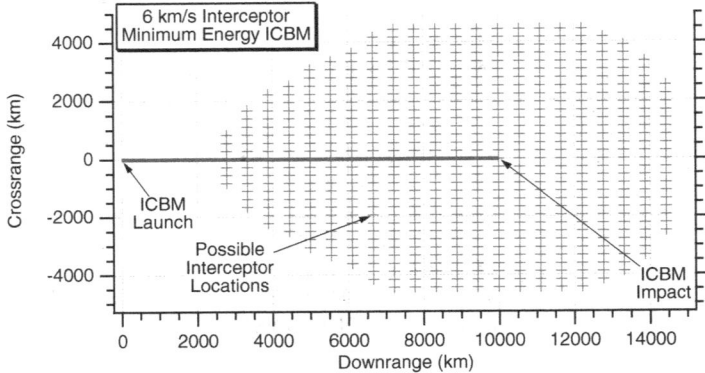

Fig. 21.7   Operational area for 6-km/s interceptor against 10,000-km minimum energy target.

that the operational area is symmetrical about the target trajectory. At first glance these results might appear to disagree with the two-dimensional results of Fig. 21.3 because the operational area of Fig. 21.7 at zero latitude extends from 3000 km to more than 14,000 km. However, recall in Fig. 21.3 we only considered three downrange cases ($z = 1000$ km, $z = 5000$ km, and $z = 9000$ km). In Fig. 21.3 it was indicated that there were solutions for a 6-km/s interceptor at $z = 5000$ km and $z = 9000$ km, which is certainly consistent with Fig. 21.7.

Figure 21.8 shows that when the maximum interceptor velocity is reduced from 6 km/s to 5 km/s, the operational area shrinks considerably. When the

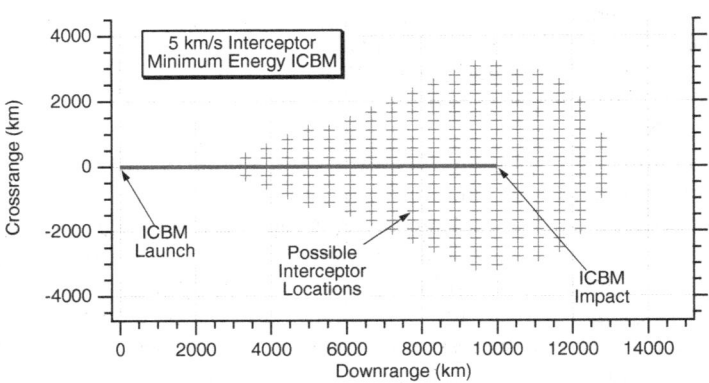

Fig. 21.8   Operational area for 5-km/s interceptor against 10,000-km minimum energy target.

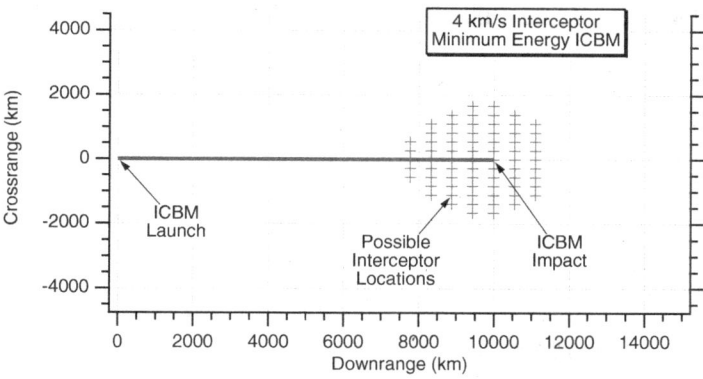

Fig. 21.9   Operational area for 4-km/s interceptor against 10,000-km minimum energy target.

interceptor had a 6 km/s capability, the interceptor could be placed from 3000 km to 14,500 km from the target launch site. However, when the interceptor velocity is reduced to 5 km/s, we can see from Fig. 21.8 that the interceptor must be placed from 3500 km to 12,500 km from the target launch site.

Figure 21.9 shows that the operational area reduces even more dramatically when the interceptor velocity is further reduced to 4 km/s. Now the interceptor must be placed 7500 km to 10,500 km from the target launch site for an intercept to be successful.

LAUNCH AREA DENIED

Successful target intercept information can be presented in a variety of ways. The operational area method of presentation, presented in the last section, assumed a fixed target trajectory that was previously represented by a thick straight line, and possible interceptor launch locations that resulted in a successful engagement were represented by plus signs. Cases can also be run in which the target aimpoint and initial interceptor locations are fixed and the initial target launch site is varied. Graphics generated in this way are known as *launch area denied plots*. The operational area code of Listing 21.3 can be modified so that launch area denied results can be generated (shown in Listing 21.4). From Listing 21.4 we can see that the interceptor is nominally placed at the equator at 60-deg longitude, and the target launch point is varied from −100 to 200 deg in longitude (in steps of 5 deg) and from −60 deg to 60 deg in latitude (in steps of 2.5 deg). Cases are ruled out if the required impulsive target velocity exceeds 7.5 km/s.

LISTING 21.4   GENERATING THREE-DIMENSIONAL LAUNCH AREA DENIED RESULTS

```
clear
count=0;
TLAUNCH=300.;
RDESKM=10000.;
ALTMKMIC=0.;
TLOFT=0.;
VBOLIM=5.;
XLONGMDEG=60.;
XLATMDEG=0.;
I=1;
for XLONGTDEG=-100:5:200,
        for XLATTDEG=-60:2.5:60,
                for TLAUNCH=300:100:1600,
                        ALTM=ALTMKMIC*3280.;
                        A=2.0926E7;
                        GM=1.4077E16;
                        XLONGFDEG=57.3*RDESKM*3280./A;
                        XLATFDEG=0.;
                        PI=3.14159;
                        SWITCH=0;
                        SWITCHM=0;
                        T=0.;
                        S=0.;
                        XLONGF=XLONGFDEG/57.3;
                        XLATF=XLATFDEG/57.3;
                        XF=A*cos(XLATF)*cos(XLONGF);
                        YF=A*cos(XLATF)*sin(XLONGF);
                        ZF=A*sin(XLATF);
                        XLONGT=XLONGTDEG/57.3;
                        XLONGM=XLONGMDEG/57.3;
                        XLATM=XLATMDEG/57.3;
                        XLATT=XLATTDEG/57.3;
                        XT=A*cos(XLATT)*cos(XLONGT);
                        YT=A*cos(XLATT)*sin(XLONGT);
                        ZT=A*sin(XLATT);
                        XM=(A+ALTM)*cos(XLATM)*cos(XLONGM);
                        YM=(A+ALTM)*cos(XLATM)*sin(XLONGM);
                        ZM=(A+ALTM)*sin(XLATM);
                        DISTFKM=distance3dkm(XF,YF,ZF,XT,YT,ZT);
                        TFTOT=252.+.223*DISTFKM-(5.44E-6)*DISTFKM*DISTFKM;
                        TFTOT=TFTOT+TLOFT;
                        for TF=(TLAUNCH+60.):50:TFTOT,
                                TGOLAM=TFTOT-T;
% CALCULATE TARGET VELOCITY REQUIRED TO REACH ITS DESTINATION
```

```
[VRX,VRY,VRZ]=LAMBERT3D(XT,YT,ZT,TGOLAM,XF,YF,ZF,SWITCH);
                        XTD=VRX;
                        YTD=VRY;
                        ZTD=VRZ;
% CALCULATE TARGET STATES AT DESIRED INTERCEPT TIME
                        T0=0.;
                        T1=TF;
                        X0(1)=XT/3280.;
                        X0(2)=YT/3280.;
                        X0(3)=ZT/3280.;
                        X0(4)=XTD/3280.;
                        X0(5)=YTD/3280.;
                        X0(6)=ZTD/3280.;
                        [X1]=KEPLER1(X0,T0,T1);
                        XTF=X1(1)*3280.;
                        YTF=X1(2)*3280.;
                        ZTF=X1(3)*3280.;
                        ALTFKM=(sqrt(XT^2+YTF^2+ZTF^2)-A)/3280.;
                        if ~(ALTFKM<50.)
%CALCULATE MISSILE VELOCITY REQUIRED TO INTERCEPT TARGET AT DESIRED
INTERCEPT TIME
                        TGOLAMM=TF-TLAUNCH;
[VRXM,VRYM,VRZM]=LAMBERT3D(XM,YM,ZM,TGOLAMM,XTF,YTF,ZTF,SWITCHM);
                        XMD=VRXM;
                        YMD=VRYM;
                        ZMD=VRZM;
                        VBOM=sqrt(XMD^2+YMD^2+ZMD^2)/3280.;
                        VBOT=sqrt(XTD^2+YTD^2+ZTD^2)/3280.;
                        if (VBOM < VBOLIM & VBOT<7.5)
                                count=count+1;
                                ArrayTF(count)=TF;
                                ArrayTLAUNCH(count)=TLAUNCH;
                                ArrayXLONGT(count)=XLONGTDEG*111.;
                                ArrayXLATT(count)=XLATTDEG*111.;
                                ArrayVBOM(count)=VBOM;
                                ArrayVBOT(count)=VBOT;
                        end
                    end
                end
            end
        end
end
figure
plot(ArrayXLONGT,ArrayXLATT,'r+'),grid
xlabel('Downrange (km)')
ylabel('Crossrange (km) ')
```

clc
output=[ArrayTF',ArrayTLAUNCH',ArrayXLONGT',ArrayXLATT',ArrayVBOM',ArrayVBOT'];
save datfil.txt output -ascii
disp 'simulation finished'
% LAMBERT3D can be found in Listing 17.2
% distance3dkm can be found in Listing 17.2
% KEPLER1 can be found in Listing 17.3

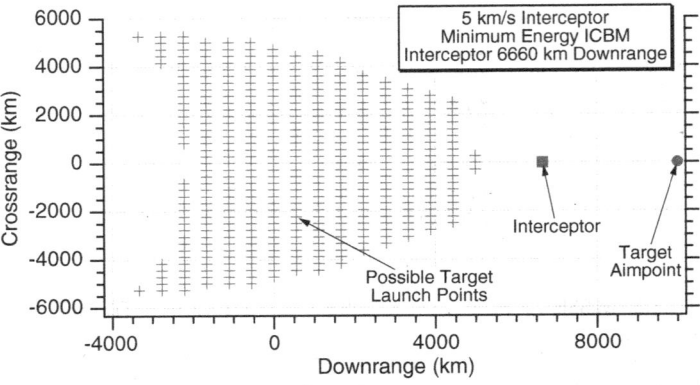

Fig. 21.10   Launch area denied for 5-km/s interceptor against minimum energy target.

The nominal case of Listing 21.4 was run for a 5-km/s interceptor against the impulsive ICBM target flying a minimum energy trajectory. Figure 21.10 displays possible target launch points that are denied, indicated by plus signs, because they

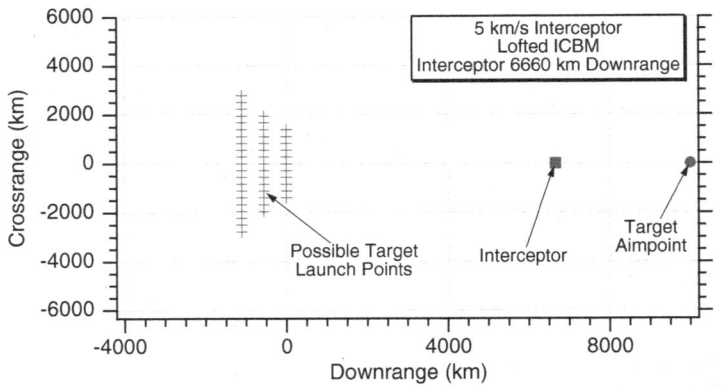

Fig. 21.11   Launch area denied for 5-km/s interceptor against a lofted target.

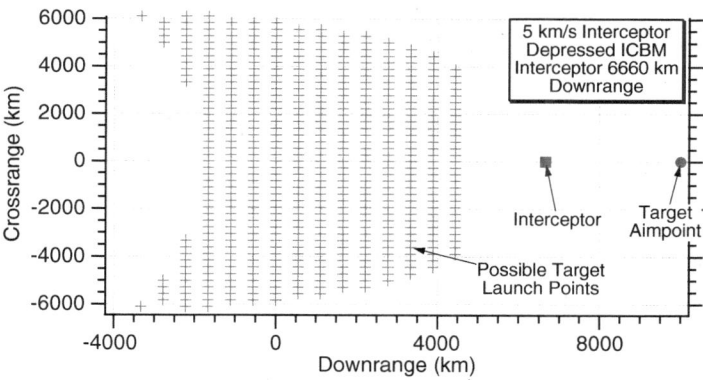

Fig. 21.12   Launch area denied for 5-km/s interceptor against a depressed target].

can be successfully intercepted by a 5-km/s interceptor whose initial launch location is 6660 km downrange (60-deg longitude). We can see that the launch area denied is very large because the interceptor is located in a favorable location. Again, it is important to note that each plus sign in Fig. 21.10 corresponds to a different interceptor launch and intercept time that results in a successful intercept. Figure 21.11 shows the same case, except this time the target trajectory is lofted (TLOFT=500). We can see that the size of the launch area denied region shrinks considerably because the target is harder to reach because it may be higher in altitude. Figure 21.12 shows that when the target trajectory is depressed (TLOFT=-200), the launch area denied region is roughly the same size as it was when the target was flying a minimum energy trajectory.

DEFENDED AREA

As was mentioned in the previous sections, successful target intercept information can be presented in a variety of ways. The operational area method of presentation assumed a fixed target trajectory that was previously represented by a thick straight line, and possible interceptor launch locations, represented by plus signs, were varied. The previous section showed that launch area denied plots can be generated by having the target aimpoint and initial interceptor locations fixed and varying the initial target launch site. Finally, cases can also be run in which the target launch point and initial interceptor locations are fixed and the final target impact point is varied. Graphics generated in this way are known as *defended area plots*. The launch area denied code of Listing 21.4 can be modified so that defended area results can be generated; this is shown in Listing 21.5. From Listing 21.5 we can see that the interceptor is nominally placed at the equator at 60-deg longitude and the target impact point is varied from 20 to 200 deg in longitude and from –60 deg to 60 deg in latitude. Cases are ruled out if the required target velocity exceeds 7.5 km/s.

LISTING 21.5   GENERATING THREE-DIMENSIONAL DEFENDED AREA RESULTS

```
clear
count=0;
TLAUNCH=300.;
RDESKM=10000.;
ALTMKMIC=0.;
TLOFT=0.;
VBOLIM=5.;
XLONGMDEG=60.;
XLATMDEG=0.;
XLONGTDEGIC=0.;
XLATTDEGIC=0.;
I=1;
for XLONGFDEG=20:5:200,
        for XLATFDEG=-60:2.5:60,
                for TLAUNCH=300:100:1600,
                        ALTM=ALTMKMIC*3280.;
                        A=2.0926E7;
                        GM=1.4077E16;
                        XLONGTDEG=XLONGTDEGIC;
                        XLATTDEG=XLATTDEGIC;
                        PI=3.14159;
                        SWITCH=0;
                        SWITCHM=0;
                        T=0.;
                        S=0.;
                        XLONGF=XLONGFDEG/57.3;
                        XLATF=XLATFDEG/57.3;
                        XF=A*cos(XLATF)*cos(XLONGF);
                        YF=A*cos(XLATF)*sin(XLONGF);
                        ZF=A*sin(XLATF);
                        XLONGT=XLONGTDEG/57.3;
                        XLONGM=XLONGMDEG/57.3;
                        XLATM=XLATMDEG/57.3;
                        XLATT=XLATTDEG/57.3;
                        XT=A*cos(XLATT)*cos(XLONGT);
                        YT=A*cos(XLATT)*sin(XLONGT);
                        ZT=A*sin(XLATT);
                        XM=(A+ALTM)*cos(XLATM)*cos(XLONGM);
                        YM=(A+ALTM)*cos(XLATM)*sin(XLONGM);
                        ZM=(A+ALTM)*sin(XLATM);
```

```
                    DISTFKM=distance3dkm(XF,YF,ZF,XT,YT,ZT);
                    TFTOT=252.+.223*DISTFKM-(5.44E-6)*DISTFKM*DISTFKM;
                    TFTOT=TFTOT+TLOFT;
                    for TF=(TLAUNCH+60.):50:(TFTOT-50.),
                        TGOLAM=TFTOT-T;
% CALCULATE TARGET VELOCITY REQUIRED TO REACH ITS DESTINATION
            [VRX,VRY,VRZ]=LAMBERT3D(XT,YT,ZT,TGOLAM,XF,YF,ZF,SWITCH);
                        XTD=VRX;
                        YTD=VRY;
                        ZTD=VRZ;
% CALCULATE TARGET STATES AT DESIRED INTERCEPT TIME
                        T0=0.;
                        T1=TF;
                        X0(1)=XT/3280.;
                        X0(2)=YT/3280.;
                        X0(3)=ZT/3280.;
                        X0(4)=XTD/3280.;
                        X0(5)=YTD/3280.;
                        X0(6)=ZTD/3280.;
                        [X1]=KEPLER1(X0,T0,T1);
                        XTF=X1(1)*3280.;
                        YTF=X1(2)*3280.;
                        ZTF=X1(3)*3280.;
                        ALTFKM=(sqrt(XTF^2+YTF^2+ZTF^2)-A)/3280.;
                        if ALTFKM<50.
                            break
                        end
% CALCULATE MISSILE VELOCITY REQUIRED TO INTERCEPT TARGET AT DESIRED
INTERCEPT TIME
                        TGOLAMM=TF-TLAUNCH;
    [VRXM,VRYM,VRZM]=LAMBERT3D(XM,YM,ZM,TGOLAMM,XTF,YTF,ZTF,SWITCHM);
                        XMD=VRXM;
                        YMD=VRYM;
                        ZMD=VRZM;
                        VBOM=sqrt(XMD^2+YMD^2+ZMD^2)/3280.;
                        VBOT=sqrt(XTD^2+YTD^2+ZTD^2)/3280.;
                        if (VBOM<VBOLIM & VBOT<7.5)
                            count=count+1;
                            ArrayTF(count)=TF;
                            ArrayTLAUNCH(count)=TLAUNCH;
                            ArrayXLONGF(count)=XLONGFDEG*111.;
                            ArrayXLATF(count)=XLATFDEG*111.;
                            ArrayVBOM(count)=VBOM;
                        end
                    end
                end
```

```
            end
end
figure
plot(ArrayXLONGF,ArrayXLATF,'r+'),grid
xlabel('Downrange (km)')
ylabel('Crossrange (km) ')
clc
output=[ArrayTF',ArrayTLAUNCH',ArrayXLONGF',ArrayXLATF',ArrayVBOM'];
save datfil.txt output -ascii
disp 'simulation finished'
% LAMBERT3D can be found in Listing 17.2
% distance3dkm can be found in Listing 17.2
% KEPLER1 can be found in Listing 17.3
```

The nominal case of Listing 21.5 was run for a 5-km/s interceptor against the impulsive ICBM target flying a minimum energy trajectory. Figure 21.13 displays possible target impact points that can be defended, indicated by plus signs, because they can be successfully intercepted by a 5-km/s interceptor located 6660 km downrange from the target launch point. Again, it is important to note that each plus sign in Fig. 21.13 corresponds to a different interceptor launch and intercept time that results in a successful intercept. Figure 21.14 shows the same case except this time the target trajectory is lofted (TLOFT=500). We can see that the size of the defended area region shrinks because the target is harder to reach because its apogee is higher than that of a minimum energy trajectory and thus harder to reach with a 5-km/s interceptor. However, the difference in results between the minimum energy and loft target trajectories is not as great as it was in the previous section because the interceptor is much

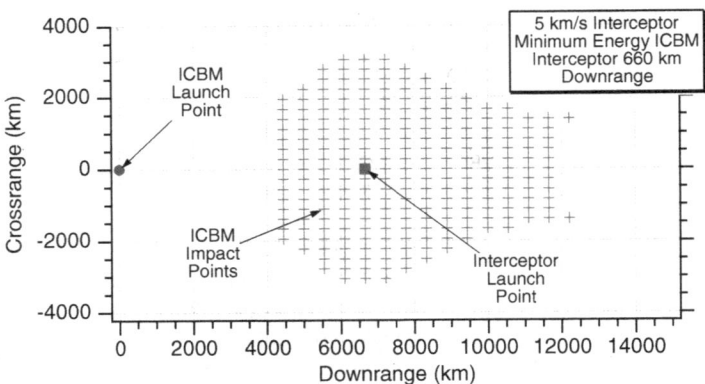

Fig. 21.13   Defended area for 5-km/s interceptor against minimum-energy trajectory target.

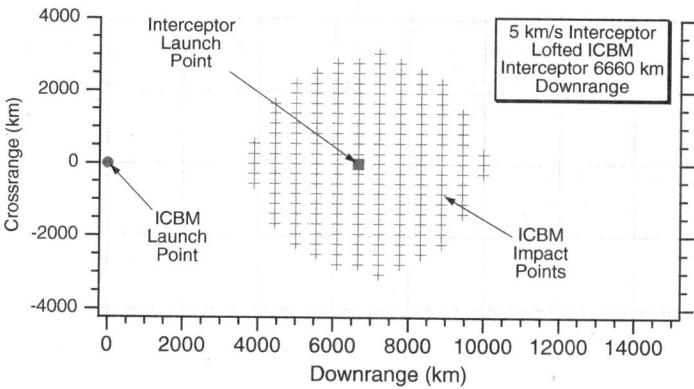

Fig. 21.14   Defended area for 5-km/s interceptor against a lofted trajectory target.

closer to the target impact point. Figure 21.15 shows that when the target trajectory is depressed (TLOFT=−200), the defended area region is roughly the same size as it was when the target was flying a minimum-energy trajectory.

SUMMARY

In this chapter we have seen how to present information on successful intercepts of an impulsive ballistic target being pursued by an impulsive interceptor. The operational area method of presentation assumes that the target trajectory is fixed and we figure out where the interceptor can be placed for a successful intercept. The launch area denied method of presentation assumes that the target

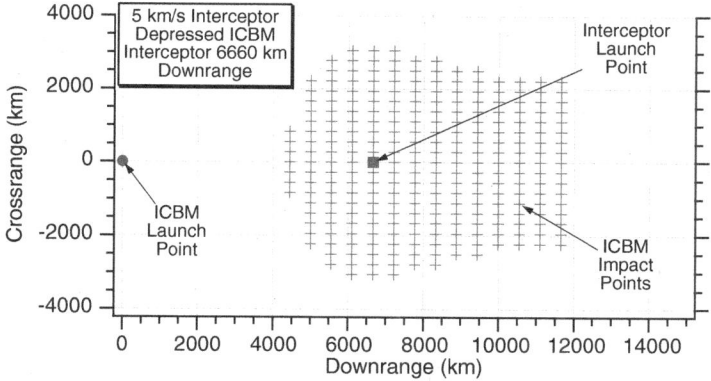

Fig. 21.15   Area for 5-km/s interceptor against a depressed trajectory target.

aimpoint and initial interceptor locations are fixed and the initial target launch site is varied. Finally, the defended area method of presentation assumes that the target launch point and initial interceptor locations are fixed and the final target impact point is varied. Using the three methods of presentation, cases were run in which the size of the area was influenced by either the missile velocity or the target trajectory type.

CHAPTER 22

Kill Vehicle Guidance and Control Sizing for Boost-Phase Intercept

INTRODUCTION

This chapter addresses some of the guidance and control issues involved in enabling an air-launched interceptor carrying a highly maneuverable kinetic kill vehicle (KKV) to perform an exoatmospheric intercept of a boosting threat target capable of traveling many thousands of kilometers [1]. The chapter takes the reader through part of the first iteration of the multi-iteration design process in order to get a preliminary estimate of how much divert and acceleration may be required by the kinetic kill vehicle to hit the target. Simplified examples are presented to indicate how conventional guidance and filtering techniques can be used as a starting point in the iterative design process for this important problem in missile defense. More advanced guidance and filtering techniques can be used in subsequent iterations to more accurately size the kinetic kill vehicle and improve system performance and robustness.

BACKGROUND

As was mentioned in Chapter 20, intercept of intercontinental ballistic missiles (ICBMs) during their boost phase has long been considered attractive. During 2003 the American Physical Society (APS) released a detailed report [2] that studied the use of surface-based interceptors for intercepting ICBMs during their boost phase. The APS report found that the surface-based interceptors required for boost-phase intercept would have to be very heavy, due to both the high required burnout velocities and lateral divert requirements of the interceptor. In addition, in the case of an Iranian ICBM launch against the United States, for example, the APS report showed that the interceptors would have to be based in countries that might present a political challenge for the United States. Thus, in general, the APS report was pessimistic about the success of a terrestrial-based boost-phase intercept system.

AIR-LAUNCHED INTERCEPTOR APPROACH

An alternative approach to boost-phase intercept involves the use of airborne interceptors and was first considered in the open literature by Wilkening [3] and then expanded upon, with considerable practical detail, by Corbett [4]. At first glance this alternative approach to boost-phase intercept might be considered to be inferior to surface-based interceptors, because airborne interceptors would have even lower burnout velocities than surface-based interceptors due to aircraft payload weight constraints. However, in this alternative approach, stealthy aircraft, which would be used as both launch and sensor platforms, initially would be manned but in the future would be unmanned using a platform such as the Naval Unmanned Combat Air System (N-UCAS) carrier [5]. Stealthiness would enable the aircraft to penetrate enemy territory to get much closer to ICBM launch sites than would be possible with surface-based interceptors located near the borders of an enemy nation. Having the defensive interceptor launch platform closer to an enemy launch site means that the required burnout velocity of an air-launched interceptor can be much less than that of a surface-launched interceptor.

The key elements in Corbett's boost-phase intercept system construct are stealthy fighter aircraft with infrared search-and-track (IRST) systems to detect and track the target and airborne interceptors with highly maneuverable kinetic kill vehicles for exoatmospheric intercepts of the enemy missiles. In this system, construct pairs of stealthy aircraft travel in oval racetracks in opposite directions over enemy territory. Their combined IRST systems have 360-deg coverage and can search for, detect, and track enemy ICBMs and intermediate range ballistic missiles (IRBMs) autonomously during their boost phase. When the IRST system of one aircraft detects a threat, it can cue another off-board IRST system to establish an additional angles-only track on the target so that the position, velocity, and acceleration of the target can be estimated. When sufficient track accuracy of the target states are obtained, a prediction is made of the target's position at the desired intercept time. This prediction will be imperfect because it is impossible to know a boosting target's future intentions. The launch aircraft turns so that it can fire its interceptor directly at the predicted intercept point (PIP). Not only is the interceptor's thrust used to increase the speed of the interceptor, but because the PIP is constantly changing, the interceptor thrust vector must also be steered in order for the interceptor to hit the latest and most refined estimate of the PIP. When the interceptor burns out, the PIP will still be in considerable error. Therefore, additional fuel and guidance are required so that a KKV, which separates from the interceptor after the interceptor burns out, will hit the target using its lateral divert engines for responding to guidance commands outside of the Earth's atmosphere.

GUIDANCE AND CONTROL ISSUES

The purpose of this chapter is to illustrate how some key guidance and control issues influence the amount of fuel and acceleration the KKV must have so that it can successfully engage both IRBMs and ICBMs during their boost phase. Sample trade-offs will be conducted using conventional guidance and filtering methods to illustrate the first step of an iterative design process that must take place for all practical designs. Subsequent steps in the design process may consider more advanced guidance and filtering techniques, which in turn might reduce the KKV divert requirements derived in this chapter.

Lambert guidance was shown in Chapter 17 of the previous volume to be an effective method of guidance when the control authority of the interceptor is in the axial direction, and augmented proportional navigation (APN) was shown in Chapter 8 of the previous volume to be appropriate when the control authority of the KKV was in the lateral direction. Therefore, Lambert guidance can be used while the interceptor is thrusting and APN can be used afterwards by the KKV's lateral divert engines. The interceptor must be sized for both adequate burnout velocity and sufficient fuel and acceleration for the KKV's divert engines. When the KKV gets close enough to the target, its seeker can acquire the target plume. The seeker software must be capable of distinguishing the target hard body from the plume and enable the KKV to hit that target's warhead.

This chapter starts out by first considering the effects of apparent target maneuver and guidance law (assuming zero PIP error) in a noise-free, one-dimensional engagement environment. Formulas will be developed showing how KKV divert requirements are related to target maneuver and KKV action or homing time. Next, engagement experiments in two dimensions are conducted in a noise-free environment to see how simulation results compare with closed-form solutions. Finally, it is demonstrated that sensor noise and filtering effects also play an important role in establishing KKV lateral divert requirements.

ONE-DIMENSIONAL MODEL FOR UNDERSTANDING GUIDANCE

Figure 22.1 presents the classical interceptor homing loop for understanding guidance. Here n_T represents the apparent target acceleration, as seen by the pursuing interceptor, of a boosting threat. That portion of the target's axial acceleration that is perpendicular to the KKV-target line of sight will appear as a target maneuver to the KKV. In this example we want to ensure that the KKV has adequate acceleration capability (does not saturate near the end of the flight) so that it can hit the target. For exoatmospheric intercepts, the kill vehicle time constants are so small that they can be neglected in a preliminary analysis. In addition, we will mainly be concerned about the amount of KKV fuel or lateral divert required for a successful intercept.

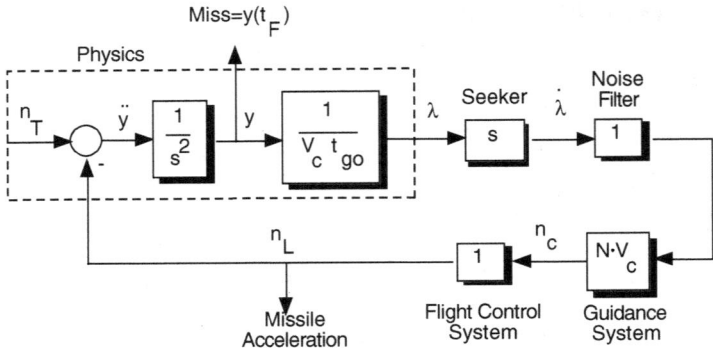

Fig. 22.1 One-dimensional guidance system model for initial analysis.

Under worst-case geometrical conditions, all of the threat's axial acceleration will be seen by the KKV as an apparent target maneuver. A typical acceleration profile for a generic one-stage, 180-s-burn IRBM is displayed as the solid curve in Fig. 22.2. Here it can be seen that the threat acceleration increases with increasing time because as the IRBM propellant burns, the weight of the IRBM decreases. In this example the maximum acceleration of the IRBM is approximately 9 g at 180 s. For academic and analytical purposes, the generic IRBM acceleration can be approximated by a parabola with zero acceleration initially and $n_{T\text{MAX}}$ acceleration finally. The parabolic approximation is given by

$$n_{T_{\text{Parabola}}} = n_{T\text{MAX}} \left(\frac{t}{t_F}\right)^2$$

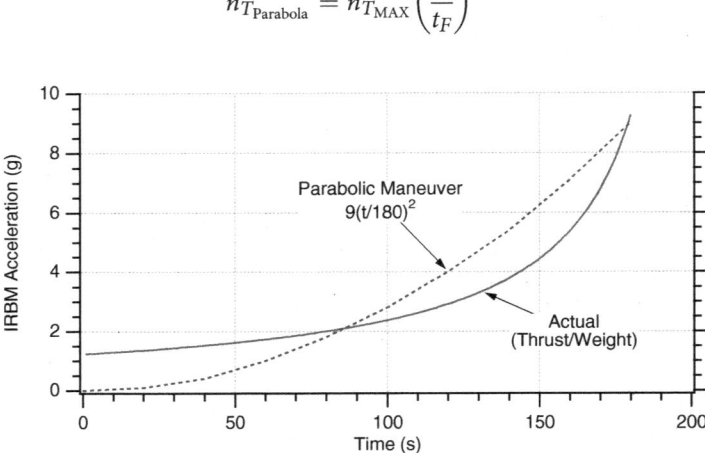

Fig. 2.2 For simplicity, the axial acceleration of an IRBM will be approximated by a parabola.

where t_F is the IRBM burnout time. It will be shown later that the real purpose of the parabolic approximation is to enable us to rapidly estimate the KKV divert and acceleration requirements on hypothetical threats when only a minimal amount of information is available.

Let us simulate the one-dimensional guidance system of Fig. 22.1, assuming the proportional navigation (PN) guidance law with an effective navigation ratio of 3, for the academic case in which the IRBM target of Fig. 22.2 and the interceptor are both launched at time zero and intercept occurs at target burnout (180 s). The linearized one-dimensional engagement simulation of Listing 2.2 of the previous volume was modified so that the effects on the guidance system due to a parabolic target maneuver could be compared to the actual apparent target maneuver (that is, thrust divided by weight). The modified code appears in Listing 22.1. The changes to the original code are highlighted in bold. Here the guidance law can be changed from proportional navigation (APN=0) to augmented proportional navigation (APN=1). In addition, the effects of the parabolic target maneuver (OPTION=0) on the guidance system can be compared to the actual apparent target maneuver (OPTION=1). The code for the IRBM's thrust and weight profiles (ITGT=1) appears with the differential equations before the FLAG=1 statement. The code also considers an ICBM threat (ITGT=2) that will be discussed later in this chapter.

LISTING 22.1   ONE-DIMENSIONAL ENGAGEMENT SIMULATION BASED ON LINEARIZED GEOMETRY

```
clear
count=0;
IOPTION=0;
ITGT=1;
XNTAV=117.6;
if ITGT==1
        TF=180.;
else
        TF=240.;
end
PRED=0.*3280.;
VM=9000.;
VC=18000.;
XNTMAX=9.*32.2;
XNCMAX=966.;
APN=0.;
HEDEG=-57.3*PRED/(VM*TF);
YD=-VM*HEDEG/57.3;
Y=0.;
```

```
XNP=3.;
T=0.;
H=.001;
S=0.;
DELV=0.;
SUM=0.;
XN=0.;
while T<(TF-.0001)
        YOLD=Y;
        YDOLD=YD;
        DELVOLD=DELV;
        STEP=1;
        FLAG=0;
        while STEP<=1
                if FLAG==1
                        STEP=2;
                        Y=Y+H*YD;
                        YD=YD+H*YDD;
                        DELV=DELV+H*DELVD;
                        T=T+H;
                end;
                TGO=TF-T+.00001;
                if ITGT==1
                        if IOPTION==0
                                XNT=XNTMAX*(T/TF)^2;
                        else
                                if T<180.
                                        WGT=-212.*T+44000.;
                                        TRST=54100.;
                                else
                                        WGT=3300.;
                                        TRST=0.;
                                end
                                        XNT=32.2*TRST/WGT;
                        end
                else
                        if IOPTION==0
                                XNT=XNTAV;
                        else
                                if T<120.
                                        WGT=-2622*T+440660.;
                                        TRST=725850.;
                                elseif T<240.
                                        WGT=-642.*T+168120.;
                                        TRST=182250.;
```

```
                    else
                          WGT=5500.;
                          TRST=0.;
                    end
                    XNT=32.2*TRST/WGT;
             end
      end
      XLAMD=(Y+YD*TGO)/(VC*TGO*TGO);
      XNC=XNP*VC*XLAMD+.5*APN*XNP*XNT;
      if XNC>XNCMAX
             XNC=XNCMAX;
      end
      if XNC<-XNCMAX
             XNC=-XNCMAX;
      end
      DELVD=abs(XNC);
      YDD=XNT-XNC;
      FLAG=1;
end
FLAG=0;
Y=.5*(YOLD+Y+H*YD);
YD=.5*(YDOLD+YD+H*YDD);
DELV=.5*(DELVOLD+DELV+H*DELVD);
S=S+H;
if S>=.09999
      S=0.;
      SUM=SUM+XNT;
      XN=XN+1.;
      count=count+1;
      ArrayT(count)=T;
      ArrayXNT(count)=XNT/32.2;
      ArrayXNC(count)=XNC/32.2;
      ArrayDELV(count)=DELV/3.28;
end
end
figure
plot(ArrayT,ArrayXNC),grid
xlabel('Missile Flight Time (s)')
ylabel('KKV Acceleration (g)')
clc
output=[ArrayT',ArrayXNC',ArrayDELV'];
save datfil.txt output -ascii
disp 'simulation finished'
Y
DELV/3.28
```

The nominal case of Listing 22.1 was run in which the parabolic target maneuver (OPTION=0) and actual target maneuver (OPTION=1) were used when the proportional navigation guidance law (APN=0) was employed. Figure 22.3 shows that in this case the actual KKV acceleration required by the PN guidance law at the end of the flight is 30 g or three times the maximum acceleration capability of the target. This KKV acceleration requirement is the same as would be the case if there was a constant target maneuver, as was shown in Chapter 2 of the previous volume. It can also be seen from Fig. 22.3 that the parabolic target maneuver approximation yields nearly identical KKV acceleration requirements at the end of flight, indicating that the parabolic target maneuver approximation might be a good approximation for a boosting target being pursued by a KKV employing PN guidance.

The one-dimensional simulation experiment was also repeated for the case in which the interceptor guidance law was changed to augmented proportional navigation (APN=1) with an effective navigation ratio of 3. The solid curve of Fig. 22.4 represents the actual acceleration required by the KKV using the augmented proportional navigation guidance law against the boosting IRBM, whereas the dashed curve represents the required KKV acceleration due to a parabolic target maneuver approximation. From the solid curve of Fig. 22.4 it can be seen that the maximum acceleration required by the KKV against the actual boosting target is now only 3 g or about one-third of the maximum axial acceleration capability of the target, whereas the parabolic approximation indicates a 4.5-g maximum KKV acceleration or one-half the maximum acceleration capability of the target. In addition, for both the actual and parabolic target acceleration models, the maximum acceleration no longer occurs at the end of flight. This means that should the KKV acceleration saturate, there is a chance that it will come out of saturation and not cause miss distance. *Thus, the maximum KKV*

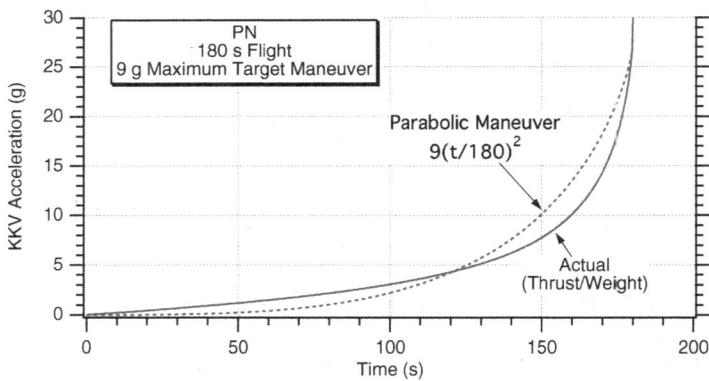

Fig. 22.3   For a PN guidance system, parabolic maneuver approximation yields accurate performance projections.

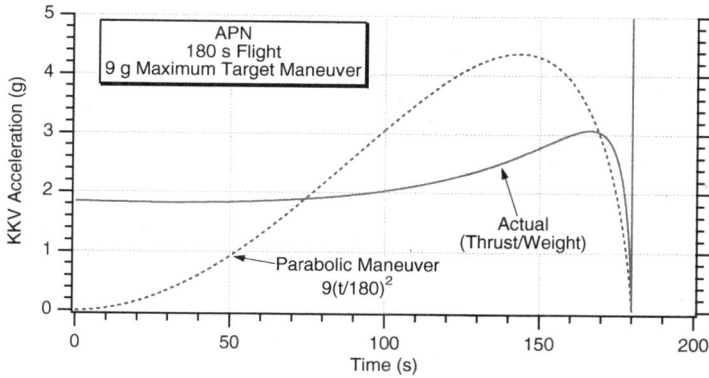

Fig. 22.4   For an augmented proportional navigation guidance system, parabolic maneuver approximation to IRBM boosting target overestimates KKV acceleration requirements.

acceleration required by the augmented proportional navigation guidance law is nearly an order of magnitude smaller than that required by the proportional navigation guidance law! This means that the performance improvement with augmented proportional navigation is so great that we shall not even consider using proportional navigation guidance against a boosting target. Figure 22.4 also indicates that the parabolic approximation to the target maneuver is not as good in predicting interceptor performance as it was in the previous example where proportional navigation guidance was used by the KKV. However, the parabolic approximation still indicates that there is a dramatic performance improvement with augmented proportional navigation guidance. It can be seen that using the parabolic approximation for the boosting IRBM tends to overestimate the KKV acceleration requirements. Therefore, for a conservative starting point, the parabolic approximation to the boosting target might be appropriate for analysis because of its simplicity.

DEVELOPING FORMULAS FOR DIVERT DUE TO BOOSTING TARGET AND PIP ERRORS

Generally speaking, an exoatmospheric interceptor has to be used against a long-range boosting target because most of the target's flight during its boost phase is outside the atmosphere. An endoatmospheric interceptor may be required against short-range ballistic missiles (SRBMs) because their apogees are very low. In this chapter only long-range IRBMs and ICBMs are considered. The KKV part of the interceptor has lateral divert engines for implementing guidance commands outside of the atmosphere. Because the divert engines burn propellant to implement the guidance law, guidance terminates when the propellant is expended because the KKV can no longer maneuver. The amount of propellant required by the KKV is related to the lateral divert ΔV through the rocket equation. Thus the amount of lateral divert required for an intercept is an important measure of interceptor performance. As was shown in Chapter 18 of the previous volume, the lateral divert is simply the integral of the absolute value of the interceptor acceleration, or

$$\Delta V = \int_0^{t_F} |n_c| \, dt$$

Figure 22.1 was evaluated, using Listing 22.1, for a 10-g parabolic target maneuver and the APN guidance law for the missile or KKV flight times ranging from 10 s to 50 s in steps of 10 s. The simulation results of Fig. 22.5 indicate that the amount of lateral divert required increases linearly with KKV flight time. We can also see from Fig. 22.5 that the simulation results can be curve fitted with a straight line and an empirical formula can be developed for the KKV lateral divert due to a parabolic target maneuver as

$$\Delta V_{MVR} = 0.25 n_{T_{MAX}} t_F$$

where n_{TMAX} is the maximum value of the parabolic maneuver in units of m/s^2 and ΔV_{MVR} is in units of m/s. The quantity t_F represents the KKV action time or the amount of time the KKV is maneuvering.

Up to this point it has been assumed that the only disturbance entering the KKV guidance system was an apparent target maneuver. Another important error source is the PIP error. This error source is due to the fact that the location of the boosting target at the desired intercept time is unknown. A prediction of the

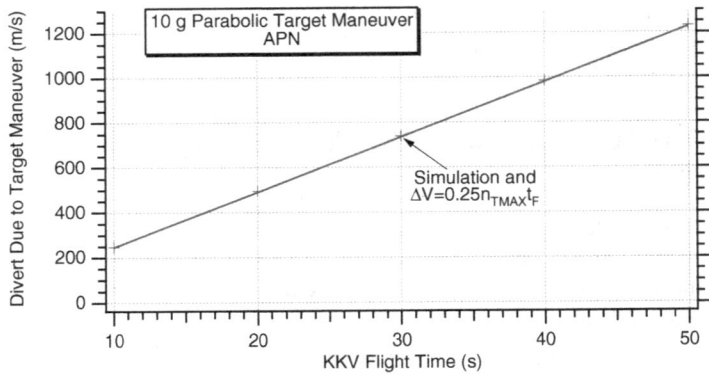

Fig. 22.5   Formula for KKV divert due to parabolic target maneuver in APN guidance system can be developed.

intercept point must be made, and this prediction will have errors. The errors will be the same whether a PN or APN guidance law is used.

It is important to note that some people believe that the PIP errors for a boosting target will be small because *a priori* information concerning the threat will be available. The point of view taken in this chapter is that even if the thrust-weight profile of the threat was known perfectly, the future direction of the acceleration vector is unknown. In other words, we do not know where the target is going or when it will get there based on past information. Therefore, the conservative point of view taken in this chapter is that the system must work when information concerning the threat is not available or denied and that the lateral divert of the KKV must be sized accordingly. It was shown in Chapter 18 of the previous volume that for the model of Fig. 22.1, the KKV lateral divert due to the PIP error for either the PN or APN guidance laws with an effective navigation ratio of 3 is given by

$$\Delta V_{\text{PIP}} = 1.5 \frac{\text{PIP}}{t_F}$$

Here it can be observed that for a given PIP error, more divert will be required for shorter KKV action times t_F. Thus the total divert required, under worst case conditions, is simply given by

$$\Delta V_{\text{TOT}_{\text{APN}}} = \frac{1.5 * \text{PIP}}{t_F} + 0.25 * n_{T\text{MAX}} t_F$$

INTERCEPTOR – IRBM ENGAGEMENTS

The previously mentioned single-stage generic IRBM model was put in a two-dimensional nonlinear simulation, assuming a round Earth and Newton's law of universal gravitation, based on Listing 17.3 of the previous volume. The simulation was modified for a sample 2000-km IRBM lofted trajectory, and the modified simulation appears in Listing 22.2. Although Listing 22.2 is a two-dimensional simulation, three-dimensional Lambert and distance routines (obtained from Chapter 17) are used by setting the z-components of various quantities to zero. As was mentioned previously, an ICBM target (ITGT=2) option also appears in Listing 22.2 and will be discussed later in this chapter.

LISTING 22.2   TARGET TRAJECTORY GENERATOR

```
clear
count=0;
RDESKM=2000.;
% 1=IRBM,2=ICBM
ITGT=1;
TLOFT=200.;
```

```
TUPT=15.;
if ITGT==1
   TPZ=180.;
else
   TPZ=240.;
end
QMIN=1;
TF=2000.;
TFINISH=3000.;
LEFT=1;
QBOOST=1;
QOOMPH=1;
CW=0;
SWITCH=0;
GAMDEG=89.99;
H=.01;
T=0.;
S=0.;
A=2.0926E7;
GM=1.4077E16;
ALT=0;
ANGDEG=0.;
ANG=ANGDEG/57.3;
XLONGM=ANG;
X=(A+ALT)*cos(ANG);
Y=(A+ALT)*sin(ANG);
Z=0;
ALT=sqrt(X^2+Y^2)-A;
XFIRST=X;
YFIRST=Y;
ZFIRST=Z;
X1=cos(1.5708-GAMDEG/57.3+ANG);
Y1=sin(1.5708-GAMDEG/57.3+ANG);
AXT=0.;
AYT=0.;
XLONGTDEG=57.3*RDESKM*3280./A;
TF=252.+.223*RDESKM-(5.44E-6)*RDESKM*RDESKM;
TF=TF+TLOFT;
XLONGT=XLONGTDEG/57.3;
XF=A*cos(XLONGT);
YF=A*sin(XLONGT);
ZF=0;
while ALT>-1
   XOLD=X;
   YOLD=Y;
   X1OLD=X1;
```

```
Y1OLD=Y1;
STEP=1;
FLAG=0;
while STEP <=1
   if FLAG==1
      STEP=2;
      X=X+H*X1;
      Y=Y+H*Y1;
      X1=X1+H*X1D;
      Y1=Y1+H*Y1D;
      T=T+H;
   end
   if ITGT==1
      if T<180.
         WGT=-212.*T+44000.;
         TRST=54100.;
      else
         WGT=3300.;
         TRST=0.;
      end
   else
      if T<120
         WGT=-2622*T+440660.;
         TRST=725850.;
      elseif T<240.
         WGT=-642.*T+168120.;
         TRST=182250.;
      else
         WGT=5500.;
         TRST=0.;
      end
   end
   AT=32.2*TRST/WGT;
   TEMBOT=(X^2+Y^2)^1.5;
   X1D=-GM*X/TEMBOT+AXT;
   Y1D=-GM*Y/TEMBOT+AYT;
   ALT=sqrt(X^2+Y^2)-A;
   FLAG=1;
end
FLAG=0;
X=(XOLD+X)/2+.5*H*X1;
Y=(YOLD+Y)/2+.5*H*Y1;
X1=(X1OLD+X1)/2+.5*H*X1D;
Y1=(Y1OLD+Y1)/2+.5*H*Y1D;
S=S+H;
if QBOOST==1
```

```
        TGOLAM=TF-T;
        [VRX,VRY,VRZ]=LAMBERT3D(X,Y,Z,TGOLAM,XF,YF,ZF,SWITCH);
        DELX=VRX-X1;
        DELY=VRY-Y1;
        DEL=sqrt(DELX^2+DELY^2);
        if T<TPZ & DEL>500
           AXT=AT*DELX/DEL;
           AYT=AT*DELY/DEL;
        elseif DEL<500
           TRST=0.;
           QBOOST=0;
           AXT=0.;
           AYT=0.;
           X1=VRX;
           Y1=VRY;
           X1OLD=X1;
           Y1OLD=Y1;
        else
           QBOOST=0;
           AXT=0.;
           AYT=0.;
        end
        if T<TUPT
           RTMAG=sqrt(X^2+Y^2);
           AXT=AT*X/RTMAG;
           AYT=AT*Y/RTMAG;
        end
     end
     if S>=.99999
        S=0.;
        DISTKM=distance3dkm(X,Y,Z,XFIRST,YFIRST,ZFIRST);
        ALTKM=(sqrt(X^2+Y^2)-A)/3280.;
        VELK=sqrt(X1^2+Y1^2)/3280.;
        count=count+1;
        ArrayT(count)=T;
        ArrayDISTKM(count)=DISTKM;
        ArrayALTKM(count)=ALTKM;
        ArrayVELK(count)=VELK;
     end
  end
end
figure
plot(ArrayDISTKM',ArrayALTKM'),grid
xlabel('Downrange (km)')
ylabel('Altitude (km) ')
clc
output=[ArrayT',ArrayDISTKM',ArrayALTKM'];
```

save datfil.txt output -ascii
disp 'simulation finished'

% LAMBERT3D can be found in Listing 17.3
% distance3dkm can be found in Listing 17.2

The nominal case of Listing 22.2 was run (TLOFT=200, TUPT=200), and the resultant IRBM trajectory is presented in Fig. 22.6. We can see that the IRBM travels 2000 km downrange with an apogee of approximately 500 km.

The boost-phase portion or the first 180 s of the IRBM trajectory is presented with 10-s time ticks in Fig. 22.7. If it is assumed that there is cloud cover until 7-km altitude, and then the IRBM can be seen by airborne IRST sensors at 80 s. Of course, on a clear day the target can be seen much sooner.

Listing 22.3 presents an engagement simulation in which an impulsively launched interceptor pursues the just-discussed IRBM. Here we have two options for calculating the PIP at the desired intercept time TF. The first option assumes that PIP is known perfectly (QPERFECT=1). With this option, the routine predict34.m integrates the target equations of motion forward to calculate the exact location of the target at the desired intercept time. Another option for calculating the PIP, assuming no *a priori* information is available, is to use a three-term Taylor series (QPERFECT=0). At the time of interceptor launch, the required velocity vector of the interceptor is calculated given the location of the interceptor, its launch time, the PIP, and desired intercept time. It is important to note that a for loop is included in Listing 22.3 to find the earliest possible intercept time for an impulsive interceptor whose speed is less than 4 km/s. Augmented proportional navigation guidance begins at a user-specified time TGUID. Again, we can see from Listing 22.3 that the three-dimensional Lambert and distance routines are used in the two-dimensional engagement simulation with the required z-component inputs set to zero.

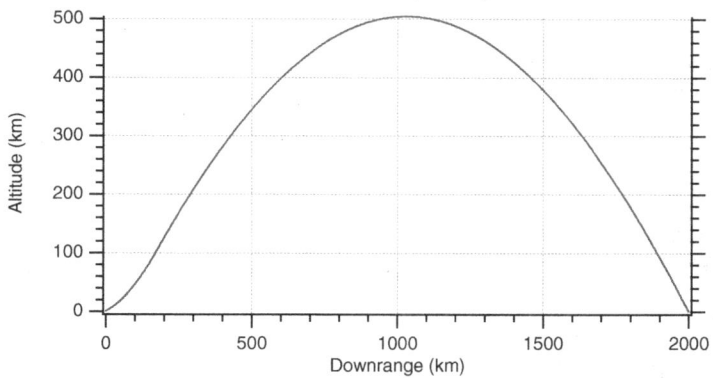

Fig. 22.6   Sample 2000-km IRBM trajectory.

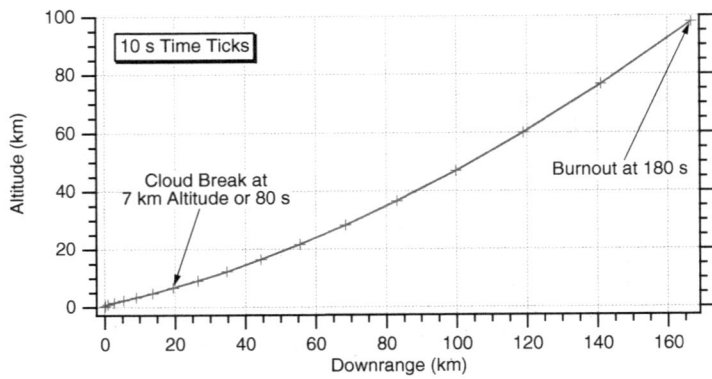

Fig. 22.7   Boost-phase portion of IRBM trajectory.

LISTING 22.3   TWO-DIMENSIONAL NONLINEAR ENGAGEMENT SIMULATION

```
clear
count=0;
TLAUNCH=90.;
TF=170.;
TS=1.;
XLONGMDEGICKM=400.;
RDESKM=2000.;
GAMDEG=89.99;
TFTOT=2000.;
TUPT=15.;
TGUID=110.;
XNCLIM=322.;
XNP=3.;
QPERFECT=1;
TLOFT=200.;
ALTMKMIC=15.;
QTAYLOR=1;
PIPERRKM=0.;
QGUID=1;
ITGT=1;
SWITCH1=0;
SWITCHM=0;
DELTF=0.;
QFIX=1;
if ITGT==1
        TPZ=180.;
else
```

```
         TPZ=240.;
end
ALTM=ALTMKMIC*3280.;
TFTOT=252.+.223*RDESKM-(5.44E-6)*RDESKM*RDESKM;
TFTOT=TFTOT+TLOFT;
A=2.0926E7;
GM=1.4077E16;
W=0.;
XLONGFDEG=57.3*RDESKM*3280./A;
XLONGMDEGIC=XLONGMDEGICKM/111.;
XLONGMDEG=XLONGMDEGIC;
QLAUNCH=0;
QFIRST=1;
XLONGTDEG=0.;
QBOOST=1;
QOOMPH=1;
T=0.;
S=0.;
AXT=0.;
AYT=0.;
ATP=0.;
XLONGF=XLONGFDEG/57.3;
XLONGF=XLONGF-W*TFTOT;
PIPERR=0.;
XF=A*cos(XLONGF);
YF=A*sin(XLONGF);
ZF=0;
XLONGT=XLONGTDEG/57.3;
XLONGM=XLONGMDEG/57.3;
XT=A*cos(XLONGT);
YT=A*sin(XLONGT);
XTINIT=XT;
YTINIT=YT;
RTINIT=sqrt(XTINIT^2+YTINIT^2);
XM=(A+ALTM)*cos(XLONGM);
YM=(A+ALTM)*sin(XLONGM);
ZM=0;
XFIRST=XT;
YFIRST=YT;
ZT=0;
ZFIRST=0;
DISTRTKMIC=distance3dkm(XT,YT,ZT,XFIRST,YFIRST,ZFIRST);
XMINIT=XM;;
YMINIT=YM;
RMINIT=sqrt(XMINIT^2+YMINIT^2);
XTD=cos(1.5708-GAMDEG/57.3);
```

```
YTD=sin(1.5708-GAMDEG/57.3);
ATP=1.;
AXM=0.;
AYM=0.;
AMP=0.;
H=.01;
ALTTKM=(sqrt(XT^2+YT^2)-A)/3280.;
XMD=0.;
YMD=0.;
RTM1=XT-XM;
RTM2=YT-YM;
RTM=sqrt(RTM1^2+RTM2^2);
VTM1=XTD-XMD;
VTM2=YTD-YMD;
VC=-(RTM1*VTM1+RTM2*VTM2)/RTM;
DELV=0.;
ACC=0.;
AXMGUID=0.;
AYMGUID=0.;
PREDERRKM=0.;
ZEM1=0.;
ZEM2=0.;
ALTMKM=(sqrt(XM^2+YM^2)-A)/3280.;
TBOT=0.;
DELVELM=0.;
PIPKMBO=0.;
ZEMPERPTOT=0.;
if QFIX==1
% FIND (EXACT LOCATION OF TARGET AT DESIRED INTERCEPT TIME EXACT PIP)
        [XTFACT,YTFACT]=predict45(T,XT,YT,XTD,YTD,TF,TFTOT,TUPT,XF,YF,ITGT);
        ZTFACT=0;
        TGOLAM=TF-TLAUNCH;
        [VRX,VRY,VRZ]=LAMBERT3D(XM,YM,ZM,TGOLAM,XTFACT,YTFACT,
                ZTFACT,SWITCHM);
        VMXRQD=VRX;
        VMYRQD=VRY;
        VMRQDKM=sqrt(VMXRQD^2+VMYRQD^2)/3280.;
else
        for TF=(TLAUNCH+30.):10:(TPZ-10.),
% FIND EXACT LOCATION OF TARGET AT DESIRED INTERCEPT TIME EXACT PIP)
        [XTFACT,YTFACT]=predict45(T,XT,YT,XTD,YTD,TF,TFTOT,TUPT,XF,YF,ITGT);
        TGOLAM=TF-TLAUNCH;
        [VRX,VRY,VRZ]=LAMBERT3D(XM,YM,ZM,TGOLAM,XTFACT,YTFACT,
                        ZTFACT,SWITCHM);
        VMXRQD=VRX;
        VMYRQD=VRY;
```

```
            VMRQDKM=sqrt(VMXRQD^2+VMYRQD^2)/3280.;
            if VMRQDKM<4.
                            break
                    end
            end
    end
    TF=TF+DELTF;
    while ~((T>(TF-10.)) & VC<0.)
            if RTM<1000
                    H=.00001;
            else
                    H=.01;
            end
            XTOLD=XT;
            YTOLD=YT;
            XTDOLD=XTD;
            YTDOLD=YTD;
            XMOLD=XM;
            YMOLD=YM;
            XMDOLD=XMD;
            YMDOLD=YMD;
            DELVOLD=DELV;
            STEP=1;
            FLAG=0;
            while STEP <=1
                    if FLAG==1
                            STEP=2;
                            XT=XT+H*XTD;
                            YT=YT+H*YTD;
                            XTD=XTD+H*XTDD;
                            YTD=YTD+H*YTDD;
                            XM=XM+H*XMD;
                            YM=YM+H*YMD;
                            XMD=XMD+H*XMDD;
                            YMD=YMD+H*YMDD;
                            DELV=DELV+H*DELVD;
                            T=T+H;
                    end
                    if ITGT==1
                            if T<180.
                                    WGT=-212.*T+44000.;
                                    TRST=54100.;
                            else
                                    WGT=3300.;
                                    TRST=0.;
                            end
```

```
            else
                  if T<120.
                        WGT=-2622*T+440660.;
                        TRST=725850.;
                  elseif T<240.
                        WGT=-642.*T+168120.;
                        TRST=182250.;
                  else
                        WGT=5500.;
                        TRST=0.;
                  end
            end
            ATP=32.2*TRST/WGT;
            TEMPBOTT=(XT^2+YT^2)^1.5;
            XTDD=-GM*XT/TEMPBOTT+AXT;
            YTDD=-GM*YT/TEMPBOTT+AYT;
            RTM1=XT-XM;
            RTM2=YT-YM;
            VTM1=XTD-XMD;
            VTM2=YTD-YMD;
            RTM=sqrt(RTM1^2+RTM2^2);
            VC=-(RTM1*VTM1+RTM2*VTM2)/RTM;
            TGO=RTM/VC;
            ACCDOTRTM=(XTDD*RTM1+YTDD*RTM2)/RTM;
            ACCPER1=XTDD-ACCDOTRTM*RTM1/RTM;
            ACCPER2=YTDD-ACCDOTRTM*RTM2/RTM;
            ACCPERPTOT=sqrt(ACCPER1^2+ACCPER2^2)/32.2;
            if T>TGUID
                  TEMPBOTM=(XM^2+YM^2)^1.5;
                  XMDDGRAV=-GM*XM/TEMPBOTM;
                  YMDDGRAV=-GM*YM/TEMPBOTM;
                  ZEM1=RTM1+VTM1*TGO+.5*(XTDD-XMDDGRAV)*TGO^2;
                  ZEM2=RTM2+VTM2*TGO+.5*(YTDD-YMDDGRAV)*TGO^2;
                  ZEMDOTRTM=(ZEM1*RTM1+ZEM2*RTM2)/RTM;
                  ZEMPER1=ZEM1-ZEMDOTRTM*RTM1/RTM;
                  ZEMPER2=ZEM2-ZEMDOTRTM*RTM2/RTM;
                  ZEMPERPTOT=sqrt(ZEMPER1^2+ZEMPER2^2)/3280.;
                  AXMGUID=XNP*ZEMPER1/(TGO^2);
                  AYMGUID=XNP*ZEMPER2/(TGO^2);
                  TGO=RTM/VC;
                  if QGUID==0
                        XNCLIM=0.;
                  end
                  if AXMGUID>XNCLIM
                        AXMGUID=XNCLIM;
                  elseif AXMGUID<-XNCLIM
```

```
                        AXMGUID=-XNCLIM;
                end
                if AYMGUID>XNCLIM
                        AYMGUID=XNCLIM;
                elseif AYMGUID<-XNCLIM
                        AYMGUID=-XNCLIM;
                end
        else
                AXMGUID=0.;
                AYMGUID=0.;
        end
        if T>TLAUNCH
                TEMPBOTM=(XM^2+YM^2)^1.5;
                XMDD=-GM*XM/TEMPBOTM+AXMGUID;
                YMDD=-GM*YM/TEMPBOTM+AYMGUID;
        else
                XMDD=0.;
                YMDD=0.;
        end
        ACCNEW=sqrt(AXMGUID^2+AYMGUID^2);
        DELVD=ACCNEW;
        ALTMKM=(sqrt(XM^2+YM^2)-A)/3280.;
        FLAG=1;
end
FLAG=0;
XT=.5*(XTOLD+XT+H*XTD);
YT=.5*(YTOLD+YT+H*YTD);
XTD=.5*(XTDOLD+XTD+H*XTDD);
YTD=.5*(YTDOLD+YTD+H*YTDD);
XM=.5*(XMOLD+XM+H*XMD);
YM=.5*(YMOLD+YM+H*YMD);
XMD=.5*(XMDOLD+XMD+H*XMDD);
YMD=.5*(YMDOLD+YMD+H*YMDD);
DELV=.5*(DELVOLD+DELV+H*DELVD);
S=S+H;
if QBOOST==1
        TGOLAM=TFTOT-T;
        [VRX,VRY,VRZ]=LAMBERT3D(XT,YT,ZT,TGOLAM,XF,YF,ZF,SWITCH1);
        VTX=VRX;
        VTY=VRY;
        DELVXT=VTX-XTD;
        DELVYT=VTY-YTD;
        VELT=sqrt(XTD^2+YTD^2);
        DELVELT=sqrt(DELVXT^2+DELVYT^2);
        if (T<TPZ & DELVELT>500.)
                AXT=ATP*DELVXT/DELVELT;
```

```
                    AYT=ATP*DELVYT/DELVELT;
            elseif DELVELT<500.
                    TRST=0.;
                    QBOOST=0;
                    AXT=0.;
                    AYT=0.;
                    XTD=VTX;
                    XTDOLD=XTD;
                    YTD=VTY;
                    YTDOLD=YTD;
                    TBOT=T;
            else
                    QBOOST=0;
                    QOOMPH=0;
                    AXT=0.;
                    AYT=0.;
                    TBOT=T;
            end
    end
    if T<TUPT
            RTMAG=sqrt(XT^2+YT^2);
            AXT=ATP*XT/RTMAG;
            AYT=ATP*YT/RTMAG;
    end
    if T>=TLAUNCH
            TGOLAMM=TF-T;
            if QPERFECT==1
                    XTF=XTFACT;
                    YTF=YTFACT;
            elseif (QPERFECT==0 & QTAYLOR==1)
                    TGOM=TF-T;
                    XTF=XT+XTD*TGOM+.5*XTDD*TGOM*TGOM;
                    YTF=YT+YTD*TGOM+.5*YTDD*TGOM*TGOM;
            end
            ZTF=0;
            QLAUNCH=1;
            PIPERR=sqrt((XTF-XTFACT)^2+(YTF-YTFACT)^2)/3280.;
    end
    TGOLAMM=TF-T;
    if (T>=TLAUNCH & QFIRST==1)
            QFIRST=0;
            TGOPZ=TF-TLAUNCH;
            [VRX,VRY,VRZ]=LAMBERT3D(XM,YM,ZM,TGOPZ,XTF,YTF,ZTF,SWITCHM);
            VMXRQD=VRX;
            VMYRQD=VRY;
            VMRQDKM=sqrt(VMXRQD^2+VMYRQD^2)/3280.;
```

```
                XMD=VMXRQD;
                XMDOLD=XMD;
                YMD=VMYRQD;
                YMDOLD=YMD;
        end
        if S>=(TS-.0001)
                S=0.;
                ALTTKM=(sqrt(XT^2+YT^2)-A)/3280.;
                DISTRTKM=distance3dkm(XT,YT,ZT,XFIRST,YFIRST,ZFIRST);
                ALTMKM=(sqrt(XM^2+YM^2)-A)/3280.;
                DISTRMKM=distance3dkm(XM,YM,ZM,XFIRST,YFIRST,ZFIRST);
                VTK=sqrt(XTD^2+YTD^2)/3280.;
                ATG=sqrt(XTDD^2+YTDD^2)/32.2;
                VMKM=sqrt(XMD^2+YMD^2)/3280.;
                VTKM=sqrt(XTD^2+YTD^2)/3280.;
                XLAM=atan2(RTM2,RTM1);
                ATPLOS=-XTDD*sin(XLAM)+YTDD*cos(XLAM);
                ATPLOSG=ATPLOS/32.2;
                ZEMPLOS=-ZEM1*sin(XLAM)+ZEM2*cos(XLAM);
                ZEMPLOSG=ZEMPLOS/3280.;
                XNCPLOSG=(-AXMGUID*sin(XLAM)+AYMGUID*cos(XLAM))/32.2;
                DELVKM=DELV/3280.;
                ACCNEWG=ACCNEW/32.2;
                AXMGUIDG=AXMGUID/32.2;
                AYMGUIDG=AYMGUID/32.2;
                if T>TLAUNCH
                        PIPKM=sqrt((XTFACT-XTF)^2+(YTFACT-YTF)^2)/3280.;
                        PIPPLOS=-(XTFACT-XTF)*sin(XLAM)+(YTFACT-YTF)*cos(XLAM);
                        PIPPLOSKM=PIPPLOS/3280.;
                else
                        PIPKM=0.;
                        PIPPLOSKM=0.;
                end
                count=count+1;
                ArrayT(count)=T;
                ArrayDISTRTKM(count)=DISTRTKM;
                ArrayALTTKM(count)=ALTTKM;
                ArrayDISTRMKM(count)=DISTRMKM;
                ArrayALTMKM(count)=ALTMKM;
                ArrayATPLOSG(count)=ATPLOSG;
                ArrayXNCPLOSG(count)=XNCPLOSG;
                ArrayPIPPLOSKM(count)=PIPPLOSKM;
        end
end
figure
plot(ArrayDISTRTKM,ArrayALTTKM,ArrayDISTRMKM,ArrayALTMKM),grid
```

```
xlabel('Downrange (km)')
ylabel('Altitude (km) ')
figure
plot(ArrayT,ArrayATPLOSG,ArrayT,ArrayXNCPLOSG),grid
xlabel('Time (s)')
ylabel('Acceleration (g) ')
clc
output=[ArrayT',ArrayDISTRTKM',ArrayALTTKM',ArrayDISTRMKM',ArrayALTMKM',...
        ArrayATPLOSG',ArrayXNCPLOSG'];
save datfil.txt output -ascii
disp 'simulation finished'
DELV/3.28

% predict45.m subroutine file
function [xtf,ytf]=predict45(tp,xtp,ytp,xtdp,ytdp,tf,tftot,tupt,xf,yf,itgt)
if itgt==1
        tpz=180;
else
        tpz=240;
end
t=tp;
switch1=0;
xt=xtp;
yt=ytp;
zt=0.;
xtd=xtdp;
ytd=ytdp;
ztd=0.;
zf=0.;
a=2.0926E7;
gm=1.4077E16;
qboost=1;
h=.01;
s=0.;
axt=0.;
ayt=0.;
ztd=0;
while t<=(tf-.00001)
        xtold=xt;
        ytold=yt;
        xtdold=xtd;
        ytdold=ytd;
        step=1;
        flag=0;
        while step <=1
                if flag==1
```

```
                    xt=xt+h*xtd;
                    yt=yt+h*ytd;
                    xtd=xtd+h*xtdd;
                    ytd=ytd+h*ytdd;
                    t=t+h;
                    step=2;
            end
            tembot=(xt^2+yt^2)^1.5;
            xtdd=-gm*xt/tembot+axt;
            ytdd=-gm*yt/tembot+ayt;
            if itgt==1
                    if t<180.
                            wgt=-212.*t+44000.;
                            trst=54100.;
                    else
                            wgt=3300.;
                            trst=0.;
                    end
            else
                    if t<120
                            wgt=-2622*t+440660.;
                            trst=725850.;
                    elseif t<240.
                            wgt=-642.*t+168120.;
                            trst=182250.;
                    else
                            wgt=5500.;
                            trst=0.;
                    end
            end
            atp=32.2*trst/wgt;
            flag=1;
    end;
    flag=0;
    xt=(xtold+xt)/2+.5*h*xtd;
    yt=(ytold+yt)/2+.5*h*ytd;
    xtd=(xtdold+xtd)/2+.5*h*xtdd;
    ytd=(ytdold+ytd)/2+.5*h*ytdd;
    if qboost==1
            tgolam=tftot-t;
            [vrx,vry,vrz]=LAMBERT3D(xt,yt,zt,tgolam,xf,yf,zf,switch1);
            vtx=vrx;
            vty=vry;
            delvxt=vtx-xtd;
            delvyt=vty-ytd;
            delvelt=sqrt(delvxt^2+delvyt^2);
```

```
        if (t<tpz & delvelt>500.)
                axt=atp*delvxt/delvelt;
                ayt=atp*delvyt/delvelt;
        elseif delvelt<500.
                trst=0.;
                qboost=1;
                axt=0.;
                ayt=0.;
                xtd=vtx;
                xtdold=xtd;
                ytd=vty;
                ytdold=ytd;
        else
                qboost=0;
                qoomph=0;
                axt=0.;
                ayt=0.;
        end
        if t<tupt
                rtmag=sqrt(xt^2+yt^2);
                axt=atp*xt/rtmag;
                ayt=atp*yt/rtmag;
        end
    end
end
xtf=xt;
ytf=yt;
% LAMBERT3D can be found in Listing 17.3
% distance3dkm can be found in Listing 17.2
```

Assuming that 10 s are required by the aircraft IRST sensors to establish a firm track on the IRBM, then the earliest an interceptor can be launched would be at 90 s. (Fig. 22.7 shows cloud break occurring at 80 s.) Figure 22.8 depicts an impulsively air-launched 3.4-km/s interceptor (that is, it takes 0 s to get up to speed being launched at 90 s and 15-km altitude) at the PIP. In this scenario an intercept is to occur at 170 s, or 10 s before the IRBM burns out. The case in which the PIP is known perfectly is done first in order to establish KKV divert requirements due to an apparent target maneuver. Next, a posttarget burnout intercept, where there is no apparent target maneuver, is investigated in order to establish KKV divert requirements due to PIP error.

Let us first examine the case in which there is no PIP error. In this exercise the interceptor would not even require a KKV because it can fly directly toward the perfect PIP. However, as part of this academic exercise, the KKV guidance system, using the APN guidance law, is turned on at 110 s (20 s after interceptor launch to account for the fact that in the real world it might take 20 s for the

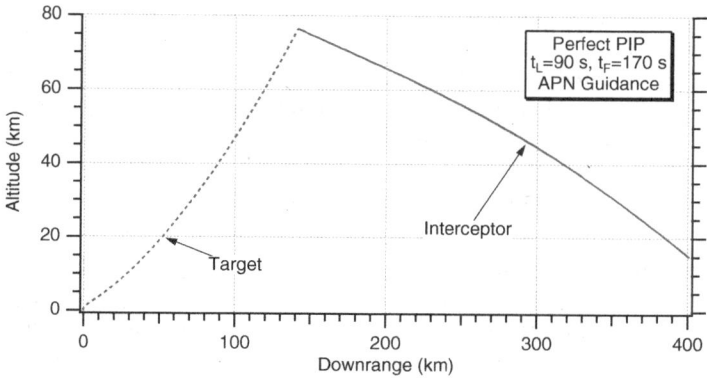

Fig. 22.8   Engagement with interceptor launched 10 s after cloud break and intercept occurring 10 s before the end of the IRBM boost phase.

interceptor to build up to speed). Because the portion of the IRBM acceleration that is perpendicular to the LOS appears as a target maneuver to the pursuing interceptor, the KKV will maneuver in order to hit the apparently maneuvering target. The KKV maneuvering will occur even though the PIP is known perfectly! The interceptor–target engagement geometry for this example is displayed in Fig. 22.8.

The required KKV acceleration to hit the target along with the target acceleration perpendicular to the LOS are displayed in Fig. 22.9. Here it can be observed that for this engagement geometry the maximum target acceleration perpendicular to the LOS is approximately 3.6 g, and the maximum KKV acceleration

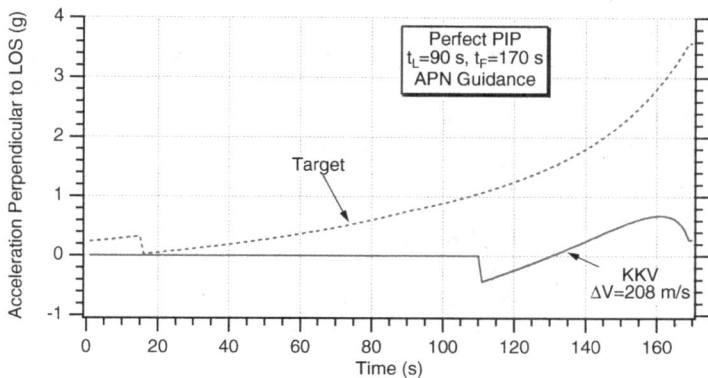

Fig. 2.9   For maximum range engagement, KKV requires much less acceleration than target.

required by the APN guidance law is approximately 0.7 g or approximately five times less acceleration than the target. The amount of lateral divert required by the KKV, as indicated by the engagement simulation, is 208 m/s in this example.

If the divert formula based on the parabolic target maneuver is utilized, it can be seen that the theoretical KKV divert predicted is 540 m/s, or

$$\Delta V_{MVR} = 0.25 n_{TMAX} t_F = 0.25 * 36 * (170 - 110) = 540 \frac{m}{s}$$

As expected, the theoretical KKV divert prediction is observed to be very conservative because it is more than twice as high as the actual KKV divert required, as indicated by the engagement simulation. Generally speaking, it was found that the KKV divert due to the apparently maneuvering one-stage IRBM target, as indicated by the engagement simulation, for different initial separations between both the interceptor and target launch points, was usually less than a few hundred m/s and is thus considered to be negligible.

A potentially more important error source against this one-stage IRBM target can be the PIP error. One simple way of predicting where the target will be at intercept is to use a three-term Taylor series based on the current position, velocity, and acceleration of the target, or in two dimensions

$$x_F = x + \dot{x} t_{go} + 0.5 \ddot{x} t_{go}^2$$

$$y_F = y + \dot{y} t_{go} + 0.5 \ddot{y} t_{go}^2$$

where t_{go} is given by

$$t_{go} = t_F - t$$

In other words, the future location of the object at some time t_F is the current target position plus the current target velocity times the time to go plus one-half the current target acceleration times the time to go squared. The Taylor series method of prediction has many faults, but its main virtue is that it does not require *a priori* information.

Figure 22.10 shows a case in which the PIP is not known as and is calculated from the preceding three-term Taylor series. In this example, the PIP error perpendicular to the line of sight is 89 km when the interceptor starts to guide. In this academic exercise, the KKV guidance system is immediately turned on after the target burns out (so that none of the KKV divert is due to the apparent target maneuver), and an intercept geometry is set up so that the desired intercept time is also after target burnout at 230 s, but the achieved intercept time in the nonlinear engagement simulation turns out to be 223 s. To accommodate the postboost intercept time and to ensure that the intercept geometry was kinematically feasible, the interceptor launch point was moved further downrange from the target launch point.

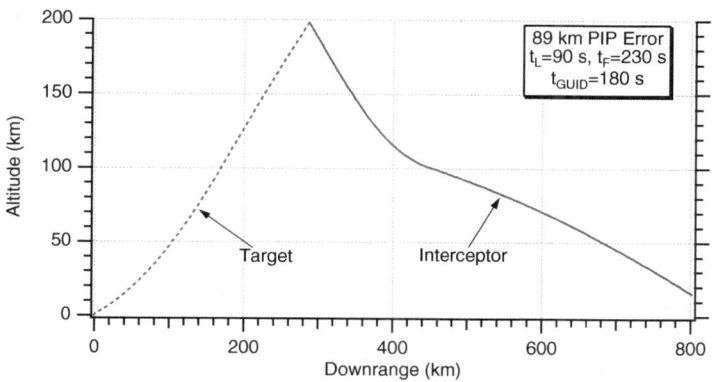

Fig. 22.10   Postboost phase engagement to test formula for PIP error.

Figure 22.11 displays the KKV acceleration profile for the engagement with the 89-km PIP error. It can be seen that when the KKV guidance system turns on at 180 s there is an immediate step in KKV acceleration. After the initial step, the KKV acceleration saturates for a brief period of time (simulation has a 10-g limit on KKV) and then linearly decreases to zero, which is in accordance with theory. The lateral divert required by the KKV, as indicated by the engagement simulation, is 2.8 km/s, and intercept occurs at 223 s. Had the KKV not saturated, the divert would have been slightly higher. The enormous amount of divert required for this hypothetical scenario was due to the fact that the interceptor was launched while the target was boosting. If the interceptor was launched after the target boost phase and the engagement was kinematically feasible, much less divert would be required for the intercept.

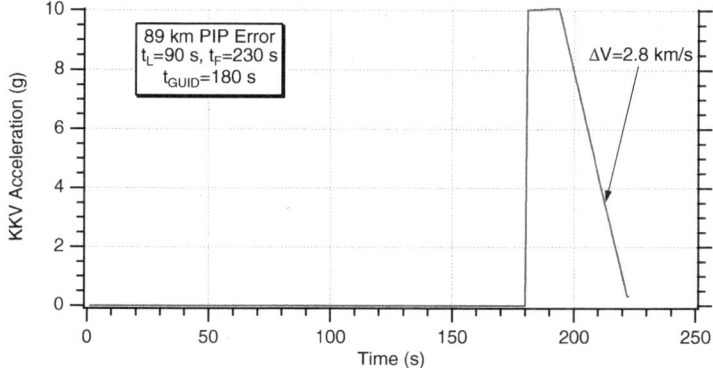

Fig. 22.11   Large initial KKV acceleration is required to take out PIP error in a short time.

TABLE 22.1   DETAILS OF PIP ERROR EXPERIMENT

Distance (km)	t_F (s)	ΔV_{Sim} (m/s)	PIP (m)	$\Delta V_{Formula}$ (m/s)
400	170	570	20,400	510
350	160	410	12,500	375
300	150	290	7200	270
250	140	210	3900	195
200	130	140	1800	135
150	120	110	700	105
100	120	110	680	102

A formula was previously provided for the KKV divert due to PIP error. For the case in which there is 89 km of PIP error and the time to take it out is 43 s (223 − 180 = 43), the theoretical divert formula indicates that 3.1 km/s of KKV lateral divert is required, or

$$\Delta V_{PIP} = 1.5 \frac{PIP}{t_F} = 1.5 \frac{89,000}{(223 - 180)} = 3.1 \frac{km}{s}$$

which is close to the engagement simulation results of Fig. 22.11.

The scenario of Fig. 22.8 was rerun for several cases where the interceptor launch point was gradually moved closer to the target launch point and intercept occurred during the target boost phase. Recall that the interceptor launch time is 90 s, and KKV guidance starts at 110 s. In each case, the earliest boost-phase intercept time was selected, assuming the interceptor speed could be no greater than 4 km/s. The simulated and calculated divert results appear along with the PIP error perpendicular to the line of sight in Table 22.1. Figure 22.12 compares both the simulated and computed divert requirements for the KKV due to PIP error. It is important to note that the simulated results include the effect of the apparent target maneuver, whereas the divert formula does not. We can see that the theoretical divert formula due to PIP error slightly underestimates the total required KKV lateral divert, as indicated by the simulation. *This means that for the one-stage IRBM example, most of the KKV divert is due to the PIP error rather than the apparent target maneuver.*

INTERCEPTOR – ICBM ENGAGEMENTS

Next, a two-stage liquid ICBM threat with a 240-s burn time, as described in [2], was considered as the target for analysis. A sample lofted 10,000-km trajectory for the ICBM was generated using Listing 22.2 (ITGT=2, TLOFT=500, TUPT=20) and is depicted in Fig. 22.13, where it can be seen that the apogee of this trajectory

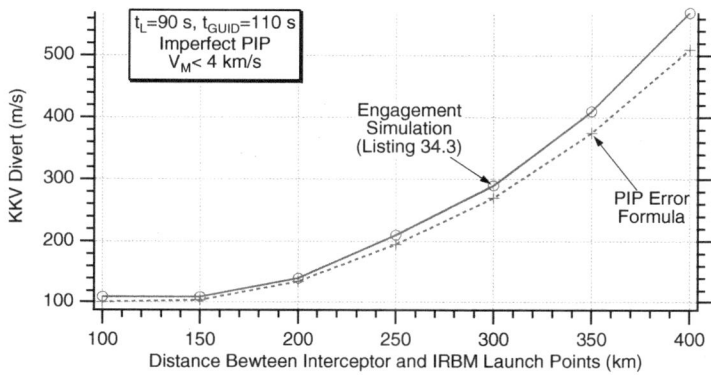

Fig. 22.12   Most of the KKV divert is due to PIP error for one-stage IRBM.

is approximately 2000 km. The boost-phase portion or first 240 s of the ICBM trajectory is displayed in Fig. 22.14 with 20-s time ticks. It can be observed that the ICBM breaks the clouds, assuming a 7-km altitude cloud cover, at approximately 60 s.

The total axial acceleration of the ICBM is depicted in Fig. 22.15. Here it can be seen that the first staging event occurs at 120 s with a maximum acceleration of 6 *g*, and the second staging event occurs at 240 s with a maximum acceleration of nearly 13 *g*. If the interceptor is launched early and the KKV guidance starts before 120 s, then the KKV guidance system will experience a large step in target acceleration at 120 s. It is apparent that this complex apparent acceleration cannot be represented by a single parabola.

An engagement was set up using Listing 22.3 in which the initial interceptor launch point is about 900 km from the target launch site. The interceptor is

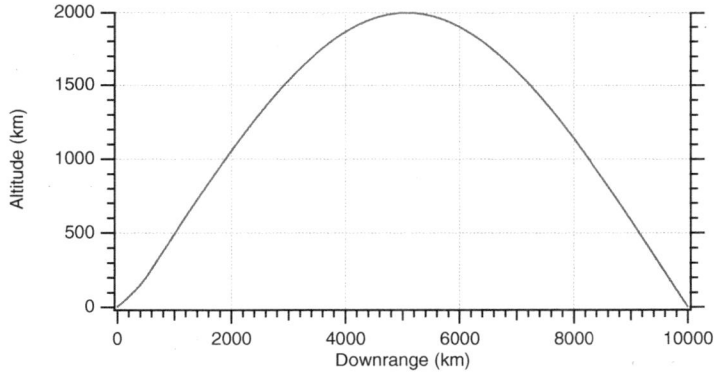

Fig. 22.13   Sample 10,000-km ICBM trajectory.

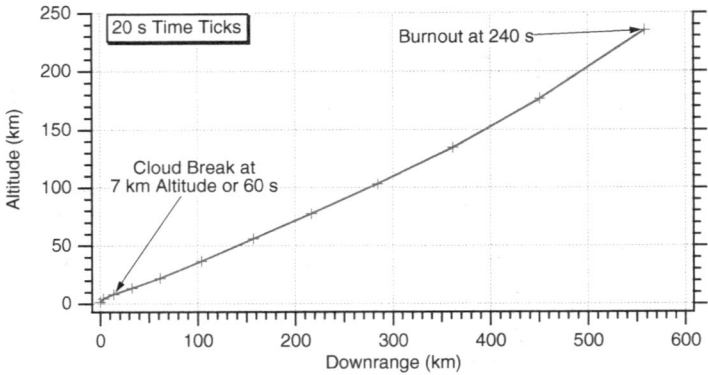

Fig. 22.14   Boost-phase portion of ICBM trajectory.

launched at 80 s, and the KKV guidance system using the APN guidance law is initiated at 100 s or about 20 s before the first target staging event. The engagement geometry is depicted in Fig. 22.16.

Figure 22.17 displays the target axial acceleration that is perpendicular to the LOS (dashed curve) for the engagement of Fig. 22.16. As was mentioned previously, this acceleration projection appears as a target maneuver to the KKV. The resultant KKV acceleration response to the apparent target maneuver (solid curve) is also displayed in Fig. 22.17. It can be seen that at first the KKV acceleration closely follows the apparent target maneuver and then changes abruptly when the target goes through the staging event at 120 s. The magnitude of the KKV acceleration after the staging event becomes a fraction of the actual target acceleration perpendicular to the LOS. The resultant lateral divert required by the KKV is shown in Fig. 22.17 to be 998 m/s.

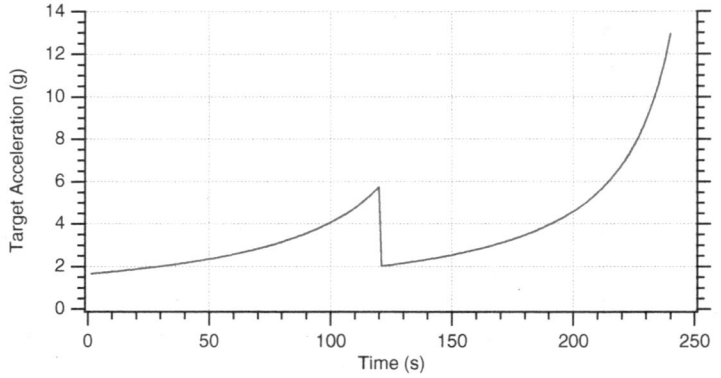

Fig. 22.15   ICBM has two staging events.

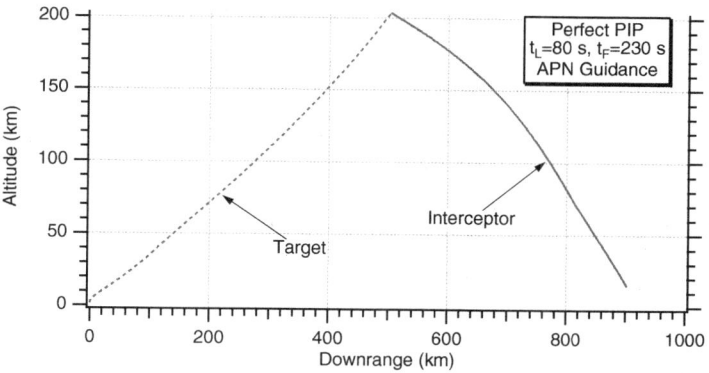

Fig. 22.16   ICBM engagement with interceptor launched 20 s after cloud break and intercept occurring 10 s before end of boost.

The average acceleration of the target maneuver while the KKV guidance system is activated can be computed from the target acceleration perpendicular to the LOS and is shown in Fig. 22.17 to be 1.3 g. If we pretend that the average target acceleration represents the complex apparent target maneuver shown in Fig. 22.17, then we can calculate the theoretical KKV divert from Chapter 18 of the previous volume (assuming APN guidance with an effective navigation ratio of 3) to be

$$\Delta V_{\text{APN}} = 0.75 n_{\text{TAV}} t_F = 0.75 * 13 * (230 - 100) = 1268 \frac{m}{s}$$

which is approximately 25% greater than the value of 998 m/s shown in Fig. 22.17.

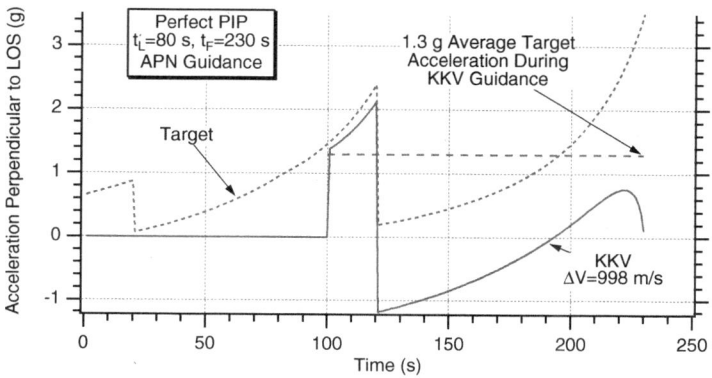

Fig. 22.17   For maximum-range ICBM engagement, apparent target acceleration is not a parabola.

TABLE 22.2   DETAILS OF BOOSTING ICBM EXPERIMENT

Downrange (km)	t_F (s)	ΔV_{SIM} (m/s)	$\Delta V_{Formula}$ (m/s)
900	220	896	1170
800	210	823	1072
700	190	727	878
600	180	689	780
500	160	557	585
400	150	463	487
300	130	177	292
200	120	141	195
100	110	2	98

An important purpose of the preceding divert formula is to qualitatively explain the simulation results and to suggest that the required KKV divert increases with increasing homing or action time. The formula also suggests that launching the interceptor earlier may increase the KKV divert requirements for a given intercept time. Thus an important purpose of the divert formula is to suggest future simulation experiments that must be conducted as part of the iterative design process.

The scenario of Fig. 22.16 was rerun for several cases where the interceptor launch point was gradually moved closer to the target launch point. Recall that the interceptor launch time is 80 s and KKV guidance starts at 100 s. *In each case, the earliest boost-phase intercept time was selected, assuming the*

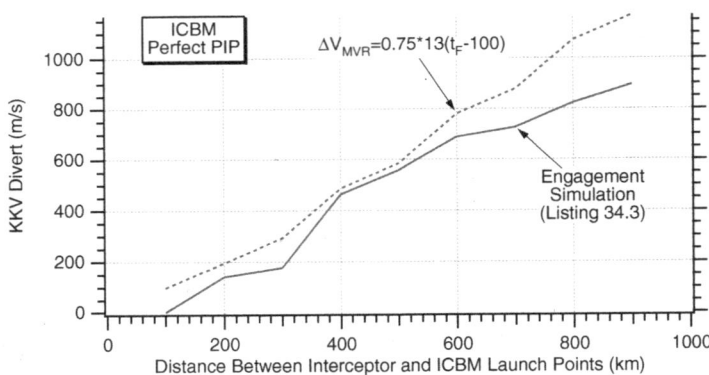

Fig. 22.18   Divert due to apparent constant target maneuver formula captures trends of actual KKV divert due to boosting two-stage ICBM.

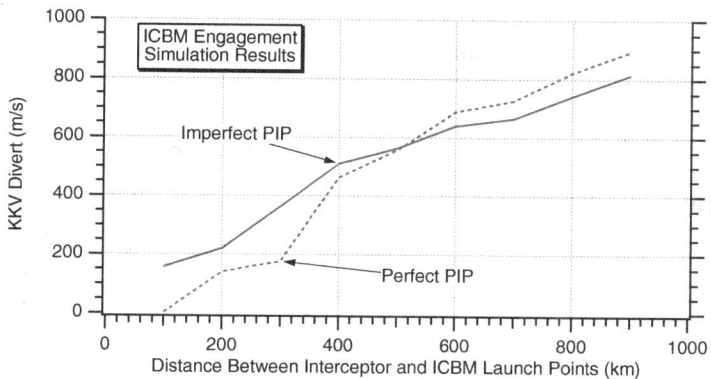

Fig. 22.19   Influence of PIP errors on KKV lateral divert for ICBM intercept is small.

interceptor speed could be no greater than 4 km/s. Therefore, for the 900-km downrange case it was found the intercept time could be reduced from 230 s (see Fig. 22.17) to 220 s. The simulated and calculated divert results appear in both Table 22.2 and Fig. 2.18. It is important to note that the calculated results are based on an average target acceleration of 1.3 g, which may not be accurate for all of the cases examined. However, it can be seen that the KKV divert trend is accurately captured with the simple divert formula appropriate for a constant average target maneuver.

Figure 22.19 shows another set of cases where the distance from the interceptor launch point to the target launch point is varied in the same way as was done in Fig. 22.18. However, this time the PIP is calculated from a three-term Taylor series rather than being perfect. It can be seen from the engagement simulation results of Fig. 22.19 that for the two-stage ICBM case the KKV divert requirements do not change significantly when PIP errors are considered. In fact, sometimes the PIP errors reduce the miss distance because of the direction of the PIP error. Thus it can be concluded that, for the ICBM case, the apparent target maneuver is the major contributor to the KKV divert requirements.

NOISE AND FILTERING

Ideally we would like to have an aircraft sensor that could measure both range and angle to the target so that we can estimate the target states required for calculating the PIP and for implementing Lambert guidance for the boosting interceptor and APN guidance for the KKV. Unfortunately, an airborne radar that can see the target at the long distances required might be too heavy for airborne applications. Similarly, an airborne LADAR (Laser Detection and Ranging) may also not work at required distances to see the target. On the other hand, an IRST sensor can see the boosting target at great distances but can only measure angle. For angle-only tracking of an unpredictable target, triangulation or stereo tracking is generally required to get target state estimates. In order to triangulate on the target, two aircraft are required, each having an IRST sensor, separated by a large distance known as a baseline. From the angle-only measurements of the two sensors, filters can be designed to estimate the position, velocity, and acceleration of the target. The triangulation of the angle measurements is also known as stereo tracking. One logical choice for filtering would be to design an extended Kalman filter (EKF) for this stereo tracking application.

Another choice might be to design an even simpler filter for the first step of the iterative design process so that initial estimates of the increase in KKV divert requirements due to sensor noise can be rapidly obtained. Such a choice might be the use of linear decoupled polynomial three-state Kalman filters using pseudo measurements. Although decoupled linear polynomial three-state Kalman filters are not optimal in this stereo tracking application, they can easily and rapidly be designed by pretending the sensors are measuring distances to the target rather than angles from the sensor to the target. Later on, during subsequent stages of the iterative design and sizing process, more complex filters such as the EKF can be considered to see if they can reduce the resultant KKV divert requirements.

The basis for the pseudo measurements for the decoupled linear polynomial three-state Kalman filters in two dimensions can be derived from Fig. 22.20. Here we see two sensors measuring angles θ_1 and θ_2. It is assumed that the location of the sensors (x_{s1}, y_{s1} and x_{s2}, y_{s2}) are known, but the location of the target (x_T, y_T) is unknown.

From Fig. 22.20 one can express the two sensor measurements of the angles θ_1 and θ_2 as

$$\theta_1 = \tan^{-1}\left(\frac{y_{s1} - y_T}{x_{s1} - x_T}\right)$$

$$\theta_2 = \tan^{-1}\left(\frac{y_{s2} - y_T}{x_{s2} - x_T}\right)$$

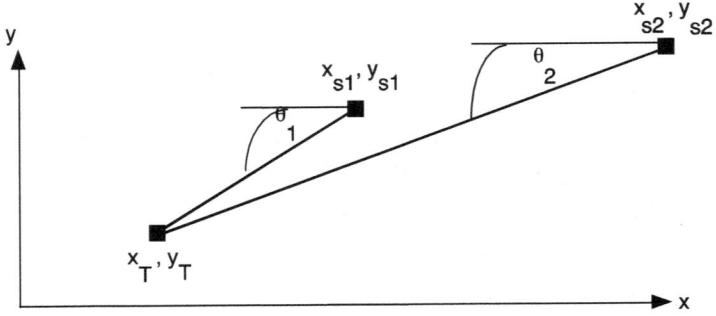

Fig. 22.20   Two angle-only sensors tracking a target.

Here we have two nonlinear equations with two unknowns. After some algebraic manipulation one can solve for the coordinates of the target in terms of the angle measurements and the sensor locations as

$$x_T^* = \frac{x_{s2} \tan(\theta_2) - x_{s1} \tan(\theta_1) + y_{s1} - y_{s2}}{\tan(\theta_2) - \tan(\theta_1)}$$

$$y_T^* = y_{s1} - x_{s1} \tan(\theta_1) + \frac{x_{s2} \tan(\theta_2) \tan(\theta_1) - x_{s1}\tan^2(\theta_1) + (y_{s1} - y_{s2}) \tan(\theta_1)}{\tan(\theta_2) - \tan(\theta_1)}$$

Thus the two pseudo measurements x_T^* and y_T^* will serve as inputs to the two decoupled three-state linear polynomial Kalman filters. One also has to develop formulas for the variance of the pseudo measurement noise to be used by the Riccati equations in the Kalman filter. The variance of the pseudo measurement noise on x_T can be found by using the chain rule from calculus. According to the chain rule,

$$\Delta x_T = \frac{\partial x_T}{\partial \theta_1} \Delta \theta_1 + \frac{\partial x_T}{\partial \theta_2} \Delta \theta_2$$

By squaring and taking expectations of both sides of the preceding equation one can express the variance of the pseudo measurement noise in terms of the variances of each of the IRST sensors as

$$\sigma_{x_T}^2 = \left(\frac{\partial x_T}{\partial \theta_1}\right)^2 \sigma_{\theta_1}^2 + \left(\frac{\partial x_T}{\partial \theta_2}\right)^2 \sigma_{\theta_2}^2$$

where the partial derivatives are evaluated as

$$\frac{\partial x_T}{\partial \theta_1} = \frac{(x_{s2} - x_{s1}) \tan(\theta_2) + y_{s1} - y_{s2}}{[(\cos(\theta_1) \tan(\theta_2) - \tan(\theta_1))]^2}$$

$$\frac{\partial x_T}{\partial \theta_2} = \frac{(x_{s1} - x_{s2}) \tan(\theta_1) + y_{s2} - y_{s1}}{[(\cos(\theta_2) \tan(\theta_2) - \tan(\theta_1))]^2}$$

The variance of the pseudo measurement noise on y_T can be found in a similar way.

A Monte Carlo simulation was set up where it was assumed that angle measurements were taken 10 times per second. The interceptor and KKV were command guided (first with Lambert guidance and then APN) until 10 s before intercept based on the Kalman filter estimates. The two linear polynomial three-state Kalman filter initial state estimates were set to zero (in other words, "cold starting" the filter) and within 10 s accurate state estimates were obtained using least squares filter techniques [6]. During the last 10 s it was assumed that homing guidance took place and that near-perfect estimates of the target states were available at a 100-Hz rate so that APN guidance could be implemented with range estimates being uplinked from the aircraft to the KKV. The code appears in Listing 22.4. Setting RUN=1 allows the simulation to run in the single flight mode; setting RUN=50 allows the simulation to run in the Monte Carlo mode.

LISTING 22.4   TWO-DIMENSIONAL MONTE CARLO ENGAGEMENT SIMULATION USING STEREO TRACKING

```
clear
count=0;
PHIS=576;
TLAUNCH=80.;
TF=230.;
TS=.1;
XLONGMDEGICKM=400.;
XLONGS2DEGKM=500.;
RDESKM=10000.;
GAMDEG=89.99;
TFTOT=2000.;
TUPT=20.;
TGUID=100.;
XNCLIM=322.;
XNP=3.;
QPERFECT=0;
TLOFT=500.;
ALTMKMIC=15.;
QTAYLOR=1;
PIPERRKM=0.;
QGUID=1;
ITGT=2;
DELTF=0.;
QFIX=0;
SIGTHET1=.00005;
THOM=10.;
BIAS1=0.;
RUN=50;
VMRQDKMIC=4.;
if ITGT==1
        TPZ=180.;
else
        TPZ=240.;
end
SIGTHET2=SIGTHET1;
XLONGS1DEGKM=XLONGMDEGICKM;
ORDER=3;
```

```
ALTM=ALTMKMIC*3280.;
TFTOT=252.+.223*RDESKM-(5.44E-6)*RDESKM*RDESKM;
TFTOT=TFTOT+TLOFT;
for JJ=1:RUN,
        SWITCHM=0;
        SWITCH1=0;
        A=2.0926E7;
        GM=1.4077E16;
        W=0.;
        XLONGFDEG=57.3*RDESKM*3280./A;
        XLONGMDEGIC=XLONGMDEGICKM/111.;
        XLONGMDEG=XLONGMDEGIC;
        XLONGS1DEG=XLONGS1DEGKM/111.;
        XLONGS2DEG=XLONGS2DEGKM/111.;
        QLAUNCH=0;
        QFIRST=1;
        XLONGTDEG=0.;
        QBOOST=1;
        QOOMPH=1;
        T=0.;
        S=0.;
        AXT=0.;
        AYT=0.;
        ATP=0.;
        XLONGF=XLONGFDEG/57.3;
        XLONGF=XLONGF-W*TFTOT;
        PIPERR=0.;
        XF=A*cos(XLONGF);
        YF=A*sin(XLONGF);
        ZF=0;
        XLONGT=XLONGTDEG/57.3;
        XLONGM=XLONGMDEG/57.3;
        XLONGS1=XLONGS1DEG/57.3;
        XLONGS2=XLONGS2DEG/57.3;
        XT=A*cos(XLONGT);
        YT=A*sin(XLONGT);
        ZT=0;
        XTINIT=XT;
        YTINIT=YT;
        ZTINIT=0;
        RTINIT=sqrt(XTINIT^2+YTINIT^2);
        XM=(A+ALTM)*cos(XLONGM);
        YM=(A+ALTM)*sin(XLONGM);
        ZM=0;
        XS1=(A+ALTM)*cos(XLONGS1);
        YS1=(A+ALTM)*sin(XLONGS1);
```

```
XS2=(A+ALTM)*cos(XLONGS2);
YS2=(A+ALTM)*sin(XLONGS2);
XFIRST=XT;
YFIRST=YT;
ZFIRST=0;
DISTRTKMIC=distance3dkm(XT,YT,ZT,XFIRST,YFIRST,ZFIRST);
XMINIT=XM;
YMINIT=YM;
RMINIT=sqrt(XMINIT^2+YMINIT^2);
XTD=cos(1.5708-GAMDEG/57.3);
YTD=sin(1.5708-GAMDEG/57.3);
ATP=1.;
AXM=0.;
AYM=0.;
AMP=0.;
H=.01;
ALTTKM=(sqrt(XT^2+YT^2)-A)/3280.;
XMD=0.;
YMD=0.;
RTM1=XT-XM;
RTM2=YT-YM;
RTM=sqrt(RTM1^2+RTM2^2);
VTM1=XTD-XMD;
VTM2=YTD-YMD;
VC=-(RTM1*VTM1+RTM2*VTM2)/RTM;
DELV=0.;
ACC=0.;
AXMGUID=0.;
AYMGUID=0.;
PREDERRKM=0.;
ZEM1=0.;
ZEM2=0.;
ALTMKM=(sqrt(XM^2+YM^2)-A)/3280.;
TBOT=0.;
DELVELM=0.;
PIPKMBO=0.;
ZEMPERPTOT=0.;
if QFIX==1
        [XTFACT,YTFACT]=predict45(T,XT,YT,XTD,YTD,TF,TFTOT,TUPT,XF,YF,ITGT);
        ZTFACT=0;
        TGOLAM=TF-TLAUNCH;
        XLONGM=atan2(YM,XM);
        XLONGT=atan2(YTFACT,XTFACT);
        [VRX,VRY,VRZ]=LAMBERT3D(XM,YM,ZM,TGOLAM,XTFACT,
                YTFACT,ZTFACT,SWITCHM);
        VMXRQD=VRX;
```

```
                    VMYRQD=VRY;
                    VMRQDKM=sqrt(VMXRQD^2+VMYRQD^2)/3280.
         else
                    for TF=(TLAUNCH+30.):10:(TPZ-10.),
                              [XTFACT,YTFACT]=predict34(T,XT,YT,XTD,YTD,TF,TFTOT,TUPT,
                                        XF,YF,ITGT);
                              ZTFACT=0;
                              TGOLAM=TF-TLAUNCH;
                              XLONGM=atan2(YM,XM);
                              XLONGT=atan2(YTFACT,XTFACT);
                              [VRX,VRY,VRZ]=LAMBERT3D(XM,YM,ZM,TGOLAM,XTFACT,
                                        YTFACT,ZTFACT,SWITCHM);
                              VMXRQD=VRX;
                              VMYRQD=VRY;
                              VMRQDKM=sqrt(VMXRQD^2+VMYRQD^2)/3280.;
                              if VMRQDKM<VMRQDKMIC
                                        break
                              end
                    end
         end
end
TF=TF+DELTF;
XH=0.;
XDH=0.;
XDDH=0.;
YH=0.;
YDH=0.;
YDDH=0.;
PHI=zeros([3,3]);
P=zeros([3,3]);
Q=zeros([3,3]);
IDNPZ=eye(3);
P(1,1)=99999999999.;
P(2,2)=99999999999.;
P(3,3)=99999999999.;
PP(1,1)=99999999999.;
PP(2,2)=99999999999.;
PP(3,3)=99999999999.;
PHI(1,1)=1;
PHI(1,2)=TS;
PHI(1,3)=.5*TS*TS;
PHI(2,2)=1;
PHI(2,3)=TS;
PHI(3,3)=1;
HMAT(1,1)=1.;
HMAT(1,2)=0.;
HMAT(1,3)=0.;
```

```
PHIT=PHI';
HT=HMAT';
Q(1,1)=PHIS*TS^5/20;
Q(1,2)=PHIS*TS^4/8;
Q(1,3)=PHIS*TS^3/6;
Q(2,1)=Q(1,2);
Q(2,2)=PHIS*TS^3/3;
Q(2,3)=PHIS*TS*TS/2;
Q(3,1)=Q(1,3);
Q(3,2)=Q(2,3);
Q(3,3)=PHIS*TS;
XN=0.;
while ~((T>(TF-10.)) & VC<0.)
        if RTM<1000
                H=.00001;
        else
                H=.01;
        end
        XTOLD=XT;
        YTOLD=YT;
        XTDOLD=XTD;
        YTDOLD=YTD;
        XMOLD=XM;
        YMOLD=YM;
        XMDOLD=XMD;
        YMDOLD=YMD;
        DELVOLD=DELV;
        STEP=1;
        FLAG=0;
        while STEP <=1
                if FLAG==1
                        STEP=2;
                        XT=XT+H*XTD;
                        YT=YT+H*YTD;
                        XTD=XTD+H*XTDD;
                        YTD=YTD+H*YTDD;
                        XM=XM+H*XMD;
                        YM=YM+H*YMD;
                        XMD=XMD+H*XMDD;
                        YMD=YMD+H*YMDD;
                        DELV=DELV+H*DELVD;
                        T=T+H;
                end
                if ITGT==1
                        if T<180
                                WGT=-212.*T+44000.;
```

```
                        TRST=54100.;
            else
                        WGT=3300.;
                        TRST=0.;
            end
      else
            if T<120.
                        WGT=-2622*T+440660.;
                        TRST=725850.;
            elseif T<240.
                        WGT=-642.*T+168120.;
                        TRST=182250.;
            else
                        WGT=5500.;
                        TRST=0.;
            end
      end
      ATP=32.2*TRST/WGT;
      TEMPBOTT=(XT^2+YT^2)^1.5;
      XTDD=-GM*XT/TEMPBOTT+AXT;
      YTDD=-GM*YT/TEMPBOTT+AYT;
      RTM1=XT-XM;
      RTM2=YT-YM;
      VTM1=XTD-XMD;
      VTM2=YTD-YMD;
      RTM=sqrt(RTM1^2+RTM2^2);
      VC=-(RTM1*VTM1+RTM2*VTM2)/RTM;
      TGO=RTM/VC;
      ACCDOTRTM=(XTDD*RTM1+YTDD*RTM2)/RTM;
      ACCPER1=XTDD-ACCDOTRTM*RTM1/RTM;
      ACCPER2=YTDD-ACCDOTRTM*RTM2/RTM;
      ACCPERPTOT=sqrt(ACCPER1^2+ACCPER2^2)/32.2;
      if (T>TGUID & TGO<THOM)
            TEMPBOTM=(XM^2+YM^2)^1.5;
            XMDDGRAV=-GM*XM/TEMPBOTM;
            YMDDGRAV=-GM*YM/TEMPBOTM;
            ZEM1=RTM1+VTM1*TGO+.5*(XTDD-XMDDGRAV)*TGO^2;
            ZEM2=RTM2+VTM2*TGO+.5*(YTDD-YMDDGRAV)*TGO^2;
            ZEMDOTRTM=(ZEM1*RTM1+ZEM2*RTM2)/RTM;
            ZEMPER1=ZEM1-ZEMDOTRTM*RTM1/RTM;
            ZEMPER2=ZEM2-ZEMDOTRTM*RTM2/RTM;
            ZEMPERPTOT=sqrt(ZEMPER1^2+ZEMPER2^2)/3280.;
            AXMGUID=XNP*ZEMPER1/(TGO^2);
            AYMGUID=XNP*ZEMPER2/(TGO^2);
            TGO=RTM/VC;
            if QGUID==0
```

```
                        XNCLIM=0.;
                end
                if AXMGUID>XNCLIM
                        AXMGUID=XNCLIM;
                elseif AXMGUID<-XNCLIM
                        AXMGUID=-XNCLIM;
                end
                if AYMGUID>XNCLIM
                        AYMGUID=XNCLIM;
                elseif AYMGUID<-XNCLIM
                        AYMGUID=-XNCLIM;
                end
        end
        if T<=TGUID
                AXMGUID=0.;
                AYMGUID=0.;
        end
        if T>TLAUNCH
                TEMPBOTM=(XM^2+YM^2)^1.5;
                XMDD=-GM*XM/TEMPBOTM+AXMGUID;
                YMDD=-GM*YM/TEMPBOTM+AYMGUID;
        else
                XMDD=0.;
                YMDD=0.;
        end
        ACCNEW=sqrt(AXMGUID^2+AYMGUID^2);
        DELVD=ACCNEW;
        ALTMKM=(sqrt(XM^2+YM^2)-A)/3280.;
        FLAG=1;
end
FLAG=0;
XT=.5*(XTOLD+XT+H*XTD);
YT=.5*(YTOLD+YT+H*YTD);
XTD=.5*(XTDOLD+XTD+H*XTDD);
YTD=.5*(YTDOLD+YTD+H*YTDD);
XM=.5*(XMOLD+XM+H*XMD);
YM=.5*(YMOLD+YM+H*YMD);
XMD=.5*(XMDOLD+XMD+H*XMDD);
YMD=.5*(YMDOLD+YMD+H*YMDD);
DELV=.5*(DELVOLD+DELV+H*DELVD);
S=S+H;
if QBOOST==1
        TGOLAM=TFTOT-T;
        XLONGM=atan2(YT,XT);
        XLONGT=atan2(YF,XF);
        [VRX,VRY,VRZ]=LAMBERT3D(XT,YT,ZT,TGOLAM,XF,YF,ZF,SWITCH1);
```

```
        VTX=VRX;
        VTY=VRY;
        DELVXT=VTX-XTD;
        DELVYT=VTY-YTD;
        VELT=sqrt(XTD^2+YTD^2);
        DELVELT=sqrt(DELVXT^2+DELVYT^2);
        if (T<TPZ & DELVELT>500.)
                AXT=ATP*DELVXT/DELVELT;
                AYT=ATP*DELVYT/DELVELT;
        elseif DELVELT<500.
                TRST=0.;
                QBOOST=0;
                AXT=0.;
                AYT=0.;
                XTD=VTX;
                XTDOLD=XTD;
                YTD=VTY;
                YTDOLD=YTD;
                TBOT=T;
        else
                QBOOST=0;
                QOOMPH=0;
                AXT=0.;
                AYT=0.;
                TBOT=T;
        end
    end
    if T<TUPT
        RTMAG=sqrt(XT^2+YT^2);
        AXT=ATP*XT/RTMAG;
        AYT=ATP*YT/RTMAG;
    end
    if T>=TLAUNCH
        TGOLAMM=TF-T;
        if QPERFECT==1
                XTF=XTFACT;
                YTF=YTFACT;
        elseif (QPERFECT==0 & QTAYLOR==1)
                TGOM=TF-T;
                XTF=XT+XTD*TGOM+.5*XTDD*TGOM*TGOM;
                YTF=YT+YTD*TGOM+.5*YTDD*TGOM*TGOM;
        end
        ZTF=0;
        QLAUNCH=1;
        PIPERR=sqrt((XTF-XTFACT)^2+(YTF-YTFACT)^2)/3280.;
    end
```

```
TGOLAMM=TF-T;
if (T>=TLAUNCH & QFIRST==1)
        QFIRST=0;
        TGOPZ=TF-TLAUNCH;
        [VRX,VRY,VRZ]=LAMBERT3D(XM,YM,ZM,TGOPZ,XTF,YTF,
                ZTF,SWITCHM);
        VMXRQD=VRX;
        VMYRQD=VRY;
        VMRQDKM=sqrt(VMXRQD^2+VMYRQD^2)/3280.;
        XMD=VMXRQD;
        XMDOLD=XMD;
        YMD=VMYRQD;
        YMDOLD=YMD;
end
if S>=(TS-.0001)
        S=0.;
        THET1=atan2(YS1-YT,XS1-XT);
        THET2=atan2(YS2-YT,XS2-XT);
        THET1NOISE=SIGTHET1*randn;;
        THET2NOISE=SIGTHET2*randn;;
        THET1S=THET1+THET1NOISE+BIAS1;
        THET2S=THET2+THET2NOISE;
        TOP1=XS2*tan(THET2S)-XS1*tan(THET1S)+YS1-YS2;
        XTS=TOP1/(tan(THET2S)-tan(THET1S));
        TOP2=XS2*tan(THET2S)*tan(THET1S)-XS1*tan(THET1S)*
                tan(THET1S)...
        +tan(THET1S)*(YS1-YS2);
        YTS=YS1-XS1*tan(THET1S)+TOP2/(tan(THET2S)-tan(THET1S));
        XTNOISE=XT-XTS;
        YTNOISE=YT-YTS;
        DXDT1=(tan(THET2)*(XS2-XS1)+YS1-YS2)/((cos(THET1)*
                (tan(THET2)...
                        -tan(THET1)))^2);
        DXDT2=(tan(THET1)*(XS1-XS2)+YS2-YS1)/((cos(THET2)*
                (tan(THET2)-...
        tan(THET1)))^2);
        SIGX=sqrt((DXDT1*SIGTHET1)^2+(DXDT2*SIGTHET2)^2);
        DYDT1=-XS1/(cos(THET1)*cos(THET1));
        DYDT1=DYDT1+(XS2*tan(THET2)*tan(THET2)-2.*XS1*
                tan(THET1)*...
                        tan(THET2)+(YS1-YS2)*tan(THET2)+XS1*
                                tan(THET1)...
                        *tan(THET1))/((cos(THET1)*(tan(THET2)...
                                -tan(THET1)))^2);
        DYDT2=(tan(THET1)*tan(THET1)*(XS1-XS2)-(YS1-YS2)*...
        tan(THET1))/((cos(THET2)*(tan(THET2)-tan(THET1)))^2);
```

```
SIGY=sqrt((DYDT1*SIGTHET1)^2+(DYDT2*SIGTHET2)^2);
XN=XN+1.;
XK1=3*(3*XN*XN-3*XN+2)/(XN*(XN+1)*(XN+2));
XK2=18*(2*XN-1)/(XN*(XN+1)*(XN+2)*TS);
XK3=60/(XN*(XN+1)*(XN+2)*TS*TS);
RMAT(1,1)=SIGX^2;
PHIP=PHI*P;
PHIPPHIT=PHIP*PHIT;
M=PHIPPHIT+Q;
HM=HMAT*M;
HMHT=HM*HT;
HMHTR=HMHT+RMAT;
HMHTRINV(1,1)=1./HMHTR(1,1);
MHT=M*HT;
K=MHT*HMHTRINV;
KH=K*HMAT;
IKH=IDNPZ-KH;
P=IKH*M;
if XN<10.
        XK1PZ=XK1;
        XK2PZ=XK2;
        XK3PZ=XK3;
else
        XK1PZ=K(1,1);
        XK2PZ=K(2,1);
        XK3PZ=K(3,1);
end
RES=XTS-XH-TS*XDH-.5*TS*TS*XDDH;
XH=XH+XDH*TS+.5*TS*TS*XDDH+XK1PZ*RES;
XDH=XDH+XDDH*TS+XK2PZ*RES;
XDDH=XDDH+XK3PZ*RES;
RMATP(1,1)=SIGY^2;
PHIPP=PHI*PP;
PHIPPHITP=PHIPP*PHIT;
MP=PHIPPHITP+Q;
HMP=HMAT*MP;
HMHTP=HMP*HT;
HMHTRP=HMHTP+RMATP;
HMHTRINVP(1,1)=1./HMHTRP(1,1);
MHTP=MP*HT;
KP=MHTP*HMHTRINVP;
KHP=KP*HMAT;
IKHP=IDNPZ-KHP;
PP=IKHP*MP;
if XN<10.
        XK1PZP=XK1;
```

```
            XK2PZP=XK2;
            XK3PZP=XK3;
    else
            XK1PZP=KP(1,1);
            XK2PZP=KP(2,1);
            XK3PZP=KP(3,1);
    end
    RESP=YTS-YH-TS*YDH-.5*TS*TS*YDDH;
    YH=YH+YDH*TS+.5*TS*TS*YDDH+XK1PZP*RESP;
    YDH=YDH+YDDH*TS+XK2PZP*RESP;
    YDDH=YDDH+XK3PZP*RESP;
    if (T>TGUID & TGO>THOM)
            RTM1H=XH-XM;
            RTM2H=YH-YM;
            RTMH=sqrt(RTM1H^2+RTM2H^2);
            VTM1H=XDH-XMD;
            VTM2H=YDH-YMD;
            VCH=-(RTM1H*VTM1H+RTM2H*VTM2H)/RTMH;
            TGOH=RTMH/VCH;
            TEMPBOTM=(XM^2+YM^2)^1.5;
            XMDDGRAV=-GM*XM/TEMPBOTM;
            YMDDGRAV=-GM*YM/TEMPBOTM;
            ZEM1H=RTM1H+VTM1H*TGOH+.5*(XDDH-XMDDGRAV)*
                    TGOH^2;
            ZEM2H=RTM2H+VTM2H*TGOH+.5*(YDDH-YMDDGRAV)*
                    TGOH^2;
            ZEMDOTRTMH=(ZEM1H*RTM1H+ZEM2H*RTM2H)/RTMH;
            ZEMPER1H=ZEM1H-ZEMDOTRTMH*RTM1H/RTMH;
            ZEMPER2H=ZEM2H-ZEMDOTRTMH*RTM2H/RTMH;
            AXMGUID=XNP*ZEMPER1H/(TGOH^2);
            AYMGUID=XNP*ZEMPER2H/(TGOH^2);
            if QGUID==0
                    XNCLIM=0.;
            end
            if AXMGUID>XNCLIM
                    AXMGUID=XNCLIM;
            elseif AXMGUID<-XNCLIM
                    AXMGUID=-XNCLIM;
            end
            if AYMGUID>XNCLIM
                    AYMGUID=XNCLIM;
            elseif AYMGUID<-XNCLIM
                    AYMGUID=-XNCLIM;
            end
    end
    if RUN==1
```

```
                        end
                  end
                  if RUN==1
                        ERRXTDD=(XTDD-XDDH)/32.2;
                        SP33X=sqrt(P(3,3))/32.2;
                        SP33XP=-SP33X;
                        ERRYTDD=(YTDD-YDDH)/32.2;
                        SP33Y=sqrt(PP(3,3))/32.2;
                        SP33YP=-SP33Y;
                        count=count+1;
                        ArrayT(count)=T;
                        ArrayERRXTDD(count)=ERRXTDD;
                        ArraySP33X(count)=SP33X;
                        ArraySP33XP(count)=SP33XP;
                  end
            end
            count=count+1
            ArrayJJ(count)=count;
            ArrayRTM(count)=RTM;
            ArrayDELV(count)=DELV/3.28;
      end
      if RUN==1
            figure
            plot(ArrayT,ArrayERRXTDD,ArrayT,ArraySP33X,ArrayT,ArraySP33XP),grid
            xlabel('Time (s)')
            ylabel('Acceleration Error (f/s) ')
            axis([0 240 -10 10])
            clc
            output=[ArrayT',ArrayERRXTDD',ArraySP33X',ArraySP33XP'];
            save datfil.txt output -ascii
      else
            figure
            plot(ArrayJJ,ArrayRTM),grid
            xlabel('Run')
            ylabel('Miss (ft) ')
            figure
            plot(ArrayJJ,ArrayDELV),grid
            xlabel('Run')
            ylabel('Divert (m/s) ')
            clc
            output=[ArrayJJ',ArrayRTM',ArrayDELV'];
            %save ('datfil.txt', 'output', '-ascii')
                  save datfil.txt output -ascii
            RTMSORT=sort(ArrayRTM)
```

```
    DELVSORT=sort(ArrayDELV)
    RTM90=RTMSORT(45)
    DELV90=DELVSORT(45)
end
disp 'simulation finished'
```

```
% LAMBERT3D can be found in Listing 17.3
% distance3dkm can be found in Listing 17.3
% predict34 can be found in Listing 22.3
```

A case was run against the ICBM target in which the shooter aircraft (with the first sensor) was 800-km downrange of the target launch point and the second sensor was 900-km downrange of the target launch point. Both sensors were 15 km in altitude, and the intercept time was set at 230 s for the case where there were no PIP errors. In addition it was assumed that the angular accuracy of both sensors was 50 μrad and that the filter sampling time was 0.1 s. Figures 22.21 and 22.22 show how the linear polynomial Kalman filter is able to estimate the target acceleration for different values of filter process noise Φ_s. Figure 22.22 shows that the filter with less process noise yields a much smoother estimate of target acceleration. However, by comparing the acceleration estimates to those of Fig. 22.21, one can see that the price paid for the smoother estimate is that the estimated target acceleration lags the actual target acceleration.

Frequently, in investigating Kalman filter performance, it is popular to look at the error in the estimates, as was shown in Chapter 9 of the previous volume. Figure 22.23 shows how the single run error in the acceleration estimate for the filter with larger process noise ($\Phi_s= 576$) compares to the theoretical predictions provided by the Riccati equations. It can be seen that the single run results fall within the theoretical bounds, which indicates that the filter is consistent. On the other hand, Fig. 22.24 shows that when the process noise is reduced ($\Phi_s= 9$), the filter is no longer consistent. However, one can also see that the errors

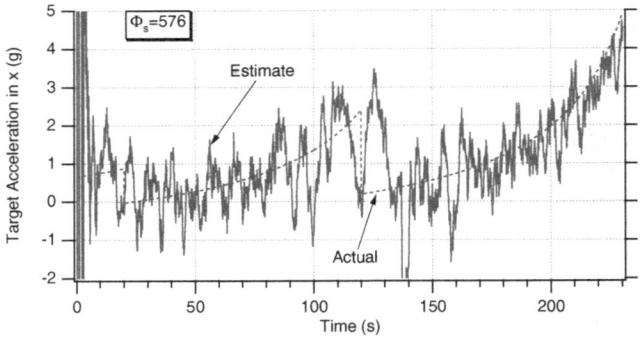

Fig. 22.21   Large Kalman filter process noise yields noisy estimates of target acceleration.

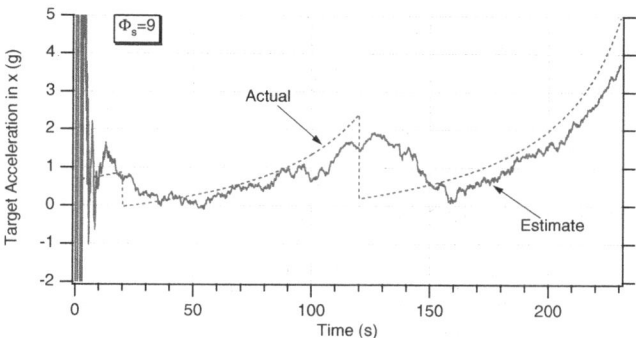

Fig. 22.22   Reducing Kalman filter process provides smoother estimates of target acceleration.

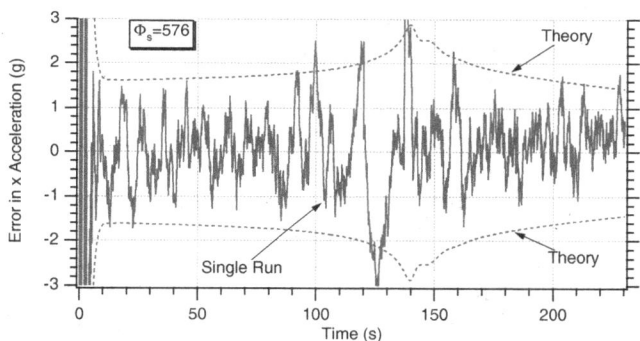

Fig. 22.23   Filter is consistent when process noise is high, but estimation errors are large.

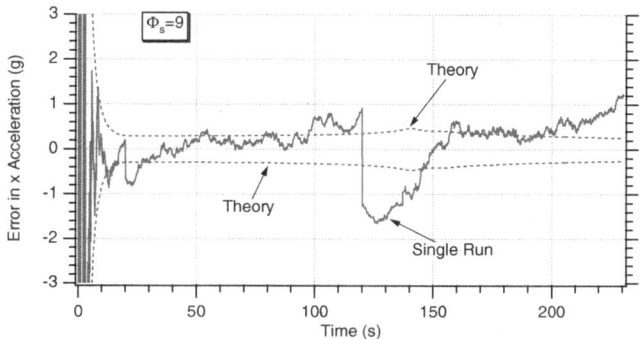

Fig. 22.24   Filter is not consistent when process noise is low, but estimation errors are much lower.

in the acceleration estimate of the filter with less process noise are smaller than when the process noise is increased. We shall soon see from a performance point of view which filter is better for this particular boost-phase intercept problem.

INTERCEPTOR ENGAGEMENTS WITH NOISE AND FILTERING

Fifty-run Monte Carlo sets were run for both the ICBM and IRBM targets to see how much the inclusion of noise and filtering increased the KKV lateral divert requirements. The 90% point (that is, 90% of the flights had required divert that was less than the amount shown in the following figures) in a Monte Carlo set was used as the figure of merit. Figure 22.25 shows that the KKV divert requirements still increase with the initial distance from the interceptor launch point to the target launch site increasing. One can also see that sensor noise may increase the KKV lateral divert requirements substantially over the case where the state estimates are perfect. Figure 22.25 also shows that the selection of the amount of process noise that is used in the filter is important. *Lower amounts of process noise may reduce the interceptor divert requirements—even though the filter is not consistent!* However, even with a small value of process noise the divert requirements can still increase more than 500 m/s over the case without noise or filtering. In addition, the inclusion of noise and filtering limits the maximum distance that the interceptor launch point can be from the target launch site. Thus, the inclusion of sensor noise and filtering is a very important part of the design process in setting KKV divert requirements. Subsequent iterations in the design and sizing process must include alternative filtering approaches such as the EKF to see if KKV divert requirements can be reduced.

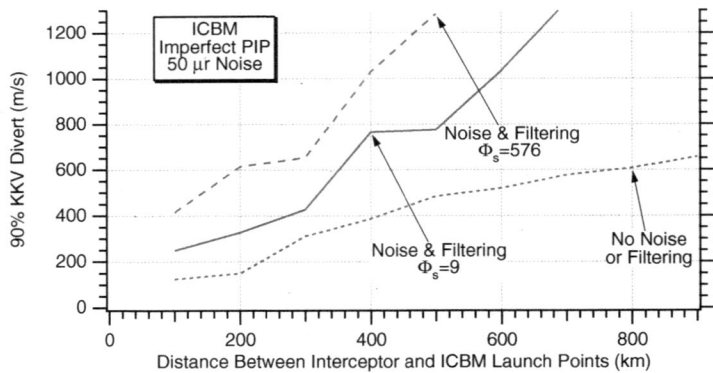

Fig. 22.25 Choice of process noise in Kalman filter can be very important in setting KKV divert requirements.

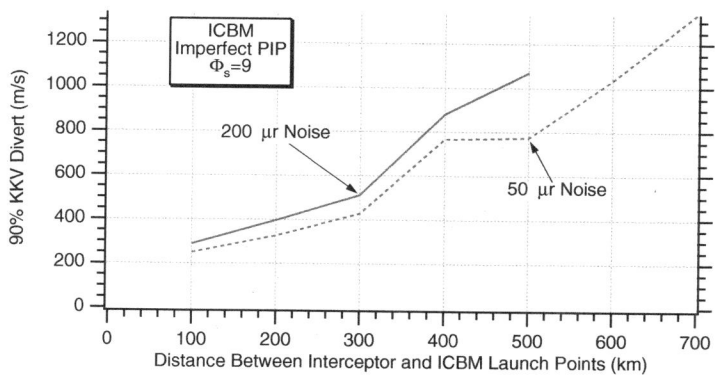

Fig. 22.26   KKV divert requirements increase with increasing measurement noise for the ICBM engagement.

The amount of sensor measurement noise is also very important. Figure 22.26 shows that if the sensor noise increases from its nominal value of 50 μrad to 200 μrad, the lateral divert requirements can sometimes increase by more than 200 m/s at the longer ranges. More importantly, the maximum distance the missile launch point can be from the target launch point is reduced when the measurement noise is increased. Thus, one can see that the guidance and control engineer must interact with the sensor designer to establish reasonable sensor requirements to achieve a balanced system design.

The IRBM engagement results were also repeated for the case in which noise and filtering were considered. One can see from Fig. 22.27 that including these realistic effects also increases the KKV lateral divert requirements. In practice, a

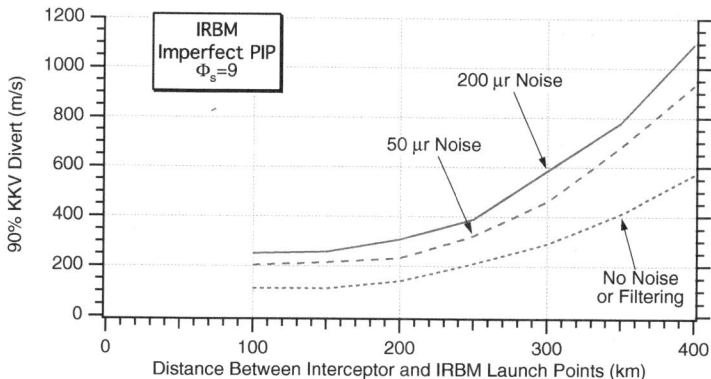

Fig. 22.27   KKV divert requirements also increase with increasing measurement noise for the IRBM engagement.

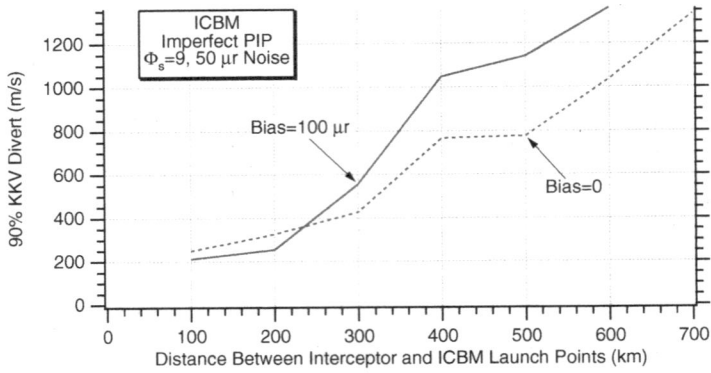

Fig. 22.28   Angle bias errors can increase KKV divert requirements or reduce maximum effective range of interceptor.

wide variety of threats and possible trajectories would have to be considered in deriving KKV lateral divert requirements.

It has been mentioned that there are many steps in the iterative design process. In later steps, other sources of error must be considered to highlight potential problems and to suggest design work that must be performed. For example, in the analysis conducted so far we have not considered measurement angle bias errors. Let us repeat the results of Fig. 22.25, where there is 50 μrad of measurement noise and no bias errors, to now include a 100-μrad angle bias error on the first IRST sensor. Figure 22.28 shows that the bias error increases the KKV lateral divert requirements. Methods for improving the KKV divert performance have to be explored. Thus alternative filter structures must also be examined in subsequent iterations of the design process to see if the influence of biases can be alleviated. Included for study in later iterations of the design process should be such traditional approaches as redesigning the Kalman filter, increasing the homing time to avoid biases, placing more stringent design requirements on the IRST sensor, or including a star tracker on the aircraft. If no acceptable solutions to the bias error can be found, we can see that the effective range of the interceptor will be diminished. It is important to point out that problems are a normal part of the design process and that the search for solutions to these problems often leads to engineering innovation if the right people are involved.

SUMMARY

This chapter illustrates, through simplified examples, how airborne interceptor guidance and filtering technology are important in determining KKV lateral divert requirements against both boosting IRBM and ICBM threats. It was shown that with 50 μrad of aircraft IRST sensor noise and 2 km/s of KKV, divert boost-phase intercepts of both IRBM and ICBM targets could be achieved if the interceptor-launching aircraft could be within 300–800 km of the target launch site. Traditional methods of guidance and filtering were employed to achieve these results. These methods were used to illustrate how sensor noise, prediction error, and apparent target maneuver work together in setting the KKV's lateral divert requirements. Simple formulas have been developed that can be used to understand and explain engagement simulation results. Alternative guidance and filtering techniques have to be explored in subsequent iterations of the design process to see if the KKV divert and acceleration requirements can be reduced.

REFERENCES

[1]   Zarchan, P., "Kill Vehicle Guidance and Control Sizing for Boost-Phase Intercept," *Journal of Guidance, Control and Dynamics*, March–April 2010, pp. 513–521.

[2]   D. Kleppner, and F. K. Lamb (eds.), *Boost-Phase Intercept Systems for National Defense*, American Physical Society, July 2003.

[3]   Wilkening, D. A., "Airborne Boost-Phase Ballistic Missile Defense," *Science and Global Security*, June 2004, pp. 1–67.

[4]   Corbett, M., and Zarchan, P., "The Role of Air Power in Active Missile Defense," *U.S. Air and Space Power Journal*, Summer 2010, pp. 57–71.

[5]   Ehrhard, T. P., and Work, R. O., "Range, Persistence, Stealth, and Networking: The Case for a Carrier-Based Unmanned Combat Air System," *Center for Strategic and Budgetary Assessments*, 2008.

[6]   Zarchan, P., *Fundamentals of Kalman Filtering: A Practical Approach*, 3rd ed, Progress in Astronautics and Aeronautics, AIAA, Reston, VA, 2009, pp. 91–128.

INDEX

Note: Page numbers followed by "t" indicate table, and "f" indicate figure.

A

Accelerometer, 13–14
Adjoints
  and blind time, 114–126, 114f, 115f,
    124f–126f
  of discrete linear Kalman filter,
    101–113, 103f, 107f,
    112f–113f
  of homing loop with three-state
    Kalman filter, 107–113, 107f,
    112f–113f
  impulse, simulation of, 126–129,
    127f
  Kalman filter homing loop with
    blind time option, 115, 115f,
    124–126, 124f, 125f, 126
  multiple sampling rate, 89–101,
    90f–91f, 92t, 96f, 100f
  of Poisson target maneuver model,
    316–319, 317f
  single-lag homing loop with
    shaping filters equivalent of
    constant target maneuver,
    140–145, 140f, 142f–144f
  of single time constant guidance
    system
    with Poisson target maneuver,
      318–319
    with shaping filter equivalent
      of uniformly distributed
      vertical-S maneuver, 449–452,
      449f, 452f
    with weaving target, 151–153,
      152f, 153f, 154–155
Aerodynamic forces, 31
Airframe
  demonstration of, 1–5, 2f, 3f

instability, 1
linearization, 10–14, 12f
review of, 1–5
rigid and flexible body rate
  parameters, 33t
simulation, 5–8, 9f
unstable, three-loop autopilot gain
  algorithm for, 14–15, 15f
Air-launched interceptor
  approach, 640
APN. *See* Augmented proportional
  navigation (APN)
Apollo guidance law, 381
A priori information, 583–584
Augmented proportional navigation
  (APN), 101, 113, 171–174,
  185, 188, 199, 257–258,
  262–263, 270, 289, 641
Autopilot design, 55
  modern control techniques to
    missile roll, 55–78, 57f, 62f,
    63f, 70f–71f, 78f

B

Ballistic targets
  direct integration *vs.* Kepler
    propagation, 513–517, 513t
  interception, kinematics of,
    611–638
  defended area, 633–637,
    636f, 637f
  launch area denied, 629–633,
    632f, 633f
  operational area, 623–629, 624f,
    628f–629f
  intercept point prediction for,
    512–517, 513t

Ballistic targets (*Continued*)
  MATLAB missile trajectory
    generator, 506–512
  trajectory generator in three
    dimensions, 503–512,
    503*f*–505*f*, 512*f*
  two-dimensional kinematic
    multiple-run engagement
    simulation, 616–621
Biased proportional navigation for
    stationary targets, 409–421,
    416*f*–420*f*
Blind range, 114
Blind time
  and adjoints, 114–126, 114*f*, 115*f*,
    124*f*–126*f*
  definition, 114
Boost-phase filtering, 583–584
  ICBM
    guidance, 584–592, 590*f*, 591*f*
    Kalman filter for tracking,
      595–604
    model, 584, 585*f*
    trajectory generator for,
      586–592, 590*f*, 591*f*
  options, 592–593
    three-state filter, 594–610,
      604*f*–609*f*
    two-state template-based filter,
      593–594

C
Cubic flight-control system
  control gains for, 360–363, 363*f*,
    364–366, 364*f*
  for guidance law, 356–362,
    357*f*, 359*f*
  alternative approach to,
    362–366, 363*f*, 364*f*
  performance comparison,
    366–374, 367*f*, 372*f*, 373*f*
  homing loop with, 358

Riccati equations for optimal
    control gains, 360–362
  transfer function, 359, 362

D
Defended area plots, 633
Differential game theory, 555–556
  guidance law, 558–559
    with bounded controls, 571–575,
      571*f*–574*f*
    practical, 575–578, 575*f*–578*f*
  pursuit-evasion games, 555, 558
Discrete Riccati equations, 542–544

E
Earth-centered coordinate system,
    503, 503*f*
EKF. *See* Extended Kalman
    filter (EKF)
Extended Kalman filter (EKF),
    211–229, 212*f*, 215*f*–219*f*,
    225*f*–229*f*, 231, 592–593

F
Filter bank methodology, 237–238
Filter residual, likelihood function
    of, 238
Five-state extended-Kalman-filter,
    231–233, 233*f*
  miss distance analysis, guidance
    system model for,
    232–233, 232*f*
  nominal system inputs for various
    studies, 232**t**
Flexible body dynamics, 31–42, 32*f*
  airframe parameters, 33**t**
  fin deflection command, response
    to, 35–37, 38*f*
  first vibrational mode shape,
    31–32, 31*f*
  notch filter simulation, 44–52, 44*f*,
    46*f*, 50*f*–53*f*

rate gyro flight control system with, 37–42, 39*f*, 42*f*
rate gyro sensing body bending and, 43–52, 43*f*, 44*f*, 46*f*, 50*f*–53*f*
simulation of, 34–35
transfer function, 34
Flight-control system, 55
numerical approach for guidance law, 351–355, 352*f*–353*f*
Rusnak and Meir technique for, 351–355, 352*f*–353*f*
single time constant, optimal guidance law for, 351–355, 352*f*–353*f*
transfer function, 351–352
Four-state weave Kalman filter, 194–207, 194*f*, 199*f*–202*f*, 237–238
miss distance performance, 235–237, 237*f*
performance, 233–237, 235*f*–237*f*

G
Guidance law
alternative approaches to, 333–374
comparison, 562–575, 562*f*–564*f*, 565t, 571*f*–574*f*
for cubic flight-control system, 356–362, 357*f*, 359*f*
alternative approach to, 362–366, 363*f*, 364*f*
performance comparison, 366–374, 367*f*, 372*f*, 373*f*
differential game, 558–559
flight-control system dynamics, numerical approach for, 351–355, 352*f*–353*f*
jerk, 289
alternate form of, 305–306
derivation of, 289–294
development, model for, 290*f*

vs. jerk proportional navigation, 295–304, 298*f*–304*f*
preliminary evaluation of, 294–304, 298*f*–304*f*
for Kalman filter, 101
to maneuvering targets, 349–351
navigation ratio for, 558, 558*f*
optimal control theory for, 333–336
single-lag flight control system, 336–342, 336*f*, 342*f*–343*f*
weaving target, 343–348, 343*f*, 344*f*, 348*f*
proportional navigation, 556–558
review, 556–558, 558*f*
single-lag flight control system for, 336–342, 336*f*, 342*f*–343*f*
trajectory shaping, 375
alternate form of, 382–383
closed-form solutions, 391–396, 395*f*, 396*f*
comparison, linear engagement simulation for, 383–391, 384*f*, 387*f*–390*f*
homing loop model for, 383, 383*f*
nonlinear results, 397–407, 398*f*, 402*f*–406*f*
problem setup, 375–377
Schwartz inequality for, 377–381
testing in linear world, 383–391, 383*f*–384*f*, 386*f*–390*f*
zero-time constant homing loop model for, 376–377, 376*f*
for weave maneuvers, brute force simulation, 176–178
for weaving target using optimal control theory, 343–348, 343*f*, 344*f*, 348*f*

H
Higher-order guidance system dynamics, 158*f*, 159*f*, 161–165, 162*f*

Homing loop
　adjoints
　　of Kalman filter with blind time
　　　option, 115, 115*f*, 124–126,
　　　124*f*, 125*f*, 126
　　of single-lag with shaping filters
　　　equivalent of constant target
　　　maneuver, 140–145, 140*f*,
　　　142*f*–144*f*
　　with three-state Kalman filter,
　　　107–113, 107*f*, 112*f*–113*f*
　　with cubic flight-control
　　　system, 358
　　for derivation of zero over pole
　　　guidance law, tail-controlled
　　　missiles, 468*f*, 478*f*
　　for Kalman filter, 186–188, 186*f*
　　with measurement noise and
　　　three-state Kalman filter, 562*f*
　　for tail-controlled missiles zero over
　　　pole guidance law, 468*f*, 478*f*
　　with three-loop autopilot and
　　　linearized airframe, 482*f*
　　for trajectory shaping guidance law,
　　　383, 383*f*
　　vertical-S target maneuvers in,
　　　452–453, 453*f*
　　zero-time constant, 376–377, 376*f*

I
ICBM. *See* Intercontinental ballistic
　　missile (ICBM)
Impulse, simulation of, 126–129, 127*f*
Infrared search-and-track (IRST)
　　systems, 640
Intercontinental ballistic missile
　　(ICBM), 584, 592–593, 612,
　　640–641
　engagements, 668–673, 669*f*–673*f*,
　　672**t**
　guidance, 584–592, 590*f*, 591*f*
　model, 584, 585*f*

　trajectory generator for, 586–592,
　　590*f*, 591*f*
　two-dimensional trajectory
　　simulation of impulsive
　　two-dimensional, 612–614
Intermediate range ballistic missiles
　　(IRBMs), 640
　engagements, 649–668, 653*f*, 654*f*,
　　665*f*, 667*f*, 668**t**
　target trajectory generator,
　　649–653
　two-dimensional nonlinear
　　engagement simulation,
　　654–664
IRBMs. *See* Intermediate range
　　ballistic missiles (IRBMs)

J
Jerk guidance law, 289
　alternate form of, 305–306
　derivation of, 289–294
　development, model for, 290*f*
　vs. jerk proportional navigation,
　　295–304, 298*f*–304*f*
　preliminary evaluation of, 294–304,
　　298*f*–304*f*
Jerk proportional navigation, 295–304
　command, 299, 299*f*
　heading error, 300–302, 300*f*–302*f*
　vs. jerk guidance law, 295–304,
　　298*f*–304*f*
　maximum acceleration, 298, 299*f*
　missile acceleration response for,
　　303–304, 303*f*, 304*f*
　performance index, 300, 300*f*
　relative trajectory, 298*f*, 300, 300*f*
　vertical-S target maneuver for, 302*f*

K
Kalman filter
　adjoints of discrete linear, 101–113,
　　103*f*, 107*f*, 112*f*–113*f*

comparison, 319–325, 326f
control gains, 101–102
extended, 211–229, 212f,
   215f–219f, 225f–229f, 231,
   592–593
four-state weave, 194–207, 194f,
   199f–202f
guidance laws for, 101
homing loop model for, 186–188,
   186f
   sinusoidal target maneuver, 194f
homing loop with blind time option
   adjoint model, 115, 115f,
     124–126, 124f, 125f, 126
   forward model, 114–115,
     116–126, 124f–126f
miss distance analysis, 208–211,
   208f, 209t, 210f–211f
Monte Carlo simulation of two
   different, 320–325
normal, process noise of, 326–331,
   326f–329f
three-state, 101, 594–610,
   604f–609f
   adjoint of homing loop with,
     107–113, 107f, 112f–113f
   forward model of homing loop
     with, 102–107, 103f
   original, review of, 185–193,
     186f, 192f, 193f
   for tracking ICBM during boost
     phase, 595–604
   for vertical-S target maneuvers,
     452–464, 453f, 461f–464f
   four-state, 461f
   homing loop in, 452–453, 453f
   six-state, 455–464, 462f–464f
   three-state, 461f
and weaving target, 202–207
Kinetic kill vehicle (KKV), 639
air-launched interceptor
   approach, 640

background, 639–640
boosting target and PIP errors,
   formulas for divert due to,
   647–649, 648f
ICBM engagements, 668–673,
   669f–673f, 672t
IRBMs engagements, 649–668,
   653f, 654f, 665f, 667f, 668t
noise and filtering, 673–690, 674f,
   688f, 689f
interceptor engagements with,
   690–692, 690f–692f
one-dimensional model, 641–647,
   641f, 646f, 647f
KKV. *See* Kinetic kill vehicle (KKV)

L
Lambert guidance, 421, 585–592, 607,
   611, 641, 673, 675
Launch area denied plots, 629
Linearization of airframe,
   10–14, 12f
Linear polynomial Kalman
   filters, 594

M
Mach number, tail-controlled
   missile, 3
Magill, D. T., 237
MATLAB ballistic missile trajectory
   generator, 506–512
Miss distance
   closed-form solutions for, 155–160,
     156f, 158f, 159f
   five-state extended-Kalman-filter,
     guidance system model for,
     232–233, 232f
   four-state weave Kalman filter,
     performance of,
     235–237, 237f
   Kalman filter, analysis of, 208–211,
     208f, 209t, 210f–211f

Miss distance (*Continued*)
Monte Carlo simulation of single
time constant guidance system
with vertical-S target
maneuvers, 445–448
for radar homing missile, 537–553,
538*f*, 544*f*, 545*f*, 550*f*–553*f*
RMS, 555
MMAE. *See* Multiple model adaptive
estimator (MMAE)
Modern control techniques, 55
classical analysis of, 78–85,
79*f*–81*f*, 83*f*–84*f*
to missile roll autopilot, 55–78, 57*f*,
62*f*, 63*f*, 70*f*–71*f*, 78*f*
for practical system design, 85–87,
85t, 87*f*
Monte Carlo simulation
for comparing guidance law
effectiveness, 565–570
of homing loop with three-state
Kalman filter, 546–553,
550*f*–553*f*
Kalman filter, two different,
320–325
miss distance, for actual vertical-S
maneuver in single time
constant guidance system,
445–448
of single-lag homing loop using
uniformly distributed random
target maneuver, 134–139
of single-time-constant guidance
system by Poisson target
maneuver model,
312–316, 313*f*
of single time constant guidance
system with vertical-S target
maneuvers
miss distance, 445–448
Multiple model adaptive estimator
(MMAE), 237

fixed, 237–238, 239–253,
249*f*–253*f*
three-filter bank for weaving target
problem, 240–249
Multiple sampling rate adjoints,
89–101, 90*f*–91*f*, 92t,
96*f*, 100*f*
system inputs for, 92t
two samplers operating at
different rates
adjoint model of homing loop
with, 95–101, 96*f*, 100*f*
forward model of homing loop
with, 91–93, 91*f*
simulation of adjoint model of
homing loop with, 96–99
simulation of forward model of
homing loop with, 93–95
two sampling rates and fading
memory filter, homing loop
with, 90*f*

N
Naval Unmanned Combat Air System
(N-UCAS), 640
Notch filter simulation, 44–52, 44*f*,
46*f*, 50*f*–53*f*
N-UCAS. *See* Naval Unmanned
Combat Air System
(N-UCAS)

O
Optimal control theory, 333–336
for single-lag flight control system
guidance law, 336–342, 336*f*,
342*f*–343*f*
single time constant flight-control
system, optimal guidance law
for, 351–355, 352*f*–353*f*
weaving target, guidance law
for, 343–348, 343*f*,
344*f*, 348*f*

P

PIP. *See* Predicted intercept point (PIP)

Poisson square wave, 312

Poisson target maneuver model, 307–308, 307*f*
  adjoint of, 316–319, 317*f*
  modeling, 312–316, 313*f*
  Monte Carlo simulation of single-time-constant guidance system by, 312–316, 313*f*
  shaping filter equivalent of, 316, 316*f*
  shaping networks for, 308–312, 309*f*
  simulation, 312–316, 313*f*

Polynomial guidance
  for impact time control, 428–435, 434*f*–435*f*
  for stationary targets, 421
    caveats, 435–436, 435*f*–436*f*
    impact time control, 428–435, 434*f*–435*f*
    impact time control problem formulation, 421–428
    line-of-sight rate, 422–427
    missile-target engagement geometry, 421, 422*f*
    simulation to test, 430–433
    theoretical formula for x2, 427–428

Position-velocity filter, 584

Predicted intercept point (PIP), 611, 640

Predictor–corrector guidance, 255
  engagement simulation with, 265–269
  proportional navigation, 276*f*
  missile flies to target with, 276*f*
  of unusual form, engagement simulation, 274–275

with roll angle/rate guidance problem, 272–288, 272*f*, 276*f*, 277*f*, 279*f*–284*f*
roll-rate profile, nominal model for, 277*f*
with surface-to-surface intercept problem, 263–271, 264*f*, 270*f*, 271*f*
unusual problem using, 284–288

Proportional navigation, 556–558, 571*f*

Pursuit-evasion games, 555, 558

R

Random telegraph signal model. *See* Poisson target maneuver model

Rate gyro flight control system, 37–42, 39*f*, 42*f*

Rate gyro sensing body bending, 43–52, 43*f*, 44*f*, 46*f*, 50*f*–53*f*

Riccati differential equation, 56–57

Rigid body dynamics, 31–42, 32*f*
  airframe parameters, 33**t**
  fin deflection command, response to, 35–37, 37*f*
  first vibrational mode shape, 31–32, 31*f*
  rate gyro flight control system with, 37–42, 39*f*, 42*f*
  simulation of, 34–35

Roll angle/rate guidance problem, 272–288, 272*f*, 276*f*, 277*f*, 279*f*–284*f*

Roll autopilot design
  four-state system
    control gains in, 65–67
    with four-state control, 68–70
    open-loop transfer function, gain of, 82–83
    plant model for, 63*f*
    with two-state control, 60–62

Roll autopilot design (*Continued*)
  modern control techniques to
    missile, 55–78, 57*f*, 62*f*, 63*f*,
    70*f*–71*f*, 78*f*
  six-state system
    control gains in, 73–75
    with four-state control, 68–70
    open-loop transfer function, gain
      of, 83–84
    plant model for, 71*f*
    with six-state control law,
      transient response of, 76–77
  two-state system
    control gains, controller for
      different sets of, 86
    control gains in, 58–60
    open-loop transfer function,
      magnitude of, 86
    plant model for, 57*f*
    with two-state control, 60–62
    with two-state control,
      open-loop transfer function,
      79–80
Rusnak and Meir technique, 351–355,
  352*f*–353*f*

S
Schwartz inequality, 173, 293, 356
  to derive guidance laws, 333, 336
  for trajectory shaping guidance,
    377–381
Second-order Runge–Kutta
  integration technique, 126
Seeker blind range, 89
Shaping filters
  background, 131
  uniformly distributed target
    maneuver
    adjoint of single-lag homing loop
      with equivalent of constant
      target maneuver, 140–145,
      140*f*, 142*f*–144*f*

Monte Carlo simulation of
  single-lag homing loop,
  134–139
proportional navigation single-
  lag homing loop, 132*f*,
  133–134
representation, 131–145, 132*f*,
  139*f*–140*f*, 142*f*–144*f*
Short-range ballistic missiles
  (SRBMs), 647
Simulation. *See also* Monte Carlo
  simulation
  adjoint model of homing loop with
    two samplers operating at
    different rates, 96–99
  adjoints impulse, 126–129, 127*f*
  airframe, 5–8, 9*f*
  of flexible body dynamics, 34–35
  homing loop with two samplers
    operating at different rates,
    forward model of, 93–95
  ICBM, two-dimensional trajectory,
    612–614
  impulse, 126–129, 127*f*
  IRBMs, two-dimensional nonlinear
    engagement, 654–664
  linear engagement simulation for
    trajectory shaping guidance
    law, comparison, 383–391,
    384*f*, 387*f*–390*f*
  notch filter, 44–52, 44*f*, 46–50, 46*f*,
    50*f*–53*f*
  two-dimensional kinematic
    multiple-run engagement,
    616–621
  weave maneuvers, brute force,
    176–178
Singer target maneuver model, 307,
  308, 308*f*
  shaping networks for,
    308–312, 310*f*
Single-lag flight control system

guidance law, optimal control
    theory for, 336–342, 336*f*,
    342*f*–343*f*
numerical approach to guidance-
    law development for,
    351–355, 352*f*–353*f*
Riccati equations for optimal
    control gains, 339–341,
    342*f*–343*f*
SRBMs. *See* Short-range ballistic
    missiles (SRBMs)
Stationary targets
    biased proportional navigation for
        controlling impact angle
        against, 409–421, 416*f*–420*f*
    closing velocity, 411
    flight-path rate, 411
    missile-target engagement
        geometry for, 410, 410*f*
    optimal guidance law against,
        410–411
    polynomial guidance for, 421
        caveats, 435–436, 435*f*–436*f*
        impact time control, 428–435,
            434*f*–435*f*.
        impact time control problem
            formulation, 421–428
        line-of-sight rate, 422–427
        missile-target engagement
            geometry, 421, 422*f*
        simulation to test, 430–433
        theoretical formula for x2,
            427–428
    two-dimensional engagement
        simulation with optimal
        trajectory shaping, 412–416
Strategic missile-target engagement
    simulation, 517–526, 525*f*
Surface-to-surface missile
    accelerations on, 255–256, 256*f*
    effectiveness, 262–263, 262*f*–263*f*
    line of sight, 256–258, 257*f*

predictor–corrector guidance with,
    263–271, 264*f*, 270*f*, 271*f*
problem of, 255–263, 256*f*–259*f*,
    262*f*–263*f*
proportional navigation guidance,
    258–259, 258*f*
target engagement simulation with
    conventional guidance,
    259–262

T
Tactical ballistic missiles (TBMs), 147
Tail-controlled missiles, 1–5, 465
    airframe
        demonstration of, 1–5
        with stable, 1–2, 2*f*
    angular acceleration on, 5
    center of gravity, 4
    forces on, 2–4, 3*f*
    hypothetical, 5–6, 6*f*
    Mach number, 3
    moment coefficient, 4
    normal acceleration on, 4
    zero over pole guidance law,
        465–467
        homing loop for derivation of,
            468*f*, 478*f*
        optimal control equation to
            confirm, 467, 468*f*, 469–474,
            475
        optimal control gains for,
            469–474, 475*f*
        performance, 475–477,
            475*f*–477*f*, 478*f*, 479–482,
            480*f*–481*f*
        transfer function, 465
Target dynamics, 578–579, 579*f*, 580*f*
Target maneuvers, 559–562,
    560*f*–562*f*
TBMs. *See* Tactical ballistic missiles
    (TBMs)
Template, 592

Three dimensions
  ballistic target trajectory generator
    in, 503–512, 503f–505f, 512f
  compensated weave guidance law
    in, 494–495
  spiraling target, tactical engagement
    simulation with, 495–500
  strategic missile-target engagement
    simulation, 517–526, 525f
  trajectory shaping guidance in,
    526–535, 532f–534f
  weave targets in, 491–502,
    500f–502f
  zero effort miss in, 492–495
Three-filter bank, 239–253, 249f–253f
Three-loop autopilot, 1
  airframe unstable, gain algorithm
    for, 14–15, 15f
  closed-loop analysis of, 17–29,
    27f, 28f
  flight-control system with, 15f
  open-loop analysis of, 15–17, 15f
  simulation of homing loop with,
    482f, 483–488
  and zero over pole guidance law,
    482–489, 482f, 483f, 488f, 489f
Three-state Kalman filter, 101,
    594–610, 604f–609f
  homing loop with
    adjoint, 107–113, 107f,
      112f–113f
    for development, 538f
    forward model, 102–107, 103f
    Monte Carlo simulation of,
      546–553, 550f–553f
    random error sources, 545f
  original, review of, 185–193, 186f,
    192f, 193f
  Poisson target maneuver with,
    563f, 564
  random constant target maneuver
    with, 563, 563f

random vertical-S target maneuver
    with, 564f, 565
  random weave target maneuver
    with, 564–565, 564f
Trajectory shaping guidance law, 375
  alternate form of, 382–383
  closed-form solutions, 391–396,
    395f, 396f
  comparison, linear engagement
    simulation for, 383–391, 384f,
    387f–390f
  homing loop model for, 383, 383f
  nonlinear results, 397–407, 398f,
    402f–406f
  problem setup, 375–377
  Schwartz inequality for, 377–381
  testing in linear world, 383–391,
    383f–384f, 386f–390f
  in three dimensions, 526–535,
    532f–534f
  zero-time constant homing loop
    model for, 376–377, 376f
Two-state template-based filter,
    593–594

U
Unstable airframe, 1

V
Vertical-S target maneuvers, 439, 440f
  analytical expression for, 442–443
  as Fourier series, 441, 444, 444f
  in homing loop, 452–453, 453f
  Kalman filter for, 452–464, 453f,
    461f–464f
    four-state, 461f
    homing loop in, 452–453, 453f
    six-state, 455–464, 462f–464f
    three-state, 461f
  shaping filter approach for,
    439–452, 444f, 445f,
    448f–449f, 452f

single time constant guidance system with, 445*f*

  adjoint diagram of, 449–452, 449*f*, 452*f*

  Monte Carlo miss distance simulation, 445–448

  shaping filter equivalent of, 448*f*, 449

W

Weave targets

  acceleration saturation, 165–168, 166*f*–167*f*

  advanced guidance techniques to improve performance, 171–179, 172*f*, 175*f*, 176*f*, 178*f*

  brute force simulation

    for generating normalized design curves, 162–164

    for guidance law evaluation, 176–178

  discrete Fourier transform to find miss caused by, 179–182, 182*f*

  higher-order guidance system dynamics, 158*f*, 159*f*, 161–165, 162*f*

  miss distance, closed-form solutions for, 155–160, 156*f*, 158*f*, 159*f*

  optimal control for, 343–348, 343*f*, 344*f*, 348*f*

  in single time constant guidance system, 147–155, 151*f*, 152*f*–153*f*

  adjoint, 151–153, 152*f*, 153*f*, 154–155

  linearized, 151, 152*f*

  nonlinear engagement simulation, 148–150

  in three dimensions, 491–502, 500*f*–502*f*

  three-filter fixed MMAE to weaving target, 240–249

  time constant reduction to improve performance, 168–170, 169*f*, 170*f*

Z

ZEM. *See* Zero effort miss (ZEM)

Zero effort miss (ZEM), 492–495

Zero over pole guidance law

  tail-controlled missiles, 465–467

    homing loop for derivation of, 468*f*, 478*f*

    optimal control equation to confirm, 467, 468*f*, 469–474, 475

    optimal control gains for, 469–474, 475*f*

    performance, 475–477, 475*f*–477*f*, 478*f*, 479–482, 480*f*–481*f*

    transfer function, 465

  and three-loop autopilot, 482–489, 482*f*, 483*f*, 488*f*, 489*f*

Zero-time constant homing loop model, 376–377, 376*f*

SUPPORTING MATERIALS

A complete listing of titles in the Progress in Astronautics and Aeronautics series is available from AIAA's electronic library, Aerospace Research Central (ARC) at arc.aiaa.org. Visit ARC frequently to stay abreast of product changes, corrections, special offers, and new publications.

AIAA is committed to devoting resources to the education of both practicing and future aerospace professionals. In 1996, the AIAA Foundation was founded. Its programs enhance scientific literacy and advance the arts and sciences of aerospace. For more information, please visit www.aiaafoundation.org.